Lecture Notes in Computer Science 630

Edited by G. Goos and J. Hartmanis

W. R. Cleaveland (Ed.)

CONCUR '92

Third International Conference
on Concurrency Theory
Stony Brook, NY, USA, August 24-27, 1992
Proceedings

Springer-Verlag
Berlin Heidelberg New York
London Paris Tokyo
Hong Kong Barcelona
Budapest

Series Editors

Gerhard Goos
Universität Karlsruhe
Postfach 69 80
Vincenz-Priessnitz-Straße 1
W-7500 Karlsruhe, FRG

Juris Hartmanis
Department of Computer Science
Cornell University
5149 Upson Hall
Ithaca, NY 14853, USA

Volume Editor

W. Rance Cleaveland
Department of Computer Science, North Carolina State University
Raleigh, NC 27695-8206, USA

CR Subject Classification (1991): F.3.2, F.3.1, F.3.3, D.3.2, F.1.2

ISBN 3-540-55822-5 Springer-Verlag Berlin Heidelberg New York
ISBN 0-387-55822-5 Springer-Verlag New York Berlin Heidelberg

Printed in Germany

Typesetting: Camera ready by author/editor
Printing and binding: Druckhaus Beltz, Hemsbach/Bergstr.
45/3140-543210 - Printed on acid-free paper

Preface

CONCUR '92 is the third in what is becoming an annual series of conferences devoted to the study of theories of concurrency. The first two CONCURs were held in Amsterdam, the Netherlands, in 1990 and 1991; 1992 marks the first year that the conference has been held outside of Europe. The goal of CONCUR is to provide a forum for reporting on advances in theories of concurrency and their applications and to facilitate exchanges of ideas and information among the several different schools of concurrency theory that have arisen over the past 20 years.

This proceedings contains 34 papers that were culled from 112 submissions, 41 more than last year. Five invited papers are also included. The number of submissions substantially exceeded expectations, and I would like especially to thank the members of the program committee, and their subreferees, for their efforts in assembling this collectio of papers.

Support for CONCUR '92 has generously been provided by the National Science Foundation (NSF grant CCR-9201450) and ESPRIT. The organizers of the conference would like to thank Nat Macon of NSF and Michel Bosco of ESPRIT for serving as the cognizant officers of these awards.

I would also like to thank the State University of New York at Stony Brook for hosting CONCUR '92 and the organizing committee for their tireless efforts in arranging the conference.

Raleigh, June 1992 Rance Cleaveland

Program Committee

R. Cleaveland (N.C. State University), *Chai*

F. de Boer (TU Eindhoven)
B. Bloom (Cornell)
S. Brookes (CMU)
N. Francez (Technion)
U. Goltz (GMD)
J. Gunawardena (Hewlett-Packard)
L. Logrippo (University of Ottawa)
M. Merritt (AT&T Bell Labs)
J. Sifakis (Université de Grenoble)
P. Sistla (University of Illinois-Chicago)
E. Stark (SUNY Stony Brook)
B. Steffen (RWTH-Aachen)
P. Wolper (Université de Liège)

Organizing Committee

S. Smolka, *Chair*

R. Cleaveland, *Program Chair*
K. Germana
F. Moller, *Tutorials Chair*
A. Zwarico, *Publicity Chair*

Steering Committee

J.C.M. Baeten, *Chair*

E. Best
K.G. Larsen
U. Montanari
P. Wolper

Table of Contents

Testing Equivalence for Mobile Processes

(Extended Abstract)

Michele Boreale and Rocco De Nicola

Università di Roma "La Sapienza"
Dipartimento di Scienze dell'Informazione
Via Salaria 113, I-00198 Roma (Italy)

Abstract. The impact of applying the testing approach to a calculus of processes with a dynamically changing structure is investigated. A proof system for the finite fragment of the calculus is introduced which consists of two groups of laws: those for strong observational equivalence and those needed to deal with τ actions. Soundness and completeness w.r.t. a testing preorder are shown. A fully abstract denotational model for the language is presented which relies on the existence of normal forms for processes.

1. Introduction

Process Algebras ([Mil89], [Hoa85], [BK89]) are generally recognized as a good formalism for describing and studying properties of distributed concurrent systems. A process algebra is often defined by specifying its syntax and the transitional semantics of its terms by means of Structured Operational Semantics [Plo81]. By now, this approach has become a standard tool for specifying basic semantics of process algebras, but it was early recognized that it does not yield extensional accounts of processes. Thus, techniques have been developed to abstract from unwanted details in systems descriptions. Many of these techniques are based on behavioural equivalences; two terms are identified if and only if no observer can notice any difference between their external behaviours.

Process description languages, such as CCS, have been (and are) thoroughly studied using equivalence notions based on bisimulations (see e.g. [Mil89]) or on testing (see e.g. [Hen88]). Complete axiomatizations have been put forward which are of fundamental importance for manipulating process expressions by means of simple axioms and inference rules and constitute the theoretical basis for a class of verification tools (see e.g. [DIN90, Hui92]).

Almost all process algebras which have been considered permit describing only systems whose subparts can interact by performing *pure synchronizations*. Lately, a language with value-passing has been investigated and a complete axiomatization of a testing-based equivalence has been performed by provided for it [HI91]. A further step toward improving the descriptive power of process algebras has been performed by adding

---

Work partially supported by "Progetto Finalizzato Sistemi Informatici e Calcolo Parallelo" of CNR, contract n. 91.00894.69 and by Istituto di Elaborazione dell'Informazione of CNR, Pisa.

The Polyadic π-calculus

Robin Milner

Laboratory for the Foundations of Computer Science
Department of Computer Science
The King's Buildings
University of Edinburgh
Edinburgh EH9 3JZ
U.K.

The π-calculus is a calculus of concurrent processes based upon the idea of *naming*. It models dynamically changing concurrent systems, has a rich algebraic theory, and contains in a precise way both functions (the λ-calculus) and data structures, represented as processes. The calculus is a generalisation of CCS, and was introduced by Robin Milner, Joachim Parrow and David Walker, based on important ideas of Mogens Nielsen and Uffe Engberg.

The way in which everything is built upon naming is this: When two processes interact, they *use* a name (which can be thought of as a channel). This name is called the *subject* of the interaction. The *object* of an interaction – its information content – is also a name; this is the *mention* of a name, not the *use* of it. To receive a name is to acquire the ability to use it, perhaps to interact with a process which was previously inaccessible. This process may indeed represent a datum, as explained above; then one can think of the datum itself having been received.

In this lecture I shall discuss the polyadic version of the π-calculus, which supports a very fruitful notion of *sort* and *sorting*, akin to simple typing in the λ-calculus. It will be seen how different sortings are appropriate for different applications. The encoding of the λ-calculus into π-calculus, and the uniform representation of data structures, are best seen in the polyadic setting with suitable sortings.

This article was processed using the LaTeX macro package with LLNCS style

primitives for expressing exchange of more complex objects such as channel names or processes themselves, see e.g. ECCS of [EN86] and CHOCS, [Tho90]. Recently an extension of CCS called π-calculus has been put forward [MPW89] that permits describing systems whose linking configuration may dynamically change by exchanging channel names (in passing, we note that this permits simulating process passing). For these higher order languages only (strong-)bisimulation-based theories have been investigated; and no *weak* (i.e. abstracting from internal moves) behavioural equivalence has been considered and e.g. equipped with a complete axiomatization.

The aim of this paper is therefore twofold: to investigate the applicability of the testing approach to higher order process description languages and to provide an axiomatization of a weak equivalence for π-calculus. The new equivalence is defined by following the general approach of [DH84, DeN87, Hen88]. It requires formally defining a *set of observers, a way of observing* and *a general criterion for interpreting the result of observations.* If we call **P** the set of systems we want to experiment upon, to apply the general setting, we need to define a set of observers, say **O**, and explain the evolution of pairs such as $<p,o> \in \mathbf{P}\times\mathbf{O}$ which will represent the interactions between p and o. Interactions may be *failing* or *successful* depending on whether particular states (which *report success*) are reached. For specific process p and observer o, one might be interested in knowing whether a successful interaction does exists, i.e. whether p **may satisfy** o, or whether *all* interactions are successful, i.e. whether p **must satisfy** o. Two *testing preorders* over **P**, naturally arise, associated with each of the above points of view:

- $p \ll_m q$ if and only if $\forall o \in \mathbf{O}$: p **may satisfy** o implies q **may satisfy** o;
- $p \ll_M q$ if and only if $\forall o \in \mathbf{O}$: p **must satisfy** o implies q **must satisfy** o.

These preorders can be naturally combined to get a third one:

- $p \ll q$ if and only $\forall o \in \mathbf{O}$: $p \ll_m q$ and $p \ll_M q$.

In [DH84] this machinery is applied to CCS and gives rise to three testing preorders which have simple axiomatic and denotational characterization communication scheme (bound names of input actions are instantiated only when the COM rule is actually used, see rules INPUT, OUTPUT, COM and CLOSE of the transitional semantics). Indeed, whereas the so-called *early instantiation* scheme of CCS with value passing ([Mil88]) permits easily determining the pairs of complementary sequences of visible actions which make an interaction possible, the same is not true for late instantiation.

Additional complications are induced by the fact that, in π-calculus, beside distinguishing between input and output actions, two distinct kinds of output actions are considered; in particular, a process can export public names, *free outputs*, or private ones, *bound outputs*.

The presence of input actions implies that, when performing a synchronization, the next state of a process might depend on the received name. The distinction between free and bound outputs gives a process different options when prompted for an output. To fully discriminate the different behaviours, observers are needed which can test whether two (received) names are the same name. This has called for an extension of the experimenter language with a sort of *if-then-else* construct; the original π-calculus has been extended with a mismatch operator which together with the original match operator gives us the needed observational power.

In this paper we concentrate on the finite fragment of the calculus. Extensions to the whole calculus are not difficult, but would require additional technicalities which would somehow obscure the main issues; generalizations to infinite processes have been fully worked out in [Bor91], on which this paper is based.

The rest of the paper is organized as follows. In Section 2 syntax and transitional semantics of π-calculus are introduced, together with a presentation of the testing equivalence approach and its application to π-calculus. In Section 3, a sound and complete (w.r.t. a testing preorder) proof system for the calculus is introduced. It consists of two groups of laws: one containing (essentially) the laws for the strong observational equivalence of [MPW89], the other needed to deal with τ actions and nondeterminism. In Section 4 a fully abstract denotational model for the language is presented. The model is based on an algebraic cpo, obtained by ideal completion from a poset of (equivalence classes of) certain canonical forms for processes. The full abstractness proof of the model is sketched. The concluding section contains a few comments about the difference between testing and bisimulation-based equivalences over dynamic processes and makes out a strong case for including a new operator (mismatch) whose presence is highly debated among the proposers of the π-calculus. Proofs will appear in a full version of the paper.

2. Applying Testing Equivalence to π-calculus

Π-calculus and its Transitional Semantics

In this subsection we give a succinct presentation of π-calculus, by introducing its syntax and transitional semantics and some standard notation. The reader is referred to [MPW89] for motivations and additional definitions. Actually, we confine ourself to *finite* agents, i.e. consider a language without recursion, and add two additional operators: *mismatch* and Ω. In spite of these changes, we shall continue to call this language π-calculus. We will argue later about the usefulness of the new operators.

Definition 2.1 ***(Syntax)***
The syntax of π-calculus is given by means of the following BNF-like grammar (each clause is followed by the name we use to refer to the construct):

p::= NIL	(null process)
$\mid \Omega$	(divergence)
$\mid \tau.p$	(silent prefix)
$\mid y(x).p$	(input prefix)
$\mid \overline{y}x.p$	(output prefix)
$\mid (y)p$	(restriction)
$\mid [x=y]p$	(match)
$\mid [x\neq y]p$	(mismatch)
$\mid p_1+p_2$	(summation)
$\mid p_1 \mid p_2$	(parallel composition)

where x, y, ... range over an infinite set **N** of *names*, called also *variables* or *labels* or *ports*. We call **P** the set of terms generated by the syntax above. The elements of **P** are called *processes*. Metavariables p, q, r,... range over **P** and α denotes generic action prefix, x(y), $\overline{x}y$, or τ. ♦

The language is essentially an extension of CCS with value passing [Mil80] which permits exchanging also communication channel names and thus to describe systems with a dynamically changing structure.

Most operators are the same as in CCS, this is the case for NIL, the terminated process; [•]+[•], the nondeterministic sum of two processes, and [•]|[•], the operator for parallel composition. Action prefix, α.[•], is only similar to that of CCS; here a distinction is introduced between input and output actions. The role of restriction, (y)[•] as a scope binder (similar in spirit to that of CCS) is rendered more explicit by the new prefix notation. The meaning of the match operators [x=y][•] , is that [x=y]p behaves like p if x is equal to y and like NIL otherwise.

The new operators we introduce are *mismatch* [x≠y][•], and the *divergence* constant Ω that in the testing approach is used to represent those agents which perform only infinite sequences of silent actions. The *mismatch* operator [x≠y][•] is complementary to match: [x≠y]p behaves like p if x is different from y and like NIL otherwise. It is worth noticing that match and mismatch together permit easily describing the if-then-else construct, indeed [x=y]p+[x≠y]q corresponds to *if* x=y *then* p *else* q.

Together with the above operators another type of action prefix will be used: the *bound output* $\overline{x}$(y).[•]. It is not a primitive operator but only a shorthand for (y)($\overline{x}$y.[•]).

Free variables, bound variables and substitutions of a process p, denoted by fv(p), bv(p) and $\sigma = \{x_1/y_1,...,x_n/y_n\}$ respectively, are defined as expected when the prefix x(y) and the restriction (y) are considered as *binders* for the name y. The notation pσ is used to denote the agent obtained from p by simultaneously replacing each y $\in$ fv(p) by yσ. *Clashes of names* are dealt with, as usual, by means of renamings.

Definition 2.2 (*Actions*)
- We define
  - the set of *input actions* as IA = {x(y) | x,y $\in$ N};
  - the set of *output actions* as OA = { $\overline{x}$y | x,y $\in$ N}$\cup${ $\overline{x}$(y) | x,y $\in$ N and x≠y};
  - the set of *visible actions* as Λ = OA $\cup$ IA; and
  - the set of all *actions* as Act = $\Lambda\cup\{\tau\}$. Symbol α, β will be used to range over Act.
- Functions fv, bv, subj (*subject*), obj (*object*) are defined over Act as follows:

	x(y)	$\overline{x}$(y)	$\overline{x}$y	τ
bv	{y}	{y}	Ø	Ø
fv	{x}	{x}	{x, y}	Ø
subj	x	x	x	—
obj	y	y	y	—

If α is an action then v(α) is the union of fv(α) and bv(α). If σ is a substitution $\alpha\sigma$ denotes the action obtained by simultaneously replacing in α each y $\in$ fv(α) by yσ. ♦

Definition 2.3 (*Transitional Semantics*)
The transitional semantics of **P** is given as a Labelled Transition System, in the SOS style, by means of the following axioms and inference rules:

$\tau.p \text{—}\tau\text{—>} p$ (TAU ACT)

$\overline{x}y.p \text{—}\ \overline{x}y\text{—>} p$ (OUTPUT ACT)

$x(y).p \text{—}x(w)\text{—>} p\{w/y\}$, *if* $w \notin fv((y)p)$ (INPUT ACT)

$p \text{—}\alpha\text{—>} p'$ *implies* $[x=x]p \text{—}\alpha\text{—>} p'$ (MATCH)

$p \text{—}\alpha\text{—>} p'$ *implies* $[x\neq y]p \text{—}\alpha\text{—>} p'$ *if* $x \neq y$ (MISMATCH)

$p_1 \text{—}\alpha\text{—>} p'_1$ *implies* $p_1+p_2 \text{—}\alpha\text{—>} p'_1$ (SUMMATION)

$p \text{—}\alpha\text{—>} p'$ *implies* $(y)p \text{—}\alpha\text{—>} (y)p'$ *if* $y \notin v(\alpha)$ (RESTRICTION)

$p \text{—}\ \overline{x}y\text{—>} p'$ *implies* $(y)p \text{—}\ \overline{x}(w)\text{—>} p'\{w/y\}$ *if* $y \neq x$, $w \notin fv((y)p')$ (OPEN)

$p_1 \text{—}\alpha\text{—>} p'_1$ *implies* $p_1|p_2 \text{—}\alpha\text{—>} p'_1|p_2$, *if* $bv(\alpha) \cap fv(p_2) = \emptyset$ (PAR)

$p_1 \text{—}\ \overline{x}y\text{—>} p'_1$ *and* $p_2 \text{—}x(z)\text{—>} p'_2$ *implies* $p_1|p_2 \text{—}\tau\text{—>} p'_1|p'_2\{y/z\}$ (COM)

$p_1 \text{—}\ \overline{x}(w)\text{—>} p'_1$ *and* $p_2 \text{—}x(w)\text{—>} p'_2$ *implies* $p_1|p_2 \text{—}\tau\text{—>} (w)(p'_1|p'_2)$ (CLOSE)

plus
the obvious symmetric rules for PAR, COM, CLOSE and SUMMATION. ♦

Notations

- We assume the following precedences between operators in agent expressions:
  {restriction, prefix, match, mismatch}>parallel composition>summation
  Moreover, substitutions will have the strongest precedence, thus e.g. $\alpha.p\sigma$ will stand for $\alpha.(p\sigma)$;
- p[+r] denotes an agent expression where the summand r is *optional*;
- [p] denotes an agent expression which may have a certain number ($n \geq 0$) of restrictions at its top level, i.e. [p] stands for $(y_1)\ldots(y_n)p$;
- $[y \notin Y][.]$ will be a shorthand for the context $[y\neq y_1]\ldots[y\neq y_k][.]$, where Y is a finite set of names, $Y=\{y_1,\ldots,y_k\}$, $k\geq 0$; if $Y=\emptyset$, $[y\notin Y]$ will denote the empty context [];
- $p=\alpha\Rightarrow q$, with $\alpha \in \Lambda$, stands for $p\text{—}\tau^*\text{—>}p'\text{—}\alpha\text{—>}q'\text{—}\tau^*\text{—>}q$;
- $p=s\Rightarrow q$, for any $s=\alpha_1\ldots\alpha_k \in \Lambda^*$, stands for $p = p_0=\alpha_1\Rightarrow p_1 \ldots p_{k-1}=\alpha_k\Rightarrow p_k = q$;
- $p|q\text{—}c\text{—>}[p'|q']$ denotes a *communication* between p and q, i.e. a transition $p|q\text{—}\tau\text{—>}[p'|q']$ whose derivation (from the inference rules) contains an application of the COM or CLOSE rule involving p and q; we will write $p|q =c\Rightarrow [p'|q']$ for $p|q \text{—}\tau^*\text{—>}\text{—}c\text{—>} [p'|q']$.
- The symbol $\equiv$ denotes syntactic equality between agents, while $\equiv_\alpha$ denote equality up to renaming of bound variables.
- /R, where R is any predicate or relation, is used to denote the negation of R, and "a R" is used to say "∃b s.t. aRb". ♦

Testing π-calculus

In this subsection, we set up a few definitions that permit applying the testing machinery to π-calculus. In particular, we will first introduce observers then experiments and finally testing preorders.

Definition 2.4 (*Observers*)
- The set **O** of observers is defined like **P**, but an additional distinct action ω is used to report success; Act' = Act$\cup\{\omega\}$, where $\omega \notin$ Act.
- We have $\Lambda' = \Lambda\cup\{\omega\}$, $bv(\omega) = fv(\omega) = \emptyset$, $subj(\omega) = obj(\omega) = -$.
The actions $\bar{x}y$ (*free output*), τ and ω are *free actions*, the remaining ones are *bound actions*.
- The operational semantics for **P** extends to **O** by adding the obvious rule: $\omega.o \text{—}\omega\text{—>} o$.
- **O** is ranged over by the metavariables o, o', ... ♦

Definition 2.5 (*Experiments*)
The set **Exp** of *experiments* is defined as: $\{[p|o] \mid p\in \mathbf{P} \text{ and } o\in \mathbf{O}\}$
Metavariables e, f,... range over **Exp**. We recall that square brackets [ ] around a term indicate that they might be prefixed by some restrictions. ♦

Henceforth, we use the word agent to mean either process or observer or experiments.

Definition 2.6 (*Interactions*)
Given an experiment [p|o], an *interaction* (or *computation*) from [p|o] is a maximal sequence of τ transitions such that:

$[p|o]\text{—}\tau\text{—>}[p_1|o_1]\text{—}\tau\text{—>}\ldots\text{—}\tau\text{—>}[p_k|o_k]$ with $k\geq 0$ and $[p_k|o_k] \not\text{—}\tau\text{—>}$. ♦

The next definition formalizes the concept of *partially divergent* agent. Here, Ω represents the "totally divergent" or "totally undefined" agent, i.e. it models the agent which computes an infinite sequence of internal actions, without ever interacting with the external environment. In spite of the fact that we do not consider recursion and infinitary behaviours, we have enclosed Ω in our language, because this will allow us to easily generalize our theory to the full language, by adding rules for recursion. Furthermore, Ω will be needed when defining denotational semantics; it will represent the *bottom* of our model.

Definition 2.7 (*Definedness predicates*)
• $\downarrow$ is the least (postfix) predicate over **O** which satisfies:
- $NIL\downarrow$, $\alpha.p\downarrow$, $[x\neq x]p\downarrow$, $x\neq y$ implies $[x = y]p\downarrow$
- $p\downarrow$ and $q\downarrow$ imply $(y)p\downarrow$, $[x = x]p\downarrow$, $[x\neq y]p\downarrow$, $p|q\downarrow$, $p+q\downarrow$.

We write $p\uparrow$ if we cannot prove $p\downarrow$.
• $\Downarrow$ is the least (postfix) predicate over **O** which satisfies:
$p\Downarrow$ iff $p\downarrow$ and for every p', $p=\varepsilon\Rightarrow p'$ implies $p'\Downarrow$.

We write $p\Uparrow$ if it is not the case that $p\Downarrow$. ♦

Definition 2.8 (*May and Must*)
• p **may** o if and only if there exists an interaction
$p|o\text{—}\tau\text{—>}\ldots\text{—}\tau\text{—>}[p_k|o_k]$ such that for some $0\leq i\leq k$, $o_i\text{—}\omega\text{—>}$.
• p **must** o if and only if for all interactions
$p|o\text{—}\tau\text{—>}\ldots\text{—}\tau\text{—>}[p_k|o_k]$ $\exists j$, $0\leq j\leq k$, such that $o_j\text{—}\omega\text{—>}$ and $\forall i\leq k$, $((p_i|o_i)\uparrow$ implies $i\geq j)$. ♦

Definition 2.9 (*Testing Preorders*)
- $p \ll_m q$ if and only if for every $o \in \mathbf{O}$ we have: p **may** o implies q **may** o
- $p \ll_M q$ if and only if for every $o \in \mathbf{O}$ we have : p **must** o implies q **must** o
- $p \ll q$ if and only if $p \ll_m q$ and $p \ll_M q$. ♦

As it might be expected $\ll_M$ (hence $\ll$) is not preserved by the + operator. A standard way of solving this problem is "closing" these relations with respect to summation (see [DH84], [Mil89]). An alternative characterization of the new testing preorders can also be given in terms of the original preorders and the initial invisible actions of agents. This leads to the following:

Definition 2.10 (*Revisited Testing Preorders*)
- $p \ll_M' q$ if and only if ($p \ll_M q$ and ($p\Uparrow$ or ($q \xrightarrow{\tau}$ implies $p \xrightarrow{\tau}$)))
- $p \ll' q$ if and only if ($p \ll_m q$ and $p \ll_M' q$) ♦

It can be easily shown that the above preorders coincides with those generated by the closure under + of the original preorders; i.e.:
- $p \ll_M' q$ if and only if for each r, $p+r \ll_M q+r$
- $p \ll' q$ if and only if for each r, $p+r \ll q+r$

Thus, the above definition provides us with preorders which are substitutive for + operator; however, differently from CCS, we have that none of the relations $\ll_m$ (which is preserved by +), $\ll_M'$, $\ll'$ is a congruence: they are not preserved by the input prefix operator. This is proved by the following counterexample.

Let x, y, u, w, z be distinct names in $p \equiv [x=y]\ \bar{x}w.NIL$ and $q \equiv [x=z]\ \bar{x}w.NIL$. We have $p \ll_m q$, because both p and q are trivially equivalent to NIL. On the other hand we have that u(y).p is not equivalent u(y).q. Indeed, if we take $o \equiv \bar{u}x.x(w).\omega.NIL$ we have $(u(y).p) \mid o \xrightarrow{\tau} [x = x]\ \bar{x}w.NIL \mid x(w).\omega.NIL \xrightarrow{\tau} NIL \mid \omega.NIL \xrightarrow{\omega}$ i.e. u(y).p **may** o. Conversely, the only possible computation from $(u(y).q) \mid o$ is: $(u(y).q) \mid o \xrightarrow{\tau} [x = z]\ \bar{x}w.NIL \mid x(w).\omega.NIL \not\xrightarrow{\tau}$ and thus u(y).q /**may** o.

Similar examples can be exhibited for $\ll_M'$ and $\ll'$; indeed the same problem arises also in [MPW89] for strong ground bisimulation. As in [MPW89], a sort of congruence law for the input prefix operator can be given. Equivalence of p and q is preserved by prefix u(y).[•] provided that it is preserved by all substitutions {w/y}, for any name $w \in fv(p, q)$. Indeed, this will be a rule of our proof system; notice that this is still an effective one, since we have a finite number of names to be tested (as fv(p,q) is finite), although not very efficient.

3. A Proof System for finite π-calculus

In this section, we provide an equational (sound and complete) theory for our preorders. For the sake of space, we only present the axiomatization of $\ll'$; the extension to $\ll_M'$ and $\ll_m'$ is only a matter of rephrasing the results. The axioms and the inference rules of the system, which we call F, are shown in Table 3.1. The following definition of *complementary actions* will be useful for stating the expansion theorem:

Definition 3.1 (*Complementary actions and communications*)

• Let α,β be actions, we have α **comp** β iff, for some names x,y,w, one of the following holds:

i) $\alpha = \overline{x}y$ and $\beta = x(w)$ ii) $\alpha = \overline{x}(y)$ and $\beta = x(y)$

iii) $\alpha = x(y)$ and $\beta = \overline{x}w$ iv) $\alpha = x(y)$ and $\beta = \overline{x}(y)$

• We call $R(\alpha,\beta,p,q,p',q')$ the result [p"|q"] of a communication transition p|q—c—>[p"|q"] from p and q, whose premises are the transitions p—α—>p' and q—β—>q', with α **comp** β. Whenever no ambiguity arises about p, q, p' and q', we write $R_{\alpha\beta}$ for $R(\alpha,\beta,p,q,p',q')$.♦

Most of the axioms and rules in Table 3.1 are taken from [MPW89] and [DH84]. In particular, the Restriction, Match and Expansion laws and the inference rules are from [MPW89]; the Omega law and the τ-laws N1, N2 and N4 are from [DH84], while the τ-law N3 and, obviously, the Mismatch laws are new. In the Expansion law, an empty summation by convention denotes NIL. By "F |— p = q" we mean that the relation p = q is provable by applying the axioms and the inference rules of the system. Sometimes we abbreviate this simply as "p = q".

The next two propositions state soundness of the inference rules and of the axioms; the actual proofs can be done along the lines of the corresponding ones for CCS; however, specific care is needed when interactions between processes and observers are analyzed, due to the exchange of names between them. Thus the proofs of the inference rules for parallel composition and restriction in C0 are significantly harder than those for CCS; rule C1 was not present in the proof system for CCS.

Proposition 3.1 (*Soundness of the inference rules*)
The inference rules α-conv, C0, C1, R and T of Table 3.1 are sound w.r.t. «'. ♦

Proposition 3.2 (*Soundness of the axioms*)
The axioms of Table 3.1 are sound w.r.t. «'. ♦

The above propositions sum up in the following:

Theorem 3.1 (*Soundness of F*)
F |— $p \leq q$ implies p «' q. ♦

The completeness proof is heavily based on the existence of certain canonical forms for processes called *head normal forms*; similar forms were already present in [DH84], though there the completeness proof was based on *normal forms*. Due to the lack of space, we will not report the full proof here, it will appear in the complete version of the paper.

Head normal forms (definition 3.3) aim at presenting processes as a non-deterministic choice among a set of *initial states* (represented by p_A, $A \in L$); each of these states is essentially described by a set (A) of *channels* through which the process can communicate with the environment, and by the corresponding possible *initial actions* (α's) and their *α-derivatives* (p_α 's). If there are different initial states (this is the case of τ-hnf's), each state (p_A) can be non-deterministically reached by performing an internal (τ-) action.

Two processes in head normal form can be syntactically compared to determine whether they are related by «'. *Saturation* of set L makes this comparison easier. It is worth noting that the notion of saturation is about *events*, i.e. input or output channels, not about actions (in [DH84] these two notions were the same).

Axioms

Sum

S 0 $p + NIL = p$

S 1 $p + p = p$

S 2 $p + q = q + p$

S 3 $p + (q + r) = (p + q) + r$

Restriction

R0 $(x)p = p$ if $x \notin fv(p)$

R1 $(x)(y)p = (y)(x)p$

R2 $(x)(p + q) = (x)p + (x)q$

R3 $(x)\alpha.p = \alpha.(x)p$ if $x \notin v(\alpha)$

R4 $(x)\alpha.p = NIL$ if $subj(\alpha)=x$

Match

M0 $[x = y]p = NIL$ if $x \neq y$

M1 $[x = x]p = p$

Mismatch

U0 $[x \neq x]p = NIL$

U1 $[x \neq y]p = p$ if $x \neq y$

Expansion

E Let $I = \{1,...,k\}$, $J = \{1,...,h\}$ with $h, k \geq 0$.
Let $p \equiv \Sigma_{i\in I}\alpha_i.p_i[+\Omega]$ and $q \equiv \Sigma_{j\in J}\beta_j.q_j[+\Omega]$.
If for each $i \in I$, $bv(\alpha_i) \cap fv(q) = \emptyset$ and for each $j \in J$, $bv(\beta_j) \cap fv(p) = \emptyset$, then

$$p|q = \Sigma_{i\in I}\alpha_i.(p_i|q) + \Sigma_{j\in I}\beta_j.(p|q_j) + \Sigma_{\alpha_i comp \beta_j}\ \tau.R_{\alpha_i\beta_j}\ [+\Omega, \text{ if } \Omega \text{ is a summand of } p \text{ or } q]$$

Omega

Ω $\Omega \leq p$

τ-laws

N1 $\alpha.p + \alpha.q = \alpha.(\tau.p + \tau.q)$

N2 $p + \tau.q \leq \tau.(p + q)$

N3 $\alpha.p + \tau.(\beta.q + r) = \tau.(\alpha.p + \beta.q + r)$ if $subj(\alpha) = subj(\beta)$ and ($\alpha,\beta \in IA$ or $\alpha,\beta \in OA$)

N4 $\tau.p \leq p$

Inference Rules

α–conv $p \equiv_\alpha q$ *implies* $p = q$

C0 $p \leq q$ *implies*:

a) $\alpha.p \leq \alpha.q$ if $\alpha \notin IA$

b) $p+r \leq q+r$

c) $p|r \leq q|r$

d) $(x)p \leq (x)q$

e) $[x = y]p \leq [x = y]q$

f) $[x \neq y]p \leq [x \neq y]q$

C1 ($p\{z/y\} \leq q\{z/y\}$ for each $z \in \{y\} \cup fv(p) \cup fv(q)$) *implies* $x(y).p \leq x(y).q$

R $p \leq p$

T $p \leq q$ and $q \leq r$ *implies* $p \leq r$

Table 3.1: The proof system F

Definition 3.2 (*Saturation*)
Let $IN = N$, $OUT = \{ \overline{x} \mid x \in N\}$, $Ev = IN \cup OUT$. Ev is said to be the set of all *events* and is ranged over by metavariable "a".
• Given a set of sets of events $L = \{A_1,...,A_n\}$, $Ev(L) = \cup\{A_i \mid 0 \leq i \leq n\}$
• A non-empty finite collection of finite sets of events L is *saturated* if for each $A \subseteq Ev$: $(\exists B \in L \text{ s.t. } B \subseteq A \subseteq Ev(L))$ *implies* $A \in L$. ♦

Definition 3.3 (*Head Normal Forms*)
- An agent p is in *head normal form (hnf)* if one of the following conditions holds:
  i. $p \equiv \Sigma_{a \in A}\, g_1(a)$, with $A \subseteq Ev$ and A finite
  ii. $p \equiv \Sigma_{A \in L}\, \tau.p_A$, with $p_A \equiv \Sigma_{a \in A}\, g_2(a)$ and L saturated

where g_1: $A \to P$ and g_2: $Ev(L) \to P$ are functions such that, for $i \in \{1, 2\}$,
  a) $a \in IN$ implies $g_i(a) = a(y).p_{a(y)}$, and
  b) $a \in OUT$ implies $g_i(a) = \Sigma_{z \in N_a} az.p_{az}$ $[+a(w).p_{a(w)}]$, with $N_a \subseteq N$, N_a finite and N_a empty implies that the optional bound output summand is present.

Functions g_1 and g_2 are called the *associated function* of p.
- If p is in hnf with associated function g:
  • $Ev(p) = domain(g)$;
  • $Init(p) = \{\alpha \in A \mid \exists a \in domain(g)$ s.t. either $\alpha \in IA$ and $g(a) = \alpha.p_\alpha$ or $\alpha \in OA$ and $\alpha.p_\alpha$ is a summand of $g(a)\}$. ♦

The counterpart of head normal form for divergent agents is the notion of *Ω-hnf*.

Definition 3.4 (*Ω-Head Normal Forms*)
- An agent p is in *Ω-head normal form (Ω-hnf)* if $p \equiv (\Sigma_{\alpha \in A}\, \alpha.(p_\alpha + \Omega)) + \Omega$, with A not containing α–equivalent actions and τ's.
- If p is a Ω-hnf
  • $Init(p) = A$;
  • $Ev(p) = \{a \in Ev \mid \exists \alpha \in Init(p)$ s.t. $\alpha = ay$ or $\alpha = a(y)$, for some $y \in N\}$ ♦

From the above definitions, it is evident that for each $\alpha \in Init(p)$ there exists a unique p_α such that $p =\alpha\Rightarrow p_\alpha$. Moreover, if p is a head normal form then the following property holds: given a port $x \in Ev(p)$, every summand p_A in p of the form $\Sigma_{a \in A}\, g(a)$, contains (as summands) *all* or *none* of the terms $\alpha.p_\alpha$, for each $\alpha \in OA \cap Init(p)$ such that $subj(\alpha) = x$.

We establish now a few facts about head normal forms. Their proofs go on along the same lines of the corresponding ones for CCS in [DH84]; additional complications are again introduced by the need of applying rule (C1) and, in the proof of proposition 3.3, by the property above, which requires the application of the new law N3.

Proposition 3.3
If $p\Downarrow$ then there exists a head normal form h(p) such that $F \vdash p = h(p)$. ♦

Proposition 3.4
If $p\Uparrow$, then there exists a Ω-head normal form Ω(p) such that $F \vdash p = \Omega(p)$. ♦

For the proof of completeness we need a special induction parameter, namely, the greatest number of communications that a process can perform with another agent. This parameter can be defined as the "length", in terms of number of visible actions, of the maximal among the observers which are satisfied by the process.

Definition 3.5 *(depth of communication)*
We define the *depth of communication* dpc as: dpc(p)= max{k | p **may** $\alpha_1 \ldots \alpha_k.\omega.NIL$, with the α_i, $0 \le i \le k$, ranging over Λ}. ♦

Theorem 3.2 *(Completeness)*
p «' q implies F |—p ≤ q.
Sketch of proof. The proof goes by induction on dpc(p) and relies on the fact that, in virtue of the above propositions, we can suppose that both p and q are either in hnf or in Ω-hnf and on the two facts below about α-derivatives of p and q:

- if $\alpha \in OA$ then p_α «' q_α ;
- if $\alpha = x(y) \in IA$ then $p_\alpha\sigma$ «' $q_\alpha\sigma$ for any name z and any substitution $\sigma=\{z/y\}$.

By applying the induction hypothesis, the above relations still hold when substituting «' by ≤. Then, rules C0 and C1 and the τ-laws suffice to prove that p ≤ q holds. ♦

4. Denotational Semantics

In this section, we sketch a denotational semantics for π-calculus and discuss its relationships with the operational semantics of Section 2. We assume familiarity with the standard algebraic semantics techniques as described in [Hen88] or [Gue81].

To avoid complications which would not provide any additional insight, we confine ourselves in searching for a model which is fully abstract w.r.t. the *must preorder* $«_M$' (rather than «'). A complete proof system can be obtained for this preorder from the one presented in Section 3 by adding the three laws:

$\Omega+p=\Omega$, $\tau p+\tau q \le p$ and $\alpha p+\beta q \le \alpha p$ if subj(α)=subj(β) and α,β∈ IA. or α,β∈ OA;
we indicate with $\le_M$ the corresponding proof-theoretic preorder.

The proof of completeness for this preorder is very similar to that presented for ≤ in Section 3, except that we can consider *strong normal form* (formally described below) instead of hnf's. By and large, strong normal forms reproduce recursively, at each level, the top structure of head normal forms, but also the law Ω+p=Ω is taken into account; therefore we have a tree structure, in which leaves are represented by NIL (a final blocked state has been reached) or Ω (a divergent state has been reached, and we are uninterested to the rest of computation).

We give the model under the form of a *natural interpretation*, a slight variant of the usual algebraic semantics introduced in [HP80] to deal with languages with explicit value passing.

A natural interpretation for (finite) π-calculus is a 4-tuple <D, F_{op}, inp, out>, where:

- D is a cpo
- F_{op} is a family of continuous functions, one for each operator different from input and output prefixes.
- inp: (N × (N—>D)) —>D is a function continuous in its second argument.
- out: (N × N × D) —>D is a function continuous in its third argument.

Given a natural interpretation, we can define a semantic function: [.]: **P** —> D as follows:

$[op(t_1,...,t_k)] = f_{op}([t_1],...,[t_k])$, for each k-ary operator different from input and output

$[x(y).t] = inp(x, \lambda z \in N.[t\{z/y\}])$

$[\ \overline{x}y.t] = out(x, y, [t])$

Since the considered calculus does not allow us to describe infinite behaviours, one might ask whether cpo's and continuous functions are really needed to model it. To obtain a *fully abstract* model, the partial ordering of the domain we use must properly reflect the testing preorder $\ll_M'$. Now, although we are only dealing with finite agents, we still have elements which are approximable by infinite chains. Consider process p = x(y).NIL; it is (syntactically) finite but can be approximated by the *infinite* chain of processes

$K = \{x(y).t_i \mid t_i = \Sigma_{z \in F_i}[y{=}z]NIL + [y \notin F_i]\Omega, i \geq 0\}$.

where $\{F_i \mid i \geq 0\}$ is a strictly increasing chain of finite set of names such that for each j, $y \notin F_j$. Indeed, we have: $x(y).t_0 \ll_M' x(y).t_1 \ll_M' ... \ll_M' x(y).NIL$. Thus we can expect that every fully abstract model will be based on a cpo rather than simply on a poset; and, in fact, the construction illustrated below naturally leads us to a cpo. It is worth noting that exactly the same cpo is however suitable for modelling the whole calculus with recursion [Bor91].

What makes non-trivial the search for such a cpo, is the proper definition of function inp(.,.,.). We will first define a poset, D', of equivalence classes of *finite strong normal forms* for processes, which can be used to approximate all processes, and easily define over D' the functions f_{op}, out and inp; then, we will extend (by *ideal completion*) D' into a cpo D and f_{op}, out and inp into continuous functions over D, obtaining a natural interpretation. Then it is straightforward to show that this natural interpretation is fully abstract.

Below, we sketch the actual construction which takes four steps:

1. Canonical form (SNF) for all processes are introduced.

Strong normal forms (snf) are defined inductively as follows :

i) Ω is a snf

ii) *(α-snf)*

$\Sigma_{a \in A} g(a)$ is a α-hnf if g:A—>**P** ($A \subseteq Ev$ and A finite) enjoys the following conditions:

a) for each input event $x \in A$, $g(x) = x(y).(\Sigma_{z \in F}[y{=}z]n_{xz} + [y \notin F]n_{x(y)})$, where:

1. $F \subseteq N$ finite, $y \notin F$
2. for each $z \in F$, $y \notin fv(n_{xz})$
3. each of the n_{xz}, $z \in N$, is either Ω or a τ-snf where, for $z \notin F$, n_{xz} is defined as $n_{x(y)}\{z/y\}$

b) for each $\overline{x} \in A$, $g(\overline{x})$ is a non-empty summation $\Sigma_{z \in N_x} \overline{x}z.n_{\overline{x}z}$, which has at most one bound output summand and where each n_{xz} is either Ω or a τ-snf.

iii) *(τ-snf)*

$\Sigma_{A \in L}\tau.\Sigma_{a \in A} g(a)$ is a τ-snf if L is saturated and g: Ev(L) —> **P** enjoys a) and b) above.

A strong normal form is either Ω or an α-snf or a τ-snf.

We call SNF the set of strong normal forms. By induction on dpc(p), it can be shown that each process p is provably equivalent to a snf. Each snf(p) is, up to obvious identifications, *unique*, thus we can think of a function *snf* which associates a snf to each process.

The compact element of our cpo will be equivalence classes of finite strong normal forms (FSNF):

A *finite strong normal form* is a snf n with associated function g, s.t.

a) $n \equiv \Omega$ or

b) $x \in$ domain(g) and $x \in$ IN implies $g(x) = x(y).(\Sigma_{z\in F}[y=z]n_{xz}+[y\notin F]\Omega)$, where: for each $z \in F$, n_{xz} is a fsnf different from Ω.

2. The set of finite strong normal form is turned into a poset.
To guarantee antisymmetry, $FSNF_{=M}$ is defined as the poset given by the set of equivalence classes of FSNF w.r.t. the kernel equivalence of the preorder $\leq_M$ (i.e. $\leq_M \cap \leq_M^{-1}$); the new partial ordering $\leq$ over $FSNF_{=M}$ is defined as follows: $[n] \leq [m]$ iff $n \leq_M m$, the least element is $[\Omega]$. Over $FSNF_{=M}$ a set of functions, corresponding to all operators, is defined.

• To define the operators different from input and output prefixes, this is done via snf; for example, for a restriction (y) we define $f_{(y)}$: $FSNF_{=M} \longrightarrow FSNF_{=M}$ as: $f_{(y)}([n]) = [snf((y)(n))]$.

• To define input prefix, we first define the poset of functions $I = \{f: N \longrightarrow FSNF_{=M} \mid$ for finitely many x, $f(x)\neq[\Omega]\}$, with the partial ordering naturally inherited from $FSNF_{=M}$. Function inp: N x I: $\longrightarrow FSNF_{=M}$ is $inp(x,f) = [x(y).(\Sigma_{z\in F}[y=z]n_z+[y\notin F]\Omega)]$ where y is fresh, $f(z) = [n_z]$ and $F=\{x \mid f(x) \neq [\Omega]\}$.

• Function out: $N x N x FSNF_{=M} \longrightarrow FSNF_{=M}$ is $out(x,y,[n]) = [\,\bar{x}y.n]$.

It is easily checked that the above functions are properly defined and monotonic in $FSNF_{=M}$ or I (in the case of inp).

3. $FSNF_{=M}$ and I are extended to *algebraic* cpo's.
Ideal completion is used to obtain $FSNF_{=M}^{\infty}$ and I^{∞} and the unique continuous extensions of our functions over them. The crucial point here is that I^{∞} is isomorphic to $N \longrightarrow FSNF_{=M}^{\infty}$, thus they can be identified. The wanted domain is then $D = FSNF_{=M}^{\infty}$, this together with the extended functions give a natural interpretation of **P**.

4. Full abstractness of our interpretation with respect to $\ll_M'$ is proven.
The proof relies on the existence, for each p, of a set of approximants App(p) of fsnf (easily definable via snf(p)), such that:

a) if n is a fsnf and $n \leq_M p$ then there exists an $n' \in$ App(p) s.t. $n \leq_M n'$

b) if for each $n \in$ App(p) $n \leq_M q$, then $p \leq_M q$

c) $[App(p)] = \{[n] \mid n \in App(p)\}$ is directed in $FSNF_{=M}^{\infty}$ and $[p] = \sup [App(p)]$.

First show that for each $n \in$ FSNF, [n]=[n]. Thus by exploiting this, a) - c) above and algebraicity of $FSNF_{=M}^{\infty}$, it is straightforward to show that for each p and q: $[p] \leq [q]$ iff $p \leq_M q$, and, hence, iff $p \ll_M' q$.

Conclusions

We conclude with a few considerations about our achievements and contrast them with previous works. We will mainly concentrate on the differences between our equivalence and bisimulation equivalence, on the role of the mismatch operator and on the weakness of our denotational model.

In [MPW89] two bisimulation based equivalences are presented, called *late* and *early* strong ground bisimulation. They differ in that, late bisimulation requires that each bound input transition is simulated by an equipotent input transition, while early bisimulation only requires that for each received name there is a simulating transition. Early strong bisimulation is considered by its proposers closer to the original idea of bisimulation and is in a sense more extensional (coarser). In spite of this, in [MPW89] only an algebraic theory for the late bisimulation is provided and it is pointed out that extra absorption laws are needed for the early bisimulation. There, it is also mentioned that a set of τ-laws similar to those for CCS should suffice to fully characterize weak observational equivalences over π-calculus but again no axiomatization is worked out for it. In our setting the question "early vs. late" does not arise, or (if one prefers) the only possible point of view is the early one. Indeed, it can be shown that within the testing approach the two views do collapse as it is suggested by our full abstractness result which relates the late operational semantics with an early denotational model.

In order to define our testing preorders we have extended the original calculus with a mismatch operator. This operator had been considered by Milner, Parrow and Walker but then abandoned for the sake of minimality. Mismatch is, indeed, unnecessary for the algebraic characterization of the strong late bisimulation; however, it is not clear whether it will be needed for axiomatizing strong early bisimulation and weak bisimulation. In our approach, mismatch is far from redundant; without it, the existence of normal forms would become problematic. Fundamental lemmata in the completeness proof can be established only thanks to the additional testing power provided by the match and mismatch operators. Indeed, in their proof, it is crucial for the observers to be able to distinguish between bound and free outputs. An observer determines whether a process has performed one kind of output by testing whether the received name belongs to a certain set of names: if a process performs a free output, the exported name is in the set of the free variable of the process, while, if the process performs a bound output, it is not. Omitting the mismatch would deeply affect the testing power of the observers and it appears that no "reasonable" testing equivalence could be obtained without it. In fact, consider a new may preorder, $<_m$, originated by observers without mismatch. Consider the two processes x(w).NIL and $\overline{x}$y.NIL. We would like them to be unrelated, and indeed, they are not related by $\ll'$, $\ll_M'$ and $\ll_m'$ because they perform different kinds of outputs. However, since they can perform an output on the same port the new may preorder would relate them, more precisely it can be proved that $\overline{x}$(w).NIL $<_m$ $\overline{x}$y.NIL.

The construction of our denotational model is not completely satisfactory, we would have preferred a more abstract characterization, for example in terms of a *powerdomain*. The main difficulty we encountered was in modelling the restriction operator, and, in particular, in giving a correct account of bound output actions. We have tried to use a kind of *generalized acceptance trees* as in [HI91], but the role bound output would play on such model is not yet clear. An attempt similar to our has been independently developed in [Hen91]. There a term model is presented which is fully abstract w.r.t. a testing preorder. The actual syntax and transitional semantics are slightly different; the operator for internal moves is replaced by a binary internal choice operator and the match-mismatch pair of operators is replaced by an if-then-else. These differences somehow hinder the relationships between the bisimulation and the testing approach and do not permit any assessment of the widely studied original calculus.

References

[BK89] Bergstra,J.A. and Klop,J.-W. Process Theory based on Bisimulation Semantics, in *Linear Time, Branching Time and Partial Orders in Logic and Models for Concurrency*, LNCS, **354**, Springer-Verlag, 1989, 50-122.

[Bor91] Boreale,M. Semantica osservazionale ed assiomatica per un'algebra di processi dinamici, Master Thesis, Università di Pisa, February, 1991.

[DH84] De Nicola,R. and Hennessy,M. Testing Equivalence for Processes, *Theoret. Comput. Sci.*, **34** (1984), 83-133.

[DeN87] De Nicola,R. Extensional Equivalences for Transition Systems, *Acta Informatica*, **24**, 1987, pp. 211-237.

[DIN90] De Nicola,R., Inverardi,P. and Nesi,M. Using Axiomatic Presentation of Behavioural Equivalences for Manipulating CCS Specifications. In *Automatic Verification Methods for Finite State Machines* (J. Sifakis, ed.) LNCS **407**, 1990; pp. 54-67.

[EN86] Engberg,U. and Nielsen,M. A Calculus of Communicating Systems with Label Passing; Int. Rep. Computer Science Dept.., Aarhus University, DAIMI-PB 208, 1986.

[Gue81] Guessarian,I. *Algebraic Semantics*, LNCS, **99**, 1981

[Hen88] Hennessy,M. *An Algebraic Theory of Processes*, MIT Press, Cambridge, 1988.

[Hen91] Hennessy,M. A Model for the π-Calculus, Internal Report No 8/91, Computer Science, University of Sussex, 1991.

[HI91] Hennessy,M. and Ingolfsdottir,A,. A Theory of Testing Equivalence with Value-passing, to appear in *Information and Computation*, 1991

[Hoa85] Hoare,C.A.R. *Communicating Sequential Process*, Prentice Hall Int., London 1985.

[HP80] Hennessy,M. and Plotkin,G. A Term Model for CCS, LNCS, **88**, Springer-Verlag, 1980

[Hui92] Huimin L., Pam: A Process Algebra Manipulator, in *Computer Aided Verification*(K.G. Larsen and A. Skou, eds) LNCS **575**, 1992.

[Mil80] Milner,R. *A Calculus of Communicating Systems*. LNCS, **92**, Springer-Verlag, 1980.

[Mil89] Milner,R. *Communicating and Concurrency*. Prentice Hall Int., London 1989.

[MPW89] Milner,R., Parrow,J. and Walker,D. A Calculus of Mobile Processes part I and II, *LFCS Report Series*, Department of Computer Science, University of Edinburgh, 1989. To appear in *Information and Computation.*

[Plo81] Plotkin, G. A Structural Approach to Operational Semantics. Technical Report Computer Science Department, Aarhus University, DAIMI FN-19; 1981).

[Tho91] Thomsen,B A Calculus of Higher Order, Proc. of POPL 1989; ACM 1989; pp. 143-154.

Testing Equivalence for Petri Nets with Action Refinement: Preliminary Report

Lalita Jategaonkar* and Albert Meyer**

MIT Laboratory for Computer Science
Cambridge, MA 02139

Abstract. A definition of "action refinement" is given for an operational model of concurrent processes based on safe Petri Nets, generalizing previous work of Vogler and van Glabbeek/Goltz. A failure-style denotational semantics is described for process nets. The semantics is fully abstract for Hennessy Testing-equivalence on nets acting as refinement *operators* as well as operands. The semantics embodies the notions of deadlock, failures and divergences found in the Hoare/CSP and Hennessy Testing-equivalence theories, as well as some of the basic ideas of "pomset runs" and "causal" partial orders of Net theory.

1 Introduction

The operation of refining atomic actions in a concurrent process suggests aspects of top-down "modular" development and also requires use of some sort of "true" concurrency process model [1, 2, 3, 5, 8, 12, 13, 14, 17]. In a previous paper [10], we developed a semantics, $[\![\cdot]\!]^{\text{MUST}}$, for certain simple "splitting" and "choice" action refinements on a Petri Net model of processes. Our semantics generalizes and simplifies a similar semantics developed in a seminal paper by Vogler [17]. These semantics support full process theories involving parallel process communication, deadlock, failures, hidden actions, and divergences (*cf.* [4, 7, 9, 11]). Their essential component consists of pomsets paired with failure sets. We observed that Vogler's semantics can equivalently be described as the restriction, $[\![\cdot]\!]^{\text{MUST}}_{\text{intvl}}$, of our general pomset semantics to interval pomsets.

Definition 1. A semantics, $[\![\cdot]\!]$, assigning to any process, N, a meaning, $[\![N]\!]$, is *compositional* for an operator on processes if semantic equality is a congruence for the operator, *i.e.*, the operator preserves semantic equality. We say that a semantics is *adequate* for an equivalence on processes if semantic equality implies process equivalence. Finally, we say that a semantics is *fully abstract* for a process equivalence *with respect to* a set of operators if the semantics is adequate for the equivalence and semantic equality is the coarsest congruence for those operators.

* Supported by the AT&T GRPW Fellowship, NSF Grant No. 8511190–DCR and ONR grant No. N00014–83–K–0125.

** Supported by NSF Grant No. 8511190–DCR and ONR grant No. N00014–83–K–0125.

Both our and Vogler's semantics are adequate for MUST-equivalence [7] of nets and are compositional for splitting and choice refinements as well as for net operations corresponding to the familiar CCS/CSP operations. It follows from [17] that $[\![\cdot]\!]^{\text{MUST}}_{\text{intvl}}$ is, in fact, *fully abstract* for MUST-equivalence with respect to splitting and choice refinements.

Vogler generalizes splitting and choice refinements to allow a large class of refinement operators corresponding to a class of "refinement nets" required to satisfy some rather technical structural and behavioral conditions [17]. Both $[\![\cdot]\!]^{\text{MUST}}$ and $[\![\cdot]\!]^{\text{MUST}}_{\text{intvl}}$ semantics are compositional with respect to each of the operators corresponding to Vogler's refinement nets. Namely, if two nets are equivalent under these semantics, then applying the *same* action refinement ρ to each of them yields semantically equivalent nets.

However, it is *not* the case that these semantics are compositional for nets as action refinement *operators*. For example, the nets a and $\tau.a$, where τ is the hidden action, are semantically equivalent as operands or *targets* of action refinement, but they behave differently when used as operators refining an action b, *viz.*,

$$[\![a]\!] = [\![\tau.a]\!],$$

but

$$[\![(b+c)[b{:=}a]]\!] \neq [\![(b+c)[b{:=}\tau.a]]\!].$$

In this paper, we resolve this problem by establishing that a surprisingly simple variation of the earlier semantics yields MUST-adequate semantics that are compositional for nets as *targets and operators* of action refinement, with the modified $[\![\cdot]\!]^{\text{MUST}}_{\text{intvl}}$ semantics being fully abstract. We similarly show how to handle MAY-equivalence; the MAY- and MUST-semantics together provide a fully abstract semantics for Testing Equivalence [7].

We begin by presenting in Section 2 our general class of *Well Terminating (WT) Nets*. These are possibly infinite, safe nets with designated transitions for signaling successful termination. We then give a definition of action refinement that allows *any* WT net to be used as a refinement operator. Our class of nets and refinements generalizes those of Vogler and van Glabbeek/Goltz [14], since both their target and refinement nets, and indeed arbitrary safe nets, can be understood as special cases of WT nets.

In addition to action refinement, we indicate how to define net operators corresponding to familiar CCS/CSP operations of prefixing ($a.$), restriction ($\backslash a$), hiding ($-a$), renaming, CSP-style sequencing (;), CSP-style parallel-composition-with-synchronization ($\|_L$), CCS-style parallel-composition-with-hiding ($|$), internal choice, and CCS-style choice ($+$).

Section 3 illustrates in some more detail the difficulties with compositionality encountered by our earlier semantics [10], and describes our modification to the semantics that repairs these problems. A discussion of other results, related work and future work is given in Section 4. Finally, in order to keep this paper self-contained, the Appendix provides the definitions of our earlier semantics [10].

2 Nets and Operations

Our class of "Well-Terminating" Nets is related to the class of CSP processes that signal successful termination by performing a distinguished action, $\surd$. In a similar manner, our well-terminating nets signal successful termination by firing any transition labeled with $\surd$. In order to ensure that the net has actually terminated, we require that all places in the net be unmarked after any $\surd$-labeled transition fires.

We wish to restrict our attention to nets with "computable behavior," and we thus impose some syntactic conditions that guarantee finite-markings and finite-branching.

Definition 2. The class of *Well-Terminating (WT) Nets* is the class of labeled, safe, possibly infinite Petri nets that satisfy the following properties:

- The initial marking is finite.
- All places have finite out-degree.
- All transitions have finite in-degree and finite out-degree.
- All places of the net are unmarked immediately after any $\surd$-labeled transition fires. This condition must be satisfied in every reachable marking of the net.

We note that the first three conditions together imply that all reachable markings are finite and that only a finite number of transitions are enabled under any reachable marking. In particular, our nets have only finite concurrency.

The condition on the $\surd$-transitions ensures that no transition (not even a $\surd$-transition) can be fired concurrently with, or following, a $\surd$-transition.

We assume for expository simplicity that all transitions have non-empty presets, and that the initial marking is non-empty. Formally, we write a WT net N as a triple $\langle S_N, T_N, Start_N\rangle$, where S_N is the set of places, T_N is the set of transitions, and $Start_N$ is the (finite) set of initially marked places. Furthermore, for every transition $t \in T_N$, we write $l_N(t)$ to refer to its label, and $pre_N(t)$ and $post_N(t)$ to refer to its preset and post-set. Similarly, for every place $s \in S_N$, we write $pre_N(s)$ and $post_N(s)$ to refer to its preset and post-set. Our syntactic conditions on nets imply that all places have finite postsets and that all transitions have finite presets and finite postsets.

We can understand the target nets of [14, 17] to be a special case of WT nets having no $\surd$-labeled transitions. The connection between WT nets and the refinement nets of [14, 17] is slightly more subtle, since the latter signal successful termination by marking some designated "accept" places rather than by firing some designated transitions. However, these nets can be easily understood as WT nets by adding a $\surd$-transition whose preset is essentially this set of "accept" places. (For technical reasons, some of the initially marked places of their refinement nets must also feed into this $\surd$-transition; we omit the details here.)

Our $\surd$-labeled transitions serve to distinguish deadlock from successful termination. We say that a net *successfully terminates* when a $\surd$-labeled transition fires, while a net is *deadlocked* exactly when no transition is enabled. We write *Succ* to denote the WT net which must immediately successfully terminate, *i.e.*, exactly one transition is enabled under its initial marking and this transition is $\surd$-labeled. For notational convenience, we simply write a to refer to the net $a.Succ$. Furthermore, we write *Dead* to denote the deadlocked process, and $a.Dead$ to refer to the net that does an a and then deadlocks.

The $\surd$-action plays a distinguished role in our theory, and we forbid prefixing with $\surd$ and renaming of other actions to $\surd$. We also forbid refinement of $\surd$-labeled transitions.

Our sequencing operator $N_1; N_2$ makes critical use of the $\surd$-transitions of N_1 by relabeling them with τ and using them as a hidden (τ-labeled) signal to transfer control to N_2. We illustrate the definition of "sequencing" through the following simple example. Suppose that we are given the WT nets $N_1 = a_1 + a_2$ and $N_2 = b_1 \|_\emptyset b_2$ of Figure 1, and we want to define $N_1; N_2$. We want the firing of either of the $\surd$-transitions of N_1 to be a hidden signal that enables both b_1 and b_2 to fire concurrently. Therefore, we relabel the $\surd$-transitions of N_1 to τ, and then have both of these τ-transitions feed into both of the start places of N_2. The resulting net $N_1; N_2$ is given in Figure 1.

We also illustrate the definition of $+$ for N_1 and N_2 of Figure 1. Clearly, we want to introduce conflicts between the a_i and the b_j but preserve the concurrency within the b_j, and so we do a simple cross product construction on the start places of both nets. We note that this causes all the $\surd$-labeled transitions to be in conflict, as desired. The resulting net is also given in Figure 1. As discussed in [15], one technical complication arises due to initially marked places that have incoming transitions, and in general, we apply a ***start-unwinding*** operator on nets [15] before doing the above construction.

We also have a CSP-style parallel composition operator $\|_L$ on WT nets, where two nets are placed in parallel, but must synchronize on all actions in the set $L \cup \{\surd\}$, where L is a set of visible labels. Our definition is essentially the same as [17] and is omitted here. Similar to [6], we also have a CCS-style parallel composition operator $|$, where two nets are placed in parallel and are allowed to perform hidden synchronizations on all complementary actions; again, they must (visibly) synchronize on the $\surd$ action. The net operators for prefixing, restriction, hiding, renaming, and internal choice are straightforward, and are omitted. The class of WT nets is closed under all of these operations.

Except for the parallel composition operators, all of our net operations are closely related to the corresponding CCS/CSP operators on labeled transition systems (*lts*'s). In particular,

Lemma 3. *For our CCS/CSP WT net-operators other than $\|_L$ and $|$, the lts of the constructed net is strongly bisimilar to the lts obtained by applying the corresponding CCS/CSP lts-operator to the lts's of the component nets. Also, lts($N_1 \|_L N_2$) is strongly bisimilar to lts(N_1)$\|_{L\cup\{\surd\}}$lts(N_2).*

The relationship between our WT net operator, $|$, and the corresponding CCS *lts* operator is slightly more complicated, but similar.

Two simple WT net operators play a significant role in our technical development. Namely, ***split refinements*** ($split_{(a,a_1,a_2)}$) replace every a-labeled by two consecutive transitions labeled a_1 and a_2, and ***choice refinements*** ($choice_{(a,a_1,a_2)}$) replace every a-labeled transition by two conflicting transitions labeled a^L and a^R. Figure 2 gives examples of these kinds of refinements.

The major new WT net operator we develop is ***action refinement***. Our action refinement operator has a rich algebraic theory. For example, the following simple

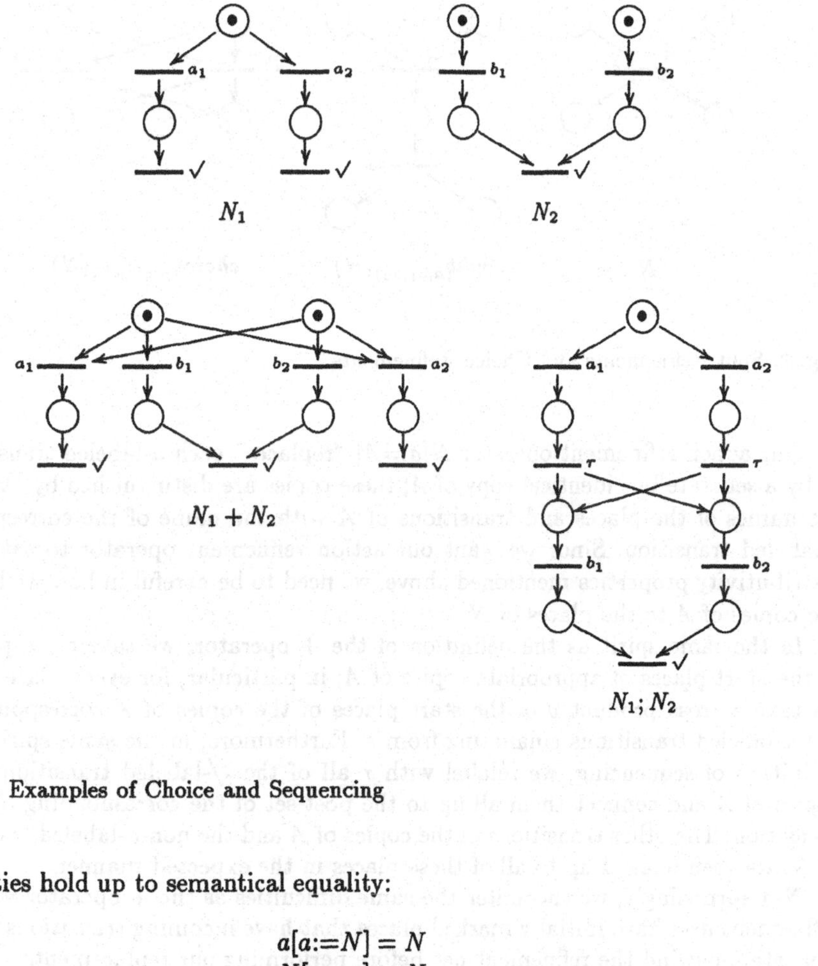

Fig. 1. Examples of Choice and Sequencing

identities hold up to semantical equality:

$$
\begin{aligned}
a[a{:=}N] &= N \\
N[a{:=}a] &= N \\
N[a{:=}\mathit{Dead}] &= N \backslash a \\
N[a{:=}\tau] &= N[a{:=}\mathit{Succ}] = N - a \\
\mathit{split}_{(a,a_1,a_2)}(N) &= N[a{:=}a_1.a_2] \\
\mathit{choice}_{(a,a^L,a^R)}(N) &= N[a{:=}a^L + a^R]
\end{aligned}
$$

Assuming that a and b are "fresh" labels, we also have:

$$
\begin{aligned}
((a+b)[a{:=}N_1])[b{:=}N_2] &= N_1 + N_2 \\
((a.b)[a{:=}N_1])[b{:=}N_2] &= N_1; N_2 \\
((a\|_\emptyset b)[a{:=}N_1])[b{:=}N_2] &= N_1 \|_\emptyset N_2
\end{aligned}
$$

For all refinements ρ, the following distributivity properties hold:

$$
\begin{aligned}
(N_1 + N_2)\rho &= N_1\rho + N_2\rho \\
(N_1; N_2)\rho &= N_1\rho; N_2\rho \\
(N_1 \|_\emptyset N_2)\rho &= N_1\rho \|_\emptyset N_2\rho
\end{aligned}
$$

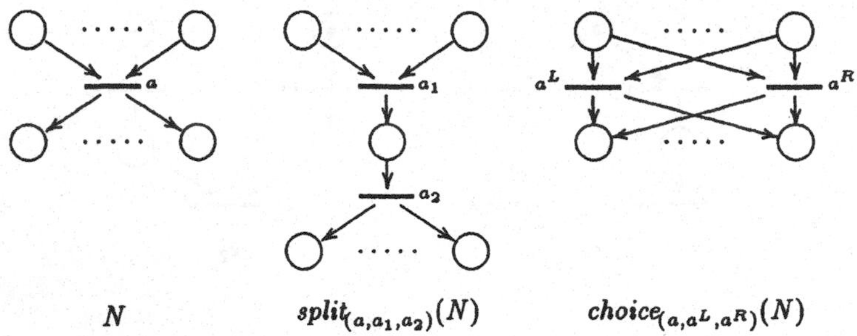

Fig. 2. Split Refinements and Choice Refinements

Our action refinement operator $N[a{:=}A]$ "replaces" each a-labeled transition in N by a separate but identical copy of A; these copies are distinguished by "tagging" the names of the places and transitions of A with the name of the corresponding a-labeled transition. Since we want our action refinement operator to satisfy the distributivity properties mentioned above, we need to be careful in how we hook up the copies of A to the places of N.

In the same spirit as the definition of the + operator, we take cross products of the start places of appropriate copies of A; in particular, for every place s in N, we take a cross product v of the start places of the copies of A corresponding to the a-labeled transitions emanating from s. Furthermore, in the same spirit as the definition of sequencing, we relabel with τ all of the $\surd$-labeled transitions of the copies of A and connect them all up to the post-set of the corresponding a-labeled transition. The other transitions of the copies of A and the non-a-labeled transitions of N are then hooked up to all of these places in the expected manner.

Not surprisingly, we encounter the same difficulties as the + operator when our refinement nets have initially marked places that have incoming transitions, and we thus start-unwind the refinement net before performing our replacements.

We now define the action refinement operator. For simplicity we assume that the refinement net A is already start-unwound; otherwise, we first start-unwind A and then carry out this construction using the start-unwound version of A rather than A itself.

Definition 4. Let N and A be WT Nets, and let a be a label other than $\surd$. Then $P = N[a{:=}A]$ is defined as:

$$\begin{aligned} S_P = \{(s,v) \mid s \in S_N \text{ and } v\colon T \rightarrow Start_A, \text{ where } T = \{t \in post_N(s) \mid l_N(t) = a\}\ \} \\ \cup\ \ \{(t,s') \mid t \in T_N,\ l_N(t) = a \text{ and } s' \in S_A - Start_A\} \end{aligned}$$

$$T_P = \{(t,*) \mid t \in T_N \text{ and } l_N(t) \neq a\} \cup \{(t,t') \mid t \in T_N,\ l_N(t) = a \text{ and } t' \in T_A\}$$

$$pre_P((t,*)) = \{(s,v) \in S_P \mid s \in pre_N(t)\}$$

$$post_P((t,*)) = \{(s,v) \in S_P \mid s \in post_N(t)\}$$

$$l_P((t,*)) = l_N(t)$$

$$pre_P((t,t')) = \{(s,v) \in S_P \mid s \in pre_N(t) \text{ and } v(t) \in pre_A(t')\} \cup \{(t,s') \in S_P \mid s' \in pre_A(t')\}$$

$$post_P((t,t')) = \begin{cases} \{(t,s') \in S_P \mid s' \in post_A(t')\} & \text{if } l_A(t') \neq \surd \\ \{(s,v) \in S_P \mid s \in post_N(t)\} & \text{otherwise} \end{cases}$$

$$l_P((t,t')) = \begin{cases} l_A(t') & \text{if } l_A(t') \neq \surd \\ \tau & \text{otherwise} \end{cases}$$

$$Start_P = \{(s,v) \in S_P \mid s \in Start_N\}$$

Our definition of refinement generalizes the definitions of refinement given by Vogler and van Glabbeek/Goltz in the sense that our refined net is semantically equivalent to their nets. In fact, there is an even tighter relationship between them, namely, these nets are equivalent up to a weaker form of history-preserving bisimulation [13] which treats τ-moves as hidden and respects concurrent divergences. We omit the definition here since it is not necessary in the development below.[3]

We note that our definition of action refinement preserves finiteness of nets, and thus, in the same spirit as our full class of WT nets, we can allow arbitrary finite WT nets to function as both target nets and refinement nets. The class of finite WT nets is also closed under all of the CCS/CSP operations discussed previously.

3 The Semantics

We presume that the reader is familiar with the theory of MUST-equivalence on labeled transition systems developed in [9, 7]. The definition of MUST-equivalence carries over directly to WT nets: two WT nets will be said to be MUST-equivalent iff their labeled transition systems are MUST-equivalent under the standard definition.

As mentioned in the introduction, all of the semantics of [10] (*cf.*, the Appendix) are compositional for WT nets as *targets* of action refinement. Namely, if two nets are semantically equivalent, then applying the *same* action refinement ρ to both of them yields semantically equivalent nets.

Theorem 5. *$[\![\cdot]\!]^{\text{MUST}}$ and $[\![\cdot]\!]^{\text{MUST}}_{\text{intvl}}$ on WT nets are adequate for* MUST*-equivalence and compositional for split refinements, choice refinements, and our CCS/CSP operators; likewise for $[\![\cdot]\!]^{\text{MAY}}$ and $[\![\cdot]\!]^{\text{MAY}}_{\text{intvl}}$. Furthermore, all four of these semantics are compositional for WT nets as* targets *of action refinement.*

[3] Since Vogler and van Glabbeek/Goltz use a cross-product construction on the "accept" places of their refinement nets rather than using τ-moves to transfer control back to the target net, our refined net is not quite *strongly* history-preserving bisimilar to their nets.

However, as explained in the introduction, neither $[\![\cdot]\!]^{\text{MUST}}$ nor $[\![\cdot]\!]^{\text{MUST}}_{\text{intvl}}$ are compositional as *operators* of action refinement. The problem with both semantics is improper accounting of "initial" τ-moves. For example, as indicated in the introduction, $[\![a]\!]^{\text{MUST}} = [\![\tau.a]\!]^{\text{MUST}}$, but the net $(b+c)[b{:=}a]$ must offer a c, while $(b+c)[b{:=}\tau.a]$ would not offer a c after a hidden move. Hence, these refined nets are MUST-distinguishable and so have different meanings.

Rather surprisingly, this is the *only* problem with the semantics, even for WT nets that can diverge. In contrast to WT nets, another problem arises with the semantics when applied to Vogler's refinement nets; namely, for these nets, the semantics does not properly detect successful termination. In particular, although Vogler's refinement nets signal successful termination by marking some designated "accept" places, the semantics is not tuned to detect whether these accept places are marked.

For example, $[\![a]\!]^{\text{MUST}} = [\![a.Dead]\!]^{\text{MUST}}$, where a is the Vogler refinement net that fires an a-transition and then successfully terminates (by marking its accept places); however, $(a;b)$ is trivially MUST-distinguishable from $((a.Dead);b)$. The net $a.Dead$ would be disallowed as a refinement net by [17], but the problem occurs even for non-deadlocking refinement nets. For example, let $N_3 = a + ab + a(b+c)$ and $N_4 = a + a(b+c)$. Then N_3 and N_4 are refinement nets in the sense of [17], and $[\![N_3]\!]^{\text{MUST}} = [\![N_4]\!]^{\text{MUST}}$. However, $(N_3;c)$ and $(N_4;c)$ are MUST-distinguishable, since $(N_3;c)$ can do an a—the "middle" one—and then refuse to do the action c, while $(N_4;c)$ is ready to do the action c after doing any a.

However, since WT nets signal successful termination by firing $\surd$-transitions (which are visible), the $[\![\cdot]\!]^{\text{MUST}}$ semantics applied to WT nets *does* properly detect successful termination. Thus, the semantics of WT nets as refinement operators can be captured by simply +'ing the net with a fresh, distinguished action γ, and taking the semantics of the resulting net. In fact, the MAY-semantics $[\![\cdot]\!]^{\text{MAY}}$ and $[\![\cdot]\!]^{\text{MAY}}_{\text{intvl}}$ of [10] are already compositional *for WT nets.*

Definition 6. For a WT net N,

$$[\![N]\!]^{\text{MUST}}_{\gamma} =_{\text{def}} [\![\gamma + N]\!]^{\text{MUST}}$$
$$[\![N]\!]^{\text{MUST}}_{\text{intvl-}\gamma} =_{\text{def}} [\![\gamma + N]\!]^{\text{MUST}}_{\text{intvl}}$$

Theorem 7. *$[\![\cdot]\!]^{\text{MUST}}_{\gamma}$ and $[\![\cdot]\!]^{\text{MUST}}_{\text{intvl-}\gamma}$ on WT nets are adequate for MUST-equivalence and compositional for split refinements, choice refinements, and our CCS/CSP operators; likewise for $[\![\cdot]\!]^{\text{MAY}}$ and $[\![\cdot]\!]^{\text{MAY}}_{\text{intvl}}$. Furthermore, all four of these semantics are compositional for WT nets as targets* and operators *of action refinement.*

Using the information provided by the γ transition as well as that provided by the $\surd$-transitions of the original net, it is fairly straightforward to prove these results for the $[\![\cdot]\!]^{\text{MUST}}_{\gamma}$ semantics and for both of the MAY-semantics. However, the fact that every interval pomset-divergence contains information about only a *single* divergence makes it much more difficult to reason compositionally about the $[\![\cdot]\!]^{\text{MUST}}_{\text{intvl-}\gamma}$ semantics when *concurrent* transitions are refined with nets that can *diverge*. Since our semantics "blurs" all information that "extends" a pomset-divergence, we have to be careful that combining such "blurred" information about these divergent nets does not somehow make finer distinctions based on concurrent divergences (*cf.* [10]).

As promised, the full abstraction properties hold of the interval semantics.

Theorem 8. *The $[\![\cdot]\!]^{\text{MUST}}_{\text{intvl-}\gamma}$ semantics on WT nets is fully abstract for* MUST*-equivalence with respect to the set of operators consisting of split refinements, choice refinements, and +; likewise for the $[\![\cdot]\!]^{\text{MAY}}_{\text{intvl}}$ except that + is not needed.*

As in [12, 17], the proof of full abstraction is based on the well-known result that every interval ordering is order-isomorphic to a set of intervals of the real line with (interval x) $<$ (interval y) iff every point in x is less than every point in y. Since split refinements and choice refinements allow us to associate unique beginnings and endings with all of the transition firings, we can use certain *sequential* observations of these beginnings and endings to fully determine the ordering of the pomset-failures and pomset-divergences in our $[\![\cdot]\!]^{\text{MUST}}_{\text{intvl-}\gamma}$ semantics. Using the + operator appropriately, we can then show that all of the semantical distinctions can be detected by *sequential* experiments.

These ideas are crystallized in the following lemma. For the sake of simplicity, we assume here that our action alphabet is finite; however, our results easily extend to infinite alphabets in the same manner as [17].

Lemma 9. *Let σ be a sequence of split refinements, one for each action in the alphabet, and for all $k \geq 1$, let ρ_k be a sequence of choice refinements mapping each action a in the alphabet to the net $a^1 + \ldots + a^k$. If N_1 and N_2 are WT nets with $[\![N_1]\!]^{\text{MUST}}_{\text{intvl-}\gamma} \neq [\![N_2]\!]^{\text{MUST}}_{\text{intvl-}\gamma}$, then there is some integer k bounded by the maximum concurrency of N_1 and N_2 (which may be infinite), such that the net $(\gamma + N_1\rho_k\sigma)$ is* MUST*-distinct from the net $(\gamma + N_2\rho_k\sigma)$.*

An analogous result holds for the $[\![\cdot]\!]^{\text{MAY}}_{\text{intvl}}$ semantics.

We now restrict our attention to *finite* WT nets. Since all finite WT nets have bounded concurrency, we can use Theorem 7 and Lemma 9 to reduce the decidability of $[\![\cdot]\!]^{\text{MUST}}_{\text{intvl-}\gamma}$-equivalence and $[\![\cdot]\!]^{\text{MAY}}_{\text{intvl}}$-equivalence for finite WT nets to the decidability of MUST-equivalence and MAY-equivalence for finite safe nets. As illustrated by Vogler [16], there are simple automata-theoretic arguments that reduce these latter problems to the equality of regular languages, which is known to be decidable. We thus have:

Corollary 10. *For finite WT nets, $[\![\cdot]\!]^{\text{MUST}}_{\text{intvl-}\gamma}$-equivalence and $[\![\cdot]\!]^{\text{MAY}}_{\text{intvl}}$-equivalence are decidable.*

We remark that Vogler [18] obtained this result for the $[\![\cdot]\!]^{\text{MUST}}_{\text{intvl}}$ semantics by using a ST-representation of interval pomsets, which he argued was more convenient for the purpose than pomsets. Our reduction of interval pomset-failures to ordinary failures via Lemma 9 indicates that this alternative representation need not be introduced.

4 Other Results, Related Work, and Future Work

All of our semantics are, in fact, fully abstract for MAY- or MUST-*approximation*, where MAY-semantics are partially ordered by set-theoretic containment, and MUST-semantics are partially ordered by component-wise reverse containment. As usual, the conjunction of MAY and MUST semantical equality corresponds to full Testing congruence [7]. It is easy to show that our decidability results extend to testing-approximation.

We expect that all of our semantical spaces form continuous partial orders, and that our action refinement and CCS/CSP operators on nets correspond to continuous semantical operations. Consequently, we expect that our theory will routinely support *arbitrary* (not merely guarded) *recursive* definitions of nets, with recursion understood as usual via least fixed points.

An important direction for further research is development of the algebra of process terms with refinement. One immediate problem to consider is finding a complete axiom system for equations between closed recursion-free CSP/CCS process terms—corresponding to the (non-divergent) isolated elements in our semantical spaces.

There is not yet a consensus on what an action refinement operator should be. For example, the action refinement operator of [3, 8, 14, 17] contrasts with the one used in [1, 2], since the operators of [3, 8, 14, 17] distribute over CCS-choice but not over a Hoare/Hennessy external choice operator, $+_H$, while the operators of [1, 2] distribute over $+_H$ but not over CCS-choice. While our $[\![\cdot]\!]^{\mathbf{MUST}}_{\mathbf{intvl}}$ semantics is compositional for all our CSP operations, including $+_H$, it is not compositional for CCS-choice, and therefore not compositional for nets as action refinement operators. However, we can define a modified action refinement operator that is tuned to $+_H$ and for which $[\![\cdot]\!]^{\mathbf{MUST}}_{\mathbf{intvl}}$ is fully abstract. We believe this $[\![\cdot]\!]^{\mathbf{MUST}}_{\mathbf{intvl}}$ semantics subsumes that of Aceto/Engberg [1], who give a fully abstract failure semantics for action refinement in a restricted framework without parallel synchronization, a restriction operator, or divergence.

A more significant contrast in approaches to action refinement is that our action refinement operator and that of [17] are tuned to a CSP-style synchronization-with-restriction, while those of [3, 8] are tuned to a CCS-style synchronization-by-hiding-complementary-actions. In this regard, an action-refinement theory closely related to ours has been proposed by Hennessy [8]. His theory incorporates an interesting, and in certain respects more powerful, action refinement operation, and he has compositionality and full abstraction results similar to ours. Unlike our action refinement operation, Hennessy's definition allows "concurrent" refinement nets to "communicate" with one another in a manner closely related to CCS-style parallel composition, where concurrent, complementary actions (*i.e.*, a and $\bar{a}$) can synchronize and perform a hidden move. However, in order for Hennessy's semantics to remain compositional for this powerful sort of action refinement, this inter-communication must in fact be quite restricted: in particular, "initial" hidden communications between refinement nets must be disallowed. As a result, Hennessy forbids some simple action refinements like $(a \mid b)[a{:=}c,\ b{:=}\bar{c}]$. The connection between Hennessy's and our theories of action refinement will be the topic of a paper now in progress.

Acknowledgments

We are grateful to Rob van Glabbeek, Ursula Goltz, Matthew Hennessy, Wolfgang Reisig, Boris Trakhtenbrot, Frits Vaandrager, Walter Vogler, and David Wald for helpful discussions. We thank Roberto Segala for proofreading previous versions of this paper.

References

1. L. Aceto and U. Engberg. Failure semantics for a simple process language with refinement. Technical report, INRIA, Sophia-Antipolis, 1991.
2. L. Aceto and M. Hennessy. Towards action-refinement in process algebras. In *Proceedings of 4th LICS*, pages 138–145. IEEE Computer Society Press, 1989.
3. L. Aceto and M. Hennessy. Adding action refinement to a finite process algebra. In *Proceedings of 18th ICALP*, volume 510 of *Lecture Notes in Computer Science*. Springer-Verlag, 1991.
4. S. D. Brookes and A. W. Roscoe. An improved failures model for communicating processes. In *Seminar on Concurrency*, volume 197 of *Lecture Notes in Computer Science*, pages 281–305. Springer-Verlag, 1984.
5. L. Castellano, G. De Michelis, and L. Pomello. Concurrency vs. interleaving: an instructive example. *Bull. Europ. Assoc. Theoretical Computer Sci.*, 31:12–15, 1987.
6. U. Goltz. CCS and petri nets. Technical report, GMD, July 1990.
7. M. C. Hennessy. *Algebraic Theory of Processes*. Series on Foundations of Computing. MIT Press, 1988. 272 pp.
8. M. C. Hennessy. Concurrent testing of processes. In *Proceedings of 3rd CONCUR*, 1992. Appears in this volume.
9. C. A. R. Hoare. *Communicating Sequential Processes*. Series in Computer Science. Prentice-Hall, Inc., 1985. 256 pp.
10. L. Jategaonkar and A. R. Meyer. Testing equivalence for Petri nets with split and choice refinements. Paper presented at the Eighth Workshop on the Mathematical Foundations of Programming Semantics, Oxford, England, Apr. 1992.
11. R. Milner. *Communication and Concurrency*. Series in Computer Science. Prentice-Hall, Inc., 1989.
12. M. Nielsen, U. Engberg, and K. S. Larsen. Fully abstract models for a process language with refinement. In *Linear Time, Branching Time and Partial Order in Logics and Models for Concurrency*, volume 354 of *Lecture Notes in Computer Science*, pages 523–548. Springer-Verlag, 1988.
13. R. van Glabbeek. *Comparative Concurrency Semantics and Refinement of Actions*. PhD thesis, CWI, 1990.
14. R. van Glabbeek and U. Goltz. Refinement of actions in causality based models. In *Stepwise Refinement of Distributed Systems: Models, Formalisms, Correctness*, volume 430 of *Lecture Notes in Computer Science*, pages 267–300. Springer-Verlag, 1990.
15. R. van Glabbeek and F. Vaandrager. Petri net models for algebraic theories of concurrency. In *Proceedings of PARLE Conference*, volume 259 of *Lecture Notes in Computer Science*, pages 224–242. Springer-Verlag, 1987.
16. W. Vogler. Failure semantics and deadlocking of modular petri nets. *Acta Informatica*, 26(4):333–348, 1989.
17. W. Vogler. Failures semantics based on interval semiwords is a congruence for refinement. *Distributed Computing*, 4:139–162, 1991.
18. W. Vogler. Is partial order semantics necessary for action refinement? Technical report, Technische Universitat Munchen, 1991.

A Appendix

In order to keep this paper self-contained, this appendix provides the definitions of our semantics [10] for safe nets. The main idea is that we first simultaneously "split" every *visible* transition t into two consecutive transitions labeled a_1 and a_2, where a is the label of t. We leave all τ-labeled transitions unsplit. We then straightforwardly extract the "pomset-failures" and "pomset-divergences" of the split net, perform some closure operations, and then restrict to interval pomsets.

We begin with the standard notions of pomsets and pomset runs of safe nets:

Definition 11. A *pomset* is a labeled partial order. The *pomset runs* of a net N are pomsets whose elements, called "events," are occurrences of transitions of N, labeled with the labels of the corresponding transitions, and partially ordered by the usual "causal" ordering defined on firings of transitions of N [15], *cf.* Figure 3.

Since pomset runs of nets may contain τ-labeled events which are unobservable, we define an operation, *visible*, on pomset runs which keeps only the visible events of the pomset run.

Definition 12. Let q be a pomset run of a net N. Then $visible(q)$ is the restriction of q to its events with visible labels. The *pomset-traces* of N are the set of $visible(q)$ such that q is a finite pomset run of N, *cf.* Figure 3.

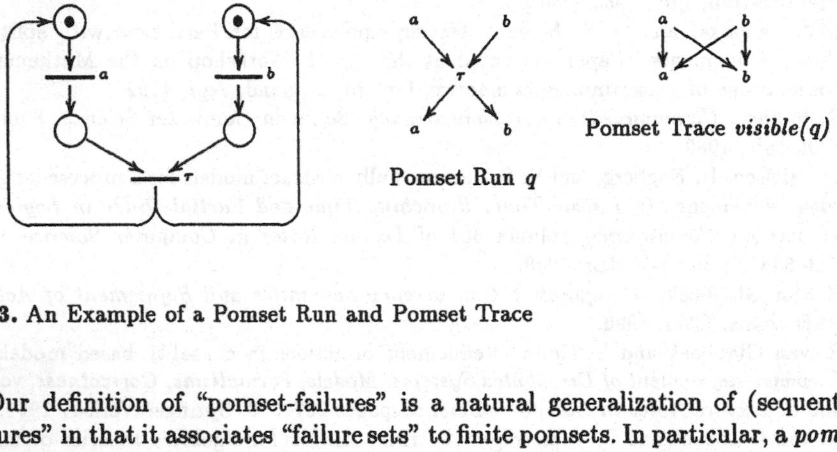

Fig. 3. An Example of a Pomset Run and Pomset Trace

Our definition of "pomset-failures" is a natural generalization of (sequential) "failures" in that it associates "failure sets" to finite pomsets. In particular, a *pomset-failure* is a pair $\langle p, F \rangle$, where p is a finite pomset with visible labels, and F is a set of visible labels.

It is well known that there is a uniquely determined final marking associated with each finite pomset run of a net; this is the marking reached after sequentially firing the events of the run in any order that is consistent with its partial order. Now, for any marking which is not divergent, we define a "failure set" of that marking as any set of *visible* actions that are not enabled under that final marking, even after further firing any finite sequence of τ-labeled transitions. (Note that this is exactly the standard definition of "failure sets" on the labeled transition system whose states are the reachable markings of the net and whose labeled transitions correspond to firings of single net transitions.)

Definition 13. The *pomset-failures* of a net N are the pairs $\langle visible(q), F \rangle$ such that q is a finite pomset run of N and F is a failure set of the marking after q.

We also wish to define a notion of "pomset-divergences" that is a natural generalization of (sequential) "divergences."

Definition 14. The *downward closure*, $\widehat{e}$, of an element e in a pomset p is $\{e' \in p \mid e' \leq_p e\}$. The downward closure, $\widehat{E}$, of a subset E of (the carrier of) a pomset p is $\bigcup\{\widehat{e} \mid e \in E\}$; E is *downward closed* iff $E = \widehat{E}$. A *pomset-divergence* is a pair $\langle p, \mathcal{D}\rangle$, where p is a finite pomset and $\mathcal{D}$ is a non-empty set of downward closed subsets of (the carrier of) p.

Given any pomset run with only a finite number of visible events, it is easy to see that any infinite chain of τ-labeled-events indicates a divergence of the net. We wish to define pomset-divergences of nets in such a way that we keep track of all the *concurrent divergences* within a pomset run while abstracting away from the τ-labeled events.

Definition 15. Let q be an infinite pomset run of a net N with a finite number of visible events. Let $\mathcal{D}$ be the family of sets of the form (carrier of) $visible(\widehat{C})$ such that C is an infinite chain of τ-labeled events of q. Then $\langle visible(q), \mathcal{D}\rangle$ is a *pomset-divergence* of N, *cf.* Figure 4.

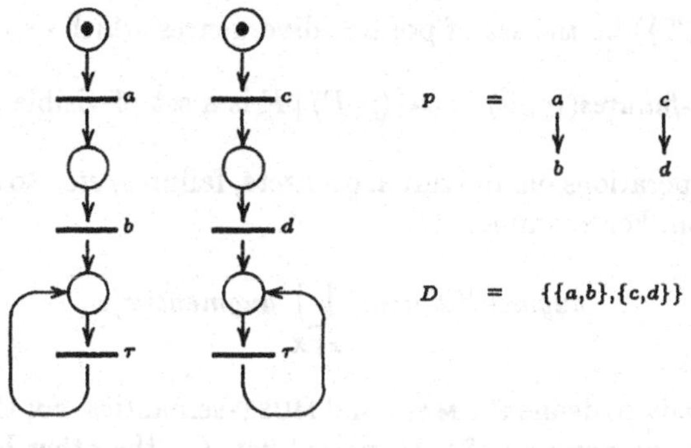

Fig. 4. An Example of a Pomset Divergence $\langle p, D\rangle$

It turns out that the semantics defined by simply taking these pomset-failures and pomset-divergences makes too many distinctions between nets, and we need to "blur" certain kinds of information from our runs. This we accomplish through various closure operations. The first such closure involves taking "augmentations" of our pomset-failures and pomset-divergences. We first restate the standard definition for pomsets, where an augmentation is simply an increase in the partial ordering.

Definition 16. Pomset p' is an *augmentation* of pomset p iff p and p' have the same set of elements with the same labels, and the partial ordering of p' contains the partial ordering of p. Let $augment(p)$ be the set of augmentations of p. The augmentations, $augment(\langle p, F\rangle)$, of a pomset-failure $\langle p, F\rangle$ is the set $\{\langle p', F\rangle \mid p' \in augment(p)\}$. For pomset-divergences, let $augment(\langle p, \mathcal{D}\rangle)$ be

$$\{\langle p', \mathcal{D}'\rangle \mid p' \in augment(p) \text{ and } \mathcal{D}' = \{d' \mid d' \text{ is the downward closure } in\ p' \text{ of some } d \in \mathcal{D}\}\}$$

Our other closure operation arises from the fact that MUST-experiments fail to yield information about the behavior of a net after a divergence. To get around this difficulty, we define below the notion of an *extension* of a pomset-divergence; the idea is that the extension is another pomset-divergence which may contain more information concerning events and divergences which "happen after" one or more divergences in the original pomset-divergence. All the information about a process after a pomset-divergence is blurred by throwing in all possible pomset-failures and pomset-divergences which extend the original pomset-divergence.

Definition 17. Pomset-divergence $\langle p', \mathcal{D}'\rangle$ *extends* pomset-divergence $\langle p, \mathcal{D}\rangle$ iff

p is the restriction of p' to a downward-closed subset of p';
for all $e \in p' - p$, there is some $d \in \mathcal{D}$ with $d \subseteq \widehat{e}$; and
for all $d' \in \mathcal{D}'$, there is some $d \in \mathcal{D}$ with $d \subseteq d'$.

Let $extend(\langle p, \mathcal{D}\rangle)$ be the set of pomset-divergences which extend it. Finally, let

$$implied\text{-}failures(\langle p, \mathcal{D}\rangle) =_{\text{def}} \{\langle p, F\rangle \mid F \text{ is a set of visible labels}\}.$$

We lift these operations on individual pomsets, failures, *etc.*, to sets of individuals by pointwise union. For example,

$$augment(X) =_{\text{def}} \bigcup_{x \in X} augment(x).$$

We are now ready to define the MAY- and MUST-semantics. For the MAY-semantics, we can work with the pomsets of the original net. On the other hand, as explained earlier, for the MUST-semantics we need to work with the pomsets of the "split" net.

Definition 18. For any safe net N,

$$\begin{aligned}
[\![N]\!]^{\text{MAY}} &=_{\text{def}} augment(pomset\text{-}traces(N)),\\
Div(N) &=_{\text{def}} extend(augment(pomset\text{-}divergences(split(N)))),\\
Fail(N) &=_{\text{def}} augment(pomset\text{-}failures(split(N))) \cup implied\text{-}failures(Div(N)),\\
[\![N]\!]^{\text{MUST}} &=_{\text{def}} \langle Fail(N), Div(N)\rangle.
\end{aligned}$$

Our definition of semantical equality implicitly equates label-preserving order-isomorphic pomsets. For finitely-marked safe nets with finite out-degree of places and transitions, the semantics satisfy the following properties:

Theorem 19. *$[\![\cdot]\!]^{\text{MUST}}$ is adequate for* MUST-*equivalence and compositional for split refinements, choice refinements, and our CCS/CSP operators; likewise for $[\![\cdot]\!]^{\text{MAY}}$.*

However, it turns out that $[\![\cdot]\!]^{\text{MUST}}$ and $[\![\cdot]\!]^{\text{MAY}}$ make more distinctions than are apparent to a single experimenter. Namely, single experimenters can only detect differences between pomsets with *interval orderings* [12, 17]. (A partial order $\leq$ is considered to be an interval ordering iff whenever both $w < x$ and $y < z$, then either $w < z$ or $y < x$.)

We define the interval-MUST- and interval-MAY-semantics by restricting the general semantics to interval pomsets:

Definition 20. Let $\mathcal{P}$ be a set of pomsets, let $\mathcal{PF}$ be a set of pomset-failures, and let $\mathcal{PD}$ be a set of pomset-divergences. Then *intervals*($\mathcal{P}$) is the set of interval pomsets $p \in \mathcal{P}$. Let *intervals*($\mathcal{PF}$) be the set of $\langle p, F\rangle \in \mathcal{PF}$ such that p is an interval pomset, and *intervals*($\mathcal{PD}$) be the set of $\langle p, \{p\}\rangle \in \mathcal{PD}$ such that p is an interval pomset. Let

$$[\![N]\!]^{\text{MUST}}_{\text{intvl}} =_{\text{def}} \langle \mathit{intervals}(\mathit{Div}(N)), \mathit{intervals}(\mathit{Fail}(N))\rangle$$
$$[\![N]\!]^{\text{MAY}}_{\text{intvl}} =_{\text{def}} \mathit{intervals}([\![N]\!]^{\text{MAY}})$$

We now have:

Theorem 21. *The $[\![\cdot]\!]^{\text{MUST}}_{\text{intvl}}$ semantics is fully abstract for* MUST*-equivalence with respect to split refinements and choice refinements; likewise for $[\![\cdot]\!]^{\text{MAY}}_{\text{intvl}}$. Furthermore, both are compositional for our CCS/CSP operators.*

The problem of "weak bisimulation up to"

Davide Sangiorgi Robin Milner

LFCS - Department of Computer Science

Edinburgh University, Edinburgh EH9 3JZ, U.K.

Abstract

"Bisimulation up to" is a technique for reducing the size of the relation needed to define a bisimulation. It works smoothly in the strong *case, where it was first introduced ([4]). But this does not directly generalize to the* weak *case, as erroneously reported in [4]. To overcome this problem, two new "up-to" techniques are proposed: They are respectively based on the use of* expansion *([1]) and of* almost-weak bisimulation. *The second solution is more general than the first one, but expansion enjoys a nicer mathematical treatment. The usefulness and generality of the solutions is motivated with non-trivial examples: two different implementations of a sorting machine.*

1 Introduction

Bisimulation has emerged as one of the more stable and mathematically natural concepts formulated in the study of concurrency over the past decade. Both in the *strong* case and in the *weak* case – distinguished by whether or not we want to abstract from internal details of systems – bisimulation seems the finest extensional or behavioural equivalence one would want to impose.

By definition, two processes are bisimilar if there exists a bisimulation relation containing them as a pair. However, in practice this definition is hardly ever followed plainly; instead, to reduce the size of the relations exhibited one prefers to define relations which are bisimulations only when closed up under some specific and privileged relation in order to relieve the proof work needed (let us call this an *"up-to" technique*). This is a general technique which allows a great variety of possibilities and which so far has not received deserved attention. For instance, one could define strong bisimulation up to strong bisimilarity [4], a weak bisimulation up to a strong equivalence, a bisimulation up to a preorder, a preorder up to an equivalence, and so on... One does not even have to limit oneself to a *single* closure relation, but different equivalences/preorders may be employed in different positions.

This variety is useful because it allows us to make each time the most convenient choice, depending upon the equilibrium we want between the size of the relation to exhibit and the fineness of the closure relation(s). However there are limits on the range of legitimate choices, and the border between what is legitimate and what is not is uncertain. In particular, care is needed when both the relation to prove and the closure relation(s) are weak, and abstract from internal behaviour. For instance, the technique of weak bisimulation up to weak bisimilarity ($\approx$), as originally proposed in [4] is not sound, in that it allows us to define relations which are not contained in $\approx$.

This opens an important problem, which is also the main concern of this paper. Weak bisimilarity is one of the most studied equivalences in process algebra and is the one where the up-to technique is most needed. We investigate here two new up-to techniques for $\approx$ based on the use of *expansions* and *almost-weak bisimulation*.

Expansion is a preorder derived from $\approx$ by, essentially, comparing the number of silent actions. It enjoys an elegant mathematical theory, explored in [1]. We show here that it also enjoys a nice up-to technique. It can be exploited for $\approx$ because expansion implies bisimilarity and because in most practical cases where $\approx$ holds, P and Q are indeed ordered by expansion.

We introduce almost-weak bisimulation not for its intrinsic merit, but purely as supporting relation for $\approx$. With almost weak bisimulation, our purpose is to define a relation as coarse as possible but still capable of providing us with a sound up-to technique for $\approx$. Almost-weak bisimulation is not very appealing on its own; for instance, it is not preserved by parallel composition. But a supporting relation does not need to be substitutive; its usefulness may come from the fact that:

1. It implies the supported substitutive relation: Then the former, if more convenient to prove and although by itself not substitutive, can even be used to derive compositional proofs of the latter, by using the former on subsystems and then lifting it up to the latter.

2. It gives rise to a powerful up-to technique to prove results for the supported relation.

The generality and usefulness of the techniques presented is tested with examples involving different implementations of a sorting machine.

NOTATION. The results we present hold in general for any transition system equipped with a special action called *silent action*. However examples/counterexamples and comparison results among relations will be presented in the setting of CCS, of which we assume some familiarity.

Let $(Pr, Act, \rightarrow)$ represent our transition system, where Pr is its domain, Act is the set of actions (labels) and $\rightarrow$ describes the possible transitions. We use P, Q, R to range over Pr; α to range over Act; ℓ to range over $Act - \{\tau\}$, where τ represents the silent action. Actions in $Act - \{\tau\}$ are usually called *visible actions*. It is important to fix a convenient notation for the arrows. We write $P \stackrel{\alpha}{\rightarrow} Q$ when $(P, \alpha, Q) \in \rightarrow$, to be interpreted as "$P$ may become Q by performing an action labelled α". We also use $P \stackrel{\hat{\alpha}}{\rightarrow} Q$ to mean:

- $P \stackrel{\alpha}{\rightarrow} Q$, if $\alpha \neq \tau$,
- $P = Q$ or $P \stackrel{\tau}{\rightarrow} Q$, if $\alpha = \tau$.

We shall often abbreviate $P \stackrel{\tau}{\rightarrow} Q$ and $P \stackrel{\hat{\tau}}{\rightarrow} Q$ with $P \rightarrow Q$ and $P \stackrel{\wedge}{\rightarrow} Q$, respectively. As usual, the "weak" arrow $\Rightarrow$ is the reflexive and transitive closure of $\rightarrow$, and then $\stackrel{\alpha}{\Rightarrow}$ stands for $\Rightarrow \stackrel{\alpha}{\rightarrow} \Rightarrow$. Finally, similarly to $P \stackrel{\hat{\alpha}}{\rightarrow} Q$, we use $P \stackrel{\hat{\alpha}}{\Rightarrow} Q$ to mean:

- $P \stackrel{\alpha}{\Rightarrow} Q$, if $\alpha \neq \tau$,
- $P \Rightarrow Q$, if $\alpha = \tau$.

2 The problem

Let us start off with a mention of the strong case, where the "bisimulation up to" technique was first introduced and where its theory works smoothly. We refer to [4] for details. *Strong bisimulation*, written $\sim$, is defined as the largest symmetric relation such that whenever $P \sim Q$ and $P \stackrel{\alpha}{\rightarrow} P'$, then for some Q' also $Q \stackrel{\alpha}{\rightarrow} Q'$ with $P' \sim Q'$. Then, a symmetric relation $\mathcal{S}$ is a *strong bisimulation up to* $\sim$, if whenever $P \mathcal{S} Q$ and $P \stackrel{\alpha}{\rightarrow} P'$, there exists Q' such that $Q \stackrel{\alpha}{\rightarrow} Q'$ and $P' \sim \mathcal{S} \sim Q'$. With a simple diagram-chasing argument one can prove that if $\mathcal{S}$ is a bisimulation up to $\sim$, then $\mathcal{S} \subset \sim$.

Unfortunately, the "bisimulation up to" technique for the strong case is not directly generalisable to the weak case. Consider the definition of *weak bisimulation*, also called *observational equivalence*:

Definition 1 *$\mathcal{S}$ is a* weak bisimulation *if $P \mathcal{S} Q$ implies, for all α:*

1. *whenever $P \stackrel{\alpha}{\rightarrow} P'$, then Q' exists s.t. $Q \stackrel{\hat{\alpha}}{\Rightarrow} Q'$ and $P' \mathcal{S} Q'$;*
2. *vice versa, i.e. whenever $Q \stackrel{\alpha}{\rightarrow} Q'$, then P' exists s.t. $P \stackrel{\hat{\alpha}}{\Rightarrow} P'$ and $P' \mathcal{S} Q'$.*

Two processes P and Q are weakly bisimilar, *written $P \approx Q$, if $P\,\mathcal{S}\,Q$, for some weak bisimulation $\mathcal{S}$.* □

If we want now to define bisimulation up to $\approx$, the experience from the strong case would suggest the following definition:

Definition 2 *$\mathcal{S}$ is a* weak bisimulation up to $\approx$ *if $P\,\mathcal{S}\,Q$ implies, for all α:*

1. *whenever $P \xrightarrow{\alpha} P'$, then Q' exists s.t. $Q \stackrel{\hat{\alpha}}{\Rightarrow} Q'$ and $P' \approx \mathcal{S} \approx Q'$;*
2. *vice versa, i.e.*
   whenever $Q \xrightarrow{\alpha} Q'$, then P' exists s.t. $P \stackrel{\hat{\alpha}}{\Rightarrow} P'$ and $P' \approx \mathcal{S} \approx Q'$. □

This is in fact the definition originally proposed in [4]. Unfortunately, it is not true in general that if $\mathcal{S}$ is a bisimulation up to $\approx$ as from Definition 2, then $\mathcal{S} \subseteq \approx$. This was proved independently by the first author of this paper and by Gunnar Sjödin and Bengt Jonsson using more or less the same counterexample, namely $\mathcal{S} = \{(\tau.a.\mathbf{0}, \mathbf{0})\}$.

One could adjust things by replacing all $\xrightarrow{\alpha}$ in Definition 2 with $\stackrel{\alpha}{\Rightarrow}$: Then the resulting condition is indeed necessary and sufficient for $\approx \mathcal{S} \approx$ to be a bisimulation, but it is expensive to check in general, even sometimes prohibitive, due to the appearance of $\stackrel{\alpha}{\Rightarrow}$ rather than $\xrightarrow{\alpha}$. Therefore it is important to look for something else.

The solution which has been used to correct the mistake in [4] replaces some of the occurrences of $\approx$ in Definition 2 by $\sim$.

Proposition 1 *Let $\mathcal{S}$ be such that $P\,\mathcal{S}\,Q$ implies, for all α:*

1. *whenever $P \xrightarrow{\alpha} P'$, then Q' exists s.t. $Q \stackrel{\hat{\alpha}}{\Rightarrow} Q'$ and $P' \sim \mathcal{S} \approx Q'$;*
2. *vice versa, i.e. whenever $Q \xrightarrow{\alpha} Q'$, then P' exists s.t. $P \stackrel{\hat{\alpha}}{\Rightarrow} P'$ and $P' \approx \mathcal{S} \sim Q'$.*

Then $\mathcal{S} \subseteq \approx$. □

However in this solution the presence of $\sim$ may represent a too heavy restriction, as the following example shows.

Example 1 For this example and the following ones which continue it, we shall use the value passing version of CCS (the extension to value passing of the results examined in this paper can be done in the usual fashion). Consider the sorting machine $Sorter_n$

described on pag.136 of [4], with $n = 2$ fixed. *Sorter*$_2$ must accept exactly two integers one by one at port *in*. Then it must deliver them one by one at port $\overline{out}$, terminated by a zero. For simplicity here we assume also that it does not resume its initial state after completing its job. The specification of *Sorter*$_2$ is given by $Spec_{2,0}$, which is described as follows:

(*)

$$\begin{aligned}
Spec_{2,0} &\stackrel{def}{=} in\,(x_1).Spec_{2,1}(x_1)\\
Spec_{2,1}(x_1) &\stackrel{def}{=} in\,(x_2).Spec_{2,2}(x_1,x_2)\\
Spec_{2,2}(x_1,x_2) &\stackrel{def}{=} \overline{out}\,\big(max\{x_1,x_2\}\big).Spec_{2,-1}\big(min\{x_1,x_2\}\big)\\
Spec_{2,-1}(x_1) &\stackrel{def}{=} \overline{out}\,(x_1).Spec_{2,-2}\\
Spec_{2,-2} &\stackrel{def}{=} \overline{out}\,(0).0
\end{aligned}$$

Sorter$_2$ is build using 2 identical cells C and a single barrier B:

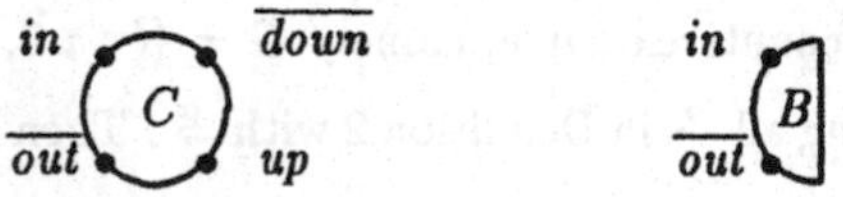

The cell C has storage capacity for two numbers, and is able to compare them. Its behaviour has two phases. In the first phase — described by the equations for C and $C'(x)$ — it receives inputs at *in* and puts them out at $\overline{down}$; but since it does not know the size of the sorter, it is ready to change at any moment to its second phase — described by the equation for $C''(x,y)$ — in which it receives inputs at *up* and (using comparisons) puts them out at $\overline{out}$.

$$\begin{aligned}
C &\stackrel{def}{=} in\,(x).C'(x)\\
C'(x) &\stackrel{def}{=} \overline{down}\,(x).C + up\,(y).C''(x,y)\\
C''(x,y) &\stackrel{def}{=} \overline{out}\,\big(max\{x,y\}\big).\ \text{if } y=0 \text{ then } \overline{out}\,(0).0\\
&\qquad\qquad \text{else } up\,(z).C''\big(z, min\{x,y\}\big)
\end{aligned}$$

The barrier cell B is fixed by the equation $B \stackrel{def}{=} Spec_{2,-2}$. Cells are then linked together through the chaining combinator $\frown$ which is defined using renaming and restriction as follows:

$$P \frown Q \stackrel{def}{=} \big(P[a/down, b/up] \mid Q[a/in, b/out]\big) \setminus \{a,b\}$$

Finally, $Sorter_2 \stackrel{def}{=} C \frown C \frown B$.

Let us consider how one could proceed in proving $Spec_{2,0} \approx Sorter_2$.

1. One could prove that $Sorter_2$ satisfies equations $(*)$ too, up to weak congruence;

2. one could try to simplify $Sorter_2$; following the inductive structure of $Sorter_n$, prove first that $Spec_{1,0} \approx Sorter_1$ and then prove that $C \frown Spec_{1,0} \approx Sorter_2$;

3. one could directly find the bisimulation containing $(Spec_{2,0}, Sorter_2)$.

Now, (3) is the least cunning method, but in many examples we would perhaps not be cunning enough to do something like (1) or (2). Then our weak bisimulation should include at least (replacing processes of the form $P \frown \mathbf{0}$ with P)

$$\begin{array}{lll} \mathcal{S} = \{ & (Spec_{2,0}\ ,\ C \frown C \frown B), & (Spec_{2,1}(x)\ ,\ C \frown C''(x,0)), \\ & (Spec_{2,2}(x,y)\ ,\ C''(x,y) \frown \overline{out}\,(0).\mathbf{0}), & (Spec_{2,-1}(x)\ ,\ C''(x,0)), \\ & (Spec_{2,-2}\ ,\ \overline{out}\,0.\mathbf{0}), & (\mathbf{0}\ ,\ \mathbf{0}) \qquad \} \end{array}$$

and in fact, preferably we would like to consider *only* $\mathcal{S}$. (The full bisimulation would have at least 11 pairs). However, we cannot apply Proposition 1: it is ok in one direction, but if you take for instance

$$C \frown C''(x,0), \quad \text{then you have}$$
$$C \frown C''(x,0) \stackrel{in(y)}{\longrightarrow} C'(y) \frown C''(x,0)$$

and the right hand-side is not strongly bisimilar to anything in the codomain of $\mathcal{S}$. It is the strong demand $\sim$ which fails us. □

3 Expansion

The expansion relation, denoted by $\lesssim$, is an asymmetric version of $\approx$ where $P \lesssim Q$ means that $P \approx Q$, but also that P achieves the same as Q with no more work, i.e. with no more τ actions. Intuitively, if $P \lesssim Q$, we can think of P as being at least as fast as Q, or more generally, we can think that Q uses at least as many resources as P. The interest of $\lesssim$ derives from the fact that in practice, most of the uses of $\approx$ are indeed instances of $\lesssim$. For example, we are rightly interested in the question $Spec \lesssim Imp$, and $Spec$ is as usual a sequential agent with no internal τ transition possible.

Definition 3 *$\mathcal{E}$ is an* expansion *if $P\mathcal{E}Q$ implies, for all α:*

1. *whenever $P \stackrel{\alpha}{\rightarrow} P'$, then Q' exists s.t. $Q \stackrel{\alpha}{\Rightarrow} Q'$ and $P'\mathcal{E}Q'$;*

2. *whenever $Q \stackrel{\alpha}{\rightarrow} Q'$, then P' exists s.t. $P \stackrel{\hat{\alpha}}{\rightarrow} P'$ and $P'\mathcal{E}Q'$;*

We say that Q expands *P, written $P \lesssim Q$, if $P\mathcal{E}Q$, for some expansion $\mathcal{E}$.* □

Relation $\lesssim$ is studied – using a different terminology – by Arun-Kumar and Hennessy: In [1] they show that $\lesssim$ is a mathematically tractable preorder in that it is preserved by all CCS operators but sum and that it has a complete proof system for finite terms based on a modification of the standard τ laws for CCS. Independently from [1], $\lesssim$ is also examined in [5], where its use to help the definition of "bisimulations up to" was first proposed. A preorder called *contraction*, very close indeed to expansion, has also been implemented in the *Concurrency Workbench* ([2]). The following two theorems are from [1]:

Theorem 1 *$\lesssim$ is preserved by all operators but sum.* □

Theorem 2 *It holds that $\sim\, \subset\, \lesssim$ and $\lesssim\, \subset\, \approx$; moreover each inclusion is strict.*

Proof. The inclusions are obvious. For the strictness, we have that $P \not\sim \tau.P$, $P \lesssim \tau.P$, and $\tau.P \not\lesssim P$, $\tau.P \approx P$. □

We show now that there is a simple and efficient way to define "expansions up to $\lesssim$", which in turn, since $\lesssim\, \subset\, \approx$, becomes an important instrument to prove results for $\approx$.

Definition 4 *$\mathcal{S}$ is an* expansion up to $\lesssim$ *if $P\,\mathcal{S}\,Q$ implies, for all α:*

1. *whenever $P \stackrel{\alpha}{\rightarrow} P'$, then Q' exists s.t. $Q \stackrel{\alpha}{\Rightarrow} Q'$ and $P' \sim \mathcal{S} \lesssim Q'$;*

2. *whenever $Q \stackrel{\alpha}{\rightarrow} Q'$, then P' exists s.t. $P \stackrel{\hat{\alpha}}{\rightarrow} P'$ and $P' \lesssim \mathcal{S} \lesssim Q'$.* □

Proposition 2 *If $\mathcal{S}$ is an expansion up to $\lesssim$, then $\mathcal{S} \subseteq \lesssim$.*

Proof. Simple diagram-chasing □

Comparing Propositions 1 and 2: since all inclusions in Theorem 2 are strict, it is not true that Proposition 1 is more general than Proposition 2, nor is it true the other way round.

Example 2 (continues Example 1) Let $\mathcal{S}$ be the relation defined in Example 1. We show $\mathcal{S} \subseteq \approx$ by proving that $\mathcal{S}$ is an expansion up to $\lesssim$. We only examine the transitions for $C \frown C \frown B$; the other cases are easier and can be treated similarly. Process $C \frown C \frown B$ can only perform the transition

$$C \frown C \frown B \overset{in(x)}{\rightarrow} C'(x) \frown C \frown B$$

This can be matched by $Spec_{2,0} \overset{in(x)}{\rightarrow} Spec_{2,1}$ if we can show that

$$C \frown C''(x,0) \lesssim C'(x) \frown C \frown B \tag{1}$$

since $(Spec_{2,1}(x),\ C \frown C''(x,0)) \in \mathcal{S}$. By simple use of the expansion law and the obvious law $P \lesssim \tau.P$, we get

$$C'(x) \frown C \ \gtrsim\ C \frown C'(x) \tag{2}$$

$$C'(x) \frown B \ \gtrsim\ C''(x,0) \tag{3}$$

Now (1) can be derived applying (2) and (3) in sequence and using the congruence properties of $\lesssim$ from Theorem 1. □

4 Almost-weak Bisimulation

We study in this section how close to Definition 2 it is possible to go without losing soundness w.r.t $\approx$. The solution we derive is based on the notion of *almost-weak bisimulation*, denoted by $\leqq$, and subsumes both Proposition 1 and Proposition 2. This however does not cancel the interest in Propositions 1 and 2 to prove results for $\approx$, since $\leqq$ does not have the same nice mathematical theory as $\sim$ and $\lesssim$. To introduce $\leqq$, let us consider again the counterexample $\mathcal{S} = \{(\tau.a.0, 0)\}$ of section 2, and the faulty diagram-chasing associated with it:

$$\begin{array}{ccc} \tau.a.\mathbf{0} & \mathcal{S} & \mathbf{0} \\ \downarrow & & \Downarrow \\ a.\mathbf{0} & \approx\ \mathcal{S} & \mathbf{0} \end{array}$$

This diagram makes $\mathcal{S}$ a bisimulation up to $\approx$ according to Definition 2, despite the fact that $\tau.a.\mathbf{0} \not\approx \mathbf{0}$. What is wrong here is that no "evolution" of $\tau.a.\mathbf{0}$ can be tested in $\mathcal{S}$.

We consider the transition $\tau.a.0 \to a.0$, but then we go back to $\tau.a.0$ without testing $a.0$. The idea for $\leqq$ is to recognise an evolution when a visible action has been produced. Once such an evolution has been met, ties can be relaxed; but as long as only τ-action are produced, we need to be restrictive and avoid any expansion.

Definition 5 *A relation $\mathcal{S}$ is an* almost-weak bisimulation *if $P\,\mathcal{S}\,Q$ implies, for all α, ℓ:*

1. *whenever $P \xrightarrow{\alpha} P'$, then Q' exists s.t. $Q \stackrel{\hat{\alpha}}{\Rightarrow} Q'$ and $P' \approx Q'$;*

2. *whenever $Q \to Q'$, then P' exists s.t. $P \stackrel{\wedge}{\to} P'$ and $P'\,\mathcal{S}\,Q'$;*

3. *whenever $Q \xrightarrow{\ell} Q'$, then P' exists s.t. $P \xrightarrow{\ell}\Rightarrow P'$ and $P' \approx Q'$.*

Two processes P and Q are almost-weak bisimilar, *written $P \leqq Q$, if $P\,\mathcal{S}\,Q$, for some almost-weak bisimulation $\mathcal{S}$.* □

Notice that the only restrictions w.r.t. the definition of $\approx$ are in clause 3, the use of $\xrightarrow{\ell}\Rightarrow$ instead of $\stackrel{\ell}{\Rightarrow}$, and in clause 2, the bound on the silent actions which P can perform together with the subsequent appearance of $\mathcal{S}$.

Indeed, there are interesting cases in which $\approx$ implies $\leqq$; we mention here informally, and omitting the proof, two of them.

- Suppose P is *stable*, i.e. it cannot perform any τ-action; then $P \approx Q$ implies $P \leqq Q$.
- Consider the contexts $a.[\,]$ and $a.a.[\,]\,|\,\overline{a}$: they have the property that some visible action has to be produced before the process in the context can become active, i.e. perform some action. Let us call *strongly guarded* this kind of contexts. Then if $C[\,]$ is a strongly guarded context, for every Q_1, Q_2, it holds that $C[Q_1] \approx C[Q_2]$ implies $C[Q_1] \leqq C[Q_2]$. (An example of a context which is not strongly guarded is $a.[\,]\,|\,\overline{a}$)

As for $\lesssim$, it is straightforward to verify that also $\leqq$ is a preorder. What makes $\leqq$ mathematically less tractable is that $\leqq$ is not a congruence over parallel composition. Consider in fact $P = \tau.a + \tau.\tau.b$ and $Q = \tau.\tau.a + \tau.b$. Then $P \approx Q$ and hence also $a.P \leqq a.Q$, since $a.P$ is stable; however $P\,|\,\overline{a} \not\leqq Q\,|\,\overline{a}$. This example proves also that inclusions $\lesssim \subset \leqq$ and $\leqq \subset \approx$ are strict, since it holds that $a.P \not\lesssim a.Q$ and $a.P|\overline{a} \approx a.Q|\overline{a}$.

Theorem 3 *The following is a chain of strict inclusions:* $\sim \subset \lesssim \subset \leqq \subset \approx$. □

We define now bisimulation up to $\leqq$, following the same idea for $\lesssim$.

Definition 6 $\mathcal{S}$ *is a* bisimulation up to $\leqq$ *if* $P \mathcal{S} Q$ *implies, for all* ℓ*:*

1. *whenever* $P \rightarrow P'$, *then* Q' *exists s.t.* , $Q \Rightarrow Q'$ *and* $P' \geqq \mathcal{S} \approx Q'$;

2. *whenever* $P \xrightarrow{\ell} P'$, *then* Q' *exists s.t.* , $Q \stackrel{\ell}{\Rightarrow} Q'$ *and* $P' \approx \mathcal{S} \approx Q'$.

3. *vice versa of (1), i.e.*

   whenever $Q \rightarrow Q'$, *then* P' *exists s.t.* , $P \Rightarrow P'$ *and* $P' \approx \mathcal{S} \leqq Q'$;

4. *vice versa of (2), i.e.*

   whenever $Q \xrightarrow{\ell} Q'$, *then* P' *exists s.t.* , $P \stackrel{\ell}{\Rightarrow} P'$ *and* $P' \approx \mathcal{S} \approx Q'$. □

If we compare it with Definition 2, the only restrictions in Definition 6 are in clauses (1) and (3), the occurrence of $\leqq$ instead of $\approx$. In fact in some cases, such as in Example 3 below, the verification of clauses of Definition 6 reduces exactly to the verification of clauses of Definition 2.

We shall prove now that if $\mathcal{S}$ is a bisimulation up to $\leqq$, then $\mathcal{S} \subseteq \approx$. For this, we need the following lemma.

Lemma 1 *Let* $\mathcal{S}$ *be a bisimulation up to* $\leqq$. *We have:*

1. *if* $P \mathcal{S} Q$ *and* $P \Rightarrow P'$, *then* Q' *exists s.t.* $Q \Rightarrow Q'$ *and* $P' \geqq \mathcal{S} \approx Q'$;

2. *if* $P \approx \mathcal{S} \approx Q$ *and* $P \Rightarrow P'$, *then* Q' *exists s.t.* $Q \Rightarrow Q'$ *and* $P' \approx \mathcal{S} \approx Q'$;

3. *if* $P \geqq \mathcal{S} \approx Q$ *and* $P \xrightarrow{\ell} P'$, *then* Q' *exists s.t.* $Q \stackrel{\ell}{\Rightarrow} Q'$ *and* $P' \approx \mathcal{S} \approx Q'$

4. *if* $P \approx \mathcal{S} \approx Q$ *and* $P \xrightarrow{a} P'$, *then* Q' *exists s.t.* $Q \stackrel{a}{\Rightarrow} Q'$ *and* $P' \approx \mathcal{S} \approx Q'$

Proof.

(1). By induction on the length of the transition $P \Rightarrow P'$. In the basic case, this length is zero and there is nothing to prove. For the inductive case, let us write $P \Rightarrow P'$ as $P \Rightarrow R \rightarrow P'$. By induction hypothesis, there exist R', T', T, such that

$$Q \Rightarrow T \quad \text{and} \quad R \geqq R' \mathcal{S} T' \approx T$$

Then, since $R \rightarrow P'$, by definitions of $\leqq$, bisimulation up to $\leqq$ and $\approx$, respectively, there exist R'', T'', Q' such that $R \overset{\wedge}{\rightarrow} R''$, $T' \Rightarrow T''$, $Q \Rightarrow Q'$ and:

$$P' \geqq R'' \geqq \mathcal{S} \approx T'' \approx Q$$

Now the assertion follows from transitivity of $\leqq$ and $\approx$.

(2). A simple diagram-chasing which uses (1).

(3). Take R, T such that $P \geqq R\,\mathcal{S}\,T \approx Q$. Then, since $P \overset{\ell}{\rightarrow} P'$ and $P \geqq R$, it is

$$R \overset{\ell}{\rightarrow} R' \Rightarrow R''$$

Since $R\,\mathcal{S}\,T$, by definition of bisimulation up to $\leqq$ we get that:

$$T \overset{\ell}{\Rightarrow} T' \quad \text{with} \quad R' \approx \mathcal{S} \approx T''.$$

Hence, since $R' \Rightarrow R''$, using (2) also

$$T' \Rightarrow T'' \quad \text{with} \quad R'' \approx \mathcal{S} \approx T''$$

Moreover, since $T \approx Q$, also $Q \overset{\ell}{\Rightarrow} Q' \approx T''$. Now, $P' \approx \mathcal{S} \approx Q'$ follows from transitivity of $\approx$.

(4). Suppose $\alpha \neq \tau$ (the case $\alpha = \tau$ is simpler) and let $P \approx R\,\mathcal{S}\,T \approx Q$. By definition of $\approx$, if $P \overset{\alpha}{\rightarrow} P'$, then

$$R \Rightarrow R' \overset{\alpha}{\rightarrow} R'' \Rightarrow R''' \approx P'$$

Now, using (1), (2) and (3), we show that T can match these moves of R: We get

$$T \Rightarrow T' \quad \text{with} \quad R' \geqq \mathcal{S} \approx T'$$

by (1),

$$T' \overset{\alpha}{\Rightarrow} T'' \quad \text{with} \quad R'' \approx \mathcal{S} \approx T''$$

by (3), and

$$T'' \Rightarrow T''' \quad \text{with} \quad R''' \approx \mathcal{S} \approx T'''$$

by (2). Summarising, we have found T''' such that $T \overset{\alpha}{\Rightarrow} T'''$ and $R''' \approx \mathcal{S} \approx T'''$. Since $Q \approx T$, also $Q \overset{\alpha}{\Rightarrow} Q' \approx T'''$. Finally, by transitivity of $\approx$, we get $P' \approx \mathcal{S} \approx Q'$. □

Proposition 3 *If S is a bisimulation up to $\lessapprox$, then $S \subseteq \approx$.*

Proof. Take $S' = \{(P,Q) \mid P \approx S \approx Q\}$. Now, use Lemma 1(4) to prove that S' is a weak bisimulation. □

Because of Theorem 3, Proposition 3 is indeed more general both of Proposition 1 and of Proposition 2. Moreover, it is possible to show that if any of the restrictions in Definition 5 or 6 were removed, Proposition 3 would fail. Therefore it really seems that with bisimulations up to $\lessapprox$ we brought ourselves as close as possible to Definition 2.

The following is an example where Proposition 3 is needed. It also shows how verification of clauses of Definition 6 may reduce to the verification of clauses of Definition 2.

Example 3 (continues Examples 1 and 2) In [4], pag. 137, exercise 7, a different implementation of $Sorter_n$ is suggested, where now the cell D used performs comparisons and exchanges during the first phase:

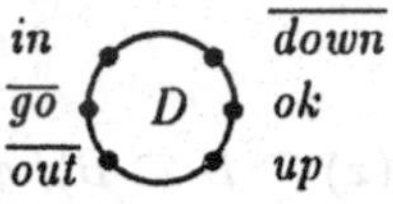

The ports go, ok implement control synchronisations between consecutive cells, which are needed if we want to limit ourselves to cells with storage capacity of two numbers. In our solution, communications along ports go and ok guarantee that $Sorter_n$ does not input more than n numbers. Behaviour of D is defined by: (constants D_1, D_2 are introduced to give a name to states which will later be referenced)

$$D \stackrel{def}{=} \overline{go}.D_1 \qquad D_1 \stackrel{def}{=} in(x).D'(x)$$

$$D'(x) \stackrel{def}{=} ok.\overline{go}.D_2(x) + up(y).D''(x,y)$$

$$D_2(x) \stackrel{def}{=} in(y).\overline{down}(min\{x,y\}).D'(max\{x,y\})$$

$$D''(x,y) \stackrel{def}{=} \overline{out}(x).\ \text{if } y=0 \ \text{ then } \overline{out}(0).0 \ \text{ else } up(z).D''(y,z)$$

The same barrier cell B as in Example 1 is used and cells are linked together with the obvious chain operator. Also, the hiding operator '$/\ go$' (described in [4] at page 122) is needed to hide go actions produced by the first cell of $Sorter_n$ to the external environment. Then

$$Sorter'_n \stackrel{def}{=} (D_1 \frown \underbrace{D \frown \dots \frown D}_{n-1} \frown B)/\,go$$

Let us take $Sorter'_2$ and consider how we could prove $Sorter_2 \approx Sorter'_2$ by exhibiting a bisimulation containing the two. The following are pairs which quite reasonably should go in the bisimulation:

$$\begin{aligned} \mathcal{S} = \{ \ & \big(C \frown C \frown B \ , \ (D_1 \frown D \frown B)/\,go\big) \ , \\ & \big(C \frown C''(x,0) \ , \ (D_2(x) \frown D_1 \frown B)/\,go\big) \ , \\ & \big(C''(x,y) \frown \overline{out}\,(0).0 \ , \ (D''(x,y) \frown \overline{out}\,(0).0)/\,go\big) \ , \\ & \big(C''(x,0) \ , \ D''(x,0)/\,go\big) \ , \ \big(\overline{out}\,(0).0 \ , \ \overline{out}\,(0).0\big) \ , \ (0 \ , \ 0) \ \} \end{aligned}$$

If we tried to prove $\mathcal{S} \subseteq \approx$ by using Proposition 1 or 2, we would fail. The reason we cannot use expansions is that $Sorter_2$ and $Sorter'_2$ are not refinements or implementations of each other. Instead, we can use Proposition 3. As all the processes in $\mathcal{S}$ are stable, verification of clauses of Definition 6 reduces to the verification of clauses in Definition 2. Hence, consider for instance $\big(C \frown C \frown B \ , \ (D_1 \frown D \frown B)/\,go\big)$ and the action

$$(D_1 \frown D \frown B)/\,go \overset{in\,(x)}{\rightarrow} (D_2(x) \frown D \frown B)/\,go$$

To match it with

$$C \frown C \frown B \overset{in\,(x)}{\rightarrow} C'(x) \frown C \frown B$$

it is enough to show

$$(D_2(x) \frown D \frown B)/\,go \approx (D_2(x) \frown D_1 \frown B)/\,go \quad \text{and}$$

$$C'(x) \frown C \frown B \approx C \frown C''(x,0)$$

This can be done similarly to what we did in Example 2 to derive equation (1). All remaining transitions of processes in $\mathcal{S}$ can be treated in analogous way. □

5 Conclusions

We have presented two new techniques for proving "weak bisimulations up to", the first based on the notion of expansion, the second one on the notion of almost-weak bisimulation. The second solution is more general than the first one, but expansions enjoy a nicer mathematical treatment and are easier to prove. We have demonstrated the usefulness of the two techniques with examples involving different implementations of a sorting machine.

We introduced almost-weak bisimulation with the purpose of defining a relation as coarse as possible but still capable of providing us with a powerful "bisimulation up to" technique for $\approx$. A plausible question is whether $\lneqq$ is the best possible relation for this. We could have made $\lneqq$ more general by following the idea of *branching bisimulation* [3] of discriminating among τ transitions: Then in Definition 5 we would have asked to recognize an evolution not only when the action performed is a visible one, but also when it represents a silent transition between non-equivalent states. However we have preferred to avoid this in order to keep the resulting relation and the proof of Lemma 1 more manageable.

Instead, if we agree to abandon some of the generality of $\lneqq$ and try to get the best of Definitions 4 and 6, a third solution is possible.

Let $\mathcal{S}$ be such that $P \mathcal{S} Q$ implies, for all α:

1. if $P \stackrel{\alpha}{\rightarrow} P'$, then then Q' exists s.t. $Q \stackrel{\hat{\alpha}}{\Rightarrow} Q'$ and $P' \gtrsim \mathcal{S} \approx Q'$;
2. if $Q \stackrel{\alpha}{\rightarrow} Q'$, then then P' exists s.t. $P \stackrel{\hat{\alpha}}{\Rightarrow} P'$ and $P' \approx \mathcal{S} \lesssim Q'$;

Then $\mathcal{S} \subset \approx$.

We think that this solution would often suffice. For instance, it is enough indeed to handle Example 3. This solution subsumes Propositions 1 and 2; it is subsumed by Proposition 3 but, again, it is easier to prove than Proposition 3 because it uses $\lesssim$ instead of $\lneqq$.

Having mentioned branching bisimulation, let us point out that it would not help to have $\approx$ meaning branching bisimulation in Definition 2: The counterexample presented would still hold.

We believe that the search of techniques capable of reducing the size of the relations to exhibit to define bisimulations, could lead to interesting and useful developments. A different direction from the one we pursue in this paper is taken in [7], with the technique of *bisimulation up to context*. The idea is to factorize out common contexts and to replace the ordinary clause in the definition of bisimulation with the clause (we consider here the strong case):

- if $P \xrightarrow{\alpha} P'$, then Q', P'' and a context $C[\,]$ exist s.t.

$$P' = C[P''],\ Q \xrightarrow{\alpha} C[Q'] \text{ and } P''SQ'$$

This idea gives rise to some nice piece of theory and in some case it allows to reduce substantially the size of $\mathcal{S}$. It would be interesting then, and we leave it for future work, to examine if and and what extent, the techniques of "bisimulation up to context" and "bisimulation up to" can be integrated each other.

6 Acknowledgement

We would like to thank Faron Moller and the anonymous referees for their comments and suggestions.

References

[1] Arun-Kumar, S., and Hennessy, M., *An efficiency preorder for Processes*, Proceedings of Theoretical Aspects of Computer Science, Tokyo, (1991), LNCS vol. 526, pp 152-175. To appear in Acta Informatica.

[2] Moller, F. *The Edinburgh Concurrency Workbench (Version 6.0)* Report No. LFCS-TN-34, Dept of Computer Science, University of Edinburgh.

[3] van Glabeek, R., and Weijland, W. P., *Branching Time and Abstraction in Bisimulation Semantics*, in Information Processing '89 (G.X. Ritter ed.), Elsevier Science (1989), pp 613-618.

[4] Milner, R., A Calculus of Communicating Systems, Prentice Hall, 1989.

[5] Milner, R., *Expansions - notes RM12*, handwritten notes, Edinburgh, march 1990.

[6] Park, D.M., *Concurrency on Automata and infinite sequences*, in Conf. on Theoretical Computer Science, (P. Deussen ed.) Springer Verlag (1981) Berlin, LNCS vol. 104.

[7] Sangiorgi, D., *Bisimulations up to context* (provisory title), in preparation, Edinburgh.

On the uniqueness of fixpoints modulo observation congruence *

Ed Brinksma

Tele-Informatics and Open Systems Group,
University of Twente
P.O. Box 217, 7500 AE Enschede, The Netherlands
brinksma@cs.utwente.nl

Abstract. We revisit the question of the uniqueness of solutions to fixpoint equations modulo observation congruence. In the literature various sufficient conditions are given for the uniqueness of such solutions for a given signature of process combinators, such as *guardedness* and *sequentiality* (CCS) or the absence of *abstraction* (ACP), *concealment* (CSP), or *hiding* (LOTOS) combinators. Our study is based on *contexts*, i.e. behaviour expressions with 'holes', whose operational semantics can be characterised by *action transductions*. Using these we define the context properties of *opacity* and *abstraction-freedom*, thus generalising the signature dependent conditions, which can be deduced from these properties. The condition of abstraction-freedom is further improved upon in two ways. First, by relativizing the notion of abstraction-freedom to essential guarding actions, allowing proofs of uniqueness in more cases. Second, by a new criterion based on context transductions that can be applied even more generally. We apply it to a fixpoint equation based on a context where a hiding combinator removes what seem to be essential guarding actions. This implies that the condition of (relativised) abstraction-freedom is generally too strict.

1 Introduction

In process algebraic calculi such as CCS [Mil80], CSP [Hoa85], and ACP [BK85], *fixpoint equations* play an important role. Such equation systems have the following form

$$\tilde{X} \approx \tilde{C}(\tilde{X}) \tag{1}$$

where $\tilde{X}$ is a vector of *process variables*, $\tilde{C}(\tilde{X})$ is a vector of *behaviour expressions* with its free variables in $\tilde{X}$, and $\approx$ is a suitable equivalence relation over process behaviour. Many characterisations of equivalent behaviour have been reported in the literature. Among the equivalences that have been employed successfully to model properties of concurrent systems are, for example, *observation congruence* [Mil89], *failure equivalence* [Hoa85], and *testing equivalence* [DNH84], to mention a few of the better known ones. In this paper we will restrict ourselves to observation

* This work has been supported in part by the CEC research programme ESPRIT (LOTOSPHERE, ref: 2303).

congruence ($\approx^c$), which allows the use of the elegant proof technique of (*rooted*) *weak bisimulations* [Mil89, BK89].

The problem of the unique solvability of (1) is not only of theoretical interest, but has also direct relevance for the specification and verification of concurrent systems in process algebraic calculi. First of all, systems like (1) can be used to specify indirectly behaviours, viz. their solutions. Unique solvability of fixpoint equations then implies the unambiguous definition of a behaviour modulo the chosen equivalence $\approx_{ob}$. Some formalisms allow the selection of a single solution from the set of all solutions by means of the definition of *process constants* or by *fixpoint operators* even in the case that unique solvability is not guaranteed [Mil83, Mil89]. In that case, as well as in the other, the unique solvability of (1) remains an important *verification* issue. A very elegant technique to show that to behaviours are equivalent is to prove that they are all solutions of the same uniquely solvable set of equations.

The best-known condition on fixpoint-equation systems to obtain unique solvability modulo observation congruence is that of being (*externally*) *guarded*. This condition is usually defined as that each occurrence of a process variable in $\tilde{C}(\tilde{X})$ appears in a subexpression of the form $a.E$ with $a \neq \tau$, i.e. an application of an *action-prefix expression* [Mil89, BK89]. This condition is in itself not sufficient as interference with other combinators may annihilate the effect of the guards a. This is the case with all combinators that can abstract observable actions into the invisible action τ, such as the *abstraction, concealment,* and *hiding* combinators in CSP, ACP, and LOTOS, respectively.

Example 1. If $/A$ is the postfix hiding operator defined in table 1 (the semantics are explained in the next section) then the fixpoint equation

$$X \approx^c a.(X/\{a\}) \tag{2}$$

has more than one solution modulo $\approx^c$ if the alphabet of actions contains at least one observable action other than a, viz. all X of the form $a.Y$ where Y is any solution of the equation $Y \approx^c Y/\{a\}$, e.g. all Y of the form b^n, i.e. the behaviour consisting of n consecutive b-actions. □

In order to take care of such interferences additional requirements are needed to guarantee the uniqueness of solutions. In languages like ACP, CSP, and LOTOS $\tilde{C}(\tilde{X})$ must be *abstraction-, concealment-,* or *hiding-free*, i.e. it must not contain applications of the corresponding operators, see e.g. [BK89]. In CCS the *composition* combinator | has abstraction-like properties, which gives rise to the requirement of *sequentiality*, i.e. the property that none of the process variables should occur within an argument of a composition application [Mil89].

The drawback of such conditions for the uniqueness of solutions is that they are language dependent and generally too strict. It would be much nicer to be able to express such conditions not in the context of a given signature, but as conditions *on* signatures, so that the results could be applied to families of process algebraic languages. It would be even better if such conditions could be related to the specific properties of the expressions $\tilde{C}(\tilde{X})$. This would allow to obtain results for fixpoint systems that contain combinators that could, in principle, interfere, but do not do so in the given case. It is the latter approach that we will pursue, making use of the fact that $\tilde{C}(\tilde{X})$ can be seen as a vector of *contexts* in the sense of Larsen [Lar90], i.e. as *action transducers* that transform actions of behaviours that can be substituted for the variables into actions of the context. We will show that it is possible to give properties in terms of the transductions of $\tilde{C}(\tilde{X})$, viz. *opacity* and *abstraction-freedom*, that guarantee the unique solvability of (1), and generalise the existing signature dependent results. We will also show how the requirement of abstraction-freedom can be weakened significantly in a number of interesting cases.

In Kranakis in [Kra87] the notion of guardedness is generalised so that it is no longer completely syntactic, but the subsequent application of this idea is only to ACP without the abstraction operator. The analysis, moreover, is given in the context of the *projective model* A^∞ of ACP, which has a theory different from that of $\approx^c$. Our work is most closely related to that of Sanderson in [San82], where the uniqueness of fixpoint equation modulo *observation equivalence* is studied for CCS. Although this work also uses action transducers, the proofs and the results are specific for the CCS signature. We are not aware of other, signature independent approaches to the unique solvability of fixpoint equations.

The structure of the rest of this paper is as follows. In section 2 we introduce the concepts and theory that we need about contexts, and recall some main facts of process algebra. In section 3 we formulate sufficient conditions on contexts that ensure unique solvability modulo $\approx^c$. In section 4 we present a weaker criterion than abstraction freedom that can also be used. In section 5 we present our conclusions.

2 Contexts

The concept of a context is central to our signature independent approach to the unique solvability of (1). Informally, contexts can be most easily imagined as behaviour expressions with a number of *holes* in them. By plugging these holes with behaviour expressions a new behaviour expression is obtained, so that a context can be seen as a kind of *generalized combinator*. In [Lar90] it is shown how contexts can be modelled quite naturally as *action transducers* that produce actions by consuming actions provided by behaviours occurring at the places of the holes. Let $\mathcal{L}$ be the alphabet of observable actions then such transductions can be denoted by

$$C \xrightarrow[\tilde{b}]{a} C'$$

where C and C' are contexts with n holes, $a \in \mathcal{L} \cup \{\tau\}$, and $\tilde{b} \in (\mathcal{L} \cup \{\tau, 0\})^n$. This should be interpreted as follows: if for each $1 \leq i \leq n$ the behaviour in the i-th hole can execute action b_i then context C can transform into context C' by excuting action a. The case that $b_i = 0$ is needed to express the possibility that the process in hole i is *idle*, i.e. does not make a transition. By convention we have $C \xrightarrow[\tilde{0}]{0} C$ for each C.

To obtain results that are general enough to deal with languages that contain *infinitary* combinators, such as the generalised sum $\sum$, we allow the contexts to have an infinite number of holes. It turns out to be convenient to number the holes with *ordinal numbers*, which we will denote with $\alpha, \beta, \gamma, \ldots$. When a context has α holes we speak of an α-context, and its transductions are like the one given above with n replaced by α everywhere. Although it is likely that in practice we will only have to deal with countable α, the results that we will present do not require such a restriction.

A convenient generalisation of contexts is to regard ordered lists or *vectors* of contexts again as contexts. Let $\tilde{C}$ be a β-dimensional vector of α_i-contexts C_i for $i \in \beta$, then the transductions of are of the form

$$\tilde{C} \xrightarrow[\tilde{b}]{\tilde{a}} \tilde{C}'$$

with $\tilde{a} \in (\mathcal{L} \cup \{\tau, 0\})^\beta$, and $\tilde{b} \in (\mathcal{L} \cup \{\tau, 0\})^\alpha$, where $\alpha = \sum_{i \in \beta} \alpha_i$. We say such to be (α, β)-contexts: in this sense an α-context is therefore an $(\alpha, 1)$-context, and an ordinary behaviour expression (without holes) is a (0,1)-context. Where dimensions are relevant we write $\tilde{C}^\beta_\alpha$ for an (α, β)-context, and C_α for an α-context. For all $i \in \beta$ we denote the ith component of $\tilde{C}^\beta_\alpha$ by $\tilde{C}^\beta_\alpha(i)$. Thus context vectors can be seen as the result of a product operation on their elements, viz. $\tilde{C}^\beta_\alpha = \prod_{i \in \beta} \tilde{C}^\beta_\alpha(i)$ based on the possibly infinitary rule

$$\frac{\forall i \in \beta \;\; \tilde{C}_i \xrightarrow[\tilde{a}_i]{b_i} \tilde{C}'_i}{\prod_{i \in \beta} \tilde{C}_i \xrightarrow[\tilde{\alpha}]{\tilde{\beta}} \prod_{i \in \beta} \tilde{C}'_i} \quad \text{with } \; \tilde{\alpha} = \prod_{i \in \beta} \tilde{a}_i, \; \tilde{\beta} = \prod_{i \in \beta} b_i$$

As usual we define the set of derivatives of a context $\tilde{C}$, denoted by *Der*($\tilde{C}$), as the smallest set of contexts that contains $\tilde{C}$ and is closed under $\xrightarrow[\tilde{b}]{\tilde{a}}$-derivations, i.e. if $\tilde{D} \in Der(\tilde{C})$ and $\tilde{D} \xrightarrow[\tilde{b}]{\tilde{a}} \tilde{E}$ for some $\tilde{a}, \tilde{b}$ then $\tilde{E} \in Der(\tilde{C})$.

The well-known generalised transitions $\xrightarrow{\sigma}$ and $\overset{\sigma}{\Rightarrow}$ for behaviour expressions (see e.g. [Mil89]) can be generalised to transductions as follows.

$$\tilde{C}^\beta_\alpha \xrightarrow[\tilde{\sigma}_1]{\tilde{\sigma}_2} \tilde{C}' \quad \text{iff } \exists n \; \exists \tilde{a}_1, \tilde{b}_1, \ldots, \tilde{a}_n, \tilde{b}_n \; \exists \tilde{C}_0, \ldots, \tilde{C}_n$$

$$\tilde{C}^\beta_\alpha = \tilde{C}_0 \xrightarrow[\tilde{b}_1]{\tilde{a}_1} \tilde{C}_1 \xrightarrow[\tilde{b}_2]{\tilde{a}_2} \cdots \xrightarrow[\tilde{b}_n]{\tilde{a}_n} \tilde{C}_n = \tilde{C}' \text{ and}$$

$$\forall i \in \alpha \; \tilde{\sigma}_1(i) = \tilde{a}_1(i) \ldots \tilde{a}_n(i)$$

$$\forall j \in \beta \; \tilde{\sigma}_2(j) = \tilde{b}_1(j) \ldots \tilde{b}_n(j)$$

$$\tilde{C}^\beta_\alpha \overset{\tilde{\sigma}_2}{\underset{\tilde{\sigma}_1}{\Rightarrow}} \tilde{C}' \quad \text{iff } \exists \tilde{v}_1, \tilde{v}_2 \;\; \tilde{C}^\beta_\alpha \xrightarrow[\tilde{v}_1]{\tilde{v}_2} \tilde{C}' \text{ and } \tilde{\sigma}_1 \lhd \tilde{v}_1, \tilde{\sigma}_2 \lhd \tilde{v}_2$$

where for $\tilde{\sigma}, \tilde{v} \in ([\mathcal{L} \cup \{\tau, 0\}]^*)^\alpha$ $\tilde{\sigma} \lhd \tilde{v}$ iff for all $i \in \alpha$ $\tilde{\sigma}(i)$ can be obtained by deleting zero or more occurrences of 0 or τ from $\tilde{v}(i)$. Note that because of $C \xrightarrow[0]{0} C$ this implies the reflexivity of the $\overset{\epsilon}{\underset{\epsilon}{\Rightarrow}}$-relation, where ϵ is the empty string.

We adopt the convention that if $\tilde{\sigma}_1$ consists only of occurrences of 0, τ, and ϵ, then it may be dropped from $\xrightarrow[\tilde{\sigma}_1]{\tilde{\sigma}_2}$ or $\overset{\tilde{\sigma}_2}{\underset{\tilde{\sigma}_1}{\Rightarrow}}$. We use the notations $\tilde{C} \xrightarrow[\tilde{\sigma}_1]{\tilde{\sigma}_2}$ and $\tilde{C} \overset{\tilde{\sigma}_2}{\underset{\tilde{\sigma}_1}{\Rightarrow}}$ to express that there exists a $\tilde{C}'$ such that $\tilde{C} \xrightarrow[\tilde{\sigma}_1]{\tilde{\sigma}_2} \tilde{C}'$ and $\tilde{C} \overset{\tilde{\sigma}_2}{\underset{\tilde{\sigma}_1}{\Rightarrow}} \tilde{C}'$, respectively.

The crucial operator for contexts is the (context) *composition* $\circ$. If $\tilde{C}$ is an (α,β)-context and $\tilde{D}$ is an (β,γ)-context, then $\tilde{D}\circ\tilde{C}$ is an (α,γ)-context, whose transductions are fully determined by the rule

$$\frac{\tilde{C}\xrightarrow[\tilde{a}]{\tilde{b}}\tilde{C}',\ \tilde{D}\xrightarrow[\tilde{b}]{\tilde{c}}\tilde{D}'}{\tilde{D}\circ\tilde{C}\xrightarrow[\tilde{a}]{\tilde{c}}\tilde{C}'\circ\tilde{D}'} \tag{3}$$

Instead of $\tilde{D}\circ\tilde{C}$ we also write $\tilde{D}[\tilde{C}]$.

In table 1 we have listed a number of process algebraic operators as they occur in languages like CCS and LOTOS [ISO89, BB87]. Instead of using the well-known inference rules on transitions their operational semantics is captured here by viewing them as contexts, and giving the corresponding transductions. By composing them we get language contexts, and using $\sum\emptyset$ for inactive behaviour (cf. **0**, **nil**, **stop** in the various process algebraic languages), also non-recursive behaviour expressions can be obtained. It is left to the reader to check how the expected transitions for behaviour expressions can be derived in this way. Examples are contained in [Lar90].

On the basis of the generalisations of the transition relation over behaviour expressions it is possible to generalise the different notions of bisimulation to contexts as well. We will not do so here, as this is not needed to obtain our results. Instead we recall the definitions of *weak bisimulation equivalence* and *observation congruence* [Mil89] for behaviour expressions. The interested reader is refered to [Lar90] for bisimulation between contexts.

Definition 1. Let $\mathcal{B}$ denote the set of behaviour expressions, i.e. (0,1)-contexts. A relation $\mathcal{R}\subseteq\mathcal{B}\times\mathcal{B}$ is a *weak bisimulation (relation)* iff for all $\langle B_1,B_2\rangle\in\mathcal{R}$ and for all $\sigma\in\mathcal{L}^*$

1. if $\exists B_1'\ B_1\overset{\sigma}{\Rightarrow}B_1'$ then $\exists B_2'\ B_2\overset{\sigma}{\Rightarrow}B_2'$ and $\langle B_1',B_2'\rangle\in\mathcal{R}$
2. if $\exists B_2'\ B_2\overset{\sigma}{\Rightarrow}B_2'$ then $\exists B_1'\ B_1\overset{\sigma}{\Rightarrow}B_1'$ and $\langle B_1',B_2'\rangle\in\mathcal{R}$

B_1 is said to be *weak bisimulation equivalent* to B_2, $B_1\approx B_2$, if there exists a weak bisimulation $\mathcal{R}$ such that $\langle B_1,B_2\rangle\in\mathcal{R}$.

Definition 2. Let $B_1,B_2\in\mathcal{B}$. A relation $\mathcal{R}\subseteq\mathcal{B}\times\mathcal{B}$ is a *rooted weak bisimulation (relation)* for B_1,B_2 iff $\mathcal{R}$ is a weak bisimulation relation with $\langle B_1,B_2\rangle\in\mathcal{R}$, and for all $a\in\mathcal{L}\cup\{\tau\}$

1. if $\exists B_1'\ B_1\overset{a}{\rightarrow}B_1'$ then $\exists B_2'\ B_2\overset{a}{\Rightarrow}B_2'$ and $\langle B_1,B_2\rangle\in\mathcal{R}$
2. if $\exists B_2'\ B_2\overset{a}{\rightarrow}B_2'$ then $\exists B_1'\ B_1\overset{a}{\Rightarrow}B_1'$ and $\langle B_1,B_2\rangle\in\mathcal{R}$

B_1 is said to be *observation congruent* to B_2, $B_1\approx^c B_2$, if there exists a rooted weak bisimulation relation for B_1,B_2.

We call an (α,β)- context $\tilde{C}$ $\approx^c$*-preserving* if $\approx^c$ is indeed a congruence with respect to substitution in $\tilde{C}$, i.e. for $\tilde{D},\tilde{E}\in\mathcal{B}^\beta$ if $\tilde{D}\approx^c\tilde{E}$ then $\tilde{C}\circ\tilde{D}\approx^c\tilde{C}\circ\tilde{E}$ (apply $\approx^c$ elementwise). No simple criterion on context transductions is known to us that is both sufficient for a context to be $\approx^c$-preserving, and liberal enough to capture most of the 'reasonable' $\approx^c$-preserving contexts (let alone necessary and sufficient conditions). We will therefore assume that we work with a *set* of contexts *Con* whose elements are built up out of combinators that do preserve $\approx^c$, as the unique solvability of (1) is indeed most relevant in that case. Useful results on the characterization of $\approx^c$-preservation of combinators on the basis of their inference rules are available. In [GV89] such characterizations are given for the preservation of strong bisimulation equivalence, together with an indication on how this result can be extended to $\approx^c$.

There is a standard way to demonstrate the *existence* of solutions to fixpoint equations, viz. by the introduction of a *fixpoint combinator*. If $\tilde{C}(\tilde{X})$ is a vector of behaviour expressions where the free process variables are all in $\tilde{X}$ of the same dimension as $\tilde{C}$, then fix $\tilde{X}.\tilde{C}(\tilde{X})$ is a vector of behaviour expressions that, as we shall see, describes a solution to the fixpoint equation. We will write $\text{fix}_i\ \tilde{X}.\tilde{C}(\tilde{X})$ for the behaviour expression that is the ith element of this vector. As we work with contexts we now get to the subtle but important difference between between a β-dimensional vector $\tilde{C}(\tilde{X})$ of behaviour expressions and and an (α,β)-context $\tilde{C}$. The process variables $\tilde{X}$ have been substituted for the holes in C, but not necessarily on a one-to-one basis, as in general $\tilde{C}(\tilde{X})$ will contain multiple occurrences of the same variable X. This multiplicity does not only arise because of the different copies of a variable X in different elements of $\tilde{C}$, but also occurs within elements, as can be seen by 'unrolling' the system.

$$\begin{array}{lcl} X \approx^c C[X,Y] & & X \approx^c C[C[X,Y],D[X,Y]] \\ & \text{into} & \\ Y \approx^c D[X,Y] & & Y \approx^c D[C[X,Y],D[X,Y]] \end{array}$$

Such unrolling plays an important role in the calculation of fixpoints, and its usual meaning is given in terms of a *substitution* procedure of the form $\tilde{D}\{\tilde{C}/\tilde{X}\}$, which is interpreted as the result of the elementwise simultaneous replacement of the occurrences of the elements of $\tilde{X}$ in $\tilde{D}(\tilde{X})$ by elements of $\tilde{C}$. To account correctly for such substitution procedures in terms of context composition we introduce the notion of *substitution factor*.

Definition 3. Let $\tilde{D}(\tilde{X})$ be a vector of behaviour expressions with α occurrences of elements of the β-dimensional vector of pairwise different process variables $\tilde{X}$, then the *substitution factor* $\theta_{\tilde{D}}$ is the unique mapping from β-vectors to α-vectors such that for all β-dimensional $\tilde{C}$

$$\tilde{D}\{\tilde{C}/\tilde{X}\} \quad \equiv \quad \tilde{D} \circ \theta_{\tilde{D}} \tilde{C}$$

Using substitution factors we may now reinterpret any system of expressions $\tilde{C}(\tilde{X})$ as a context (vector) $\tilde{C}[\theta_{\tilde{C}}\tilde{X}]$. Where this introduces no ambiguity we will drop the subscript and simplify it to $\tilde{C}[\theta\tilde{X}]$.

Having introduced this new concept we can transpose the usual definition of the fixpoint operators (as in [Mil89]) to the world of contexts. Let $\tilde{C}^\beta_\alpha$ be a context, $\tilde{X}$ an β-dimensional vector of process variables, and θ a corresponding substitution factor then the transitions (transductions) of fix $\tilde{X}.\tilde{C}^\beta_\alpha(\tilde{X})$ are completely characterized by the rule

$$\frac{\tilde{C}^\beta_\alpha \circ \theta \ \text{fix}\ \tilde{X}.\tilde{C}^\beta_\alpha(\tilde{X}) \xrightarrow{\tilde{a}} \tilde{C}'}{\text{fix}\ \tilde{X}.\tilde{C}^\beta_\alpha(\tilde{X}) \xrightarrow{\tilde{a}} \tilde{C}'}$$

The well-known existence theorem now follows.

Theorem 4. *Let $\tilde{C}^{\beta}_{\alpha}$ be a context, and $\tilde{X}$ an β-dimensional vector of process variables, then*

$$\tilde{X} \approx^{c} \tilde{C}^{\beta}_{\alpha}(\tilde{X}) \tag{4}$$

has a solution.

Proof. Standard. A solution is fix $\tilde{X}.\tilde{C}^{\beta}_{\alpha}(\tilde{X})$, which can be proved by showing that

$$\{\langle \text{fix}_i\ \tilde{X}.\tilde{C}^{\beta}_{\alpha}(\tilde{X}), \tilde{C}^{\beta}_{\alpha}(i)\{\text{fix}\ \tilde{X}.\tilde{C}^{\beta}_{\alpha}(\tilde{X})/\tilde{X}\}\rangle \mid i \in \beta\} \cup Id_{\mathcal{B}}$$

where $Id_{\mathcal{B}}$ is the identity relation on $\mathcal{B}$, is a rooted weak bisimulation for all the pairs contained in it (it is in fact a *strong bisimulation relation*, see [Mil89]). □

3 Fixpoints of abstraction-free contexts

Our first aim is to find a semantic property of contexts that corresponds to the syntactic notion of guardedness. Informally, a behaviour expression is externally guarded when a process variable cannot be 'reached' without first executing an observable action. This property can be expressed quite directly in terms of transductions, where we require that transductions cannot become dependent upon a context argument, before some observable action has been produced that does not consume actions from any of the context arguments. We refer to this property as *opacity*.

Definition 5. Let C_α be a context.

1. C_α is *transparent* if $C_\alpha \xrightarrow[\tilde{b}]{a}$ with $\tilde{b} \neq \tilde{0}$.
2. C_α is *opaque* if $C_\alpha \stackrel{\epsilon}{\Rightarrow} C'_\alpha$ implies C'_α is not transparent.

   A context $\tilde{C}^{\beta}_{\alpha}$ is opaque if for all $i \in \beta$ the $\tilde{C}^{\beta}_{\alpha}(i)$ are opaque.

   Note that opacity is a stronger property than non-transparancy.

Proposition 6. *Let C_α be an opaque context and $C_\alpha \stackrel{\epsilon}{\Rightarrow} C'_\alpha$ then C'_α is opaque.*

Proof. Follows directly from the definition of opacity. □

Proposition 7. *Let C_β be an opaque context and $\tilde{D}^{\beta}_{\alpha}$ a context, then for all $\sigma \in \{\tau^n, \tau^n a \mid n \in \omega\}$ $C_\beta \circ \tilde{D}^{\beta}_{\alpha} \stackrel{\sigma}{\rightarrow} E$ implies that there exists a C'_β $C_\beta \stackrel{\sigma}{\rightarrow} C'_\beta$ with $E \equiv C'_\beta \circ \tilde{D}^{\beta}_{\alpha}$.*

Proof. By induction on n; omitted. □

The common idea behind all additional syntactic requirements that must accompany guardedness to ensure unique solvability, is that no contexts are created in which the protective power of the guards can be undone by making them unobservable. Again, this idea can expressed in terms of transductions quite naturally. We call this additional property *abstraction-freedom*.

Definition 8. Let $AF(\alpha) \subseteq Con$ be the largest set of α-contexts satisfying that for all $C_\alpha \in AF(\alpha)$

1. $C_\alpha \xrightarrow[\tilde{b}]{\tau}$ implies $\tilde{b} \in \{0, \tau\}^\alpha$
2. $Der(C_\alpha) \subseteq AF(\alpha)$

C_α is called *abstraction-free* if $C_\alpha \in AF(\alpha)$.
$\tilde{C}^\beta_\alpha$ is called *abstraction-free* if for all $i \in \beta$ the $\tilde{C}^\beta_\alpha(i)$ are abstraction-free.

Proposition 9. *Let $\tilde{C}^\gamma_\beta$ and $\tilde{D}^\beta_\alpha$ be abstraction-free contexts then $\tilde{C}^\gamma_\beta \circ \tilde{D}^\beta_\alpha$ is an abstraction-free context.*

Proof. Omitted, see [Bri92]. □

Proposition 10. *Let $\tilde{C}^\gamma_\beta$ be an abstraction-free context and $\tilde{D}^\beta_\alpha$ an opaque context then $\tilde{C}^\gamma_\beta \circ \tilde{D}^\beta_\alpha$ is an opaque context.*

Proof. Omitted, see [Bri92]. □

The above proposition and proposition 7 are instrumental in the proof of our main theorem.

Theorem 11. *Let $\tilde{C}^\beta_\alpha$ be an opaque, and abstraction-free context, and $\tilde{X}$ a β-dimensional vector of process variables, then*

$$\tilde{X} \approx^c \tilde{C}^\beta_\alpha(\tilde{X})$$

has a unique solution modulo $\approx^c$.

Proof. We give a proof sketch only; a detailed proof can be found in [Bri92]. By theorem 4 it suffices to prove the uniqueness modulo $\approx^c$. Let $\theta_{\tilde{C}}$ be the substitution factor of $\tilde{C}^\beta_\alpha$, i.e. $\tilde{C}^\beta_\alpha(\tilde{X}) \equiv \tilde{C}^\beta_\alpha[\theta_{\tilde{C}}\tilde{X}]$. Let $\tilde{O}_1$ and $\tilde{O}_2$ be solutions of (4), then

$$R \quad = \quad \{\langle D[\theta_D\tilde{O}_1], D[\theta_D\tilde{O}_2]\rangle \mid D = D_\gamma \text{ opaque, abstraction-free}\}$$

is a weak bisimulation relation upto $\approx$ (see e.g. [Mil83]).
To see this it is sufficient to check the bisimulation conditions for $\overset{\epsilon}{\Rightarrow}$ and $\overset{\epsilon}{\Rightarrow}\overset{a}{\rightarrow}$ derivations with $a \in \mathcal{L}$, as the $\overset{\sigma}{\Rightarrow}$ cases follow by induction on $|\sigma|$. Moreover, we have to verify only that $D[\theta_D\tilde{O}_2]$ in each step simulates $D[\theta_D\tilde{O}_1]$, as the other direction follows by a symmetrical argument.

- $D[\theta_D\tilde{O}_1] \overset{\epsilon}{\Rightarrow} F$: this case follows by factorizing $D[\theta_D\tilde{O}_1] \equiv D \circ \theta_D\tilde{O}_1$ and applying proposition 7, using the preservation of opacity and abstraction-freedom under $\overset{\epsilon}{\Rightarrow}$ derivations (proposition 6 and definition 8).
- $D[\theta_D\tilde{O}_1] \overset{\epsilon}{\Rightarrow}\overset{a}{\rightarrow} F$: applying lemma 7 again, we find a D' with $D \overset{\epsilon}{\Rightarrow}\overset{a}{\rightarrow} D'$ and $F \equiv D' \circ \theta_D\tilde{O}_1 \equiv D'[\theta_D\tilde{O}_1]$. It follows that $D[\theta_D\tilde{O}_2] \overset{\epsilon}{\Rightarrow}\overset{a}{\rightarrow} D'[\theta_D\tilde{O}_2]$. Now we use that $\tilde{O}_1$ and $\tilde{O}_2$ are solutions of (4). This implies that $D'[\theta_D\tilde{O}_1] \approx^c D'[\theta_D\tilde{C}[\theta_{\tilde{C}}\tilde{O}_1]] \equiv (D' \circ \theta_D\tilde{C}) \circ \theta^*\tilde{O}_1$, with $\theta^* = \theta_{D'\{\tilde{C}^\beta_\alpha(\tilde{X})/\tilde{X}\}}$, and similarly that $D'[\theta_D\tilde{O}_2] \approx^c D'[\theta_D\tilde{C}[\theta_{\tilde{C}}\tilde{O}_2]] \equiv (D' \circ \theta_D\tilde{C}) \circ \theta^*\tilde{O}_2$. Defining $H = D' \circ \theta_D\tilde{C}$ we thus have $D'[\theta_D\tilde{O}_1] \approx^c H[\theta^*\tilde{O}_1]$ and $D'[\theta_D\tilde{O}_2] \approx^c H[\theta^*\tilde{O}_2]$. It follows by proposition 10 that H is opaque, and by proposition 9 that H is abstraction-free, so that $\langle H[\theta^*\tilde{O}_1], H[\theta^*\tilde{O}_2]\rangle \in R$.

It is easy to show that $\approx R \approx$ is a rooted bisimulation relation for all pairs in R. With this result and the fact that for all $i \in \beta$ $\langle C_\alpha^\beta(i)[\theta_{\tilde{C}}\tilde{O}_1], C_\alpha^\beta(i)[\theta_{\tilde{C}}\tilde{O}_2]\rangle \in R$ it follows that $\tilde{O}_1 \approx^c \tilde{C}_\alpha^\beta[\theta_{\tilde{C}}\tilde{O}_1] \approx^c \tilde{C}_\alpha^\beta[\theta_{\tilde{C}}\tilde{O}_2] \approx^c \tilde{O}_2$, i.e. all solutions are identical modulo $\approx^c$. □

This theorem immediately gives us the well-known syntactic results.

Example 2. The language *CCS* [Mil89] essentially consists of the combinators *action-prefix*, *summation*, *composition*, *restriction*, and *renaming* from table 1, and a fixpoint combinator. A behaviour expression $B(\tilde{X})$ is called *guarded* if every process variable in B occurs in a subexpression of the form $a.E$ with $a \in \mathcal{L}$. A behaviour expression $B(\tilde{X})$ is called *sequential* if every free process variable in B occurs only in the scope of action-prefix or summation combinators. By straightforward induction on the construction of guarded, sequential CCS contexts it can be proved that these are opaque and abstraction-free. It follows that fixpoint equations in terms of such contexts have unique solutions.

The language *Basic LOTOS* [BB87] consists of the combinators *action-prefix*, *summation*, *synchronization*, *hiding*, *disruption*, and *renaming* and a fixpoint combinator. The definition of guardedness is as with CCS. A behaviour expression $B(\tilde{X})$ is called *hiding-free* and *fixpoint-free* if no free process variable of B occurs in the scope of a hiding operator, or a fixpoint combinator, respectively. Again, one proves quite easily that guarded, hiding-free, and fixpoint-free LOTOS contexts are opaque and abstraction-free. □

Although the above shows that we have successfully generalised the well-known syntactic restrictions to properties of contexts, there is still room for considerable improvement. A context need not be abstraction-free for all actions, but only for those that matter, viz. the guarding actions. Intuitively, it should suffice that there remains at least one guarding actions for each variable, as in the equation

$$X \approx^c a.(b.X \mid \text{fix}Y.a.Y) \tag{5}$$

We revise our definitions as follows.

Definition 12. Let $L \subseteq \mathcal{L}$ and let C_α be a context.
C_α is *L-opaque* if for all $\sigma \in (\mathcal{L}\backslash L)^*$ $C_\alpha \stackrel{\sigma}{\Rightarrow} C'_\alpha$ implies that C'_α is not transparent.
$\tilde{C}_\alpha^\beta$ is *L-opaque* if for all $i \in \beta$ the $\tilde{C}_\alpha^\beta(i)$ are L-opaque.

Definition 13. Let $L \subseteq \mathcal{L}$.
$AF(L, \alpha) \subseteq Con$ is the largest set of α-contexts satisfying that for all $C_\alpha \in AF(L, \alpha)$

1. $C_\alpha \xrightarrow[\tilde{a}]{b} \wedge \exists i \in \alpha\ \ \tilde{a}(i) \in L$ implies $b \in L$
2. $Der(C_\alpha) \subseteq AF(L, \alpha)$

C_α is called *L-abstraction-free* if $C_\alpha \in AF(L, \alpha)$.
$\tilde{C}_\alpha^\beta$ is called *L-abstraction-free* if for all $i \in \beta$ the $\tilde{C}_\alpha^\beta(i)$ are L-abstraction-free.

Theorem 14. *If $\tilde{C}_\alpha^\beta$ is an L-opaque, L-abstraction-free context and $\tilde{X}$ is a β-dimensional vector of process variables then*

$$\tilde{X} \approx^c \tilde{C}_\alpha^\beta(\tilde{X})$$

has a unique solution modulo $\approx^c$.

Proof. Note that proposition 7 also holds for L-opacity. It is not difficult to check that propositions 9 and 10 also hold if the conditions of abstraction-freedom and opacity are replaced everywhere by those of L-abstraction-freedom and L-opacity, respectively. The proof of theorem 11 can now be adapted by carrying out the same replacement in the definition of R. □

Example 3. The context $a.(b.[\,] \mid \text{fix}Y.a.Y)$ is $\{b\}$-opaque and $\{b\}$-abstraction-free. It follows that equation (5) has a unique solution modulo $\approx^c$.

This in fact an instance of a result by Sanderson in [San82] that implies that all CCS fixpoint of the format $X \approx^c G[\lambda.C[X]]$ have unique solutions if neither C nor G is of the form $H[B|D[X]]$ with λ in the action sort of B, and λ is not relabelled anywhere. We invite the reader to check that this is a special case of our last theorem. □

4 Fixpoints in abstraction contexts

In this section we look at the uniqueness of fixpoints when the condition of abstraction-freedom is not fulfilled. To convince him/herself that the influence of abstraction contexts on the uniqueness of fixpoints is a subtle issue, we invite the reader to consider the solvability of the following equations.

$$X \approx^c a.(X/\{a\}) + b.X \tag{6}$$
$$X \approx^c a.(X/\{b\}) + b.X \tag{7}$$
$$X \approx^c a.(X/\{b\}) + b.(X/\{a\}) \tag{8}$$

Here, all of the contexts $C[X]$ have derivatives $D[X]$ such that $D[C[X]]$ is transparent, thus undermining the essential step in the proofs of theorems 11 and 14. In view of our previous results one could conjecture that this would be a necessary and sufficient condition for the existence of multiple solutions.

It is no so difficult to see that (6) has infinitely many solutions, as the case is quite close to (2). It is a simple exercise to prove that every solution of X in

$$X \approx^c a.(Y/\{a\}) + b.X$$
$$Y \approx^c \tau.Y + b.X$$

is a solution of (6). Applying some τ-laws for $\approx^c$ ($a.\tau.B \approx^c a.B$ and $\tau.(B+C) \approx^c \tau.(B+C)+C$) one can check that if O_Y is a solution of the equation for Y then so is $\tau.(O_Y + B)$ for arbitrary B.

A similar argument cannot be devised for (7). We can, in fact, prove that (7) has a unique solution modulo $\approx^c$. This contradicts the sufficiency of the above conjecture, and implies that the conditions of theorems 11 and 14 are generally too strict.

Although the proof can be given directly with the use of *Koomen's fair abstraction rule* [BK89], we want to prove this through a general theorem for (1,1)-contexts in terms of transduction properties. We will write C^n for the (1,1)-context that is the n-fold composition of C with itself. We also need the assumption that the context and all its derivatives preserve $\approx$. In terms of contexts of calculi like CCS this boils down to the requirement that the arguments of summation-like operators are *weakly guarded* [Mil89] (i.e. τ may also be used as a guard; in [San82] this context property is referred to as *firmly guardedness*).

Theorem 15. *Let C be a guarded (1,1)-context over a finite action alphabet, and let* $\mathrm{Der}(C)$ *be $\approx$-preserving. If*

$$X \approx^c C[X] \tag{9}$$

has more than one solution modulo $\approx^c$ then there exist $n \in \omega$, $\sigma \in \mathcal{L}^*$, and a C' such that

$$\sigma \neq \epsilon \;\wedge\; C^n \overset{\sigma}{\Rrightarrow} C' \underset{\sigma}{\overset{\epsilon}{\Rrightarrow}} \tag{10}$$

Proof. Let O_1 and O_2 be different solutions to (9) and let $\mathcal{R} \subseteq \mathcal{B} \times \mathcal{B}$ be defined by

$$\mathcal{R} = \{\langle D[O_1], D[O_2]\rangle \mid \exists\, n > 0\; D = C^n \;\vee\; C^n \overset{\sigma}{\rightarrow} D\}$$

$\approx \mathcal{R} \approx$ cannot be a weak bisimulation, since it is rooted in $\langle C[O_1], C[O_2]\rangle$ because of the guardedness of C, which would imply that $O_1 \approx^c C[O_1] \approx^c C[O_2] \approx^c O_2$. Without loss of generality we may now assume that there exist $\langle D[O_1], D[O_2]\rangle \in \mathcal{R}$ and $\sigma \in \mathcal{L}^*$ with $D[O_1] \overset{\sigma}{\Rrightarrow} E$ so that there is *no* F with $D[O_2] \overset{\sigma}{\Rrightarrow} F$ and $E \approx \mathcal{R} \approx F$. Using the construction of $\mathcal{R}$ and the fact that O_1 and O_2 are solutions of (9), we may conclude that

$$\exists \sigma_1, O_1^{(1)}\; O_1 \overset{\sigma_1}{\Rrightarrow} O_1^{(1)} \text{ and } \nexists O_2^{(1)}\; O_2 \overset{\sigma_1}{\Rrightarrow} O_2^{(1)} \wedge O_1^{(1)} \approx \mathcal{R} \approx O_2^{(1)} \tag{11}$$

Using that $O_1 \approx C[O_1]$ we find that there must be a maximal prefix σ_{11} of σ_1 that can be produced independently of O_1 by transductions of the form $C^n \underset{0^*}{\overset{\sigma_{11}}{\Rrightarrow}}$. Let n_1 be the smallest number with this property. This gives us that there exist $n_1, C^{(1)}, C'^{(1)}, \sigma_2, O_1^{(2)}$

$$C^{n_1} \underset{0^*}{\overset{\sigma_{11}}{\Rrightarrow}} C^{(1)} \underset{\sigma_2}{\overset{\sigma_{12}}{\Rrightarrow}} C'^{(1)} \tag{12}$$

with $\sigma_1 = \sigma_{11}\sigma_{12} \;\wedge\; \sigma_{11} \neq \epsilon \;\wedge\; C^{(1)}$ transparent $\wedge\, O_1 \overset{\sigma_2}{\Rrightarrow} O_1^{(2)} \;\wedge$
$C'^{(1)}[O_1^{(2)}] \approx O_1^{(1)}$

Such a transparent derivative of C^{n_1} must exist, as a similar derivation would otherwise exist for O_2, which contradicts (11). It follows from this transparency and from the guardedness of C that $\sigma_{11} \neq \epsilon$.
We claim that $\nexists O_2^{(2)}\; O_2 \overset{\sigma_2}{\Rrightarrow} O_2^{(2)} \wedge O_1^{(2)} \approx \mathcal{R} \approx O_2^{(2)}$, i.e. the situation of (11) holds again, but with the appropriate indices increased by one. For, if we assume otherwise, it would follow that also

$$C^{n_1}[O_2] \overset{\sigma_1}{\Rrightarrow} C'^{(1)}[O_2^{(2)}]$$

and combining this with the $\approx$-preservation of $C'^{(1)}$ we get $C'^{(1)}[O_2^{(2)}] \approx \mathcal{R} \approx C'^{(1)}[O_1^{(2)}] \approx O_1^{(1)}$, i.e. there exists an $O_2^{(1)} \approx C'^{(1)}[O_2^{(2)}]$ contradicting (11).
Repeating the above process we obtain also a variant op (12) with increased indices, viz. there exist $n_2, C^{(2)}, C'^{(2)}, \sigma_3, O_1^{(3)}$

$$C^{n_2} \overset{\sigma_{21}}{\underset{0^*}{\Rrightarrow}} C^{(2)} \overset{\sigma_{22}}{\underset{\sigma_3}{\Rrightarrow}} C'^{(2)} \tag{13}$$

with

$$\sigma_2 = \sigma_{21}\sigma_{22} \ \wedge\ \sigma_{21} \neq \epsilon \ \wedge\ C^{(2)} \text{ transparent } \wedge\ O_1 \overset{\sigma_3}{\Rrightarrow} O_1^{(3)} \ \wedge$$
$$C'^{(2)}[O_1^{(3)}] \approx O_1^{(2)}$$

We can conclude additionally that $C^{(1)} \overset{\epsilon}{\underset{\sigma_{21}}{\Rrightarrow}}$, as the maximality property of σ_{11} would be contradicted otherwise.
By recursively applying the argumentation process given above we obtain for all $i > 1$ $n_i, \sigma_{i1}, C^{(i)}$ such that, simplifying by putting $\sigma_i = \sigma_{i1}$,

$$C^{n_i} \overset{\sigma_i}{\Rrightarrow} C^{(i)} \overset{\epsilon}{\underset{\sigma_{i+1}}{\Rrightarrow}} \tag{14}$$

It follows from the finiteness of the action alphabet of C that there exist $1 < i < j$ such that σ_i is a prefix of σ_j: let T be the tree consisting of the nodes formed by the elements of $N = \{\sigma \mid \exists k\ \sigma \text{ a prefix of } \sigma_k\}$ and edges running from between $\sigma, \sigma a \in N$ for $a \in \mathcal{E}(C)$, then T is a finitely branching tree of infinite depth because $\{\sigma_k\}_{k>1}$ contains σ_k of arbitrary length. By König's lemma there exists an infinite branch in T, which must therefore be covered by infinitely many σ_k. Choose $i = k$ for the shortest σ_k on the branch then, because k has only finitely many predecessors, there must be a suitable σ_j on this branch with $i < j$, which extends σ_i by construction. Putting $n = \sum_{k=i}^{j} n_k$ and combining (14) for i through j we find some C' so that

$$C^n \overset{\sigma_i}{\Rrightarrow} C' \overset{\epsilon}{\underset{\sigma_j}{\Rrightarrow}}$$

and, as σ_i is a prefix of σ_j, our theorem follows. □

We apply the criterion of the above theorem to (7) in the next example.

Example 4. Although the context in (7) has two occurrences of X we can still treat it like a (1,1)-context. This is so because the two copies of X are never needed concurrently as they are in the different alternatives of the binary summation combinator $+$. The initial transductions are $a.([\ ]_0/\{b\}) + b.[\ ]_0 \xrightarrow{a} [\ ]_0/\{b\}$ and $a.([\ ]_0/\{b\}) + b.[\ ]_0 \xrightarrow{b} I_1$. As both alternatives are guarded all derivatives of $a.([\ ]_0/\{b\}) + b.[\ ]_0$ preserve $\approx$, and the context is guarded. Now suppose that

$$(a.([\ ]_0/\{b\}) + b.[\ ]_0)^n \overset{\sigma}{\Rrightarrow} C' \overset{\epsilon}{\underset{\sigma}{\Rrightarrow}}$$

then C' is transparent, which implies that σ must contain a occurrence of a. But since the only abstraction operator in C is $/\{b\}$, it is the only abstraction operator in C', so that $C' \overset{\epsilon}{\underset{\sigma}{\Rrightarrow}}$ is impossible. It follows that (7) has a unique solution modulo $\approx^c$.

It is easy to see that the argument of the above example fails for equation (8), because here the context can abstract from both a and b. The reader is invited to show that (8) indeed allows multiple solutions modulo $\approx^c$.

5 Conclusions

We have studied the uniqueness modulo observation congruence of solutions of fixpoint equations in process algebraic calculi from a new angle. We have shown how syntax-dependent conditions that are known to be sufficient for the uniqueness of solutions, such as external guardedness, sequentiality, and the absence of abstraction combinators, can be generalised to syntax-independent properties in terms of the transductions of the contexts that appear in the equations. These properties are opacity and abstraction-freedom.

The advantage of such a syntax-independent characterisation is that it gives context-specific results for all combinator signatures that respect observation congruence. The conditions of opacity and abstraction-freedom can be determined staticly only by syntax-dependent requirements that imply the characterisations in the operational transduction semantics. Thus our work can be seen as a typical meta-result that serves as a basis for deriving syntactic properties of equation systems that guarantee a unique solution in a given signature. We have shown examples of this.

We have studied two ways to improve on the condition of abstraction-freedom. First, we relativised the notions of opacity and abstraction-freedom to essential guarding actions. This was shown to give uniqueness results in a number of cases where the old definitions did not suffice. Next, we demonstrated that the relativised notion of abstraction-freedom is not necessary, by giving an equation in which a seemingly essential guard is removed by an abstraction operator, but which nevertheless has a unique solution. From this case we were able to derive an interesting result, which we formulated only for guarded (1,1)-contexts: if an equation has more than one solution then there is an observable sequence of actions that transforms the corresponding context to a transparent context that can absorb the same sequence, i.e. the sequence when produced by the context's argument can be transformed into the unobservable ϵ-sequence. The contraposition of this result gives a criterion for unique solvability.

An interesting question is whether it is possible to obtain a criterion that supplies us with a necessary and sufficient condition in the presence of guardedness. So far, we have not been able to obtain such a result. It should also be remarked that from a pragmatic viewpoint such results are less interesting than the formulation of widely applicable sufficient conditions. We have not discussed the necessity of guardedness in this paper. Our research has shown us, however, that there is not much to be gained here: only in rather unusual cases does non-opacity go together with uniqueness of solutions.

Finally, do our results generalise to fixpoint equations modulo observation equivalence? As we need the derived contexts to preserve this equivalence – which is straightforward for *congruences* – this would seem to require a signature independent characterisation of firmly guardedness as in [San82] for contexts. We have not found an obvious way to do this.

Acknowledgements This paper has benefited from discussions with several members from the Formal Methods Group, especially Rom Langerak and Arend Rensink, at the various stages in producing conjectures, proofs, and counterexamples.

References

[BB87] Tommaso Bolognesi and Ed Brinksma. Introduction to the ISO specification language LOTOS. *Computer Networks and ISDN Systems*, 14:25–59, 1987.

[BK85] J. A. Bergstra and J. W. Klop. Algebra of communicating processes with abstraction. *Theoretical Computer Science*, 37(1):77–121, 1985.

[BK89] J. A. Bergstra and J. W. Klop. Process theory based on bisimulation semantics. In J. W. de Bakker, Willem-Paul de Roever, and Grzegorz Rozenberg, editors, *Linear Time, Branching Time and Partial Order in Logics and Models for Concurrency*, volume 354 of *Lecture Notes in Computer Science*, pages 50–122. Springer-Verlag, 1989.

[Bri92] Ed Brinksma. On the uniqueness of fixpoints modulo observation congruence. Memoranda informatica, University of Twente, 1992. Forthcoming.

[DNH84] Rocco De Nicola and Matthew Hennessy. Testing equivalences for processes. *Theoretical Computer Science*, 34:83–133, 1984.

[GV89] Jan Friso Groote and Frits W. Vaandrager. Structured operational semantics and bisimulation as a congruence. In G. Ausiello, M. Dezani-Ciancaglini, and S. Ronchi Della Rocca, editors, *Automata, Languages and Programming*, volume 372 of *Lecture Notes in Computer Science*, pages 423–438. Springer-Verlag, 1989.

[Hoa85] C. A. R. Hoare. *Communicating Sequential Processes*. Prentice-Hall, 1985.

[ISO89] ISO. Information processing systems — open systems interconnection — LOTOS — a formal description technique based on the temporal ordering of observational behaviour. International Standard 8807, ISO, Geneva, February 1989. 1st Edition.

[Kra87] E. Kranakis. Fixed point equations with parameters in the projective model. *Information and Control*, 75(3), 1987.

[Lar90] Kim G. Larsen. Compositional theories based on an operational semantics of contexts. In J. W. de Bakker, W. P. de Roever, and Grzegorz Rozenberg, editors, *Stepwise Refinement of Distributed Systems — Models, Formalisms, Correctness*, volume 430 of *Lecture Notes in Computer Science*, pages 487–518. Springer-Verlag, 1990.

[Mil80] R. Milner. *A Calculus of Communicating Systems*, volume 92 of *Lecture Notes in Computer Science*. Springer-Verlag, 1980.

[Mil83] R. Milner. Calculi for synchrony and asynchrony. *Theoretical Computer Science*, 25:267–310, 1983.

[Mil89] R. Milner. *Communication and Concurrency*. Prentice-Hall, 1989.

[San82] Michael Thomas Sanderson. *Proof Techniques for CCS*. PhD thesis, University of Edinburgh, November 1982.

This article was processed using the LaTeX macro package with LLNCS style

Inaction	$\tilde{C}^{\beta}_{\alpha} \xrightarrow[\tilde{0}]{\tilde{0}} \tilde{C}^{\beta}_{\alpha}$	**for all** $\tilde{C}^{\beta}_{\alpha}$
Action-prefix	$a.[\ ]_0 \xrightarrow[0]{a} I_1$	
Summation	$\sum_{i\in\alpha}[\ ]_i \xrightarrow[j(a)]{a} \Pi^{\alpha}_j$	**for all** $j \in \alpha$, $j(a)$ **an** α**-dimensional vector with** j^{th} **element is** a **and other elements are 0**
Composition	$[\ ]_0 \mid [\ ]_1 \xrightarrow[a^*]{a} [\ ]_0 \mid [\ ]_1$	**for** $a^* \in \{\langle a,0\rangle, \langle 0,a\rangle\}$
	$[\ ]_0 \mid [\ ]_1 \xrightarrow[\langle a,\bar{a}\rangle]{\tau} [\ ]_0 \mid [\ ]_1$	
Synchronization	$[\ ]_0 \mid_G [\ ]_1 \xrightarrow[a^*]{a} [\ ]_0 \mid_G [\ ]_1$	**for** $a^* \in \{\langle a,0\rangle, \langle 0,a\rangle\}$ **and** $a \notin G$
	$[\ ]_0 \mid_G [\ ]_1 \xrightarrow[\langle a,a\rangle]{a} [\ ]_0 \mid_G [\ ]_1$	**for** $a \in G$
Restriction	$[\ ]_0 \backslash A \xrightarrow[a]{a} [\ ]_0 \backslash A$	**for** $a \notin A$
Hiding	$[\ ]_0 / A \xrightarrow[a]{a} [\ ]_0 / A$	**for** $a \notin A$
	$[\ ]_0 / A \xrightarrow[a]{\tau} [\ ]_0 / A$	**for** $a \in A$
Disruption	$[\ ]_0 [> [\ ]_1 \xrightarrow[\langle a,0\rangle]{a} [\ ]_0 [> [\ ]_1$	
	$[\ ]_0 [> [\ ]_1 \xrightarrow[\langle 0,a\rangle]{a} \Pi^2_1$	
Renaming	$[\ ]_0[S] \xrightarrow[a]{S(a)} [\ ]_0[S]$	**for** $S : \mathcal{L} \to \mathcal{L}$ **a relabelling mapping**
Identity	$I_\alpha \xrightarrow[\tilde{a}]{\tilde{a}} I_\alpha$	**for** $\tilde{a} \in (\mathcal{L} \cup \{\tau, 0\})^\alpha$
Projection	$\Pi^{\alpha}_j \xrightarrow[j(a)]{a} \Pi^{\alpha}_j$	**for** $j \in \alpha$ **and** $j(a)$ **defined as above**

Table 1. Some combinators as contexts

Verification of Parallel Systems via Decomposition*

Jan Friso Groote[†]
CWI, Amsterdam

Faron Moller[‡]
University of Edinburgh

Abstract

Recently, Milner and Moller have presented several decomposition results for processes. Inspired by these, we investigate decomposition techniques for the verification of parallel systems. In particular, we consider those of the form

$$p_1 \parallel p_2 \parallel \cdots \parallel p_m \quad = \quad q_1 \parallel q_2 \parallel \cdots \parallel q_n \qquad (\star)$$

where p_i and q_j are (finite-) state systems, and $\parallel$ denotes parallel composition. We provide a decomposition procedure for all p_i and q_j and give criteria that must be checked on the decomposed processes to see whether $(\star)$ does or does not hold. We analyse the complexity of our procedure and show that it is polynomial in n, m and the sizes of p_i and q_j if there is no communication. We also show that with communication the verification of $(\star)$ is co-NP hard, which makes it very unlikely that a polynomial complexity bound exists. But by applying our decomposition technique to Milner's cyclic scheduler we show that verification can become polynomial in space and time for practical examples, where standard techniques are exponential.

1 Introduction

Most common techniques for the automated verification of parallel systems are based on some kind of state-space exploration. Contemporary computer technology limits exploration to state spaces of about 10^7 states. However, state spaces of most parallel systems are substantially larger.

This problem is identified by many researchers, and various solutions have been proposed. For instance one may apply minimisation techniques when constructing state spaces [2], one may represent the state space using hash techniques [12], or one may restrict the state space using some additional information [7]. A more successful approach seems to be the smart encoding of state spaces, employing the regularity that is often present in the state spaces of parallel systems. In particular, the results based on binary decision diagrams (BDD's) seem more than promising [3]. An argument that one could raise against BDD's is that it is not directly based on notions inherent to processes, such as amount of communication, the structure of processes or the structure of communication, etc. This may obscure the true causes of the success of BDD's, and it may hinder further developments and a proper understanding of applicability.

Recently, some interesting decomposition results have emerged in process theory [17, 18]. Inspired by these results, we study whether decomposition techniques can be applied in order to obtain alternative means for the verification of parallel systems. Basically, the idea is as follows: Consider processes $p = \|_{i=1}^{m} p_i$ and $q = \|_{j=1}^{n} q_j$. We want to establish whether $p = q$ where '=' represents some reasonable process equivalence. In order to do so, we decompose each p_i into $p_{i1} \ldots p_{in}$ and each q_j into $q_{j1} \ldots q_{jm}$ according to some particular decomposition rules. Then we must verify whether $p_{ij} = q_{ji}$ for all i and j. The method is beneficial if the combination

*Research supported by ESPRIT BRA Grant No. 3006 — CONCUR
[†]Present address: Dept of Philosophy, University of Utrecht, Box 80126, 3508 TC Utrecht, The Netherlands.
[‡]Dept of Computer Science, University of Edinburgh, The King's Buildings, Edinburgh EH9 3JZ, Scotland.

of performing the decompositions of the p_i's and q_j's along with checking each $p_{ij} = q_{ji}$ is considerably more efficient than checking $p = q$ directly. We show that this is indeed so in particular cases, but we show also that it is very unlikely to be true in general.

This paper first presents the decomposition scheme (after some preliminaries). Then we analyse what we have actually gained. It turns out that when there is no communication, verification via decomposition has a polynomial time and space complexity in the number and sizes of the processes p_i and q_j. In the case where communication is allowed, we provide a straightforward proof that verification is co-NP hard even in the case where the p_i's and q_j's are finite and determinate. More results of this kind can be found in [19]. Hence, polynomial verification is rather unlikely in this case.

In order to understand whether this intractability result rules out application of our techniques, we consider an example. This is Milner's scheduler [15], which is generally used as a benchmark for verification tools [6, 10, 13], due to its simple description, and its exponentially growing state spaces that it generates (in the number of 'cyclers' from which the scheduler is constructed). Verification via decomposition uses only polynomial time and linear space. The largest intermediate state space that is used in the verification has size $3k$ where k is the number of cyclers in the scheduler.

Our conclusions from the complexity analysis is that decomposition can indeed be a good technique for the verification of parallel systems. When there is little communication, i.e. in the case where the system has been adequately structured, the benefits of this technique may be especially high.

2 Preliminaries

In this paper we do not employ a particular process language. Rather, it turns out to be handy to work in a setting where processes are viewed as (possibly infinite) transition systems.

Definition 2.1. A *transition system* (*TS*) $p = (S_p, \alpha_p, \longrightarrow_p, s_p)$ is a four tuple, where

- S_p is a set of *states*;
- α_p is a set of *actions*;
- $\longrightarrow_p \subseteq S_p \times \alpha_p \times S_p$ is a *transition relation*; and
- $s_p \in S_p$ is the *initial state* of the transition system.

We use p, q, r to range over transition systems, and α to range over sets of actions. We often write $(t, a, t') \in \longrightarrow_p$ as $t \xrightarrow{a}_p t'$. We also write $t \xrightarrow{a_1 \cdots a_n}_p t'$ for $t \xrightarrow{a_1}_p \cdots \xrightarrow{a_n}_p t'$. A function α gives the set of actions of a transition system, e.g. $\alpha\Big((S_p, \alpha_p, \longrightarrow_p, s_p)\Big) = \alpha_p$. The TS p is *finite-state* if S_p is finite, and it is *finite* if there is no infinite sequence $t_1 \xrightarrow{a_1}_p t_2 \xrightarrow{a_2}_p \cdots \xrightarrow{a_{i-1}}_p t_i \xrightarrow{a_i}_p t_{i+1} \cdots$.

Definition 2.2. A TS $p = (S, \alpha, \rightarrow, s)$ is called *determinate* with respect to some equivalence relation $\sim$ iff for all $t \in S$ and $a \in \alpha$, $t \xrightarrow{a} t_1$ and $t \xrightarrow{a} t_2$ implies $t_1 \sim t_2$. In general it will be clear which equivalence relation is meant, in which case we will simply say that p is determinate.

Definition 2.3. Let α be a set of actions. We have the following 'standard' transition systems.

- The *willing* process on α is the process that can always do an action from α:

$$W_\alpha \stackrel{def}{=} \Big(\{s\}, \alpha, \longrightarrow, s\Big) \quad \text{where} \quad \longrightarrow = \Big\{\langle s, a, s\rangle \mid a \in \alpha\Big\}.$$

- The *nil* process is not willing to do anything: $nil \stackrel{def}{=} W_\emptyset$.

Definition 2.4. Let $p = (S_p, \alpha_p, \longrightarrow_p, s_p)$ and $q = (S_q, \alpha_q, \longrightarrow_q, s_q)$ be TS's. We can define the following useful operations on TS's.

- For an action a the *a-prefix* of p is the TS
$$a{:}p \stackrel{def}{=} \left(S_p \cup \{s\}, \alpha_p \cup \{a\}, \longrightarrow_p \cup \{(s, a, s_p)\}, s\right) \qquad \text{for } s \notin S_p.$$

- Assuming (without loss of generality) $S_p \cap S_q = \emptyset$, the *sum* or *choice* of p and q is the TS
$$p + q \stackrel{def}{=} \left(S_p \cup S_q \cup \{s_{p+q}\}, \alpha_p \cup \alpha_q, \longrightarrow_{p+q}, s_{p+q}\right) \qquad \text{for } s_{p+q} \notin S_p \cup S_q,$$
where $\longrightarrow_{p+q} = \longrightarrow_p \cup \longrightarrow_q \cup \left\{\langle s_{p+q}, a, s'\rangle \mid s_p \xrightarrow{a}_p s' \text{ or } s_q \xrightarrow{a}_q s'\right\}$.

- The *parallel composition* or *synchronisation merge* [11] of p and q is the TS
$$p \parallel q \stackrel{def}{=} \left(S_p \times S_q, \alpha_p \cup \alpha_q, \longrightarrow_{p\parallel q}, \langle s_p, s_q\rangle\right)$$
$$\text{where } \langle s_1, s_2\rangle \xrightarrow{a}_{p\parallel q} \langle s_1', s_2'\rangle \text{ iff } \begin{cases} s_1 \xrightarrow{a}_p s_1' \text{ and } s_2 \xrightarrow{a}_q s_2', \text{ or} \\ s_1 \xrightarrow{a}_p s_1',\ s_2 = s_2' \text{ and } a \notin \alpha_q, \text{ or} \\ s_2 \xrightarrow{a}_q s_2',\ s_1 = s_1' \text{ and } a \notin \alpha_p. \end{cases}$$
The synchronisation merge thus forces common actions to synchronise. We often write $\|_{i=1}^{n} p_i$ for $p_1 \parallel p_2 \parallel \cdots \parallel p_n$ and $\|_{i=1, i\neq k}^{n}$ for $p_1 \parallel \ldots \parallel p_{k-1} \parallel p_{k+1} \parallel \ldots \parallel p_n$. It is clear from the definition that the associativity of the composition operator is immaterial.

- Let α_1, α_2 be two sets of actions (which in our applications will naturally, but not necessarily, satisfy $\alpha_1 \subseteq \alpha_2$). The (α_1, α_2)-*projection* of p is the TS
$$\upharpoonright_{\alpha_2}^{\alpha_1}(p) \stackrel{def}{=} \left(S_p, \alpha_2 \cap \alpha_p, \xrightarrow{a}_{\upharpoonright_{\alpha_2}^{\alpha_1}(p)}, s_p\right)$$
$$\text{where } s \xrightarrow{a}_{\upharpoonright_{\alpha_2}^{\alpha_1}(p)} s' \text{ iff } \begin{cases} s \xrightarrow{b_1\cdots b_n a}_p s' \text{ with } b_i \notin \alpha_2 \ \&\ a \in \alpha_1 \cap \alpha_2, \text{ or} \\ s \xrightarrow{a}_p s' \text{ for } a \in \alpha_2. \end{cases}$$
The projection operator $\upharpoonright$ is also used for traces: $(a_1 \cdots a_n) \upharpoonright_\alpha$ is the trace $a_1 \cdots a_n$ from which the actions $a_i \notin \alpha$ are removed.

Remark 2.5. The projection operator $\upharpoonright_{\alpha_2}^{\alpha_1}$ has, as far as we know, not appeared in the literature. In this article, it is solely introduced for the purpose of defining the decompositions. For an idea how this operator works, consider the process p given in the diagram. This represents a transition system with actions a, b and c, states s_1, s_2, s_3, s_4 and s_5, initial state s_1, and a transition relation as suggested by the arrows. Clearly p is the result of composing $p_1 = b{:}a{:}nil$ and $p_2 = c{:}a{:}nil$ in parallel. Using the projection operator $\upharpoonright_{\alpha_2}^{\alpha_1}$ we can project p onto its parallel components, where α_1 contains those actions through which the components communicate and α_2 contains all the actions of that component. That is, $p_1 = \upharpoonright_{\{a,b\}}^{\{a\}}(p)$ and $p_2 = \upharpoonright_{\{a,c\}}^{\{a\}}(p)$. In the composition, the actions a and b appear in p_1, a and c appear in p_2, and a is the action through which p_1 and p_2 communicate. Note that when calculating p_1 and p_2, the possibility of extending actions backwards is essentially used. Also note that if we take $\alpha_1 = \emptyset$, then the projection operator $\upharpoonright_{\alpha_2}^{\emptyset}(p)$ behaves as the encapsulation operator $\partial_{\alpha(p)\setminus\alpha_2}(p)$ from ACP [1] and the restriction operator $p\backslash\left(\alpha(p) \setminus \alpha_2\right)$ from CCS [16].

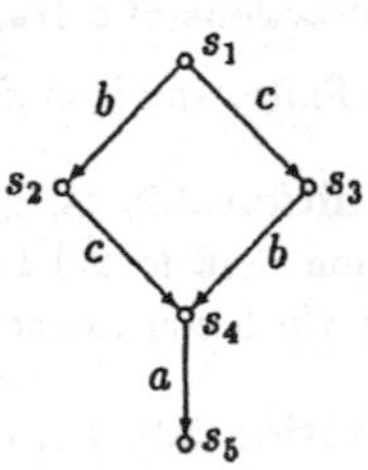

Remark 2.6. We now have three ways of specifying transition systems. We can describe them explicitly, we can write them down algebraically using the operators that have just been introduced, or we can draw a diagram. In this paper, we also specify transition systems by simple

recursive equations containing only choice, action prefix and a single variable. A construction that is sufficient for the examples in this paper is the following. Consider an equation $X = e(X)$ where e consists of action prefixes and choices only. Define the self-loop TS $r = \big(\{s\}, \{\star\}, \{(s, \star, s)\}, s\big)$ where $\star \notin \alpha\big(e(nil)\big)$. Construct the TS $e(r) = (S, \alpha, \longrightarrow, t)$. The TS defined by the equation is then the TS $p = (S, \alpha \setminus \{\star\}, \longrightarrow_p, t)$ where

$$\longrightarrow_p = \Big(\longrightarrow \cap \, (S \times \alpha(e(nil)) \times S)\Big) \cup \Big\{\langle t_1, a, t_2\rangle \mid t_1 \xrightarrow{\star} t_1 \text{ and } t \xrightarrow{a} t_2\Big\}.$$

For the examples in this paper, this definition coincides with the generally accepted interpretation of equations.

Remark 2.7. We can give operational characterisations of the above operators. We do not go into this any further except to list them as follows, and refer the interested yet uninitiated reader to e.g. [9] for understanding in interpreting these.

$$W_{\alpha\cup\{a\}} \xrightarrow{a} W_{\alpha\cup\{a\}} \qquad \frac{q \xrightarrow{a} q'}{p+q \xrightarrow{a} q'} \qquad \frac{p \xrightarrow{a} p' \quad q \xrightarrow{a} q'}{p \parallel q \xrightarrow{a} p' \parallel q'}$$

$$a{:}p \xrightarrow{a} p \qquad \frac{p \xrightarrow{a} p'}{p \parallel q \xrightarrow{a} p' \parallel q}\ \big(a \notin \alpha(q)\big) \qquad \frac{p \xrightarrow{a} p'}{\restriction^{\alpha_1}_{\alpha_2}(p) \xrightarrow{a} \restriction^{\alpha_1}_{\alpha_2}(p')}\ \big(a \in \alpha_2\big)$$

$$\frac{p \xrightarrow{a} p'}{p+q \xrightarrow{a} p'} \qquad \frac{q \xrightarrow{a} q'}{p \parallel q \xrightarrow{a} p \parallel q'}\ \big(a \notin \alpha(p)\big) \qquad \frac{p \xrightarrow{b} p' \quad \restriction^{\alpha_1}_{\alpha_2}(p') \xrightarrow{a} p''}{\restriction^{\alpha_1}_{\alpha_2}(p) \xrightarrow{a} p''}\ \big(a \in \alpha_1, b \notin \alpha_2\big)$$

3 Basic axioms

We will prove our results using axioms for $\parallel$, $\restriction$ and W only. In this section we introduce these. The axioms hold in strong bisimulation semantics, and therefore in most other reasonable semantics as well.

Definition 3.1. Let $p = (S_p, \alpha_p, \longrightarrow_p, s_p)$ and $q = (S_q, \alpha_q, \longrightarrow_q, s_q)$ be TS's. We call a relation $\mathcal{R} \subseteq S_p \times S_q$ a *(p, q)-bisimulation relation* iff $t\mathcal{R}u$ implies

1. if $t \xrightarrow{a}_p t'$ then $u \xrightarrow{a}_q u'$ for some $u' \in S_q$ with $t'\mathcal{R}u'$; and
2. if $u \xrightarrow{a}_q u'$ then $t \xrightarrow{a}_p t'$ for some $t' \in S_p$ with $t'\mathcal{R}u'$.

Two states $t \in S_p$ and $u \in S_q$ are *(p, q)-bisimilar*, written $t \leftrightarrow_{p,q} u$, iff there is a (p, q)-bisimulation relation $\mathcal{R}$ relating t and u. We abbreviate $\leftrightarrow_{p,p}$ by $\leftrightarrow_p$. The two TS's p and q are bisimilar, written $p \leftrightarrow q$, if $\alpha(p) = \alpha(q)$ and $s_p \leftrightarrow_{p,q} s_q$.

Lemma 3.2 (Congruence). *$\leftrightarrow$ is a congruence with respect to action prefix, choice, parallel composition and (α_1, α_2)-projection.*

Proof. Standard. □

The axioms that we use are presented in table 1. We do not strive for completeness of the axiomatisation. Rather, the axioms need only be sufficiently complete to satisfy our goal.

Lemma 3.3 (Soundness). *The axioms in table 1 are sound with respect to $\leftrightarrow$.*

$\Vert_1$.	$p \parallel (q \parallel r) = (p \parallel q) \parallel r$	
$\Vert_2$.	$p \parallel q = q \parallel p$	
R_1.	$p = \lceil^{\alpha}_{\alpha(p)} (p)$	
R_2.	$\lceil^{\alpha_1}_{\alpha_2} (p) = \lceil^{\alpha_1}_{\alpha_2 \cap \alpha(p)} (p)$	
R_3.	$\lceil^{\alpha_1}_{\alpha_2} (p) = \lceil^{\alpha_1 \cap \alpha_2}_{\alpha_2} (p)$	
R_4.	$\lceil^{\alpha_1}_{\alpha_2} (p) = \lceil^{\alpha_1}_{\alpha_2} \left(\lceil^{\alpha_1 \cup \alpha}_{\alpha_2 \cup \alpha} (p)\right)$	if $\alpha_2 \cap \alpha = \emptyset$
R_5.	$\lceil^{\alpha_1}_{\alpha_2} (p \parallel q) = \lceil^{\alpha_1}_{\alpha_2} (p) \parallel \lceil^{\alpha_1}_{\alpha_2} (q)$	if $\alpha_1 \subseteq \alpha(p) \cap \alpha(q) \subseteq \alpha_2$
R_6.	$p = p \parallel \lceil^{\alpha}_{\alpha} (p)$	if $\lceil^{\alpha}_{\alpha} (p)$ is determinate
R_7.	$\lceil^{\alpha}_{\emptyset} (p) = nil$	
W_1.	$p \parallel W_{\alpha(p)} = p$	
W_2.	$W_{\alpha_1} \parallel W_{\alpha_2} = W_{\alpha_1 \cup \alpha_2}$	
W_3.	$\lceil^{\alpha_1}_{\alpha_2} (W_{\alpha}) = W_{\alpha_2 \cap \alpha}$	

Table 1: Basic axioms for operators

Proof. For each axiom, we must construct an appropriate bisimulation relation. Let $p = (S_p, \alpha_p, \longrightarrow_p, s_p)$ and $q = (S_q, \alpha_q, \longrightarrow_q, s_q)$. We present proofs only for axioms R_4, R_5 and R_6.

R_4. We will show that the identity relation on S_p is a $\left(\lceil^{\alpha_1}_{\alpha_2} (p), \lceil^{\alpha_1}_{\alpha_2} (\lceil^{\alpha_1 \cup \alpha}_{\alpha_2 \cup \alpha} (p))\right)$-bisimulation. Suppose then that $s \xrightarrow{a}_{\lceil^{\alpha_1}_{\alpha_2}(p)} s'$. We will show that $s \xrightarrow{a}_{\lceil^{\alpha_1}_{\alpha_2}(\lceil^{\alpha_1 \cup \alpha}_{\alpha_2 \cup \alpha}(p))} s'$. We know that $a \in \alpha_2$. We distinguish between the following two cases.

(a) $a \notin \alpha_1$. Then, $s \xrightarrow{a}_p s'$ and thus $s \xrightarrow{a}_{\lceil^{\alpha_1 \cup \alpha}_{\alpha_2 \cup \alpha}(p)} s'$. Therefore $s \xrightarrow{a}_{\lceil^{\alpha_1}_{\alpha_2}(\lceil^{\alpha_1 \cup \alpha}_{\alpha_2 \cup \alpha}(p))} s'$.

(b) $a \in \alpha_1$. Then $s \xrightarrow{b_1 \cdots b_n a}_p s'$ for some $b_i \notin \alpha_2$. Thus $s \xrightarrow{(b_1 \cdots b_n) \lceil_\alpha a}_{\lceil^{\alpha_1 \cup \alpha}_{\alpha_2 \cup \alpha}(p)} s'$. Therefore we have that $s \xrightarrow{a}_{\lceil^{\alpha_1}_{\alpha_2}(\lceil^{\alpha_1 \cup \alpha}_{\alpha_2 \cup \alpha}(p))} s'$.

Now suppose that $s \xrightarrow{a}_{\lceil^{\alpha_1}_{\alpha_2}(\lceil^{\alpha_1 \cup \alpha}_{\alpha_2 \cup \alpha}(p))} s'$. We will show that $s \xrightarrow{a}_{\lceil^{\alpha_1}_{\alpha_2}(p)} s'$. We know that $a \in \alpha_2$. We distinguish between the following two cases.

(a) $a \notin \alpha_1$. Then $s \xrightarrow{a}_{\lceil^{\alpha_1 \cup \alpha}_{\alpha_2 \cup \alpha}(p)} s'$. From the side condition that $\alpha_2 \cap \alpha = \emptyset$ we know that $a \notin \alpha$. Therefore $s \xrightarrow{a}_p s'$ and hence $s \xrightarrow{a}_{\lceil^{\alpha_1}_{\alpha_2}(p)} s'$.

(b) $a \in \alpha_1$. Then $s \xrightarrow{b_1 \cdots b_n a}_{\lceil^{\alpha_1 \cup \alpha}_{\alpha_2 \cup \alpha}(p)} s'$ for some $b_i \notin \alpha_2$. So $b_i \in \alpha$ and therefore $s \xrightarrow{c^1_1 \cdots c^1_{m_1} b_1}_p \ldots \xrightarrow{c^n_1 \cdots c^n_{m_n} b_n}_p \xrightarrow{c^{n+1}_1 \cdots c^{n+1}_{m_{n+1}} a}_p s'$ with $c^i_j \notin \alpha_2 \cup \alpha$. Hence $s \xrightarrow{a}_{\lceil^{\alpha_1}_{\alpha_2}(p)} s'$.

R_5. We will show that the identity relation on $S_p \times S_q$ is a $\left(\lceil^{\alpha_1}_{\alpha_2} (p \parallel q), \lceil^{\alpha_1}_{\alpha_2} (p) \parallel \lceil^{\alpha_1}_{\alpha_2} (q)\right)$-bisimulation. Suppose then that $(s_1, s_2) \xrightarrow{a}_{\lceil^{\alpha_1}_{\alpha_2}(p \parallel q)} (s_1', s_2')$. We will show that $(s_1, s_2) \xrightarrow{a}_{\lceil^{\alpha_1}_{\alpha_2}(p) \parallel \lceil^{\alpha_1}_{\alpha_2}(q)} (s_1', s_2')$. We know that $a \in \alpha_2$. We distinguish between the following two cases:

(a) $a \notin \alpha_1$. Then $(s_1, s_2) \xrightarrow{a}_{p\|q} (s_1', s_2')$. We distinguish between the following three subcases:

i. $a \in \alpha_p \cap \alpha_q$. Then $s_1 \xrightarrow{a}_p s_1'$ and $s_2 \xrightarrow{a}_q s_2'$. Therefore $s_1 \xrightarrow{a}_{\lceil^{\alpha_1}_{\alpha_2}(p)} s_1'$ and $s_2 \xrightarrow{a}_{\lceil^{\alpha_1}_{\alpha_2}(q)} s_2'$. Hence $(s_1, s_2) \xrightarrow{a}_{\lceil^{\alpha_1}_{\alpha_2}(p)\|\lceil^{\alpha_1}_{\alpha_2}(q)} (s_1', s_2')$.

ii. $a \notin \alpha_q$. Then $s_1 \xrightarrow{a}_p s_1'$ and $s_2 \equiv s_2'$. Thus $s_1 \xrightarrow{a}_{\lceil^{\alpha_1}_{\alpha_2}(p)} s_1'$ and again we have that $(s_1, s_2) \xrightarrow{a}_{\lceil^{\alpha_1}_{\alpha_2}(p)\|\lceil^{\alpha_1}_{\alpha_2}(q)} (s_1', s_2')$.

iii. $a \notin \alpha_p$. This case is symmetric to the case above.

(b) $a \in \alpha_1$. Then $(s_1, s_2) \xrightarrow{b_1 \cdots b_n a}_{p\|q} (s_1', s_2')$ for some $b_i \notin \alpha_2$. As $a \in \alpha_1$, we know from the side condition that $\alpha_1 \subseteq \alpha_p \cap \alpha_q$ that $a \in \alpha_p \cap \alpha_q$. Hence $s_1 \xrightarrow{(b_1 \cdots b_n)\lceil_{\alpha_p} a}_p s_1'$ and $s_2 \xrightarrow{(b_1 \cdots b_n)\lceil_{\alpha_q} a}_q s_2'$. Thus $s_1 \xrightarrow{a}_{\lceil^{\alpha_1}_{\alpha_2}(p)} s_1'$ and $s_2 \xrightarrow{a}_{\lceil^{\alpha_1}_{\alpha_2}(q)} s_2'$ and we therefore have $(s_1, s_2) \xrightarrow{a}_{\lceil^{\alpha_1}_{\alpha_2}(p)\|\lceil^{\alpha_1}_{\alpha_2}(q)} (s_1', s_2')$.

Now suppose $(s_1, s_2) \xrightarrow{a}_{\lceil^{\alpha_1}_{\alpha_2}(p)\|\lceil^{\alpha_1}_{\alpha_2}(q)} (s_1', s_2')$. We will show $(s_1, s_2) \xrightarrow{a}_{\lceil^{\alpha_1}_{\alpha_2}(p\|q)} (s_1', s_2')$. We know that $a \in \alpha_2$. We distinguish between the following three cases.

(a) $a \in \alpha_p \cap \alpha_q$. Then $a \in \alpha(\lceil^{\alpha_1}_{\alpha_2}(p))$ and $a \in \alpha(\lceil^{\alpha_1}_{\alpha_2}(q))$. Hence $s_1 \xrightarrow{a}_{\lceil^{\alpha_1}_{\alpha_2}(p)} s_1'$ and $s_2 \xrightarrow{a}_{\lceil^{\alpha_1}_{\alpha_2}(q)} s_2'$. Therefore $s_1 \xrightarrow{b_1 \cdots b_n a}_p s_1'$ and $s_2 \xrightarrow{c_1 \cdots c_m a}_q s_2'$ for some $b_i, c_j \notin \alpha_2$. From the side condition that $\alpha_p \cap \alpha_q \subseteq \alpha_2$, we know that $b_i, c_j \notin \alpha_p \cap \alpha_q$. Therefore $(s_1, s_2) \xrightarrow{b_1 \cdots b_n c_1 \cdots c_m a}_{p\|q} (s_1', s_2')$. Hence $(s_1, s_2) \xrightarrow{a}_{\lceil^{\alpha_1}_{\alpha_2}(p\|q)} (s_1', s_2')$.

(b) $a \notin \alpha_q$. Then $s_2 \equiv s_2'$ and $s_1 \xrightarrow{a}_{\lceil^{\alpha_1}_{\alpha_2}(p)} s_1'$. From the side condition $\alpha_1 \subseteq \alpha_q$, we know $a \notin \alpha_1$. Therefore $s_1 \xrightarrow{a}_p s_1'$. Thus $(s_1, s_2) \xrightarrow{a}_{p\|q} (s_1', s_2')$ and so again $(s_1, s_2) \xrightarrow{a}_{\lceil^{\alpha_1}_{\alpha_2}(p\|q)} (s_1', s_2')$.

(c) $a \notin \alpha_p$. This case is symmetric to the previous case.

R_6. It is sufficient to prove that the relation

$$\mathcal{R} \stackrel{def}{=} \left\{ \langle s, (t, u) \rangle \mid s \leftrightarrows_p t \ \&\ \exists\, v \leftrightarrows_{\lceil^{\alpha}_{\alpha}(p)} u \text{ such that } v \xrightarrow{b_1 \cdots b_n}_p s \text{ for some } b_i \notin \alpha \right\}$$

is a $\left(p, p \,\|\, \lceil^{\alpha}_{\alpha}(p)\right)$-bisimulation relation. Suppose then that $s\mathcal{R}(t, u)$ and $s \xrightarrow{a}_p s'$. We must show that $(t, u) \xrightarrow{a}_{p\|\lceil^{\alpha}_{\alpha}(p)} (t', u')$ where $s'\mathcal{R}(t', u')$. We know that $t \xrightarrow{a}_p t'$ where $s' \leftrightarrows_p t'$ and that there is some $v \leftrightarrows_{\lceil^{\alpha}_{\alpha}(p)} u$ such that $v \xrightarrow{b_1 \cdots b_n a}_p s'$ for some $b_i \notin \alpha$. We distinguish between the following two cases.

(a) $a \in \alpha$. Then $v \xrightarrow{a}_{\lceil^{\alpha}_{\alpha}(p)} s'$ and thus $u \xrightarrow{a}_{\lceil^{\alpha}_{\alpha}(p)} u'$ where $s' \leftrightarrows_{\lceil^{\alpha}_{\alpha}(p)} u'$. Hence $(t, u) \xrightarrow{a}_{p\|\lceil^{\alpha}_{\alpha}(p)} (t', u')$ and $s'\mathcal{R}(t', u')$.

(b) $a \notin \alpha$. Then $(t, u) \xrightarrow{a}_{p\|\lceil^{\alpha}_{\alpha}(p)} (t', u)$ and $s'\mathcal{R}(t', u)$.

Now suppose that $s\mathcal{R}(t, u)$ and $(t, u) \xrightarrow{a}_{p\|\lceil^{\alpha}_{\alpha}(p)} (t', u')$. We must show that $s \xrightarrow{a}_p s'$ where $s'\mathcal{R}(t', u')$. We know that there is some $v \leftrightarrows_{\lceil^{\alpha}_{\alpha}(p)} u$ such that $v \xrightarrow{b_1 \cdots b_n}_p s$ for some $b_i \notin \alpha$. Furthermore, $t \xrightarrow{a}_p t'$ so $s \xrightarrow{a}_p s'$ where $s' \leftrightarrows_p t'$. It remains to show that $s'\mathcal{R}(t', u')$. We distinguish between the following two cases.

(a) $a \in \alpha$. Then $v \xrightarrow{a}_{\lceil^{\alpha}_{\alpha}(p)} s'$ and $u \xrightarrow{a}_{\lceil^{\alpha}_{\alpha}(p)} u'$, so $v \xrightarrow{a}_{\lceil^{\alpha}_{\alpha}(p)} v'$ where $u' \leftrightarrows_{\lceil^{\alpha}_{\alpha}(p)} v'$. From the side condition that $\lceil^{\alpha}_{\alpha}(p)$ is determinate (with respect to $\leftrightarrows_{\lceil^{\alpha}_{\alpha}(p)}$), we have that $s' \leftrightarrows_{\lceil^{\alpha}_{\alpha}(p)} v'$. Hence $s' \leftrightarrows_{\lceil^{\alpha}_{\alpha}(p)} u'$ and thus $s'\mathcal{R}(t', u')$.

(b) $a \notin \alpha$. Then $u \equiv u'$, so again we have that $s'\mathcal{R}(t', u)$.

□

Example 3.4. The following examples show why the conditions in R_4, R_5 and R_6 of the last theorem are necessary. For the condition in R_4, observe that

$$\lceil^{\emptyset}_{\{b\}}\ (a{:}b{:}nil) \leftrightarrows nil_b \quad \text{whereas} \quad \lceil^{\emptyset}_{\{b\}} \left(\lceil^{\{b\}}_{\{b\}}\ (a{:}b{:}nil) \right) \leftrightarrows b{:}nil.$$

By nil_b, we mean the TS *nil* with alphabet $\{b\}$, which can be defined by $\lceil^{\emptyset}_{\{b\}}\ (a{:}b{:}nil)$. For the first condition in R_5, observe that

$$\lceil^{\{c\}}_{\{b,c\}} \left((a{:}nil + b{:}nil) \parallel c{:}nil \right) \leftrightarrows b{:}nil \parallel c{:}nil\ +\ c{:}nil \quad \text{whereas}$$

$$\lceil^{\{c\}}_{\{b,c\}}\ (a{:}nil + b{:}nil)\ \parallel \lceil^{\{c\}}_{\{b,c\}}\ (c{:}nil) \leftrightarrows b{:}nil \parallel c{:}nil.$$

For the second condition in R_5, observe that

$$\lceil^{\{a\}}_{\{a\}} \left(b{:}a{:}nil \parallel (a{:}nil + b{:}nil) \right) \leftrightarrows nil_a \quad \text{whereas}$$

$$\lceil^{\{a\}}_{\{a\}}\ (b{:}a{:}nil)\ \parallel \lceil^{\{a\}}_{\{a\}}\ (a{:}nil + b{:}nil) \leftrightarrows a{:}nil.$$

For the condition in R_6, observe that for $p = a{:}b{:}a{:}nil + a{:}b{:}b{:}nil$, $p\ \parallel \lceil^{\{a\}}_{\{a\}}\ (p) \leftrightarrows p\ +\ a{:}b{:}nil$.

4 Verification via decomposition

In this section we formulate our main result which explains how the verification of an equation $p = q$ with $p = \parallel_{i=1}^{m} p_i$ and $q = \parallel_{j=1}^{n} q_j$ can be performed via decomposition. In theorem 4.4 we describe the decomposition and we give some conditions that must be checked in order for the method to be applicable. In the theorem, we use p and q on both the left and right hand sides, so that nothing is apparently gained by applying the theorem. However in remark 4.6 we show how p and q can be eliminated from the right hand side.

We begin with some straightforward lemmata that are used in the proofs to follow.

Lemma 4.1. *If $\alpha \subseteq \alpha(p)$ then $p = p \parallel W_\alpha$. In particular, $p = p \parallel nil$.*

Proof. $p \overset{W_1}{=} p \parallel W_{\alpha(p)} \overset{W_2}{=} p \parallel W_{\alpha(p)} \parallel W_\alpha \overset{W_1}{=} p \parallel W_\alpha$. □

Lemma 4.2. *If $p = p_1 \parallel p_2$, $\alpha \subseteq \alpha(p)$ and $\lceil^{\alpha}_{\alpha}\ (p_2 \parallel W_\alpha)$ is determinate, then $p = p \parallel \lceil^{\alpha}_{\alpha}\ (p_2 \parallel W_\alpha)$.*

Proof. $p \overset{\text{lemma 4.1}}{=} p_1 \parallel p_2 \parallel W_\alpha \overset{R_6}{=} p_1 \parallel p_2 \parallel W_\alpha \parallel \lceil^{\alpha}_{\alpha}\ (p_2 \parallel W_\alpha) \overset{\text{lemma 4.1}}{=} p \parallel \lceil^{\alpha}_{\alpha}\ (p_2 \parallel W_\alpha)$. □

Lemma 4.3. *If $\alpha \cap \beta = \emptyset$ and $\lceil^{\alpha}_{\alpha}\ (p)$ is determinate, then $\lceil^{\alpha}_{\alpha\cup\beta}\ (p) \parallel \lceil^{\alpha}_{\alpha}\ (p) = \lceil^{\alpha}_{\alpha\cup\beta}\ (p)$.*

Proof.

$$\begin{aligned} \lceil^{\alpha}_{\alpha\cup\beta}\ (p) &\overset{R_2,R_3,R_6}{=} \lceil^{\alpha\cap\alpha(p)}_{(\alpha\cup\beta)\cap\alpha(p)} \left(p \parallel \lceil^{\alpha\cap\alpha(p)}_{\alpha\cap\alpha(p)}\ (p) \right) \\ &\overset{R_5}{=} \lceil^{\alpha\cap\alpha(p)}_{(\alpha\cup\beta)\cap\alpha(p)}\ (p) \parallel \lceil^{\alpha\cap\alpha(p)}_{(\alpha\cup\beta)\cap\alpha(p)} \left(\lceil^{\alpha\cap\alpha(p)}_{\alpha\cap\alpha(p)}\ (p) \right) \overset{R_1,R_2,R_3,R_4}{=} \lceil^{\alpha}_{\alpha\cup\beta}\ (p) \parallel \lceil^{\alpha}_{\alpha}\ (p). \end{aligned}$$

□

Theorem 4.4 (Verification via decomposition). *Let* $p = \|_{i=1}^{m} p_i$ *and* $q = \|_{j=1}^{n} q_j$. *Let* α *consist of the synchronous (communicating) actions of* p *and* q. *That is,*

$$\alpha \stackrel{def}{=} \bigcup_{1\le i<j\le m} \Big(\alpha(p_i) \cap \alpha(p_j)\Big) \cup \bigcup_{1\le i<j\le n} \Big(\alpha(q_i) \cap \alpha(q_j)\Big).$$

Assume that $\upharpoonright^{\alpha}_{\alpha} (p_i \parallel W_\alpha)$ *and* $\upharpoonright^{\alpha}_{\alpha} (q_j \parallel W_\alpha)$ *are determinate for all* i *and* j. *Then*

$$p = q \qquad \text{iff} \qquad \begin{cases} p_{ij} = q_{ji} & \text{for } 1 \le i \le m,\ 1 \le j \le n, \\ \upharpoonright^{\alpha}_{\alpha\cup\alpha(p_i)} (p) = \mathop{\|}\limits_{j=1}^{n} p_{ij} & \text{for } 1 \le i \le m, \text{ and} \\ \upharpoonright^{\alpha}_{\alpha\cup\alpha(q_j)} (q) = \mathop{\|}\limits_{i=1}^{m} q_{ji} & \text{for } 1 \le j \le n, \end{cases}$$

where $p_{ij} \stackrel{def}{=} \upharpoonright^{\alpha}_{\alpha\,\cup\,(\alpha(p_i)\cap\alpha(q_j))} (p)$ *and* $q_{ji} \stackrel{def}{=} \upharpoonright^{\alpha}_{\alpha\,\cup\,(\alpha(p_i)\cap\alpha(q_j))} (q)$

Proof.

($\Leftarrow$) For each $1 \le i \le m$ we can prove that $p \stackrel{lemma\ 4.2}{=} p \parallel \mathop{\|}\limits_{k=1,k\ne i}^{m} \upharpoonright^{\alpha}_{\alpha} (p_k \parallel W_\alpha)$. Thus

$$\begin{aligned}
p &\stackrel{lemmas\ 4.2,4.1}{=} \Big(p \parallel W_\alpha\Big) \parallel \Big(\mathop{\|}\limits_{i=1}^{m} \mathop{\|}\limits_{k=1,k\ne i}^{m} \upharpoonright^{\alpha}_{\alpha} (p_k \parallel W_\alpha)\Big) \\
&\stackrel{\|_1,\|_2,W_2}{=} \mathop{\|}\limits_{i=1}^{m} \Big(p_i \parallel W_\alpha\Big) \parallel \Big(\mathop{\|}\limits_{i=1}^{m} \mathop{\|}\limits_{k=1,k\ne i}^{m} \upharpoonright^{\alpha}_{\alpha} (p_k \parallel W_\alpha)\Big) \\
&\stackrel{\|_1,\|_2}{=} \mathop{\|}\limits_{i=1}^{m} \Big(\Big(p_i \parallel W_\alpha\Big) \parallel \Big(\mathop{\|}\limits_{k=1,k\ne i}^{m} \upharpoonright^{\alpha}_{\alpha} (p_k \parallel W_\alpha)\Big)\Big) \\
&\stackrel{R_1,R_2}{=} \mathop{\|}\limits_{i=1}^{m} \Big(\upharpoonright^{\alpha}_{\alpha\cup\alpha(p_i)} (p_i \parallel W_\alpha) \parallel \Big(\mathop{\|}\limits_{k=1,k\ne i}^{m} \upharpoonright^{\alpha}_{\alpha\cup\alpha(p_i)} (p_k \parallel W_\alpha)\Big)\Big) \\
&\stackrel{\|_1,\|_2}{=} \mathop{\|}\limits_{i=1}^{m} \mathop{\|}\limits_{k=1}^{m} \upharpoonright^{\alpha}_{\alpha\cup\alpha(p_i)} (p_k \parallel W_\alpha) \stackrel{R_5}{=} \mathop{\|}\limits_{i=1}^{m} \upharpoonright^{\alpha}_{\alpha\cup\alpha(p_i)} \Big(\mathop{\|}\limits_{k=1}^{m} (p_k \parallel W_\alpha)\Big) \\
&\stackrel{lemma\ 4.1}{=} \mathop{\|}\limits_{i=1}^{m} \upharpoonright^{\alpha}_{\alpha\cup\alpha(p_i)} (p) \stackrel{assumption}{=} \mathop{\|}\limits_{i=1}^{m} \mathop{\|}\limits_{j=1}^{n} p_{ij}
\end{aligned}$$

In the same way, we can deduce that $q = \|_{j=1}^{n} \|_{i=1}^{m} q_{ji}$. Hence from the assumption that $p_{ij} = q_{ji}$ for each $1 \le i \le m$ and $1 \le j \le n$, we can deduce that $p = q$.

($\Rightarrow$) First it is clear that $p = q$ immediately implies that $p_{ij} = q_{ji}$. So we now prove that $p = q$ implies the second condition of the theorem. (The third condition can be deduced in the same way.) For each i we can compute the following.

$$\begin{aligned}
\mathop{\|}\limits_{j=1}^{n} p_{ij} &= \mathop{\|}\limits_{j=1}^{n} \upharpoonright^{\alpha}_{\alpha\cup(\alpha(p_i)\cap\alpha(q_j))} (p) \\
&\stackrel{lemma\ 4.1}{=} \mathop{\|}\limits_{j=1}^{n} \upharpoonright^{\alpha}_{\alpha\cup(\alpha(p_i)\cap\alpha(q_j))} (q \parallel W_\alpha) \\
&= \mathop{\|}\limits_{j=1}^{n} \upharpoonright^{\alpha}_{\alpha\cup(\alpha(p_i)\cap\alpha(q_j))} \Big(\mathop{\|}\limits_{k=1}^{n} (q_k \parallel W_\alpha)\Big) \\
&\stackrel{R_5}{=} \mathop{\|}\limits_{j=1}^{n} \mathop{\|}\limits_{k=1}^{n} \upharpoonright^{\alpha}_{\alpha\cup(\alpha(p_i)\cap\alpha(q_j))} (q_k \parallel W_\alpha) \\
&\stackrel{R_2}{=} \mathop{\|}\limits_{j=1}^{n} \mathop{\|}\limits_{k=1}^{n} \upharpoonright^{\alpha}_{\alpha\cup(\alpha(p_i)\cap\alpha(q_j)\cap\alpha(q_k))} (q_k \parallel W_\alpha) \\
&\stackrel{R_2}{=} \mathop{\|}\limits_{j=1}^{n} \Big(\mathop{\|}\limits_{k=1,k\ne j}^{n} \upharpoonright^{\alpha}_{\alpha} (q_k \parallel W_\alpha) \parallel \upharpoonright^{\alpha}_{\alpha\cup\alpha(p_i)} (q_j \parallel W_\alpha)\Big) \\
&\stackrel{lemma\ 4.3}{=} \mathop{\|}\limits_{j=1}^{n} \upharpoonright^{\alpha}_{\alpha\cup\alpha(p_i)} (q_j \parallel W_\alpha) \stackrel{R_5}{=} \upharpoonright^{\alpha}_{\alpha\cup\alpha(p_i)} \Big(\mathop{\|}\limits_{j=1}^{n} (q_j \parallel W_\alpha)\Big) \\
&\stackrel{lemma\ 4.1}{=} \upharpoonright^{\alpha}_{\alpha\cup\alpha(p_i)} (q) = \upharpoonright^{\alpha}_{\alpha\cup\alpha(p_i)} (p)
\end{aligned}$$

□

Remark 4.5. One may wonder whether it is enough simply to check $p_{ij} = q_{ji}$ in theorem 4.4. This would be a substantial optimisation. Unfortunately, this is not possible, as shown by the following example. Consider $p = (a{:}nil + b{:}nil) \parallel c{:}nil$ and $q = a{:}nil \parallel (b{:}nil + c{:}nil)$. One may try to verify that $p = q$ by applying theorem 4.4. In this case $\alpha = \emptyset$, so the determinacy constraints are easily satisfied. Calculating each p_{ij} and q_{ji} yields the following.

$$p_{11} = q_{11} = a{:}nil \qquad p_{21} = q_{12} = b{:}nil$$
$$p_{12} = q_{21} = nil \qquad p_{22} = q_{22} = c{:}nil$$

So clearly $p_{ij} = q_{ji}$ for all i and j, but $p \neq q$.

Remark 4.6. The right hand side of theorem 4.4 can be calculated using the following.

$$p_{ij} = \restriction^{\alpha}_{\alpha\cup(\alpha(p_i)\cap\alpha(q_j))} \Big(\mathop{\Big\|}_{k=1}^{n} p_k\Big) \overset{lemma\ 4.1,\ R_5}{=} \mathop{\Big\|}_{k=1}^{n} \restriction^{\alpha}_{\alpha\cup(\alpha(p_i)\cap\alpha(q_j))} (p_k \parallel W_\alpha).$$

We can calculate $\restriction^{\alpha}_{\alpha\cup\alpha(p_i)} (p)$ using the following:

$$\restriction^{\alpha}_{\alpha\cup\alpha(p_i)} (p) = \restriction^{\alpha}_{\alpha\cup\alpha(p_i)} \Big(\mathop{\Big\|}_{j=1}^{n} (p_j \parallel W_\alpha)\Big) = \mathop{\Big\|}_{j=1}^{n} \restriction^{\alpha}_{\alpha\cup\alpha(p_i)} (p_j \parallel W_\alpha).$$

Of course this also applies to q_{ji} and $\restriction^{\alpha}_{\alpha\cup\alpha(q_j)} (q)$.

In section 6 we give an application of the above technique which takes advantage of the preceding remark. However we first analyse the verification problem to demonstrate the benefit of the technique.

5 On the complexity of verification by decomposition

In this section we consider the complexity of verification through decomposition. We do this in the setting of bisimulation equivalence, as the verification of trace based equivalences is generally intractable on finite-state systems [14]. We show that in the case where there is no communication between the components, the verification is polynomial. In the case where there is communication between the components, we show that the verification is co-NP hard, and hence inherently intractable. The proof that we give is a simplified variant of those given in [19]. From these observations we draw the conclusion that verification via decomposition is especially worthwhile when there are relatively many asynchronous or non-communicating actions, and that its use is rather limited if almost every action is used for communication. But it is exactly the former case that leads to enormous state graphs, while in the latter case state graphs remain relatively small, and therefore, they can be more readily handled by existing means.

We start out by reformulating theorem 4.4, but now with the restriction that there are no communication actions among the component processes, which means that $\alpha = \emptyset$. For convenience, we write $\restriction_\beta$ for $\restriction^{\emptyset}_\beta$.

Corollary 5.1. *Let $p = \|_{i=1}^{m} p_i$ and $q = \|_{j=1}^{n} q_j$ with $\alpha(p_i) \cap \alpha(p_j) = \emptyset$ whenever $1 \leq i < j \leq m$ and $\alpha(q_i) \cap \alpha(q_j) = \emptyset$ whenever $1 \leq i < j \leq n$. Then*

$$p = q \quad \text{iff} \quad \begin{cases} p_{ij} = q_{ji} & \text{for } 1 \leq i \leq m,\ 1 \leq j \leq n, \\ p_i = \mathop{\Big\|}_{j=1}^{n} p_{ij} & \text{for } 1 \leq i \leq m, \text{ and} \\ q_j = \mathop{\Big\|}_{i=1}^{m} q_{ji} & \text{for } 1 \leq j \leq n, \end{cases}$$

where $p_{ij} \overset{def}{=} \restriction_{\alpha(q_j)} (p_i)$ and $q_{ji} \overset{def}{=} \restriction_{\alpha(p_i)} (q_j)$.

Proof. From R_1, R_2, R_7, lemma 4.1 and remark 4.6, we can show that $p_i = \restriction_{\alpha(p_i)} (p)$ and $q_j = \restriction_{\alpha(q_j)} (q)$, and from R_2, R_7, lemma 4.1 and remark 4.6, we can show that $\restriction_{\alpha(p_i)\cap\alpha(q_j)} (p) = \restriction_{\alpha(q_j)} (p_i)$ and $\restriction_{\alpha(p_i)\cap\alpha(q_j)} (q) = \restriction_{\alpha(p_i)} (q_j)$. The result then follows directly from theorem 4.4. □

Equality	Time complexity Space complexity
$p_{ij} = q_{ji} \quad (1 \leq i \leq m,\ 1 \leq j \leq n)$	$O\Big(m\,n\,(\max_{i,j}(\mid\longrightarrow_{p_i}\mid + \mid\longrightarrow_{q_j}\mid))\log\,(\max_{i,j}(\mid S_{p_i}\mid + \mid S_{q_j}\mid))\Big)$ $O\Big(\max_{i,j}(\mid\longrightarrow_{p_i}\mid + \mid\longrightarrow_{q_j}\mid)\Big)$
$p_i = \Vert_{j=1}^{n} p_{ij} \quad (1 \leq i \leq m)$	$O\Big(m\,n\,\max_i \mid\longrightarrow_{p_i}\mid \log(\max_i \mid S_{p_i}\mid)\Big)$ $O\Big(\max_i \mid\longrightarrow_{p_i}\mid\Big)$
$q_i = \Vert_{i=1}^{m} q_{ji} \quad (1 \leq j \leq n)$	$O\Big(m\,n\,\max_j \mid\longrightarrow_{q_j}\mid \log(\max_j \mid S_{q_j}\mid)\Big)$ $O\Big(\max_j \mid\longrightarrow_{q_j}\mid\Big)$
$p = q$	$O\Big(m\,n\,(\max_{i,j}(\mid\longrightarrow_{p_i}\mid + \mid\longrightarrow_{q_j}\mid))\log\,(\max_{i,j}(\mid S_{p_i}\mid + \mid S_{q_j}\mid))\Big)$ $O\Big(\max_{i,j}(\mid\longrightarrow_{p_i}\mid + \mid\longrightarrow_{q_j}\mid)\Big)$

Table 2: Complexities of deciding bisimulation in non-communicating processes

In order to verify that $p = q$, we must check the three identities at the right hand side of the curly bracket in corollary 5.1. In table 2 we have put the complexities for each step and the complexity for the total calculation. Here, S_r and $\longrightarrow_r$ represent the sets of states and transitions, respectively, of TS r. We assume that the number of states of our TS's is smaller than the number of transitions, as it is reasonable to assume that all states are reachable. The complexities in table 2 are motivated as follows.

1. In order to calculate p_{ij}, we take p_i and remove all transitions labelled with actions not in $\alpha(q_j)$. Then we remove all unreachable states, along with their outgoing transitions. This takes $O(\mid\longrightarrow_{p_i}\mid)$ time and space. In the same way we construct q_{ji}. In order to calculate $p_{ij} = q_{ji}$, we apply a standard bisimulation algorithm [14], which takes $O\Big((\mid\longrightarrow_{p_i}\mid + \mid\longrightarrow_{q_j}\mid)\log(\mid S_{p_i}\mid + \mid S_{q_j}\mid)\Big)$ time and $O(\mid\longrightarrow_{p_i}\mid + \mid\longrightarrow_{q_j}\mid)$ space. As this must be repeated for each $1 \leq i \leq m$ and $1 \leq j \leq n$, we obtain the complexities as given in table 2.
2. We obtain the second complexity measures via the following (easily proved) observation.

Lemma 5.2. *Let $r_0 = (S_{r_0}, \alpha_{r_0}, \longrightarrow_{r_0}, s_{r_0})$ and $r_1 = (S_{r_1}, \alpha_{r_1}, \longrightarrow_{r_1}, s_{r_1})$ with $\alpha_{r_0} \cap \alpha_{r_1} = \emptyset$. For all $u, u' \in S_{r_0}$ and $v, v' \in S_{r_1}$, $u \leftrightarrows_{r_0} u'$ and $v \leftrightarrows_{r_1} v'$ iff $\langle u, v\rangle \leftrightarrows_{r_0 \Vert r_1} \langle u', v'\rangle$.*

Reading this lemma from right to left, it says that if $r_0 \Vert r_1$ is not minimised with respect to bisimulation, i.e. it contains different states that are bisimilar, then this is due to the fact that either r_0 or r_1 was not minimal with respect to bisimulation. Reversing this reasoning says that if we ensure that r_0 and r_1 are minimal, then $r_0 \Vert r_1$ will also be minimal.

We use this observation as follows in constructing $\Vert_{j=1}^{n} p_{ij}$. First construct p_{i1} as indicated above. This takes $O(\mid\longrightarrow_{p_i}\mid)$ time and space. Minimise p_{i1} with respect to bisimulation, obtaining $\widehat{p}_{i1}$. Using the ordinary bisimulation algorithms, this takes $O\Big(\mid\longrightarrow_{p_i}\mid\log(\mid S_{p_i}\mid)\Big)$ time and $O(\mid\longrightarrow_{p_i}\mid)$ space. Now construct p_{i2} and its minimised variant $\widehat{p}_{i2}$ likewise. Then calculate $\widehat{p}_{i1} \Vert \widehat{p}_{i2}$, but stop if the number of states of the result exceed those of p_i. As $\widehat{p}_{i1}$ and $\widehat{p}_{i2}$ are minimal w.r.t. bisimulation, $\widehat{p}_{i1} \Vert \widehat{p}_{i2}$ is minimal. Hence if the number of states of $\widehat{p}_{i1} \Vert \widehat{p}_{i2}$ exceed the number of states of p_i, then p_i cannot be bisimilar to $\Vert_{j=1}^{n} p_{ij}$.

The complexity of calculating $\hat{p}_{i1} \parallel \hat{p}_{i2}$ is therefore $O(|\longrightarrow_{p_i}|)$. We thus calculate $\|_{j=1}^{n} p_{ij}$ by stepwise adding $p_{i3}, p_{i4}, \ldots, p_{in}$ in the same way. This takes $O\big(n|\longrightarrow_{p_i}|\log(|S_{p_i}|)\big)$ time and $O(|\longrightarrow_{p_i}|)$ space. The verification of $p_i = \|_{j=1}^{n} p_{ij}$ can then be done without increasing the time and space complexities. The steps above must be repeated for each $1 \le i \le m$. So we obtain the figures in table 2.

3. The analysis in this case is the same as in case 2, using q instead of p.

4. Combining the above gives these complexities for calculating $p \leftrightarrows q$.

The procedure sketched above is rather wasteful, e.g. p_{ij} and q_{ji} are calculated rather often. We have not investigated optimisations, as we expect that they will not improve the time and space complexities. However, the example in section 6 gives the impression that by using the regularity of processes p_i and q_j, substantial improvements can be expected.

In the case where there is communication between the processes, then the verification of $\|_{i=1}^{m} p_i = \|_{j=1}^{n} q_j$ becomes co-NP hard for each process equivalence between trace and bisimulation equivalence. We give a straightforward proof of this fact, actually showing that in the case that p_i and q_j are all finite and determinate, this verification is co-NP complete. In [19] it is shown that this verification becomes P-space hard if p_i and q_j are finite-state. It also gives an EXPSPACE completeness result in case abstraction of actions is allowed.

The proof technique in this section is a straightforward reduction from 3SAT [4]: Let $x_1, \ldots, x_k$ be variables and $l_{ij} \in \{x_1, \ldots, x_k, \neg x_1, \ldots, \neg x_k\}$. The problem of determining whether or not $\bigwedge_{i=1}^{n}(l_{i1} \vee l_{i2} \vee l_{i3})$ is satisfiable is well-known to be NP-complete. There is a straight-forward polynomial way of reducing an instance of 3SAT to an instance of 3SAT such that $k_{i1} < k_{i2} < k_{i3}$ where l_{ij} refers to a variable $x_{k_{ij}}$[1]. So 3SAT with this restriction is still NP-complete.

Lemma 5.3. *Determining $\|_{i=1}^{n} p_i = \|_{j=1}^{m} q_j$ is co-NP complete for finite determinate p_i and q_j.*

Proof. First we note that the problem is in co-NP. To determine inequality, we simply guess a trace which is in one side but not in the other, which can be easily checked by examining the component processes. This suffices, since for determinate processes, trace and bisimulation equivalences coincide [16].

Next we show co-NP hardness by reducing from 3SAT with the ordering restriction to the question whether $(\|_{i=1}^{n} p_i) \parallel p = p'$, for finite determinate p_i, p and p', does not hold. Consider the following instance of 3SAT with restriction over variables $x_1, \ldots, x_k$:

$$\bigwedge_{i=1}^{n}(l_{i1} \vee l_{i2} \vee l_{i3}). \tag{1}$$

The processes p_i, p and p' are constructed as in figure 1. Process p_i has actions l_{i1}, l_{i2}, l_{i3}, $\neg l_{i1}$, $\neg l_{i2}$, $\neg l_{i3}$ and $\surd$. Here $\neg l_{ij}$ stands for $\neg x$ if $l_{ij} \equiv x$ and for x if $l_{ij} \equiv \neg x$. A step l_{ij} corresponds to considering a valuation σ that assigns true to l_{ij}, and a step $\neg l_{ij}$ corresponds to considering a valuation σ that assigns false to l_{ij}. Clearly, p_i can perform a $\surd$ step iff $\sigma(l_{i1} \vee l_{i2} \vee l_{i3})$ is true.

The process p is used to guarantee that in $(\|_{i=1}^{n} p_i) \parallel p$, first a step corresponding to x_1 must be performed, then one corresponding to x_2 etc. It has actions $x_1, \ldots, x_k$, $\neg x_1, \ldots, \neg x_k$ and $\surd$. The process p' is equal to p with the only difference being that it has no $\surd$ step at the end.

We have the following fact, from which our co-NP hardness result follows immediately.

[1] First remove all clauses $l_{i1} \vee l_{i2} \vee l_{i3}$ that contain a variable occurring both with and without negation. Next remove double occurrences of variables in the clauses. Finally, introduce two new variables x_{k+1} and x_{k+2} and add these to incomplete clauses.

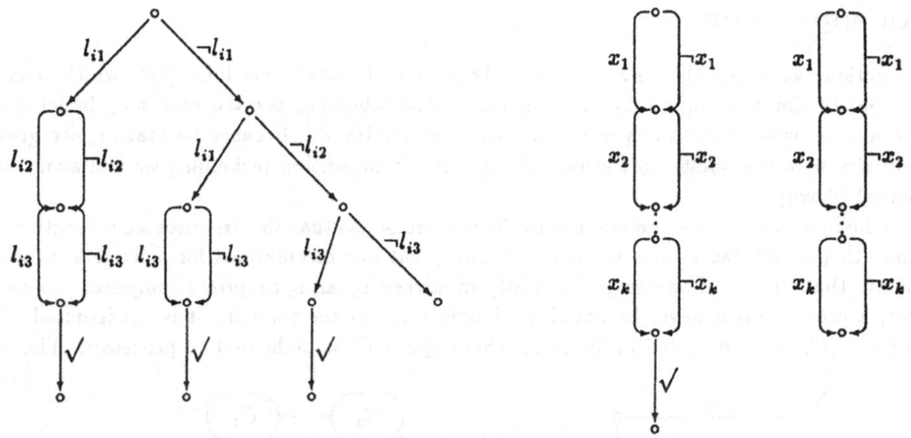

Figure 1: The processes p_i, p and p'

$$\bigwedge_{i=1}^{n}(l_{i1} \vee l_{i2} \vee l_{i3}) \text{ is satisfiable} \qquad \text{iff} \qquad \mathop{\|}_{i=1}^{n} p_i \ \| \ p = p' \text{ does not hold.}$$

Here '=' represents any equivalence between trace and bisimulation equivalence [8]. We now prove this fact:

($\Rightarrow$) Let σ be a valuation satisfying (1). Then $(\|_{i=1}^{n} p_i) \| p$ can perform the trace $a_1 \cdots a_k \surd$ where

$$a_i = \begin{cases} x_i & \textit{if } \sigma(x_i) = \textit{true}, \\ \neg x_i & \textit{if } \sigma(x_i) = \textit{false}. \end{cases}$$

Clearly, such a trace cannot be performed by p'. So, $\|_{i=1}^{n} p_i \ \| \ p$ and p' are not trace equivalent.

($\Leftarrow$) If $\|_{i=1}^{n} p_i \ \| \ p$ can perform a trace $a_1 \cdots a_k \surd$, then the assignment σ defined as:

$$\sigma(x_i) = \begin{cases} \textit{true} & \textit{if } a_i = x_i, \\ \textit{false} & \textit{if } a_i = \neg x_i. \end{cases}$$

is clearly a satisfying truth assignment for (1). Thus if (1) is not satisfiable, then $\|_{i=1}^{n} p_i \ \| \ p$ cannot perform traces ending in $\surd$. So exactly the traces $a_1 \cdots a_k$ with $a_i \equiv x_i$ or $a_i \equiv \neg x_i$ can be performed by both $\|_{i=1}^{n} p_i \ \| \ p$ and p', and hence they are trace equivalent. As all processes are determinate, $\|_{i=1}^{n} p_i \ \| \ p$ and p' are also bisimulation equivalent [5].

□

It is not difficult to extend the proof above to include only two-way communication (see [19]) or to use only two actions. However this is outside the setting of this paper, and it complicates matters slightly.

6 An application

In this section, we apply the decomposition theorem to Milner's scheduler [15], which is constructed out of simple components, called cyclers. The scheduler is often used as a benchmark for programmes which calculate process equivalences [6, 10, 13], because its state space grows exponentially with the number of cyclers. Using our decomposition technique, we can avoid this exponential blowup.

The scheduler schedules k processes in cyclic succession, so that the first process is reactivated after the kth process has been activated. We carry out our calculations for a fixed k in order to illustrate that these can be straightforwardly mimicked by an appropriate computer program. However, a process must never be reactivated before it has terminated. It is constructed of k cyclers $C_0, \ldots, C_{k-1}$, as depicted in figure 2, where cycler C_i is dedicated to process i. The left

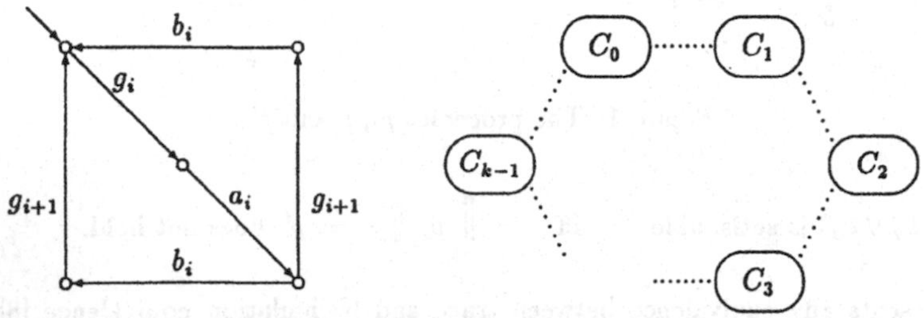

Figure 2: A cycler and a scheduler

part of the figure shows the transition system for cycler C_i, while the right part depicts the architecture of the scheduler. The dotted lines indicate where the cyclers synchronise. Cycler C_i first synchronises on a signal g_i which indicates that it may start. It then activates process i via an action a_i. Next, it waits for termination of process i, indicated by b_i, and in parallel, using g_{i+1}, activates the next cycler. Here, the indices are taken mod k, so that $g_k = g_0$. It then returns to its initial state. The cycler C_i is described by:

$$\begin{aligned} C_0 &= a_0{:}(b_0{:}g_1{:}g_0{:}C_0 + g_1{:}b_0{:}g_0{:}C_0), \\ C_i &= g_i{:}a_i{:}(b_i{:}g_{i+1}{:}C_i + g_{i+1}{:}b_i{:}C_i) \quad \text{for} \quad 1 \leq i < k. \end{aligned}$$

The first cycler is assumed to have been initiated. The complete scheduler for k processes is thus described by:

$$Sched_k = C_0 \parallel C_1 \parallel \cdots \parallel C_{k-1}.$$

A correctness criterion for the scheduler has been formulated in [15]. The a_i and b_i actions must happen alternately, and the a_i actions must happen cyclically. For the purposes of this example, we are also interested in the precise relationship between the synchronisation actions g_i and the actions b_j. Therefore we prove the scheduler $Sched_k$ equal to the specification $Correct_k$ from which the behaviour of the scheduler can easily be understood. The process $Correct_k$ is defined by

$$Correct_k = D_0 \parallel D_1 \parallel \cdots \parallel D_{k-1} \parallel BB_k, \ where \quad \begin{aligned} &D_0{=}a_0{:}b_0{:}g_0{:}D_0, \\ &D_i{=}g_i{:}a_i{:}b_i{:}D_i \quad \text{for } 1 \leq i < k, \\ &BB_k{=}a_0{:}g_1{:}a_1{:}\cdots{:}g_{k-1}{:}a_{k-1}{:}g_0{:}BB_k. \end{aligned}$$

The letters BB in BB_k stand for 'backbone'. It is easy to see that $Correct_k$ satisfies the correctness criteria as given by Milner. This can be shown formally by applying hiding, but as this is rather standard, we do not prove that here. For an idea of the proof, see the verification of the scheduler in [15].

We wish to apply theorem 4.4 to verify that $Sched_k = Correct_k$. We thus let $p_i = C_i$ for $0 \leq i < k$, and define $q_j = D_j$ for $0 \leq j < k$ and $q_k = BB_k$.

First note that $\alpha = \{a_i, g_i \mid 0 \leq i < k\}$. A small calculation tells us that $\lceil^{\alpha}_{\alpha} (p_i \parallel W_\alpha)$ is bisimilar to $E_i \parallel W_\alpha$ and that $\lceil^{\alpha}_{\alpha} (q_j \parallel W_\alpha)$ is bisimilar to $F_j \parallel W_\alpha$, where E_i and F_j are defined by

$$E_0 = a_0{:}g_1{:}g_0{:}E_0, \qquad F_0 = a_0{:}g_0{:}F_0,$$
$$E_i = g_i{:}a_i{:}g_{i+1}{:}E_i \quad \text{for} \quad 1 \leq i < k; \qquad F_j = g_j{:}a_j{:}F_j \quad \text{for} \quad 1 \leq j < k,$$
$$F_k = BB_k.$$

Obviously these are all determinate, so theorem 4.4 is applicable. We use remark 4.6 to calculate p_{ij}, q_{ji}, $\lceil^{\alpha}_{\alpha\cup\alpha(p_i)} (p)$ and $\lceil^{\alpha}_{\alpha\cup\alpha(q_j)} (q)$. For $i \neq j$, we find that

$$p_{ij} = \Big\Vert_{l=0}^{k-1} \lceil^{\alpha}_{\alpha\cup(\alpha(p_i)\cap\alpha(q_j))} (p_l \parallel W_\alpha) = \Big\Vert_{l=0}^{k-1} \lceil^{\alpha}_{\alpha} (p_l \parallel W_\alpha) = \Big\Vert_{l=0}^{k-1} E_l \parallel W_\alpha = BB_k, \quad \text{and}$$

$$q_{ji} = \Big\Vert_{l=0}^{k} \lceil^{\alpha}_{\alpha\cup(\alpha(p_i)\cap\alpha(q_j))} (q_l \parallel W_\alpha) = \Big\Vert_{l=0}^{k} \lceil^{\alpha}_{\alpha} (q_l \parallel W_\alpha) = \Big\Vert_{l=0}^{k} F_l \parallel W_\alpha = BB_k.$$

For $i = j$, we find that

$$p_{ii} = \Big\Vert_{l=0}^{k-1} \lceil^{\alpha}_{\alpha\cup(\alpha(p_i)\cap\alpha(q_i))} (p_l \parallel W_\alpha) = \Big\Vert_{l=0}^{k-1} \lceil^{\alpha}_{\alpha\cup\{b_i\}} (p_l \parallel W_\alpha) \tag{2}$$

$$= \Big\Vert_{l=0, l\neq i}^{k-1} E_l \parallel C_i \parallel W_\alpha$$

$=$

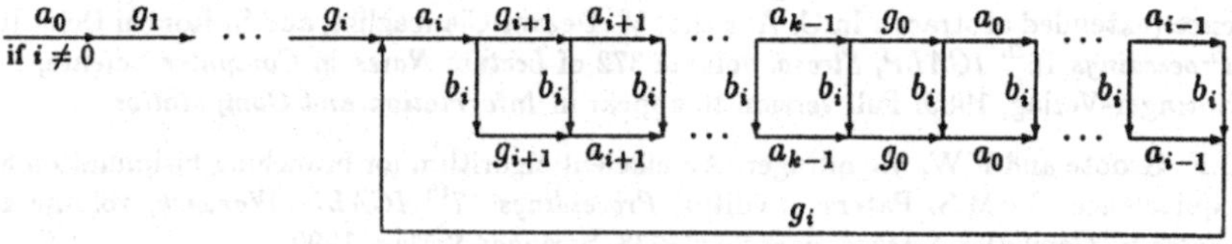

and

$$q_{ii} = \Big\Vert_{l=0}^{k-1} \lceil^{\alpha}_{\alpha\cup(\alpha(p_i)\cap\alpha(q_i))} (q_l \parallel W_\alpha) = \Big\Vert_{l=0}^{k-1} \lceil^{\alpha}_{\alpha\cup\{b_i\}} (q_l \parallel W_\alpha) \tag{3}$$

$$= \Big\Vert_{l=0}^{k} F_l \parallel D_i \parallel W_\alpha$$

$=$

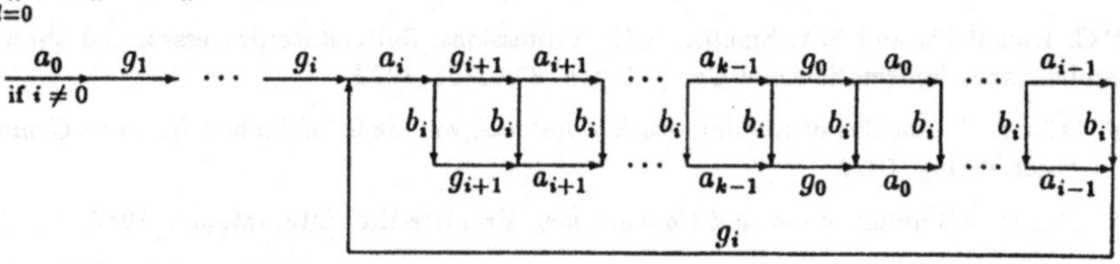

In the diagrams, the first transition $\xrightarrow{a_0}$ only appears for the cases when $i \neq 0$; for the case when $i = 0$, the TS's start with the $\xrightarrow{g_1}$ transition.

Obviously, p_{ij} and q_{ji} are thus equivalent. Note that the number of states of each intermediate term is always smaller than $3k$, i.e. linear in k.

Now note that $p_{ii} \parallel BB_k = p_{ii}$ and hence $\Vert_{j=0}^{k} p_{ij} = p_{ii}$. Similarly, $\Vert_{i=0}^{k-1} q_{ji} = q_{jj}$. Hence

$$\lceil^{\alpha}_{\alpha\cup\alpha(p_i)} (p) \overset{remark\ 4.6}{=} \Big\Vert_{j=0}^{k-1} \lceil^{\alpha}_{\alpha\cup\alpha(p_i)} (p_j \parallel W_\alpha) = \Big\Vert_{j=0}^{k-1} \lceil_{\alpha\cup\{b_i\}} (p_j \parallel W_\alpha) \overset{(2)}{=} \Big\Vert_{j=0}^{k} p_{ij}.$$

Equally, from remark 4.6 and (3) we have that $\lceil^{\alpha}_{\alpha\cup\alpha(q_j)} (q) = \Vert_{i=0}^{k-1} q_{ji}$. So by theorem 4.4, $p = q$.

References

[1] J.C.M. Baeten and W.P. Weijland. *Process Algebra.* Cambridge Tracts in Theoretical Computer Science 18. Cambridge University Press, 1990.

[2] A. Bouajjani, J.-C. Fernandez and N. Halbwachs. Minimal model generation. Draft, 1991.

[3] J.R. Burch, E.M. Clarke, K.L. McMillan, D.L. Dill, and L.J. Hwang. Symbolic model checking 10^{20} states and beyond. In *Proceedings* 5th *Annual Symposium on Logic in Computer Science, Philadelphia, USA*, pages 428–439, 1990.

[4] S.A. Cook. The complexity of theorem-proving procedures. In *Proceedings of the* 3rd *Annual ACM Symposium on Theory of Computing, Shaker Heights, Ohio*, pages 151–158, 1971.

[5] J. Engelfriet. Determinacy → (observation equivalence = trace equivalence). *Theoretical Computer Science*, 36(1):21–25, 1985.

[6] J.-C. Fernandez. An implementation of an efficient algorithm for bisimulation equivalence. *Science of Computer Programming*, 13:219–236, 1989/1990.

[7] J.-C. Fernandez and L. Mounier. "On the fly" verification of behavioural equivalences and preorders. In K.G. Larsen, editors, *Proceedings CAV'91, Aalborg*, pages 238–250. 1991.

[8] R.J. van Glabbeek. The linear time – branching time spectrum. In J.C.M. Baeten and J.W. Klop, editors, *Proceedings CONCUR'90, Amsterdam*, volume 458 of *Lecture Notes in Computer Science*, pages 278–297. Springer-Verlag, 1990.

[9] J.F. Groote and F.W. Vaandrager. Structured operational semantics and bisimulation as a congruence (extended abstract). In G. Ausiello, M. Dezani-Ciancaglini, and S. Ronchi Della Rocca, editors, *Proceedings* 16th *ICALP, Stresa*, volume 372 of *Lecture Notes in Computer Science*, pages 423–438. Springer-Verlag, 1989. Full version to appear in *Information and Computation.*

[10] J.F. Groote and F.W. Vaandrager. An efficient algorithm for branching bisimulation and stuttering equivalence. In M.S. Paterson, editor, *Proceedings* 17th *ICALP, Warwick*, volume 443 of *Lecture Notes in Computer Science*, pages 626–638. Springer-Verlag, 1990.

[11] C.A.R. Hoare. *Communicating Sequential Processes.* Prentice-Hall International, 1985.

[12] G.J. Holzmann. *Design and Validation of Computer Protocols.* Prentice-Hall International, 1991.

[13] H. Qin. Efficient verification of determinate processes. In J.C.M. Baeten and J.F. Groote, editors, *Proceedings CONCUR'91, Amsterdam*, volume 527 of *Lecture Notes in Computer Science*, pages 471–494. Springer-Verlag, 1991.

[14] P.C. Kanellakis and S.A. Smolka. CCS expressions, finite-state processes, and three problems of equivalence. *Information and Computation*, 86:43–68, 1990.

[15] R. Milner. *A Calculus of Communicating Systems*, volume 92 of *Lecture Notes in Computer Science.* Springer-Verlag, 1980.

[16] R. Milner. *Communication and Concurrency.* Prentice-Hall International, 1989.

[17] R. Milner and F. Moller. Unique decomposition of processes. *Bulletin of the European Association for Theoretical Computer Science*, 41:226–232, 1990.

[18] F. Moller. *Axioms for concurrency.* PhD thesis, Report CST-59-89, Department of Computer Science, University of Edinburgh, 1989.

[19] A. Rabinovich. Checking equivalences between concurrent systems of finite agents. Draft, 1991.

Interface Refinement in Reactive Systems
(extended abstract)

Rob Gerth*, Ruurd Kuiper*, John Segers
Eindhoven University of Technology†

May, 1992

Abstract

Suppose one has a system that has a *synchronous* interface with its environment. Now, suppose that one refines this system and changes its interface to an *asynchronous* one. Whatever is meant here by refinement, it cannot be standard (process) refinement since the interface actions have changed; nor is it action refinement in the sense that a process is substituted for an action, as the intention presumably is to allow the system to proceed without having to wait until the environment is willing to synchronize.

In this paper we propose a notion of *interface refinement* of which changing synchronous to asynchronous communication is an instance; as is in fact the reverse change. This notion of interface refinement is quite powerful; it generalizes all existing methods w.r.t. the class of interface changes that it allows.

The major part of the paper is concerned with developing proof rules with which to verify interface refinement. We use (linear) temporal logic as specification language and an adaptation of the Manna–Pnueli verification framework. The method is illustrated by verifying an interface change in which synchronous communication is replaced by asynchronous send and receive. Proofs of the various theorems and lemma's are delegated to an appendix.

Keywords refinement, interface change, temporal logic, leads to property, concurrency, transition systems, ∀-automata.

1 Introduction

The position that a system's development starts with a high level, abstract, descriptive specification which is gradually changed through a sequence of refinement steps into a low level, concrete, prescriptive specification that can be readily implemented hardly needs motivation. There are many theories that formalize such systems' development [Lam91, Lam86, Mil89, Hoa86, BJO91, Jon90]. In most of them, control structure can change during a refinement step; in some of them, data representation can change as well [Lam91, Jon90]; few of them allow interface actions to change during development [Lam86, BJO91]; none of these methods allow all three types of change to take place.

*The author is currently working in and partially supported by ESPRIT project P3096: "Formal Methods and Tools for the Development of Distributed and Real-Time Systems (SPEC)".

†Department of Computing Science, Eindhoven University of Technology, P.O. Box 513, 5600 MB Eindhoven, The Netherlands. Email: robg@win.tue.nl.

Yet, surely, the argument for being able to change control and data representation during refinement is equally valid for allowing interface changes: the more detailed and low-level a specification becomes, the more concrete data representation becomes; the more detailed program control is described; and also the more specific the system's interface is defined.

As a concrete example, consider a network of server tills that interfaces with customers. At an abstract level it might be convenient to describe a banking transaction at some till as a single *atomic* interface event that checks the banking card and PIN code, dispenses the requested amount and charges the account. During subsequent refinement steps, such events will be broken up. Inserting the card, checking the PIN code, obtaining the amount to be dispensed need to appear as separate and new interface events. More interesting, the need for a centralized database of client accounts will have to be recognized, which has an *asynchronous* interface with the tills[1]. Consequently, although from the point of view of a client his transaction ends with the money being dispensed, the system will view the transaction's end only when the account has been charged. However, because these updates occur *asynchronously* it is now possible that a client, possibly the same, initiates a second transaction before the first one is completed. Clearly, there is a big difference between the computations that are described by the top level specification and the corresponding low-level ones. If one looks at what happens at a single server till, then the high-level behavior will be a sequence of atomic transaction events and the state in between events will always describe properly balanced accounts. The low-level behavior looks different: first of all, the low-level events that 'implement' some high-level transaction appear distributed along the sequence of events; moreover, the states in between events may now show unbalanced accounts because money may have been dispensed without the account yet showing it.

If we agree that this example makes sense, then there is a problem: to develop a notion of refinement with which the above refinement steps can be formalized. Before we proceed, we want to discuss two objections that might be levied against the claim that there is a problem to be solved at all.

1. If a system S is to be shown a refinement of T, S *ref* T, then what one should show in case the interfaces differ, is $P(S)$ *ref* T, where $P(\cdot)$ encapsulates S and translates S's interface to T's. Thus, there is no need to develop new theories. In fact, [BJO91] starts from essentially the above suggestion.

   Ignoring questions as to whether the direction of translation is the most appropriate in all circumstances, we posit that this is not so much a solution as a suggestion for an adaptation of the notion of refinement. Changes in the interface will often be quite standard, for instance involving standard communication protocols. Instead of proving $P(S)$ *ref* T for various S's, where part of the proof effort is wasted in repeatedly showing that $P(\cdot)$ correctly translates between the interfaces, it is more desirable in such cases to prove $S\ ref_P\ T$ where ref_P has 'absorbed' the interface change and imposes conditions on the behavior of S itself.

2. Even if one concedes the previous argument, there are still many 'interface' changes that actually are internal changes between the interfaces of components within the system; these are of no concern as they involve internal behavior which is not visible to the system's environment. The introduction of the centralized database in the above example is a case in point[2].

[1] At least if one aims for a standard implementation.

[2] Actually in this case, a client might detect differences, depending on the precise form of asynchrony.

This point of view is embodied in most existing refinement theories. The problem is that, taken literally, it tends to place architectural design decisions beyond the reach of these theories; at least those decisions that do not change the behavior on the interface with the system's environment. Since such design decisions are pack and parcel in 'real life' systems specification, it seems unreasonable to not at least aspire to permitting them first class citizenship when formalizing the specification and development process.

The process refinement relation that we propose here, generalizes standard process refinement as can be found in for instance [Lam91, LT87]. Such process refinements can often be viewed as semantic inclusion:

$$S \; ref \; T \text{ iff } \mathrm{Beh}(S) \subseteq \mathrm{Beh}(T) \, ,$$

for appropriate notions of system behavior, Beh(·). We rewrite this to

$$S \; ref \; T \text{ iff } \forall\, s \in \mathrm{Beh}(S) \; \exists t \in \mathrm{Beh}(T) \;\; s = t \, .$$

This is a trivial change, but what it makes explicit is the use of *equality* to compare behaviors. We propose that interface refinement means substituting equality by some relation that describes how the interface changes. More specifically, given some notion of behavior, Beh(·), and a relation C between low and high level behaviors, define

$$S \; ref_C \; T \text{ iff } \forall\, s \in \mathrm{Beh}(S) \; \exists t \in \mathrm{Beh}(T) \;\; s \, C \, t \, . \qquad (1)$$

This is a straightforward but quite general definition. The definition applies to any notion of refinement that is based on inclusion of behavior. Also, the relation C as it stands is unconstrained; i.e., any pair of low and high level behaviors can be related by judiciously choosing a C. Far from being a vice, we consider this generality to be a virtue as it indicates that our proposal does not place arbitrary a priori constraints on what can and what cannot be viewed as an interface refinement. Note, however, that the refinement relation is parametrized with C. In other words, it is up to the user to choose appropriate interface changes; just as it is already up to him to choose a sensible refinement relation.

In this paper we intend to develop this proposal of interface refinement. First, the basic definition (1) is refined and formally defined in Section 2. Defining interface changes in terms of complete behaviors is awkward and it is more natural to define such changes in terms of the individual actions that are changed. We give a semantical definition in terms of action sequences, after which the definition is adapted to the use of temporal logic for defining action sequences and then to include a store of variables that is changed by such actions. After introducing in Section 3 the necessary terminology of transition systems, the specification language and Manna–Pnueli temporal logic, Section 4 reduces the problem of proving interface refinement to proving a number of temporal logic formulae and a standard process refinement. Subsequently an example is presented in Section 5. Section 6 formulates some conclusions and the appendix contains proofs of the theorems and lemma's. Finally, a more detailed example of interface refinement can be found in [SGK92]. That example is worked out using an earlier version of the theory than the one described in this paper.

2 The interface refinement relation

We restate the definition:

$$S \; ref_C \; T \text{ iff } \forall\, s \in \mathrm{Beh}(S) \; \exists t \in \mathrm{Beh}(T) \;\; s \, C \, t \, .$$

The question is how to define C. Describing C directly as a relation on behaviors is awkward, since this does not correspond to how interface changes are usually conceived of. In the service tills example one would like to specify the interface changes by describing how a high level transaction event splits into a number of separate, not necessarily consecutive, events or how a database update event divides into an update request and a subsequent update. Using this description, we proceed to define what related behaviors are: every transaction or database update event in the high level behavior should be seen to split up in the related low level behavior.

We shall illustrate this on the concrete example of an interface change in which synchronous communication is replaced by asynchronous send and receives. Ignore for the moment that values are being communicated; also forget about the FIFO nature of the asynchronous communication. Let $\mathbf{c}$ stand for the communication action, $\mathbf{s}$ for the send and $\mathbf{r}$ for the receive action. If we take the high level behavior, $\mathbf{cc}$, consisting of two communication events (i.e., occurrences of actions) then this can be related to either one of the low level sequences $\mathbf{ssrr}$ and $\mathbf{srsr}$; it should not be related to, e.g., $\mathbf{srrs}$ or $\mathbf{ssr}$ for obvious reasons. If we now think about defining C, then this example offers three observations. First, C cannot relate just the *set* of $\mathbf{s}$ and $\mathbf{r}$ actions to $\mathbf{c}$, since then it is hard to see how we can have $\mathbf{srsr}$ as a legal 'implementation', while disallowing $\mathbf{srrs}$. Second, C must allow the $\mathbf{s}$ and $\mathbf{r}$ actions to not be consecutive, since otherwise we cannot have $\mathbf{ssrr}$ as an implementation. Third, in showing that $\mathbf{cc}$ is related to, say, $\mathbf{ssrr}$ we need to *partition* the events in both sequences and show that the classes of these partitions can be C-related, because if we do not, we cannot relate $\mathbf{cc}$ to $\mathbf{ssrr}$ but disallow it to be related to $\mathbf{ssr}$: $\mathbf{cc}$ can be related to $\mathbf{ssrr}$ because we can partition the sequences as $\mathbf{c}^1\mathbf{c}^2$ and $\mathbf{s}^1\mathbf{s}^2\mathbf{r}^1\mathbf{r}^2$ (the superscripts identify the partition classes); moreover, if we look at what we call the *chunks* of 1-labeled events then these are C-related and this is true as well for the chunks of 2-labeled events. No such partition exists for $\mathbf{cc}$ and $\mathbf{ssr}$. We conclude that C must relate sequences (not sets) of events; it must be able to state whether events must be consecutive or not; and because of the need to partition, it must separate the events that are part of the same chunk from events that are not part of the chunk but just happen to occur in between chunk events.

Thus we motivate defining C as a relation over $(\mathbf{A} \times \{E, C\})^\omega$, the set of infinite sequences of *tagged* actions. The tags indicate whether the action belongs to the chunk—tagged with C—or is an environment action—tagged with E. We usually write $\mathbf{a}^C$ and $\mathbf{a}^E$ instead of $(\mathbf{a}, C)$ and $(\mathbf{a}, E)$.

For the above example, we might define a relation

$$C \;=\; \{(us^C vr^C w,\ sc^C t) \mid s, u, v \in (\mathbf{A} \times \{E\})^*,\ t, w \in (\mathbf{A} \times \{E\})^\omega\}\,. \tag{2}$$

If in addition we want to express the FIFO nature of the asynchronous communications, then we should add the condition that $u\#\mathbf{s} = u\#\mathbf{r} + v\#\mathbf{r}$, where $s\#\mathbf{a}$ stands for the number of a-actions in s. This possibility is an additional bonus of distinguishing chunk from environment events. We should perhaps remark that this interface change is beyond the reach of [BJO91] because the asynchronous chunks are not definable by finite state automata.

Next, partitioning the events in a computation will be done by labeling events with natural numbers; events with the same label belong to the same partition class. So, if $s \in \mathbf{A}^\omega$ is some behavior and g is some function of type $\omega \rightarrow \omega$ then $s \times g \in (\mathbf{A} \times \omega)^\omega$ is the behavior in which every event s_i is replaced by the event $(s_i, g(i))$. Again, we usually write $\mathbf{a}^3$ instead of $(\mathbf{a}, 3)$. With such labelings go projections: define $\mathbf{a}^i{\downarrow}j = \mathbf{a}^C$ if $i = j$ and $\mathbf{a}^E$ otherwise. Projection extends pointwise to behaviors. Projecting a behavior on a label gives a tagged sequence; this is the partition class—and hopefully chunk—determined by this label.

This leads to the following definition of interface refinement:

2–1 Definition (semantic interface relation). *For behaviors $s, t \in \mathbf{A}^\omega$ and a chunk relation $C \subseteq (\mathbf{A} \times \{E, C\})^\omega \times (\mathbf{A} \times \{E, C\})^\omega$ define*

$$s\, C\, t \quad \text{iff} \quad \exists g, h{:}\, \omega \to \omega\ \forall i \in \text{ran}(g) \cup \text{ran}(h) \ \ (s \times g){\downarrow}i\ C\ (t \times h){\downarrow}i$$

Using this definition, we can formally show that $\mathtt{ssrr}\ C\ \mathtt{cc}$:[3] define g and h so that $\mathtt{ssrr} \times g = \mathtt{s}^1\mathtt{s}^2\mathtt{r}^2\mathtt{r}^1$ and $\mathtt{cc} \times h = \mathtt{c}^1\mathtt{c}^2$; then $\mathtt{s}^1\mathtt{s}^2\mathtt{r}^2\mathtt{r}^1{\downarrow}1\ C\ \mathtt{c}^1\mathtt{c}^2{\downarrow}1$ because $\mathtt{c}^1\mathtt{c}^2{\downarrow}1 = \mathtt{c}^C\mathtt{c}^E$ and $\mathtt{s}^1\mathtt{s}^2\mathtt{r}^2\mathtt{r}^1{\downarrow}1 = \mathtt{s}^C\mathtt{s}^E\mathtt{r}^E\mathtt{r}^C$ and likewise for the projection on 2.

Interface refinement satisfies the following evident properties:

> for any interface refinement relations C and C', ref_C is a pre order and ref_C is transitive in the sense that if $S\ ref_C\ T$ and $T\ ref_{C'}\ U$ then also $S\ ref_{C;C'}\ U$ where ';' is relational composition.

2.1 A syntactic variant

In the sequel we shall specify refinement relations by temporal logic formulae. E.g., we replace (2) by the specification }

$$\begin{aligned} C \;=\; \{ & (\Diamond(s \wedge \mathsf{C} \wedge \Diamond(r \wedge \mathsf{C})) \wedge \Box(\mathsf{C} \wedge \text{past}\Diamond\mathsf{C} \to \bigcirc\Box\mathsf{E}) \,, \\ & \Diamond(c \wedge \mathsf{C}) \wedge \Box(\mathsf{C} \to \bigcirc\Box\mathsf{E}))\} \end{aligned}$$

The FIFO nature of asynchronous communication would then be expressed by replacing $c \wedge \mathsf{C}$ by $c \wedge \mathsf{C} \wedge \mathsf{h}\#\mathsf{c} = n$; by replacing $s \wedge \mathsf{C}$ with $s \wedge \mathsf{C} \wedge \mathsf{h}\#\mathsf{s} = n$; and by replacing $r \wedge \mathsf{C}$ with $r \wedge \mathsf{C} \wedge \mathsf{h}\#\mathsf{r} = n$. Here, h is the history of events and C has become a set of pairs of formulae; a pair for every n.

These formulae are interpreted on tagged sequences. For $\mathsf{a} \in \mathbf{A}$, a proposition a is true just in case the last action was a; the proposition C (E) is true if the previous action was (was not) a chunk action; and h is valuated at some point by the sequence of events that have occurred up to that point. Action propositions are not essential since a proposition a can always be replaced by the atom 'last(h) = a'.

If a refinement relation is specified as such sets of pairs of formulae, $\{(\phi_i, \psi_i) \mid i \in I\}$, then its extension to a relation on behaviors takes the following form:

$$s\, C\, t \quad \text{iff} \quad \exists g, h{:}\, \omega \to \omega\ \forall i \in \text{ran}(g) \cup \text{ran}(h)\ \ \exists j \in I\ \ (s \times g){\downarrow}i \models \phi_j \ \&\ (t \times h){\downarrow}i \models \psi_j \ .$$

2.2 Stores

Although we have discussed interface refinement in terms of sequences of uninterpreted actions that do not change some store, the definitions immediately extend to this more general case: if Σ is some set of stores, $\sigma, \tau, \ldots$, then we just replace the action set $\mathbf{A}$ by $\bar{\mathbf{A}} = \Sigma \times \mathbf{A} \times \Sigma$. The intention is that every new 'action' $(\sigma, \mathsf{a}, \tau) \in \bar{\mathbf{A}}$ describes the execution of action a in the store σ, which causes the store to change to τ; i.e., (σ, τ) is in the input-output semantics of a. Although we shall not provide details we mention that this extension allows data type refinement to be interpreted as interface refinement as well.

---

[3] Assuming the definition to extend to finite computations.

An implementation of synchronous value communication by asynchronous sends and receives might now be described by $C = \{ (\psi_x(n,m),\ \phi_x(n,m)) \mid n, m \in \mathbb{N},\ x \in \mathsf{Var} \}$, with

$$\begin{aligned}\phi_x(n,m) &= \Diamond(c?x \| c!n \wedge \mathsf{C} \wedge x = n \wedge h\#c = m) \wedge \Box(\mathsf{C} \to \bigcirc\Box\mathsf{E}) \text{ and} \\ \psi_x(n,m) &= \Diamond(s!n \wedge \mathsf{C} \wedge \mathsf{h}\#\mathsf{s} = m \wedge \Diamond(r?x \wedge \mathsf{C} \wedge x = n \wedge \mathsf{h}\#\mathsf{r} = m)) \wedge \\ &\qquad \Box(\mathsf{C} \wedge \mathrm{past}\Diamond\mathsf{C} \to \bigcirc\Box\mathsf{E})\end{aligned}$$

Here, $c?x \| c!n$ is the synchronized communication transferring the value n to x; $s!n$ and $r!x$ have the expected meanings. The expression $\mathsf{h}\#\mathsf{c}$ now stands for the number of communications and likewise for $\mathsf{h}\#\mathsf{s}$ and $\mathsf{h}\#\mathsf{r}$. We have not stated that variables other than x are not changed by these actions. In fact, we do not have to since this is enforced by the semantics of these actions.

3 Transition systems, Manna–Pnueli temporal logic

In the remainder of the paper we shall use labeled, interpreted transition systems to describe systems and the Manna–Pnueli proof techniques for linear temporal logic. This section is a quick introduction to these and establishes some notation.

Transition systems π have the form $(\mathcal{V}, \mathcal{E}, I, \mathcal{L}, \mathcal{T})$. $(\mathcal{V}, \mathcal{E})$ is a directed graph and every edge $e \in \mathcal{E}$ has a source vertex ${}^\bullet e \in \mathcal{V}$ and a target vertex $e^\bullet \in \mathcal{V}$; $I \in \mathcal{V}$ is the initial vertex; to any edge e, $\mathcal{L}(e)$ associates an action and $\mathcal{T}(e)$ associates a description of the store change that traversing this edge would entail, often called the *transition relation*. As an example, the program `while true do x := x - 1` would be written as $(\{v\}, \{e\}, v, \mathcal{L}, \mathcal{T})$ with ${}^\bullet e = v$, $e^\bullet = v$, $\mathcal{L}(e) = \mathtt{x} := \mathtt{x} - 1$ and $\mathcal{T}(e) \equiv \mathtt{x}' = \mathtt{x} - 1 \wedge \bigwedge_{\mathtt{y} \not\equiv \mathtt{x}} \mathtt{y}' = \mathtt{y}$. Only $\mathcal{T}(e)$, or $\mathcal{T}_e$ as we shall often write, needs explaining: store changes are defined using 1st order logic; primed (program) variables refer to the variables values in the resulting stores. So $\mathcal{T}_e$, indeed, expresses the expected semantics of the assignment $\mathtt{x} := \mathtt{x} - 1$: the new value, $\mathtt{x}'$, of $\mathtt{x}$ is one less than the old value and the values of the other variables remain the same[4]. Observe that we associate the formulae $\mathcal{T}_e$ with the edges rather than with the actions $\mathcal{L}(e)$ labeling the edges. This small generalization will pay off in the formulation of Lemma 4–2 below. The above notwithstanding we shall use uninterpreted transition systems whenever they suffice and shall write, e.g., $a(b+c)$ instead of $(\{u, v, w\}, \{e, f, g\}, u, \mathcal{L}, \mathcal{T})$ with $u = {}^\bullet e$, $v = e^\bullet = {}^\bullet f = {}^\bullet g$, $w = f^\bullet = g^\bullet$, $\mathcal{L}(e) = a$, $\mathcal{L}(f) = b$, $\mathcal{L}(g) = c$ and $\mathcal{T}$ everywhere, say, true.

Given a transition system π, we can define the set of its *computation sequences* $\mathrm{Seq}(\pi)$ as those finite or infinite sequences $(\sigma_i \xrightarrow{e_i} \sigma_{i+1})_{i<\alpha}$ for which $e_i^\bullet = {}^\bullet e_{i+1}$ and every pair of successive stores σ_i and σ_{i+1} satisfies $\mathcal{T}(e_i)$. The *behaviors* of π, $\mathrm{Beh}(\pi)$ is the subset of computation sequences $(\sigma_i \xrightarrow{e_i} \sigma_{i+1})_{i<\alpha}$ that are of maximal length and for which ${}^\bullet e_0 = I$ holds. We will say that π is *closed* if any infinite sequence $s = (\sigma_i \xrightarrow{e_i} \sigma_{i+1})_{i<\omega}$ such that there are arbitrary long prefixes of s that agree with an initial part of some behavior of π is in fact a behavior of π. Thus, a closed transition system does not have any justice or fairness constraints [MP89a] imposed upon its behaviors.

Temporal logic We shall use a first order, linear temporal logic with the standard future modalities, $\mathcal{L}an$. Given a transition system π, $\mathcal{L}an$ is interpreted on frames determined by the behaviors of π. The underlying predicate logic, $\mathcal{A}ss$, has constants C, E and h as hinted at earlier and

[4] Formally speaking we also should extend π with the set of variables that it may use so as to limit the quantification over $y \not\equiv x$ to a finite set

constants v and e which valuate to the vertex control resides at and to the edge last traversed; it has variables of edge-type over which variables may be quantified and rigid constants[5] for the actions, vertices and edges; it also has rigid functions $^\bullet.$ and $.^\bullet$ valuated by the corresponding function in π (; hence, we always have $\mathsf{v} = \mathsf{e}^\bullet$ and by convention $I = {}^\bullet i = i^\bullet$ and in every initial store $\mathsf{v} = i^\bullet$ for some rigid constant i whose interpretation differs from that of all other edge constants.). We allow projection of h on a set of actions *aset*, $\mathsf{h}\lceil_{aset}$. Finally, we assume rigid constants $z, \ldots$, of appropriate types unequal to program variables, $x, y, \ldots$. Rather than fully defining $\mathcal{A}ss$ we shall let it be implicitly determined by whatever is needed for chunk definitions and in proof rules. We feel justified doing so because the definition of syntax and satisfaction is completely standard. Only the valuation of the history constant h deserves to be mentioned: at point j of a behavior $(\sigma_i \xrightarrow{e_i} \sigma_{i+1})_{i<\alpha} \in \mathrm{Beh}(\pi)$, h valuates to the sequence $(\sigma_i, e_i, \sigma_{i+1})_{i<j}$. In other words, the last executed action might be 'accessed' as $\mathcal{L}(\mathsf{last}(\mathsf{h}))$ (and also as $\mathcal{L}(\mathsf{e})$). Furthermore, $\mathsf{h}[i]$ valuates to the ith triple $(\sigma_i, a_i, \sigma_{i+1})$ in h; and $\mathsf{h}[i](x)$ valuates to the value of x in state σ_i. The temporal logic will be the anchored version as defined in [MP89a]. Hence, $\pi \models \phi$ with $\phi \in \mathcal{L}an$ holds if $s \models \phi$ for every $s \in \mathrm{Beh}(\pi)$. Satisfaction $s \models \phi$ is standard and assumed known.

The Manna–Pnueli proof method is based on the possibility of reducing proofs of temporal logic formulae to proving so-called *leads to* properties $p \overset{e}{\leadsto}_\pi q$ defined as

$$\models p[{}^\bullet e/\mathsf{v}] \wedge T_e \rightarrow q[\vec{x}'/\vec{x},\ e^\bullet, e, \mathsf{h}\hat{\ }e/\mathsf{v}, \mathsf{e}, \mathsf{h}]$$

where $p, q \in \mathcal{A}ss$ and $\vec{x} = \mathrm{FV}(q)$. For example, to prove $\pi \models p \Rightarrow \Diamond q$ $(p, q \in \mathcal{A}ss)$, which means $\pi \models \Box(p \rightarrow \Diamond q)$, one might use the following rule, paraphrased from [MP89a]

$$\frac{\begin{array}{c}\models (p \rightarrow \exists \alpha\, r(\alpha)) \wedge (r(0) \rightarrow q) \wedge (r(\alpha+1) \rightarrow en(\pi)) \\ \forall e \in \mathcal{E}\ \ r(\alpha) \overset{e}{\leadsto}_\pi \exists \beta < \alpha\, r(\beta)\end{array}}{\pi \models p \Rightarrow \Diamond q},$$

where α and β are ordinals and $\sigma \models en(\pi)$ holds if at least one transition of π is enabled in σ. Instead of $\forall e \in \mathcal{E}\ p \overset{e}{\leadsto}_\pi q$, one often writes $p \leadsto_\pi q$.

In the sequel we shall need the notion of *comes from* properties, dual to leads to properties: $p \overset{e}{\hookleftarrow}_\pi q$ is defined as $\models q[\vec{x}'/\vec{x},\ e^\bullet, e, \mathsf{h}\hat{\ }e/\mathsf{v}, \mathsf{e}, \mathsf{h}] \wedge T_e \rightarrow p[{}^\bullet e/\mathsf{v}]$. Also, $p \hookleftarrow_\pi q$ has the expected meaning.

4 Verifying interface refinement

The problem of verifying that $\pi^0\ ref_C\ \pi^1$ (with $C = \{(\phi_i, \psi_i) \mid i \in I\}$) will be tackled in two steps. The first step is to partition the behaviors of π^0 and π^1 so that every partition of π^0 (π^1) satisfies a TL formula ϕ (ψ). Given such partitions, the second step is to relate the partitions of π^0 and π^1 so that the TL formulae, ϕ and ψ, that two related partitions satisfy are in C: $(\phi, \psi) \in C$.

4.1 Partitioning into chunks

Behaviors are partitioned by labeling the events after which every partition class is shown to be a chunk. We first concentrate on associating labels with events.

---

[5] I.e., constants whose interpretation is the same in every world of the frame. In fact, in this case the interpretation will be the same across all the frames determined by π as well.

4.1.1 Partitioning

Since we have transition systems that generate behaviors instead of the behaviors themselves, it seems natural to label the events on each behavior by associating labels with the edges of the transition system π that generates them. This we do by defining a function $\lambda: \mathcal{E} \rightarrow \mathcal{A}ss$. Each assertion $\lambda(e)$ may use the constant L and the intention is that if $\lambda(e) \equiv \mathsf{L} = 3$ then the edge e receives the label 3. Note that L is *not* interpreted in the stores of π. Rather, if we define the formula $\mathsf{S}\lambda(e)$ by $\{\ell \mid \lambda(e)[\ell/\mathsf{L}]\}$, then the set of labels, S, associated with edge e in a store σ is determined by $\sigma \models S = \mathsf{S}\lambda(e)$. Since $\mathcal{A}ss$ allows edge constants, λ itself is an assertion and $\lambda(e)$ stands for $\lambda[e/\mathbf{e}]$.

There are two remarks to be made. First, since an edge may be traversed many times during a computation, the label to associate with an edge will depend in general on the computation history. E.g., if the chunk is a^C then in the computation of the transition system $a^\dagger$, every *occurrence* of the edge a should be in a separate partition and we might define $\lambda(a) \equiv \mathsf{L} = |\mathsf{h}|$. Second, an assertion $\lambda(e) \equiv \mathsf{L} = 3 \vee \mathsf{L} = 5$ clearly allows edge e to be labeled with more than one label (within the same store). This means that the action $\mathcal{L}(e)$ can be part of both the partition determined by 3 as well as the one determined by 5. There is a good reason for allowing this, which we shall discuss next. It also causes problems, which we shall point out and solve subsequently.

Consider a transition system $a(b+c)$ on whose computations we want to recognize the chunks $a^C b^C$ and $a^C c^C$. It is clear that the computations can be partitioned into such chunks. But if we attempt to label the edges we run into a problem. We could define $\lambda(b) \equiv \mathsf{L} = 1$ and $\lambda(c) \equiv \mathsf{L} = 2$ but what should $\lambda(a)$ be? Its label depends on the sequence on which the a event occurs, but this is determined only *after* the a transition has been taken. In other words, the state and the computation history do not determine the correct label of a. One might try and associate labels relative to assumptions about the future, were it not for the difficulty of discharging such assumptions in general. Instead, we propose to leave the choice of labels unresolved and to take in this case $\lambda(a) \equiv \mathsf{L} = 1 \vee \mathsf{L} = 2$. A concrete example of this branching phenomenon is a small extension of our running example: communications may fail. On the computations in the asynchronous systems we then might want to distinguish chunks $\mathtt{s}^C\mathtt{r}^C$ from $\mathtt{s}^C\mathtt{f}^C$; the latter representing a failed communication.

The problem is now the following. Remember that our verification strategy is a reduction of the proof obligations to linear TL formulae. In other words, the proof obligations are reduced to formulae that should be satisfied on *every* behavior. Yet, when verifying whether, e.g., the partition determined by the 1-labeled actions is a chunk we definitely should not attempt to prove this for the sequence ac as only the a-action can be labeled with a 1. To resolve this dilemma, observe that since the proof obligations ultimately reduce to leads to properties, it would suffice to trivialize proving such properties for the 'wrong' transitions; i.e., for those transitions that cannot occur on a behavior containing this chunk.

We introduce a second function $\delta: \mathcal{E} \rightarrow \mathcal{A}ss$. Each assertion $\delta(e)$ may use the constant D and the intention is that if $\delta(a) \equiv \mathsf{D} = 1 \vee \mathsf{D} = 2$ then the edge a may be part of sequences on which partition classes of both 1 and 2-labeled actions may occur. In the example we would take $\delta(a)$ as above, $\delta(b) \equiv \mathsf{D} = 1$ and $\delta(c) \equiv \mathsf{D} = 2$. Leads to properties $p \overset{e}{\leadsto}_\pi q$ that would be used to prove that the ℓ-labeled actions constitute a chunk, are now *weakened* to $p \overset{e}{\leadsto}_\pi (\delta(e)[\ell/\mathsf{D}] \rightarrow q)$. Hence, the property trivializes in case $\delta(e)$ does not 'allow' the label ℓ. Note that $\delta(e)$ (as well as $\lambda(e)$) is valuated in the state produced by the e-transition. Here as well, δ is an assertion.

Since nothing prevents defining $\delta \equiv \mathsf{false}$, which effectively trivializes every leads to property,

There is a $\rho \in \mathcal{A}ss$ with $\mathsf{D}, \mathsf{L} \notin \mathrm{FV}(\rho)$ such that for all $\ell, \bar{\ell} \in \mathbb{N}$ with $\ell \neq \bar{\ell}$, λ and δ satisfy conditions (1), ..., (5) below. Define $\mathsf{S}\lambda = \{\ell \mid \lambda[\ell/\mathsf{L}]\}$ (hence $\mathsf{S}\lambda(e) = \{\ell \mid \lambda(e)[\ell/\mathsf{L}]\}$); define $\mathsf{S}\delta$ likewise.

1. $\models \mathsf{h} = \Lambda \wedge \mathsf{v} = I \to \rho$ and $\rho \rightsquigarrow_\pi \rho$
2. $\models \rho \wedge \mathsf{v} = \mathsf{e}^\bullet \to 0 < |\mathsf{S}\lambda| < \aleph_0$, $\quad \models \rho \wedge \mathsf{v} = \mathsf{e}^\bullet \wedge \lambda \to \delta[\mathsf{L}/\mathsf{D}]$
3. $z \subseteq \mathsf{S}\delta \leftarrow\!\sim_\pi \rho \wedge z = \mathsf{S}\delta$
4. $\pi \models \Box(\, (\rho \wedge z = \mathsf{S}\lambda) \to \Box\, z \cap \mathsf{S}\delta \neq \emptyset\,)$
5. $\pi \models \Box(\, (\rho \wedge \lambda[\ell/\mathsf{L}] \wedge \lambda[\bar{\ell}/\mathsf{L}]) \to \Diamond\, \neg(\delta[\ell/\mathsf{D}] \wedge \delta[\bar{\ell}/\mathsf{D}])\,)$

Figure 1: Conditions on λ and δ

there is a question to be answered. Namely, when and in what sense do a pair λ and δ determine a proper labeling of the edges of a transition system? The intended use of λ and δ suggests that the following property should hold:

for any behavior $(\sigma_i \xrightarrow{e_i} \sigma_{i+1})_{i<\alpha} \in \mathrm{Beh}(\pi)$ *and any* $j < \alpha$*:*

$$\exists!\, \ell \quad \sigma_{j+1} \models \lambda(e_j)[\ell/\mathsf{L}] \;\&\; \forall i < \alpha \;\; \sigma_{i+1} \models \delta(e_i)[\ell/\mathsf{D}]\,. \tag{3}$$

So, on every behavior and for every edge on that behavior there is precisely one label with which to label the edge, that δ always allows along the behavior. Uniqueness of such labels is not required at this point but will be needed in the sequel in Theorem 4–3. Verifying that λ and δ satisfy this property is reducible to proving a number of temporal logic and leads to properties:

4–1 LEMMA. *Let* $\pi = (\mathcal{V}, \mathcal{E}, I, \mathcal{L}, \mathcal{T})$ *be a labeled transition system; let* $\lambda, \delta \in \mathcal{A}ss$ *with* $\mathsf{D} \notin \mathrm{FV}(\lambda)$ *and* $\mathsf{L} \notin \mathrm{FV}(\delta)$ *satisfy the Conditions (1), ...,(5) in Figure 1. Then* π *satisfies Property (3).*

We give a quick proof sketch. First note that Condition (1) makes ρ a *reachability function* which characterizes the reachable states. So, if we fix a computation, then ρ will be satisfied in every state. By Condition (5) there can be at most one ℓ satisfying Property (3). Next, take some event $(\sigma \xrightarrow{e} \sigma')$ on the computation. By Condition (2) there is a label ℓ, with which to label e, that is allowed by $\lambda(e)$ (i.e., $\sigma' \models \lambda(e)[\ell/\mathsf{L}]$) and which is also allowed by $\delta(e)$. Repeated application of Condition (3) gives that there is an ℓ that is allowed by δ for all earlier events in the computation as well. Condition (4) allows the extension to the future: there is an ℓ that is also allowed by δ for all later events in the computation and, hence, that is allowed everywhere. The full proof can be found in the Appendix. We remark that a proof of Property (3) need not follow the strategy of Lemma 4–1; indeed, it is often more advantageous to give a semantical argument than to prove Conditions (1), ..., (5).

Thus is established that, given λ and δ for some transition system π that satisfy Property (3), we can uniquely partition its behaviors $s = (\sigma_i \xrightarrow{e_i} \sigma_{i+1})_{i<\alpha} \in \mathrm{Beh}(\pi)$ by $s_{\lambda,\delta} = (\sigma_i \xrightarrow{e_i,\, \ell_i} \sigma_{i+1})_{i<\alpha}$ where ℓ_j $(j < \alpha)$ is the unique label satisfying $\sigma_{j+1} \models \lambda(e_j)[\ell_j/\mathsf{L}]$ and $\forall i < \alpha \;\sigma_{i+1} \models \delta(e_i)[\ell_j/\mathsf{D}]$. Write $\mathrm{Beh}_{\lambda,\delta}(\pi)$ for the resulting set of behaviors. Observe that $s_{\lambda,\delta}$ is a labeled sequence in the sense of Section 2, page ; we just preferred to write $s_{\lambda,\delta}$ instead of $s \times g$ with $g(i) = \ell_i$.

4.1.2 Proving partitions to be chunks

The next step is to verify that every partition corresponds to a chunk specified by some temporal logic formula. Let $\Phi\colon \omega \to \mathcal{L}an$ associate a chunk definition to every label. We define

> λ, δ *chunk partitions* π *by* Φ iff δ and λ satisfy Property (3) and $s{\downarrow}\ell \models \Phi(\ell)$ holds for any $s = (\sigma_i \xrightarrow{e_i, \ell_i} \sigma_{i+1})_{i<\alpha} \in \mathrm{Beh}_{\lambda,\delta}(\pi)$ and for all ℓ such that $\sigma_{i+1} \models \delta(e_i)[\ell/\mathsf{D}]$ for every $i < \alpha$.

The thing to realize is, that when proving that λ, δ *chunk partitions* π *by* Φ, it is *not* necessary to know the precise label to be associated to some occurrence of a transition along a computation. Indeed, the strategy is that when verifying that, say, the 1-labeled events constitute a chunk, the δ function will trivialize the proof obligation for any transition that cannot occur on a computation that 'contains' this chunk. In other words, we can just assume that the label to associate with some transition e is any label allowed by $\lambda(e)$.
The implication of all this is the following

4–2 Lemma. *Let $\Phi\colon \omega \to \mathcal{L}an$ and assume that π only admits infinite behaviors. Then λ, δ chunk partitions π by Φ if λ and δ satisfy Property (3) and $\forall j \;\; (\pi, \lambda, \delta, j) \models' \Phi(j)$*

In this lemma, $(\pi, \lambda, \delta, j)$ is a transition system derived from π, and proving $(\pi, \lambda, \delta, j) \models' \Phi(j)$ now falls within the domain of the Manna–Pnueli TL approach. The definition of $(\pi, \lambda, \delta, j)$ is surprisingly simple: it is π except that for any edge e its old transition formula $\mathcal{T}_e$ is replaced by

$$\mathcal{T}'_e \equiv \mathcal{T}_e \wedge \delta(e)[j/\mathsf{D},\ \vec{x}'/\vec{x}] \wedge (\mathsf{C} \leftrightarrow \lambda(e)[j/\mathsf{L},\ \vec{x}'/\vec{x}]) \wedge (\mathsf{E} \leftrightarrow \neg\mathsf{C}) ,$$

where $\vec{x} = \mathrm{FV}(\lambda(e), \delta(e))$. The validity relation $\models'$ is such that $\pi \models' \phi$ holds if ϕ is satisfied on every *infinite* behavior of π. This makes sense because by assumption π only has infinite computations, so that a finite computation of $(\pi, \lambda, \delta, j)$ constitutes a computation on which $\delta(e)[j/\mathsf{D}]$ becomes unsatisfiable, meaning that no j-chunk occurs on this computation. This means that $(\pi, \lambda, \delta, j) \models' \Phi(j)$ can be verified using the Manna-Pnueli temporal logic rules, except that the premisses in these rules that ensure the enabledness of transitions as long as the formula to be proven is not yet established may be ignored. For example, the 'BASIC RESPONSE RULE' of [MP89a],

$$\frac{\models p \Rightarrow (q \vee \phi), \quad \phi \rightsquigarrow_\pi q, \quad \models \phi \rightarrow en(\pi)}{\pi \models p \Rightarrow \Diamond q} ,$$

becomes a rule for proving $\pi \models' p \Rightarrow \Diamond q$ by omitting the third premiss.

4.2 Relating behaviors

We are given two transition systems π^0 and π^1 and an interface relation $C = \{(\phi_i, \psi_i) \mid i \in I\}$ such that π^0 ref_C π^1 is valid. Suppose λ^i, δ^i *chunk partitions* π^i *by* Φ^i for $i = 0, 1$ has been established. By judicious choice of Φ^i we can have $(\Phi^0(i), \Phi^1(i)) \in C$ for every i. Validity of π^0 ref_C π^1 guarantees this possibility. At this point we know that the behaviors of π^0 and π^1 can be partitioned into chunks and that, moreover, an ℓ-labeled chunk in π^0 correctly implements or refines the ℓ-labeled chunk in π^1, since $(\Phi^0(\ell), \Phi^1(\ell)) \in C$. What is as yet unknown is whether for every 'sequence' of chunks that constitutes a behavior of π^0 there is a 'sequence' of related chunks that is allowed as a behavior of π^1. This last question can be resolved using the δ^0 and δ^1 functions.

Given π and δ, for any $s = (\sigma_i \xrightarrow{e_i} \sigma_{i+1})_{i<\alpha} \in \mathrm{Beh}(\pi)$ define $s_\delta = (\sigma_i \xrightarrow{D_i} \sigma_{i+1})_{i<\alpha}$ where $\sigma_{i+1} \models D_i = \mathrm{S}\delta(e_i)$; define $s^0_{\delta^0} =_D s^1_{\delta^1}$ by $\bigcap_{i<\alpha^0} D^0_i = \bigcap_{i<\alpha^1} D^1_i$ and define $\pi^0\ ref_{\delta^0,\delta^1}\ \pi^1$ by $\forall\, s^0 \in \mathrm{Beh}(\pi^0)\ \exists\, s^1 \in \mathrm{Beh}(\pi^1)\ \ s^0_{\delta^0} =_D s^1_{\delta^1}$. If δ is defined properly, then such an intersection contains the labels of chunks that appear on the behavior. This observation immediately yields

4–3 THEOREM. *Let* $C = \{(\phi^0_i, \phi^1_i) \mid i \in I\}$. *Suppose there are* $\Phi^j : \omega \to \mathcal{L}an\ (j = 0,1)$ *such that*

1. *for every* $i < \omega$ $(\Phi^0(i), \Phi^1(i)) \in C$,
2. λ^0, δ^0 *chunk partitions* π^0 *by* Φ^0 ,
3. λ^1, δ^1 *chunk partitions* π^1 *by* Φ^1 *and*
4. $\pi^0\ ref_{\delta^0,\delta^1}\ \pi^1$.

Then $\pi^0\ ref_C\ \pi^1$.

This theorem reduces the proof of $\pi^0\ ref_C\ \pi^1$ to proving a number of 'standard' properties. Partitioning transition systems reduces to proving a number of leads to and linear temporal logic properties, thanks to Lemma 4–1. The remaining condition of Theorem 4–3 is implied by:

$$\forall\, (\sigma^0_i \xrightarrow{e^0_i} \sigma^0_{i+1})_{i<\alpha^0} \in \mathrm{Beh}(\pi^0)\ \exists\, (\sigma^1_i \xrightarrow{e^1_i} \sigma^1_{i+1})_{i<\alpha^1} \in \mathrm{Beh}(\pi^1)\ \forall n\ \forall D\ \exists i_n, j_n \tag{4}$$
$$i_{n-1} < i_n \wedge j_{n-1} < j_n \wedge \sigma^0_{i_n+1} \models D = \mathrm{S}\delta^0(e^0_{\min(i_n,\alpha^0)}) \Longleftrightarrow \sigma^1_{j_n+1} \models D = \mathrm{S}\delta^1(e^1_{\min(j_n,\alpha^1)})\ .$$

This is a more or less classical 'process refinement' question in which for every computation another computation has to be constructed that agrees with the first w.r.t. to some condition on individual transitions. For such, there exist various proof principles; most of them variations of Milner-simulation. We choose to adapt Lynch's technique of *multivalued possibilities mappings* [Lyn89].

Semantically, the idea is to define a *set valued* function, f, from the vertices and stores of π^0 to sets of such of π^1 satisfying that initial vertices are mapped onto the same and that for every transition $\sigma^0 \xrightarrow{e^0} \bar\sigma^0$ in π^0 and every $(v^1, \sigma^1) \in f({}^\bullet e^0, \sigma^0)$ there is a 'corresponding' transition in π^1, $\sigma^1 \xrightarrow{e^1} \bar\sigma^1$, with $v^1 = {}^\bullet e^1$ and $(e^{1\bullet}, \bar\sigma^1) \in f(e^{0\bullet}, \bar\sigma^0)$. Using these conditions, we can stepwise construct with every sequence in π^0 a corresponding one in π^1 (provided π^1 is closed).

To use this technique in proving $\pi^0\ ref_{\delta^0,\delta^1}\ \pi^1$, introduce notation $v, \sigma \xrightarrow[\pi,\delta]{D} v', \sigma'$ meaning that there is a sequence $(\sigma_i \xrightarrow{e_i} \sigma_{i+1})_{i<n} \in \mathrm{Seq}(\pi)$ such that $\sigma_0 = \sigma$, ${}^\bullet e_0 = v$, $\sigma_n = \sigma'$, $e^\bullet_{n-1} = v'$ and $\sigma_n \models D = \mathrm{S}\delta(e_{n-1})$.

4–4 DEFINITION. *Call* f a multivalued D-possibility mapping *between* (π^0, δ^0) *and* (π^1, δ^1), $(\pi^0, \delta^0) \xrightarrow{f} (\pi^1, \delta^1)$, *if* $f : \mathcal{V}^0 \times \Sigma \to 2^{\mathcal{V}^1 \times \Sigma}$ *and there is a* $d \subseteq \mathcal{V}^0 \times \Sigma$ *such that*

1. $(I^0, \sigma[\Lambda/h]) \in d$ *for any* $\sigma \in \Sigma$ *and for any* $(\sigma^0_i \xrightarrow{e^0_i} \sigma^0_{i+1})_{i<\alpha^0} \in \mathrm{Beh}(\pi^0)$ *and any* $i < \alpha^0$ *there is a* $j \geq i$ *such that* $({}^\bullet e^0_j, \sigma^0_j) \in d$,

2. $\forall \sigma^0 \in \Sigma\ \exists \sigma^1 \in \Sigma\ (I^1, \sigma^1[\Lambda, I^1, i/\mathsf{h}, \mathsf{v}, \mathsf{e}]) \in f(I^0, \sigma^0[\Lambda, I^0, i/\mathsf{h}, \mathsf{v}, \mathsf{e})])$ and

3. if $(v^1, \sigma^1) \in f(v^0, \sigma^0)$, $(v^0, \sigma^0) \in d$, $v^0, \sigma^0 \xrightarrow[\pi^0, \delta^0]{D} \bar{v}^0, \bar{\sigma}^0$ and $(\bar{v}^0, \bar{\sigma}^0) \in d$

   then $\exists \bar{v}^1, \bar{\sigma}^1\ v^1, \sigma^1 \xrightarrow[\pi^1, \delta^1]{D} \bar{v}^1, \bar{\sigma}^1$ and $(\bar{v}^1, \bar{\sigma}^1) \in f(\bar{v}^0, \bar{\sigma}^0)$

Note that if we view d as a state assertion, then 4–4(1) translates as $\pi^0 \models (\mathsf{h} = \Lambda \rightarrow d) \wedge \Box\Diamond d$. However, we stick to the 'tradition' of defining simulations semantically. The function of d is to define states in which the sets of potential chunk labels (D) in corresponding behaviors should coincide.

4–5 LEMMA. *Assume that π^0 and π^1 do not admit finite computations; and suppose π^1 is closed. Then π^0 ref_{δ^0,δ^1} π^1 if there is a mapping f such that $(\pi^0, \delta^0) \xrightarrow{f} (\pi^1, \delta^1)$*

The restriction to infinite computations can be relaxed for the price of a more complex Definition 4–4. The closedness condition is quite essential.

With this Lemma we have completed the reduction of interface refinement proofs to verifying standard properties. A proof of π^0 ref_C π^1 starts with defining λ^i and δ^i functions for both systems. These functions must be shown to determine unique partitions for every behavior of the respective systems as formalized by Property (3) (page). This can be done by proving the conditions of Figure 1 (page). The second step is to show that each class in these partitions satisfy a chunk definition given as a TL formula. Lemma 4–2 shows that for this the Manna-Pnueli proof technique applies although on a slightly modified transition system. The third and final step—according to Theorem 4–3—is the proof of a standard process refinement for which possibility mappings can be used as shown in Lemma 4–5.

4.3 Completeness

We do not claim completeness for our criterion. There are a number of reasons for this. A trivial source of incompleteness is the use of possibility mappings in proving π^0 ref_{δ^0,δ^1} π^1. It is well-known that such mappings do not exists if π^0 is more deterministic than π^1 is. This source is easily quenched, either by using history and prophesy variables [AL91] or by using subset simulation [Ger90, Jon91]. However, even if we make such changes it is Condition (4) that we prove and this is stronger than required by π^0 ref_{δ^0,δ^1} π^1: the fact that the intersections of all D-sets along related computations agree does not imply that there is even one coinciding pair of such D-sets. Completeness here can be bought with a more complex simulation criterion; one in which the cardinality of the set-difference of the D-sets along related sequences should be shown to converge to 0. Of course, this approach only works with finite label sets and one should ask whether this is always possible. This question is part of a larger problem, namely whether our λ-δ approach to defining unique labelings of behaviors applies in all circumstances. Although we think it does, we have no proof of this as yet.

5 An example

In Section 2 we used the refinement of synchronous communication actions into asynchronous send and receive actions as an example. Now we apply the criterion formulated in Section 4 to a (high level) program with synchronous communication and a (low level) program with asynchronous communication.

5.1 The problem

In Figures 2 and 3 two programs are defined. The are abstracted producer-consumer systems; the first program, P^1, uses synchronous communications, while the second one, P^0, uses asynchronous communications. The entry locations of P^1 are l_{10} and l_{20}; the entry locations of P^0 are m_{10} and m_{20}.

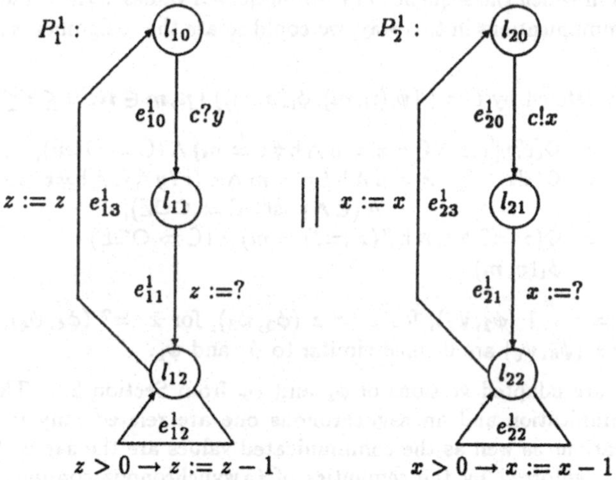

Figure 2: Program P^1.

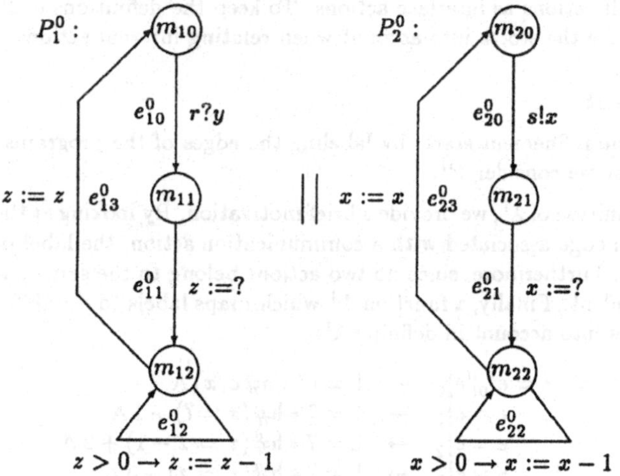

Figure 3: Program P^0.

These programs can easily be transformed into transition systems as described in Section 3 by taking the Cartesian product of the nodes and adding edges in the obvious way, taking into account that the actions $c!x$ and $c?y$ can only be executed in parallel. We assume that the asynchronous communication in P^0 is buffered. The assignments have their expected meaning. The actions other than the communications abstract 'internal' computations of arbitrary, finite length.

We impose weak fairness constraints (see [MP89a]) $(\mathcal{T}_{P_1^0}, \mathcal{T}_{P_1^0})$ and $(\mathcal{T}_{P_2^0}, \mathcal{T}_{P_2^0})$ on P^0 to prevent P_1^0 from never being scheduled. No fairness constraints on P^1 are imposed since in Section 3 we already assumed that behaviors are maximal.

To prove that P^0 refines P^1, we have to define a chunk relation which expresses what *refines* means here. We consider P^0 to be a refinement of P^1 because for every behavior of P^0 a behavior of P^1 exists in which the sequence of communicated values is the same. (In fact, apart from relating the communications in this way, we could relate the assignments in the same way as well.)

The chunk relation is defined by $C = \{ (\psi_i(n,m), \phi_i(n,m)) \mid n, m \in \mathbb{N},\ 0 \leq i \leq 6 \}$. We have

$$\begin{aligned}
\phi_0(n,m) &= \Diamond(c?y \| c!x \wedge \mathsf{C} \wedge x = n \wedge \mathsf{h}\#\mathsf{c} = m) \wedge (\mathsf{C} \Rightarrow \bigcirc\Box\mathsf{E}),\\
\psi_0(n,m) &= \Diamond(s!x \wedge \mathsf{C} \wedge x = n \wedge \mathsf{h}\#\mathsf{s} = m \wedge \Diamond(r?y \wedge \mathsf{C} \wedge \mathsf{h}\#\mathsf{r} = m))\\
&\qquad \wedge (\mathsf{C} \wedge \mathsf{past}\Diamond\mathsf{C} \Rightarrow \bigcirc\Box\mathsf{E}),\\
\phi_1(n,m) &= \Diamond(z :=? \wedge \mathsf{C} \wedge \mathsf{h}\#(z :=?) = m) \wedge (\mathsf{C} \Rightarrow \bigcirc\Box\mathsf{E}),\\
\psi_1(n,m) &= \phi_1(n,m)\ .
\end{aligned}$$

The formulae for $z := z - 1$ (ϕ_2, ψ_2), for $z := z$ (ϕ_3, ψ_3), for $x :=?$ (ϕ_4, ψ_4), for $x := x - 1$ (ϕ_5, ψ_5) and for $x := x$ (ϕ_6, ψ_6) are defined similar to ϕ_1 and ψ_1.

Formulae ϕ_0 and ψ_0 are adapted versions of ϕ_x and ψ_x from Section 2.2. They express that a synchronous communication and an asynchronous one are related only if the numbers of preceding communications as well as the communicated values are the same. Equality of sent and received values is enforced by the semantics of (a)synchronous communication. Hence, we do not have to add conjuncts '$y = n$' to ϕ_0 and ψ_0. The other formulae express that the remaining actions do not change in the refinement. We need these formulae because in this example we regard all actions as interface actions. To keep the definitions of δ^0 and δ^1 a little simpler, we do not take the stores into account when relating internal actions.

5.2 The refinement

The verification of the refinement starts by labeling the edges of the programs by means of λ- and δ-functions. First we consider P^1.

Before giving the definition of λ^1, we provide a brief motivation. By looking at the chunk relation C, we see that for an edge associated with a communication action, the label must include the communicated value. Furthermore, since no two actions belong to the same partition class, λ^1 will be history dependent. Finally, a function Φ^1 which maps labels to chunk formulae must be defined. We take this into account in defining λ^1:

$$\begin{aligned}
\lambda^1 = \quad & \mathsf{e} = e^1_{10} \| e^1_{20} &\rightarrow\ & \mathsf{L} = (7 * \mathsf{h}\#\mathsf{c}, x) \wedge\\
& \mathsf{e} = e^1_{11} &\rightarrow\ & \mathsf{L} = 7 * \mathsf{h}\#(z :=?) + 1 \wedge\\
& \mathsf{e} = e^1_{12} &\rightarrow\ & \mathsf{L} = 7 * \mathsf{h}\#(z := z - 1) + 2 \wedge\\
& \mathsf{e} = e^1_{13} &\rightarrow\ & \mathsf{L} = 7 * \mathsf{h}\#(z := z) + 3 \wedge\\
& \mathsf{e} = e^1_{21} &\rightarrow\ & \mathsf{L} = 7 * \mathsf{h}\#(x :=?) + 4 \wedge\\
& \mathsf{e} = e^1_{22} &\rightarrow\ & \mathsf{L} = 7 * \mathsf{h}\#(x := x - 1) + 5 \wedge\\
& \mathsf{e} = e^1_{23} &\rightarrow\ & \mathsf{L} = 7 * \mathsf{h}\#(x := x) + 6.
\end{aligned}$$

We use pairs to label the edges associated with communications. Obviously such pairs can be represented by integers.

There is only one interesting point in the definition of δ^1 given below. If in the ith communication a value n is communicated, then label $(7i, n)$ should always be admitted by δ^1, and any other label $(7i, n')$ should be split off. Thus we have to 'look up' the communicated values in h. We arrive at the following definition for δ^1:

$$\begin{aligned}\delta^1 \quad = \quad & \exists\, i \;\; 1 \le i < \mathsf{h\#c} \wedge \mathsf{D} = (7i, \mathsf{h}\lceil_{\{\mathsf{c}\}}[i-1](x)) \\ & \vee \exists\, i, n \;\; i > \mathsf{h\#c} \wedge n \ge 0 \wedge \mathsf{D} = (7i, n) \\ & \vee \exists\, i \;\; i \ge 7 \wedge i \bmod 7 \ne 0 \wedge \mathsf{D} = i.\end{aligned}$$

It is straightforward to show that λ^1 and δ^1 satisfy Property (3) (see page). Their definition shows that on any behavior $(\sigma_i^1 \xrightarrow{e_i^1} \sigma_{i+1}^1)_{i<\alpha^1}$ of P^1 all edges are labeled with unique labels. Furthermore, if for some $i < \alpha^1$ and for some label l we have that $\sigma_{i+1} \models \lambda^1(e_i^1)[l/\mathsf{L}]$, then $\forall\, j < \alpha^1 \;\; \sigma_j^1 \models \delta^1(e_j^1)[l/\mathsf{D}]$ holds by definition of δ^1: labels of the form i are never split off, and labels of the form $(7i, n)$ are split off only if the ith communicated value is unequal to n.

With λ^1 and δ^1 defined, it is not difficult to define their counterparts for P^0:

$$\begin{aligned}\lambda^0 \; = \; & \mathsf{e} = e_{10}^0 \;\rightarrow\; \mathsf{L} = (7 * \mathsf{h\#}x, y) \wedge \\ & \mathsf{e} = e_{11}^0 \rightarrow \mathsf{L} = 7 * \mathsf{h\#}(z :=?) + 1 \wedge \\ & \mathsf{e} = e_{12}^0 \rightarrow \mathsf{L} = 7 * \mathsf{h\#}(z := z-1) + 2 \wedge \\ & \mathsf{e} = e_{13}^0 \rightarrow \mathsf{L} = 7 * \mathsf{h\#}(z := z) + 3 \wedge \\ & \mathsf{e} = e_{20}^0 \rightarrow \mathsf{L} = (7 * \mathsf{h\#s}, x) \wedge \\ & \mathsf{e} = e_{21}^0 \rightarrow \mathsf{L} = 7 * \mathsf{h\#}(x :=?) + 4 \wedge \\ & \mathsf{e} = e_{22}^0 \rightarrow \mathsf{L} = 7 * \mathsf{h\#}(x := x-1) + 5 \wedge \\ & \mathsf{e} = e_{23}^0 \rightarrow \mathsf{L} = 7 * \mathsf{h\#}(x := x) + 6, \text{ and}\end{aligned}$$

$$\begin{aligned}\delta^0 \quad = \quad & \exists\, i \;\; 1 \le i < \mathsf{h\#s} \wedge \mathsf{D} = (7i, \mathsf{h}\lceil_{\{\mathsf{s}\}}[i-1](x)) \\ & \vee \exists\, i, n \;\; i > \mathsf{h\#s} \wedge n \ge 0 \wedge \mathsf{D} = (7i, n) \\ & \vee \exists\, i \;\; i \ge 7 \wedge i \bmod 7 \ne 0 \wedge \mathsf{D} = i.\end{aligned}$$

λ^0 and δ^0 also satisfy Property (3).

The functions Φ^0 and Φ^1 that we use in the proof are defined as follows:

$$\begin{aligned}\Phi^0(m,n) &= \psi_0(n, m \operatorname{div} 7) && \text{for } m \ge 7,\ m \bmod 7 = 0, \\ \Phi^0(m) &= \psi_{m \bmod 7}(0, m \operatorname{div} 7) && \text{for } m \ge 7,\ m \bmod 7 \ne 0, \\ \Phi^1(m,n) &= \phi_0(n, m \operatorname{div} 7) && \text{for } m \ge 7,\ m \bmod 7 = 0, \\ \Phi^1(m) &= \phi_{m \bmod 7}(0, m \operatorname{div} 7) && \text{for } m \ge 7,\ m \bmod 7 \ne 0.\end{aligned}$$

For the value 0 occurring in the second and fourth formulae of the above definition, we might as well have chosen any other value, since for $i \neq 0$ chunk formulae ψ_i and ϕ_i are constant in their first arguments anyway. We immediately have that $(\Phi^0(l), \Phi^1(l)) \in C$ for every label l.

Now we have to show that λ^i, δ^i *chunk partitions* π^i *by* Φ^i for $i = 0, 1$. Since λ^i, δ^i satisfy Property (3) and the programs do not admit of finite behaviors, we only have to prove that $\forall\, l \;\; (P^i, \lambda^i, \delta^i, l) \models' \Phi^i(l)$ for $i = 0, 1$. Because of space limitations we shall not supply the proof.

There remains to prove that Condition 4–3(4) holds. We apply Lemma 4–5 to show that $\pi^0 \; ref_{\delta^0,\delta^1} \; \pi^1$. The possibility mapping is the following:

$$\begin{aligned}f((m_{10}, m_{20}), \sigma^0[\Lambda/\mathsf{h}]) &= \{ ((l_{10}, l_{20}), \sigma^1[\Lambda/\mathsf{h}]) \mid \sigma^1(x) = \sigma^0(x) \} && \text{for } \sigma^0 \in \Sigma, \\ f((m_{1i}, m_{21}), \sigma^0) &= \{ ((l_{11}, m_{21}), \sigma^1) \mid \sigma^1(x) = \sigma^0(x) \wedge \sigma^1(\mathsf{h\#s}) = \sigma^0(\mathsf{h\#c}) \} \\ & \qquad \text{for } 0 \le i \le 2 \text{ and } \sigma^0 \text{ such that } \mathcal{L}(\sigma^0(\mathsf{e})) = s!x,\end{aligned}$$

and for any other argument f is defined arbitrarily. If we choose $d = \{ ((m_{10}, m_{20}), \sigma[\Lambda/\mathsf{h}]) \mid \sigma \in \Sigma \} \cup \{ ((m_{1i}, m_{21}), \sigma) \mid 0 \leq i \leq 2 \wedge \mathcal{L}(\sigma(\mathsf{e})) = s!x \}$ then it is straightforward to verify that f is a multivalued possibility mapping between (π^0, δ^0) and (π^1, δ^1).

With this, we have established that $P^0 \ ref_C \ P^1$.

6 Conclusions

We have introduced a quite general notion of interface refinement and have developed a verification criterion that reduces the proof of $\pi^0 \ ref_C \ \pi^1$ to showing that the systems satisfy a number of linear temporal logic specifications and showing an ordinary process refinement. Interface changes have been left completely free (but for the constraint that it be expressible in terms of behaviors). In the absence of any consensus about what constitutes a sensible interface change, any restriction would appear arbitrary and ad hoc. In any case, we have shown that the unrestricted case can be dealt with. In [SGK92] we work out a refinement of synchronous to asynchronous communication in some more detail than in Section 5 and we are currently extending that to the server tills example of Section 1, in which we first reduce the granularity of the user transactions and subsequently introduce asynchronous update of the accounts. We are also working on a correctness proof of a two-phase locking protocol [EGLT76] to serialize database transactions, in which the interface refinement relation C is used to formalize the condition of serializability. This is an interesting example because serializability is not usually construed of as changing the interface. Yet, we find that our formalism supports, in our opinion, a consise and straightforward statement of serializability and an easy serializability proof; at least as compared to [PKP91] which to our knowledge is the only paper that gives a (syntactic) serializability proof.

The generality that we allow interface refinements to have, does have its effects. We expect the problem to be Σ_1^1-hard, and the complexity of the auxiliary assertions that are needed to carry out a refinement proof will reflect this. Also, we do not yet claim relative completeness. It is an interesting and relevant question what restrictions to place on interface refinements in order to simplify the criterion. Specifically, are there restrictions that would fundamentally simplify our approach? The need to partition the events on behaviors is arguably the main source of complexity. However, it seems hard to defend a notion of interface change in which one cannot express that a low-level event belongs to the implementation of precisely one high-level event; at least the interface change that substitutes asynchronous sends and receives for synchronous communications will then become inexpressible. So, we would expect the overall set-up to remain essentially the same. This does not mean that there is no room for change, though. For instance, an alternative to introducing δ is to allow edge labelings to depend on the *future*; i.e., to allow λ to be a TL formula. It might very well be easier to specify such a λ than it is to specify old style λ and δ, although showing that such a λ defines unique labelings for the behaviors might very well become harder. Also, the use of different formalisms—transition systems and temporal logic—can be questioned. In fact, we already have switched to using $\forall$-automata [MP89b] to specify interface relations. An advantage of this formalism is that there is a single uniform proof rule, which compares favorably with the various temporal logic proof rules that depend on the specific forms of the formulae to be verified. Also, chunk definitions often involve *sequences* of actions and such are easier to specify with automata.

One topic that we have purposely ignored is that of compositionality. The generality of our framework as compared to that of Brinksma et al's [BJO91] is set off against the compositionality of that paper's framework since ours is not. It is an open problem whether this paper's framework can be extended so as to yield a notion of interface change that is a pre-congruence for at least the parallel composition of systems.

References

[AL91] M. Abadi and L. Lamport. The existence of refinement mappings. *Theoretical Computer Science*, 82(2):253–285, May 1991.

[BJO91] E. Brinksma, B. Jonsson, and F. Orava. Refining interfaces of communicating systems. In S. Abramsky and T. Maibaum, editors, *Proceedings of the Colloquium on Combining Paradigms for Software Development*, volume 494 of *Lecture Notes in Computer Science*, pages 297–312. Springer-Verlag, 1991.

[EGLT76] K. P. Eswaran, J. N. Gray, R. A. Lorie, and I. L. Traiger. The notions of consistency and predicate locks in a relational database system. *Communications of the ACM*, 33:151–178, 1976.

[Ger90] R. Gerth. Foundations of compositional program refinement: safety properties. In J. W. de Bakker, W.-P. de Roever, and G. Rozenberg, editors, *Proceedings of the NFI/REX Workshop on Stepwise Refinement of Distributed Systems*, volume 430 of *Lecture Notes in Computer Science*, pages 777–809. Springer-Verlag, 1990.

[Hoa86] C. A. R. Hoare. *Communicating Sequential Processes*. International series on computer science. Prentice-Hall, 1986.

[Jon90] C.B. Jones. *Systematic Software Development using VDM*. Prentice-Hall, second edition, 1990.

[Jon91] B. Jonsson. Simulations between specifications of distributed systems. In J. C. M. Baeten and J. F. Groote, editors, *Proceedings of the 2nd International Conference on Concurrency Theory (CONCUR'91)*, volume 527 of *Lecture Notes in Computer Science*, pages 346–361. Springer-Verlag, 1991.

[Lam86] L. Lamport. On interprocess communication, Part I: Basic formalism. *Distributed Computing*, 1(2):77–85, 1986.

[Lam91] L. Lamport. The temporal logic of actions. Technical Report 79, DEC Systems Research Center, Palo Alto, California, December 1991.

[LT87] N.A. Lynch and M.R. Tuttle. Hierarchical correctness proofs for distributed algorithms. In *Proceedings of the Fourteenth ACM Symposium on the Principles of Programming Languages*, pages 137–151, Munich, Germany, 1987.

[Lyn89] N. A. Lynch. Multivalued possibilities mappings. In J.W. de Bakker, W.-P. de Roever, and G. Rozenberg, editors, *Linear Time, Branching Time, and Partial Order in Logics and Models for Concurrency*, volume 354 of *Lecture Notes in Computer Science*, pages 519–544. Springer-Verlag, 1989.

[Mil89] R. Milner. *Communication and concurrency*. International series on computer science. Prentice-Hall, 1989.

[MP89a] Z. Manna and A. Pnueli. The anchored version of the temporal framework. In J.W. de Bakker, W.-P. de Roever, and G. Rozenberg, editors, *Linear Time, Branching Time, and Partial Order in Logics and Models for Concurrency*, volume 354 of *Lecture Notes in Computer Science*, pages 201–284. Springer-Verlag, 1989.

[MP89b] Z. Manna and A. Pnueli. Specification and verification of concurrent programs by ∀-automata. In B. Banieqbal, H. Barringer, and A. Pnueli, editors, *Proc. Temporal Logic in Specification*, volume 398 of *Lecture Notes in Computer Science*, pages 124–165. Springer-Verlag, 1989.

[PKP91] D. Peled, S. Katz, and A. Pnueli. Specifying and proving serializability in temporal logic. IEEE, July 1991.

[SGK92] J. Segers, R. Gerth, and R. Kuiper. Implementing synchronous by asynchronous communication. an example of interface refinement. Technical report, Department of Computing Science, Eindhoven University of Technology, 1992. Deliverable of ESPRIT/BRA project 3096 (SPEC).

Concurrent Testing of Processes ★ (Extended Abstract)

M. Hennessy

University of Sussex

Abstract. We develop a noninterleaving semantic theory of processes based on testing. We assume that all actions have a non-zero duration and the allowed tests take advantage of this assumption. The result is a semantic theory in which concurrency is differentiated from nondeterminism.
We show that the semantic preorder based on these tests is preserved by so-called "stable" action refinement and may be characterised as the largest such preorder contained in the standard testing preorder.

1 INTRODUCTION

In recent years there has been much research into semantic theories of processes which distinguish nondeterminism from concurrency. See for example [DD89b, DNM90, vGV87], [BC89]. Most of these are based on some variation of bisimulation equivalence, [Mil89], a widely used "interleaving theory" of processes.

Another well-established "interleaving" theory of processes is based on testing, [DH84]. Here processes are said to be equivalent if they are guaranteed to pass exactly the same tests. Although the framework of testing equivalence is quite general it has only been applied, apart from [MP91], to generate "interleaving theories". The purpose of this paper is to use this framework to develop a "non-interleaving theory" of processes.

In the standard theory a test, which is usually itself a process, is applied to a process by running both together in parallel. A particular run is considered to be successful if the test reaches a certain designated successful state and the process guarantees the test if every run is successful. The test and the process under observation interact by communicating with each other or synchronising. In most process algebras synchronisation is modelled as the simultaneous occurrence of complementary actions although there is a variety of definitions of complementation. In *CCS*, it is assumed that the set of actions is divided in two, Λ, a basic set of actions and $\overline{\Lambda} = \{\overline{\lambda} \mid \lambda \in \Lambda\}$, the set of complements of basic actions. Then synchronisation consists of the simultaneous occurrence of some pair of actions λ and $\overline{\lambda}$. But regardless of the precise definition of complementation the actions which comprise the synchronisations are considered to be instantaneous and indivisible. This of course

★ This work has been supported by SERC and the ESPRIT/BRA CEDISYS project

is an idealisation and abstraction from reality but it has proved to be most useful as it has lead to a range of elegant mathematical theories of processes. Here we relax this restriction. Now we will assume that the synchronisations take a non-zero but indefinite amount of time. This is still an abstraction from reality as we are not saying exactly what form the interaction takes; only that it takes time. For example we could have in mind the rendez-vous mechanism of ADA or the existence of some non-trivial communication medium connecting the tester and the process. Under these assumptions we can see how the process performing a and b in parallel can be differentiated from its sequential counterpart which performs the actions in either order. Consider the test which requests a synchronisation via the action a and then will succeed only if it can successfully initiate a second synchronisation via the action b before the first synchronisation has terminated. The first process will always pass this test whereas the second will always fail.

Let us now discuss how this intuitive idea of non-instantaneous actions can be formalised. The simplest consequence is that each action a has a distinct beginning and end, $s(a)$ and $f(a)$ respectively. If we are thinking of synchronisation over a communication network $s(a)$ corresponds to the sending of a synchronisation signal while $f(a)$ corresponds to the receipt of a confirmation that the signal has been received. The elements of Λ may be viewed as virtual synchronisation channels or ports which are supported by the underlying communication network. But only using these two sub-events we do not capture the idea that we have individual actions with duration, particularly if two distinct occurrences of an action a can be active at the same time. In terms of the communication network we have to assume that it is sufficiently intelligent to distinguish distinct instances of the same virtual synchronisation channel. In this way we are lead to view the behaviour of processes in terms of their ability to produce *labeled beginning and endings of actions.* Such an operational semantics has been given for a process algebra in [AH91b, AH91a] and a similar semantics had previously been developed in [vGV87] for Petri nets where it is used to define a variation on bisimulation equivalence called *ST-bisimulation.* So we will call this type of semantics *ST-operational semantics* and using it we can apply the standard framework of testing to define a behavioural equivalence, or more generally a preorder, between processes which captures the intuitions of concurrent testing described above: $p \sqsubseteq_c q$ if q guarantees every concurrent test guaranteed by p and $p \eqsim_c q$ if they guarantee exactly the same set of concurrent tests. A concurrent test will be a test which can demand of a process either the beginning of a synchronisation along a specific instance of a communication channel or wait for the end of such a synchronisation, i.e. await confirmation that the synchronisation has been successfully completed.

This theory is developed for a process algebra which is a slight extension to *CCS*, essentially the language studied in [AH92]. It contains all of the operators of *CCS* together with sequential composition and recursion. The first major result of the paper is that $\sqsubseteq_c$ is preserved by all the operators in the language apart from, as usual, the choice operator $+$ from *CCS*. This is proved using an alternative characterisation in terms of so-called st-sequences and Acceptance sets which is a generalisation of the alternative characterisation of the standard testing preorder, [Hen88]. St-sequences are simply sequences of labeled beginning and ending of actions where the actual labels involved are not important; they are only a mechanism for describing sequences which may contain a number of distinct occurrences of the sub-actions $s(a)$ or $f(a)$. They have been used previously in a number of papers, such as [Ace91, Vog91a, AH91b, vGV87].

However the main result of the paper concerns *action refinement*. We add to the language a new operator which allows the refinement of an action by a process: an action refinement is a mapping ρ from actions to processes and intuitively the process $p\rho$ is supposed to act like the process p where each action a is replaced by the process $\rho(a)$. We first make this notion precise and then show that $\underset{\sim}{\sqsubset}{}_c^c$, the closure of $\underset{\sim}{\sqsubset}{}_c$ with respect to all $+$ contexts, is also preserved by a class of refinements which we call *stable*. That is if $p \underset{\sim}{\sqsubset}{}_c^c p'$ and ρ, σ are two stable refinements such that $\rho(a) \underset{\sim}{\sqsubset}{}_c^c \sigma(a)$ for every action a then $p\rho \underset{\sim}{\sqsubset}{}_c^c p'\sigma$.
The final result in the paper shows that $\underset{\sim}{\sqsubset}{}_c^c$ may in fact be characterised using action refinement and the standard testing preorder which we call $\underset{\sim}{\sqsubset}{}_s$. The latter is be defined using the standard operational semantics where actions are instantaneous and it generates an "interleaving theory". We show that $\underset{\sim}{\sqsubset}{}_c^c$ is the largest preorder contained in $\underset{\sim}{\sqsubset}{}_s^c$ which is preserved by stable action refinements.

The development of *non-interleaving* semantic theories for concurrent processes has recently received much attention in the literature. See for example [DNM90], [TV87], [vGV87], [BC89], [Vog91a]. However most of the semantic equivalences investigated in these papers are based on modifying in some way the basic definition of *bisimulation equivalence*, [Mil89]. There are exceptions such as [TV87] and [Vog91a] but as far as the author is aware the present paper is the first attempt at applying the paradigm of testing equivalence, [DH84], to process algebras so as to obtain a non-interleaving semantic theory. However the properties of this equivalence, or preorder, have not been fully investigated and the main contribution is the refinement theorem and the characterisation theorem. The research literature on the subject of action refinements is also quite extensive. The novelty of the results on this topic in the present paper is in the conjunction of three characteristics:

- the semantic theory and the definition of action refinement applies directly to the syntax of the process algebra, rather than some intermediate intensional model such as *event structures* or *causal trees*
- the semantic theory used is based on testing equivalence, rather than the much finer equivalence *bisimulation equivalence*
- the class of allowed action refinements is reasonably general, as is the manner in which these refinements are applied to processes.

Nevertheless our restriction to stable refinements is quite strong and it would be of interest to find a more general behavioural equivalence based on testing which is preserved by a wider class of refinements such as that considered in [AH91a].

In papers such as [DD90], [DD89a], [vG90] [Vog90], restricted forms of action refinements are defined on *event structures, causal trees* or *synchronisation trees* and related refinement theorems are proved with respect to *causal bisimulation* or *st-bisimulation*. The last reference also contains characterisation theorems for a variety of equivalences based on bisimulation. In [AH91a] refinements are defined directly

on a process algebra and a refinement theorem and characterisation theorem is given with respect to weak bisimulation. But in the language considered there is no recursion so all processes are necessarily finite. In [AE91], failures equivalence is used and there are similar results although the language used is very restricted; all processes are finite, there is no communication and no restriction. Nevertheless the results of the present paper may be viewed as a direct generalisation of the results of [AE91] to a full process algebra. (However we have not been able to generalise their results on fully-abstract models).
The existing work which appears closest to our results is [Vog91a] and [JM92]. In [Vog91a] the author deals with *safe Petri Nets* and *failure equivalence.* A restricted form of refinement theorem is proved for a generalisation of *failure equivalence* based on *interval semi-words* and there is a characterisation with respect to the standard failures equivalence. As the author points out in a separate paper, [Vog91b], *interval semi-words* are more or less equivalent to st-sequences; it therefore follows (from the results of section three) that this equivalence is closely related to $\eqsim_c$. However the notion of action refinement used is more restrictive than what we allow. In particular when a refinement ρ is applied to a process there can be no interaction between occurrences of $\rho(a)$ and $\rho(b)$ in the refined process. In [JM92] this work is extended to a more general class of Petri Nets but the restrictions on the type of action refinements remain.

2 THE LANGUAGE AND ITS SEMANTICS

The language is parameterised on a set of basic actions or *channel names* Λ and we use $\overline{\Lambda}$ to denote the set of its complements, $\overline{\Lambda} = \{\, \overline{\lambda} \mid \lambda \in \Lambda \,\}$. Let *Act* denote the set of actions, $\Lambda \cup \overline{\Lambda}$ which is ranged over by $a, b, \ldots$. We also assume a special action symbol, τ, different from all symbols in *Act* and a set of recursion variables X ranged over by x. Then the syntax of the language is given in the usual way using a finite set of process operators and a mechanism for recursive definitions:

$$\begin{aligned} t ::= {} & \Omega \mid nil \mid \delta \mid a \mid \tau \mid x \\ & \mid t+t \mid t \,|\, t \mid t\backslash\lambda \\ & \mid t;t \mid rec\,x.\,t \end{aligned}$$

We have two kinds of terminated processes, *nil* the successfully terminated process and δ the deadlocked process, while Ω denotes a process which can only diverge internally and $rec\,x.\ t$ stands for the process defined recursively by the equation $x = t$. We use the choice, parallel and restriction operators from *CCS*, $|, +$ and $\backslash$, while ; represents sequential composition. Let BP_Λ denote the set of *closed* terms, i.e. those terms with no free variables, which we often refer to as *processes*; this set is ranged over by p, q. Much of the development does not depend on the set of actions Λ and so we usually abbreviate BP_Λ to BP.

The st-operational semantics of the language is given in terms of a next state relation with respect to labelled beginning and ending of actions. We also need a relation to define the effect of an internal move or communication between sub-processes, which is represented in the language by the special action τ. One way of giving rise to such an action is by the simultaneous occurrence of two complementary *complete* actions and therefore it will convenient to define in addition a next state relation with respect to complete actions also although in principle these are derivable from those for sub-actions. Let L be an infinite set of labels, ranged over by l, $LSAct$ denote the set of labelled sub-actions $\{ s(a_l), f(a_l) \mid a \in Act, l \in L \}$ and $LAct$ denote the union of all the external actions, $LSAct \cup Act$. For any set S we use S_τ to denote $S \cup \{\tau\}$. So for example $LSAct_\tau$, Act_τ, denote the sets $LSAct \cup \{\tau\}$, $Act \cup \{\tau\}$ respectively. We let μ range over the set of all possible actions $LAct_\tau$, α over the set of external actions $LAct$, a over the set of complete actions Act and finally e over $LSAct$, the set of (labelled) sub-actions. The execution of the sub-actions will often lead to states of processes where actions have started and not yet terminated and therefore we have to enrich the language in order to define such states. We call the more general terms *configurations*, C, and they are defined by

$$c ::= p \mid a_l \mid c \mid c \mid c; p \mid c \backslash \lambda$$

where we assume that in $c\backslash\lambda$ c contains no occurrences of any $\lambda_l, \overline{\lambda}_l$ and more importantly that every occurrence of a labelled action a_l is unique; an occurrence of a_l is meant to denote that there is an a action active and since we use the labels to distinguish different occurrences it is important that there is no duplication of labels. So for example the configuration $a_l; p \mid b_l; q \mid a_k; r$ describes a process which has three subprocesses, two of which are performing an a action and one a b action. The next state relations $\xrightarrow{\mu}$, for each $\mu \in LAct_\tau$ are given in Figure 1. Many of the rules are straightforward and do not require comment. The obvious symmetric components of the rules ($O2$) and ($O3$) are omitted and the predicate admits used in (O4) has the obvious definition: λ admits μ unless μ has one of the forms $\lambda, \overline{\lambda}, s(\lambda_l)$ or $f(\lambda_l)$. The main nonstandard rule is (O1) which allows the basic process a to perform not only the complete action a but also simply start the action by performing the move $s(a_l)$ for any label l to arrive at the state a_l: this indicates that an instance of the action a is running. The side condition in ($O3$) ensures that labels continue to be unique so that the actual labels used in applications of (O1) will be restricted by the labels not already occurring in the configurations which contain the process to which this rule is applied. The rule for sequential composition uses an auxilary predicate of "proper termination", $\surd$, which is defined in the normal way:

Definition 1. Let $\surd$ be the least relation over configurations which satisfies

1. $nil\surd$
2. $p\surd$, $q\surd$ implies $p+q\surd$
3. $p\surd$, $c\surd$ implies $p;c\surd$
4. $c\surd$, $c'\surd$ implies $c \mid c'\surd$
5. $c\surd$ implies $c\backslash\lambda\surd$
6. $t[rec\, x.\, t/t]\surd$ implies $rec\, x.\, t\surd$

The last two rules of Figure 1 are concerned with the derivation of internal moves of which ($O7$) is the most important. It says that an internal move may occur because of a communication between two subprocesses. The remaining rules are

(O1) $\mu \xrightarrow{\mu} nil$

$a \xrightarrow{s(a_l)} a_l$ for every label l

$a_l \xrightarrow{f(a_l)} nil$

(O2) $p \xrightarrow{\mu} c$ implies $p+q \xrightarrow{\mu} c$

(O3) $c_1 \xrightarrow{\mu} c_1'$ implies $c_1 \mid c_2 \xrightarrow{\mu} c_1' \mid c_2$
provided $c_1' \mid c_2$ is in C

(O4) $c \xrightarrow{\mu} c'$ implies $c\backslash\lambda \xrightarrow{\mu} c'\backslash\lambda$
provided λ admits μ

(O5) $c \xrightarrow{\mu} c'$ $c;p \xrightarrow{\mu} c';p$

$c\surd,\ p \xrightarrow{\mu} c'$ implies $c;p \xrightarrow{\mu} c'$

(O6) $t[rec\,x.\ t/x] \xrightarrow{\mu} q$ implies $rec\,x.\ t \xrightarrow{\mu} q$

(O7) $c_1 \xrightarrow{a} c_1',\ c_2 \xrightarrow{\bar{a}} c_2'$ implies $c_1 \mid c_2 \xrightarrow{\tau} c_1' \mid c_2'$

(O8) $\Omega \xrightarrow{\tau} \Omega$

Fig. 1. Operational semantics

straightforward; Ω can only perform internal moves and the moves of a recursive definition are determined by its body.

We have need for one further relation over terms as in this language there can be processes whose behaviour is "under-specified" in an informal sense although they can not perform an infinite sequence of internal τ moves; a typical example is $rec\,x.\ x$. This "under-specification" is captured by the predicate $\uparrow$ although it is easier to define its converse which we denote by $\downarrow$:

Definition 2. Let $\downarrow$ be the least relation over configurations which satisfies

1. $nil\downarrow$ and $\delta\downarrow$
2. $p\downarrow,\ q\downarrow$ implies $p+q\downarrow$
3. $p\downarrow,\ c\downarrow$ implies $p;c\downarrow$
4. $c\downarrow,\ c'\downarrow$ implies $c \mid c'\downarrow$
5. $c\downarrow$ implies $c\backslash\lambda\downarrow$
6. $t[rec\,x.\ t/t]\downarrow$ implies $rec\,x.\ t\downarrow$

The reason for developing this st-operational semantics is to formalise the concurrent testing discussed in the introduction. However in order to be able to describe the appropriate tests we need a language which is strictly more expressive than BP_{Λ}. This is because with this form of testing we need to be able to test the ability of processes to initiate new synchronisations before other previously initiated synchronisations have terminated. Such tests are not possible in the basic language BP_{Λ}. So we include in our set of tests processes which can perform as fully-fledged actions the subactions of the language BP_{Λ}. Specifically we use as the set of basic actions

$$\Delta = \{ s(a_l), f(a_l) \mid a \in \Lambda \cup \overline{\Lambda} \} \cup \Lambda.$$

There is another problem which can not be resolved by mimicing the framework of testing developed in [Hen88]. Here we have two forms of termination, in *nil* and δ, and it is difficult to see how they can be differentiated by purely computation means. Accordingly we introduce into the testing language the ability to recognise proper termination; this takes the form of a special action *term* which the test can execute only when the process under observation is properly terminated. So we use as the set of tests, *Tests*, closed terms in the language BP_{Φ} where Φ is the set of basic actions

$$\Delta \cup \{term, \omega\}.$$

Here the action ω will be used to report the successful completion of an experiment. Of course there are many tests in this language which will not be used but, at least here, it is not of great importance to characterise the collection of meaningful tests.

We now define formally how tests and processes, or more generally configurations, interact. An *experimental state* takes the form $e \parallel c$ where e is a test and c a configuration. An experiment proceeds by moving from state to state and this is defined using a transition relation of the form $e \parallel c \ \longmapsto\ e' \parallel c'$.

Definition 3. Let $\longmapsto$ be the least relation between experimental states which satisfies

1. $e \xrightarrow{\alpha} e',\ c \xrightarrow{\alpha} c'$ implies $e \parallel c \ \longmapsto\ e' \parallel c'$ for every α in $LAct$
2. $e \xrightarrow{\tau} e'$ implies $e \parallel c \ \longmapsto\ e' \parallel c$
3. $c \xrightarrow{\tau} c'$ implies $e \parallel c \ \longmapsto\ e \parallel c'$
4. $e \xrightarrow{term} e',\ c\surd$ implies $e \parallel c \ \longmapsto\ e' \parallel c$. □.

We can now define in the standard way when processes guarantee tests. A *computation* from the state $e \parallel p$ is a maximal sequence of the form

$$e \parallel p \equiv e_0 \parallel c_0 \ \longmapsto\ e_1 \parallel c_1 \ldots \longmapsto\ e_n \parallel c_n \ \longmapsto\ \ldots$$

that is, it is either infinite or has a maximal element $e_m \parallel c_m$ from which no further derivation can be made. A state $e_m \parallel c_m$ in such a computation is successful if e_m can perform the action ω and $p_n \downarrow$ for every $n < m$. We say p *guarantees* e or p *must* e if every computation from $e \parallel p$ is successful. Finally we write $p \sqsubseteq_c q$ if for every test e p *must* e implies q *must* e. The associated equivalence relation, the kernel of $\sqsubseteq_c$, is denoted by $\approx_c$.

One can check that, for example, the processes *nil* and δ are incomparable as they can be distinguished by the tests $term;\omega$ and $\tau;\omega + term$. The processes $a \mid b$ and $a;b+b;a$ are also incomparable; $a \mid b$ guarantees the test $s(a_l);s(b_l);\omega$ which can be failed by $a;b+b;a$ while the converse is true of the test $s(a_l);(\tau;\omega + s(b_l))$.

Because of the ability to test for termination this theory of testing, even when restricted to purely sequential proceses, is a little different than that of [Hen88]. There, for example, $a + a;(b+c) \sqsubseteq a + a;b + a;c$ but here they can be distinguished by the test $a;(term;\omega + b;\omega)$.

We end this section by outlining the properties of processes which determine their ability to to guarantee tests. They are essentially the st-sequences they can produce and their acceptance sets after these sequences. These acceptance sets are finite collections of finite subsets of *Act* but because of the ability to test for proper termination they have to be defined slightly differently than in [Hen88]. First some notation.

For each $s \in LAct^*_\tau$ we define the relation $\stackrel{s}{\Longrightarrow}$ between configurations in the standard way. For any configuration c let $S(c) = \{\, a \in Act \mid c \stackrel{a}{\Longrightarrow} \,\}$; $S(c)$ only contains complete actions which are by definition unlabelled. We say c is *stable* if $c \downarrow$ and $c \stackrel{\tau}{\longrightarrow} c'$ for no c', it is *live* if $c\surd$ is not true i.e. if it has not properly terminated and it *converges*, written $c \Downarrow$, if it has no infinite τ derivations $c \stackrel{\tau}{\longrightarrow} c_1 \stackrel{\tau}{\longrightarrow} \ldots$ and $c' \downarrow$ whenever $c \stackrel{\varepsilon}{\Longrightarrow} c'$. The set of *acceptance sets of p after s*, for $s \in LAct^*$, is defined by

$$\mathcal{A}(p,s) = \{\, S(c) \mid p \stackrel{s}{\Longrightarrow} c,\ c\ \textit{stable} \text{ and } \textit{live} \,\}.$$

In this definition the requirement that c be live is crucial. Acceptance sets are compared as in [Hen88], using a slight variation on subset inclusion; We write $\mathcal{A}(p,s) \ll \mathcal{A}(q,s)$ if for every $A \in \mathcal{A}(q,s)$ there is a $B \in \mathcal{A}(p,s)$ such that $B \subseteq A$.

To obtain the alternative characterisation of $\underset{\sim}{\sqsubset}_c$ we need to parameterise both the proper termination and the convergence predicates to sequences of actions from *LAct*. We say *c (weakly) terminates with respect to* s, $c\surd s$, if there exists a c' such that $c \stackrel{s}{\Longrightarrow} c'$ and $c'\surd$. The generalisation of $\Downarrow$ is much stronger; $c \Downarrow s$ guarantees that c will never diverge when performing any subsequence of s; It is defined by induction on s:

1. $c \Downarrow \varepsilon$ if $c \Downarrow$
2. $c \Downarrow \alpha.s$ if $c \Downarrow$ and for every c' such that $c \stackrel{\alpha}{\Longrightarrow} c,\ \ c' \Downarrow s$.

Definition 4. For processes p, q we write $p \ll_{st} q$ if for every $s \in LAct^*$

$$p \Downarrow s \text{ implies } \begin{array}{ll} i) & q \Downarrow s \\ ii) & q\surd s \text{ implies } p\surd s \\ iii) & \mathcal{A}(p,s) \ll \mathcal{A}(q,s) \end{array}$$

The main result of this section is that $\underset{\sim}{\sqsubset}_c$ and $\ll_{st}$ coincide on processes:

Theorem 5. *(Alternative Characterisation) For all processes* p, q $\;p \underset{\sim}{\sqsubset}_c q$ *if and only if* $p \ll_{st} q$. □

The preorder $\underset{\approx_c}{\sqsubset}$ is not preserved by $+$ for the usual reasons: $a \underset{\approx_c}{\sqsubset} \tau; a$ but $a+b$ guarantees the test $b;\omega$ which may be failed by $\tau; a+b$. However it is very easy to adapt it so that it is preserved by $+$; Let $p \underset{\approx_c}{\sqsubset}^c q$ if

1. $p \underset{\approx_c}{\sqsubset} q$
2. p *stable* implies q *stable*

and let $\approx_c^c$ be the associated equivalence.

Proposition 6. *The relation* $\underset{\approx_c}{\sqsubset}^c$ *is the largest relation contained in* $\underset{\approx_c}{\sqsubset}$ *which is preserved by all the operators in the language.* □

In the definition of $\ll_{st}$ there is a tremendous amount of redundant information. This is principally because many of the sequences in $LAct^*$ can not be generated by processes and the fact the the individual identity of the labels are unimportant. By eliminating some of this redundancy we can show that st-sequences are closely related to particular kinds of labelled partial orders, called *interval orders* which have been extensively studied in [Vog91b]. The details may be found in the full version of the paper.

3 ACTION REFINEMENT

In this section we investigate the circumstances under which the behavioural preorder $\underset{\approx_c}{\sqsubset}$ is preserved by action refinement and we use the approach and notation from [AH91a]. An *action refinement* is a mapping

$$\rho: Act \longmapsto BP_{\Lambda'}.$$

It associates with each action a a process $\rho(a)$. The application of an action refinement ρ to the process p will be denoted by $p\rho$. This is to be considered as a process which behaves like the process p where the action a has been replaced by the process $\rho(a)$. We will not give a direct operational semantics to $p\rho$; instead we will simply define a substitution operator *sub* which gives the effect of substituting $\rho(a)$ for a and then say that $p\rho$ behaves in exactly the same way the process $sub(\rho, p) \in BP$. The definition of substitution is straightforward except that we wish to consider the restriction operator $\backslash\lambda$ as a binding operator on actions. This is reasonable because, for example, the behaviour of $(\lambda; p \mid a; q)\backslash\lambda$ does not depend on the particular action λ; it has the same behaviour as $(\beta; p \mid a; q)\backslash\beta$ assuming λ, β does not appear in $p, a; q$. We would also expect $(\lambda; p \mid a; q)\backslash\lambda\ \rho$ to have the same behaviour as $(\beta; p \mid a; q)\backslash\beta\ \rho$. Now consider a ρ which maps a to $rec\, x.\ \lambda; x$. If we simply used syntactic substitution we would require that $(\lambda; p \mid (rec\, x.\ \lambda; x); q)\backslash\lambda$ and $(\beta; p \mid (rec\, x.\ \lambda; x); q)\backslash\beta$ have the same behaviour, and they are obviously different; the latter can perform a λ move which the former can not. The problem occurs because in the former the free occurrence of λ in $\rho(a)$ is captured by the restriction in $(\lambda; p \mid a; q)\backslash\lambda$. In order to avoid this we need to rename the actions being restricted so that no capturing occurs. So assuming that $\backslash\lambda$ acts as a *channel name binder* we obtain the usual notions of free and bound channels occurring in terms whose definitions we omit.

Definition 7. For each action refinement ρ and term t let $sub(\rho, t)$ be the term defined by

1. $sub(\rho, a) = \rho(a)$
2. $sub(\rho, (t\backslash\lambda)) = sub(\rho[\lambda \mapsto \lambda'], t)\backslash\lambda'$ where λ' is different than all free channel names in $\rho(a)$ for each a occurring free in t.
3. $sub(\rho, op(\ldots, t, \ldots)) = op(\ldots, sub(\rho, t), \ldots)$
4. $sub(\rho, x) = x$
5. $sub(\rho, rec\, x.\ t) = rec\, x.\ sub(\rho, t)$.

For each process p and action refinement ρ $sub(\rho, p)$ is also a process and we take the behaviour of $p\rho$ to be that of $sub(\rho, p)$. Via this reduction we may view $p\rho$ as a process in BP and in what follows we will take this for granted.
Now let $\rho \sqsubseteq_c^c \sigma$ if $\rho(a) \sqsubseteq_c^c \sigma(a)$ for every a. Then we wish to know under what conditions

$$p \sqsubseteq_c^c q \text{ and } \rho \sqsubseteq_c^c \sigma \text{ imply } p\rho \sqsubseteq_c^c q\sigma,$$

or speaking more strictly $p \sqsubseteq_c^c q$ and $\rho \sqsubseteq_c^c \sigma$ imply $sub(\rho, p) \sqsubseteq_c^c sub(\sigma, q)$.

In [AH91a] a similar problem was posed for bisimulation equivalence and it was shown to be true for *standard* refinements.

Definition 8. An action refinement ρ is *standard* if it satisfies

1. for each a $\rho(a) \mid \rho(\overline{a}) \stackrel{\varepsilon}{\Longrightarrow} r$ for some r such that $r\surd$
2. for each a not $\rho(a)\surd$

The first condition says that after the refinement the resulting process should be able to mimic the complete synchronisation between a and $\overline{a}$ and the second says that an action can not be refined to a process which is properly terminated. These conditions are also necessary if we wish $\sqsubseteq_c$ to be preserved by refinements.

Example 1. Let p be $(a + \lambda) \mid \overline{\lambda}; b$ and q be $p + \tau; b$. Then $p \approx_c^c q$ but if $\rho(\lambda) = c$, $\rho(\overline{\lambda}) = d$ then $p\rho \not\approx_c q\rho$; the former guarantees the test $a; \omega$. Note that this refinement does not satisfy the first requirement of *standard*.

Example 2. Let p, q be $((c + a; \overline{\lambda}) \mid \lambda; b)\backslash\lambda$, $c + a; b$ respectively. Then $p \approx_c^c q$ but if $\rho(a) = nil$, a refinement which does not satisfy the second condition, then $p\rho \not\approx_c q\rho$.

However more restrictions are necessary as can be seen from the next example.

Example 3. Let p be $a \mid (b; c) + a + b$, q be $a + b$ and ρ a standard refinement such that $\rho(a) = d$ and $\rho(b) = \overline{d}$. Then $p \sqsubseteq_c q$ but $p\rho$ guarantees the test $c\omega$ whereas $q\rho$ fails it.

Definition 9. An standard action refinement is called *stable* if

1. $\rho(a) \stackrel{\tau}{\not\longrightarrow}$ for every action a
2. if $\rho(a) \mid \rho(b) \stackrel{\tau}{\longrightarrow}$ implies $a = \overline{b}$

The example just discussed violates the second condition but we conjecture that the first is unnecessary.

Theorem 10. *For every pair of stable refinements ρ, σ, $p \underset{\sim c}{\sqsubset}^{c} q$ and $\rho \underset{\sim c}{\sqsubset}^{c} \sigma$ imply $p\rho \underset{\sim c}{\sqsubset}^{c} q\sigma$.*

Proof. (Outline) Suppose $p \underset{\sim c}{\sqsubset}^{c} q$. Using the Alternative Characterisation of $\underset{\sim c}{\sqsubset}^{c}$ we can show $p\rho \underset{\sim c}{\sqsubset}^{c} q\sigma$ by establishing, under the assumption $p\rho \Downarrow s$, the following three statements

1. $q\sigma \Downarrow s$
2. $\mathcal{A}(p\rho, s) \ll \mathcal{A}(q\sigma, s)$
3. $q\sigma\surd s$ implies $p\rho\surd s$.

Let us look briefly at the last two , in reverse order.

3. Let A be an arbitrary element in $\mathcal{A}(q\sigma, s)$; we must show that there exists a $B \in \mathcal{A}(p\rho, s)$ such that $B \subseteq A$. We know that $A = S(c)$ for some live and stable c such that

$$q\sigma \stackrel{s}{\Longrightarrow} c. \qquad (*)$$

Let us examine the composition of this computation. Underlying it is a contribution from q in the form of a sub-computation say $q \stackrel{u}{\Longrightarrow} \overline{c}$ for some configuration $\overline{c}$. There are also contributions from σ and these take two forms:

- for each a_l occurring in c there is a sub-computation from $\sigma(a)$ of the form, say, $\sigma(a) \stackrel{u_l}{\Longrightarrow} c_l$
- during the computation $(*)$ there may have been sub-computations from $\sigma(a)$ which were initiated and have already terminated. These can be captured in a *termination set* T which consists of a set of pairs of the form (a, s); each such (a, s) represents the fact that $\sigma(a)$ can perform an s computation to a properly terminated state, i.e. $\sigma(a)\surd s$.

In fact for each termination set T we can define a *merge predicate* $\mathcal{M}_T$ which has three components:

$$< u, \mathbf{r}, s >$$

where

1. $u \in LAct^*$ records the contribution from the process being refined, in this case q
2. $s \in LAct^*$ is the sequence produced by the refined process, in this case $q\sigma$
3. $\mathbf{r}$ is a partial function from $Act \times N$ to $LAct^+_\tau$
   $\mathbf{r}(a_i)$ records the contribution of $\sigma(a)$ to the computation of s which may be a non-empty sequence of elements from $LAct$ or simply τ, representing the some unknown internal activity.

If we extend refinements so that they act on labelled actions then we can characterise the relevance of $\mathcal{M}_T$ as follows: $q\sigma$ can perform the sequence of moves s, $q\sigma \stackrel{s}{\Longrightarrow} c$, if and only if c has the form $\overline{c}\sigma'$ and there is a triple $< u, \mathbf{r}, s > \in \mathcal{M}_T$ such that

- $q \stackrel{u}{\Longrightarrow} \overline{c}$
- $\sigma(a)$ can perform the sequence of actions $\mathbf{r}(a_i)$ to become $\sigma(a_l)$ for each a_i occurring in c.

– for every $(a, s) \in T$, $\sigma(a)\surd s$.

Furthermore we can assume that the pair $c\sigma'$ is *proper*, i.e. if a_l occurs in c then $\sigma(a_l)$ has not terminated.

Now ignoring potential problems with divergence, $p \underset{\sim}{\sqsubset}_c^c q$ and $\rho \underset{\sim}{\sqsubset}_c^c \sigma$ implies we can get similar sub-computations from p to $\overline{d}$ and from ρ to ρ' which can be recombined to obtain a computation $p\rho \stackrel{s}{\Longrightarrow} \overline{d}\rho'$ where the latter is assumed to be proper. One can now argue, because of the relationship between c and d and ρ' and σ', that $d\rho'$ is live and and stable and finally that $S(d\rho') \subseteq A$.

2. Here we prove that if $q\sigma \stackrel{s}{\Longrightarrow} c$ where $c\surd$ then there exists a d such that $p\rho \stackrel{s}{\Longrightarrow} d$ and $d\surd$. The argument proceeds in much the same way as in the previous case by decomposing the computation $q\sigma \stackrel{s}{\Longrightarrow} c$ into the contributions matching the merge predicate $\mathcal{M}_T$ from q and σ, finding matching contributions from p and ρ which are in turn composed to obtain a computation from $p\rho$ to a properly terminated state. The extra fact required is that if $d\rho$ is proper then $d\surd$ if and only if $d\rho\surd$.

Full accounts of these arguments may be found in the complete version of the paper.

This theorem leaves open the problem of finding a testing based behavioural preorder which is preserved by arbitrary standard refinements. We should also point out that it is very dependent on syntax of the language used. For example the theorem is not true for the language *CSP*, [Jat92].

4 CHARACTERISING ST-TESTING

The more standard notion of testing, based on complete, non-divisible actions as in [Hen88] may also be applied to our language; we denote the resulting preorder by $\underset{\sim}{\sqsubset}_s$; the formal definition may be found in the full version of the paper. We will not develop to any great extent the theory of this behavioural preorder; it is similar to that in [DH84] except that here it is applied to a slightly more general language which has two kinds of termination and sequential composition and we can test for the proper termination of processes. However it can be characterised in terms of acceptance sets along the lines of the alternative characterisation of $\underset{\sim}{\sqsubset}_c$ given in §2; it is sufficient to restrict attention to sequences from *Act*. Let us call the resulting preorder $\ll$ and we can prove that it coincides with $\underset{\sim}{\sqsubset}_s$.

In this section we examine the relationship between $\underset{\sim}{\sqsubset}_c$ and $\underset{\sim}{\sqsubset}_s$. Let $\underset{\sim}{\sqsubset}_s^c$ be defined in the obvious way from $\underset{\sim}{\sqsubset}_s$:

$$p \underset{\sim}{\sqsubset}_s^c q \text{ if } p \underset{\sim}{\sqsubset}_s q \text{ and } p \ \textit{stable} \text{ implies } q \ \textit{stable}.$$

It follows trivially that $\underset{\sim}{\sqsubset}_c^c \subseteq \underset{\sim}{\sqsubset}_s^c$.

The main result is that $\underset{\sim}{\sqsubset}_c^c$ may be characterised by $\underset{\sim}{\sqsubset}_s^c$ and action refinement. Specifically $\underset{\sim}{\sqsubset}_c^c$ satisfies the properties

1. it is contained in $\underset{\sim}{\sqsubset}_s^c$

2. it is preserved by stable action refinements.

We prove that $\underset{\approx_c}{\sqsubseteq}^c$ is the largest relation with these two properties.

We know that $\underset{\approx_c}{\sqsubseteq}^c$ is preserved by stable action refinements and therefore the result will follow if we can construct a particular stable refinement σ with the property that for all processes p, q, $p\sigma \ll q\sigma$ implies $p \ll_{st} q$. We define σ as a mapping

$$\sigma\colon Act \longmapsto BP_{\Delta}$$

where Δ is the set of basic actions defined in §2; it contains as basic actions all the elements of Λ and all labelled begin and end actions. Then σ is defined by

$$\sigma(a) = a + \sum\{\, s(a_k); f(a_k) \mid k \in K \,\} \text{ for } K \subseteq L \text{ a sufficiently large finite set.}$$

Note that σ is a stable refinement. The use of the alphabet Δ leads to some blurring of the distinction between processes and configurations as a configuration for the process language BP_{Λ} is a process for the language BP_{Δ}. However this is the basic reason why the concurrent behaviour of processes in BP_{Λ} can be simulated by the sequential behaviour of processes in BP_{Δ}.

Theorem 11. *$p\sigma \underset{\approx_s}{\sqsubseteq} q\sigma$ implies $p \underset{\approx_c}{\sqsubseteq} q$.*

Proof. In the full version of the paper.

As a corollary we have the main result of the section:

Corollary 12. *The relation $\underset{\approx_c}{\sqsubseteq}^c$ is the largest preorder contained in $\underset{\approx_s}{\sqsubseteq}^c$ which is preserved by stable action refinements.* □

Acknowledgements: The author would like to thank L. Jategoankar who pointed out a number of errors in previous versions of this paper.

References

[Ace91] L. Aceto. Full abstraction for series-parallel-pomsets. In *Proceedings of CAAP*, volume 493 of *Lecture Notes in Computer Science*, pages 1–25. Springer–Verlag, 1991.

[AE91] L. Aceto and U. Engberg. Failure semantics for a simple process language with refinement. Technical report, INRIA, Sophia-Antipolis, 1991.

[AH91a] L. Aceto and M. Hennessy. Adding action refinement to a finite process algebra. In *Proceedings of* 18th *ICALP*, Lecture Notes in Computer Science. Springer–Verlag, 1991.

[AH91b] L. Aceto and M. Hennessy. Towards action refinement in process algebras. *Information and Computation*, 1991. to appear.

[AH92] L. Aceto and M. Hennessy. Termination, deadlock and divergence in process algebras. *Journal of the ACM*, 39(1):147–187, 1992.

[BC89] G. Boudol and I. Castellani. Permutation of transitions: an event structure semantics for CCS and SCCS. In *Proceedings of Linear Time, Branching Time and Partial Order in Logics and Models for Concurrency*, number 354 in Lecture Notes in Computer Science, pages 411–427, 1989.

[DD89a] Ph. Darondeau and P. Degano. About semantic action refinement. Technical Report 11/89, Dipartimento di Informatica, Università di Pisa, 1989. To appear in *Fundamenta Informaticae*.

[DD89b] P. Degano and P. Darondeau. Causal trees. In *Proceedings of ICALP 89*, number 372 in Lecture Notes in Computer Science, pages 234–248. Springer–Verlag, 1989.

[DD90] P. Darondeau and P. Degano. Event structures, causal trees and refinements, 1990. Submitted to *Theoretical Computer Science*.

[DH84] R. DeNicola and M. Hennessy. Testing equivalences for processes. *Theoretical Computer Science*, 24:83–113, 1984.

[DNM90] P. Degano, R. De Nicola, and U. Montanari. A partial ordering semantics for CCS. *Theoretical Computer Science*, 75:223–262, 1990.

[Hen88] M. Hennessy. *An Algebraic Theory of Processes*. MIT Press, 1988.

[Jat92] L. Jategaonkar. Personal communication. 1992.

[JM92] L. Jategoankar and A. Meyer. Testing equivalence for petri nets with action refinement. Technical report, MIT, 1992.

[Mil89] R. Milner. *Communication and Concurrency*. Prentice-Hall, 1989.

[MP91] D. Murphy and D. Pitt. Testing, betting and true concurrency. In *Proceedings of Concur 91*, number 527 in Lecture Notes in Computer Science, 1991.

[TV87] D. Taubner and W. Vogler. The step failures semantics. In F.J. Brandenburg et. al., editor, *Proceedings of STACS 87*, number 247 in Lecture Notes in Computer Science, pages 348–359. Springer–Verlag, 1987.

[vG90] R.J. van Glabbeek. The refinement theorem for ST-bisimulation. In *Prooceedings IFIP Working Group, Sea of Galilee*, Lecture Notes in Computer Science. Springer–Verlag, 1990.

[vGV87] R.J. van Glabbeek and F.W. Vaandrager. Petri net models for algebraic theories of concurrency. In J.W. de Bakker, A.J. Nijman, and P.C. Treleaven, editors, *Prooceedings PARLE conference*, number 259 in Lecture Notes in Computer Science, pages 224–242. Springer–Verlag, 1987.

[Vog90] W. Vogler. Bisimulation and action refinement. Technical report, Technische Universität München, 1990.

[Vog91a] W. Vogler. Failure semantics based on interval semiwords is a congruence for refinement. *Distributed Computing*, 4:139–162, 1991.

[Vog91b] W. Vogler. Is partial order semantics necessary for action refinement ? Technical report, Technische Universität München, 1991.

This article was processed using the LaTeX macro package with LLNCS style

A Theory of Processes with Localities (Extended Abstract)*

G. Boudol, I. Castellani, INRIA, Sophia-Antipolis,
M. Hennessy, CSAI, University of Sussex,
A. Kiehn, TUM, Munich.

Abstract

We study a notion of observation for concurrent processes which allows the observer to see the distributed nature of processes, giving explicit names for the location of actions. A general notion of bisimulation related to this observation of distributed systems is introduced. Our main result is that these bisimulation relations, particularized to a process algebra extending CCS, are completely axiomatizable. We discuss in details two instances of location bisimulations, namely the location equivalence and the location preorder.

1 Introduction

A distributed system may be described as a collection of computational activities spread among different sites or *localities*, which may be physical or logical. Such activities are viewed as being essentially independent from each other, although they may require to synchronize or communicate at times. It has been argued in previous work [CH89,Cas88,Kie89,BCHK91a] that the standard interleaving approach to the semantics of concurrent systems may not be adequate to model such distributed computations: more precisely, it may not be able to express naturally properties of distributed systems which depend on their inherent distribution in space, like e.g. a local deadlock, that is a deadlock in a specific site of the system.

Most noninterleaving semantics proposed so far in the literature for algebraic languages such as CCS [Mil80,Mil89] are based on the notion of *causality* between actions, or on the complementary notion of causal independence or concurrency. Here we pursue the different approach of [CH89,Cas88,Kie89,BCHK91a], which focusses more specifically on the distributed aspects of systems. At first sight, the concepts of causality and *distribution in space* may appear as dual notions, which should give rise to the same kind of noninterleaving semantics for distributed systems. In fact this is not the case, essentially because communication may introduce causal dependencies between activities at different locations.

In this paper we develop a general semantic theory for CCS which takes the distributed nature of processes into account rather than their causal structure. As in [BCHK91a], we shall deal with processes with explicit localities, or *locations*, extending CCS with a construct of *location prefixing*, $l :: p$, which denotes the process p residing at location l. Let us illustrate our approach with a concrete example. We may describe in CCS a simple protocol, tranferring data one at a time from one port to another, as follows:

*This work has been supported by the ESPRIT/BRA CEDISYS project.

$$\begin{aligned} Sys &\Leftarrow (Sender \mid Receiver)\backslash\alpha,\beta \\ Sender &\Leftarrow in.\bar{\alpha}.\beta.Sender \\ Receiver &\Leftarrow \alpha.out.\bar{\beta}.Receiver \end{aligned}$$

where α represents transmission of a message from the sender to the receiver, and β is an acknowledgement from the receiver to the sender, signalling that the last message has been processed. In the standard theory of *weak bisimulation equivalence*, usually noted $\approx$, one may prove that this system is equivalent to the following specification:

$$Spec \Leftarrow in.out.Spec$$

That is to say, $Spec \approx Sys$. The reader familiar with the *causal* weak bisimulation of [DD89], which we denote $\approx_c$, could note that $Spec \approx_c Sys$: intuitively, this is because the synchronizations on α, β in *Sys* create "cross-causalities" between its visible actions *in* and *out*, constraining them to happen alternately in sequence. On the other hand *Spec* will be distinguished from *Sys* in our theory, because *Spec* is completely sequential and thus performs the actions *in* and *out* at the same location l, what can be represented graphically as follows:

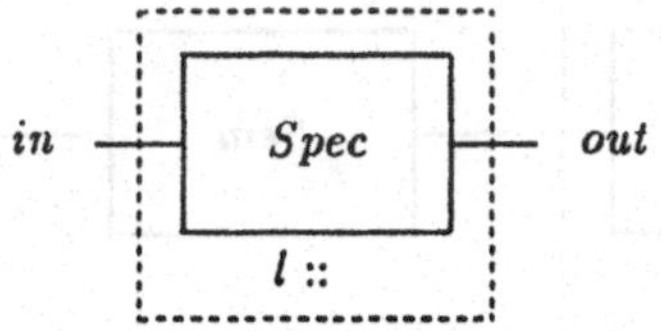

while *Sys* is a system distributed among two different localities l_1 and l_2, with the actions *in* and *out* occurring at l_1 and l_2 respectively. Thus *Sys* may be represented as:

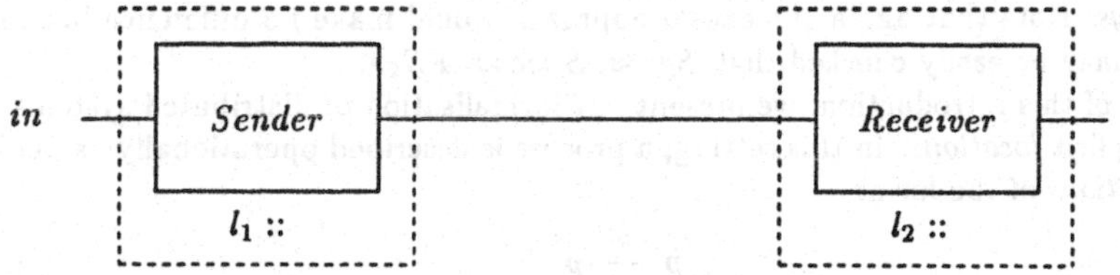

Here the unnamed link represents the communication lines α, β, which are private to the system. Although *Spec* and *Sys* will not be equated in our theory, we will be interested in relating them by a weaker relation, a *preorder* that orders processes according to their degree of distribution.

Consider another example, taken from [BCHK91], describing the solution to a simple mutual exclusion problem. In this solution, two processes compete for a device, and a semaphore is used to serialize their accesses to this device:

$$\begin{aligned} Proc &\Leftarrow \bar{p}.enter.exit.v.Proc \\ Sem &\Leftarrow p.\bar{v}.Sem \\ Sys &\Leftarrow (Proc \mid Sem \mid Proc)\backslash\{p,v\} \end{aligned}$$

Take now a variant of the system *Sys*, where one of the processes is faulty and may deadlock after exiting the critical region (the deadlocked behaviour is modelled here as *nil*). This system, *FSys*, may be defined by:

$$FProc \Leftarrow \overline{p}.\,enter.\,exit.(v.\,FProc + v.\,nil)$$
$$FSys \Leftarrow (Proc \mid Sem \mid FProc)\backslash\{p, v\}$$

In the standard theory of weak bisimulation the two systems *Sys* and *FSys* are equivalent. In fact they are both equivalent to the sequential specification:

$$Spec \Leftarrow enter.\,exit.\,Spec$$

that is $Sys \approx Spec \approx FSys$. Note that both *Sys* and *FSys* are globally deadlock-free. On the other hand *FSys* has the possibility of entering a *local deadlock* in its faulty component, which has no counterpart in *Sys*. More precisely, consider the following distributed representation of *Sys*:

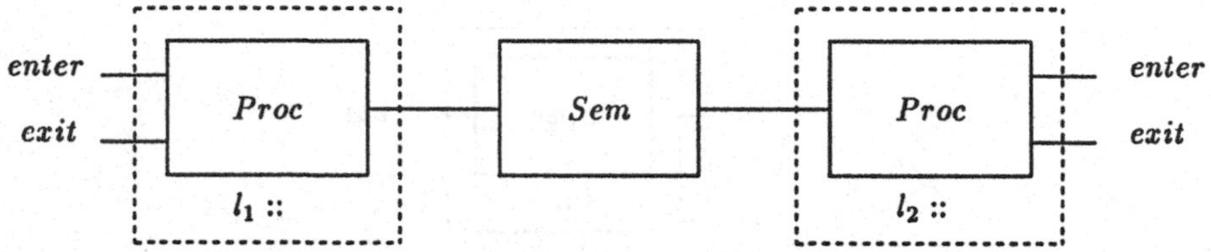

The faulty system *FSys* has a similar representation, with *FProc* in place of the second occurrence of *Proc*. In this distributed view *Sys* and *FSys* have different behaviours, because *FSys* may reach a state in which no more actions can occur at location l_2, while this is not possible for *Sys*. Note that again the causal approach would make no difference between *Sys* and *FSys*: it may be easily checked that $Sys \approx_c Spec \approx_c FSys$.

In the rest of this introduction, we present our formalisation of distributed systems as processes with explicit *locations*. In this setting, a process is described operationally as performing *location transitions* of the form:

$$p \xrightarrow[u]{a} p'$$

which differ from the standard transitions of CCS in that any (observable) action a has associated with it a particular location u. To deal with the dynamic evolution of the distributed structure, we will take locations to be words $u, v, \ldots$ over atomic locations $l, l', \ldots$. The essence of our semantics is expressed by the transition rules for the constructs of action prefixing $a.\,p$ and location prefixing $u :: p$. The rules for the remaining operators of CCS are formally identical to the standard ones (with $\xrightarrow[u]{a}$ replacing $\xrightarrow{a}$). Locations are introduced into processes by means of the rule for action prefixing:

$$a.\,p \xrightarrow[l]{a} l :: p \qquad \text{for any atomic location } l$$

This says that the process $a.\,p$ may be observed to perform an action a at any atomic location l. All subsequent actions of the process will be observed within this location: this is expressed by the fact that the residual of $a.p$ is $l::p$, the process p residing at location l. The rule for the location prefixing operator $u::p$ is now:

$$p \xrightarrow[v]{a} p' \quad \Rightarrow \quad u::p \xrightarrow[uv]{a} u::p'$$

Thus any action of $u::p$ is observed at a sublocation of u. Note that the process $u::p$ retains the location u throughout its execution, in other words location prefixing is a static construct.

We will mainly be interested in the *weak location transition system* associated with the transitions $\xrightarrow[u]{a}$. We assume that the unobservable τ-actions have also unobservable locations: thus the corresponding transitions will have the usual form $\xrightarrow{\tau}$, and simply pass over existing locations without introducing any new ones.

Then we define a general notion of bisimulation on location transition systems, what we call *parameterized location bisimulation*. A parameterized location bisimulation *(plb)* is a relation $\mathcal{B}(R)$ on processes with locations, parameterized on a relation R on locations. Roughly speaking, two processes are related by $\mathcal{B}(R)$ if they can perform the same actions at locations u, v related by R. Our main result is a complete *axiomatisation*, over the set of finite CCS processes, of parameterized location bisimulations satisfying some general conditions, which we call *sensible*. This is achieved by introducing an auxiliary prefixing construct $<a \text{ at } ux>.t$. Intuitively, the construct $<a \text{ at } ux>.p$ prefixes the term t by an "action with locality". Here u represents the access path to the component performing the action a, while x is a location variable that is instantiated to some actual location l when the action is performed. The operational behaviour of such a process is given by the rule:

$$<a \text{ at } ux>.p \xrightarrow[ul]{a} p[l/x]$$

This prefixing construct is used to define *normal forms*, that are terms of the form $\sum_{i \in I} <a_i \text{ at } u_i x_i>.p_i + \sum_{j \in J} \tau.q_j$, and an essential part of our proof system for sensible *plb*'s consists of laws for converting terms into normal forms.

We shall study in some detail two instances of sensible parameterized location bisimulation, the *location equivalence* $\approx_\ell$ and the *location preorder* $\sqsubseteq_\ell$. Location equivalence is obtained by taking the relation R to be the identity relation on locations: two processes are equivalent if and only if they can perform the same actions at the same locations. The equivalence $\approx_\ell$ formalises the idea that two processes are bisimilar, in the classical sense, and moreover have the same parallel structure. The other example of sensible *plb* we shall consider, the location preorder $\sqsubseteq_\ell$, relates two processes when they are bisimilar but one is less distributed than the other. Thus $\sqsubseteq_\ell$ is weaker than $\approx_\ell$, in the sense that $\approx_\ell \subseteq \sqsubseteq_\ell$. For instance, looking back at the protocol example of p. 2, we will have the following relations between the specification *Spec* and the protocol *Sys*:

$$Spec \not\approx_\ell Sys \qquad \text{but} \qquad Spec \sqsubseteq_\ell Sys$$

whereas the two systems Sys and $FSys$ in the mutual exclusion example of p.3 are distinguished by both $\approx_\ell$ and $\sqsubseteq_\ell$.

In the full version of the paper [BCHK92], we give a detailed comparison with noninterleaving theories based on causality, and with other approaches based on "distribution in space". We also give the proofs of the results stated in this extended abstract.

2 Parameterized Location Bisimulations

We introduce a new kind of transition system, called the *location transition system*, to specify processes whose actions may occur at different locations. We assume an infinite set of atomic locations *Loc*, ranged over by $k, l, m \ldots$; we then take general locations, ranged over by $u, v, w \ldots$, to be sequences of Loc^*. As usual we denote concatenation by uv, and the empty word by ε. The set of non-empty locations is Loc^+. Processes will have transitions $p \xrightarrow[u]{a} p'$, where a is an action and u is the location where it occurs, as well as unobservable τ-transitions; the locations of τ-transitions are themselves considered to be unobservable, so these transitions will have the usual form $p \xrightarrow{\tau} p'$.

Definition 2.1 A *Location Transition System* is of the form

$$(S, A, Loc, \{ \xrightarrow[u]{a} \mid a \in A,\ u \in Loc^*\}, \xrightarrow{\tau})$$

where S is a set of *processes with locations*, A is a set of *actions*, *Loc* is the set of atomic locations and each $\xrightarrow[u]{a}$, $\xrightarrow{\tau}$ is a subset of $(S \times S)$, called an action relation. The union of action relations forms the transition relation over S. □

Based on the transitions $p \xrightarrow[u]{a} p'$ and $p \xrightarrow{\tau} p'$, we define the weak transitions $p \overset{\varepsilon}{\Rightarrow} p'$ and $p \overset{a}{\underset{u}{\Longrightarrow}} p'$ in the standard way. Let $\overset{\varepsilon}{\Rightarrow} = (\xrightarrow{\tau})^n$, $n \geq 0$. We will also use $\overset{\tau}{\Rightarrow}$ to denote $(\xrightarrow{\tau})^n$, $n \geq 1$. Then the $\overset{a}{\underset{u}{\Longrightarrow}}$ are given by:

$$p \overset{a}{\underset{u}{\Longrightarrow}} p' \Leftrightarrow_{\text{def}} \exists q, q' .\ p \overset{\varepsilon}{\Rightarrow} q \xrightarrow[u]{a} q' \overset{\varepsilon}{\Rightarrow} p'$$

On the resulting (weak) location transition system we define now the notion of parameterized location bisimulation (plb). A *plb* is a relation on processes with locations, parameterized on a relation R on locations. Informally, two processes are related if they can perform the same actions, at locations u, v related by R.

Definition 2.2 Let $R \subseteq (Loc^* \times Loc^*)$ be a relation on locations. A relation $G \subseteq (S \times S)$ is a *parameterized location bisimulation (plb)* parameterized on R, or R-location bisimulation, iff $G \subseteq C_R(G)$, where $(p, q) \in C_R(G)$ iff

(i)	$p \overset{\tau}{\Rightarrow} p'$	implies	$q \overset{\varepsilon}{\Rightarrow} q'$	for some $q' \in S$ such that $(p', q') \in G$
(ii)	$q \overset{\tau}{\Rightarrow} q'$	implies	$p \overset{\varepsilon}{\Rightarrow} p'$	for some $p' \in S$ such that $(p', q') \in G$
(iii)	$p \overset{a}{\underset{u}{\Longrightarrow}} p'$	implies	$q \overset{a}{\underset{v}{\Longrightarrow}} q'$	for some $q' \in S$ and $v \in Loc^*$ such that $(u, v) \in R$ and $(p', q') \in G$
(iv)	$q \overset{a}{\underset{v}{\Longrightarrow}} q'$	implies	$p \overset{a}{\underset{u}{\Longrightarrow}} p'$	for some $p' \in S$ and $u \in Loc^*$ such that $(u, v) \in R$ and $(p', q') \in G$.

□

The function C_R is monotonic and therefore, from standard principles, it has a maximal fixpoint which we denote by $\mathcal{B}(R)$. As usual $\mathcal{B}(R) = \bigcup\{G \mid G \subseteq C_R(G)\}$ Other properties of $\mathcal{B}(R)$ depend on corresponding properties of the underlying relation R. For instance we have:

Property 2.3 *If R is reflexive (resp. symmetric, transitive) then so is $\mathcal{B}(R)$.*

Also, it should be clear that if $R \subseteq R'$ then any R-location bisimulation is also an R'-location bisimulation, therefore:

Property 2.4 $\quad R \subseteq R' \;\Rightarrow\; \mathcal{B}(R) \subseteq \mathcal{B}(R')$

If for instance we take R to be the universal relation $U = Loc^* \times Loc^*$, we obtain an equivalence relation, $\mathcal{B}(U)$, which is the largest parameterized location bisimulation. Intuitively, letting $R = U$ amounts to ignore the information on locations. In the next section we will see that indeed for the location transition system associated with the language CCS the relation $\mathcal{B}(U)$ coincides with the standard *weak bisimulation equivalence* of [Mil89].

Another important instance of parameterized location bisimulation is $\mathcal{B}(Id)$, where Id is the identity relation on locations. Again this is an equivalence relation, which we shall call *location equivalence* and denote by $\approx_\ell$. This equivalence will be studied in detail in sections 4 and 5. We shall see that in some sense location equivalence is the strongest "reasonable" parameterized location bisimulation. In section 5 we will discuss another example of *plb*, the *location preorder* $\sqsubseteq_\ell$, formalising the idea that a process is less distributed than another.

3 Language and Operational Semantics

We propose now a location transition system semantics for an extension of Milner's language CCS, and discuss the resulting parameterized location bisimulations. We assume the reader to be familiar with CCS, and in particular with its usual notational conventions (see [Mil80]). The language we consider is essentially CCS, with some additional constructs to deal with locations. The first new construct is *location prefixing* $u :: p$ (already introduced in [BCHK91a]) which represents an agent p residing at the location u. Moreover we shall assume, for axiomatization purposes, an infinite set of *location variables* $LVar$, ranged over by $x, y \ldots$, and introduce a new form of prefixing, $< a \text{ at } \sigma x >.p$, where σ is a location word possibly containing variables, that is $\sigma \in (Loc \cup LVar)^*$. Intuitively, the construct $< a \text{ at } \sigma x >.p$ prefixes a term by an "action with locality". Because of location variables, we will need a more general location construct of the form $\sigma :: p$, where $\sigma \in (Loc \cup LVar)^*$; thus $u :: p$ will be a particular case of $\sigma :: p$. Our language $\mathbb{L}$ is given by:

$$\begin{array}{rcl} p & ::= & nil \;\mid\; \mu.p \;\mid\; p + p \;\mid\; p \mid p \;\mid\; p[f] \;\mid\; p\backslash\alpha \\ & \mid & P \;\mid\; rec\, P.\, p \\ & \mid & \sigma :: p \;\mid\; < a \text{ at } \sigma x >.p \end{array}$$

We will use $p, q, \ldots$ to denote terms of our language. As usual $rec\, P.\, p$ is a binding operator for process variables (ranged over by $P, Q \ldots$). Similarly, $< a \text{ at } \sigma x >.p$ is a *binding operator* for location variables, which binds all free occurrences of the variable x in p. However, x may still occur free in σ. We will use the notation $p[\rho]$ to denote an instantiation of both process and location variables in p, and $\sigma[\rho]$ to represent an instantiation of an "open" location word σ. In general we will be only interested in *closed terms*, where all occurrences of both kinds of variables are bound. We take $\mathbb{P}$ to denote the set of such closed terms, also called *processes* in the following. For any process p, we shall denote by $loc(p)$ the set of location names $l \in Loc$ occurring in p. The set of *finite* processes, that is those not involving the recursion construct, will be denoted $\mathbb{P}_f$.

For each $a \in Act$ let $\xrightarrow[u]{a} \subseteq (\mathbb{P} \times \mathbb{P})$ be the least binary relation satisfying the following axioms and rules.

(LT1)	$a.p \xrightarrow[l]{a} l :: p \quad l \in Loc$		
(LT2)	$< a \text{ at } ux >.p \xrightarrow[ul]{a} p[l/x] \quad l \in Loc$		
(LT3)	$p \xrightarrow[u]{a} p'$	implies	$v :: p \xrightarrow[vu]{a} v :: p'$
(LT4)	$p \xrightarrow[u]{a} p'$	implies	$p + q \xrightarrow[u]{a} p'$ $q + p \xrightarrow[u]{a} p'$
(LT5)	$p \xrightarrow[u]{a} p'$	implies	$p \mid q \xrightarrow[u]{a} p' \mid q$ $q \mid p \xrightarrow[u]{a} q \mid p'$
(LT6)	$p \xrightarrow[u]{a} p'$	implies	$p[f] \xrightarrow[u]{f(a)} p'[f]$
(LT7)	$p \xrightarrow[u]{a} p'$	implies	$p\backslash\alpha \xrightarrow[u]{a} p'\backslash\alpha,\ a \notin \{\alpha, \bar{\alpha}\}$
(LT8)	$p[rec\,P.\,p/P] \xrightarrow[u]{a} p'$	implies	$rec\,P.\,p \xrightarrow[u]{a} p'$

Figure 1: Location transitions for $\mathbb{P}$

We define now the location transition system for $\mathbb{P}$, specifying its operational semantics. The transition rules are given in Figure 1. As we said in the previous section, the idea is that actions are observed at particular locations. Initially, some locations may be present in processes because of the location construct $u :: p$. Subsequently, when an action is performed by a component at some location u, an atomic location l is created, which is appended to u to form the new location ul. The word u may then be understood as the *access path* to the component performing the action. For the prefixing operator $a.p$ of CCS the "access path" is empty, and we have the following transition rule:

$$a.p \xrightarrow[l]{a} l :: p \quad \text{for any atomic location } l \in Loc$$

Here the action a may be observed at an arbitrary location $l \in Loc$. For the new prefixing construct $< a \text{ at } ux >.p$ the access path is given by u, while x is a variable which is replaced by an arbitrary location l when a is executed. The rules of Figure 1 are modelled on the standard ones for CCS. Apart from the rules for the prefixing constructs, they are exactly the same as those in [BCHK91a]. Note for instance that $u :: p$ has all the moves of p with locations prefixed by u. By inspecting the rules one can easily check the following property:

$$p \xrightarrow[v]{a} p' \;\Rightarrow\; \exists u \in loc(p)^* \; \exists l \in Loc. \quad v = ul$$

In what follows we will often write transitions explicitly in the form $p \xrightarrow[ul]{a} p'$, and refer to u as the "access path", and to l as the "actual location" of the action a.

The transitions $p \xrightarrow{\tau} p'$, whose location is not observable, are defined through a simple adaptation of the standard transition system for CCS to our extended language. The definition of the τ-moves also requires to adapt the transitions $p \xrightarrow{\mu} p'$. In fact the only new rules are the ones for the constructs $u :: p$ and $< a \text{ at } ux > .t$; in these rules the locations are in fact ignored. In particular we have for $u :: p$

$$p \xrightarrow{\mu} p' \quad \text{implies} \quad u :: p \xrightarrow{\mu} u :: p'$$

and for $< a \text{ at } ux > .t$

$$< a \text{ at } ux > .t \xrightarrow{a} t[\varepsilon/x]$$

For lack of space, we do not enter into the details here, and refer to the full version of this paper [BCHK92]. The weak transitions $p \overset{a}{\underset{u}{\Longrightarrow}} p'$ (or in fact $p \overset{a}{\underset{ul}{\Longrightarrow}} p'$) are then derived as explained in the previous section.

We may now instantiate the definition of parameterized location bisimulation to obtain a family of relations $\mathcal{B}(R)$ over $\mathbb{P}$. These relations are extended to open terms in the standard way: for terms p, q involving process and location variables we set $p\,\mathcal{B}(R)\,q$ if $p[\rho]\,\mathcal{B}(R)\,q\,[\rho]$ for every closed instantiation ρ of both process and location variables.

We already mentioned in the previous section the case where R is U, the universal relation on locations. In $\mathcal{B}(U)$ the locations are completely ignored and therefore one expects it to coincide with the usual (*weak*) *bisimulation equivalence* $\approx$. The bisimulation equivalence $\approx$ may be defined on our extended language using the weak transitions $\overset{\mu}{\Longrightarrow}$ associated with the standard transitions. We may then show the following:

Proposition 3.1 *For all processes p, q:* $\quad (p, q) \in \mathcal{B}(U)$ *if and only if* $p \approx q$.

We have seen in the previous section that for any R, the relation $\mathcal{B}(R)$ is included into $\mathcal{B}(U)$. Therefore:

Corollary 3.2 *For any relation R and processes p, q:* $\quad (p, q) \in \mathcal{B}(R)$ *implies* $p \approx q$.

We also mentioned the *plb* obtained by taking $R = Id$, the identity relation on locations. This relation, the *location equivalence* $\approx_\ell$, will be studied in detail in Section 5. Let us see an example showing that $\approx_\ell$ is strictly finer that $\approx$. Let p and q denote respectively the CCS processes $(a.\alpha.c \mid b.\bar{\alpha}.d)\backslash\alpha$ and $(a.\alpha.d \mid b.\bar{\alpha}.c)\backslash\alpha$. Since in p the actions a and c are in the same parallel component we have $p \overset{a}{\underset{l}{\Longrightarrow}} \overset{b}{\underset{k}{\Longrightarrow}} \overset{c}{\underset{u}{\Longrightarrow}} p' \;\Rightarrow\; u = ll'$ for some l', whereas this is not the case for q. Therefore $p \not\approx_\ell q$, while it is easy to check that $p \approx q$.

In the last section, we shall consider another example of *plb*, which is a preorder but not an equivalence. Both this location preorder and the location equivalence will be shown to be completely axiomatizable. In general, if we want $\mathcal{B}(R)$ to have a reasonable algebraic theory then R must enjoy certain properties. For instance we have the following:

Proposition 3.3 *If R is reflexive and compatible with concatenation on the left, that is $u\,R\,v \Rightarrow wu\,R\,wv$, then $\mathcal{B}(R)$ is preserved by all the operators in the language except $+$.*

Proposition 3.4 *If R is compatible with concatenation on the right, that is $u\,R\,v \Rightarrow uw\,R\,vw$, then $\mathcal{B}(R)$ satisfies the property: $uRv \Rightarrow u :: p\ \mathcal{B}(R)\ v :: p$*

In order to develop an equational theory for parameterized location bisimulations, we need to turn them into substitutive relations, that is relations which are preserved by all the operators of the language. This is done in the standard way. For any *plb* $\mathcal{B}(R)$ we define $\mathcal{B}^c(R)$ to be the closure of $\mathcal{B}(R)$ w.r.t. all contexts:

Definition 3.5 $t\ \mathcal{B}^c(R)\ t'$ if and only if for every term context $C[\,]$: $C[t]\ \mathcal{B}(R)\ C[t']$. □

4 Axiomatisation

We have just seen that some interesting features of parameterized location bisimulations $\mathcal{B}(R)$ depend on specific properties of the underlying relation on locations R, as for example reflexivity, transitivity, and compatibility with concatenation. In this section we propose an axiomatisation, over the set $\mathbb{L}_f$ of finite terms of $\mathbb{L}$, for parameterized location bisimulations $\mathcal{B}(R)$, or more accurately for $\mathcal{B}^c(R)$, based on particular relations R that we call *sensible*. The definition of sensible relations involves the operation of *location renaming*. A location renaming is determined by a mapping π from *Loc* to *Loc**, which is extended to words in the obvious way: $\pi(\varepsilon) = \varepsilon$ and $\pi(lu) = \pi(l)\pi(u)$. Further, π is transferred homomorphically to a mapping between processes: for example we have $\pi(u :: p) = \pi(u) :: \pi(p)$ and $\pi(<a \text{ at } ux>.t) = <a \text{ at } \pi(u)x>.\ \pi(t)$.

Definition 4.1 A relation R on locations is called *sensible* if and only if it is of the form $R = \{(ul, vl) \mid u\,\hat{R}\,v,\ l \in Loc\}$ for some relation on locations $\hat{R}$ satisfying:

1. $\hat{R}$ is a preorder
2. $\hat{R}$ is compatible with concatenation on the left and on the right:
$$u\,\hat{R}\,v \Rightarrow wu\,\hat{R}\,wv \text{ and } uw\,\hat{R}\,vw$$
3. $\hat{R}$ is compatible with location renaming:
$$u\,\hat{R}\,v \Rightarrow \pi(u)\,\hat{R}\,\pi(v) \text{ for any } \pi : Loc \to Loc^*$$

□

The prerequisite that R be of the form $\{(ul, vl) \mid u\hat{R}v,\ l \in Loc\}$ essentially translates into a requirement for the resulting *plb* $\mathcal{B}(R)$, namely that R-bisimilar processes should mark corresponding actions with the same location name l. We will see that this requirement is essential for defining the location preorder $\sqsubseteq_\ell$ in section 5.

As an example, the identity relation *Id* on non-empty locations is obviously sensible, and is the strongest sensible relation. Similarly, the relation U_ℓ given by

$$u\,U_\ell\,v \Leftrightarrow \exists l \in Loc\ \exists u', v'.\ u = u'l \ \&\ v = v'l$$

is clearly a sensible relation, and in fact the weakest one. It gives rise to the same equivalence as U, that is bisimulation equivalence $\approx$.

Since a sensible relation R is a preorder, the corresponding parameterized location bisimulation $\mathcal{B}(R)$ is also a preorder, and $\mathcal{B}^c(R)$ is a *precongruence*; they will then be denoted respectively by $\sqsubseteq_R$ and $\sqsubseteq^c_R$. However, we shall maintain the notation $\approx_\ell$ for $\mathcal{B}(Id)$. The proof system we introduce now for $p \sqsubseteq^c_R q$ consists of a set of inequations of the form $p \sqsubseteq q$, together with some inference rules. We shall use equations $p = q$ to stand for the pair of inequations $p \sqsubseteq q$ and $q \sqsubseteq p$. For terms involving location variables, we have an inference rule expressing a generalisation property:

S1. If $Lvar(p) \cup Lvar(q) \subseteq \{x_1, \ldots, x_n\}$ and $k_1, \ldots, k_n$ are distinct location names not occurring in p and q then: $p[k_1/x_1, \ldots, k_n/x_n] \sqsubseteq q[k_1/x_1, \ldots, k_n/x_n] \Rightarrow p \sqsubseteq q$

The first step of the axiomatisation consists as usual in reducing processes to *normal forms*, which are essentially notations for the transition systems used in the operational semantics. Here the normal forms will be terms built with $+$ and the prefixing construct $< a \text{ at } \sigma x >.p$. They are in fact a special kind of *head normal form*. More precisely:

Definition 4.2 A *head normal form* is a term (defined modulo axioms A1, A2, A3, see Fig. 2) of the form:

$$p = \sum_{i \in I} < a_i \text{ at } \sigma_i x_i >. p_i \; + \; \sum_{j \in J} \tau . p'_j$$

By convention this head normal form is *nil* if $I = \emptyset = J$. A *normal form* is a head normal form whose subterms are again normal forms. □

We introduce now the axioms that will allow us to transform terms of $\mathbb{L}_f$ into normal forms. From now on, the laws will be given for closed terms; by virtue of S1, these laws can then be turned into similar statements on open terms. The basic transformation, replacing ordinary prefixing by the new prefixing construct, is the following:

L1. $a.p = < a \text{ at } x >. x :: p$

The law L1 introduces a new location variable in front of the subterm p. We give now a set of laws to push locations through subterms.

L2. $u :: < a \text{ at } vx >. p = < a \text{ at } uvx >. u :: p$
L3. $u :: \tau . p = \tau . u :: p$
L4. $u :: nil = nil$
L5. $u :: (p + q) = u :: p + u :: q$

Using laws L2, S1, we may infer for instance $y :: < a \text{ at } \sigma x >. p = < a \text{ at } y \sigma x >. y :: p$. Note however that this only holds for $y \neq x$, since the variable x is bound in $< a \text{ at } \sigma x >. p$. Indeed we need a kind of α-conversion rule:

S2. $< a \text{ at } ux >. p = < a \text{ at } uy >. p[y/x]$, y not free in p

$$
\begin{array}{lrcl}
\text{(A1)} & p+(q+r) & = & (p+q)+r \\
\text{(A2)} & p+q & = & q+p \\
\text{(A3)} & p+nil & = & p \\
\text{(A4)} & p+p & = & p \\
\text{(R1)} & nil\backslash\alpha & = & nil \\
\text{(R2)} & (<a \text{ at } ux>.p)\backslash\alpha & = & \begin{cases} <a \text{ at } ux>.(p\backslash\alpha) & \text{if } a\neq\alpha,\bar{\alpha} \\ nil & \text{otherwise} \end{cases} \\
\text{(R3)} & (\tau.p)\backslash\alpha & = & \tau.(p\backslash\alpha) \\
\text{(R4)} & (p+q)\backslash\alpha & = & p\backslash\alpha + q\backslash\alpha \\
\text{(U1)} & nil[f] & = & nil \\
\text{(U2)} & (<a \text{ at } ux>.p)[f] & = & <f(a) \text{ at } ux>.p[f] \\
\text{(U3)} & (\tau.p)[f] & = & \tau.(p[f]) \\
\text{(U4)} & (p+q)[f] & = & p[f]+q[f]
\end{array}
$$

(EXP) Let $p = \sum_{i\in I} <a_i \text{ at } u_i x_i>.p_i + \sum_{j\in J} \tau.p'_j$ and $q = \sum_{k\in K} <b_k \text{ at } v_k y_k>.q_k + \sum_{l\in L} \tau.q'_l$

Then:

$$
\begin{aligned}
p \mid q = \; & \sum_{i\in I} <a_i \text{ at } u_i x_i>.(p_i \mid q) + \sum_{k\in K} <b_k \text{ at } v_k y_k>.(p \mid q_k) + \\
& \sum_{a_i=\bar{b}_k} \tau.(p_i[\varepsilon/x_i] \mid q_k[\varepsilon/y_k]) + \sum_{j\in J} \tau.(p'_j \mid q) + \sum_{l\in L} \tau.(p \mid q'_l)
\end{aligned}
$$

Figure 2: Equations $\mathcal{E}$, standard expansion laws.

$$
\begin{array}{lrcl}
\text{(L1)} & a.p & = & <a \text{ at } x>.\, x :: p \\
\text{(L2)} & u :: <a \text{ at } vx>.p & = & <a \text{ at } uvx>.\, u :: p \\
\text{(L3)} & u :: \tau.p & = & \tau.\, u :: p \\
\text{(L4)} & u :: nil & = & nil \\
\text{(L5)} & u :: (p+q) & = & u :: p \, + \, u :: q
\end{array}
$$

Figure 3: Equations $\mathcal{L}$, location laws.

$$
\begin{array}{lrcl}
\text{(T1)} & p+\tau.p & = & \tau.p \\
\text{(T2)} & <a \text{ at } ux>.\, p & = & <a \text{ at } ux>.\, \tau.p \\
\text{(T2)}' & \tau.p & = & \tau.\tau.p \\
\text{(T3)} & <a \text{ at } ux>.\, (p+\tau.q) & = & <a \text{ at } ux>.\, (p+\tau.q) + <a \text{ at } ux>.\, q
\end{array}
$$

Figure 4: Equations $\mathcal{T}$, the τ-laws.

(GEN_R) If $(u,v) \in \hat{R}$ then: $<a \text{ at } ux>.\, p \sqsubseteq \; <a \text{ at } vx>.\, p$

Figure 5: GEN_R, the parametric law.

So far we have seen how prefixing $< a \text{ at } \sigma x >.p$ and location variables are introduced. In order to obtain normal forms, we also need to get rid of the static operators occurring in terms. The idea is as usual to eliminate the parallel operator by means of an *expansion theorem* which is given in Fig. 2. The other static operators will be taken care of by standard laws, listed as R1–R4, U1–U4 in Fig. 2.

The equations $\mathcal{E}$ are more or less the standard expansion laws, adapted to account for the new prefixing construct. Similarly, the laws $\mathcal{T}$ in figure 4 are an adaptation of Milner's τ-laws to our language. To deal with the particular sensible relation $R = \{(ul, vl) \mid u\hat{R}v, l \in Loc\}$ on which the parameterized location precongruence $\sqsubseteq^c_R$ is based, we have in addition a parametric inequation GEN$_R$; note that this is the only place where R intervenes in the axiomatisation.

Let now $\mathcal{I}_R$ be the set of all the laws and rules considered so far, including S1, S2. We write $p \sqsubseteq_R q$ if $p \sqsubseteq q$ is provable in this proof system, and similarly for $p =_R q$. Our aim is to establish that on terms of $\mathbb{L}_f$ the parameterized location precongruence $\sqsubseteq^c_R$ coincides with $\sqsubseteq_R$. A first half of this result expresses the soundness of the laws $\mathcal{I}_R$ with respect to $\sqsubseteq^c_R$.

Proposition 4.3 (*Soundness of the laws*) *The laws $\mathcal{I}_R$ are sound for the parameterized location precongruence $\sqsubseteq^c_R$, that is $p \sqsubseteq_R q \Rightarrow p \sqsubseteq^c_R q$.*

Let us sketch now the main points in the proof of completeness, that is $p \sqsubseteq^c_R q \Rightarrow p \sqsubseteq_R q$. We establish first that any term of $\mathbb{L}_f$ may be transformed into a head normal form, and then into a normal form, using the laws $\mathcal{E}$ and $\mathcal{L}$ (indeed the reduction to normal forms is independent of the choice of the relation R).

Proposition 4.4 (*Normalisation*) *For each term p of $\mathbb{L}_f$, there exists a normal form $\mathrm{nf}(p)$ such that $p =_R \mathrm{nf}(p)$.*

The proof of completeness requires in addition two *absorption lemmas*, similar to those used for weak bisimulation in [HM85].

Lemma 4.5 (*τ-absorption lemma*) *If p is a closed normal form then:*

$$p \stackrel{\tau}{\Rightarrow} p' \quad \textit{implies} \quad p + \tau . p' =_R p$$

Lemma 4.6 (*General absorption lemma*) *If p is a closed normal form then:*

$$p \underset{ul}{\stackrel{a}{\Rightarrow}} p' \quad \textit{implies} \quad \exists p''.\ p =_R p \ + < a \text{ at } ux >.\ p'' \textit{ and } p''[l/x] = p'$$

Using these intermediary properties, one may establish the announced completeness theorem, which is our main result:

Theorem 4.7 (*Completeness*) *For any terms $p, q \in \mathbb{L}_f$: $p \sqsubseteq^c_R q \Rightarrow p \sqsubseteq_R q$.*

This concludes our axiomatization for parameterized location bisimulations $\mathcal{B}(R)$ based on a sensible relation R. In the next section we will examine two particular instances of *plb's* axiomatizable in this way, namely the location equivalence $\approx_\ell$ and the location preorder $\sqsubseteq_\ell$.

5 The Location Equivalence and Preorder

In this section we first discuss the generalized location bisimulation $\mathcal{B}(R)$ obtained by instantiating R as the identity relation Id. We recall that this is an equivalence relation, called *location equivalence* and denoted $\approx_\ell$. Clearly the identity relation on locations is a sensible relation, therefore our axiomatization result of the previous section holds for $\approx_\ell$, or more accurately for the associated congruence $\approx_\ell^c$. Note that the parametric absorption law GEN_R is trivial in this case.

We already saw that location equivalence is strictly finer than bisimulation equivalence $\approx$. The example we gave, namely $p = (a.\alpha.c \mid b.\bar{\alpha}.d)\backslash\alpha$ and $q = (a.\alpha.d \mid b.\bar{\alpha}.c)\backslash\alpha$, also shows that location equivalence is different from Darondeau and Degano's *weak causal bisimulation* [DD89]: denoting by $\approx_c$ the causal weak bisimulation, we have $p \approx_c q$ since, roughly speaking, both c and d causally depend on a and b in p and q. However $p \not\approx_\ell q$ since in p the d action is not spatially dependent upon the a action. We can also give examples not involving the restriction operator to show that the two equivalences $\approx_\ell$ and $\approx_c$ are incomparable: let $r = (a.\alpha + b.\beta \mid \bar{\alpha}.b + \bar{\beta}.a)$. Then

$$\begin{array}{ccccc} r + (a \mid b) & \approx_\ell & r & \not\approx_\ell & r + a.b \\ & \not\approx_c & & \approx_c & \end{array}$$

Note also that

$$(a \mid b) \quad \approx_\ell \quad (r\backslash\alpha,\beta) \quad \approx_c \quad a.b + b.a$$

These absorption phenomena, resp. of $(a \mid b)$ in r w.r.t. $\approx_\ell$, and of $a.b$ in r w.r.t. $\approx_c$, clearly show the difference between the two equivalences: the former equates processes with the same parallel structure, while the latter equates processes with the same causal structure. We refer to [Kie91] for a precise study of the relation between causal and location equivalences.
Let us see another example; one can see that the two terms

$$p = (l :: \alpha \mid \bar{\alpha}.b)\backslash\alpha \quad \text{and} \quad q = (l :: (\alpha + b) \mid \bar{\alpha}.b)\backslash\alpha$$

have different normal forms, namely $\tau.<b \text{ at } x>.nil$ and $<b \text{ at } ly>.nil + \tau.<b \text{ at } x>.nil$. Indeed we have $p \not\approx_\ell q$, and this also implies that if we let r and s be the CCS terms $(a.\alpha \mid \bar{\alpha}.b)\backslash\alpha$ and $(a.(\alpha + b) \mid \bar{\alpha}.b)\backslash\alpha$, then $r \not\approx_\ell s$. Intuitively we want to distinguish these two processes since in the first the b action is not spatially dependent upon the a action. This shows that our location equivalence is different from what is also called "location equivalence" in [BCHK91]. In fact the definition of the two equivalences is formally the same, but the underlying location transition systems are slightly different. In the full version of the paper [BCHK92] it is shown that $\approx_\ell$ is (strictly) finer than the equivalence of [BCHK91], although they coincide on the set of finite restriction-free CCS processes. As a consequence of this and a result of [BCHK91], our equivalence coincides on this set of processes with *distributed bisimulation equivalence*, introduced in [CH89] and [Cas88], and further studied in [Kie89].

In [BCHK91] we said that "introducing locations adds discriminations between processes only as far as their distributed aspect is concerned". This is still true here: let $\mathrm{CCS}_{\mathrm{seq}}$ be the set of sequential processes of CCS, that is processes built without the parallel operator. Then one can show that on sequential processes, all location bisimulations induced by a reflexive relation R collapse to weak bisimulation:

Lemma 5.1 *For any reflexive relation R on locations, and any processes $p, q \in \mathrm{CCS_{seq}}$*

$$p\mathcal{B}(R)q \Leftrightarrow p \approx q$$

In the rest of this section we discuss another instance of $\mathcal{B}(R)$, giving rise to a preorder on processes which takes into account their degree of parallelism. Roughly speaking, we seek a relation R such that if $p\ \mathcal{B}(R)\ q$ then p and q have similar behaviour but p is possibly more sequential or less distributed than q. For example, let p, q be the processes:

$$p = a.a.a.nil \qquad\qquad q = a.a.nil \mid a.nil$$

Intuitively, p is a sequential shuffle of q. If we try to relate the behaviours of these two processes using our location transition semantics, we can see that the *superword* relation on locations could be an appropriate relation to choose as $\widehat{R}$. Intuitively, if s is a sequentialized version of r, this means that some component of s corresponds to a group of components of r; then, provided at each step the same locations are introduced for corresponding actions, the location u of a component of s will always be a *shuffle* of the locations of the corresponding components of r (and thus a superword of each individual location v).

Let $\gg$ denote the *superword* relation on Loc^*. This is the inverse of the *subword* relation, which we note $\ll$. Recall that v is a subword of u if $v = v_1 \dots v_k$ and $u = w_1 v_1 \dots w_k v_k w_{k+1}$, for some collection of words v_i, w_j. Now it is easy to check that the relation R generated by $\widehat{R} = \gg$, which we denote $\gg_\ell$, is a sensible relation on locations, and therefore is a suitable candidate for our theory. Since $\mathcal{B}(\gg_\ell)$ is a preorder, we will call it the *location preorder* and denote it by $\sqsubseteq_\ell$. By virtue of the results of the previous section we have a complete axiomatisation of the location precongruence $\sqsubseteq_\ell^c$ over finite terms. Let us examine some examples.

Example 5.2 For any processes p, q

$$a.(p \mid b.q) + b.(a.p \mid q) \sqsubseteq_\ell a.p \mid b.q$$

Example 5.3 If α is different from a and does not appear in p, q then

$$a.(p \mid q) \sqsubseteq_\ell (a.\alpha.p \mid \overline{\alpha}.q)\backslash\alpha$$

Example 5.4 Let us see why using $\gg_\ell$ instead of $\gg$ is important. This will explain why in the definition of sensible relations R we require that if $u\,R\,v$ then u and v end on the same location name (this was also used in the Completeness Theorem of Section 4). Let $p = a.((b.\alpha \mid \bar{\alpha}.c)\backslash\alpha + b.c)$ and $q = (a.\alpha \mid \bar{\alpha}.b.c)\backslash\alpha$. Then $p \not\sqsubseteq_\ell q$ since the only possibility for q to match, up to $\gg_\ell$, the sequence of moves $p \overset{a}{\underset{l}{\Longrightarrow}} \overset{b}{\underset{lk}{\Longrightarrow}} l :: (k :: nil \mid c)\backslash\alpha \approx_\ell l :: c$ is $q \overset{a}{\underset{l}{\Longrightarrow}} \overset{b}{\underset{k}{\Longrightarrow}} (l :: nil \mid k :: c)\backslash\alpha \approx_\ell k :: c$, and clearly $l :: c \not\approx_\ell k :: c$ if $l \neq k$. On the other hand, q can match the moves of p up to $\gg$, since $q \overset{a}{\underset{l}{\Longrightarrow}} \overset{b}{\underset{l}{\Longrightarrow}} (l :: nil \mid l :: c)\backslash\alpha$. Indeed we have $p\mathcal{B}(\gg)q$, but intuitively we do not want to regard p as less parallel than q since in p the action c is not necessarily spatially dependent on b.

Having introduced the preorder $\sqsubseteq_\ell$, it is natural to consider the associated equivalence, i.e. the kernel $\simeq_\ell =_{\mathrm{def}} \sqsubseteq_\ell \cap \sqsupseteq_\ell$. Clearly $\approx_\ell \subseteq \simeq_\ell$ by Property 2.4. On the other hand, as could be expected, the kernel of the preorder is weaker than location equivalence, that is $\simeq_\ell \not\subseteq \approx_\ell$. An example is:

$$a.a.a + (a \mid a \mid a) \quad \text{versus} \quad a.a.a + a.a \mid a + (a \mid a \mid a)$$

These two processes are obviously not equivalent w.r.t. $\approx_\ell$, but they are equivalent w.r.t. $\simeq_\ell$ because $a.a.a \sqsubseteq_\ell a.a \mid a \sqsubseteq_\ell a \mid a \mid a$.

We have seen in the previous section that on finite *sequential* CCS processes all the *plb*'s based on sensible relations collapse to weak bisimulation. For the preorder $\sqsubseteq_\ell$ we may prove a stronger result, similar to that given in [Ace89], namely that $p \sqsubseteq_\ell q$ implies $p \approx q$ already when p is sequential:

Proposition 5.5 *If* $p \in \mathrm{CCS_{seq}}$ *and* $q \in \mathrm{CCS}$, *then* $p \sqsubseteq_\ell q \;\Leftrightarrow\; p \approx q$.

We refer to the full version of the paper for more details and examples.

References

[Ace89] L. Aceto. On relating concurrency and nondeterminism. Report 6/89, Computer Science, University of Sussex, Brighton, 1989.

[BCHK91] G. Boudol, I. Castellani, M. Hennessy, and A. Kiehn. Observing localities. Report 4/91, Sussex University, and INRIA Res. Rep. 1485, 1991. To appear in TCS.

[BCHK92] G. Boudol, I. Castellani, M. Hennessy, and A. Kiehn. A theory of processes with localities. Report 13/91, Sussex University, and INRIA Res. Rep. 1632, 1992.

[Cas88] I. Castellani. *Bisimulations for Concurrency.* Ph.d. thesis, University of Edinburgh, 1988.

[CH89] I. Castellani and M. Hennessy. Distributed bisimulations. *JACM*, 10(4):887–911, 1989.

[DD89] Ph. Darondeau and P. Degano. Causal trees. In *Proc. ICALP 88*, volume 372 of *LNCS*, pages 234–248, 1989.

[HM85] M. Hennessy and R. Milner. Algebraic laws for nondeterminism and concurrency. *JACM*, 32(1):137–161, 1985.

[Kie89] A. Kiehn. Distributed bisimulations for finite CCS. Report 7/89, University of Sussex, 1989.

[Kie91] A. Kiehn. Local and global causes. Report 342/23/91, Technische Universität München, 1991.

[Mil80] R. Milner. *A Calculus of Communicating Systems*, volume 92 of *Lecture Notes in Computer Science.* 1980.

[Mil89] R. Milner. *Communication and Concurrency.* Prentice-Hall, 1989.

Model Checking for Context-Free Processes

Olaf Burkart and Bernhard Steffen

Lehrstuhl für Informatik II
RWTH Aachen
Ahornstraße 55
W-5100 Aachen
GERMANY

Abstract. We develop a model-checking algorithm that decides for a given *context-free* process whether it satisfies a property written in the alternation-free modal mu-calculus. The central idea behind this algorithm is to raise the standard iterative model-checking techniques to *higher order*: in contrast to the usual approaches, in which the set of formulas that are satisfied by a certain state are iteratively computed, our algorithm iteratively computes a *property transformer* for each state class of the finite process representation. These property transformers can then simply be applied to solve the model-checking problem. The complexity of our algorithm is linear in the size of the system's representation and exponential in the size of the property being investigated.

1 Introduction

Model-checking provides a powerful tool for the automatic verification of behavioural systems. The corresponding standard algorithms fall into two classes: the iterative algorithms (cf. [EL86,CES86,CS92,CS91]) and the tableaux-based algorithms (cf., e.g. [Bra91,Cle90,Lar88,SW89,Win89]). Whereas the former class usually yields higher efficiency, the latter allows local model checking (cf. [SW89]). This observation has been exploited by Bradfield and Stirling [Bra91,BS91] for the automatic verification of infinite-state systems. However, their method does not guarantee termination. Another recently proposed approach to the verification of infinite state systems is deciding bisimilarity of context-free processes (cf. [BBK87,Cau90,HS91,Gro91,HT92]).

In this paper we develop an iterative algorithm that decides the alternation-free modal mu-calculus for *context-free* processes, i.e. for processes that are given in terms of a context-free grammar, or equivalently, as mutually recursive systems of procedural process graphs.[1]

The point of our algorithm is to consider a *higher order* variant of the standard iterative model-checking techniques, which instead of determining properties, determines property transformers. Altogether the model-checker has the following two step structure:

---

[1] Note that the (alternation-free) modal mu-calculus only allows to express structurally restricted properties. In fact, our algorithm decides essentially "regular" properties of context-free processes.

1. iterative computation of a ***property transformer*** for each state class of the finite process representation. These property transformers give semantics to the procedural components (or the non-terminals of the context-free representation). Intuitively they describe the set of formulas that are valid before the procedure calls (non-terminals) relative to the set of formulas that are valid afterwards. This computation is effective because we can restrict our attention to the subformulas of the property currently being investigated.

2. decision of the model-checking problem: having the predicate transformers at hand, the model-checking problem can be solved, simply by checking whether the formula under consideration is a member of the set of formulas that results from applying the property transformer associated with the initial state class of the "main procedure" to the set of formulas that are valid for a deadlocking or terminating state.

The complexity of the resulting algorithm is linear in the size of the system's representation and exponential in the size of the property.[2]

The remainder of the paper develops along the following lines. Section 2 describes context-free process systems which model context-free processes, Sect. 3 presents our logic, and Sect. 4 develops our model-checking algorithm. The final section contains our conclusions and directions for future research.

2 Modelling Behaviour

In this section we introduce the notion of a *context-free process system*. It will be used to finitely represent infinite state behaviours, which themselves are modelled by *process graphs*:

Definition 1 [HT92]. A *process graph* is a quadruple $G = \langle \mathcal{S}, Act, \rightarrow, s_0 \rangle$, where:

- $\mathcal{S}$ is a set of *states*;
- *Act* is a set of *actions*;
- $\rightarrow \subseteq \mathcal{S} \times Act \times \mathcal{S}$ is the *transition relation*; and
- $s_0 \in \mathcal{S}$ is a distinguished element, the "start state".

Intuitively, a process graph encodes the operational behaviour of a process. The set $\mathcal{S}$ represents the set of states the process may enter, *Act* the set of actions the process may perform and $\rightarrow$ the state transitions that may result upon execution of the actions. Unrooted process graphs, i.e. process graphs without a specified start state, are also called *labelled transition systems*.

In the remainder of the paper we use $s \xrightarrow{a} s'$ in lieu of $\langle s, a, s' \rangle \in \rightarrow$, and we write $s \xrightarrow{a}$ when there is an s' such that $s \xrightarrow{a} s'$. If $s \xrightarrow{a} s'$ then we call s' an *a-derivative* of s. Finally a process graph is said to be *finite-state*, when $\mathcal{S}$ and *Act* are finite.

[2] In fact, it is only exponential in the weight of the property, see Sect. 4.4.

2.1 Context-Free Process Systems

In this paper we represent *context-free processes*, which may have infinitely many states, by means of *context-free process systems*. These systems are essentially sets of named process graphs whose set of actions contain the names of the system's process graphs. Intuitively, a transition labelled with such a name represents a call to the denoted process graph. Thus the names of the process graphs correspond to the non-terminals and the atomic actions to the terminals of a context-free grammar.

Here we prefer the more dynamic procedural point of view, because it better reflects the transformational character of process graphs, which is essential for our algorithm presented in Sect. 4.

Definition 2. A *procedural process graph* (PPG) is defined as a quintuple $P = \langle \Sigma_P, Trans, \rightarrow_P, \sigma_P^s, \sigma_P^e \rangle$, where:

- Σ_P is a set of *state classes*;[3]
- $Trans =_{df} Act \cup \mathcal{N}$ is a set of *transformations*, where Act is a set of *actions* and $\mathcal{N}$ is a set of *names*;
- $\rightarrow_P = \rightarrow_P^{Act} \cup \rightarrow_P^{\mathcal{N}}$ is the *transition relation*, where $\rightarrow_P^{Act} \subseteq \Sigma_P \times Act \times \Sigma_P$ and $\rightarrow_P^{\mathcal{N}} \subseteq \Sigma_P \times \mathcal{N} \times \Sigma_P$; and
- $\sigma_P^s \in \Sigma_P$ is a class of *"start states"* and $\sigma_P^e \in \Sigma_P$ is a class of *"end states"*.

Additionally a procedural process graph P must satisfy the following two constraints:

1. The class of "end states" σ_P^e must be *terminating* in P, i.e. there must not exist any $\alpha \in Trans$ and $\sigma' \in \Sigma_P$ with $\langle \sigma_P^e, \alpha, \sigma' \rangle \in \rightarrow_P$; and
2. P must be *guarded*, i.e. all initial transitions of P must be labelled with atomic actions.

P is called *simple*, if it does not contain any calls, or equivalently, if it is a process graph with a distinguished end state. The set of all simple PPGs is denoted by $\mathcal{G}$. Moreover a PPG P is said to be *finite* if Σ_P and $Trans$ are finite.

The definition of *context-free process systems* is now straightforward.

Definition 3. A *context-free process system* (CFPS) is defined as a quadruple $\mathcal{P} = \langle \mathcal{N}, Act, \Delta, P_0 \rangle$, where

- $\mathcal{N} = \{N_0, \ldots, N_{n-1}\}$ is a set of *names*;
- Act is a set of *actions*;
- $\Delta =_{df} \{ N_i = P_i \mid 0 \leq i < n \}$ is a finite set of *PPG definitions* where the P_i are finite PPGs with names in $\mathcal{N}$ and
- P_0 is the *"main" PPG*.

The encoding machine of Fig. 1 is an example of a CFPS with only one component. We will see that this example is already sufficient to illustrate the essential features of our model-checking algorithm.

In the remainder of the paper the union of all state classes of a CFPS $\mathcal{P}$ is denoted by $\Sigma =_{df} \bigcup_{i=0}^{n-1} \Sigma_{P_i}$ and the union of all transition relations by $\rightarrow =_{df} \bigcup_{i=0}^{n-1} \rightarrow_{P_i}$.

[3] As will be explained, members of Σ_P represent classes or sets of states in the usual sense.

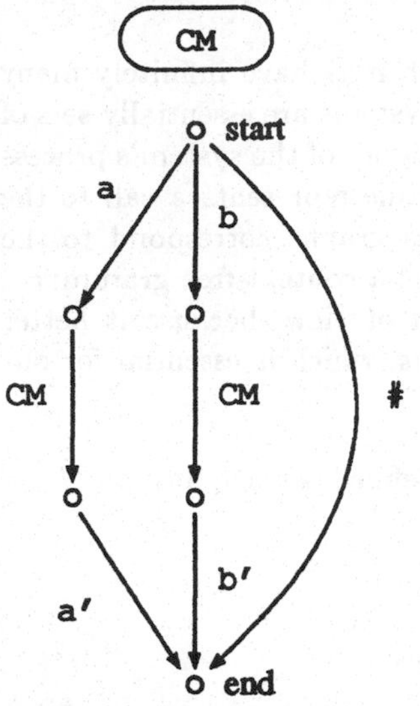

Example: The encoding machine is a directly recursive PPG. It reads a word over the alphabet $\{a, b\}$ until it finds the word delimiter #. Then, in its second phase, it outputs the corresponding sequence of codes in reverse order.

Fig. 1. The Encoding Machine

Moreover, as before for process graphs, we use $\sigma \xrightarrow{\alpha} \sigma'$ instead of $\langle \sigma, \alpha, \sigma' \rangle \in \rightarrow$, and we write $\sigma \xrightarrow{\alpha}$ when there is a σ' such that $\sigma \xrightarrow{\alpha} \sigma'$.

A CFPS $\mathcal{P}$ serves as a finite representation of the *complete expansion* of P_0, which is defined as follows.

Definition 4. Let P be a PPG of a CFPS $\mathcal{P}$. The *complete expansion* of P with respect to $\mathcal{P}$ is the simple PPG which results from successively replacing in P each transition $\sigma \xrightarrow{P_i} \sigma'$ by a copy of the corresponding PPG P_i, while identifying σ with $\sigma^s_{P_i}$ and σ' with $\sigma^e_{P_i}$. We denote the complete expansion of P with respect to $\mathcal{P}$ by $Exp_{\mathcal{P}}(P)$.

Figure 2 illustrates the stepwise expansion for the encoding machine presented in Fig. 1.

Given some state $s \in \mathcal{S}$ of a complete expansion $Exp_{\mathcal{P}}(P)$, we say that s belongs to the state class $\sigma \in \Sigma_{P_i}$ if it arose as a copy of σ during the expansion. Thus a state class stands for a possibly infinite set of states of the corresponding complete expansion.

Finally we introduce some notations which concern PPGs. If $P = \langle \Sigma_P, Trans, \rightarrow_P, \sigma^s_P, \sigma^e_P \rangle$ is a PPG and σ is a state class of P, then we denote with $P^{(\sigma)}$ the PPG $\langle \Sigma_P, Trans, \rightarrow_P, \sigma, \sigma^e_P \rangle$, which is essentially the PPG P where the initial state class changed from σ^s_P to σ. If P_1 and P_2 are two PPGs, then we define their sequential composition $P_1; P_2$ as the PPG

$$\langle \Sigma_{P_1} \cup \Sigma_{P_2} \setminus \{\sigma^s_{P_2}\}, Trans_{P_1} \cup Trans_{P_2}, \rightarrow_{P_1} \cup \rightarrow_{P_2'}, \sigma^s_{P_1}, \sigma^e_{P_2} \rangle$$

where $\rightarrow_{P_2'}$ denotes the transition relation which results from substituting in $\rightarrow_{P_2}$ all occurrences of $\sigma^s_{P_2}$ by $\sigma^e_{P_1}$.

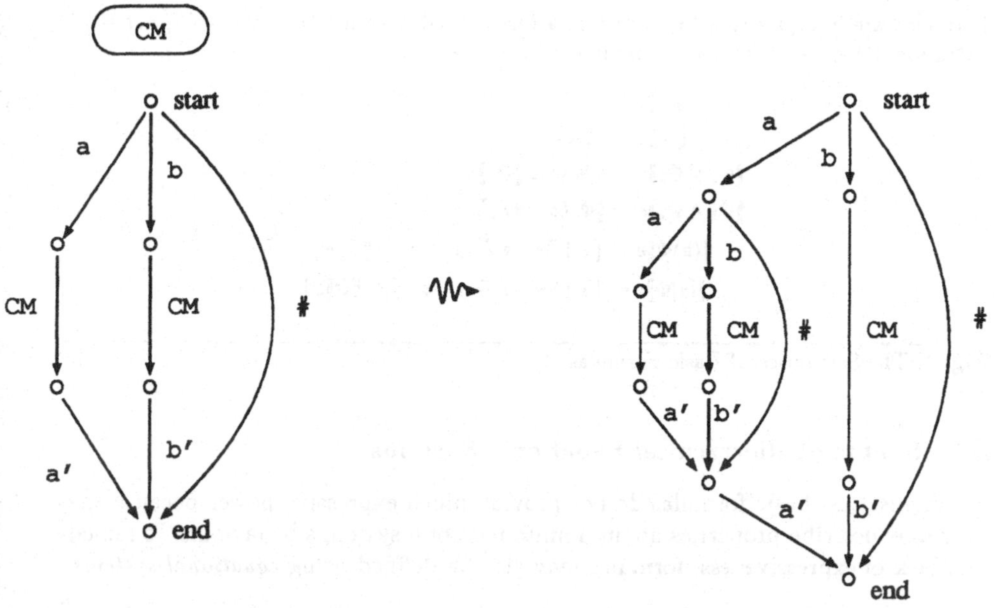

Fig. 2. Expanding a Procedural Process Graph

3 Specifying Behavioural Properties

The logic we consider is essentially the alternation-free fragment of the modal mu-calculus [Koz83]. For technical reasons, however, we represent formulas by means of *hierarchical (mutually recursive) equational systems* (cf. [CS92,CS91]).

3.1 Syntax and Semantics of Basic Formulas

Let *Var* be a (countable) set of variables, $\mathcal{A}$ a set of atomic propositions, and *Act* a set of actions. In what follows, X will range over *Var*, A over $\mathcal{A}$, and a over *Act*. Then the syntax of *basic* formulas is given by the following grammar.

$$\Phi ::= A \mid X \mid \Phi \vee \Phi \mid \Phi \wedge \Phi \mid \langle a \rangle \Phi \mid [a]\Phi$$

The formal semantics appears in Fig. 3. It is given with respect to a labelled transition system $\langle \mathcal{S}, Act, \rightarrow \rangle$, a valuation $\mathcal{V}$ mapping atomic propositions to subsets of $\mathcal{S}$, and an environment e mapping variables to subsets of $\mathcal{S}$.

Intuitively, the semantic function maps a formula to the set of states for which the formula is "true". Accordingly, a state s satisfies $A \in \mathcal{A}$ if s is in the valuation of A, while s satisfies X if s is an element of the set bound to X in e. The propositional constructs are interpreted in the usual fashion: s satisfies $\Phi_1 \vee \Phi_2$ if it satisfies one of the Φ_i and $\Phi_1 \wedge \Phi_2$ if it satisfies both of them. The constructs $\langle a \rangle$ and $[a]$ are *modal operators*; s satisfies $\langle a \rangle \Phi$ if it has an a-derivative satisfying Φ, while s satisfies $[a]\Phi$ if each of its a-derivatives satisfies Φ.

Formulas are interpreted with respect to a fixed labelled transition system $\langle \mathcal{S}, Act, \rightarrow \rangle$, a valuation $\mathcal{V} : \mathcal{A} \rightarrow 2^{\mathcal{S}}$, and an environment $e : Var \rightarrow 2^{\mathcal{S}}$.

$$
\begin{aligned}
[\![A]\!]e &= \mathcal{V}(A) \\
[\![X]\!]e &= e(X) \\
[\![\Phi_1 \vee \Phi_2]\!]e &= [\![\Phi_1]\!]e \cup [\![\Phi_2]\!]e \\
[\![\Phi_1 \wedge \Phi_2]\!]e &= [\![\Phi_1]\!]e \cap [\![\Phi_2]\!]e \\
[\![\langle a \rangle \Phi]\!]e &= \{\, s \mid \exists s'.\, s \xrightarrow{a} s' \wedge s' \in [\![\Phi]\!]e \,\} \\
[\![[a]\Phi]\!]e &= \{\, s \mid \forall s'.\, s \xrightarrow{a} s' \Rightarrow s' \in [\![\Phi]\!]e \,\}
\end{aligned}
$$

Fig. 3. The Semantics of Basic Formulas.

3.2 Syntax of Hierarchical Equational Systems

By themselves, basic formulas do not provide much expressive power because they only may describe properties about a finite part of a system's behaviour. To remedy this lack of expressiveness, formulas may also be defined using *equational systems.*

Definition 5. An *(equational) block* has one of two forms, $min\{E\}$ or $max\{E\}$, where E is a list of (mutually recursive) equations

$$\langle X_1 = \Phi_1, \ldots, X_n = \Phi_n \rangle$$

in which each Φ_i is a basic formula and the X_i are all distinct. An *equational system* $\mathcal{B} = \langle B_1, \ldots, B_m \rangle$ is a list of equational blocks where the variables appearing on the left-hand sides of the blocks are all distinct.

Intuitively, a block defines n mutually recursive propositions, one per variable. Several blocks may be used to define formulas, and the right-hand sides of an equation in one block may refer to variables serving as the left-hand sides of equations in other blocks. In this paper, however, we restrict our attention to *hierarchical equational systems.* This ensures that there are no *alternating fixed points* [EL86].

Definition 6. An equational system $\mathcal{B} = \langle B_1, \ldots, B_m \rangle$ is *hierarchical* if the existence of a left-hand-side variable of a block B_j appearing in a right-hand-side formula of a block B_i implies $i \leq j$.

Consider the equational systems of Fig. 4 for illustration. Equational system 1 is hierarchical; X_2 appears in B_1 which is listed *before* B_2 and X_1 does not appear in B_2. System 2, however, fails our syntactic restriction, since X_1 appears in B_2.

3.3 Semantics of Hierarchical Equational Systems

To define the semantics of a hierarchical equational system $\mathcal{B}$, we first define the semantics of an individual block. Let E be the list of equations

$$\langle X_1 = \Phi_1, \ldots, X_n = \Phi_n \rangle.$$

$$
\begin{array}{llcll}
1. & \{ B_1 & \equiv & min\ \{ X_1 = X_2 \wedge [a]X_1 & \} \\
 & B_2 & \equiv & max\ \{ X_2 = A \wedge [a]X_2 \wedge \langle a\rangle tt & \}\} \\
2. & \{ B_1 & \equiv & max\ \{ X_1 = X_2 & \} \\
 & B_2 & \equiv & min\ \{ X_2 = [a](A \wedge X_1) \vee X_2 & \}\}
\end{array}
$$

Fig. 4. Examples of Hierarchical Equational Systems.

Then, given a fixed environment ρ, we may build a function $f_E^\rho : (2^S)^n \to (2^S)^n$ as follows. Let $\overline{S} = \langle S_1, \ldots, S_n \rangle \in (2^S)^n$, and let $\rho_{\overline{S}} = \rho[X_1 \mapsto S_1, \ldots, X_n \mapsto S_n]$ be the environment that results from ρ by updating the binding of X_i to S_i. Then

$$f_E^\rho(\overline{S}) = \langle [\![\Phi_1]\!]\rho_{\overline{S}}, \ldots, [\![\Phi_n]\!]\rho_{\overline{S}} \rangle.$$

$(2^S)^n$ forms a complete lattice, where the ordering, join and meet operations are the pointwise extensions of the set-theoretic inclusion $\subseteq$, union $\cup$ and intersection $\cap$, respectively. Moreover, for any equation system E and environment ρ, f_E^ρ is monotonic with respect to this lattice and therefore, according to Tarski's fixed-point theorem [Tar55], has both a *greatest* fixed point, νf_E^ρ, and a *least* fixed point, μf_E^ρ. In general, these may be characterized as follows.

$$\nu f_E^\rho = \bigcup \{ \overline{S} \mid \overline{S} \subseteq f_E^\rho(\overline{S}) \}$$
$$\mu f_E^\rho = \bigcap \{ \overline{S} \mid f_E^\rho(\overline{S}) \subseteq \overline{S} \}$$

Blocks $max\{E\}$ and $min\{E\}$ are now interpreted as *environments* in the following fashion.

$$[\![max\{E\}]\!]\rho = \rho_{\nu f_E^\rho}$$
$$[\![min\{E\}]\!]\rho = \rho_{\mu f_E^\rho}$$

So $max\{E\}$ represents the "greatest" fixed point of E and $min\{E\}$ the least.

Given ρ, we define the relative semantics of a hierarchical equational system $\mathcal{B} = \langle B_1, \ldots, B_m \rangle$ in terms of the following sequence of environments.

$$\rho_m = [\![B_m]\!]\rho, \quad \ldots, \quad \rho_1 = [\![B_1]\!]\rho_2$$

Then $[\![\mathcal{B}]\!]\rho = \rho_1$.

The definition of what it means for a state in a labelled transition system to satisfy a formula whose variables are "bound" by a hierarchical equational system is now straightforward. First, we say that a basic proposition Φ is *closed* with respect to an equational system $\mathcal{B}$ if every variable in Φ appears on the left-hand side of some equation in some block in $\mathcal{B}$. We also refer to an equational system $\mathcal{B}$ as closed if each right-hand side in each block in $\mathcal{B}$ is closed with respect to $\mathcal{B}$.

It turns out that $[\![\mathcal{B}]\!]\rho$ does not depend on ρ for closed equational systems $\mathcal{B}$, and that it is sufficient to deal with the case where Φ is a variable:

Theorem 7. *Let $\mathcal{B}$ be a closed hierarchical equational system, and Φ be a (basic) proposition that is closed with respect to $\mathcal{B}$. Then we have*

1. $[\![\Phi]\!]_{[\mathcal{B}]\rho} = [\![\Phi]\!]_{[\mathcal{B}]\rho'}$ *for any environments ρ and ρ'.*
2. *There exists a closed hierarchical equational system $\mathcal{B}'$ having X' as the first variable of its first block, such that $[\![\Phi]\!]_{[\mathcal{B}]} = [\![X']\!]_{[\mathcal{B}']}$.*

The proof for 1. is straightforward, and for 2. it is sufficient to define $\mathcal{B}'$ by prepending $\mathcal{B}$ with a new *min* block $min\{ X' = \Phi \}$.

When $\mathcal{B}$ is closed Theorem 7 allows us to omit reference to ρ, and we abbreviate $s \in [\![X_1]\!]_{[\mathcal{B}]}$ with $s \models \mathcal{B}$, where X_1 is the first variable in the first block of $\mathcal{B}$. The set of all closed hierarchical equational systems is henceforth denoted with $\mathcal{F}$.

To illustrate how properties may be formulated using equational systems, consider the following list containing two blocks.[4]

$$\begin{aligned} < B_1 &\equiv max\{X_1 = X_2 \wedge [a]X_1\}, \\ B_2 &\equiv min\{X_2 = P \vee (\langle a\rangle tt \wedge [a]X_2)\} > \end{aligned}$$

Intuitively, the proposition X_1 of $\langle B_1, B_2\rangle$ represents the CTL formula $AG\,AF\,P$ — "it is always the case that, eventually, P will hold" — for process graphs in which $Act = \{a\}$.

3.4 The Alternation-Free Modal Mu-Calculus

In this section we establish that hierarchical equational systems are equally expressive as the alternation-free modal mu-calculus $L\mu_1$(cf. [EL86]), which uses the greatest and least fixpoint operators ν and μ in order to specify unbounded behaviour. Emerson and Lei define the notion of alternation depth of a formula in the modal mu-calculus which, intuitively, refers to the "level" of mutually recursive greatest and least fixed-point operators. When no such mutual recursion exists, the alternation depth is one, and the formula is said to be *alternation-free*. We have the following (cf. [CDS]).

Theorem 8 (Expressivity).
Let $\mathcal{T}$ be a labelled transition system, and ρ be a corresponding environment. Then we have:

1. *Every formula Φ in $L\mu_1$ can be translated in time proportional to the size of Φ into a hierarchical equational system $\mathcal{B}$ with $[\![\Phi]\!]_\rho = [\![X]\!]_{[\mathcal{B}]\rho}$ for some left-hand-side variable X of $\mathcal{B}$.*
2. *For every hierarchical equational system $\mathcal{B}$ and every variable X there is a formula Φ in $L\mu_1$ satisfying $[\![X]\!]_{[\mathcal{B}]\rho} = [\![\Phi]\!]_\rho$.*

[4] Here tt is an atomic proposition that holds of every state in every process graph.

4 Verifying Behavioural Properties

The heart of our algorithm consists of determining *property transformers* for each state class $\sigma \in \Sigma_{P_i}$ of the finite process representation under consideration: given a set of formulas M as an argument the property transformer for σ yields the set of formulas which are valid at σ *under the assumption* that all formulas of M are valid at the end state of P_i. That this approach provides a sound basis for model-checking follows from the higher-order formulation of the semantics of (simple) PPGs in terms of its valid formulas, together with its extension to CFPSs (see Sect. 4.1).

Our method is effective, as soon as we restrict ourselves to atomic propositions that respect state classes, i.e. they are either satisfied by none or all members of a state class. This is a consequence of the fact that it suffices to deal with the part of a property transformer which concerns the subformulas of the property to be investigated (Sect. 4.2 & 4.3).

4.1 Higher Order Semantics

The "higher order" semantics interprets a simple PPG as the *property transformer* that yields the set of formulas which are valid at the start state, relative to the assumption that the set of propositions serving as the argument are valid at the end state. Its formal definition is as follows.

Definition 9. Let $P = \langle \mathcal{S}, Act, \rightarrow_P, \sigma_P^s, \sigma_P^e \rangle$ be a simple PPG. Then we interpret P as the function $[\![P]\!] : 2^{\mathcal{F}} \longrightarrow 2^{\mathcal{F}}$ which is defined as follows.

$$[\![P]\!](M) = \{ \mathcal{B}' \in \mathcal{F} \mid \forall P' \in \mathcal{G}.\ \ \sigma_P^e \models_{P;P'} M \implies \sigma_P^s \models_{P;P'} \mathcal{B}' \}$$

That the higher order semantics is consistent with the usual semantics of simple PPGs in terms of its valid formulas, is a consequence of the following theorem.

Theorem 10 (Consistency of Higher Order Semantics).
Let P be a simple PPG and $\mathcal{B}$ be a closed hierarchical equational system. Then we have:

$$\sigma_P^s \models \mathcal{B} \quad \iff \quad \mathcal{B} \in [\![P]\!](\mathcal{F}_{deadlock}),$$

where $\mathcal{F}_{deadlock}$ is the set of all propositions that are "true" at a "deadlocked state", i.e.:

$$\mathcal{F}_{deadlock} =_{\mathrm{df}} \{ \mathcal{B} \mid s \models \mathcal{B} \quad in\ \mathcal{T} = \langle \{ s \}, Act, \emptyset \rangle \}.$$

The extension of the higher order semantics for simple PPGs to CFPSs is straightforward. We simply associate with each state class σ of some P_i the property transformer which is induced by the complete expansion of $P_i^{(\sigma)}$. This leads to the following definition.

Definition 11. Let $\mathcal{P} = \langle \mathcal{N}, Act, \Delta, P_0 \rangle$ be a CFPS and $\sigma \in \Sigma_{P_i}$ be a state class of P_i. Then we define: $\quad [\![\sigma]\!]^{\mathcal{P}} =_{\mathrm{df}} [\![Exp_{\mathcal{P}}(P_i^{(\sigma)})]\!]$.

When the CFPS $\mathcal{P}$ is clear from the context we simply write $[\![\sigma]\!]$ for the property transformer associated with σ.

4.2 The Methodology of the Algorithm

Given a CFPS our model-checker iteratively computes $[\![\sigma]\!]$ for every state class σ, and subsequently applies these property transformers to solve the model-checking problem. Thus it has the following two step structure:

1. iterative computation of a ***property transformer*** for each state class of the finite process representation. This computation works block-wise. In the case of a *min*-block it successively updates the initial "minimal" predicate transformers until the smallest fixpoint is reached. The handling of *max*-blocks is completely dual, and the global algorithm, which deals with a hierarchy of blocks works hierarchically as described in [CS92].
   The iteration mechanism as such is essentially the same as for usual iterative algorithms, except for:
   - the handling of the procedural identifiers (non-terminals). They are dealt with by applying the currently valid approximation for their defining PPG.
   - the domain for the iteration, which is higher order here, i.e. it consists of property transformers.

   This computation is effective because we can restrict our attention to the subformulas of the property currently being investigated.
2. decision of the model-checking problem: having the predicate transformers at hand, the model-checking problem can be solved, simply by checking whether the formula under consideration is a member of the set of formulas that results from applying the property transformer associated with the initial state class of the "main procedure" to the set of formulas that are valid for a deadlocking or terminating state.

4.3 The Model-Checking Algorithm

In this section we present the algorithm for *min*-blocks. This is sufficient, because the general algorithm can be obtained straightforwardly along the lines of [CS92]:

- the handling of *max*-blocks is completely dual.
- the analysis of a hierarchical equational system must be done in a hierarchical fashion according to the innermost evaluation strategy.

Preparations. To simplify the algorithm, we assume the following structure for the (*min*-)block B under consideration:

1. the right-hand sides of blocks consist of ***simple*** basic formulas, which are characterized by the following grammar.

$$\Phi^{simple} ::= A \mid X \vee X \mid X \wedge X \mid \langle a \rangle X \mid [a]X$$

2. the graph of the ***unguarded dependencies*** on the variables of B, which is defined by having an arc $X_i \rightarrow X_j$ iff X_j appears unguardedly in the right-hand side of the equation for X_i, is acyclic.

These assumptions do not impose any restrictions, because every block can be transformed into an equivalent block of the same size, satisfying these requirements. The first transformation simply introduces a new variable for each complex subexpression, and the transformation for the second requirement, which is a bit tedious, is illustrated in Fig. 5. Note that X_1 is unguarded in all right-hand side expressions.

$$\begin{aligned}
min\{X_1 = X_1 \vee \Phi\} &\leadsto min\{X_1 = \Phi\}\\
max\{X_1 = X_1 \vee \Phi\} &\leadsto \mathtt{tt}\\
min\{X_1 = X_1 \wedge \Phi\} &\leadsto \mathtt{ff}\\
max\{X_1 = X_1 \wedge \Phi\} &\leadsto max\{X_1 = \Phi\}
\end{aligned}$$

Fig. 5. Transformation for the Second Requirement.

The Algorithm. Let B be a closed equational *min*-block with left-hand side variables $\mathcal{X} = \{\, X_i \mid 1 \le i \le r \,\}$ where X_1 represents the property to be investigated. Moreover, let

- $\mathcal{D} =_{\mathrm{df}} 2^{\mathcal{X}} \rightarrow 2^{\mathcal{X}}$ be the set of all functions from $2^{\mathcal{X}}$ to $2^{\mathcal{X}}$,
- $PT_{\emptyset} \in \mathcal{D}$ be the function which maps every $M \subseteq \mathcal{X}$ to the empty set, and
- PT_{Id} be the identity on $\mathcal{D}$.

Given a CFPS $\mathcal{P}$ we now associate with each $\sigma \in \Sigma$ a function $PT_\sigma \in \mathcal{D}$ and with each transition $\xrightarrow{\alpha}$ a function $PT_{[\alpha]} \in \mathcal{D}$. The function $PT_{[\alpha]}$ represents the property transformer of the process graph $\sigma^s \xrightarrow{\alpha} \sigma^e$. It is defined as follows.

case 1: $\alpha \equiv a \in Act$: Let $M \subseteq \mathcal{X}$ and $X_i = \Phi_i$ be some equation of B.

$$\text{Then } X_i \in PT_{[a]}(M) \text{ iff } \begin{cases} \Phi_i = \langle a\rangle X_j \text{ and } X_j \in M\\ \Phi_i = [a]X_j \text{ and } X_j \in M\\ \Phi_i = [b]X_j \text{ and } b \neq a \end{cases}$$

case 2: $\alpha \equiv P_i$: $\quad PT_{[P_i]} =_{\mathrm{df}} PT_{\sigma^s_{P_i}}$

Now let PT_i for $i \in \{1, \ldots, k\}$ be a property transformer in $\mathcal{D}$ and $\sigma \in \Sigma$. Then the function $\Diamond^{\sigma}_{i=1,\ldots,k} PT_i$ is defined as follows. Given $M \subseteq \mathcal{X}$ and the equation $X_j = \Phi_j$ of B we define $M' =_{\mathrm{df}} (\Diamond^{\sigma}_{i=1,\ldots,k} PT_i)(M)$ by

$$X_j \in M' \text{ iff } \begin{cases} \Phi_j = A & \text{and } \sigma \in \mathcal{V}(A)\\ \Phi_j = X_{j_1} \wedge X_{j_2} & \text{and } X_{j_1} \in M' \text{ and } X_{j_2} \in M'\\ \Phi_j = X_{j_1} \vee X_{j_2} & \text{and } (X_{j_1} \in M' \text{ or } X_{j_2} \in M')\\ \Phi_j = \langle a\rangle X' & \text{and there exists } i \in \{1, \ldots, k\} \text{ with } X_j \in PT_i(M)\\ \Phi_j = [a]X' & \text{and } X_j \in PT_i(M) \text{ holds for all } i \in \{1, \ldots, k\} \end{cases}$$

Note that this definition is well-founded because of our assumption that the graph of unguarded dependencies must be acyclic. Our model-checking algorithm is now given by the procedure **solve** which is defined in Fig. 6.

- Initialize each PT_σ for $\sigma \in \Sigma \setminus \{ \sigma^e_{P_i} \mid 1 \leq i \leq n \}$ with $PT_\emptyset$ and each $PT_{\sigma^e_{P_i}}$ for $1 \leq i \leq n$ with PT_{Id}, respectively.
- Update the functions PT_σ according to the procedure **update** in Fig. 7.
- Check whether $X_1 \in PT_{\sigma^s_P}(\mathcal{B}_{deadlock})$, where $\mathcal{B}_{deadlock}$ is the set of all variables of $\mathcal{B}$ valid at a "deadlock" state. It is defined as follows.

$$\mathcal{B}_{deadlock} =_{df} \{ X_i \in \mathcal{X} \mid s \in [\![X_i]\!]_{[\mathcal{B}]} \quad \text{in } \mathcal{T} = \langle \{ s \}, Act, \emptyset \rangle \}.$$

Fig. 6. The procedure **solve**.

```
workset := { PT_σ | σ ∈ Σ }
WHILE workset ≠ ∅ DO
    LET PT_σ ∈ workset
        BEGIN
        workset := workset \ {PT_σ}
        PT_σ.old := PT_σ
        LET { σ -a_j-> σ_j | j ∈ {1,...,n} } the set of all available transformations for σ
            PT_σ := ◇^σ_{j=1,...,n} PT_[a_j] ∘ PT_σ_j
        IF PT_σ ≠ PT_σ.old THEN
            IF σ = σ^s_{P_i} for some i ∈ I THEN
                workset := workset ∪ { PT_σ' | σ' -Pi->, σ' ∈ Σ }
            ELSE
                workset := workset ∪ { PT_σ' | σ' -a-> σ }
            FI
        FI
        END
OD
```

Fig. 7. The procedure **update**.

4.4 Correctness and Complexity

The procedure **solve** consists of an initialization, a call of **update** and a computation of $PT_{\sigma^s_P}(\mathcal{B}_{deadlock})$. It always terminates, since the number of state classes is finite and the procedure **update** is monotonic on the finite lattice $\mathcal{D} = 2^{\mathcal{X}} \rightarrow 2^{\mathcal{X}}$. Moreover, upon termination of **update**, a single application of $PT_{\sigma^s_P}$ yields the set of formulas that are valid at the start state of the "main procedure".[5]

Theorem 12 (Correctness).
Let $\mathcal{P} = \langle \mathcal{N}, Act, \Delta, P_0 \rangle$ be a CFPS and $\mathcal{B}$ be a closed hierarchical equational system. Then we have upon termination of the algorithm

$$\forall \sigma \in \Sigma. \quad PT_\sigma = [\![\sigma]\!]_{\downarrow \mathcal{B}}$$

where $[\![\sigma]\!]_{\downarrow \mathcal{B}}$ is the restriction of the property transformer $[\![\sigma]\!]$ to the variables of $\mathcal{B}$.

[5] Note that the atomic propositions are required to preserve state classes.

In particular, the theorem yields

$$PT_{\sigma^s_{P_0}}(\mathcal{B}_{deadlock}) = [\![\sigma^s_{P_0}]\!]_{\downarrow\mathcal{B}}(\mathcal{B}_{deadlock})$$

Thus the algorithm computes the set of variables of $\mathcal{B}$ which are "true" at the start state in the complete expansion of P_0.

The proofs of the two inclusions that are necessary to establish Theorem 12 can be sketched as follows:

1. $PT_\sigma \sqsubseteq [\![\sigma]\!]_{\downarrow\mathcal{B}}$ can be proved by straightforward induction on the number of iterations during the execution of the procedure **update**.
2. $PT_\sigma \sqsupseteq [\![\sigma]\!]_{\downarrow\mathcal{B}}$ follows by induction on the depth of the "proof tree" which is necessary to establish the validity of a certain formula at a given state class.

Let us now consider the complexity of our algorithm. Here it turns out that the *weight* of a property is essential. Given a hierarchical equational system $\mathcal{B}$, we define the weight of $\mathcal{B}$ denoted by $w(\mathcal{B})$ as the number of distinct variables that occur after a modality on the right-hand-side in some equation of $\mathcal{B}$. We have:

Theorem 13 (Complexity). *Let $\mathcal{P} = \langle \mathcal{N}, Act, \Delta, P_0 \rangle$ be a CFPS and $\mathcal{B}$ be a closed hierarchical equational system. Then the worst-case time complexity of* **solve** *is*

$$O(\ |\rightarrow_P| * |\mathcal{B}|^2 * 4^{|w(\mathcal{B})|}\ )$$

Sketch of the Proof: The worst case time complexity of the procedure **update** can be estimated by the product of the number of times the while loop is executed and the maximal cost for executing the body of the while loop. Whereas the first factor can be estimated by the product of the number of states classes and the maximal chain length in the lattice of property transformers, the second factor is characterized by the costs for computing $\diamond^\sigma_{j=1,\ldots,k} PT_{[\alpha_j]} \circ PT_{\sigma_j}$, because all the other computations in the body of the while loop are constant-time.

As we can restrict our attention to the variables that count as a weight for the property under investigation the maximal chain length of the lattice of property transformers can be estimated by $|\mathcal{B}| * 2^{|w(\mathcal{B})|}$. Moreover, the costs for the pointwise computation of $\diamond^\sigma_{j=1,\ldots,k} PT_{[\alpha_j]} \circ PT_{\sigma_j}$ can be estimated by $O(2^{|w(\mathcal{B})|} * |\mathcal{B}| * k)$.

Thus, $O(\ |\Sigma| * k_{max} * |\mathcal{B}|^2 * 4^{|w(\mathcal{B})|}\ )$, where k_{max} is the maximal number of transitions that leave a state class, is an upper approximation of the desired worst case time complexity. However, applying standard techniques, which can be found e.g. in CS1 or CS2, the factor $|\Sigma| * k_{max}$ can be reduced to $|\rightarrow_P|$, which essentially reflects the fact that all state classes can be updated once by considering all transitions once.
□

Continuation of the Example

Consider the encoding machine of Fig. 1, and suppose that we are interested in proving that this machine reads the full (finite) sequence of input symbols, before it outputs the first code. As long as we restrict ourselves to finite input sequences terminated by #, this property can be reformulated by saying that after outputting an encoded character a' or b' the machine will always refuse to read #. The following figure expresses this property in terms of a *max*-block where X_1 represents the specified property.

$$max \left\{ \begin{array}{llll} X_1 = X_2 \wedge X_3 & X_6 = [\#]X_1 & Y_1 = Y_2 \wedge Y_3 & Y_6 \;= [\#]Y_7 \\ X_2 = [a]X_1 & X_7 = X_8 \wedge X_9 & Y_2 = [a]Y_1 & Y_7 \;= \texttt{ff} \\ X_3 = X_4 \wedge X_5 & X_8 = [a']Y_1 & Y_3 = Y_4 \wedge Y_5 & Y_8 \;= Y_9 \wedge Y_{10} \\ X_4 = [b]X_1 & X_9 = [b']Y_1 & Y_4 = [b]Y_1 & Y_9 \;= [a']Y_1 \\ X_5 = X_6 \wedge X_7 & & Y_5 = Y_6 \wedge Y_8 & Y_{10} = [b']Y_1 \end{array} \right\}$$

Note that, although this *max*-block is rather large, its weight is just 3, because only the variables X_1, Y_1 and Y_7 occur after a modality on the right-hand-side in some equation of $\mathcal{B}$.

5 Conclusions and Future Work

We have presented an algorithm that decides the alternation-free modal mu-calculus for *context-free* processes, i.e. for processes resulting from an extension of standard finite-state process graphs by introducing *recursive procedures* or, alternatively, a *recursive action refinement mechanism*. The point of this algorithm is its *higher orderedness*: it iteratively determines a property transformer for the procedural components of the (infinite-state) system being analyzed. This idea leads to a model checking algorithm for infinite state systems, which is linear in the size of the system's representation and exponential in the size of the property.

The complexity result about our algorithm is encouraging, because in practice the process systems are extremely large, however, the weights of the properties to be investigated are often very small. For example for deadlock checking the weight would be one. Thus the exponential factor is often uncritical. The practicability of our approach may also be supported by representing the property transformer in terms of BDD's (cf. [Bry86]). We are going to implement this algorithm in the Edinburgh Concurrency Workbench (cf. [CPS89]), which currently can only deal with finite state systems, in order to validate this statement from the practical point of view. Moreover, we are working on an extension of the algorithm that covers the full modal mu-calculus.

Acknowledgements

We would like to thank Bengt Jonsson for his constructive comments.

References

[BBK87] J.C.M. Baeten, J.A. Bergstra, and J.W. Klop. Decidability of Bisimulation Equivalence for Processes Generating Context-Free Languages. In *PARLE '87, LNCS 259*, pages 94–113. Springer, 1987.

[Bra91] J.C. Bradfield. *Verifying Temporal Properties of Systems with Applications to Petri Nets.* PhD thesis, University of Edinburgh, 1991.

[Bry86] R.E. Bryant. Graph-Based Algorithms for Boolean Function Manipulation. *IEEE Transactions on Computers*, C-35(8):677–691, 1986.

[BS91] J.C. Bradfield and C. Stirling. Local Model Checking for Infinite State Spaces. Technical Report ECS-LFCS-90-115, LFCS, June 1991.

[Cau90] D. Caucal. Graphes Canoniques de Graphes Algébriques. *RAIRO*, 24(4):339–352, 1990.

[CDS] R. Cleaveland, M. Dreimüller, and B. Steffen. Faster Model Checking for the Modal Mu-Calculus. Accepted for CAV '92.

[CES86] E.M. Clarke, E.A. Emerson, and A.P. Sistla. Automatic Verification of Finite State Concurrent Systems Using Temporal Logic Specifications. *ACM Transactions on Programming Languages and Systems*, 8(2):244–263, 1986.

[Cle90] R. Cleaveland. Tableau-Based Model Checking in the Propositional Mu-Calculus. *Acta Informatica*, 27:725–747, 1990.

[CPS89] R. Cleaveland, J. Parrow, and B. Steffen. The Concurrency Workbench. In *Proceedings of the Workshop on Automatic Verification Methods for Finite-State Systems, LNCS 407*, pages 24–37. Springer, 1989.

[CS91] R. Cleaveland and B. Steffen. Computing Behavioural Relations, Logically. In *ICALP '91, LNCS 510*. Springer, 1991.

[CS92] R. Cleaveland and B. Steffen. A Linear-Time Model-Checking Algorithm for the Alternation-Free Modal Mu-Calculus. In *CAV '91, LNCS 575*, pages 48–58. Springer, 1992.

[EL86] E.A. Emerson and C.-L. Lei. Efficient Model Checking in Fragments of the Propositional Mu-Calculus. In *Proc. 1th Annual Symp. on Logic in Computer Science*, pages 267–278. IEEE Computer Society Press, 1986.

[Gro91] J.F. Groote. A Short Proof of the Decidability of Bisimulation for Normed bpa-Processes. Technical Report CS-R9151, CWI, December 1991.

[HS91] H. Hüttel and C. Stirling. Actions Speak Louder than Words: Proving Bisimularity for Context-Free Processes. In *Proc. 6th Annual Symp. on Logic in Computer Science*, pages 376–386. IEEE Computer Society Press, 1991.

[HT92] D.T. Huynh and L. Tian. Deciding Bisimilarity of Normed Context-Free Processes is in Σ_2^p. Technical Report UTDCS-1-92, University of Texas at Dallas, 1992.

[Koz83] D. Kozen. Results on the Propositional μ-Calculus. *Theoretical Computer Science*, 27:333–354, 1983.

[Lar88] K.G. Larsen. Proof Systems for Hennessy–Milner Logic with Recursion. In *CAAP '88, LNCS 299*, pages 215–230. Springer, 1988.

[SW89] C. Stirling and D. Walker. Local Model Checking in the Modal Mu-Calculus. In *TAPSOFT '89, LNCS 351*, pages 369–383. Springer, 1989.

[Tar55] A. Tarski. A Lattice-Theoretical Fixpoint Theorem and its Applications. *Pacific Journal of Mathematics*, 5:285–309, 1955.

[Win89] G. Winskel. A Note on Model Checking the Modal Mu-Calculus. In *Automata, Languages and Programming, LNCS 372*, pages 761–772. Springer, 1989.

This article was processed using the LaTeX macro package with LLNCS style

Bisimulation Equivalence is Decidable for all Context-Free Processes

Søren Christensen* Hans Hüttel† Colin Stirling*

1 Introduction

Over the past decade much attention has been devoted to the study of process calculi such as CCS, ACP and CSP [11]. Of particular interest has been the study of the behavioural semantics of these calculi as given by labelled transition graphs. One important question is when processes can be said to exhibit the same behaviour, and a plethora of *behavioural equivalences* exists today. Their main rationale has been to capture behavioural aspects that language or trace equivalences do not take into account.

The theory of finite-state systems and their equivalences can now be said to be well-established. There are many automatic verification tools for their analysis which incorporate equivalence checking. Sound and complete equational theories exist for the various known equivalences, an elegant example is [16].

One may be led to wonder what the results will look like for *infinite-state* systems. Although language equivalence is decidable for finite-state processes, it is undecidable when one moves beyond finite automata to context-free languages. For finite-state processes all known behavioural equivalences can be seen to be decidable. In the setting of process algebra, an example of infinite-state systems is that of the transition graphs of processes in the calculus BPA (Basic Process Algebra) [3]. These are recursively defined processes with nondeterministic choice and sequential composition.

A special case is that of *normed BPA* processes. A process is said to be normed if it can terminate in finitely many steps at any point during the execution. Even though normed BPA does not incorporate all regular processes, systems defined in this calculus can in general have infinitely many states. A recent result shows that strong bisimulation equivalence for normed BPA processes is *decidable*. A number of proofs of this result exists, the original proof by Baeten, Bergstra and Klop [1], another due to Caucal [7] (see [9] too), and a third due to Hüttel and Stirling [14] (with more details in [13]) which appeals to tableaux. The tableau based approach supports a sound and complete equational theory for normed BPA. On the other hand, Huynh and Tian

*Laboratory for Foundations of Computer Science, James Clerk Maxwell Building, University of Edinburgh, Edinburgh EH9 3JZ, Scotland.

†Department of Mathematics and Computer Science, Aalborg University Centre, Fredrik Bajersvej 7E, 9220 Aalborg Ø, Denmark.

[15] and Groote and Hüttel [10] have proved that all other standard equivalences are undecidable for normed BPA and thus for BPA in general.

One remaining question to be answered is whether bisimulation equivalence is decidable for the *full* BPA language. In this paper we answer this question in the affirmative, using a technique inspired by Caucal's proof of the decidability of language equivalence for simple algebraic grammars [5]. Section 2 contains preliminary definitions. Section 3 describes the main result, namely that the maximal bisimulation of any BPA transition graph is generable from a finite *self-bisimulation* relation.

2 BPA processes

The class of guarded recursive BPA (Basic Process Algebra) processes [1, 3] is defined by the following abstract syntax

$$E ::= a \mid X \mid E_1 + E_2 \mid E_1 \cdot E_2$$

Here a ranges over a set of atomic actions *Act*, and X over a family of variables. The operator $+$ is nondeterministic choice while $E_1 \cdot E_2$ is the sequential composition of E_1 and E_2 – we usually omit the '$\cdot$'. A BPA process is defined by a finite system of recursive process equations

$$\Delta = \{X_i \stackrel{def}{=} E_i \mid 1 \leq i \leq k\}$$

where the X_i are distinct, and the E_i are BPA expressions with free variables in $Var_\Delta = \{X_1, \ldots, X_k\}$. One variable (generally X_1) is singled out as the *root*. Usually one considers relations within the transition graph for a single Δ. This can be done without loss of generality, since we can let Δ be the disjoint union of any pair Δ_1 and Δ_2 that we wish to compare (with suitable renamings of variables, if required).

We restrict our attention to *guarded* systems of recursive equations.

Definition 2.1 *A BPA expression is* guarded *if every variable occurrence is within the scope of an atomic action. The system* $\Delta = \{X_i \stackrel{def}{=} E_i \mid 1 \leq i \leq k\}$ *is guarded if each* E_i *is guarded for* $1 \leq i \leq k$.

We use $X, Y, \ldots$ to range over variables in Var_Δ and Greek letters $\alpha, \beta, \ldots$ to range over elements in Var_Δ^*. In particular, ϵ denotes the empty variable sequence.

Definition 2.2 *Any system of process equations Δ defines a labelled transition graph. The transition relations are given as the least relations satisfying the following rules:*

$$a \xrightarrow{a} \epsilon, a \in Act \qquad\qquad \frac{E \xrightarrow{a} E'}{X \xrightarrow{a} E'}\; X \stackrel{def}{=} E \in \Delta$$

$$\frac{E \xrightarrow{a} E'}{E + F \xrightarrow{a} E'} \qquad\qquad \frac{F \xrightarrow{a} F'}{E + F \xrightarrow{a} F'}$$

$$\frac{E \xrightarrow{a} E'}{EF \xrightarrow{a} E'F}\; E' \neq \epsilon \qquad\qquad \frac{E \xrightarrow{a} \epsilon}{EF \xrightarrow{a} F}$$

Example 2.1 Consider the system $\Delta = \{X \stackrel{def}{=} a + bXY;\;\; Y \stackrel{def}{=} c\}$. By the transition rules in Definition 2.2 X generates the transition graph in Figure 1. □

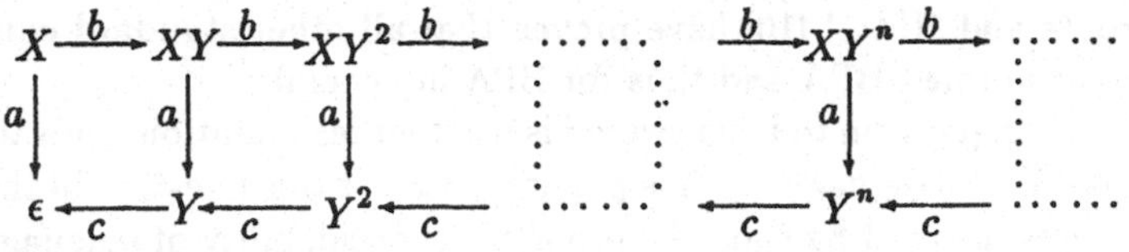

Figure 1: Transition graph for $X \stackrel{def}{=} a + bXY$; $Y \stackrel{def}{=} c$ (Example 2.1)

2.1 Bisimulation equivalence and Greibach Normal Form

Definition 2.3 *A relation R between processes is a* bisimulation *if whenever pRq then for each $a \in Act$*

1. $p \xrightarrow{a} p' \Rightarrow \exists q' : q \xrightarrow{a} q'$ *with* $p'Rq'$
2. $q \xrightarrow{a} q' \Rightarrow \exists p' : p \xrightarrow{a} p'$ *with* $p'Rq'$

Two processes p and q are *bisimulation equivalent*, written as $p \sim q$, if there is a bisimulation relation R such that pRq. The relation $\sim$ is an equivalence, and moreover it is a congruence relation with respect to the operators $+$ and $\cdot$, [3]. An alternative characterization of $\sim$ is via a sequence of approximations.

Definition 2.4 *The sequence of bisimulation approximations $\{\sim_n\}_{n=1}^{\omega}$ is defined inductively as follows.*

- $p \sim_0 q$ *for all processes p and q,*
- $p \sim_{n+1} q$ *iff for each $a \in Act$*
  - $p \xrightarrow{a} p'$ *implies* $q \xrightarrow{a} q'$ *and* $p' \sim_n q'$ *for some* q'
  - $q \xrightarrow{a} q'$ *implies* $p \xrightarrow{a} p'$ *and* $p' \sim_n q'$ *for some* p'

It is a standard result, see [17] for instance, that for any image-finite labelled transition graph (that is, where for each p and a the set $\{q \mid p \xrightarrow{a} q\}$ is finite):

$$\sim \, = \bigcap_{n=0}^{\omega} \sim_n$$

Clearly, the transition graph for any family Δ of guarded BPA processes is image-finite.

Any system Δ of guarded BPA equations has a unique solution up to bisimulation equivalence [2]. Moreover, in [1] it is shown that any such system can be effectively presented in a normal form

$$\{X_i \stackrel{def}{=} \sum_{j=1}^{n_i} a_{ij}\alpha_{ij} \mid 1 \leq i \leq m\}$$

such that bisimilarity is preserved. From the transition rules we see that if $X_i \xrightarrow{w} E$ then E is just a sequence α of variables. The normal form is called Greibach Normal Form, GNF, by analogy with context-free grammars (without the empty production) in GNF (see e.g. [12]). There is an obvious correspondence with grammars in GNF: process variables correspond to non-terminals, the root is the start symbol, actions correspond to terminals, and each equation $X_i \stackrel{def}{=} \sum_{j=1}^{n_i} a_{ij}\alpha_{ij}$ can be viewed as the family of productions $\{X_i \rightarrow a_{ij}\alpha_{ij} \mid 1 \leq j \leq n_i\}$.

3 Decidability of bisimulation equivalence

Assume a fixed system Δ of BPA equations in GNF whose variable set is Var. The bisimulation equivalence problem is whether or not $\alpha \sim \beta$ when α and β are sequences of variables drawn from Var. In the case that these are finite-state processes, a very naive decision procedure consists of enumerating all binary relations over the finite state space generated by α and β using the rules for transitions and determining if there is a relation among them which is a bisimulation containing the pair (α, β). But of course BPA processes are not generally finite-state, and therefore bisimulations can now be infinite.

On the other hand for any n, the n-bisimulation equivalence problem (whether or not $\alpha \sim_n \beta$) is decidable. This means that bisimulation inequivalence is semi-decidable via the simple procedure which seeks the least i such that $\alpha \not\sim_i \beta$. Therefore we just need to establish the semi-decidability of bisimulation equivalence. The proof of this (inspired by [5, 6, 7]) relies on showing that there is a *finite self-bisimulation* relation which generates the bisimulation equivalence.

3.1 Self-bisimulations

The notion of self-bisimulation was introduced by Didier Caucal in [7] (originally published as [6]). Here the notion of a least congruence is essential.

Definition 3.1 *For any binary relation R on Var^*, $\underset{R}{\rightarrow}$ is the least precongruence w.r.t. sequential composition that contains R, $\underset{R}{\longleftrightarrow}$ the symmetric closure of $\underset{R}{\rightarrow}$ and $\underset{R}{\longleftrightarrow}^*$ the reflexive and transitive closure of $\underset{R}{\longleftrightarrow}$ and thus the least congruence w.r.t. sequential composition containing R.*

A self-bisimulation is then simply a bisimulation up to congruence w.r.t. sequential composition.

Definition 3.2 *A relation $R \subseteq Var^* \times Var^*$ is called* a self-bisimulation *iff $\alpha R \beta$ implies that for each $a \in Act$*

1. *if $\alpha \xrightarrow{a} \alpha'$ then $\beta \xrightarrow{a} \beta'$ for some β' with $\alpha' \underset{R}{\longleftrightarrow}^* \beta'$*
2. *if $\beta \xrightarrow{a} \beta'$ then $\alpha \xrightarrow{a} \alpha'$ for some α' with $\alpha' \underset{R}{\longleftrightarrow}^* \beta'$*

The following lemma, due to Didier Caucal, shows that a self-bisimulation is a witness for bisimulation equivalence.

Lemma 3.1 *[7] If R is a self-bisimulation then $\overset{*}{\underset{R}{\longleftrightarrow}} \subseteq \sim$.*

Corollary 3.1 *$\alpha \sim \beta$ iff there is a self-bisimulation R such that $\alpha R \beta$.*

3.2 Decompositions

Our aim is to show that bisimulation equivalence on Var^* is generable from a finite self-bisimulation. To do this we must find techniques for decomposing bisimilar sequences of variables α and β into "smaller" subsequences $\alpha_1 \ldots \alpha_n$ and $\beta_1 \ldots \beta_n$ with $\alpha_i \sim \beta_i$ for each i in such a way that there are only "finitely" many pairs α and β that can not be decomposed. Extra definitions and some preliminary results are needed to achieve this.

A process $\alpha \in Var^+$ is *normed* if there is a $w \in Act^+$ such that $\alpha \xrightarrow{w} \epsilon$. When α is normed we let the *norm* of α, written as $|\alpha|$ following [1], be defined as:

$$|\alpha| = \min\{length(w) \mid \alpha \xrightarrow{w} \epsilon, w \in Act^+\}$$

By convention we also assume that $|\epsilon| = 0$. Clearly α is normed just in case each variable occurring in it has a norm. We divide the variable set Var into disjoint subsets $V_0 = \{X \in Var \mid X \text{ is normed}\}$ and $V_1 = Var \setminus V_0$. The system of equations, example 2.1, only contains normed variables with $|X| = |Y| = 1$, so $V_0 = \{X, Y\}$ and $V_1 = \emptyset$. Example 3.1 contains an unnormed X so in this case $V_0 = \{Y\}$ and $V_1 = \{X\}$.

Example 3.1 In the system of equations $\Delta = \{X \overset{def}{=} aX;\ \ Y \overset{def}{=} c + aX\}$ the variable X is not normed since there is no w such that $X \xrightarrow{w} \epsilon$ whereas $|Y| = 1$. □

A straightforward consequence of the definition of having a norm is the following:

$$\text{if } X \in V_1 \text{ then } \alpha X \beta \sim \alpha X$$

Therefore we can assume that our fixed system of BPA equations in normal form $\Delta = \{X_i \overset{def}{=} \sum_{j=1}^{n_i} a_{ij}\alpha_{ij} \mid 1 \le i \le m\}$ has the property that each $\alpha_{ij} \in (V_0^* V_1) \cup V_0^*$.

The next definition stipulates what we mean by decomposition:

Definition 3.3 *When $X\alpha \sim Y\beta$ we say that the pair $(X\alpha, Y\beta)$ is decomposable if $X, Y \in V_0$ and there is a γ such that*

- *$\alpha \sim \gamma\beta$ and $X\gamma \sim Y$ if $|X| \le |Y|$*
- *$\gamma\alpha \sim \beta$ and $X \sim Y\gamma$ if $|Y| \le |X|$.*

In the case of normed processes (where each variable in Var is normed) the important property underpinning decidability of $\sim$ is that any bisimilar pair $(X\alpha, Y\beta)$ is decomposable (see [5]). Assuming that $|X| \leq |Y|$ and that β is not empty this means that there is a decomposition of $X\alpha$ into the two smaller (with respect to norm) subsequences $X\gamma$ and β with $X\gamma \sim Y$. Consequently bisimulation equivalence is then generable from a finite self-bisimulation consisting of pairs of the form (X, α).

However, in the presence of unnormed variables the situation is much more complex, as there can be bisimilar pairs $(X\alpha, Y\beta)$ which are not decomposable. We therefore need to show that in some sense there are only finitely many of them. A special class of pairs have the form $(\alpha, X\gamma\alpha)$. The following lemma provides some information about them.

Lemma 3.2 *If $\alpha \sim X\gamma\alpha$ and $\beta \sim X\gamma\beta$ then $\alpha \sim \beta$.*

PROOF: If $\alpha \sim X\gamma\alpha$ and $\beta \sim X\gamma\beta$ then both α and β are solutions to the same (guarded) equation. As any system of guarded equations has a unique solution up to bisimulation equivalence [1] it follows that $\alpha \sim \beta$. □

This leads to a crucial finiteness lemma which appeals to the notion of the degree of Δ, $deg(\Delta)$, defined as the size of the largest set $\{\alpha | \ X \xrightarrow{a} \alpha, \ a \in Act\}$ when $X \in Var$: for instance, both systems of equations in Examples 2.1 and 3.1 have degree 2.

Lemma 3.3 *For any $\alpha, \beta \in Var^*$, if $\alpha \not\sim_n \beta$ then there are at most $(deg(\Delta))^{n-1}$ different ϕ up to $\sim$ with the feature $\alpha\phi \sim \beta\phi$.*

PROOF: Induction on n using the previous lemma. For the base case if $\alpha \not\sim_1 \beta$ then without loss of generality $\alpha \xrightarrow{a}$ but $\beta \not\xrightarrow{a}$ for some a. But there can not be a ϕ giving $\alpha\phi \sim \beta\phi$, unless $\beta = \epsilon$. By Lemma 3.2 there is only one ϕ up to $\sim$ such that $\alpha\phi \sim \phi$. If $\alpha \not\sim_{n+1} \beta$ then without loss of generality $\alpha \xrightarrow{b} \alpha'$ and for all β' such that $\beta \xrightarrow{b} \beta'$ it is the case that $\alpha' \not\sim_n \beta'$. Now suppose $\alpha\phi \sim \beta\phi$. Any transition $\alpha\phi \xrightarrow{b} \alpha'\phi$ can be matched by a transition $\beta\phi \xrightarrow{b} \beta'\phi$ with $\alpha'\phi \sim \beta'\phi$. By the induction hypothesis there are only $(deg(\Delta))^{n-1}$ distinct ϕ_1 such that $\alpha'\phi_1 \sim \beta'\phi_1$. Let $S = \{a_j\alpha'_j \mid \alpha \xrightarrow{a_j} \alpha'_j, \forall\beta' : \beta \xrightarrow{a_j} \beta' \Rightarrow \alpha_j \not\sim_n \beta'\}$. S can have at most $deg(\Delta)$ distinct elements. We can write $\alpha \sim \sum_{a_j\alpha_j \in S} a_j\alpha_j + \sum_{b_j\beta_j \notin S} b_j\beta_j$, and from this we see that since for each $a_j\alpha_j \in S$ there are at most $(deg(\Delta))^{n-1}$ distinct ϕ such that $\alpha_j\phi \sim \beta'\phi$, there are all in all at most $(deg(\Delta))^n$ distinct ϕ such that $\alpha\phi \sim \beta\phi$. □

We say that the pairs $(X\alpha, Y\beta)$ and $(X\alpha_1, Y\beta_1)$ are *distinct* when $\alpha \not\sim \alpha_1$ or $\beta \not\sim \beta_1$. The next surprising result shows that there are only finitely many interesting pairs $(X\alpha, Y\beta)$ that are not decomposable.

Lemma 3.4 *For any $X, Y \in Var$ any set R of the form*

$$\{(X\alpha, Y\beta) \mid X\alpha, Y\beta \in (V_0^* V_1) \cup V_0^*, \ X\alpha \sim Y\beta, \ (X\alpha, Y\beta) \text{ is not decomposable}\}$$

such that all pairs are distinct is finite.

PROOF: First, if both X and Y belong to V_1 then R contains just one member. Otherwise assume one of them is in V_1, without loss of generality let this be X. As Y is normed let $|Y| = n$. Therefore $Y \xrightarrow{w} \epsilon$ for some w of length n. But there are only finitely many γ such that $X \xrightarrow{w} \gamma$. If R were infinite containing pairs $(X, Y\beta_i)$ for all i, we would have that for some j $\beta_j \sim \beta_k$ for infinitely many k which contradicts distinctness. Now, assume that both $X, Y \in V_0$ and without loss of generality let $|X| \leq |Y|$ with $|X| = n$. Consider a $w = a_1 \cdots a_n$ such that $X \xrightarrow{w} \epsilon$. Since $X\alpha_i \sim Y\beta_i$ for all $(X\alpha_i, Y\beta_i) \in R$ we must have $Y \xrightarrow{w} \gamma$ for some γ. But then consider the set $B = \{\gamma_j \mid \exists u : Y \xrightarrow{u} \gamma_j, length(u) = n\}$. This set is finite and has at most $(deg(\Delta))^n$ elements. But then for some $\gamma \in B$, since $\alpha_i \sim \gamma\beta_i$, it must be the case that for infinitely many $(X\alpha_i, Y\beta_i) \in R$ we have $X\gamma\beta_i \sim Y\beta_i$. But this is impossible by Lemma 3.3, as $X\gamma \not\sim Y$ follows from the assumption that the pairs are not decomposable. □

3.3 Finite representability of $\sim$

We are now almost in a position to prove our main theorem, which relies on an induction on size, defined for every $\alpha \in (V_0^*V_1) \cup V_0^*$ and denoted by $s(\alpha)$:

$$s(\alpha X) = \begin{cases} |\alpha X| & \text{if } X \in V_0 \\ |\alpha| & \text{otherwise} \end{cases}$$

We let $\sqsubseteq$ be the well-founded ordering on $(V_0^*V_1) \cup V_0^* \times (V_0^*V_1) \cup V_0^*$ given by $(\alpha_1, \alpha_2) \sqsubseteq (\beta_1, \beta_2)$ if $\max\{s(\alpha_1), s(\alpha_2)\} \leq \max\{s(\beta_1), s(\beta_2)\}$.

Theorem 3.1 *There is a finite relation R on $(V_0^*V_1) \cup V_0^*$ such that $\sim = \underset{R}{\longleftrightarrow}^*$.*

PROOF: We define R as the union of two finite relations R_1 and R_2. R_1 is a largest relation of the form

$$\{(X, \alpha) \mid X, \alpha \in V_0^*,\ X \sim \alpha\}$$

and R_2 is a largest relation of the form

$$\{(X\alpha, Y\beta) \mid X\alpha, Y\beta \in (V_0^*V_1) \cup V_0^*,\ X\alpha \sim Y\beta,\ (X\alpha, Y\beta) \text{ is not decomposable}\}$$

such that each pair $(X\alpha, Y\beta)$, $(X\alpha', Y\beta')$ in R_2 is distinct. Moreover, we assume *minimal* elements w.r.t. $\sqsubseteq$, i.e. if $(X\alpha, Y\beta) \in R_2$ is *not* distinct from $(X\alpha', Y\beta')$ then $(\alpha, \beta) \sqsubseteq (\alpha', \beta')$. Notice that both R_1 and R_2 are finite; the finiteness of R_1 follows from the fact that there are only finitely many elements of V_0^* with a given finite norm and the finiteness of R_2 follows from lemma 3.4. Thus R is finite.

We now want to show that $\sim = \underset{R}{\longleftrightarrow}^*$. As $R \subseteq \sim$ and $\sim$ is a congruence w.r.t. sequential composition we immediately have $\underset{R}{\longleftrightarrow}^* \subseteq \sim$. So we consider proving $\sim \subseteq \underset{R}{\longleftrightarrow}^*$ and proceed by induction on $\sqsubseteq$. Let $X\alpha \sim Y\beta$. There are two cases:

- Suppose that $(X\alpha, Y\beta)$ is not decomposable. Then by the maximality of R_2 we have $(X\alpha', Y\beta')$ in R_2 such that $(\alpha', \beta') \sqsubseteq (\alpha, \beta)$ with $\alpha \sim \alpha'$ and $\beta \sim \beta'$. If $X, Y \in V_0$ then clearly $(\alpha', \beta') \sqsubseteq (\alpha, \beta) \sqsubset (X\alpha, Y\beta)$ from which it follows that $(\alpha, \alpha') \sqsubset (X\alpha, Y\beta)$ and $(\beta, \beta') \sqsubset (X\alpha, Y\beta)$. By the induction hypothesis we now conclude that $\alpha \xleftrightarrow[R]{}{}^{*} \alpha'$ and $\beta \xleftrightarrow[R]{}{}^{*} \beta'$ from which we get $X\alpha \xleftrightarrow[R]{}{}^{*} Y\beta$ as desired. If $X \in V_1$ and $Y \in V_0$ we get $\alpha = \alpha' = \epsilon$ and therefore $X \sim Y\beta$. As $Y \in V_0$ we have $s(\beta') \leq s(\beta) < s(Y\beta)$ hence $(\beta, \beta') \sqsubset (X, Y\beta)$ which by the induction hypothesis implies $\beta \xleftrightarrow[R]{}{}^{*} \beta'$. But then $X \xleftrightarrow[R]{}{}^{*} Y\beta' \xleftrightarrow[R]{}{}^{*} Y\beta$ as desired. Finally, if $X, Y \in V_1$ then $\alpha = \alpha' = \epsilon$ and also $\beta = \beta' = \epsilon$. Hence we have $(X, Y) \in R_2$ from which $X \xleftrightarrow[R]{}{}^{*} Y$ follows.

- Suppose $(X\alpha, Y\beta)$ is decomposable. By the definition of decomposability it follows that $X, Y \in V_0$. Assume without loss of generality that we have γ such that $\gamma\alpha \sim \beta$ and $X \sim Y\gamma$. As X is normed and $X \sim Y\gamma$ clearly $s(\gamma\alpha) < s(X\alpha)$. Similarly as Y is normed we also have $s(\beta) < s(Y\beta)$ and therefore $(\gamma\alpha, \beta) \sqsubset (X\alpha, Y\beta)$ from which $\gamma\alpha \xleftrightarrow[R]{}{}^{*} \beta$ follows by the induction hypothesis. As $X \sim Y\gamma$ with $X \in V_0$ we have $(X, Y\gamma) \in R_1$ from the maximality of R_1. But then $X\alpha \xleftrightarrow[R]{}{}^{*} Y\gamma\alpha \xleftrightarrow[R]{}{}^{*} Y\beta$ as desired.

This completes the proof. □

Thus, Corollary 3.1 can be strengthened to: $\alpha \sim \beta$ iff there is a *finite* self-bisimulation R such that $\alpha R \beta$. We now show that this is sufficient for semi-decidability of $\sim$. For given a finite relation R on $(V_0^* V_1) \cup V_0^*$ it is semi-decidable whether it is a self-bisimulation. The procedure consists in defining a derivation or proof system: the axioms are the pairs in R, and the rules are congruence rules for sequential composition together with the usual equivalence rules. Consequently, for each n let $D_n(R)$ be the finite set of pairs (α, β) which are derivable within n steps of the proof system.

Definition 3.4 *A finite relation R on $(V_0^* V_1) \cup V_0^*$ is an* n-self-bisimulation *iff $\alpha R \beta$ implies that for all $a \in Act$*

1. *if $\alpha \xrightarrow{a} \alpha'$ then $\beta \xrightarrow{a} \beta'$ for some β' with $(\alpha', \beta') \in D_n(R)$*

2. *if $\beta \xrightarrow{a} \beta'$ then $\alpha \xrightarrow{a} \alpha'$ for some α' with $(\alpha', \beta') \in D_n(R)$*

For each n clearly it is decidable whether a finite relation R on $(V_0^* V_1) \cup V_0^*$ is an n-self-bisimulation. Moreover, if R is a finite self-bisimulation then for some n it is an n-self-bisimulation.

We now complete the proof that bisimulation equivalence is semi-decidable using a dovetailing technique (compare [5]). Let $R_0 \ldots R_i \ldots$ be an effective enumeration list of all finite relations on $(V_0^* V_1) \cup V_0^*$ and let $g : N^2 \to N$ be an effective bijection. To check whether $\alpha \sim \beta$, for each $n \geq 0$ in turn consider the pair $(i, j) = g^{-1}(n)$: if $\alpha R_i \beta$ then test if R_i is a j-self-bisimulation. Consequently, if $\alpha \sim \beta$ this must be established at the nth stage of this procedure for some n. The decidability result is now established.

Theorem 3.2 *Bisimulation equivalence is decidable for all guarded BPA processes.*

4 Conclusion

We have shown that bisimulation equivalence is decidable for BPA. As the proof involves two semi-decision procedures it is not obvious how to determine the complexity of solving this problem. Moreover it does not provide us with an intuitive technique for deciding bisimilarity as does the tableau method in [14, 13] which also has the advantage of providing us with a way of extracting a complete axiomatization for normed BPA processes. A similar result for full BPA would be a proper extension of Milner's axiom system for regular processes [16].

More generally this work addresses the area of infinite-state processes. Besides deciding equivalences there is also the question of model checking: a recent result [4] shows decidability for fragments of the modal mu-calculus in the case of normed BPA. There is also the question of pushdown automata processes (which generate a richer family of transition graphs than BPA processes). [18] contains a very elegant characterization of their graphs.

Of more interest to concurrency theory are process languages with parallel combinators. Although bisimulation equivalence is undecidable for ACP, CCS, and CSP it is unclear if this must be true of all parallel models with full Turing power (especially those that lack abstraction mechanisms). Moreover there may be finer useful equivalences which permit general decidability results: for instance in [8] it is shown that distributed bisimulation equivalence is decidable for a recursive fragment of CCS with parallel.

Acknowledgements

The authors would like to thank Didier Caucal and Robin Milner for useful discussions and insights.

References

[1] J.C.M. Baeten, J.A. Bergstra, and J.W. Klop. Decidability of bisimulation equivalence for processes generating context-free languages. Technical Report CS-R8632, CWI, September 1987.

[2] J.A. Bergstra and J.W. Klop. Process algebra for synchronous communication. *Information and Control*, 60:109–137, 1984.

[3] J.A. Bergstra and J.W. Klop. Process theory based on bisimulation semantics. In J.W. de Bakker, W.P de Roever, and G. Rozenberg, editors, *LNCS 354*, pages 50–122. Springer-Verlag, 1988.

[4] O. Burkart and B. Steffen. Model checking for context-free processes. In these proceedings.

[5] D. Caucal. Décidabilité de l'egalité des langages algébriques infinitaires simples. In *Proceedings of STACS 86, LNCS 210*, pages 37–48. Springer-Verlag, 1986.

[6] D. Caucal. Graphes canoniques de graphes algébriques. Rapport de Recherche 872, INRIA, Juillet 1988.

[7] D. Caucal. Graphes canoniques de graphes algébriques. *Informatique théorique et Applications (RAIRO)*, 24(4):339–352, 1990.

[8] S. Christensen. Distributed bisimularity is decidable for a class of infinite state-space systems. In these proceedings.

[9] J. F. Groote. A short proof of the decidability of bisimulation for normed BPA-processes. Tech. Report Utrecht University 1992.

[10] J.F. Groote and H. Hüttel. Undecidable equivalences for basic process algebra. Technical Report ECS-LFCS-91-169, Department of Computer Science, University of Edinburgh, August 1991.

[11] C.A.R. Hoare. *Communicating Sequential Processes.* Prentice-Hall, 1988.

[12] J. Hopcroft and J.D. Ullman. *Introduction to Automata Theory, Languages, and Computation.* Addison-Wesley, 1979.

[13] H. Hüttel. *Decidability, Behavioural Equivalences and Infinite Transition Graphs.* PhD thesis, University of Edinburgh, December 1991.

[14] H. Hüttel and C. Stirling. Actions speak louder than words: Proving bisimilarity for context-free processes. In *Proceedings of 6th Annual Symposium on Logic in Computer Science (LICS 91)*, pages 376–386. IEEE Computer Society Press, 1991.

[15] Dung T. Huynh and Lu Tian. On deciding readiness and failure equivalences for processes. Technical Report UTDCS-31-90, University of Texas at Dallas, September 1990.

[16] R. Milner. A complete inference system for a class of regular behaviours. *Journal of Computer and System Sciences*, 28:439–466, 1984.

[17] R. Milner. *Communication and Concurrency.* Prentice-Hall International, 1989.

[18] D. Muller and P. Schupp. The theory of ends, pushdown automata, and second-order logic. *Theoretical Computer Science*, 37:51–75, 1985.

Distributed Bisimularity is Decidable for a Class of Infinite State-Space Systems

Søren Christensen[1]
Laboratory for Foundations of Computer Science
University of Edinburgh,
Edinburgh EH9 3JZ,
Scotland

Abstract

In general, the possibility of having parallelism within recursion will lead to systems with infinite state spaces. For instance, a language $\mathcal{P}$ consisting of recursively defined processes over a signature of *prefixing*, *non-deterministic choice* and *merge* will contain infinite state-space systems; the solution to $X = aX|b$ is such an example. Whether *bisimulation* [10] is decidable on $\mathcal{P}$ is an open problem. However, in this paper we show that *distributed bisimulation* [2] is decidable on the language $\mathcal{P}$. The proof of decidability relies on a *tableau decision method* as utilised by Hüttel and Stirling in [6].

1 Introduction

Recently, much effort has been devoted to the investigation of decidability questions for process equivalences such as *bisimularity*, *branching bisimularity* and *failure equivalence* on classes of processes encompassing systems with infinite state spaces. For instance, in [1] and later in [3, 6] it is shown that bisimularity is decidable for *normed* BPA processes, a class of processes corresponding to that of irredundant context-free grammars. In [7] that result has been extended to branching bisimularity. In [8] it is shown that the *readiness* and *failures* equivalences on BPA processes are undecidable and finally, in [5] it has been shown that *all* equivalences below bisimulation in the linear/branching time hierarchy [4] are undecidable for BPA processes.

The class of BPA processes is given by recursively defined expressions over a signature consisting of *sequencing* and *non-deterministic choice*. An important restriction in connection with the decidability results is that of *normedness* [6] which implies that the class of regular processes (such as $X = aX$) is not included.

An interesting question is whether the decidability result w.r.t. bisimulation still holds if the signature is extended to include other process constructions such as *parallelism* or *restriction*. Including parallelism will in general lead to processes with infinite state spaces even if sequencing is replaced by the simpler notion of *prefixing* because of the possibility to have parallelism within recursion. For instance, the solution to $X = aX|b$ will have infinite state space.

[1]The author gratefully acknowledges financial support from Aarhus University (Daimi) and from the Danish Research Academy.

The decidability question w.r.t. bisimulation cannot be answered affirmatively if the signature is extended to include all the process constructions of CCS. In [10] it is remarked that neither weak bisimulation, nor bisimulation is decidable for CCS.

In this paper we show decidability of ***distributed bisimularity*** on a process language consisting of recursively defined processes over a signature including prefixing, non-deterministic choice and merge. Distributed bisimulation as defined by Castellani in [2] is a congruence that takes into account the distribution of processes in space. It is finer than bisimulation, e.g. it does not obey the expansion law.

The only restriction we are going to assume on our recursively defined processes is that of ***guardedness***, a rather natural restriction often imposed on such languages in order to ensure uniqueness of solutions to recursive equations [10]. Thus we do not impose normedness and therefore our language includes regular processes.

The proof of decidability makes use of a ***tableau decision method*** involving goal directed rules as utilised by Hüttel and Stirling in [6]. It turns out that there is an upper bound on the size of every tableau and that is the essential property in providing the decision procedure; an immediate procedure for deciding distributed bisimularity ($\sim_d$) between processes p and q consists in constructing all possible tableaux for p and q followed by a search for a witness for $p \sim_d q$. By showing ***soundness*** and ***completeness*** of the tableau decision method the procedure is guaranteed to be correct.

2 Preliminaries

In the following we describe the language and the behavioural equivalence to be considered. The language will be a subset of CCS [10], i.e. we have recursively defined processes over a signature including ***prefixing***, ***non-deterministic choice*** and ***merge***. However, in order to be able to construct the decision procedure we also include a combinator called ***left merge***. It is an operator very similar to merge but the left operand has precedence over the right. The semantical equivalence over this language will be ***distributed bisimulation*** as defined by Castellani in [2].

2.1 The Language

Let A be a non-empty set of ***atomic actions*** and V a non-empty set of ***variables***. We shall let $a, b, c, \ldots$ range over A and $x, y, z, \ldots$ over V. Define $\mathcal{P}$ to be the least set of ***terms*** satisfying:

(i) $nil \in \mathcal{P}$,

(ii) $x \in \mathcal{P}$ and

(iii) if $p, q \in \mathcal{P}$ then $a.p, p+q, p|q, p\lfloor q \in \mathcal{P}$.

We shall let $p, q, r, \ldots$ range over $\mathcal{P}$. The operator $(a.)$ is called ***prefix***, the operator $(+)$ ***sum*** and finally the operators $(|)$ and $(\lfloor)$ are called ***merge*** and ***left merge*** respectively. We assume some rules in order to improve readability of terms. The term $a.p$ shall be abbreviated to ap and nil shall often be omitted. Furthermore, prefix has precedence over merge and left merge which in turn both have precedence over sum.

The ***processes*** we are interested in is given by systems of term equations. A process is defined by a finite family of recursive equations

$$\Delta = \{x_i \stackrel{\text{def}}{=} p_i \mid 1 \leq i \leq m\},$$

where $x_1, \ldots, x_m$ all are distinct. Each p_i is a term containing at most the variables $x_1, \ldots, x_m$.

A1	$x+(y+z)$	$=$	$(x+y)+z$
A2	$x+y$	$=$	$y+x$
A3	$x+nil$	$=$	x
P1	$x\vert y$	$=$	$x\lfloor y + y\lfloor x$
P2	$(x+y)\lfloor z$	$=$	$x\lfloor z + y\lfloor z$
P3	$(x\lfloor y)\lfloor z$	$=$	$x\lfloor(y\vert z)$
P4	$x\lfloor nil$	$=$	x
P5	$nil\lfloor x$	$=$	nil

Table 1: Axiom System $\mathcal{E}$.

The class of processes thus defined is rather rich. Because of the possibility to define processes containing parallelism within recursion the class comprises infinite state-space processes. The only restriction we impose on systems of term equations is *guardedness*. But that is a natural restriction often imposed on such systems in order to ensure uniqueness of solutions to recursive equations.

Suppose $p \in \mathcal{P}$ is a term containing the variable x. Then an occurrence of x in p is *guarded* if p has a subterm of the form aq where q contains this occurrence of x. Otherwise the occurrence of x is called *unguarded.* A term $p \in \mathcal{P}$ is called *guarded* if all occurrences of all variables in p are guarded. From now on we restrict attention to systems of term equations $\Delta = \{x_i \stackrel{\text{def}}{=} p_i \mid 1 \le i \le m\}$ where all the terms p_i are guarded; we shall call such systems guarded systems of term equations.

The decision procedure to be described in section 3 only works for terms in so-called normal form. A term $p \in \mathcal{P}$ is defined to be in *normal form* if

$$p \equiv \sum_{i=1}^{n} p_i \lfloor q_i$$

where for each i, q_i is in normal form and p_i is either a variable x_i, called a *front variable*, or a prefix $a_i r_i$ with r_i again being in normal form. By convention, the empty sum denotes the term *nil*. Furthermore, a term $p \in \mathcal{P}$ is defined to be in *guarded normal form* if

$$p \equiv \sum_{i=1}^{n} a_i p_i \lfloor q_i$$

where for each i, p_i is in normal form and q_i is in guarded normal form. Note that the concept of normal form is a generalisation of that of guarded normal form, i.e. terms in guarded normal form are also in normal form. All the normal forms are of course defined modulo the axioms A1 and A2 of table 1. That is, we shall ignore the ordering of terms in sums.

The decision procedure only works for guarded systems of term equations. Moreover, each rule in the decision procedure is only applicable to terms in normal form. At certain points in the tableau decision procedure terms might not be in normal form because of unfolding of variables. Thus we must provide a system to *transform* any term into an equivalent one (w.r.t. distributed bisimulation) in normal form. This is the purpose of the axiom system $\mathcal{E}$ of table 1.

The axiom system $\mathcal{E}$ or more correctly the equational theory $\mathcal{E}$ consists of the axioms of table 1 together with laws for equational reasoning, i.e. the laws of reflexivity, symmetry and transitivity in addition to the laws of congruence w.r.t. all the process constructions. If p and q are terms of $\mathcal{P}$ then by $\mathcal{E} \vdash p = q$ or equivalently $p =_{\mathcal{E}} q$ we shall denote the fact that p and q are *provably equal*

$\delta(p)$	$<$	$\delta(ap)$	$\delta(q)$	$<$	$\delta(ap\lfloor q)$
$\delta(p)$	$<$	$\delta(p+q)$	$\delta(p\lfloor r)$	$<$	$\delta((p+q)\lfloor r)$
$\delta(q)$	$<$	$\delta(p+q)$	$\delta(q\lfloor r)$	$<$	$\delta((p+q)\lfloor r)$
$\delta(p\lfloor q)$	$<$	$\delta(p\vert q)$	$\delta(p\lfloor(q\lfloor r))$	$<$	$\delta((p\vert q)\lfloor r)$
$\delta(q\lfloor p)$	$<$	$\delta(p\vert q)$	$\delta(q\lfloor(p\lfloor r))$	$<$	$\delta((p\vert q)\lfloor r)$
$\delta(p)$	$<$	$\delta(ap\lfloor q)$	$\delta(p\lfloor(q\lfloor r))$	$<$	$\delta((p\lfloor q)\lfloor r)$

Table 2: Conditions on the complexity measure.

using the equational theory $\mathcal{E}$.

We end this section by showing that any term p of $\mathcal{P}$ can be proved equal to a term $q \in \mathcal{P}$ in normal form. In order for our inductive arguments to proceed correctly we need a complexity measure on the structure of terms.

Definition 2.1.1 *Let $\alpha : \mathcal{P} \to \omega$ be defined as follows:*

(i) $\alpha(nil) = \alpha(x) = 1,$

(ii) $\alpha(ap) = 1 + \alpha(p)$ *and*

(iii) $\alpha(p+q) = \alpha(p|q) = \alpha(p\lfloor q) = \alpha(p) + \alpha(q).$

Furthermore, let $\beta : \mathcal{P} \to \omega$ be defined as follows:

$$\beta(p) = \begin{cases} \alpha(q) & \text{if } p \equiv q\lfloor r \\ \alpha(p) & \text{otherwise} \end{cases}$$

Finally, let the complexity measure $\delta : \mathcal{P} \to \omega \times \omega$ be

$$\delta(p) = (\alpha(p), \beta(p)),$$

and let $<$ be the lexicographical ordering on $\omega \times \omega$. □

As will become clear in the proof of the next lemma the properties which our complexity measure δ must satisfy are as given in table 2. It is readily checked that in each case the relationship as indicated is true. For instance, $\delta(p\lfloor q) < \delta(p|q)$ because although $\alpha(p\lfloor q) = \alpha(p|q)$ we have $\beta(p\lfloor q) = \alpha(p) < \alpha(p|q) = \beta(p|q)$.

Not only shall the complexity measure δ assure that the inductive arguments proceed properly; δ will also make it evident that normal forms can be constructed within finite time. That is important in order for the decision procedure to work.

Lemma 2.1.2 *Suppose $p \in \mathcal{P}$. Then a term $q \in \mathcal{P}$ in normal form can be constructed such that $p =_{\mathcal{E}} q$.*

PROOF: **The proof proceeds by induction on $\delta(p)$ and by analysing the cases for p.**

(i) **Suppose $p \equiv nil$. Then let q be nil. By equational reasoning $p =_{\mathcal{E}} q$ and by convention q is in normal form.**

(ii) **Suppose $p \equiv x$. Then define q to be $x\lfloor nil$. By axiom P4 we have $p =_{\mathcal{E}} q$. Moreover, q is in normal form.**

(iii) Suppose $p \equiv ap'$. By induction, let q' be in normal form such that $p' =_{\mathcal{E}} q'$. Then define q to be $aq' \lfloor nil$. By axiom P4 and equational reasoning we have $p =_{\mathcal{E}} q$. Moreover, q is in normal form.

(iv) Suppose $p \equiv p_1 + p_2$. By induction, let q_i be in normal form such that $p_i =_{\mathcal{E}} q_i$ for $i = 1, 2$. Then define q to be $q_1 + q_2$. By equational reasoning we have $p =_{\mathcal{E}} q$. Moreover, q is in normal form (modulo axiom A3).

(v) Suppose $p \equiv p_1 | p_2$. By axiom P1 we have $p =_{\mathcal{E}} p_1 \lfloor p_2 + p_2 \lfloor p_1$. Hence, by induction let q_1 and q_2 be terms in normal form such that $p_1 \lfloor p_2 =_{\mathcal{E}} q_1$ and $p_2 \lfloor p_1 =_{\mathcal{E}} q_2$. Then define q to be $q_1 + q_2$. By equational reasoning we have $p =_{\mathcal{E}} q$. Moreover, q is in normal form (modulo axiom A3).

(vi) Suppose $p \equiv p_1 \lfloor p_2$. The proof proceeds by analysing the cases for p_1.

(a) Suppose $p_1 \equiv nil$. Then define q to be nil. By axiom P5, $p =_{\mathcal{E}} q$ and q is in normal form.

(b) Suppose $p_1 \equiv x$. By induction let q_2 be a term in normal form such that $p_2 =_{\mathcal{E}} q_2$. Then define q to be $x \lfloor q_2$. By definition q is in normal form. Moreover, by equational reasoning we have $p =_{\mathcal{E}} q$.

(c) Suppose $p_1 \equiv ar$. Then by induction let q_1 and q_2 be terms in normal form with $r =_{\mathcal{E}} q_1$ and $p_2 =_{\mathcal{E}} q_2$. Define q to be $aq_1 \lfloor q_2$. By equational reasoning $p =_{\mathcal{E}} q$ and q is in normal form.

(d) Suppose $p_1 \equiv r_1 + r_2$. By axiom P2, $p =_{\mathcal{E}} r_1 \lfloor p_2 + r_2 \lfloor p_2$. Then by induction let q_1 and q_2 be terms in normal form with $r_1 \lfloor p_2 =_{\mathcal{E}} q_1$ and $r_2 \lfloor p_2 =_{\mathcal{E}} q_2$ and define q to be $q_1 + q_2$. By equational reasoning $p =_{\mathcal{E}} q$ and q is in normal form (modulo axiom A3).

(e) Suppose $p_1 \equiv r_1 | r_2$. Then by axiom P1, P2 and equational reasoning we have $p =_{\mathcal{E}} r_1 \lfloor (r_2 | p_2) + r_2 \lfloor (r_1 | p_2)$. By induction, let q_1 and q_2 be terms in normal form such that $r_1 \lfloor (r_2 | p_2) =_{\mathcal{E}} q_1$ and $r_2 \lfloor (r_1 | p_2) =_{\mathcal{E}} q_2$. Define q to be $q_1 + q_2$ and we have $p =_{\mathcal{E}} q$ by equational reasoning. Moreover, q is in normal form (modulo axiom A3).

(f) Suppose $p_1 \equiv r_1 \lfloor r_2$. Then by axiom P3 we have $p =_{\mathcal{E}} r_1 \lfloor (r_2 | p_2)$. By induction let q be a term in normal form such that $r_1 \lfloor (r_2 | p_2) =_{\mathcal{E}} q$ and hence also $p =_{\mathcal{E}} q$ by equational reasoning.

□

Corollary 2.1.3 *If $p \in \mathcal{P}$ is guarded then a term $q \in \mathcal{P}$ in guarded normal form can be constructed such that $p =_{\mathcal{E}} q$.*

PROOF: Follows readily from lemma 2.1.2 □

Of course the transformation of terms into normal form must preserve our semantical equivalence, viz. distributed bisimulation ($\sim_d$). As shall become evident in the following the axiom system $\mathcal{E}$ is sound and therefore for processes p and q, if $p =_{\mathcal{E}} q$ then $p \sim_d q$.

2.2 Distributed Bisimulation

The behavioural semantics of the process language described in the previous section shall be defined via so-called ***distributed transition systems*** and the equivalence between processes will be ***distributed bisimulation*** as defined by Castellani in [2]. In this section we shall briefly describe the notion of distributed bisimulation and refer to [2] for a thorough explanation.

Distributed bisimulation is defined on a class of transition systems in which states are augmented with information about the distribution of processes in space. That is, in observing an event the resulting state will contain process components representing distribution of the system.

A transition has the form $p \xrightarrow{a} \langle p', p'' \rangle$ where a is the atomic action observed. The state $\langle p', p'' \rangle$ is called a *compound residual* and it contains information about the distribution of processes on observing the action a. The process p' is called the *local residual* and p'' the *concurrent residual*. Intuitively, p' is the local process at which the action a was performed whereas p'' is the part of the process p which behaves independently of the local residual.

The operational semantics of guarded systems of term equations is given via a distributed transition system with the transition relations defined as follows.

Definition 2.2.1 *Let $\Delta = \{x_i \stackrel{def}{=} p_i \mid 1 \leq i \leq m\}$ be a guarded system of term equations. Then Δ defines a distributed transition system with the transition relations given as the least relations satisfying the following rules.*

$$\frac{}{ap \xrightarrow{a} \langle p, nil \rangle}$$

$$\frac{p \xrightarrow{a} \langle p', p'' \rangle}{p+q \xrightarrow{a} \langle p', p'' \rangle} \qquad \frac{q \xrightarrow{a} \langle q', q'' \rangle}{p+q \xrightarrow{a} \langle q', q'' \rangle}$$

$$\frac{p \xrightarrow{a} \langle p', p'' \rangle}{p|q \xrightarrow{a} \langle p', p''|q \rangle} \qquad \frac{q \xrightarrow{a} \langle q', q'' \rangle}{p|q \xrightarrow{a} \langle q', p|q'' \rangle}$$

$$\frac{p \xrightarrow{a} \langle p', p'' \rangle}{p \lfloor q \xrightarrow{a} \langle p', p''|q \rangle}$$

$$\frac{p \xrightarrow{a} \langle p', p'' \rangle}{x \xrightarrow{a} \langle p', p'' \rangle} \quad where\ x \stackrel{def}{=} p \in \Delta$$

□

Although we have defined the behaviour of terms of Δ via the distributed transition system above it is not at all clear how a term computes; it seems as if only the first step of terms can be observed. In order to be able to talk about computations of terms we extend the transition relation by adding the following rules for pairs $\langle p, q \rangle$ of terms.

$$\frac{p \xrightarrow{a} \langle p', p'' \rangle}{\langle p, q \rangle \xrightarrow{a} \langle p', p''|q \rangle} \qquad \frac{q \xrightarrow{a} \langle q', q'' \rangle}{\langle p, q \rangle \xrightarrow{a} \langle q', p|q'' \rangle}$$

According to these rules, after each transition $p \xrightarrow{a} \langle p', p'' \rangle$ the execution resumes with the composition of the two residuals; the pair $\langle p', p'' \rangle$ has exactly the same behaviour as the term $p'|p''$.

Example 2.2.2 Let Δ consists of $x \stackrel{def}{=} ax\lfloor b+c$. Then some of the computation steps of the process x are as indicated in figure 1. In the figure nil is represented by 0 and b^n is an abbreviation for n copies of b in parallel. Also we have omitted nil components of compound residuals whenever possible. □

The definition of distributed bisimulation is closely related to that for bisimulation as for instance utilised by Milner in [10]. However, the relation will now be based on the information contained in the individual residuals. Thus the distributed bisimulation shall relate local and concurrent residuals separately.

Definition 2.2.3 *A relation R between processes is a* distributed bisimulation *provided for all $(p, q) \in R$ the following is satisfied:*

(i) $p \xrightarrow{a} \langle p', p'' \rangle$ implies $q \xrightarrow{a} \langle q', q'' \rangle$ such that $(p', q') \in R$ & $(p'', q'') \in R$, and

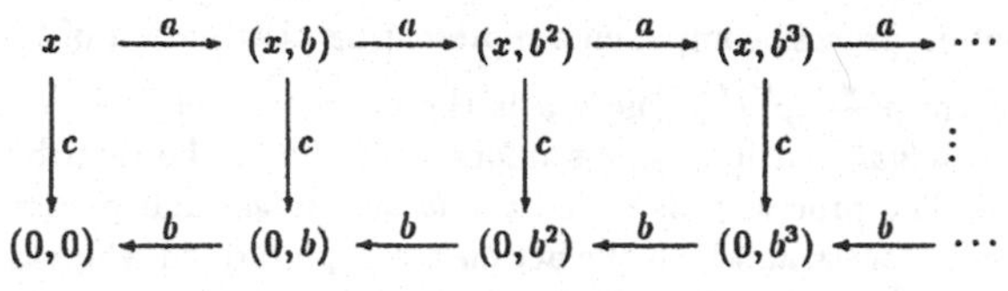

Figure 1: Part of transition graph for $x \stackrel{\text{def}}{=} ax\lfloor b + c$.

(ii) $q \xrightarrow{a} \langle q', q'' \rangle$ *implies* $p \xrightarrow{a} \langle p', p'' \rangle$ *such that* $(p', q') \in R$ *&* $(p'', q'') \in R$.

□

Definition 2.2.4 *Let* $p, q \in \mathcal{P}$. *Then* $p \sim_d q$ *iff there exists a distributed bisimulation* R *such that* $(p, q) \in R$. *If* $p \sim_d q$ *we call* p *and* q *distributed bisimulation equivalent.* □

Example 2.2.5 Let Δ consists of $x \stackrel{\text{def}}{=} a(x\lfloor x)$, $y \stackrel{\text{def}}{=} az$ and $z \stackrel{\text{def}}{=} az\lfloor az$. Then $x \sim_d y$ because $\{(x, y), (x\lfloor x, z), (x, az), (nil, nil)\}$ is easily checked to be a distributed bisimulation. □

In [2] it is shown that $\sim_d$ is an equivalence relation and also a congruence w.r.t. all the process constructors prefix, sum, left merge and merge. Moreover, it follows easily that all the axioms of table 1 are valid w.r.t. distributed bisimulation. Thus the equational theory $\mathcal{E}$ is sound and therefore for processes p and q, if $p =_{\mathcal{E}} q$ then also $p \sim_d q$. Of course the theory $\mathcal{E}$ does not form a *complete* axiomatisation. First of all, the idempotency of sum needs to be added but, more importantly, some notion of *fixed point induction* must be included in order to prove equations involving recursively defined processes. In the conclusion we will briefly discuss the possibility of obtaining a complete axiomatisation and indeed suggest an appropriate induction principle.

3 The Tableau Decision Procedure

In this section we fix a family of guarded term equations $\{x_i \stackrel{\text{def}}{=} p_i \mid 1 \le i \le m\}$ where by lemma 2.1.2 we can assume wlog. that each term p_i is in guarded normal form. We shall let x, y be arbitrary variables over $\{x_1, \ldots, x_m\}$.

First we describe a tableau method for testing $x \sim_d y$. We show that the method is sound and complete and that each tableau is finite. Finally, we show that the tableau method provides a decision procedure by proving that there are only finitely many tableaux for testing $x \sim_d y$.

3.1 The Tableau Method

The decision procedure we now present to test whether $x \sim_d y$ is a *tableau system*, i.e. a goal directed proof system. The rules of the tableau system are built around equations $p = q$ where p and q are terms in normal form. Each rule has the form

$$\frac{p = q}{p_1 = q_1 \quad \ldots \quad p_n = q_n}$$

possible with side conditions. The premise of the rule represents the goal to be achieved (that $p \sim_d q$) whereas the consequents are the subgoals to be proved.

A *tableau* for $x = y$ is a maximal finite proof tree whose root is labelled $x = y$ and where the labelling of immediate successors of a node are determined according to the rules of the tableau system presented in table 3. At a given node the tableau rule to apply is found by *matching* the

REC1	$$\frac{x\lfloor p' + p'' = q}{p\lfloor p' + p'' = q}$$	where $x \stackrel{\mathrm{def}}{=} p \in \Delta$
REC2	$$\frac{p = y\lfloor q' + q''}{p = q\lfloor q' + q''}$$	where $y \stackrel{\mathrm{def}}{=} q \in \Delta$
SUM	$$\frac{\sum_{i=1}^{n} a_i p_i \lfloor p'_i = \sum_{j=1}^{m} b_j q_j \lfloor q'_j}{\{a_i p_i \lfloor p'_i = b_{f(i)} q_{f(i)} \lfloor q'_{f(i)}\}_{i=1}^{n} \quad \{b_j q_j \lfloor q'_j = a_{g(j)} p_{g(j)} \lfloor p'_{g(j)}\}_{j=1}^{m}}$$ where $f : \{1, \ldots, n\} \rightarrow \{1, \ldots, m\}$, $g : \{1, \ldots, m\} \rightarrow \{1, \ldots, n\}$	
PREFIX	$$\frac{ap\lfloor p' = aq\lfloor q'}{p = q \qquad p' = q'}$$	

Table 3: Rules of the tableau system. All terms are assumed to be in normal form.

label of the node against the premise of the rule considered. The procedure of matching shall be performed modulo axiom A1, A2 and A3 of table 1. For instance, a node with label $p' + x\lfloor p'' = q$ will match premise of rule (REC1).

Normally, the rules shall be applied according to the following pattern: possible front variables are unfolded by a succession of applications of the rules (REC1) and (REC2) possibly followed by a use of (SUM) followed by a use of (PREFIX) whereupon the whole pattern is repeated again.

The rules are only applicable to labels with terms in normal form. Hence, strictly speaking, the root of a tableau should be represented as $x\lfloor nil = y\lfloor nil$. More importantly, terms of consequents of the rule (REC1) or (REC2) are not necessarily in normal form. Thus in building tableaux it is assumed that terms are transformed into expressions in normal form whenever it is required. By lemma 2.1.2 that is possible using the equational theory $\mathcal{E}$. For instance, a node labelled $x|y = ap\lfloor q$ with p and q already being in normal form shall be transformed into $x\lfloor y + y\lfloor x = ap\lfloor q$ before the rule (REC1) can be used.

Instead of transforming terms into normal form whenever required the tableau system could include rules to perform the transformation. Those rules would to a certain extent mimic the axioms of table 1. However, we have chosen to keep the number of rules of the tableau system as small as possible because of the analysis on tableaux to come.

The consequent $p = q$ in the rule (PREFIX) will be called the left consequent of the prefix rule in the following. Similarly, $p' = q'$ will be called the right consequent of the prefix rule. Later we shall see that the property of being able to *split* a goal of the form $ap\lfloor p' = aq\lfloor q'$ into two goals, viz. $p = q$ and $p' = q'$, is essential in providing finite tableaux.

In building tableaux the rules are only applied to nodes that are not *terminal*. A terminal node can either be *successful* or *unsuccessful*. A successful terminal node is one labelled either $p = p$ or $p = q$

$$x = y$$

Rec1,Rec2 — $\overline{}$

$$a(x\lfloor x) = az$$

Prefix — $\overline{}$

$$x\lfloor x = z$$

Rec1,Rec2 — $\overline{}$

$$a(x\lfloor x)\lfloor x = az\lfloor az$$

Prefix — $\overline{}$

$$x\lfloor x = z \qquad\qquad x = az$$

$$\overline{a(x\lfloor x) = az} \qquad \text{Rec1}$$

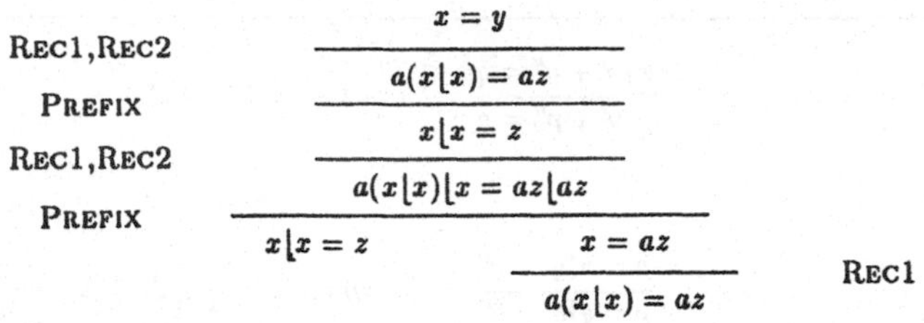

Figure 2: Successful tableau for $x = y$ where $x \stackrel{\text{def}}{=} a(x\lfloor x)$, $y \stackrel{\text{def}}{=} az$ and $z \stackrel{\text{def}}{=} az\lfloor az$.

where there is another node above in the tableau (and an application of rule (Prefix) in between) also labelled $p = q$. An unsuccessful terminal node is one labelled either $ap\lfloor p' = bq\lfloor q'$ with $a \neq b$, $ap\lfloor p' = nil$ or $nil = bq\lfloor q'$. A tableau is successful iff all terminal nodes are successful. Otherwise it is unsuccessful.

Lemma 3.1.1 *Every tableau for $x = y$ is finite.*

PROOF: Let $T(x = y)$ be a tableau with root labelled $x = y$. It can only be infinite if it has an infinite path as every node has finite branching degree. Hence suppose π is an infinite path starting at the root. The path can only be infinite if it contains unfolding of variables infinitely often, i.e. π must contain instances of (Rec1) or (Rec2) infinitely often.

Note that, π being infinite it cannot contain terminal nodes. Now, if π passes through the left consequent of the rule (Prefix) infinitely often then π will also contain a terminal node infinitely often. This follows from the fact that the expressions ap and aq of labels of nodes matching the premise of rule (Prefix) along the path will be *syntactical subexpressions* of the guarded term equations Δ. Hence there are only finitely many possibilities for terms of the form ap and aq and therefore only finitely many possibilities for instances of the left consequent of the rule (Prefix).

Thus from a point on in π the path will never go via the left consequent of the prefix rule again. But then π cannot contain infinitely many instances of the rule (Rec1) or (Rec2) because each time one of the rules is applied to a node one of the *finitely* many front variables contained in that node will be removed because of unfolding. And the only way to provide "fresh" unguarded variables is via the left consequent of rule (Prefix) as every equation of Δ is guarded.

Hence π must be finite and therefore also $T(x = y)$ is finite. □

Example 3.1.2 In figure 2 we give a successful tableau for $x = y$ where x and y are defined as in example 2.2.5. Note that not all terms are presented in normal form and therefore that only the essential part of the tableau is given in figure 2. For instance, at the first occurrence of rule (Prefix) there should be a branch with label $nil = nil$. □

Lemma 3.1.3 *(Completeness) If $x \sim_d y$ then there exists a successful tableau with $x = y$ as root.*

PROOF: Suppose $x \sim_d y$. If we can build a tableau $T(x = y)$ for $x = y$ with the property that the label of any node of $T(x = y)$ contains terms being distributed bisimular then by lemma 3.1.1 that construction must terminate and each terminal will be successful. Thus the tableau shall be successful.

Now, $T(x = y)$ can be constructed once we verify that each rule of the tableau system is *forward sound* in the sense that if the antecedent relates distributed bisimular processes then it is possible to find a set of consequents relating distributed bisimular processes.

It is readily seen that all rules of the tableau system are forward sound in the above sense. We only consider rule (PREFIX). Assume $ap\lfloor p' \sim_d aq\lfloor q'$. We must show $p \sim_d q$ and $p' \sim_d q'$. By the operational semantics $ap\lfloor p' \xrightarrow{a} \langle p, p'|nil\rangle$ which must be matched by $aq\lfloor q'$. But the only possibility for $aq\lfloor q'$ to match is by $aq\lfloor q' \xrightarrow{a} \langle q, q'|nil\rangle$ and hence $p \sim_d q$ and $p'|nil \sim_d q'|nil$. And the latter readily implies $p' \sim_d q'$. □

The proof of soundness makes use of an alternative characterisation of distributed bisimulation, viz. as a sequence of approximations.

Definition 3.1.4 *The sequence of distributed bisimulation approximations* $\{\sim_d^n\}_{n=0}^{\omega}$ *is defined inductively as follows:*

(i) $p \sim_d^0 q$ *for all processes* p *and* q,

(ii) $p \sim_d^{n+1} q$ *iff*

(a) $p \xrightarrow{a} \langle p', p''\rangle$ *implies* $q \xrightarrow{a} \langle q', q''\rangle$ *for some* q' *and* q'' *such that* $p' \sim_d^n q'$ *and* $p'' \sim_d^n q''$,

(b) $q \xrightarrow{a} \langle q', q''\rangle$ *implies* $p \xrightarrow{a} \langle p', p''\rangle$ *for some* p' *and* p'' *such that* $p' \sim_d^n q'$ *and* $p'' \sim_d^n q''$.

□

Lemma 3.1.5 *For image-finite, i.e. finitely branching, transition graphs we have*

$$\sim_d = \bigcap_{n=0}^{\omega} \sim_d^n .$$

PROOF: Follows from standard arguments. See for instance [10]. □

Lemma 3.1.6 *(Soundness) If there is a successful tableau for* $x = y$ *then* $x \sim_d y$.

PROOF: Suppose $T(x = y)$ is a successful tableau for $x = y$ but $x \not\sim_d y$. As the transition graphs for x and y are image-finite there is a least $n \in \omega$ such that $x \not\sim_d^n y$ and $x \sim_d^m y$ for all $m < n$. Notice that all the tableau rules of table 3 are backwards sound w.r.t. $\sim_d^n$. Significantly, in the case of rule (PREFIX) we can strengthen this property as $p \sim_d^n q$ and $p' \sim_d^n q'$ implies $ap\lfloor p' \sim_d^{n+1} aq\lfloor q'$. Hence there is a path π from the root $x = y$ to some leaf in the tableau such that every node $p_i = q_i$ along π satisfy $p_i \not\sim_d^{n_i} q_i$ for some n_i. For each i chose n_i such that it is the least with this property. The leaf of π cannot be of the form $p = p$ as $\sim_d$ is reflexive. Thus the leaf must be of the form $p = q$ where there is a node above it also labelled $p = q$. Moreover, between those two nodes there must be an instance of the rule (PREFIX). But then we arrive at a contradiction because the leaf cannot satisfy $p \not\sim_d^{n_i} q$ as the node above it labelled $p = q$ must have the property $p \sim_d^{n_i} q$. □

3.2 The Decision Procedure

If we can prove that there are only finitely many different tableaux for $x = y$ then we immediately have a decision procedure for $x \sim_d y$. And in showing that there are only finitely many different tableaux for $x = y$ it is sufficient to provide an uniform upper bound on the size of any tableau. The uniform bound will of course depend on the set of term equations Δ. In order to prepare the ground we define a few constants parameterised by Δ.

Definition 3.2.1 *Let $p \equiv \sum_{i=1}^{n} p_i \lfloor q_i$ be a term in normal form. Then the* depth, *denoted $d(p)$, of p is defined as follows:*

$$d(p) = \begin{cases} 1 + max\{d(q_i) \mid 1 \le i \le n\} & \text{if there exists } q_i \not\equiv nil \\ 0 & \text{otherwise} \end{cases}$$

Also, define the sum degree, *denoted $s(p)$, of p to be:*

$$s(p) = max\{n, s(p_i), s(q_i) \mid 1 \le i \le n\}.$$

By convention, we have $s(nil) = 0$. □

Definition 3.2.2 *Let $\Delta = \{x_i \stackrel{def}{=} p_i \mid 1 \le i \le m\}$ be a finite set of guarded term equations in normal form. Then define*

(i) n_Δ to be the number of different expressions ap being subexpressions of the terms p_i of Δ,

(ii) d_Δ to be the maximal depth of any guarded normal form p being a subexpression of the terms p_i of Δ,

(iii) ud_Δ to be the maximal depth of any normal form p such that ap, for some $a \in A$, is a subexpression of terms p_i of Δ and

(iv) s_Δ to be the maximal sum degree of any term p_i of Δ, i.e.

$$s_\Delta = max\{s(p_i) \mid 1 \le i \le m\}$$

□

Lemma 3.2.3 *The sum degree of any term being the left hand side or right hand side of the label of any node in any tableau for $x = y$ is bound by an uniform constant.*

PROOF: Let $T(x = y)$ be a tableau with $x = y$ as root. We shall chase all possible nodes of the tableau and show that each label contains terms with sum degree being uniformly bounded.

First, consider any node along the path starting at the root $x = y$ and never passing through the left consequent of the rule (PREFIX). Labels of nodes along that path will always contain terms being *subexpressions* of the terms p_i of Δ. Hence the sum degree of any term along that path cannot exceed s_Δ.

Now, consider nodes along a path emanating from the left consequent of the rule (PREFIX) and which never passes through the left consequent of the prefix rule again. Let π be such a path. It will start at a node labelled $p = q$ where ap and aq are subexpressions of terms p_i of Δ. Hence p and q can at most have sum degree s_Δ.

The rules applied along the path π will consist of unfoldings (via (REC1) and (REC2)), (SUM) and (PREFIX). In going from the premise to the consequents of the two rules (SUM) and (PREFIX) the sum degree of the terms constituting the labels will not increase. Hence only when unfolding variables will terms increase in sum degree.

If a front variable x of a term is unfolded it gets replaced by p where $x \stackrel{def}{=} p \in \Delta$. As p is a term of Δ its sum degree cannot exceed s_Δ. Hence it is not hard to verify that the sum degree of a term by unfolding one of its front variables (and then transforming into normal form) cannot increase by more that s_Δ. Thus we are going to find an upper bound on the number of possible unfoldings along π because then we immediately have an upper bound of the sum degree of terms.

Any term along π can at most contain s_Δ front variables. Hence at most s_Δ successions of unfoldings of variables can be performed on a term along π. Each of those unfoldings increases the sum degree by at most s_Δ. As the root $p = q$ of π satisfy that the depth of p and q cannot exceed ud_Δ there can at most be ud_Δ "groups" of unfoldings of say the left hand side of labels along π each of which with at most s_Δ applications of (REC1) in succession. Hence after at most $ud_\Delta s_\Delta$ unfoldings of variables of the left hand side (or the right hand side) of terms along π the resulting term will be guarded and further unfolding is impossible. But then the sum degree of any term along π is uniformly bounded by $ud_\Delta s_\Delta^2$.

Now, we have chased all possible nodes of the tableau $T(x = y)$ and therefore the uniform bound on the sum degree of any term is $ud_\Delta s_\Delta^2$. □

Corollary 3.2.4 *There exists an uniform upper bound on the branching degree of any node of any tableau for $x = y$.*

PROOF: From lemma 3.2.3 it follows that the branching degree of any node with label matching the rule (SUM) cannot exceed $2ud_\Delta s_\Delta^2$ □

Lemma 3.2.5 *Let $T(x = y)$ be a tableau with $x = y$ as root. A path starting at the root and never passing through the left consequent of the prefix rule will be bound in length by $2 + 2d_\Delta$.*

PROOF: Let π be a path starting at the root of $T(x = y)$. Moreover, assume that π never goes via the left consequent of the rule (PREFIX). Apart from the root and its immediate successor all nodes along the path will have labels matching either the rule (SUM) or the rule (PREFIX). In the worst case, i.e. in case π is longest, the labels will alternate between matching the two rules. Each time the path goes from the premise to the right consequent of the rule (PREFIX) the terms constituting the labels will decrease in depth by one. Hence along π there can at most be d_Δ instances of the rule (PREFIX). Therefore, π is bound by $2 + 2d_\Delta$ where the 2 originates from unfolding of the variables contained in the root. □

Lemma 3.2.6 *There exists an uniform upper bound on the length of any path in any tableau for $x = y$.*

PROOF: Let $T(x = y)$ be an arbitrary tableau with root $x = y$. Suppose π is a path starting at an instance of the left consequent of the rule (PREFIX). Moreover, assume that π never goes via the left consequent of the prefix rule again. We want to provide an upper bound on the length of π.

Notice that, the path π emanates from a node labelled $p = q$ where ap and aq, for some $a \in A$, are subexpressions of terms p_i of Δ. Hence the depths of p and q cannot exceed ud_Δ. In the analysis performed in lemma 3.2.3 we saw that at most ud_Δ "groups" of at most s_Δ unfoldings can be carried out on p (or q). Hence within $2ud_\Delta s_\Delta$ unfoldings (via (REC1) and (REC2)) along π the resulting label will contain *guarded* terms and therefore no further unfolding is possible. Together with those unfoldings we have at most ud_Δ instances of rule (SUM) and (PREFIX) along π. Thus within length $(2s_\Delta + 2)ud_\Delta$ all nodes along π will be labelled by guarded terms (in normal form).

Now, each time we have unfolded as far as possible, i.e. within $2s_\Delta$ occurrences of (REC1) and (REC2) in succession the resulting label (possibly after transformation into normal form) will contain terms having increased in depth by at most d_Δ. Let the first node along π only containing guarded terms have label $r = s$. Then $d(r) \leq ud_\Delta(d_\Delta - 1)$ and similarly with s. The (-1) part of the inequality comes from the fact that we also have ud_Δ instances of the rule (PREFIX) along π from the root until the node labelled $r = s$ each of which decreases the depth of terms by one. From the node labelled $r = s$ on, the rules along π will in case the length of π is longest alternate between instances of (SUM) and (PREFIX). From the proof of lemma 3.2.5 it follows that the path will be bound by $2ud_\Delta(d_\Delta - 1)$. Hence the path π will be uniformly bound by $2ud_\Delta(d_\Delta + s_\Delta)$.

Any path in the tableau $T(x = y)$ starting at the root $x = y$ can at most pass through the left consequent of the rule (PREFIX) n_Δ^2 times because of the possible nodes labelled with $p = q$ where ap and aq, for some $a \in A$, are subexpressions of terms p_i of Δ. Hence first time a path goes via the left consequent of the rule (PREFIX) it will be bound from that point on by $2n_\Delta^2 ud_\Delta(d_\Delta + s_\Delta)$. Starting from the root $x = y$ in $T(x = y)$, the only path that does never pass through the left consequent of instances of the rule (PREFIX) will be bound by $2 + 2d_\Delta$ (follows from lemma 3.2.5). All other paths will go via the left consequent of the rule (PREFIX) within $2+2d_\Delta$ steps. Hence any path in $T(x = y)$ starting at the root will be uniformly bound by $2 + 2d_\Delta + 2n_\Delta^2 ud_\Delta(d_\Delta + s_\Delta)$ □

From corollary 3.2.4 and lemma 3.2.6 we now conclude that there are only finitely many tableaux for $x = y$. Hence a naive decision procedure for $x \sim_d y$ could consists of first constructing all possible tableaux for $x = y$ and then search for a successful one. The answer to the question $x \sim_d y$ is then "yes" if there exists a successful tableau. Otherwise the answer is "no". By completeness and soundness of the tableau system (lemma 3.1.3 and lemma 3.1.6) we know that the decision procedure is always giving the correct answer.

Observe that in constructing tableaux we have to transform terms into normal form as outlined by axiom system $\mathcal{E}$. From lemma 2.1.2 we know that the transformation is possible within finite time. Hence is is possible to construct the tableaux within finite time. It also follows that the decision procedure works for *any* guarded system of term equations; all that is required is a transformation into guarded terms in normal form before the tableau rules can be applied.

4 Further Work

In this paper we have seen that distributed bisimulation is decidable on a language of recursively defined processes over a signature of prefixing, non-deterministic choice and merge. There are a number of interesting questions in connection with the presented work, all of which currently constitute active line of research.

First of all, is distributed bisimulation still decidable if the language is extended by replacing merge with *parallel composition* [10], i.e. allowing processes to communicate and thus introducing silent actions. Having introduced silent actions another question would be whether *weak* distributed bisimulation [2] is decidable on the language.

Yet another question would be whether we have decidability if the language is extended with *restriction* [10]. However, the definition of distributed bisimulation is not tailored towards restriction so before asking decidability we need to work out the appropriate definition of $\sim_d$ in case of restriction. On solving the problems a good starting point would perhaps be [9] where distributed bisimulation on a language including restriction has been considered.

As mentioned in section 2.2 the equational theory $\mathcal{E}$ does not form a complete axiomatisation for distributed bisimulation on the language of this paper. Significantly, some notion of fixed point induction must be included in order to prove equations involving recursively defined processes. In [6] a complete axiomatisation for bisimulation on normed BPA is obtained via the tableau decision procedure. But soundness of the axiomatisation relies on processes being in so-called Greibach Normal Form. At present we do not see how a sound and complete axiomatisation for distributed bisimulation using the technique of [6] can be obtained. However we believe that it is possible to obtain a sound and complete axiomatisation for distributed bisimulation on recursively defined terms by adding Milner's fixpoint axioms [11] to Castellani's axioms for distributed bisimulation [2]. Details will follow in a forthcoming paper.

Finally, it would be interesting to know whether bisimulation is decidable on the language studied in this paper. Notice that the only rule of the tableau system (table 3) not being forward sound w.r.t. bisimulation is (PREFIX). In case of bisimulation it is not possible to split a goal of the form

$ap\lfloor p' = aq\lfloor q'$ into the two goals $p = q$ and $p' = q'$. Therefore w.r.t. bisimulation the tableaux will not be finite.

Acknowledgement

I would like to thank Colin Stirling for many inspiring discussions on the issues addressed in this paper.

References

[1] J.C.M. Baeten, J.A. Bergstra and J.W. Klop. *Decidability of Bisimulation Equivalence for Processes Generating Context-Free Languages.* LNCS 259, pp 93–114, 1987.

[2] I. Castellani. *Bisimulation for Concurrency.* Ph.D. Thesis, Edinburgh University, CST-51-88, 1988.

[3] D. Caucal. *Graphes Canoniques de Graphes Algebriques.* Report de Recherche 972, INRIA, Juillet 1988.

[4] R.J. van Glabbeek. The Linear Time-Branching Time Spectrum. CONCUR90, LNCS 458, pp 278–297, 1990.

[5] J.F. Groote and H. Hüttel. *Undecidable Equivalences for Basic Process Algebra.* University of Edinburgh, LFCS Report Series, ECS-LFCS-91-169, 1991.

[6] H. Hüttel and C. Stirling. *Actions Speak Louder that Words: Proving Bisimularity for Context-Free Processes.* In Proceedings of 6th Annual Symposium on Logic in Computer Science (LICS 91), pp 376–386, IEEE Computer Society Press, 1991.

[7] H. Hüttel. *Silence is Golden: Branching Bisimularity is Decidable for Context-Free Processes.* In Proceedings of the 3rd International Workshop on Computer-Aided Verification (CAV91), 1991.

[8] D.T. Huynh and L. Tian. *On deciding Readiness and Failure equivalences for Processes.* Technical Report UTDCS-31-90, University of Texas at Dallas, 1990.

[9] A. Kiehn. *Distributed Bisimulations for Finite CCS*, University of Sussex, Dept. of Computer Science, Report no. 7/89, 1989.

[10] R. Milner. *Communication and Concurrency.* Prentice Hall, 1989.

[11] R. Milner. *A Complete Axiomalisation for Observational Congruence of Finite-State Behaviours.* Journal of Information and Computation, Vol. 81, No. 2, May 1989.

How Vital is Liveness? Verifying Timing Properties of Reactive and Hybrid Systems*

Amir Pnueli
Weizmann Institute of Science†

Extended Abstract

Abstract. This extended abstract discusses the importance of the notion of liveness properties and their verification. The main observation is that they provide a most useful abstraction on the qualitative (non-quantitative) level of modeling. As we construct more refined models that take real-time into account, the importance of liveness and fairness decrease, and many important properties move to the safety class.

In the talk to be presented, we propose a framework for the formal specification and verification of *timed* and *hybrid* systems. For timed systems we propose a specification language that refers to time only through *age* functions which measure the length of the most recent time interval in which a given formula has been continuously true.

We then consider hybrid systems, which are systems consisting of a non-trivial mixture of discrete and continuous components, such as a digital controller that controls a continuous environment. The proposed framework extends the temporal logic approach which has proven useful for the formal analysis of discrete systems such as reactive programs. The new framework consists of a semantic model for hybrid time, the notion of *phase transition systems*, which extends the formalism of discrete transition systems, an extended version of Statecharts for the specification of hybrid behaviors, and an extended version of temporal logic that enables reasoning about continuous change.

The talk is based on extensive collaboration with Z. Manna, T. Henzinger, O. Maler, and Y. Kesten, whose results are reported in [HMP91], [MMP92], and [KP92].

Keywords: Liveness, fairness, justice, Real-time, timed transitions system, hybrid systems, discrete and continuous systems, Statecharts.

---

*This research was supported in part by the European Community ESPRIT Basic Research Action Project 3096 (SPEC) and by the France-Israel project for cooperation in Computer Science.

†Department of Applied Mathematics, Weizmann Institute, Rehovot, Israel

Introduction

A common situation to all sciences that use mathematics is that there exists a phenomenon out there in the real physical world, for which we wish to construct a mathematical model and to apply to it mathematical techniques in order to infer and predict properties of the real thing. Philosophically, this approach is doomed for failure, since inescapably the mathematical model can only provide an approximation to the real thing, and one can never absolutely verify the extent to which the properties of the mathematical model are valid for the physical system.[1] Pragmatically, in spite of this philosophical criticism, this approach has been successfully applied for the last three thousands years with spectacular results.

The reason for this past success and a lesson to be learned for the future is that a successful scientific discipline is one that not only perfects and improves its ability to analyze existing mathematical models but also continuously assesses the validity of the existing models for capturing important properties of the physical system under study and keeps developing more refined models that manage to capture more faithfully these properties. Ideally, such a discipline should produce a hierarchy of models (infinitely many of them, given infinite time), each refining and improving the modeling accuracy of its predecessors. This set of ever more accurate models need not be linearly ordered. Partial ordering, in which some unrelated models may even be incompatible, is quite acceptable. As an example, physicists are quite happy with the two incompatible models of light as waves and light as particles. Each has to be used in the context where the phenomena it represents better are more relevant. Also, a more refined model does not necessarily invalidates its less refined version. For example, Einsteinian physics does not void the utility of Newtonian physics. Both are useful approximations that may be beneficially used for modeling and analysis of systems that fall into well defined ranges, where the main distinguishing parameter is how close are the typical speeds in the system to the speed of light.

Let us apply this freshman observation to our own humble discipline which is the application of formal methods to the specification, analysis, and correct construction of reactive programs and systems. The physical system studied is obviously the concrete computerized systems plus any elements of the environment (possibly non-digital) that are necessary in order to identify a closed context in which the computerized system operate. Various mathematical models and formalisms have been proposed over the years for approximating the behavior of such systems.

Safety and Liveness

Experimental evidence and additional contemplation led Lamport [Lam77] to make the important distinction between *safety* and *liveness* properties of reactive systems. At that time, no formal definition of the two classes was given, but the offered intuition was that safety properties claim that nothing bad ever happens while liveness properties claim that something good eventually happen. A compatible intuition for safety can be that some

[1]The only exception to this sweeping description is possibly Mathematics itself which, self-contemplatively, may use one branch of mathematics to obtain a perfect model of structures in another branch. This is perhaps why some people tend to regard Mathematics as not being a science.

state property, i.e., a property that can be evaluated based merely on the current state, is continuously maintained. According to this interpretation, safety is closely related to invariance.

An example of a safety property is that, while the elevator cage is in motion, its doors are never open. Identifying two state predicates *moving* and *doors_open*, we can view this property as stating that the bad event $moving \wedge doors\_open$ never happens, or, equivalently, that the state formula $\neg(moving \wedge doors\_open)$ is continuously maintained.

Another example of a safety property is that of *partial correctness*. Let P be a program and φ its post-condition, i.e., a state formula that is expected to hold on all terminal states in the execution of P. For example, if P is expected to compute in variable y the factorial function of an input x, then its posto-condition could be $\varphi : y = x!$. Program P is defined to be ***partially correct with respect to*** φ if whenever it terminates, its final state satisfies φ. Note that partial correctness does not imply termination. For example, a program P may be partially correct with respect to $y = x!$ even if it diverges on all negative inputs. Partial correctness can be viewed as a safety property by observing that it requires that the bad event $term \wedge \neg\varphi$ never occurs, where *term* is a state predicate characterizing final states of the program.

Many liveness properties ensure response to requests. We may identify two events (or state predicates) *req* and *res* and require that any occurrence of *req* is eventually responded to by an occurrence of *res*. We refer to such properties as *response properties*. One example of a response property is program termination, where *req* stands for the event of starting the execution of a program and *res* stands for the state predicate *term* characterizing termination.

Another example, taken from the elevator world, is when *req* stands for the event of a passenger standing at some floor pushing one of the request buttons for the elevator. An appropriate response *res* is the event that the elevator cage is at the passenger's floor with its doors open.

An immediate consequence of the safety-liveness distinction was the recognition that the two complement one another, and each tells only half the story. For example, the program that immediately enters an infinite loop is partially correct with respect to *all* post-conditions. It is only the liveness requirement of termination that forces it to do something useful. On the other hand, another trivial program, the one that terminates immediately without computing anything, obviously satisfies the liveness property of termination, but not that of partial correctness.

In the elevator world, an elevator that immediately proceeds to the fifth floor and remains there forever with its doors open, probably satisfies all the safety properties an elevator should satisfy, but fails miserably in fulfilling any of its liveness requirements.

Thus, one of the conclusions of Lamport's paper is that both safety and liveness properties are of utmost importance and any reasonable specification and verification methods must represent both and will be seriously incomplete if they catered only to one of these classes.

Indeed, several of the formal schools inspired by these observations, in particular the temporal logic school, adhered most devoutly to the recommendation that safety and liveness should be equally represented.

Liveness Necessitates Fairness

When starting to implement Lamport's recommendation, it became apparent that there is some price to be paid. Consider the trivial program PARA, presented in Fig. 1.

$$x,\ y\text{: integer where } x = 0,\ y = 0$$

$$P_1 :: \begin{bmatrix} \ell_0\text{: while } x = 0 \text{ do} \\ \quad \ell_1\text{: } y := y + 1 \end{bmatrix} \quad \| \quad P_2 :: \begin{bmatrix} m_0\text{: } x := 1 \end{bmatrix}$$

Fig. 1. PARA: A Simple Parallel Program.

This program consists of two processes P_1 and P_2. Process P_1 is ready to loop endlessly as long as x is kept at 0. Process P_2 consists of a single assignment that sets x to 1. Obviously, if we run this program on a system where each process is assigned a separate processor, in a (very short) finite time, x will be set to 1 and the complete program will terminate. The same will hold for execution of this program on a single processor system which emulates a multi-processor system by multi-programming and any reasonable scheduling policy. What should be our semantic mathematical model so that all executions according to this model will eventually terminate, similar to the behavior of real systems on this program?

The basic ingredients of such semantic model have already been laid down by Dijkstra in [Dij65] and [Dij68]. The postulates of this model are:

1. Concurrency is represented by interleaving of "atomic" actions of the concurrent processes.

2. Nothing may be assumed about the relative speeds of the N computers; we may not even assume their speed to be constant in time [Dij65]. This postulate allows us to consider computations in which the distance between two consecutive occurrences of actions of a particular process gets increasingly larger beyond any bound.

3. "We now request that a process, once initiated, will — sooner or later — get the opportunity to complete its actions" [Dij68]. This disallows the distance between two consecutive actions from the same process to become infinite, unless the process is deadlocked or has terminated. That is, as long as a process has an action which is continuously ready, this action must eventually be completed.

The last postulate incorporates into the model the requirement of *weak fairness* to which we will refer (adopting the terminology of [LPS81] and [MP91b]) as *justice*.[2] Assuming that all atomic actions of processes are modeled by *transitions* in a *transition system*, this requirement can be stated as

> It is not the case that a transition becomes continuously enabled beyond some position p in a computation without being taken beyond p.

[2]Thus, we should regard Dijkstra as being the father of fairness; or at least, the introducer of its simplest and most feasible manifestation.

Later, additional brands of fairness have been considered and analyzed (see, for example, [Par80], [LPS81], [Par83], and [Fra86]). Here we restrict our attention to the simplest fairness notion, that of justice.

Following Dijkstra's recommendation, most of the formal methods that incorporated liveness as part of their specification and verification capabilities, in particular those using temporal logic, have adopted at least the assumption of justice into the semantics associated with concurrent programs.

Different Proof Rules for Safety and Liveness

Closely following the distinction between safety and liveness properties, came the realization that different proof rules are necessary in order to verify properties of the two different classes. In view of the kinship between safety and partial correctness, on one hand, and liveness and termination, on the other hand, this did not come as a complete surprise.

It is well known from the verification theory of sequential programs that proving partial correctness only requires the establishment of appropriate *invariants*, while proving termination requires, in addition to invariants, also *variants* which are state-functions measuring the progress towards (equivalently, the distance away from) termination. A related view is that, while partial correctness uses *implicit induction* inducting on the progress of the computation (therefore also called *computational induction*), termination must rely on *explicit induction*, where the variant functions progress along a well-founded domain.

Indeed, generalizing into the domain of reactive (usually concurrent) programs, we can offer two main rules for the establishment of safety and liveness properties.

A Rule for Safety

For safety properties, we propose rule SAFE:

SAFE	S1.	$\Theta \rightarrow p$	
	S2.	$\rho_\tau \wedge p \rightarrow p'$	for every $\tau \in \mathcal{T}$
		$\Box p$	

Several notions and notations are necessary in order to understand this rule. We assume that our system (program) is modeled by a *fair transition system* whose main components are V, a finite set of *state variables*, $\mathcal{T}$, a finite set of *transitions*, and Θ a state formula over V, called the *initial condition* which characterizes all states that are acceptable as initial states for computations of the system. *States* of the system, denoted by Σ, are interpretations of the variables V.

Each transition $\tau \in \mathcal{T}$ is associated with a *transition relation* $\rho_\tau(V, V')$, relating the state variables in two consecutive states, s and its τ-successor s'.

For example, we may consider a program that can be modeled by a transition system T with state variables $V = \{x, y\}$, and initial condition $\Theta : (x = 0 \wedge y = 0)$. An assignment $x := x + y$ may be modeled by a transition τ whose transition relation is $\rho_\tau : (x' = x + y \wedge y' = y)$. This relation claims that the state s' is a τ-successor of a state s iff the value of y in s' equals the value of $x + y$ in s and the value of y in s' is the same as the value of y in s. Assume that τ is the only transition that T has. In this case, we can use rule SAFE to prove that the safety property $\Box(x = 0 \wedge y = 0)$ is valid for T, that is, all states in all computations of T satisfy $x = y = 0$. Obviously, we take the formula p appearing in the rule to be $x = 0 \wedge y = 0$. Premise S1 then assumes the form

$$(x = 0 \wedge y = 0) \quad \rightarrow \quad (x = 0 \wedge y = 0)$$

which is obviously valid. Premise S2 assumes the form

$$(x' = x + y \wedge y' = y) \wedge (x = 0 \wedge y = 0) \quad \rightarrow \quad (x' = 0 \wedge y' = 0)$$

which demonstrates that the formula p' appearing in the rule is obtained by replacing every (free) variable $v \in V$ appearing in p by its *primed version* v'. Clearly, this second premise is also valid. This establishes that $x = y = 0$ is an invariant of the trivial transition system T.

The version of the rule displayed above covers the case that p is a state formula. A more general rule, in which p may be an arbitrary past temporal formula is presented in [MP91a] and shown there to be sound and complete. Since all safety properties expressible by temporal logic can be expressed by a formula $\Box p$, where p is a past formula, this shows that this rule, based on the computational induction principle, is fully adequate for proving safety properties (expressible within TL).

A Rule for Liveness

As is to be expected, rules for establishing liveness properties are more complicated, at least to the extent that they have to introduce their own explicit mechanism of induction. Indeed, the rule we are about to introduce assumes a well-founded binary relation $(\mathcal{A}, \succ)$, where $\mathcal{A}$ is a set of elements and $\succ$ is a binary relation on $\mathcal{A}$, such that there does not exists an infinite descending sequence

$$a_1 \succ a_2 \succ a_3 \cdots \qquad \text{where } a_i \in \mathcal{A} \text{ for all } i > 0.$$

The rule also uses a *ranking function* $\delta : \Sigma \mapsto \mathcal{A}$ mapping each state s to an element $\delta(s) \in \mathcal{A}$. We expect that as the computation progresses the rank of the encountered states will keep decreasing until some goal is achieved.

Assume that the transition of the studied system are given by $\mathcal{T} = \{\tau_1, ldots, \tau_m\}$. The rule uses intermediate *assertions* (state formulas), $\varphi_1, \ldots, \varphi_m$, each corresponding to one of the transitions. Assertion φ_i is intended to characterize all the states in which taking the transition τ_i is guaranteed to decrease the rank. We describe this situation by saying that τ_i is a *helpful transition* at these states.

Consider, for example, program PARA of Fig. 1, which will serve as our running example for the rule, where the liveness property to be established is that of termination. In the initial phases of any computation of this program, only transition τ_{m_0} (i.e., the transition associated with statement m_0) is helpful, while the transitions associated with ℓ_0 or ℓ_1 make no contribution towards termination. Once, statement m_0 is executed (terminating process P_2), the transitions associated with ℓ_1 and ℓ_0 become helpful, since they are the ones responsible for the termination of P_1.

We denote the disjunction $\varphi_1 \vee \cdots \vee \varphi_m$ by φ. Rule RESP may be used to establish liveness properties of the form $p \Rightarrow \Diamond q$, i.e., the response property claiming that every occurrence of p must be eventually followed by an occurrence of q, for state formulas p and q.

$$\begin{array}{l} \textbf{RESP} \\ \begin{array}{lrcl} \text{R1.} & p & \rightarrow & q \vee \varphi \\ \text{R2.} & \rho_i \wedge \varphi_i & \rightarrow & q \vee (\varphi \wedge \delta' \succ \delta) \\ \text{R3.} & \rho_j \wedge \varphi_i & \rightarrow & q \vee (\varphi \wedge \delta' \succ \delta) \vee (\varphi_i \wedge \delta' = \delta) \\ \text{R4.} & \varphi_i & \rightarrow & q \vee En(\tau_i) \end{array} \left.\begin{array}{l} \\ \\ \\ \\ \end{array}\right\} \text{for } i,j = 1,\ldots,m \\ \hline p \Rightarrow \Diamond q \end{array}$$

In the rule, we use ρ_i and ρ_j as abbreviations for ρ_{τ_i} and ρ_{τ_j} for $i, j = 1, \ldots, m$. The rule also uses the formula $En(\tau)$ which is defined as $\exists V'.\rho_\tau(V, V')$ characterizing all states s on which τ is *enabled*, i.e., s has some τ-successor.

Premise R1 requires that every p-state, either already satisfies the goal assertion q, or satisfies φ which means one of φ_i, for $i = 1, \ldots, m$.

Premise R2 describes the effect of taking transition τ_i from a state s for which τ_i is helpful. It claims that the state s' resulting from taking τ_i, either already satisfies q or still satisfies φ and has a rank δ' lower than that of s (given by δ.

Premise R3 describes the effect of taking transition τ_j from a state s for which τ_i is helpful where, possibly, $i \neq j$. The premise claims that either real progress is achieved, i.e., q becomes true or the rank decreases, or we obtain a successor state s' for which still the same transition τ_i is helpful, and whose rank is equal to that of s. It can be interpreted as stating that taking an unhelpful transition, while possibly not improving the situation, at least does not make it worse.

Premise R4 requires that τ_i is enabled on all non-q states on which it is considered helpful.

It is clear that, starting at a p-state, we cannot keep taking helpful steps forever, since this will violate the well-foundedness of $(\mathcal{A}, \succ)$. On the other hand, if from a certain point on we keep taking unhelpful steps that do not decrease the rank, the transition τ_i which is helpful there remains helpful forever, and due to R4, remains continuous forever without being taken. This violates the requirement of justice. Thus, for all legal computations that satisfy the justice requirement, every p-state must eventually be followed by a q-state.

Let us illustrate this rule on program PARA of Fig. 1, proving that it terminates. To refer to the location of control in the states of this program we use the notation at_ℓ_i, $i = 0, 1$, to denote the location of control within process P_1, and at_m_0 to denote that control within process P_2 is at location m_0. We also use the notation $at\_\ell_{0,1}$ as abbreviation for $at\_\ell_0 \vee at\_\ell_1$. The property expressing termination can be represented by the response formula

$$at\_\ell_0 \wedge at\_m_0 \wedge x = 0 \wedge y = 0 \quad \Rightarrow \quad \neg at\_\ell_{0,1} \wedge \neg at\_m_0$$

This formula states that, starting at a state recognizable as the initial state of the program, we are guaranteed to eventually reach a state in which control of P_1 is nowhere within $\{\ell_0, \ell_1\}$ and control of P_2 is not at m_0. This is obviously a terminated state.

As a well-founded relation we choose $(\mathsf{N}, >)$, where N is the set of natural numbers (including 0), and $>$ is the usual greater-than relation. This relation is obviously well-founded.

The ranking function δ is defined by

$$\begin{array}{rllll} \delta & = & \text{if} \;\; at\_m_0 & \text{then } 3 & \text{else} \\ & & \text{if} \;\; at\_\ell_1 & \text{then } 2 & \text{else} \\ & & \text{if} \;\; at\_\ell_0 & \text{then } 1 & \text{else } 0 \end{array}$$

The transition system for this program consists of four transitions: τ_{m_0}, τ_{ℓ_1}, $\tau_{\ell_0}^{\mathrm{T}}$, and $\tau_{\ell_0}^{\mathrm{T}}$, corresponding, respectively, to the execution of statements m_0 and ℓ_1 and to the two different steps statement ℓ_0 can take, depending on whether x is found to be zero (the true case) or not (the false case).

Consequently, we define four assertions, characterizing the states for which these transitions are helpful.

$$\begin{array}{llll} \varphi_{m_0}: & x = 0 \wedge at\_m_0 & \varphi_{\ell_1}: & x > 0 \wedge \neg at\_m_0 \wedge at\_\ell_1 \\ \varphi_{\ell_0}^{\mathrm{T}}: & \mathrm{F} & \varphi_{\ell_0}^{\mathrm{F}}: & x > 0 \wedge \neg at\_m_0 \wedge at\_\ell_0 \end{array}$$

We observe, for example, that there is no state on which $\tau_{\ell_0}^{\mathrm{T}}$ is considered helpful.

It is not difficult to see that these constructs satisfy all the premises of rule RESP, which constitutes a formal proof that the program PARA always terminates.

Unlike the case of safety properties, the question of coverage of all liveness properties is more involved. On one hand, rule RESP can also be used to establish some non-liveness properties. For example, the same premises are sufficient to establish also the property $p \Rightarrow \varphi \mathcal{U} q$ which, according to the official definition given in [AS85], is not, in general a liveness property. On the other hand, there are liveness properties, such as those expressed by the *reactivity formula* $\Box\Diamond p \rightarrow \Box\Diamond q$ which cannot be phrased as response formulas, and therefore may need stronger (and more complex) rules for their establishment. We refer the interested reader to [MP91a], where these issues are discussed in full detail.

The Price Paid for Fairness

A closer examination of the implications of incorporating fairness (even in its weakest form - justice) into the semantic model reveals that there is a price to be paid for the ability to prove reasonably liveness properties in the presence of fairness. The added complexity can be described in several ways that talk, of course, about the same phenomena but may use different terminologies for doing so.

The Topological View

Let us consider an extremely simplistic view by which the semantics of a transition system (and therefore of the program modeled by them) consists of the set of possible computations that can be generated by them. Computations, in this simplistic view, are maximal finite or infinite sequences of states $s_1, s_2, s_3 \ldots$, such that s_1 satisfies the initial condition Θ and each state s_{i+1} can be obtained by applying some transition to its predecessor s_i. Maximality means that a finite sequence is accepted as a computation only if its last state s_k is *terminal*, i.e., no transition in $\mathcal{T}$ is enabled on s_k.

The set of such (finite and infinite) sequences can be considered as a metric space where the distance between two sequences $s_1, s_2, \ldots$ and $s'_1, s'_2, \ldots$ can be defined as $1/j$

where j is the minimal index such that $s_j \neq s'_j$ and, in the special case that one is a (finite) prefix of the other, j is the length of the shorter one plus 1.

If we consider the set of all computations of a transition system without requiring fairness, it can be showed that the set of all computations is always a *closed set* in the topological sense. One way of describing the effect of fairness is that it destroys the property of closure. Consider, for example, some of the computations of the program of Fig. 1. In the following we display the states belonging to these computations as pairs consisting of the current values of x and y.

$\langle 0,0\rangle$, $\langle 1,0\rangle$, $\langle 1,0\rangle$.
$\langle 0,0\rangle$, $\langle 0,0\rangle$, $\langle 0,1\rangle$, $\langle 1,1\rangle$, $\langle 1,1\rangle$.
$\langle 0,0\rangle$, $\langle 0,0\rangle$, $\langle 0,1\rangle$, $\langle 0,1\rangle$, $\langle 0,2\rangle$, $\langle 1,2\rangle$, $\langle 1,2\rangle$.
$\langle 0,0\rangle$, $\langle 0,0\rangle$, $\langle 0,1\rangle$, $\langle 0,1\rangle$, $\langle 0,2\rangle$, $\langle 0,2\rangle$, $\langle 0,3\rangle$, $\langle 1,3\rangle$, $\langle 1,3\rangle$.

Note that it takes P_1 two steps to update the value of y, and one more step after P_2 sets x to 1 to respond and terminate too.

If we take the topological limit of this sequence of computations we obtain an infinite sequence in which y keeps increasing forever, while x stays zero forever. This is accepted as a computation of program PARA if we give up the requirement of fairness, but is not a legitimate computation when fairness is adopted. Thus, under the requirement of justice, the set of computations of this program is not a closed set.

This has some severe repercussions. As a general result it disrupts Scott's grand thesis by which every computable function is continuous, and in defining and reasoning about the semantics of program it is sufficient to deal with infinite objects through their finite approximations, and the infinite objects will take care of themselves through closure and continuity.

To illustrate why this scheme is broken by fairness, consider program NONDET presented in Fig. 2.

$$x,\ y\text{: integer where } x = 0,\ y = 0$$

$$P :: \left[\begin{array}{l} \ell_0\text{: while } x = 0 \text{ do} \\ \quad \left[\begin{array}{c} \ell_1\text{: } y := y + 1 \\ \text{or} \\ \ell_2\text{: } x := 1 \end{array}\right] \end{array}\right]$$

Fig. 2. NONDET: A Simple Nondeterministic Program.

Program NONDET is very similar to program PARA of Fig. 1 except that the critical statement $x := 1$ is not performed by a parallel process but as a result of a nondeterministic choice. Since no fairness is assumed about nondeterministic selection, the semantics of this program differs from that of program PARA. The semantic of NONDET includes, in addition to all the finite computations admitted by PARA, also their limit, the infinite computation

$$\langle 0,0\rangle,\ \langle 0,0\rangle,\ \langle 0,1\rangle,\ \langle 0,1\rangle,\ \langle 0,2\rangle,\ \langle 0,2\rangle,\ \langle 0,3\rangle,\ \langle 0,3\rangle,\ \langle 0,4\rangle,\ \ldots,$$

which never sets x to 1 and keeps increasing y forever. Thus, the two compared programs differ by the presence or absence of this infinite computation from their semantics.

The standard approach of assigning *finite trace* semantics to such programs, based on Scott's grand thesis, is to take the finite prefixes all maximal computations. Unfortunately, in the case above, this leads to identical semantics for the two programs, even though when assuming justice, we wish to distinguish between the two since One always terminates and the other does not.

This shows that, in the presence of fairness, it is not always possible to distinguish two different programs by the finite prefixes of their maximal computations. We may describe this as losing the property of *finite distinguishability* which is guaranteed as long as the semantics of programs are closed sets.

The implication of this is that our semantics must be expressed in terms of full computations that may be infinite objects.

The Automata-Theoretic View

Another insight into the more complex situation caused by fairness is provided by restricting our attention to finite-state systems. In many cases, properties of finite-state systems can be studied through automata-theoretic means. When dealing with unfair systems the appropriate model is that of a finite-state automaton. Thus, for any finite-state unfair program P, we can construct a finite-state automaton that accepts precisely all the prefixes of the computations of P. Since, in the unfair case, the language of these prefixes uniquely identifies the set of all computations (by taking its closure), we can check, for example, whether two finite-state programs are equivalent by testing their corresponding automata for equivalence, a well known and understood procedure.

In the fair case, we can also construct a finite-state automaton that will recognize all the computations of a given finite-state program. Here, however, we have to take into account maximal infinite computations, and therefore the automaton will no longer be one over finite inputs but, instead, is an ω-finite-state automaton of the Büchi, Rabin, or Streett type, that accepts inifinitary languages, i.e., sets of infinite words. There is a corresponding theory for manipulating these automata and applying to them the standard procedures of determinization, complementation, and checking for equivalence. However, as anyone who studied this theory knows, it is considerably more complicated and difficult to apply. One of the things we still do not know how to do for these automata is to compute the minimal automaton equivalent to a given one, a standard procedure for normal finite-input automata.

A related and very relevant observation is that, while there exists a reasonable and complete axiomatization by which two regular expressions can be proven equivalent, no such axiomatization exists (to the best of our knowledge) for ω-regular expressions which is the infinitary analog of regular expressions.

The Ordinal View

Another symptom of the complications introduced by fairness is the question of the nature of the well-founded domain necessary to support proofs of liveness. There are concurrent programs whose termination can be proven without assuming any fairness. For these, we can have a much simpler rule than RESP. This rule will also use the basic idea of well-founded induction lifted from proofs of termination of sequential programs. However, there is a major difference in the diversity of the well-founded domains that must be used in these proofs.

For proofs of termination of sequential and concurrent but unfair programs, it is always possible to use the canonical well-founded relation $(\mathsf{N}, >)$ which can also be described as using the domain ω. For proofs of termination of fair programs one must go to considerably higher ordinals. In fact, for every countable ordinal α, there exists a terminating fair program, whose proof of termination must use as a well-founded domain α (equivalently, the set of all ordinal smaller than α in much the same way that ω stands for N).

Some of these properties have been established in [AP86], where the authors attribute them not necessarily to fair concurrent programs but to countable but unbounded non-determinism. However, it is a well established fact that unbounded (but countable) non-determinism is equivalent in expressive power to fairness. In fact, program PARA is an example of an implementation of the statement **choose**(y) which assigns an arbitrary natural number to variable y. The *choose* statement is one of the canonical examples of an unboundedly nondeterministic statement.

What is Being Done in other Schools?

The description in the preceding subsections covered the formal approach to the specification and verification of reactive systems which embraces fairness into its semantic models and uses as specification language a language that, similar to temporal logic, can specify properties of infinite computations. Some people successfully use ω-automata as the specification language.

This, of course, is only one of the proposed viable approaches to concurrency. Another very important school is the one that bases its modeling of concurrent systems on processes that communicate by synchronous message passing. This school includes a variety of process algebras such as CCS, ACP, Meije, and also CSP. With due apologies to all involved, I will refer to this important and central school as the *algebraic school*.

Many attempts have been made to identify the main differences between the algebraic school and, say, the temporal logic school. For example, the algebraic method studies processes that communicate by message-passing while the temporal logic approach traditionally studied communication by shared variables. However, this difference is not really essential. Already in [BKP85] it has been shown that the temporal approach can be extended to cover synchronous and asynchronous message-passing. The interested reader may find a detailed such extension described in [MP91b].

Since the central question asked in the algebraic school is that of equivalence between two given processes, and the rules of the game are that this must be answerable by axiomatic or algebraic rules, or on the semantic level by observing local changes, the notion of fairness which requires comparing complete infinite computations always seemed incongruous with the algebraic approach. The price as described above was deemed to be too high. There were some valiant attempts to incorporate fairness into the algebraic approach but they never seriously caught on.

The main point of the discussion here, however, is not to criticize the algebraic school for not adopting the concept of fairness. Instead, I would like to reiterate an interesting argument that was offered in defense of the decision not to adopt fairness.

Of course, the argument goes, we are equally interested in eventually establishing liveness properties of real systems. However, what you call liveness is only an abstraction of the real thing. What good is the guarantee that the proper response *eventually* arrives to the implementor or to the customer, who may die of old age waiting for the elevator which eventually arrived after 150 years? The real useful property is not qualitative (non-quantitative) response which is the property you can prove in your proof system but *bounded response*, i.e., response within a given time-bound. Both of us will eventually have to construct the next more refined model in the hierarchy. The one that addresses real-time and provides quantitative measures to liveness. The difference between us is that you already considered some abstraction of liveness properties at the first level of modeling, while we will consider liveness properties more accurately but for the first time only at the second modeling level. The price you pay for the earlier consideration of liveness properties is the disruption of Scott's thesis, the resulting more complex mathematics, and the need to use higher ordinals.

The Next Models in the Hierarchy

Indeed, the more refined modeling of reactive systems is obtained by taking real-time into account. This will enable us to study the class of *timed systems* which are reactive systems for which actions are associated with timing information. In this extended abstract, we will only outline in the next section the two main extensions applied to the main constituents of the temporal methodology: the computational model of systems, and the specification logics. The computational model of *fair transitions system* is replaced by that of *timed transitions system*. The temporal specification language is extended by the introduction of bounded temporal operators such as $\Diamond_{\leq 5}$ meaning eventually within 5 seconds, explicit reference to the real-time clock T, and the *age* function $\Gamma(p)$ measuring how long p has been continuously true.

One of the observations to be made when designing our proof system for this model, is that, in the presence of timing information, fairness and unbounded liveness lose much of their significance, and many properties move from the class of liveness to the class of safety. This should not come as a big surprise. As we traced the origins of the notion of fairness, it was born under the supposition that at the first, qualitative, level of modeling we make *no assumption* about the relative speeds of processes. In the presence of timing information, we *are* provided with assumptions about the relative speeds of processes, if not directly with respect to one another then with respect to the clock which is the great synchronizer that no process can ignore.

As the next extension leading to the 3rd level of modeling, we are interested in the concept of *hybrid systems*. These are systems that consists of a tightly coupled mixture of continuous and discrete components such as a digital computer controlling a mechanical or chemical system. Here, the introduced computational model is that of *phase transition system* whose computations consist of alternating phases of discrete and continuous activity. A discrete phase consists of several transitions taken in an interleaving manner but taking no time. The continuous phase is controlled by a set of differential equations that are the ones active in this phase. Time is consumed only in the continuous phases. The specification language is not extended by much. It is still temporal logic equipped with the ability of referring to real-time and to the age of formulas. However, in some of the hybrid models it is interpreted over dense domains rather than over discrete ones.

In this extended abstract, we will not provide additional technical details about these two models. They will be described in the talk. The interested reader may consult [MMP92] for more details of these models.

References

[AP86] K.R. Apt and G.D. Plotkin. Countable nondeterminism and random assignment. *J. ACM*, 33,4:724–767, 1986.

[AS85] B. Alpern and F.B. Schneider. Defining liveness. *Info. Proc. Lett.*, 21:181–185, 1985.

[BKP85] H. Barringer, R. Kuiper, and A. Pnueli. A compositional temporal approach to a *csp*-like language. In E.J. Neuhold and G. Chroust, editors, *Formal Models of Programming*, pages 207–227. IFIP, North Holland, 1985.

[Dij65] E.W. Dijkstra. Solution of a problem in concurrent programming control. *Comm. ACM*, 8(9):569, 1965.

[Dij68] E.W. Dijkstra. Cooperating sequential processes. In F. Genuys, editor, *Programming Languages*, pages 43–112, New York, 1968. Academic Press.

[Fra86] N. Francez. *Fairness*. Springer-Verlag, 1986.

[HMP91] T. Henzinger, Z. Manna, and A. Pnueli. Temporal proof methodologies for real-time systems. In *Proc. 18th ACM Symp. Princ. of Prog. Lang.*, pages 353–366, 1991.

[KP92] Y. Kesten and A. Pnueli. Timed and hybrid statecharts and their textual representation. In J. Vytopil, editor, *Formal Techniques in Real-Time and Fault-Tolerant Systems*, pages 591–619. Lec. Notes in Comp. Sci. 571, Springer-Verlag, 1992.

[Lam77] L. Lamport. Proving the correctness of multiprocess programs. *IEEE Trans. Software Engin.*, 3:125–143, 1977.

[LPS81] D. Lehmann, A. Pnueli, and J. Stavi. Impartiality, justice and fairness: The ethics of concurrent termination. In *Proc. 8th Int. Colloq. Aut. Lang. Prog.*, pages 264–277. Lec. Notes in Comp. Sci. 115, Springer-Verlag, 1981.

[MMP92] O. Maler, Z. Manna, and A. Pnueli. From timed to hybrid systems. In J.W. de Bakker, C. Huizing, W.P. de Roever, and G. Rozenberg, editors, *Proceedings of the REX Workshop "Real-Time: Theory in Practice"*, volume 600 of *LNCS*, Berlin, 1992. Springer Verlag.

[MP91a] Z. Manna and A. Pnueli. Completing the temporal picture. *Theor. Comp. Sci.*, 83(1):97–130, 1991.

[MP91b] Z. Manna and A. Pnueli. *The Temporal Logic of Reactive and Concurrent Systems: Specification.* Springer Verlag, New York, 1991.

[Par80] D. Park. On the semantics of fair parallelism. In *Abstract Software Specification*, pages 504–524, Berlin, 1980. Lec. Notes in Comp. Sci. 86, Springer-Verlag.

[Par83] D. Park. The fairness problem and nondeterministic computing netrworks. In J.W. de Bakker and J. Van Leeuwen, editors, *Foundations of Computer Science IV, Distributed Systems*, pages 133–161. Mathematical Centre Tracts 159, Center for Mathematics and Computer Science (CWI), Amsterdam, 1983.

Preserving Specific Properties in Program Development

–How to Debug Programs–
(Conference Version)

F. A. Stomp

Christian Albrechts Universität
Institut für Informatik und Praktische Mathematik II
2300 Kiel, Germany
Email: fst@informatik.uni-kiel.dbp.de

Abstract. The problem of preserving *specific temporal properties* in program development is addressed. A new relation between programs, or, more generally, between specifications, is defined. This relation is parameterized by a (finite) collection of temporal properties which will be preserved by that relation. Such a relation will, in general, *not* preserve *all* temporal properties. It is proved, however, that for a particular choice of the parameter the new relation coincides with Abadi and Lamport's notion of *implements* [1], which preserves all (externally visible) temporal properties. As a consequence, the approach for program development proposed in the current paper is *as least as powerful* as those which use *implements* as their basic refinement relation. Examples of the latter approaches are those proposed by Back [4], by Lynch and Tuttle [27], and by Lam and Shankar [25]. It is argued that the approach in program development which preserves only certain temporal properties is preferable to those approaches which preserve all temporal properties.
It is shown that the approach advocated here is, except for formally developing programs, also applicable for *debugging* programs, i.e., for correcting programs which do not satisfy their required specification. This provides –for the first time– a formal basis for "correcting code".

1 Introduction

Stepwise refinement [7, 10, 20, 31], see [11] for an overview on current trends, is nowadays one of the most important approaches for developing correct (sequential and concurrent) programs. The underlying idea of stepwise refinement is to start with a certain initial specification of the problem to be solved, and derive by *correctness preserving* transformations an executable, and hopefully efficient, program.
Depending on the notion of correctness, transformations are, in general, constrained by so-called *refinement relations* between programs, or, more generally, between specifications. There appear to be two main refinement relations explicitly defined in the literature: one which preserves *all total-correctness* specifications [7, 6, 30], and one which preserves *all* (*stutter-free linear-time*) *temporal* specifications [4, 5, 21, 25, 26, 27]. (A relation $\preceq$ between programs preserves property ϕ if the following is true for all programs S and S': If S satisfies ϕ, i.e., if every execution sequence of S satisfies ϕ, and if $S \preceq S'$ holds, then S' satisfies ϕ, too.) Sanders [32] and Singh

[36] define refinement relations in the context of **UNITY** [10]. Their relations may be considered to belong to the class of relations preserving all temporal properties. Both kinds of relations have been applied successfully on many occasions, see [7, 4, 26, 33, 42] to mention a few articles in which case-studies can be found.

The refinement relation which preserves total-correctness specifications makes sense only in case the specifications constrain input-output behaviors. It is, e.g., not applicable for deriving programs which are meant to run forever or which are required to satisfy additional (temporal) constraints, such as an invariant. (Examples are programs for garbage collection [8, 13, 41], and programs for stability detection [9, 14, 16].) As a consequence, the relation which preserves all temporal properties has been studied extensively. (It is indeed applicable to all the above-mentioned programs.) It is argued later in this section, however, that this relation exhibits serious problems in the context of program development.

For this reason, *the present paper introduces a new relation for developing programs.* It is parameterized by a (finite) collection of temporal properties. All properties in this collection are preserved by that relation; other temporal properties are not necessarily preserved. One does not lose, however, the possibility to preserve all temporal properties, if required: it is proved that this is achieved by chosing a particular parameter for the relation.

Suppose that at a certain stage during a development, some program S has been obtained which satisfies each of the temporal properties $\phi_1, \phi_2, \cdots, \phi_n$. In order to preserve each of these properties (separately) during the next derivation step, it suffices to preserve their conjunction $\bigwedge_{i=1}^{n} \phi_i$ during this step. Consequently, in the rest of this paper, preservation of one temporal property, rather than a finite number of them, will be considered during a derivation step; hence, the relation will be parameterized by one property. If temporal property ϕ is the parameter then the relation (defined in this paper) is referred to as a ϕ-derivation. Program T is a ϕ-derivative of program S iff the following are true:

(a) Every execution of program S satisfying ϕ is an execution of program T. Thus, the set of executions satisfying ϕ is not reduced when program S is transformed into program T.

(b) Every execution of program T satisfying $\neg\phi$ is an execution of program S. Thus, the set of executions satisfying $\neg\phi$ is not enlarged when program S is transformed into program T.

For each temporal property ϕ, the relation ϕ-derivation is a pre-order, i.e., it is reflexive and transitive. (If programs are identified with the collection of executions they generate, then it is even a partial order.) This property, and the fact that a certain temporal property is preserved, makes it feasible to apply such a relation to stepwise refinement [18].

Now, consider the relation *implements* [1], also studied by Diepstraten and Kuiper [12], by Klarlund [24], and by Meritt [29], which preserves all temporal properties. Formally, for programs S and T, T *implements* S iff the following is true for all temporal properties ϕ: If S satisfies ϕ, i.e., if for all execution sequences of program S, ϕ holds, then T satisfies ϕ, too.[1]

[1] Abadi and Lamport [1] identify a program with the collection of execution sequences generated by that program, and characterize such a collection by a temporal property.

What is the drawback of methods based on this relation in program development? Consider an arbitrary problem which is to be formally specified. Let S be some formal (initial) specification of that problem. (S could be a program [4].) Let, furthermore, S' be some program which solves the above-mentioned problem. Consequently, program S' satisfies specification S. Now, suppose that one wishes to formally derive some program S' from specification S. One then starts with specification S and tries to obtain program S' by stepwise refinement employing relation *implements* in every derivation step. In order to eventually obtain program S', one has to ensure that during this development no program S'' is ever derived where S'' satisfies a certain temporal property ϕ independent of S which contradicts some property of S'. *Why is that so?* The answer is simple: Property ϕ would be preserved by any transformation step. Since it contradicts some property satisfied by program S', it would be impossible to obtain S' (from program S''). This implies that one has to consider in advance all temporal properties satisfied by program S' in addition to specification S – *in general, an immense and impossible task.* (Deriving large and complex programs as such may be even impossible, because these programs are often *overspecified* [31].)

This drawback emerged when I applied Back's *reactive refinement relation* [4], an example of a relation which preserves all temporal properties, to derive a sequential program and a concurrent program for dynamic cycle prevention, i.e., programs for maintaining a finite, directed, and acyclic graph in the presence of additions of edges to and deletions of edges from that graph. The reactive refinement relation, as defined in [4], seems to be *too strong* for program development, because it does not take into account the environment in which the program is to be executed. Therefore, I applied a modified relation, taking the respective environments into account, in order to derive a sequential program and a concurrent program for dynamic cycle prevention. To obtain a sequential program I started with a specification using some array A to record elements $<+, v, w>$ (request to add edge $<v, w>$ to the graph) and elements $<-, v, w>$ (request to delete edge $<v, w>$ from the graph), for nodes v and w. Due to the sequential nature of the problem, it seemed obvious to incorporate in the initial specification that requests were processed one after another. After having derived a sequential program, I concentrated on the concurrent case. The goal was to derive Katz and Shmueli's program reported in [23]. In this program, however, many requests may be processed concurrently. As a consequence, the order in which edges are added to and deleted from the graph as a result of granting requests is not necessarily the same as in the sequential case. I was able to formulate a (stutter-free, linear-time) temporal property expressing that requests are treated one after another; and prove that it was implied by the initial specification for the sequential case; and prove that this property is not true for Katz and Shmueli's program (because there requests are processed concurrently). Consequently, another initial specification had to be formulated for the concurrent case.

Consequently, *implements* is defined in [1] as a relation between temporal properties: Ψ' *implements* Ψ iff $\Psi' \Rightarrow \Psi$ holds (for temporal properties Ψ and Ψ'). It is easy to see that the latter is equivalent to: For all temporal properties ϕ, $\Psi \Rightarrow \phi$ implies $\Psi' \Rightarrow \phi$. Thus, the definition of *implements* in terms of programs is consistent with Abadi and Lamport's definition.

Intuitively, this is due to the following: When sequential programs are refined into concurrent ones, more interleavings and more states will occur. As a result, fewer temporal properties will be true: simply consider a property true for the original states but not true for all the new ones. Now, Back [4], and Chandy and Misra [10] have argued that initial development of concurrent programs should take place at a level of sequential programming. With this in mind, the discussion above indicates that the reactive refinement relation is not the appropriate relation for developing concurrent programs. This result is not restricted to Back's reactive refinement method. It generalizes, as argued above, to all other methods using *implements* as their basic relation.

How does the parameterized relation introduced in this paper avoid this drawback? The drawback is avoided by preserving only temporal properties which are *relevant* for the problem under consideration and properties found *relevant* during the development. As discussed earlier this is achieved by the parameter of the relation.

Except for (formally) developing programs, the approach advocated in this paper is also shown to be applicable for *debugging* programs. More precisely, assume that someone has written a certain program S which has to satisfy (temporal) specification ϕ. It is common –in particular for larger programs– that program S admits a collection of execution sequences for which ϕ does not hold. Instead of writing another program (from scratch, or by applying formal techniques), one usually tries to *correct* program S. Now, ϕ-derivation will preserve the correct sequences and will not add any incorrect ones. Based on this, a rule is formulated which allows one to modify program S so that certain incorrect sequences are replaced by correct ones.

The remainder of this paper is organized as follows: The class of programs considered in this paper is introduced in Section 2. It is shown that each such program may be viewed as a linear time temporal property [28]. This result, due to Shengzong [34, 35], enables one to use a single formalism, viz., linear time temporal logic, for describing both programs and specifications. The parameterized relation is defined Section 3. It is shown that this relation can be applied for (formally) developing programs. It is also shown that the same relation can be applied for debugging programs. In Section 4, I demonstrate the applicability of the parameterized relation. It is envisaged that this relation can be applied to all refinement steps in [10]. (Chandy and Misra have not defined any refinement relation between programs. They use a mixed formalism to express programs and properties of programs. They justify refinement steps by proving that a program satisfies its required properties.) Section 5 contains some conclusions. Examples demonstrating various applications of this relation can be found in [39]. In another paper an example of how to formally debug programs (employing the parameterized relation) will be presented.

2 Preliminaries

The class $\mathcal{P}rog$ of programs employed in this paper is, in essence, the class of UNITY-programs [10]. Class $\mathcal{P}rog$ is discussed in Subsection 2.1. Each program in $\mathcal{P}rog$ may be viewed as a linear time temporal formula. This result, due to Shengzong [34, 35], is discussed in Subsection 2.2. Consequently, a single formalism, viz., linear time temporal logic, suffices to describe both programs and specifications.

2.1 Programs

A program consists of three parts:

- An *assignment-segment*. This part consists of a finite, strictly positive number of guarded atomic multiple-assignment statements. Each such statement is of the form $b \rightarrow x := e$, where x is some non-empty finite list of variables, e is some list of expressions having the same length as x, and b (the statement's guard) is some boolean condition. It is assumed that expression e has a unique, well-defined value when b is true.
- An *initial-segment*. This part consists of a state formula, the *precondition*, and constrains the initial values of some (not necessarily all) variables occurring in the program.
- An *always-segment*. This part consists of a state formula, and defines certain variables in the program in terms of other, possibly fresh, variables.

A statement a of the form $b \rightarrow x := e$ is *enabled* in state[2] s if b holds in s; otherwise, a is *disabled* in s. The effect of executing an enabled statement a as above in state s is another state s' defined by $s'=s[s(e)/x]$, where $s[s(e)/x]$ denotes, as usual, the state variant of s:[3] $s[s(e)/x](y) = \begin{cases} s(e) \text{ if } x \equiv y \\ s(y) \text{ otherwise.} \end{cases}$

The semantics of programs is defined by means of execution sequences. An execution sequence of a program as above is an infinite sequence $s_0 \rightarrow s_1 \rightarrow s_2 \rightarrow \cdots$ of states s_n, $n \geq 0$, such that the following are all satisfied:[4]

(a) State s_0 satisfies the precondition.

(b) For each natural number n,
  - either there exists some statement a enabled in state s_n, and state s_{n+1} is the result of executing a in s_n, or
  - all statements are disabled in state s_n, and $s_{n+1}=s_n$ holds.

(c) The state formula in the always-segment holds in every state s_n, $n \geq 0$.

(d) If some statement a is infinitely often enabled in the sequence, then the effect of a is infinitely often visible in sequence (strong-fairness, cf. [17]).[5]

2.2 Programs viewed as Temporal Formulae

The specification language used in this paper is Manna and Pnueli's linear time temporal logic [28]. It is assumed that this logic contains the *always*-operator $\Box$, the *eventual*-operator $\Diamond$, and the *next*-operator O. Every program S can be viewed as

---

2 A state is a total mapping assigning values to all variables which occur in some program.

3 Recall the x is, in general, a list of variables. For ease of exposition, the definition of a state variant treats x as a single variable.

4 For brevity, the possibility of stuttering [1] is ignored.

5 Thus, it is not required that statement a itself is infinitely often executed in the sequence. It suffices that in every execution sequence one discerns infinitely often the effect of such a statement, as if that statement is infinitely often executed. The reason for doing so is that only the effect of a statement is *relevant*; and that the actual statement responsible for that effect is *irrelevant*. This fairness assumption differs from the one in [10], where weak-fairness is assumed.

a linear time temporal property. This has been shown in [34, 35]. In this subsection, I briefly discuss how a program can be translated into a temporal formula.[6]

Consider any program S as in Subsection 2.1. Let $a_1, \cdots, a_n$ ($n \geq 1$) be program S's statements. Denote statement a_i by $b_i \rightarrow x_i := e_i$. Let y_i denote the list, in an arbitrary order, of S's programming variables different from the ones constituting list x_i. (Thus, y_i consists of variables occurring in program S, which cannot be changed during execution of statement a_i.) Assume that z denotes the list, again in an arbitrary order, of all program S's programming variables. Now, if statement a_i is executed in a certain state, then in the next state x_i has the value of e_i in the present state, and y_i is unchanged. Thus, $step(a_i) \equiv b_i \wedge (e_i = O x_i) \wedge (y_i = O y_i)$ holds in such a case. Consequently, executing any of the enabled actions in a state is expressed by the temporal property $step(S) \equiv \bigvee_{i=1}^{n} step(a_i)$.
For any execution of program S it is always the case that either an enabled action is executed, or that all the actions are disabled. Thus, the following holds for every execution sequence of program S: $steps(S) \equiv \Box(step(S) \vee (\bigwedge_{i=1}^{n} \neg b_i \wedge z = O z))$.
Incorporating that every execution sequence of program S starts in a state satisfying S's precondition pre, and that for all these sequences state formula ρ in S's always-segment continuously holds, leads to the following: For all execution sequences of program S, $Unfair\_Exec(S)$ defined by $Unfair\_Exec(S) \equiv pre \wedge steps(S) \wedge \Box\rho$ holds. It remains to take into account the fairness constraint that whenever some action is infinitely often enabled, then its effect is infinitely often visible. This is, obviously, expressed by $Fair(S)$ defined as follows: $Fair(S) \equiv \bigwedge_{i=1}^{n}((\Box\Diamond b_i) \Rightarrow (\Box\Diamond step(a_i)))$.

Program S as above can be identified with the temporal formula $Trans(S)$, the translation of S, defined by $Trans(S) \equiv Unfair\_Exec(S) \wedge Fair(S)$.

3 Preserving a Specific Temporal Property

In this section a relation $\sqsubseteq_\phi$ between specifications, for some given temporal property ϕ, is defined. Its definition implies that relation $\sqsubseteq_\phi$ ***preserves*** property ϕ; in general, it does ***not preserve*** properties different from ϕ. It is shown that such a relation can be applied for formally developing programs and for debugging (incorrect) programs. It is also shown that for a particular choice of parameter ϕ, Abadi and Lamport's notion of ***implements*** is equivalent to $\sqsubseteq_\phi$.
All proofs of the theorems presented in this section can be found in the [40].

3.1 On Designing Programs

As argued in the previous section, programs may be viewed as temporal properties. Consequently, any relation between temporal properties induces a relation between programs. Therefore, the focus will be on relations between temporal properties (rather than between programs).

For a given temporal property ϕ –the property to be preserved, a relation $\sqsubseteq_\phi$ between temporal properties, also called ***specifications***, is defined. Intuitively, $\Psi_1 \sqsubseteq_\phi \Psi_2$ holds iff the following are true:

[6] Essentially the same transformation, using the strong until-operator from linear time temporal logic, can be applied to cope with the phenomenon of stuttering.

- If some sequence satisfies both Ψ_1 and ϕ, then this sequence also satisfies Ψ_2. Thus, the set of sequences already satisfying ϕ is not reduced.
- If some sequence satisfies both Ψ_2 and $\neg\phi$, then this sequence already satisfies Ψ_1. Thus, the set of sequences satisfying $\neg\phi$ is not enlarged.

Definition 3.1 Let ϕ be an arbitrary temporal property. Define the relation $\sqsubseteq_\phi$ between temporal properties as follows: $\Psi_1 \sqsubseteq_\phi \Psi_2$ is true iff $(\Psi_1 \land \phi) \Rightarrow \Psi_2$ and $(\Psi_2 \land \neg\phi) \Rightarrow \Psi_1$ are both satisfied. These implications are interpreted over all infinite sequences of states. ■

Relation $\sqsubseteq_\phi$ will be referred to as a ϕ-derivation. If $\Psi_1 \sqsubseteq_\phi \Psi_2$ holds, for temporal properties Ψ_1, Ψ_2, and ϕ, then Ψ_2 is called a ϕ-derivative of Ψ_1; and Ψ_2 is said to be ϕ-derived from Ψ_1. For convenience, if for temporal properties Ψ and ϕ, $\Psi \Rightarrow \phi$ holds, then Ψ is said to satisfy ϕ.

Any ϕ-derivation preserves property ϕ. That is, if specification Ψ_2 is ϕ-derived from specification Ψ_1, and if Ψ_1 satisfies property ϕ, then Ψ_2 also satisfies ϕ:

Theorem 3.1 If $\Psi_1 \sqsubseteq_\phi \Psi_2$ and $\Psi_1 \Rightarrow \phi$ are true, then $\Psi_2 \Rightarrow \phi$ is also true. ■

For each temporal property ϕ, relation $\sqsubseteq_\phi$ is reflexive, anti-symmetric, and transitive. This is the subject of the next theorem. Consequently, it is possible to find a ϕ-derivative of Ψ_1 by means of a sequence of ϕ-derivatives starting with Ψ_1.

Theorem 3.2 For each temporal property ϕ, relation $\sqsubseteq_\phi$ is a partial ordering on temporal properties. ■

The previous two theorems justify the following *basic method* for developing some program which has to satisfy a certain temporal property ϕ: Start with some specification Ψ_1 which satisfies ϕ, and derive a sequence of specifications $\Psi_1, \Psi_2, \cdots, \Psi_n$ $(n \geq 1)$ such that Ψ_{i+1} is a ϕ-derivative of Ψ_i $(i=1,\cdots,n-1)$. Thus, $\Psi_1 \sqsubseteq_\phi \Psi_2 \sqsubseteq_\phi \cdots \sqsubseteq_\phi \Psi_n$ holds. From the Theorems 3.1 and 3.2, it follows that specification Ψ_n satisfies ϕ. Now, if Ψ_n is a property equivalent to $Trans(S)$, for some program S, as described in Subsection 2.2, then one may find some program from class $\mathcal{P}rog$ which corresponds to Ψ_n (and which, therefore, satisfies specification ϕ).

In general, the basic method does not allow a designer to, e.g., add new variables during a development. As an example, if ϕ denotes the temporal formula $\Box(0 \leq x \leq 3)$ and if Ψ_1 denotes $\Box(x=1 \lor x=2)$, then Ψ_2 defined by $\Psi_2 \equiv \Box(x=1 \lor x=2) \land \Diamond y=3$ is *not* a ϕ-derivative of specification Ψ_1. I will discuss in Subsection 3.2 a more liberal method, which does enable a designer to modify a collection of variables during a development. That method also enables a designer to "backtrack" during a development, when it has observed that some sequence of derivation steps leads to an undesirable result.

For some fixed property ϕ, ϕ-derivation does not necessarily preserve properties different from ϕ. In general, one does not even have monotonicity nor anti-monotonicity of ϕ-derivations w.r.t. argument ϕ. The validity of these claims is shown in the following:

Example 3.1 Let z be a variable ranging over the natural numbers; let predicate $even(z)$ assert that variable z's value is an even natural number.

(a) It is ***not true*** that ϕ-derivations are monotone in ϕ. I.e., it is not the case that $\phi \Rightarrow \phi'$ and $\Psi_1 \sqsubseteq_\phi \Psi_2$ imply $\Psi_1 \sqsubseteq_{\phi'} \Psi_2$. Consider, e.g., $\phi \equiv (\Box z{=}0) \vee (\Box z{=}2)$, $\phi' \equiv \Box even(z)$, $\Psi_1 \equiv \Box z{=}4$, and $\Psi_2 \equiv \Box z{=}0$.

(b) It is ***not true*** that every ϕ-derivation preserves all temporal properties. I.e., it is not the case that $\Psi_1 \Rightarrow \phi'$ and $\Psi_1 \sqsubseteq_\phi \Psi_2$ imply $\Psi_2 \Rightarrow \phi'$. Take, e.g., $\phi \equiv \Box z{=}0$, $\phi' \equiv \Box z{=}1$, $\Psi_1 \equiv \Box z{=}1$, and $\Psi_2 \equiv \Box z{=}0$.

(c) It is ***not true*** that ϕ-derivations are anti-monotone in ϕ. I.e., it is not the case that $\phi' \Rightarrow \phi$ and $\Psi_1 \sqsubseteq_\phi \Psi_2$ imply $\Psi_1 \sqsubseteq_{\phi'} \Psi_2$. If every ϕ-derivation would be anti-monotone in ϕ, then every ϕ-derivation would preserve all temporal properties. (Choose $\phi' \equiv$ *false*; and apply theorem 3.3 below.) ■

Example 3.1(b) above shows that ***not all temporal properties*** are preserved by every ϕ-derivation. This is the key difference between the method proposed here and the methods advocated in, e.g., in [4, 5, 25, 26, 27], which preserve ***all temporal properties*** during a derivation step. If required, however, all temporal properties may be preserved during a derivation step using the method presented in this paper:

Theorem 3.3 For specifications Ψ_1 and Ψ_2, $\Psi_1 \sqsubseteq_{false} \Psi_2$ and Ψ_2 ***implements*** Ψ_1 are equivalent.
Proof: By Definition 3.1, $\Psi_1 \sqsubseteq_{false} \Psi_2$ is equivalent to $\Psi_2 \Rightarrow \Psi_1$. ■

The method advocated above for developing some program which has to satisfy property ϕ (see the discussion after Theorem 3.2) will always preserve sequences already satisfying ϕ. In fact, as the goal is to find some end-product that satisfies ϕ, there is no reason to avoid replacing sequences satisfying ϕ by other ones (in a derivation step) as long as property ϕ is preserved. This observation is formalized in the theorem below. There, ϕ_1 and ϕ_2 are temporal properties both stronger than ϕ. The theorem shows that one can eliminate sequences satisfying ϕ_1, and replace them by sequences which satisfy ϕ_2 (without destroying property ϕ).

Theorem 3.4 Assume that specification Ψ_1 satisfies temporal property ϕ_1. Let ϕ_2 and ϕ be temporal properties such that $(\phi_1 \vee \phi_2) \Rightarrow \phi$ holds. If Ψ_2 is a ϕ_2-derivative of specification Ψ_1, then Ψ_2 satisfies ϕ. ■

This theorem allows one to replace, during a derivation step (in a derivation of a program which has to satisfy property ϕ), ***certain*** or even ***all*** sequences satisfying ϕ by other ones. Examples can be found in [39].
Theorem 3.4 has several important consequences:

(a) It allows a more liberal method for developing programs than the basic method. The more liberal method is discussed in Subsection 3.2. It does not require more and more concrete programs during a development; rather, a development will be represented by a finite, directed graph in which one can "backtrack" if desired. This corresponds to the idea that one may resume a development at stage passed earlier, even the initial one, if a sequence of derivation steps leads to an undesirable result.

(b) It allows a designer to modify the collection of (free) variables in a specification. This is discussed in Subsection 3.2.1.

3.2 Another Derivation Method

Theorem 3.4 suggests a less rigid method STRAT than the basic method for developing programs which have to satisfy property ϕ: Start with a certain specification Ψ_1 which satisfies ϕ, and derive specifications $\Psi_1, \Psi_2, \cdots, \Psi_n$ ($n \geq 1$) such that Ψ_{i+1} is a ϕ_i-derivative of Ψ_i for a certain temporal property ϕ_i. The choice of the properties ϕ_i, for $i=1, \cdots, n-1$, is not arbitrary; these properties have to be selected such that the following is true: If specification Ψ_i satisfies ϕ then specification Ψ_{i+1} also satisfies ϕ ($i=1,\cdots,n-1$). In this case, specification Ψ_n will, obviously, satisfy the required property ϕ. Thus, *the derivation relation may be applied in a development with different arguments.*

Derivations employing this method may be represented by a finite, directed, acyclic graph in which all nodes are labeled by some temporal property. The graph can be viewed as having a root labeled by ϕ, the initial specification, from which every other node can be reached along some directed path. Labels of nodes different from the root are labeled by $\Psi_1, \Psi_2, \cdots, \Psi_n$, the specifications obtained during the development. It is convenient to label some of the edges, too: an edge from a node labeled Ψ_i to a node labeled Ψ_{i+1} is labeled by ϕ_i, where $\Psi_i \sqsubseteq_{\phi_i} \Psi_{i+1}$ has been proved to hold ($i=1,\cdots,n-1$); the edge from the root labeled ϕ to the node labeled Ψ_1 is then labeled by *false* to represent that $\phi \sqsubseteq_{false} \Psi_1$ is true. Apart from these labeled edges, there also exist some edges without any label. If an edge in the graph, labeled ϕ_i is the only incoming edge of Ψ_{i+1}, then the corresponding derivation step is viewed as leading from some specification Ψ_i to the more concrete specification Ψ_{i+1} ($1 \leq i \leq n$). If, except for the edge labeled ϕ_i, specification Ψ_{i+1} has another incoming edge, which is unlabeled, from say specification Ψ_j, $1 \leq j < i$, then Ψ_{i+1} is viewed to be obtained from Ψ_i without being more concrete than Ψ_i, but more concrete than Ψ_j. This is what has been called "backtracking" earlier: It is as if the derivation is resumed at another, more abstract, level.

Two particular relations different from $\sqsubseteq_\phi$ that can be used in the development of a program, which has to satisfy property ϕ, follow from the Theorems 3.3 and 3.4. Another one is described in Theorem 3.5 below. At first, I informally motivate that theorem.

Suppose that at a certain stage during some development of some program which has to satisfy property ϕ, one has started with some specification satisfying ϕ and has derived specification Ψ_1, e.g., using ϕ-derivations. Specification Ψ_1 might satisfy some property ϕ' different from ϕ, which the designer wishes to preserve. One reason for this could be that property ϕ' allows the designer to derive an *efficient* program satisfying ϕ; another reason for the above-mentioned objective could be that property ϕ' guides the designer to some specific implementation. In order to continue the development, the designer could continue with a $\phi \wedge \phi'$-derivation: It will preserve both properties ϕ and ϕ'; and in this case the $\phi \wedge \phi'$-derivation is a ϕ-derivation, as shown in the next theorem.

Theorem 3.5 If specification Ψ_1 satisfies temporal property ϕ', and if Ψ_2 is a $\phi \wedge \phi'$-derivative of Ψ_1, then Ψ_2 satisfies ϕ', and Ψ_2 is a ϕ-derivative of Ψ_1. ■

This theorem shows that during a certain development one may *strengthen* the property to be preserved. It also allows one to *structure* temporal properties satisfied

by programs. Consider, for example, a derivation of programs which all satisfy some invariant $\Box I$. (This invariant might be a consequent of the initial specification.) Now, if the last program in this derivation satisfies, apart from invariant $\Box I$, also some invariant $\Box J$, then the designer could decide to preserve $\Box I \wedge \Box J$ during subsequent derivation steps. The programs then obtained will satisfy a stronger invariant than $\Box I$; furthermore $\Box J$ describes additional properties of the execution sequences of programs, which may bias the designer to some particular implementation.
A development in which invariants have been strengthened and structured during refinement steps can be found in [38, 37], where a simple broadcasting protocol has been derived.

3.2.1 Modifying the collection of variables

Modifying the collection of variables occurring free in a specification can be obtained by means of data-refinements [19, 2]. In this subsection it is demonstrated how to formulate this in the framework of the present paper.

Consider properties ϕ, Ψ_1, and Ψ_2 defined by $\phi \equiv \Box(0 \leq x \leq 3)$, $\Psi_1 \equiv \Box(x=1 \vee x=2)$, and $\Psi_2 \equiv \Box(x=1 \vee x=2) \wedge \Diamond y=3$. Applying the basic method does not allow one to transform Ψ_1 into Ψ_2 in a derivation of some program which has to satisfy property ϕ. (This has been argued in Subsection 3.1.) Such a transformation is allowed, however, when the method discussed in Subsection 3.2 is applied. This is an immediate consequence of Theorem 3.4. (Choose in that theorem $\phi_1 \equiv \phi$ and $\phi_2 \equiv \phi \wedge \Diamond y=3$.) Thus, one can add new variables to a specification during a development.)

It is next shown that one may even replace some of the variables by (a collection) of other variables. More precisely, it is now discussed how to replace a certain data-structure by other ones in a specification (or program) without invalidating that specification.

Assume that $\phi(x)$ is some specification which depends on variable x. Let S be a program satisfying $\phi(x)$, such that S operates on variable x, i.e., x occurs in at least one statement of S. The objective is to find some program S' which also satisfies specification $\phi(x)$, and which operates on fresh variables $x_1, \cdots, x_n$ ($n \geq 1$), but not on variable x. It is assumed that there exists some correspondence, a state formula, $C(x, x_1, \cdots, x_n)$ between the variables $x, x_1, \cdots, x_n$.
Let $\Psi_1(x)$ be the translation of program S as given in Subsection 2.2. [In general, variable x occurs free in specification $\Psi_1(x)$.] First, consider specification $\Psi_1(x) \wedge \Box C(x, x_1, \cdots, x_n)$. It corresponds to a program obtained from S by including state formula $C(x, x_1, \cdots, x_n)$, as a conjunct, to S's always-segment. Thereafter, find some specification $\Psi_2(x_1, \cdots, x_n)$, depending on the variables $x_1, \cdots, x_n$, but not on x, such that $\Psi_1(x) \sqsubseteq_{\phi(x) \wedge \Box C(x, x_1, \cdots, x_n)} \Psi_2(x_1, \cdots, x_n)$ holds. If $\Psi_2(x_1, \cdots, x_n)$ is the translation of some program S' as described in Subsection 2.2, then specification $\Psi_2(x_1, \cdots, x_n) \wedge \Box C(x, x_1, \cdots, x_n)$ corresponds to the program obtained from S' by including $C(x, x_1, \cdots, x_n)$, as a conjunct, to S''s always-segment. Observe that $\Psi_2(x_1, \cdots, x_n) \wedge \Box C(x, x_1, \cdots, x_n)$ satisfies property ϕ. This transformation is formulated in the following:

Theorem 3.6 Let $\Psi_1(x)$ be a specification in which x occurs free. Assume that relation $C(x, x_1, \cdots, x_n)$, a state formula, holds between variable x and fresh variables $x_1, \cdots, x_n$. Let $\Psi_2(x_1, \cdots, x_n)$ be some specification which depends on the variables $x_1, \cdots, x_n$, but not on x, such that $\Psi_1(x) \sqsubseteq_{\phi(x) \wedge \Box C(x, x_1, \cdots, x_n)} \Psi_2(x_1, \cdots, x_n)$ holds. Then $\Psi_1(x) \wedge \Box\ C(x, x_1, \cdots, x_n) \sqsubseteq_{\phi} \Psi_2(x_1, \cdots, x_n) \wedge \Box\ C(x, x_1, \cdots, x_n)$ hold, too. ■

3.3 Debugging Programs

It is next demonstrated that the approach advocated in the previous subsection can, apart from developing (correct) programs, be applied for correcting programs. That is, assume that someone has written a certain program S which has to satisfy property ϕ. It might be the case -this is even common for larger programs- that program S admits execution sequences for which property ϕ does not hold at all. From a formal point of view, program S is incorrect. Despite this fact, program S might be applied in some system in which failures of S are tolerable. Now, instead of developing a new program, usually program S will be corrected. Below in Theorem 3.7 a rule is formulated along which lines such a correction can take place.

To describe the rule intuitively, assume that some temporal specification Ψ_1, identified with a certain program S, should satisfy property ϕ. If specification Ψ_1 does not satisfy property ϕ, then Ψ_1 admits a collection of sequences, characterized by temporal property χ, for which ϕ does not hold. (This is expressed in condition (1) in Theorem 3.7.) These "incorrect" sequences might be removed and replaced by a collection of, possible new, sequences characterized by temporal property γ, all satisfying property ϕ. Although, this transformation may not result in a specification that satisfies property ϕ, one has come "closer" to a correct solution. The technical formulation of this rule is as follows:

Theorem 3.7 Let Ψ_1, χ, γ, ϕ be temporal properties. Assume that (1) $\chi \Rightarrow \neg\phi$ and (2) $\gamma \Rightarrow \phi$ hold. Then, for $\Psi_2 \equiv (\Psi_1 \wedge \neg\chi) \vee \gamma$, (a) Ψ_2 is a ϕ-derivative of Ψ_1 and (b) $\Psi_2 \Rightarrow \neg\chi$ are both satisfied. ■

In practice, one would like to ensure that Ψ_2 in Theorem 3.7 is the translation of some program. An example of how to apply this theorem will be given in another paper.

4 An Example

In this section it is shown how to apply the parameterized relation in program development. Because of the space limitations, a small example is considered. The emphasis has been placed on refining atomicity, see also [3].

Let x, y be two variables ranging over the natural numbers. It is required to compute the sum of x and y, and to record this sum in a variable *sum*. The initial values of the variables x, y, and *sum* are irrelevant. It must also be ensured that some boolean variable *done* is *true* when the value of variable *sum* equals the sum of (the initial values of) x and y. Formally, it is required to find a program which satisfies the following specification ϕ: ϕ is the conjunction of the initial condition

pre defined by $pre \equiv \neg done$, and the temporal properties $\Diamond\Box done$ and $\forall m, n.(x=m \wedge y=n) \Rightarrow (\Box(done \Rightarrow sum=m+n))$.
(Here, *m* and *n* denote two variables ranging over the natural numbers.)

A possible implementation of specification ϕ is easy. Consider the program S defined by $S \equiv$ *initial-segment*: *pre*
always-segment: *true*
assignment-segment: $\neg done \rightarrow sum, done := x+y, true$.

The translation of program S, *Trans*(S), as defined in Subsection 2.2, is equivalent to Ψ_1 defined by

$$\Psi_1 \equiv pre \wedge \Box \quad (\neg done \wedge (x, y, true, x+y)=O(x, y, done, sum)) \vee (done \wedge (x, y, done, x+y)=O(x, y, done, sum)).$$

It is straightforward to prove that $\Phi_1 \Rightarrow \phi$ holds.

Program S solves the problem specified by property ϕ in one single atomic step. When it is required, for some reason or another, to refine this atomic step the following program could be considered:

$T \equiv$ *initial-segment*: $pre \wedge t=0$
always-segment: *true*
assignment-segment: $t=0 \rightarrow sum, t := x+y, 1$
$t=1 \rightarrow done, t := true, 2$.

Note that not all (stutter-free) temporal properties satisfied by program S are satisfied by program T. As an example, consider the following property: $sum \neq x+y \Rightarrow \Box(done \Leftrightarrow sum=x+y)$. It expresses that, if in the initial state the value of *sum* differs from the sum of (the values of) x and y, then in any state *done* is true iff *sum* has its required value. This temporal property is satisfied for program S. Yet, however, it is not satisfied for program T, because it does not hold in the state following the initial state in any execution sequence of T in which $sum=x+y$ is true, but *done* is not.

In the setting of this paper, it can be proved that program T may result from program S in a derivation of a program which has to satisfy the initial specification ϕ. In order to show this, define temporal formula Ψ_2 defined by

$$\Psi_2 \equiv pre \wedge t=0 \wedge \Box \quad (t=0 \wedge (x, y, done, x+y, 1)=O(x, y, done, sum, t)) \vee (t=1 \wedge (x, y, true, sum, 2)=O(x, y, done, sum, t)) \vee (t=2 \wedge (x, y, done, sum, t)=O(x, y, done, sum, t)).$$

Then Ψ_2 is equivalent to the translation of program T, *Trans*(T).

Now, define ϕ_1 by $\phi_1 \equiv \Psi_1$; and ϕ_2 by $\phi_2 \equiv \phi \wedge (\neg done \Rightarrow O(\neg done \wedge sum=x+y))$. Thus, ϕ_2 is the conjunction of ϕ and the temporal property expressing the following: if in the initial state *done* is not true, then in the second state of any execution *done* is not true and *sum* has the value of x and y. For this choice of temporal properties, Theorem 3.4 can be applied. Therefore, program T satisfies specification ϕ.

In program T the sum of the variables x and y is computed and assigned to variable *sum* in a single atomic action. This action is next split up into two actions. The first one computes the sum of x and y, and records the result in variable x; the second one assigns the new value of x to variable *sum*. The refinement of program T, T', is shown below. Note, again, that the transformation from T to T' does not preserve all temporal properties: The (stutter-free) temporal property "variable x has a constant value", expressed by $\forall n.(x=n \Rightarrow \Box x=n)$, holds for any execution sequence of program T. Obviously, this property is not necessarily true for all execution sequences of program T'.

$T' \equiv$ *initial-segment*: $pre \wedge t{=}0$

always-segment: *true*

assignment-segment: $t{=}0 \rightarrow x, t := x{+}y, 1$

$$t{=}1 \rightarrow sum, t := x, 2$$

$$t{=}2 \rightarrow done, t := true, 3.$$

Program T can be identified with the temporal formula Ψ_3 defined by

$$\begin{aligned}\Psi_3 \equiv pre \wedge t{=}0 \wedge \square \quad & (t{=}0 \wedge (x{+}y, y, done, sum, 1){=}O(x, y, done, sum, t)) \\ & \vee (t{=}1 \wedge (x, y, done, x, 2){=}O(x, y, done, sum, t)) \\ & \vee (t{=}2 \wedge (x, y, true, sum, 3){=}O(x, y, done, sum, t)) \\ & \vee (t{=}3 \wedge (x, y, done, sum, t){=}O(x, y, done, sum, t)).\end{aligned}$$

In order to show that program T' can be obtained from program T in a development of a program which has to satisfy property ϕ, Theorem 3.4 is applied, again. Consider, e.g., $\phi' \equiv \Psi_2$ and $\phi'' \equiv \phi \wedge \square(done \Leftrightarrow t{=}3)$. This concludes the example, and, hence, this section.

5 Conclusions

A new method STRAT for deriving (sequential or concurrent) programs has been presented. The underlying idea is that at any point during a development a *certain* collection of *temporal properties*, viz., those of interest at that point of the development, have to be preserved; and that it is *not necessary*, in general, to preserve the collection of *all temporal properties*. This is the main difference between STRAT and those methods which use Abadi and Lamport's notion *implements* as their basic refinement relation. Examples of the latter methods are those described in [4, 5, 25, 26, 27].

In order to formulate method STRAT, a relation between specifications which is parameterized by a temporal property has been defined. If ϕ is the argument, then the relation has been denoted by $\sqsubseteq_\phi$; relation $\sqsubseteq_\phi$ has been referred to as a ϕ-derivation. It has been shown that if $S \sqsubseteq_\phi T$ holds for specifications (or programs) S and T, then the following are true:

1. If S satisfies property ϕ, then T satisfies ϕ, too. Thus, property ϕ is preserved by a ϕ-derivation.
2. In general, specification S satisfies some temporal property not satisfied by specification T. Thus, it is not always the case that all temporal properties are preserved by a ϕ-derivation.

It has also been shown that for a particular parameter ϕ, ϕ-derivation coincides with the relation *implements*. This implies that method STRAT is as least as powerful as those methods which use *implements* as their basic relation. It has been argued, that it is preferable to apply method STRAT rather than those other methods.

Method STRAT enables a designer to apply the relation defined in this paper with various arguments during a development in order to obtain an executable implementation, which satisfies the designer's initial requirement. This allows for more flexibility than other methods, which require a single refinement relation to be applied in every derivation step.

To the best of my knowledge, existing derivation techniques are always demonstrated by showing only a complete, often streamlined, derivation D of some program. Although the designer might have applied a (*correct*) transformation at some stage of the development, resulting in, say program S, from which the desired end-product cannot be derived, S will never be recorded in derivation D. This occurs, e.g., when program S satisfies some temporal property which contradicts a certain (temporal) property in the end-product. As a result, several design-decisions made during an actual derivation will not be represented in derivation D. With method STRAT, however, a program S as above may be recorded in the derivation (since the designer has derived S by means of a certain transformation). Thereafter, Theorem 3.4 could be applied to continue the derivation and to obtain the desired end-product.

The applicability of method STRAT has been demonstrated in Section 4. More examples can be found in the full paper [39]. It is believed that the parameterized relation is applicable to all derivations given in [10]. Future work will be carried out to apply method STRAT to larger examples. Future research will also be carried out to prove a soundness and completeness result of the parameterized relation in the style of Abadi and Lamport [1].

The relation proposed here has been shown to be, apart from developing programs, applicable also for debugging programs, i.e., for correcting programs. An example will be given in another paper.

The present paper has concentrated on the preservation of a collection of linear time temporal properties. It is envisaged that, without too much effort, the same kinds of ideas apply in the context of branching time temporal logic [15], total correctness [7], and partial orders [22].

Acknowledgements: I would like to thank T. Långbacka for lending me his ear when I tried to organize my ideas, and N. Francez for pointing out some recent and relevant references. I am indebted to S. Katz for his constructive criticism and suggestions, which led to many improvements of the paper.

References

1. M. Abadi and L. Lamport. The existence of refinement mappings. In *Proc. of the 3rd IEEE Conference on Logic in Computer Science*, pages 165–175, 1988.
2. R. J. R. Back. Changing data representation in the refinement calculus. In *21st Hawaii International Conference on System Sciences*, 1989.
3. R. J. R. Back. A method for refining atomicity in parallel algorithms. In *PARLE89*, pages 199–216, 1989.
4. R. J. R. Back. Refinement calculus, part ii: Parallel and reactive programs. In J. W. de Bakker, W. P. de Roever, and G. Rozenberg, editors, *Stepwise Refinement of Distributed Systems: Models, Formalisms, Correctness*, volume 430 of *LNCS*, pages 67–93, 1990.
5. R. J. R. Back and R. Kurki-Suonio. Superposition and fairness in action system refinement. Unpublished Manuscript, 1991.
6. R. J. R. Back and K. Sere. Stepwise refinement of action systems. In *Proc. of Math. of Program Construction*, volume 375 of *LNCS*, 1989.
7. R.J.R. Back. A calculus of refinements for program derivations. *Acta Informatica*, 25:593–624, 1988.
8. M. Ben-Ari. Algorithms for on-the-fly garbage collection. *Transactions of Programming Languages and Systems*, 6(3):333–344, 1984.

9. K. M. Chandy and L. Lamport. Distributed snapshots: Determining global states of distributed systems. *ACM Trans. on Comp. Syst.*, 3(1):63–75, 1985.
10. K. M. Chandy and J. Misra. *Parallel Program Design: A Foundation*. Addison-Wesley Publishing Company, Inc., 1988.
11. J. W. de Bakker, W. P. de Roever, and G. Rozenberg, editors. *Stepwise Refinement of Distributed Systems: Models, Formalisms, Correctness*, volume 430 of *LNCS*. Springer Verlag (Berlin), 1990.
12. E. Diepstraten and R. Kuiper. Abadi & Lamport and Stark: Toward a proof theory for stuttering, dense domains and refinement mappings. In J. W. de Bakker, W. P. de Roever, and G. Rozenberg, editors, *Stepwise Refinement of Distributed Systems: Models, Formalisms, Correctness*, volume 430 of *LNCS*, pages 208–238, 1990.
13. E. W. Dijkstra, L. Lamport, A. J. Martin, C. S. Scholten, and E. F. M. Steffens. On the fly garbage collection: An exercise in cooperation. *Communications of the ACM*, 21(11):966–975, 1987.
14. E. W. Dijkstra and C. S. Scholten. Termination detecting for diffusing computations. *Information Processing Letters*, 11(1):1–4, 1980.
15. E. A. Emerson and Srinivasan J. Branching time temporal logic. In J. W. de Bakker, W. P. de Roever, and G. Rozenberg, editors, *Linear Time, Branching Time, and Partial Order in Logics and Models for Concurrency*, volume 354 of *LNCS*, pages 123–172, 1988.
16. N. Francez. Distributed termination. *Transactions of Programming Languages and Systems*, 2(1):42–55, 1980.
17. N. Francez. *Fairness*. Springer Verlag, 1986.
18. R. Gerth. Foundations of compositional program refinement. In J. W. de Bakker, W. P. de Roever, and G. Rozenberg, editors, *Stepwise Refinement of Distributed Systems: Models, Formalisms, Correctness*, volume 430 of *LNCS*, pages 777–808, 1990.
19. C. A. R. Hoare. Proofs of correctness of data representations. *Acta Informatica*, 1(4):271–281, 1972.
20. C. B. Jones. *Software Development*. Prentice-Hall International, 1980.
21. B. Jonsson. *Compositional Verification of Distributed Systems*. PhD thesis, Department of Computer Science, Uppsala University, Uppsala, Sweden, 1987.
22. S. Katz and D. Peled. Interleaving set temporal logic. In *Proc. of the 6th ACM Symposium on Distributed Computing*, pages 178–190, 1987.
23. S. Katz and O. Shmueli. Cooperative distributed algorithms for dynamic cycle prevention. *IEEE Transactions on Software Engeneering*, 13(5):540–552, 1987.
24. N. Klarlund. *Progress Measures and Finite Arguments for Infinite Computations*. PhD thesis, Department of Computer Science, Cornell University, Ithaca, New York, USA, 1990. Also published as Technical Report 90-1153, Cornell University.
25. S. S. Lam and A. U. Shankar. Refinement and projection of relational projections. In J. W. de Bakker, W. P. de Roever, and Rozenberg G., editors, *Stepwise Refinement of Distributed Systems: Models, Formalisms, Correctness*, volume 430 of *LNCS*, pages 454–487, 1990.
26. L. A. Lynch. Multivalued possibility mappings. In J. W. de Bakker, W. P. de Roever, and G. Rozenberg, editors, *Stepwise Refinement of Distributed Systems: Models, Formalisms, Correctness*, volume 430 of *LNCS*, pages 519–543, 1990.
27. L. A. Lynch and M. Tuttle. Hierarchical correctness proofs for distributed algorithms. In *Proc. of the 6th ACM Symposium on Distributed Computing*, pages 137–151, 1987.
28. Z. Manna and A. Pnueli. Verification of concurrent programs: A temporal proof system. foundations of computer science iv, part 2. *MC-tracts*, 159, 1983.

29. R. Meritt. Completeness theorems for automata. In J. W. de Bakker, W. P. de Roever, and Rozenberg G., editors, *Stepwise Refinement of Distributed Systems: Models, Formalisms, Correctness*, volume 430 of *LNCS*, pages 544–560, 1990.
30. C. Morgan. The specification statement. *Transactions of Programming Languages and Systems*, 10(3):403–419, 1988.
31. H. Partsch. *Specification and Transformation of Programs – A Formal Approach to Software Development*. Springer Verlag (Berlin), 1990.
32. B. Sanders. Stepwise refinement of mixed specifications of concurrent programs. In M. Broy and C. B. Jones, editors, *Proceedings of the IFIP TC2 Conference on Programming Concepts and Methods*. Elsevier Science Publishers B. V., 1990.
33. K. Sere. *Stepwise Refinement of Parallel Programs*. PhD thesis, Department of Computer Science, Åbo Akademi, Turku, Finland, 1990.
34. Z. Shengzong. Compositional temporal logic specifications. Technical Report SFB 124 - 07/1991, Universität des Saarlandes, Deparment of Computer Science, Saarbrücken, Germany, 1991.
35. Z. Shengzong. UNITY-like logic programming. Technical Report SFB 124 - 06/1991, Universität des Saarlandes, Deparment of Computer Science, Saarbrücken, Germany, 1991.
36. A. K. Singh. Program refinement in fair transition systems. In E. H. L. Aarts, J. van Leeuwen, and M. Rem, editors, *Parallel Architectures and Languages Europe 1991*, volume 506 of *LNCS*, pages 128–147, 1991.
37. F. A. Stomp. A derivation of a broadcasting protocol using sequentially phased reasoning. In J. W. de Bakker, W. P. de Roever, and G. Rozenberg, editors, *Stepwise Refinement of Distributed Systems: Models, Formalisms, Correctness*, volume 430 of *LNCS*, pages 696–730, 1990.
38. F. A. Stomp. A derivation of a broadcasting protocol using sequentially phased reasoning (extended abstract). In L. Logrippo, R. L. Probert H., and Ural, editors, *Protocol Specification, Testing, and Verification, X*. Elsevier Science Publishers B. V. (North-Holland), 1990.
39. F. A. Stomp. Preserving specific properties in program development -How to debug programs- (journal version). In preparation, 1992.
40. F. A. Stomp. Preserving specific properties in program development -How to debug programs-. Technical report, Christian Albrechts Universität, Kiel, Germany, 1992. To appear.
41. J. van de Snepscheut. "Algorithms for on-the-fly garbage collection" revisited. *Information Processing Letters*, 24(4):211–216, 1987.
42. J. L. Welch, L. Lamport, and N. A. Lynch. A lattice-structured proof of a minimum spanning tree algorithm. In *Proc. of the 7th ACM Symposium on Distributed Computing*, pages 28–43, 1988.

Sometimes 'Some' is as Good as 'All'*

Doron Peled
Department of Computer Science
University of Warwick
Coventry CV4 7AL, UK
e-mail: doron@dcs.warwick.ac.uk

Abstract. The representation of partial order semantics as an equivalence relation on interleaving sequences was shown to extend the expressive power of interleaving semantics. A specification formalism called *existential specification* is introduced in which a formula is interpreted over equivalence classes of sequences by asserting that *some* (at least one but not necessarily all) sequences from each equivalence class satisfy a given property. It differs from the more common *universal specification*, which is interpreted over *all* sequences in all classes. Its advantage over other formalisms that deal with partial order executions lies in its simplicity: any syntax that is defined over interleaving sequences, e.g., linear temporal logic, can be adopted. It is shown how under an appropriate semantical construction, an exact existential specification of a program (i.e., each property of the program expressed using the same formalism is a consequence of this specification) can be given. Moreover, under such a construction, no information about the program is lost by choosing exact existential specification rather than exact universal specification; it is possible to generalise, by means of a proof system, an exact existential specification into an exact universal specification. Applications of these results to achieving compositional proof rules are shown.

1 Introduction

There has been a long lasting debate about the use of partial order semantics versus using interleaving semantics to model concurrent programs. Supporters of interleaving semantics often claim that this model includes all the information needed to be expressed about a program, and that partial order semantics adds no important extra information. Supporters of the partial order model show how certain properties are better described and proved using partial order semantics.

Representing partial order execution as an equivalence class of interleaving sequences (or equivalently, traces) provides a way to connect the two models. Using this model, there is a common paradigm of properties that requires that *at least one interleaving sequence from each equivalence class satisfies some conditions*. We shall call this *existential specification*, as opposed to the more common *universal specification*. Using existential specification is sometimes more convenient than asserting over *all* the interleaving sequences, as some interleaving sequences represent the desired property better than others.

One example is serializability of database programs [23], where each execution sequence is supposed to be equivalent to a sequence in which the transitions are executed one after the other. Other examples include specifying algorithms such as Chandy and Lamport's global snapshot algorithm [5], recovery from faults based on this algorithm [24], and layering of concurrent programs [8]. Program verification methods can exploit this paradigm by allowing proofs of the properties of programs using intermediate assertions

*Supported in part by SERC research grant GR/F57960

that apply only to the states of these representatives [16, 29], while in model checking it is used to reduce the size of the checked structure [30, 11]. Since existential specification covers a wide range of properties and can use a simple formalism such as linear temporal logic (LTL), it can be considered as an alternative to more powerful specification languages for partial order semantics [27, 15, 26].

Expressing properties of representatives rather than all the execution sequences was recognised in [15]. There, it was described as a class of properties embedded in the more powerful logic **ISTL***. In this paper, an alternative representation of this class of properties using a simpler framework that uses the syntax of LTL is suggested, and is further studied.

This paper deals with some basic questions about representing properties using existential specification, and its interrelation with the more familiar universal specification. The ability to specify *exactly* all the executions of a given program using existential specification is studied. Such a specification is desirable as all the properties of the program that are expressible in this formalism are consequences of it. An exact specification is often used for completeness proofs [19] and is especially important for compositional proof methods [4]. It is shown that provided that the semantical construction satisfy certain requirements, an exact existential specification can be given.

We prove that it is possible to conclude from an existential specification properties of *all* the sequences: this arises when one wants to *generalise* or conclude from an existential property a universal property of the program. It is shown that under an appropriate construction, if an exact existential specification of a program is given, it is possible to conclude from it an exact universal specification. This is especially beneficial for proof rules that exploits existential specification in some intermittent stages to achieve compositionality. Another motivation is that often the property or the program specified is best presented or proved using existential specification, but universal properties of the program ought to be provable from it.

In Section 2, some preliminary definitions concerning modeling and asserting about programs are reviewed. Then, existential satisfaction is formally presented. In Section 3 some examples of using existential specification for compositional verification are given. In Section 4, exact existential specification of a program is defined and shown to be sensitive to the choice of representing the states. Section 5 provides conditions for semantical constructions that allow exact existential specifications of programs. A construction of such structures is also presented. In Section 6, proof rules are given for generalising from an existential specification into a universal specification. It is shown that using these rules, an exact existential specification can be generalised into an exact universal specification.

2 Universal and Existential Satisfaction

2.1 Interleaving Semantics and Linear Temporal Logic

A program P is a finite set of named operations T, a finite set of variables V, and an initial condition Θ (a first order predicate). Each operation τ is a triple $\langle l_\tau, en_\tau, f_\tau \rangle$ such that $l_\tau \subseteq V$ is the set of τ's *local* variables, i.e., the only variables that τ can examine and change, en_τ is its enabling condition, i.e., a first order predicate with free variables from l_τ which determines when τ can be executed, and f_τ is a transformation on τ's local variables (which is naturally extended to transform the entire set of program variables V by not changing the variables which are not in l_τ). An operation τ will be denoted by $en_\tau \longrightarrow l_\tau := f_\tau(l_\tau)$. The assignment will not mention explicitly variables from l_τ that are not changed, e.g., writing $x = y \longrightarrow x := x + 1$ instead of $x = y \longrightarrow (x, y) := (x + 1, y)$.

A state is an assignment (interpretation) of values to the variables. In program verification, the states are usually interpretations of the set of program variables. However, by adding auxiliary variables [7] to the program one can add additional information to each state such as the sequence of operations executed so far. An *interleaving sequence* is a finite or infinite sequence of states. A program P generates a set of interleaving sequences $\xi = \xi_0 \xi_1 \xi_2 \ldots$ such that ξ_0 satisfies Θ, and for each $0 < i <|\xi|$ ($|\xi| = \omega$ if the sequence is infinite), there exists some operation $\alpha_i \in T$ such that ξ_{i-1} satisfies en_{α_i}, and $\xi_i = f_{\alpha_i}(\xi_{i-1})$. This sequence is finite iff its last state ξ_n does not satisfy en_τ for any $\tau \in T$. The sequence of operations $\alpha_1 \alpha_2 \ldots$ corresponding to ξ is called the *generating sequence* of ξ.

Let $I \subseteq T \times T$ be a symmetric and irreflexive relation called the *independence* relation, satisfying that $(\alpha, \beta) \in I$ iff $l_\alpha \cap l_\beta = \phi$. (Notice that $l_\alpha \cap l_\beta = \phi$ is usually a necessary condition for α and β to be independent but not always a sufficient condition. However, one can always force any two such operations to be dependent by adding to both sets l_α and l_β a fresh common dummy variable.) If $(\tau_1, \tau_2) \in I$, we say that τ_1 and τ_2 are *independent*, otherwise they are *dependent*. The independence relation I identifies when it is possible to commute operations.

Requirements from execution sequences called 'justice' and 'fairness' [22] are often imposed, so that only those interleaving sequences that satisfy them are considered as representing the executions of a program. We choose the following constraint:

Definition 2.1 *The* execution sequences *of a program P are exactly the interleaving sequences whose states are interpretations of P's variables and satisfy the following* justice *property: if from some state in the sequence, an operation α is enabled, then an operation that is dependent on α (possibly α itself) will occur eventually in the sequence. The set of execution sequences of P is denoted $\tilde{\mathcal{E}}[\![P]\!]$.*

Linear temporal logic [28] is a formalism that can be used to specify concurrent programs. As usual in LTL, $\Box\varphi$ means that φ holds in every future state in a sequence, $\Diamond\varphi$ means that φ will hold in some future state, $\varphi\mathcal{U}\psi$ means that φ will continue to hold until ψ holds, $\varphi \mathrm{U} \psi$ means that if either φ holds in every future state or until ψ holds, and $\bigcirc\varphi$ means that φ holds starting with the next state. Past temporal modals that correspond to the above future modals also exist. For example, $\varphi\mathcal{S}\psi$ (which is the past version of $\varphi\mathcal{U}\psi$) means that φ holds continuously since a state in the past in which ψ held. The connectives '$\neg$', '$\vee$', '$\wedge$', '$\rightarrow$', '$\leftrightarrow$' and the (rigid, i.e., state-independent) quantifiers '$\forall$' and '$\exists$' are also used. The predicate T holds always, while F never holds for every state or sequence.

In [4], a binary modal operator $\mathcal{C}$ called *chop* was added to LTL. The formula $\varphi\mathcal{C}\psi$ is satisfied by an interleaving sequence if it can be partitioned into a prefix satisfying φ and a suffix satisfying ψ or if the entire sequence satisfies φ. An LTL formula φ is interpret over a pair of an interleaving sequence ξ and an index i that designate the starting state in the sequence from which the formula needs to be satisfied. This is denoted $(\xi, i) \models \varphi$. An interleaving sequence ξ satisfies an LTL formula φ iff $(\xi, 0) \models \varphi$. This is denoted also as $\xi \models \varphi$. An LTL formula is *valid* if it holds for each sequence of states. A *past temporal formula* ψ is a formula that does not contain future modals. It is said to be *end-satisfied [6]* by a finite interleaving sequence if it holds in its last state, i.e., $(\xi, |\xi| - 1) \models \psi$.

2.2 Trace Semantics

Partial order semantics can be represented in various ways. One of the most simple representations is Mazurkiewicz's trace semantics [21].

A *history* of a program P is a pair $h = \langle J_0, v\rangle$ where J_0 is a state called the *initial interpretation of* h, and $v \in T^*$ satisfies that there exists a finite prefix of an execution sequence $\xi = J_0 J_1 \ldots J_n$ such that $v = \alpha_0\alpha_1 \ldots \alpha_n$ is its generating sequence. For each history $h = \langle J, v\rangle$, let the nth state J_n in the above sequence (which is a function of J_0 and v) be denoted by fin_h. This is called the *final interpretation* of h.

Two histories $h = \langle J, v\rangle$ and $h' = \langle J, w\rangle$ are *equivalent* (denoted $h \equiv h'$) if it is possible to obtain w from v by repeatedly commuting adjacent independent operations. That is, if there exists a sequence of histories $\langle J, v_1\rangle, \langle J, v_2\rangle, \ldots, \langle J, v_n\rangle$ with $v_1 = v$ and $v_n = w$, and for each $1 \le i < n$ there exist $u, \bar{u} \in T^*$, $(\alpha, \beta) \in I$ such that $v_i = u\alpha\beta\bar{u}$, $v_{i+1} = u\beta\alpha\bar{u}$. For example, if $T = \{\alpha, \beta, \gamma\}$ and $I = \{(\alpha, \beta), (\beta, \alpha), (\gamma, \beta), (\beta, \gamma)\}$, then $\langle J, \alpha\alpha\beta\beta\gamma\gamma\rangle \equiv \langle J, \alpha\beta\alpha\gamma\beta\gamma\rangle \equiv \langle J, \beta\alpha\alpha\gamma\gamma\beta\rangle$, but $\langle J, \alpha\alpha\beta\gamma\gamma\rangle \not\equiv \langle J, \gamma\gamma\beta\alpha\alpha\rangle$. A *trace* is an equivalence class of histories, denoted by $[J, w]$, where J is the common initial interpretation and $\langle J, w\rangle$ is a member of the equivalence class.

Obviously, if $h = \langle J, u\alpha\beta\bar{u}\rangle$ is a history of P, and $(\alpha, \beta) \in I$, then $h' = \langle J, u\beta\alpha\bar{u}\rangle$ is also a history of P. Moreover, $fin_h = fin_{h'}$. Therefore, the final interpretations of all the histories in a trace are identical. This permits defining fin_σ as fin_h for any $h \in \sigma$. Concatenation between two traces $\sigma_1 = [J_1, v]$ and $\sigma_2 = [J_2, w]$, denoted $\sigma_1\sigma_2$, is defined when $fin_{\sigma_1} = J_2$, and is given by $[J_1, vw]$. The relation '$\sqsubseteq$' between traces is defined as $\sigma_1 \sqsubseteq \sigma_2$ iff there exists some σ_3 such that $\sigma_1\sigma_3 = \sigma_2$. It is then said that σ_1 is *subsumed* by σ_2 or that σ_1 is a *prefix* of σ_2. If, in addition, σ_3 contains a single operation, σ_2 is an *immediate successor* of σ_1.

An *accessible* trace of a program P is any trace obtained as the equivalence class of some of the histories of P with an initial interpretation satisfying Θ. The traces *generated by a program P* are traces σ such that for some ρ, $\rho\sigma$ is an accessible trace of P. In the sequel we consider only traces generated by programs.

Two traces σ_1, σ_2 are *consistent*, denoted $\sigma_1 \Uparrow \sigma_2$, iff there exists σ_3 such that $\sigma_1 \sqsubseteq \sigma_3$ and $\sigma_2 \sqsubseteq \sigma_3$. A *run* (partial order execution) Π of P is a maximal set of pairwise consistent accessible traces. The set of all runs of a program P is denoted R_P. A set of traces Π is *directed* if for each two traces $\sigma, \rho \in \Pi$, a trace subsuming both traces exists in Π. It is not difficult to show that each run Π is also prefix closed and maximal consistent.

For each run Π, an *interleaving sequence of traces* is a sequence whose states are traces $\sigma_0\,\sigma_1\,\sigma_2 \ldots$ of Π, such that $\sigma_0 = [J, \epsilon]$ (an empty trace) for some $J \models \Theta$, and for each $i \geq 0$, $\sigma_i \sqsubset_{im} \sigma_{i+1}$.

Definition 2.2 *An* observation *of Π is an interleaving sequence where for each $\sigma \in \Pi$, there exists some σ_i, $i \geq 0$ such that $\sigma \sqsubseteq \sigma_i$.*

The set of observations of Π is denoted $obs(\Pi)$. Denote the fact that o, o' are both in $obs(\Pi)$ for some $\Pi \in R_P$ by $o \approx o'$. This is obviously an equivalence relation on the observations of each program P. Denote $\Delta_P = \bigcup_{\Pi \in R_P} \Pi$, i.e., the set of all the traces of the program P. If Σ is a set of traces, denote $\downarrow\Sigma = \{\sigma \mid \exists\rho \in \Sigma(\sigma \sqsubseteq \rho)\}$ (i.e., all the traces subsumed by traces of Σ).

It is shown in [17, 25] that the set of observations of a program are exactly the interleaving sequences of traces satisfying the justice condition of Definition 2.1. Another important characterisation of the observations of P is given by the following Lemma.

Lemma 2.3 *An interleaving sequence of traces $\xi = \sigma_0\sigma_1\sigma_2 \ldots$ of P is an observation iff for each trace $\sigma \in \Delta_P$, there exists some trace σ_i on ξ, such that either $\sigma \sqsubseteq \sigma_i$ or $\sigma_i \not\Uparrow \sigma$. In particular, this holds for $\Sigma = \Delta_P$. Moreover, this still holds if Δ_P is replaced by any set of traces Σ such that $\downarrow\Sigma = \Delta_P$.*

Proof. Follows from the definition of runs and observations. ∎

In order to reason about sequences of traces using temporal logic, it is necessary to map traces into states, such that state predicates and functions symbols on state variables can be used to assert about traces, and temporal operators can then be added to formalise sequence assertions. These state predicates and functions can refer to information extracted from traces. For example, in specification and verification of concurrent programs, often only the value of the program variables according to fin_σ (i.e., the values of the variables after executing operations according to any history in σ) are of interest.

Definition 2.4 *A* representation of traces *is a mapping $f : \Gamma \mapsto \mathcal{R}$ from a class of traces Γ to a set of elements (the domain of the representation) $\mathcal{R}$. The function f is naturally generalised to sequences and sets of sequences. Sequences of $\mathcal{R}$-elements will be called $\mathcal{R}$-sequences.*

For trace semantics, $\mathcal{E}[\![P]\!] = \{f(obs(\Pi)) \mid \Pi \in R_P\}$ represents not only the set of observations of P, but also the grouping of them into sets. Hence, if f maps traces into interpretations of program variables, then $\tilde{\mathcal{E}}[\![P]\!] = \bigcup_{\Pi \in R_P} \mathcal{E}[\![P]\!]$. That is, $\tilde{\mathcal{E}}[\![P]\!]$ is a single set containing all the sequences of $\mathcal{E}[\![P]\!]$ (but not the partitioning). Denote by $\xi \Subset \mathcal{E}[\![P]\!]$ the fact that $\xi \in \bigcup_{\Pi \in R_P} \mathcal{E}[\![P]\!]$.

2.3 Universal and Existential Satisfaction

Definition 2.5 *An LTL formula φ is* universally satisfied *by a set of sequences M, denoted by $M \models \varphi$, iff for each sequence $\xi \in M$, $\xi \models \varphi$. This is extended to a set of sets of sequences $\mathcal{A}$, denoting $\mathcal{A} \models \varphi$ iff $M \models \varphi$ for each $M \in \mathcal{A}$. (Notice that the meaning of the relation '$\models$' depends on the type of its first argument.)*

Definition 2.6 *An LTL formula φ is* existentially satisfied *by a set of sequences M, denoted by $M \models^{\exists} \varphi$, iff there exists some sequence $\xi \in M$ such that $\xi \models \varphi$. This is extended to a set of sets of sequences $\mathcal{A}$, denoting $\mathcal{A} \models^{\exists} \varphi$ iff $M \models^{\exists} \varphi$ for each $M \in \mathcal{A}$.*

Thus, $\mathcal{E}[\![P]\!] \models^{\exists} \varphi$ means that φ is satisfied by at least one representative observation from any equivalence class of $\mathcal{E}[\![P]\!]$, while $\mathcal{E}[\![P]\!] \models \varphi$ ignores the equivalence relation by requiring that a formula is satisfied by all the (representations of the) observations of P (i.e., all the sequences $\xi \in\in \mathcal{E}[\![P]\!]$ satisfy the formula φ). Existential satisfaction is weaker than universal satisfaction with respect to the same formula, because it demands that only *representatives* of the equivalence classes satisfy a formula, rather than all of them. Denote by $P \vdash \mu$ the fact that one can prove using some proof system that $\mathcal{E}[\![P]\!] \models \mu$. Similarly, $P \vdash^{\exists} \mu$ means that $\mathcal{E}[\![P]\!] \models^{\exists} \mu$ is provable.

The same LTL formula can be interpreted in both universal and existential ways. Henceforth, we will mention explicitly the way a formula is interpreted over the semantic models. The expressive power [9, 18] of existential and universal satisfaction is incomparable. This follows from the fact that if N and M are two sets of sequences such that $N \subseteq M$, then there exists no formula which is satisfied existentially by N but not by M (even in the cases where there exists such a universally satisfied formula), and no formula which is satisfied universally by M but not by N (even in the cases where there exists such an existentially satisfied formula).

3 Applicability of Existential Specification

Existential specification is more appropriate than universal specification when some execution sequences of a program correspond better than others to the property under consideration. In these cases, it is preferable to specify the behaviour of at least one sequence out of each equivalence class.

Composing a program out of smaller segments often results in a program whose behaviour is best described existentially. The reason for this is that by composing, concurrency is enhanced and the executions of the segments are interleaved. However, some of the concurrent activities of different parts are independent of each other. Thus, sequences in which a pair of activities are considered to execute separately one after the other in one of either orders are equivalent to all the sequences in which occurrences of operations from both activities are interleaved (and thus can be chosen as representatives to this larger set of sequences).

In order to achieve compositionality, we use *program segments* rather than programs. A program segment differs from a program by not having its own initial condition. The pair $\langle P, \Theta_P \rangle$ of a program segment and a state predicate Θ_P is treated similarly to a program with an initial condition Θ_P, except that Θ_P is not restricted to be satisfied by representations of empty traces. Thus, the execution of such a pair can be modeled by runs that can start with a non-empty trace.

A trivial example is the composition of two terminating processes (segments) P and Q which do not interact, i.e., all the operations from P are independent of all the operations of Q. Then, $\mathcal{E}[\![\langle P \parallel Q, \Theta_{PQ} \rangle]\!]$ includes all the interleavings of the execution sequences from P and from Q. However, each execution sequence of $\langle P \parallel Q, \Theta_{PQ} \rangle$ is equivalent to an execution sequence that consists of some (finite) execution sequence of $\mathcal{E}[\![\langle P, \Theta_{PQ} \rangle]\!]$ followed by an execution sequence of $\mathcal{E}[\![\langle Q, \Theta_Q \rangle]\!]$. If $\langle P, \Theta_{PQ} \rangle \vdash^{\exists} \varphi \wedge \Diamond(\theta_P \wedge \bigcirc \mathrm{F})$ (i.e., φ holds when P is executed, and P terminates with a state satisfying the state predicate θ_P) and $\langle Q, \Theta_Q \rangle \vdash^{\exists} \psi$, and $\theta_P \rightarrow \Theta_Q$ (i.e., the condition Θ_Q allows executing Q from any state in which the execution of the segment P can terminate), then $\langle P \parallel Q, \Theta_{PQ}\rangle \models^{\exists} \varphi \mathcal{C} \psi$. This can be formalised as a proof rule:

$$\frac{\begin{array}{l}\langle P, \Theta_{PQ}\rangle \models^{\exists} \varphi \wedge \Diamond(\theta_P \wedge \bigcirc \mathrm{F}) \\ \langle Q, \Theta_Q\rangle \models^{\exists} \psi \\ \theta_P \rightarrow \Theta_Q\end{array}}{\langle P \parallel Q, \Theta_{PQ}\rangle \models^{\exists} \varphi \mathcal{C} \psi} \tag{1}$$

The existential '$\models^{\exists}$' in the antecedents of this proof rule can be replaced with a universal '$\vdash$', but not in the consequence.

Although the above example may be considered simple and untypical, the following are examples of general cases to which similar verification techniques can be applied. First, consider the composition of CSP processes in the partial correctness proof rules of [2]. There, it is observed that composing the segment $S_1; \alpha; S_2$ in one process with $S_3; \bar{\alpha}; S_4$ in the second, where α and $\bar{\alpha}$ are matching send and receive communication commands and $S_1 \ldots S_4$ are local segments, behaves as $S_1; S_3; \alpha \parallel \bar{\alpha}; S_2; S_4$. The soundness of the proof rules in [2] relies on the fact the set of executions in which the segments are executed in this order contains enough representatives equivalent to all other executions, and that if at least one sequence from each equivalence class satisfies the partial correctness property, then all the other sequences also satisfy it.

Another way to compose programs is sequentially as *communication closed layers* [8]. There, two layers $S = [S_1 \parallel \ldots \parallel S_n]$ and $M = [M_1 \parallel \ldots \parallel M_n]$ are composed into a program $S; M = [S_1; M_1 \parallel \ldots \parallel S_n; M_n]$. If there is no possible communication between any S_i and M_j for $1 \leq i, j \leq n$, the following compositional proof rule can be formulated:

$$\frac{\begin{array}{l}\langle S, \Theta_S\rangle \models^{\exists} \eta \wedge \Diamond(\theta_S \wedge \bigcirc \mathrm{F}) \\ \langle M, \Theta_M\rangle \models^{\exists} \delta \\ \theta_S \rightarrow \Theta_M\end{array}}{\langle S; M, \Theta_S\rangle \models^{\exists} \eta \mathcal{C} \delta} \tag{2}$$

This rule reflects the fact that for each equivalence class of $\langle S; M, \Theta_S\rangle$, there exists a representative sequence in which M is executed entirely after S [15]. It is interesting to observe that although the layered composition and the concurrent composition are rather different, the structure of their proof rules is identical. Here as before, the existential '$\models^{\exists}$' in the antecedents of the proof rule can be replaced with a universal '$\vdash$', but not in the consequence. This simple compositional proof rule (compare [4]) allows one to compose provably correct programs from layers. This rule does not restrict the formulas η and δ (but it requires that S will terminate and that the layers are compatible to execute one after the other). This extends proof rules for layered programs that deal only with partial correctness [8] and total correctness [16]. Moreover, using the results of Section 6, compositional completeness [31] of the rules (1) and (2) can be shown, i.e., that every universal property of the program can be deduced from properties of S and M using this rule and pure temporal logic reasoning.

The last example is to consider transforming a program which was designed to operate in a fault-free environment into a fault-tolerant program that is able to recover from some expected faults [32]. A transformation exists that adds recovery handlers to each such basic program. These handlers are based upon taking snapshots [5] of the global state of the program from time to time and retracting upon the occurrence of a failure to the last snapshot taken. For such a program transformation, a *specification transformation* can be formalised, converting properties of the basic program into properties of its fault-tolerant version. Thus, verification of the fault-tolerant version can be done compositionally, by proving properties of the basic program, applying the specification transformation and doing some pure temporal verification. There is no need to actually verify that the property is satisfied by using the code of the fault-tolerant version of the program directly. Moreover, the verification of the specification transformation is done only once for all possible programs [24]. The snapshot taken in some execution sequence does not necessarily correspond to a global state that occurred in the past of the same sequence. However, there always exists an equivalent sequence in which this is true. Thus, the outcome of the formula transformation can be conveniently given by an existentially satisfied formula.

4 Exact Universal and Existential Specification of Programs

4.1 Exact Specification of a Program

Our aim is to determine the ability of a specification formalism to describe the structures (be it execution sequences or equivalence classes of them) modeling the behaviour of each given program P. This means that for any program P, there exists a formula φ such that

- all the structures modeling P satisfy φ, and
- all the structures that satisfy φ represent executions of P.

We call such a formula φ an *exact specification* of P (with respect to the class of structures used to model P) [19]. An exact specification φ of a program P in some formalism $\mathcal{L}$ has the property that every formula $\psi \in \mathcal{L}$ that holds for P is a consequent of φ, i.e., each structure that satisfies φ satisfies also ψ. This is denoted $\varphi \models \psi$.

This is especially important for achieving completeness of compositional verification methods [4, 31], i.e., proofs in which segments of a program are verified separately and then the proofs are combined (instead of proving each property with respect to the entire program). Completeness proofs of compositional methods often use the following conditions:

1. It is known that for each pair of programs (or program segments) P_1 and P_2 there exist exact specifications, say, φ_1 and φ_2, respectively.
2. It is possible to verify from φ_1 and φ_2 using some compositional proof rule [4] that ψ holds for the program P that is combined from P_1 and P_2.
3. ψ is an exact specification of P (i.e., the composition of the specifications preserves exactness).

Given the above conditions, the compositional proof method under discussion is compositional complete relative to verification in $\mathcal{L}$: since every property η of the combined program P is a consequence of ψ, it can be proved that P satisfies η by properties of P_1 and P_2 as follows:

(i) $P \vdash \psi$ (with ψ an exact specification of P) is proved using the compositional proof rule from exact specifications of P_1 and P_2.
(ii) $\psi \rightarrow \eta$ is proved in $\mathcal{L}$.

Then, $P \vdash \eta$ follows by a simple deduction rule from (i) and (ii).

For universal specification we define exact specification over sets of sequences by ignoring the partitioning into sets.

Definition 4.1 *A formula Φ_P is an* exact universal specification of a program P *if for each interleaving sequence ξ, $\xi \models \Phi_P$ iff $\xi \Subset \mathcal{E}[\![P]\!]$.*

It was shown in [19] that for interleaving semantics, for each program P, there exists a temporal formula that is satisfied exactly by its set of executions (i.e., universal satisfaction).

4.2 Exact Existential Specification

For existential specification, similarly to exact universal specification, for each program P, there should exist a formula that is satisfied exactly by equivalence classes of sequences of P. However, this is not as straightforward to achieve as in the universal case. Consider the following definition:

Definition 4.2 (first attempt) *A formula* Υ_P *is an* exact existential specification *of a program* P *iff for each set of sequences* M, $M \models^{\exists} \Upsilon_P$ *iff* $M \in \mathcal{E}[\![P]\!]$.

According to this definition, exact existential specification does not exist for any program. This is because one can add arbitrary new sequences to any set of sequences M such that $M \models^{\exists} \Upsilon_P$, obtaining $M' \notin \mathcal{E}[\![P]\!]$ such that $M' \models^{\exists} \Upsilon_P$ (by Definition 2.6, since $M' \subseteq M$). Thus, there is a need to limit the sets of sequences under consideration.

Definition 4.3 (second attempt) *A formula* Υ_P *is an* exact existential specification of a program P *iff for each set of sequences* M *obtained as an equivalence class of sequences of* some *program,* $M \models^{\exists} \Upsilon_P$ *iff* $M \in \mathcal{E}[\![P]\!]$.

Again, it can be shown that it is not true that under all representations for each program there exists an exact existential specification.

Example. Consider the most obvious representation, where each trace σ is represented by the values of the program variables in fin_σ. We will show that there is a program P_1 for which there is no exact existential specification. Consider the following two programs P_1 and P_2: The program P_1 has two interdependent operations $\alpha : x = y \longrightarrow x := x + 1$, and $\beta : x > y \longrightarrow y := y + 1$. The program P_2 has two independent operations $\alpha : true \longrightarrow x := x + 1$, and $\beta : true \longrightarrow y := y + 1$. Both programs are initiated with $x = 0$ and $y = 0$.

The programs have a mutual interleaving sequence, namely $(x = 0, y = 0) \xrightarrow{\alpha} (x = 1, y = 0) \xrightarrow{\beta} (x = 1, y = 1) \xrightarrow{\alpha} (x = 2, y = 1) \xrightarrow{\beta} \ldots$, denoted according to its generating sequence as $(\alpha\beta)^\omega$, where u^ω means u occurring infinitely many times. That is, x and y are incremented in turn, one after the other. However, for P_1, there is only one interleaving sequence, while for P_2, the operations can be selected to repeat arbitrarily and all the sequences $(\beta^*\alpha\alpha^*\beta)^\omega$ are equivalent. Hence, any formula such as

$$(\Box 0 \leq x - y \leq 1) \wedge \forall z \, (\Diamond x > z) \wedge$$
$$\Box \forall z \, \forall t \, (\, (x = z \wedge y = t) \rightarrow \bigcirc (\, (x = z + 1 \wedge y = t) \vee (x = z \wedge y = t + 1)\,)\,),$$

that is satisfied exactly by the the single interleaving sequence of P_1, is not an exact existential specification of P_1. This follows from the fact that P_2's executions constitute a single equivalence class that properly contains P_1's single interleaving sequence, and thus also existentially satisfies any such formula. ⊣

Another problem of the attempted Definition 4.3 is that the phrase "set of sequences M obtained as the equivalence class of sequences of *some* program" refers to either adding some program-dependent information to the sequences, or the ability of showing that a set of sequences is obtainable as an equivalence class of some program (by axiomatising the properties of such classes). Both of these are undesirable: adding additional information about the program defies the very idea of giving the exact specification of a program by a formula; alternatively, proving that an equivalence class corresponds to some program would require additional effort.

Instead, using some appropriate representation, it should be possible to define a condition on sets of representations sequences that limits their scope. Such a condition will later help to guarantee that if one of the sequences in a set is obtained as the representation of an execution sequence of some program, the rest of the sequences are exactly all the representations of sequences equivalent to it.

Definition 4.4 *Let B be a condition (expressed in some predefined formalism) on sets of $\mathcal{R}$-sequences. Such a condition is called a* bounding condition. *Each set of $\mathcal{R}$-sequences satisfying B is called a B-set.*

A bounding condition B can be seen as a program-independent restriction, limiting the scope of sets of possible $\mathcal{R}$-sequences. (It may allow B-sets that contain sequences that do not represent execution sequences of programs.) It is now possible to redefine exact existential specification using a bounding condition.

Definition 4.5 *A formula Υ_P is an* exact existential specification of a program P with respect to a bounding condition B *iff for each B-set M, $M \models^{\exists} \Upsilon_P$ iff $M \in \mathcal{E}[\![P]\!]$.*

5 Obtaining Exact Existential Specification

In this section, we define some general conditions which guarantee the existence of a bounding condition that facilitates exact existential specification. Then we construct a representation of traces that satisfies these conditions. The following goals are considered:

- No additional information about the modeled program is needed other than the existential specification (e.g., the independence relation).
- The construction should be obtainable using a syntactical transformation of the program, and then applying a simple familiar semantical construction (such as in [21] or [17]). Such a transformation should be as simple as adding to the program a set of appropriate auxiliary variables [7]. This will allow existing verification methods to be adopted.
- Generalisation, i.e., inferring a universal specification from an existential specification, must be possible. This is useful for the cases where an exact specification of a program (e.g., a result of superimposing programs) is best given in existential form, but some interesting properties of the superimposed program are of universal form.

5.1 Requirements from the Representation

We consider now some requirements that are imposed on any domain $\mathcal{R}$ of elements that represent a class of traces.

R1 The relations '$\sqsubseteq$' and '$\sqsubset_{im}$' between traces must be isomorphically definable on the representations. That is, for each pair of traces σ_1, σ_2 of a program P, $\sigma_1 \sqsubseteq \sigma_2 \Leftrightarrow subsumed(f(\sigma_1), f(\sigma_2))$, where $subsumed(s, t)$ is the corresponding relation defined among representations. Similarly, $\sigma_1 \sqsubset_{im} \sigma_2 \Leftrightarrow im\_subsumed(f(\sigma_1), f(\sigma_2))$.

Thus, a representation function f that identifies all the traces with the same final interpretation by assigning them to the same $\mathcal{R}$-element is inappropriate, as a triple $\sigma_1 \sqsubseteq \sigma_2 \sqsubseteq \sigma_3$, $\sigma_1 \neq \sigma_3$, with σ_1 and σ_3 having the same representation must not exist.

Since no additional information about the program need to be given together with an existential specification we require:

R2 The predicates $subsumed$ and $im\_subsumed$, defined over the representation elements, must be program-independent. That is, if $subsumed(s, t)$, then for *every* program P with traces σ and ρ such that $f(\sigma) = s$ and $f(\rho) = t$ it must be that $\sigma \sqsubseteq \rho$. (Notice that '$\sqsubseteq$' *is* program-dependent.) Moreover, when $subsumed(s, t)$ holds for some $t = f(\sigma)$, where $\sigma \in \Delta_P$, then $s = f(\rho)$ for some $\rho \sqsubseteq \sigma$. The same must hold for the relation '$\sqsubset_{im}$' and the corresponding relation $im\_subsumed(s, t)$ among $\mathcal{R}$-elements.

If X is a set of $\mathcal{R}$-elements, denote $\downarrow X = \{s \mid \exists t \in X \wedge subsumed(s, t)\}$. Observe that by **R2**, for any set of traces Σ, $f(\downarrow\Sigma) = \downarrow f(\Sigma)$. Denote by $\langle\langle\xi\rangle\rangle$ the set of elements that appear on the sequence ξ.

In the sequel, we will fix B as the following condition on sets of $\mathcal{R}$-sequences M: let ξ be some $\mathcal{R}$-sequence of M. Then, B is the condition that M includes *exactly all* the sequences of the form $s_0\, s_1\, s_2 \ldots$ that satisfy the following conditions:

B1 $\neg\exists s\,(subsumed(s, s_0) \wedge (s \neq s_0))$ *(i.e., s_0 represents an empty trace),*

B2 *for each* $0 \leq i <|\xi'|$, $s_i \in \downarrow \langle\langle\xi\rangle\rangle$,

B3 *for each* $0 < i <|\xi'|$, $im\_subsumed(s_{i-1}, s_i)$, *and*

B4 *for each* $s \in \langle\langle\xi\rangle\rangle$, *there exists some* $0 \leq i <|\xi'|$ *such that* $subsumed(s, s_i)$.

Notice that B already contains an appropriate restriction on adjacent elements.

Lemma 5.1 *If M is a B-set that contains an $\mathcal{R}$-sequence $\xi = f(o)$ for some observation o of P, then M is exactly the set of $\mathcal{R}$-sequences $\{\xi' \mid \xi' = f(o') \wedge o \approx o'\}$ (i.e., M is the set of $\mathcal{R}$-sequences representing all the observations of the same run as o).*

Proof. Let ξ be an $\mathcal{R}$-sequence representing some observation o of some run $\Pi \in R_P$. Then, it is easy to check that by **R1** and **R2**, $f(\downarrow \langle\langle o\rangle\rangle) = \downarrow \langle\langle\xi\rangle\rangle = f(\Pi)$. Thus by B1–B4 and using the definition of observations 2.2, the sequences in M are exactly the $\mathcal{R}$-sequences representing the observations of Π (i.e., the observations that are equivalent to o). ∎

For convenience, we fix some variable c to refer to the 'current state' when appearing in a temporal formula. The following notation will be used in the sequel:

- $consistent(s, t)$ – holds iff $\exists r\,(subsumed(s, r) \wedge subsumed(t, r))$. Notice that if $\sigma_1 \Uparrow \sigma_2$ for *some* program P, then $consistent(f(\sigma_1), f(\sigma_2))$.
- $[\![\varphi]\!]$ – the set of elements satisfying φ.
- $Subsumed_\varphi$ – a predicate transformer returning a predicate that is satisfied by exactly the elements subsumed by elements satisfying φ, i.e., $[\![Subsumed_\varphi]\!] = \downarrow [\![\varphi]\!]$. Both φ and $Subsumed_\varphi$ are state predicates (i.e., do not contain temporal modals). $Subsumed_\varphi$ can be defined as $\exists t\,(subsumed(s, t) \wedge \varphi(t))$ (where s is its free variables that stands for an $\mathcal{R}$-element, i.e., $Subsumed_\varphi(s)$ is a predicate of s).
- $Maximal$ – The temporal formula

$$\Box\forall s\,(im\_subsumed(\mathrm{c}, s) \rightarrow \Diamond(subsumed(s, \mathrm{c}) \vee \neg consistent(s, \mathrm{c}))).$$

The requirements **R1** and **R2** assures that the relations $\sqsubseteq$, $\sqsubset_{im}$ between traces are correspondingly definable among their representations. The third requirement allows defining maximal sets of $\mathcal{R}$-elements that correspond to the runs of programs. It is not true that this is already guaranteed by the requirements **R1** and **R2**: although the relation $subsumed$ between representations corresponds to '$\sqsubseteq$' (and the maximality of the runs is based upon the relation '$\sqsubseteq$'), the representations of the traces Δ_P of a single run are embedded in $\mathcal{R}$ with representations of other runs.

R3 If $\Pi \in R_P$ for some program P, then the set of $\mathcal{R}$-elements $f(\Pi)$ satisfies that any representation which immediately subsumes an element that belongs to $f(\Pi)$, is either itself in $f(\Pi)$ or is inconsistent with some element of $f(\Pi)$. This is formally written as:

$$\forall t\,((t \notin f(\Pi) \wedge \exists r\,(im\_subsumed(r, t) \wedge r \in f(\Pi))) \rightarrow \exists s \in f(\Pi)\, \neg consistent(t, s)).$$

Lemma 5.2 *A sequence of elements $\xi = s_0\, s_1 \ldots$ from $f(\Delta_P)$ such that $\forall i, 0 \leq i <|\xi|, subsumed(s_i, s_{i+1})$ is a representation of an observation of P (i.e., $\xi = f(o)$ for some $o \in obs(\Pi)$, $\Pi \in R_P$) iff $\xi \models Maximal$.*

Proof. Assume first that $\xi = f(o)$ for some observation o of Π. Let $t \in \langle\langle\xi\rangle\rangle$ be an element satisfying $im\_subsumed(t, s)$. If $s = f(\sigma)$ for $\sigma \in \Pi$ then by Definition 2.2, there exists a trace $\rho \in \langle\langle o\rangle\rangle$ that subsumes σ, and thus $f(\sigma) \in \langle\langle\xi\rangle\rangle$, and $subsumed(s, f(\rho))$. Otherwise, by **R3** there exists some element $\rho \in \Pi$ such that $\neg consistent(t, f(\rho))$. Since o is an observation, there exists a trace $\rho' \in \langle\langle o\rangle\rangle$ that subsumes ρ. It is easy to see that since $\neg consistent(t, f(\rho)) \wedge subsumed(f(\rho), f(\rho'))$, then $\neg consistent(t, f(\rho'))$.

To prove the other direction, assume that $\xi \models Maximal$. Suppose that some operations α is enabled in P after some trace σ of $\langle\langle o\rangle\rangle$. Then, $\sigma[\alpha] \in \Delta_P$. According to $Maximal$, there exists some element $t = f(\rho), \rho \in \langle\langle\xi\rangle\rangle$ such that either $subsumed(f(\sigma[\alpha]), t)$ or $\neg consistent(f(\sigma[\alpha]), t)$, and thus in the latter case, $\sigma[\alpha] \not\Uparrow \rho$. From properties of traces it can be shown that both cases correspond to the fact that an operation that is dependent on α (including α itself, in the former case) is executed in ξ (and is contained in $f(\rho)$). This condition on interleaving sequences of traces of P was shown in [17, 25] to be equivalent to the definition of observations. Thus, ξ represents an observation. ∎

The following theorem provides a useful form of exact existential specification:

Theorem 5.3 *If $[\![\varphi]\!] \subseteq f(\Delta_P)$, and $\mathcal{E}[\![P]\!] \stackrel{\exists}{\models} Maximal \wedge \Box\varphi$, then $Maximal \wedge \Box\varphi$ is an exact existential specification of P with respect to B.*

Proof. Let $\eta = Maximal \wedge \Box\varphi$. Let M be an arbitrary B-set such that $M \stackrel{\exists}{\models} \eta$. Then there exists an $\mathcal{R}$-sequence $\xi \in M$ satisfying η. Since $[\![\varphi]\!] \subseteq f(\Delta_P)$, all the elements in $\langle\langle\xi\rangle\rangle$ are $\mathcal{R}$-elements representing traces of P. From B3, for each adjacent elements $s_i, s_{i+1} \in \langle\langle\xi\rangle\rangle$, $im\_subsumed(s_i, s_{i+1})$. Thus, from Lemma 5.2 it follows that ξ is an observation of P. It follows from Lemma 5.1 that all the sequences in M represent observations of P as well. ∎

For each program P, a predicate φ_P that includes exactly the elements of $f(\Delta_P)$ is obtainable by a standard construction (it is used as a strongest invariant of a program in standard completeness proofs, e.g., [20]). This predicate satisfies the conditions of Theorem 5.3. Thus, an exact existential specification *always exists*. However, this predicate φ_P is not necessarily the best choice (or the only one) to be used for exact existential specification, as for this choice of φ_P, $Maximal \wedge \Box\varphi_P$ is both an exact existential and an exact universal specification for P.

In addition to the above requirements, it is also required that the representations must contain in a retrievable form any information that is needed for formulating the intended program properties. (In particular, the initial condition of the program must be interpretable over the representations.) For example, if the assertion language specifies properties of a program by asserting on its variables, then the value of these variables must be retrievable from the trace representation such that relation and functions on these values can be defined. Alternatively, one might be interested only in formulating properties of the sequences of operations occurred [13].

5.2 A Construction of a Representation

A construction of representations for traces that satisfy the requirements **R1**, **R2** and **R3** will now be presented. Its main purpose is to demonstrate that these conditions are satisfiable. There is no claim that this is the best or most efficient choice of representation for any other purpose.

It is sometimes convenient to represent a trace by denoting one of its histories. This is followed here, when histories, rather than traces, are actually the objects that are represented directly. Predicates on history representations, such as $subsumed$ and $consistent$, can treat $\mathcal{R}$-elements as traces, by using in their definition an equivalence relation between $\mathcal{R}$-elements which is defined for this purpose. Hence, there can exist multiple representations for each trace and observation. This can be easily avoided by assigning unique weights to the $\mathcal{R}$-elements, and then representing each trace by the minimal $\mathcal{R}$-element corresponding to a history of a given trace.

A *snapshot* is finite set of pairs (v, a), where v is a variable and a is a value. It represents a valuation of a set of variables. An *event* is an occurrence of some operation $\tau \in T$. It is represented as a pair containing an operation name τ and a snapshot that includes a pair (v, a), for each $v \in l_\tau$. For each such pair, a is the value of v just after executing τ. (Notice that the name of the variable can be represented by a string or an integer.)

An $\mathcal{R}$-element is then a triple $\ll J, \underline{E}, U \gg$, where J is a snapshot, E is a finite sequence of events (underlined for convenience of reading), and U is a set of pairs, each one containing the name τ of the operation that is enabled after the occurrence of the events in E and its set of variables l_τ. The following restrictions guarantee that each element in $\mathcal{R}$ represents a history of some program: (1) the set of variables of the snapshot J must include at least all the variables that appear in the events E, and (2) each two events with the same name have snapshots of the same variables. This agreement on the variables must also hold for any event and an element of U with the same name of operation.

As an example, the program P_1 that appear in the example of Section 4 has an $\mathcal{R}$-element

$$\ll \{(x,0), (y,0)\}, \underline{\langle\alpha, \{(x,1), (y,0)\}\rangle \langle\beta, \{(x,1), (y,1)\}\rangle \langle\alpha, \{(x,2), (y,1)\}\rangle}, \{\langle\beta, \{x,y\}\rangle\} \gg, \quad (3)$$

while P_2 has

$$\ll \{(x, 0), (y, 0)\}, \underline{\langle\alpha, \{(x, 1)\}\rangle \langle\beta, \{(y, 1)\}\rangle \langle\alpha, \{(x, 2)\}\rangle}, \{\langle\alpha, \{x\}\rangle, \langle\beta, \{y\}\rangle\} \gg. \quad (4)$$

Both $\mathcal{R}$-elements correspond to the execution of three events: executing α, β and then α again.

The independence between pairs of operations can be easily translated using this representation into a corresponding independence of their events: Two events are independent iff they have no variable in common. For example, the occurrences of α and β in (3) are dependent, as they both have in common the variables x and y, while the occurrences of α and β in (4) are independent.

It is thus possible to define the relations *subsumed* and *im_subsumed* using only the information that is in the $\mathcal{R}$-elements without additional information about the programs they represent. The third component U_s of an $\mathcal{R}$-element $s = f(\sigma)$ is the set of operations enabled immediately after σ. The relation *im_subsumed* (and similarly *subsumed*) must reflect this fact by allowing *im_subsumed*(s, t) only if t is obtained from s (or an element equivalent to s up to repeatedly commuting independent events) by the occurrence of an additional event with the same operation name and the same set of variables as a member of U_s. This use of the third component is essential to guarantee that the requirement **R3** holds.

6 Generalisation of Existential Specification

In this section, we show how under an appropriate representation of traces that satisfy the requirements **R1**, **R2** and **R3** of Section 5 (such as the one given in Section 5.2), it is possible to generalise from an existential specification into a universal specification. The proof system presented in this section also guarantees that one can infer an *exact* universal specification of a program P from an *exact* existential specification of P. This provides a framework that augment compositional proof rules such as (1) and (2) in Section 3 and guarantees compositional completeness.

Lemma 6.1 *If φ is a temporal property, then there exists a past temporal formula η that end-satisfies a finite sequence ξ iff ξ is a prefix of an interleaving sequence that satisfies φ.*

Proof. It is possible to construct for every property φ that is expressible in LTL an equivalent formula of the form $Safe\varphi \wedge Live\varphi$, where $Safe\varphi$ is of the form $\Box\eta$, with η satisfying the above requirements. This is the well known separation into safety and liveness [1]. The proof of this is very similar to [6, Section 4.11], where a related property is proved. This is based on using Gabbay's separation theorem [10]. ∎

We next assume that the first order logic we use for state formulas allows encoding finite sequences [3, 12]. Specifically, we assume that the following functions and predicate can be expressed:

$Pref(\chi, i)$ the prefix of length i of χ.
$Len(\chi)$ the length of χ.
$Seq(\chi, i, s)$ holds when s is the ith element in the sequence χ (and $0 \leq i < Len(\chi)$).

Lemma 6.2 *If φ is a temporal property, then there exists a first order formula $States_\varphi(s)$ that holds exactly for the $\mathcal{R}$-elements s that appear in interleaving sequences that satisfy φ.*

Proof. Let η be the formula constructed from φ in Lemma 6.1. Then define a translation $\mathcal{T}(\kappa, \chi)$ that transforms a temporal formula κ into a first order formula that is applied to the encoded sequence χ. The translation can be defined inductively on the structure of κ. For example, if $\kappa = \kappa_1 \mathcal{S} \kappa_2$, then

$$\mathcal{T}(\kappa, \chi) = \exists i\, (0 < i \leq Len(\chi) \wedge \mathcal{T}(\kappa_2, Pref(\chi, i)) \wedge \forall j\, (i < j \leq Len(\chi) \rightarrow \mathcal{T}(\kappa_1, Pref(\chi, j))))$$

The details of how to translate other past modals are omitted. Then $States_\varphi(s)$ can be defined as

$$\exists n \exists \chi\, (Len(\chi) = n \wedge \mathcal{T}(\eta, \chi) \wedge Seq(\chi, n-1, s)).$$

∎

Consider now the following simple deduction rule for existential specification.

$$\frac{\begin{array}{c}\varphi \rightarrow \psi \\ P \models^{\exists} \varphi\end{array}}{P \models^{\exists} \psi} \tag{5}$$

and the following rule for generalisation, where μ is a state formula

$$\frac{P \models^{\exists} \Box \mu}{P \vdash Maximal \wedge \Box Subsumed_\mu} \tag{6}$$

The soundness of (6) is guaranteed by the following lemma:

Lemma 6.3 *If M is a B-set that contains a representation of an observation of some run Π and $M \models^{\exists} \Box \mu$ such that μ is a state formula, then $M \models Maximal \wedge \Box Subsumed_\mu$.*

Proof. From Lemma 5.1, if M contains an observation of some run $\Pi \in \mathcal{R}_P$, then all the sequences of M are exactly $f(obs(\Pi))$. Since $M \models^{\exists} \Box \mu$, there exists a sequence ξ in M such that $\xi = f(o)$, where $o \in obs(\Pi)$ and $\xi \models \Box \mu$. Thus, $\langle\langle \xi \rangle\rangle \subseteq [\![\mu]\!]$. Consider now any other sequence ξ' of M. According the conditions of B, $\langle\langle \xi' \rangle\rangle \subseteq [\![Subsumed_\mu]\!]$ and therefore $\xi' \models \Box Subsumed_\mu$. Each sequence in M, is a representation of an observation of Π, and thus from Lemma 5.2, it satisfies $Maximal$. ∎

The above rules can be used to infer a universal property of P from an existential property φ.

1. Prove in LTL that $\varphi \rightarrow \Box \mu$ for some state formula μ.
2. Use the proof rule (5) to infer that $\Box \mu$ is an existential specification of P.
3. Use the rule (6) to infer that $Maximal \wedge \Box Subsumed_\mu$ is a universal specification of P.

Furthermore, if φ is an exact existential specification of P, then an exact universal specification of P can be obtained from it. This is done by choosing μ as $States_\varphi$. By Lemma 6.2, $\varphi \rightarrow \Box States_\varphi$ is valid, and thus can be proved in LTL. The rule (5) can be used to prove that $P \models^{\exists} \Box States_\varphi$. Then (6) can be used to prove that $P \models^{\exists} Maximal \wedge \Box Subsumed_{States_\varphi}$. Since φ is an exact existential specification, $[\![Subsumed_{States_\varphi}]\!] = \downarrow [\![States_\varphi]\!] = f(\Delta_P)$ ($\downarrow [\![States_\varphi]\!] \subseteq f(\Delta_P)$ since each sequence satisfying φ is a representation of an observation of P, and $\downarrow [\![States_\varphi]\!] \supseteq f(\Delta_P)$ since for each run there exists at least one representation of an observation that satisfies φ). Thus, by Lemma 5.2, $Maximal \wedge \Box Subsumed_{States_\varphi}$ is satisfied exactly by all the representations of observations of P.

Recall that, as was shown in Section 4.1, any universal property of P can be deduced from an exact universal specification such as $Maximal \wedge \Box Subsumed_{States_\varphi}$ relative to proving assertions in LTL.

7 Conclusions and Further Research

Representing properties of concurrent programs using existential specification was introduced. It was shown that given an appropriate representation, an exact existential specification Υ_P of a program P can be given such that all the properties of P that are expressed within this formalism are consequences of Υ_P. Moreover, it is possible to transform such an exact existential specification to an exact universal specification efficiently using a proof system.

A representation of trace semantics that allows exact existential specification and generalisation was demonstrated. Alternative constructions, based on a direct representation of partially ordered sets of events can be given under the same requirements. Moreover, when the class of programs dealt with is more constraint, simpler constructions can be made. In particular, the requirement **R1** of Section 5 does not allow a finite representation of finite state programs. It is interesting to check if weakening **R1** is possible or is possible only in some limited cases. The results of this paper are not confined to the temporal logic formalism and can also be adapted to other formalisms such as Büchi automata.

It was shown that existential specification is convenient for compositionality which involves composing concurrent and layered segments. The ability to generalise exact existential specifications is important for achieving completeness of such compositional methods. Thus, existential specification is suggested for compositional program construction, complementing algebraic methods such as [14].

Finally, notice that the generalising proof rule (6) is not optimal in the following sense: it does not guarantee that the strongest universal property ψ that holds for a program that satisfies a given existential property φ can be inferred from φ (however, it does guarantee this when φ is exact). Thus, although useful for achieving compositional completeness of proof rules such as those presented in Section 3, additional proof rules for generalising existential properties are sought.

Acknowledgements

I would like to thank Mathai Joseph and Nissim Francez for their helpful comments.

References

[1] B. Alpern, F.B. Schneider, Defining Liveness, Information Processing Letters 21, 1985, 181–185.

[2] K. R. Apt, N. Francez, W.P. de Roever, A proof system for communicating sequential processes, ACM Transactions on Programming Languages and Systems, Vol 2, 1980, 359–385.

[3] J. W. deBakker, Mathematical Theory of Program Correctness, Prentice–Hall, Englewood Cliffs, N. J, 1980.

[4] H. Barringer, R. Kuiper, A. Pnueli, Now You May Compose Temporal Logic Specification, Proceedings of 16th ACM Symposium on Theory of Computing, 1984, 51–63.

[5] K. M. Chandy, L. Lamport, Distributed Snapshots: determining the global state of distributed systems, ACM Transactions on Computer Systems 3, 1985, 63–75.

[6] E. Chang, Z. Manna, A. Pnueli, The Safety–Progress Classification, Manuscript 1992.

[7] M. Clint, Program proving: Coroutines, Acta Informatica 2, 1973, 50–63.

[8] Tz. Elrad, N. Francez, Decomposition of Distributed Programs into Communication–Closed Layers, Science of Computer Programming 2 (1982), 155–173

[9] E. A. Emerson, J. Y. Halpern, "Sometimes" and "Not Never" Revisited: On Branching versus Linear Time Temporal Logic, Journal of the ACM 33 (1986), 151–178.

[10] D. Gabbay, The declarative past and imperative future, in: B. Banieqbal, H. Barringer, A. Pnueli (editors), Temporal Logic in Specification, LNCS 398, Springer–Verlag, 1987, 407–448.

[11] P. Godefroid, P. Wolper, Using partial orders for the efficient verification of deadlock freedom and

safety properties, Proceedings of Computer–Aided Verification, Aalborg, Denmark, 1991.

[12] D. Harel, First order Dynamic Logic, Lecture Notes in Computer Science 68, Springer–Verlag, 1979.

[13] C. A. R. Hoare, Communicating Sequential Processes, Prentice–Hall, 1985.

[14] W. Janssen, J. Zwiers, Protocol Design by Layered Decomposition, A compositional Approach, 2nd Symposium on Formal Techniques in Real-Time and Fault-Tolerant Systems, Nijmegen, The Netherlands, 1992, LNCS 571, Springer–Verlag, 307–326.

[15] S. Katz, D. Peled, Interleaving Set Temporal Logic, Theoretical Computer Science 75(1990),21–43.

[16] S. Katz, D. Peled, Verification of Distributed Programs using Representative Interleaving Sequences, to appear in Distributed Computing.

[17] M. Z. Kwiatkowska, Fairness for Non–interleaving Concurrency, Phd. Thesis, Faculty of Science, University of Leicester, 1989.

[18] "Sometime" is Sometimes "Not Never" – On the Temporal Logic of Programs, Proceedings of the 7th ACM symposium on Principles of Programming Languages, 1980, 174–185.

[19] Z. Manna, A. Pnueli, How to Cook a Temporal Proof System for Your Pet Language. Proceedings of the Symposium on Principles on Programming Languages, Austin, Texas, 1983, 141–151.

[20] Z. Manna, A. Pnueli, Completing the temporal picture, Theoretical Computer Science 83(1991), 97–130.

[21] A. Mazurkiewicz, Traces, Histories, Graphs: Instances of a process monoid, in M. Chytil (Ed.), Mathematical Foundation of Computer Science, LNCS 176, Springer-Verlag, 1984, 115–133.

[22] D. Park, On the Semantics of Fair Parallelism, in D. Biorner (ed.), Proceedings on Abstract Software Specification, LNCS 86, Springer–Verlag, 1979, 504–526.

[23] D. Peled, S. Katz, and A. Pnueli, Specifying and Proving Serializability in Temporal Logic, LICS 91', Amsterdam, The Netherlands, July 1991, 232–245.

[24] D. Peled, M. Joseph, A Compositional Approach to Fault Tolerance Using Specification Transformation, Manuscript, 1992.

[25] D. Peled, A. Pnueli, Proving partial order liveness properties, in M.S. Paterson (ed.), Proceedings of the 17th ICALP, LNCS 443, Springer–Verlag, 1990, 553–571.

[26] W. Penczek, A Concurrent Branching Time Temporal Logic, Conference on Computer Science Logic, LNCS 440, Springer–Verlag, 1989, 337–354.

[27] S. Pinter, P. Wolper, A Temporal Logic for Reasoning about Partially Ordered Computations, Proceedings of the 3rd ACM Symposium on Principles of Distributed Computing, Vancouver, B. C., August 1984, 23–27.

[28] A. Pnueli, The Temporal Logic of Programs, Proceedings of the 18th Symposium on Foundation of Computer Science, IEEE, Providence, 1977, 46–57.

[29] F. A. Stomp, W. P. deRoever, Designing Distributed Algorithms by Means of Formal Sequentially Phased Reasoning, Proceedings of the 3rd International Workshop on Distributed Algorithms, LNCS 392, Springer–Verlag, 1989, 242–253.

[30] A. Valmari, Stubborn Sets for Reduced State Space Generation, 10th International Conference on Application and Theory of Petri Nets, Germany, 1989, Vol 2, 1–22.

[31] J. Zwiers, Compositionality, Concurrency and Partial Correctness, LNCS 321, Springer–Verlag, 1987.

[32] L. Zhiming , M. Joseph, Transformations of programs for fault–tolerance, to appear in Formal Aspects of Computing.

The Weakest Compositional Semantic Equivalence Preserving Nexttime-less Linear Temporal Logic

Roope Kaivola
University of Helsinki
Department of Computer Science
Teollisuuskatu 23, SF-00510 Helsinki, Finland
tel. +358-0-708 4163
fax. +358-0-708 4441
email rkaivola@cc.helsinki.fi

Antti Valmari
Technical Research Centre of Finland
Computer Technology Laboratory
PO Box 201, SF-90571 Oulu, Finland
tel. +358-81-551 2111
fax. +358-81-551 2320
email ava@tko.vtt.fi

Abstract. *Temporal logic model checking is a useful method for verifying properties of finite-state concurrent systems. However, due to the state explosion problem modular methods like compositional minimisation based on semantic congruences are essential in making the verification task manageable. In this paper we show that the so-called CFFD-equivalence defined by initial stability, infinite traces, divergence traces and stable failures is exactly the weakest compositional equivalence preserving nexttime-less linear temporal logic with an extra operator distinguishing deadlocks from divergences. Furthermore, a slight modification of CFFD, called the NDFD-equivalence, is exactly the weakest compositional equivalence preserving standard nexttime-less linear temporal logic.*

1 Introduction

Many important correctness considerations of concurrent systems lend themselves to representing the system by a finite-state model, and consequently, to automatic verification. However, due to the state-explosion problem caused by the customary way of modelling concurrency by arbitrary interleavings, such models of realistic systems tend to be too large for the verification task to be feasible.

One approach to tackling this problem is using semantic equivalences between models: if an equivalence preserves the property to be verified, we may try to find a smaller model equivalent to the original one, and verify that it has the required property, instead. What is more, if the equivalence is a congruence with respect to the syntactic operations used in constructing the systems, i.e. if $P_i \approx Q_i$ implies $f(P_1, \ldots, P_n) \approx f(Q_1, \ldots, Q_n)$ for all syntactic operators f, we may replace the components of the system by smaller equivalent ones before composition. This can mean a dramatic reduction in the size of the system [GS90] [Val90] [KV91].

The process algebra literature contains a multitude of various equivalence notions. What properties should an equivalence have in order to be useful in this compositional minimisation approach? Two immediate requirements are, as already stated, that it preserves the class of properties to be verified and is congruent with respect to the syntactic operators. In order to be usable in practice, the equivalence should also have a reduction algorithm. Finally, the weaker the equivalence is the more potential it has for savings, since the probability of finding a small equivalent model naturally increases with the number of models the equivalence considers as equal.

In this paper we are interested in using this approach to verify properties expressed in linear-time temporal logic without nexttime. Many established semantic equivalences do not preserve such properties, and are therefore not usable. E.g. trace-equivalence and observation-equivalence [Mil89] do not pay attention to divergences, and CSP-equivalence [OH86] does not keep track of events after a potential divergence, both of which features mean that the equivalences do not necessarily preserve liveness-properties. Strong bisimulation equivalence [Mil89] is truth-preserving, but is usually too strong to allow significant reduction. A temporal logic preserving equivalence between Kripke-models is presented in [BCG87], but compositionality issues are not addressed there.

In [VT91] a novel *chaos-free failures divergences* or CFFD-equivalence was introduced, its compositionality with respect to standard LOTOS-operators [BB89] established, and a reduction algorithm for it described. In [KV91] the present authors showed that CFFD is truth-preserving with respect to nexttime-less linear temporal logic LTL'. The paper also demonstrated compositional verification of a mutex protocol and a version of the AB-protocol.

In this paper we show that a slight modification of CFFD, called *nondivergent failures divergences* or NDFD-equivalence, is exactly the *weakest* compositional equivalence that is truth-preserving with respect to LTL'. Furthermore, CFFD is exactly the weakest compositional equivalence that is truth-preserving with respect to LTL' augmented with an extra operator $\overset{\infty}{\Diamond}$, *infinitely often in the proper future.*

These results mean that the NDFD and CFFD-equivalences seem to be ideal for verification using compositional minimisation. If any other equivalence, like e.g. divergence bisimulation [Wal87], is both compositional and LTL' (LTL^{∞}) preserving, it is always a refinement of NDFD (CFFD). Consequently, any reduction algorithm preserving such an equivalence preserves NDFD (CFFD) too, but may preserve some additional unnecessary features as well.

The paper proceeds as follows: we first recall some standard definitions of process algebras and define the CFFD and NDFD-equivalences. In Section 3 we discuss how to interpret temporal logics over labelled transition systems. After that, the truth-preservation results are presented. In Section 5 it is shown that the equivalences are the weakest possible, and in Section 6 an equivalence-preserving reduction algorithm is outlined.

2 CFFD and NDFD-equivalences

In this section we recall some basic concepts of process algebras and give the definitions of CFFD and NDFD-equivalences. For a more detailed discussion of these equivalences and the intuitions behind them please see [VT91] and [VT92].

Definition 2.1 A *transition alphabet* is a countably infinite set Σ not containing the *empty transition label* τ. We write Σ_τ for $\Sigma \cup \{\tau\}$, and Σ^* (Σ^∞) for the set of all finite (infinite) strings consisting of elements of Σ. The symbol ϵ is used to denote the empty string. If $\sigma \in (\Sigma^* \cup \Sigma^\infty)$ and $n \geq 1$, we write σ_n for the n:th element of σ and $\sigma^{(n)}$ for the string obtained by leaving the first n elements out of σ. If $\sigma, \pi \in (\Sigma^* \cup \Sigma^\infty)$, $\sigma \cdot \pi$ is used to denote the catenation of σ and π, $\sigma \prec \pi$ to denote that σ is a prefix of π, and $|\sigma|$ to denote the length of σ. If $\rho \in \Sigma_\tau^* \cup \Sigma_\tau^\infty$, $vis(\rho)$ is used to denote the string obtained by removing all τ-symbols from ρ, and $\Sigma(\rho)$ to denote the set of elements of ρ. □

Definition 2.2 A *labelled transition system (lts)* is a triple $L = (S, s, \Delta)$, where S is the set of states, $s \in S$ is the unique initial state, and $\Delta \subseteq S \times \Sigma_\tau \times S$ is the transition relation. The *alphabet of* L, $\Sigma(L)$, is the set $\Sigma(L) = \{l \in \Sigma \mid \exists s, s' : (s, l, s') \in \Delta\}$.

The alphabet of any lts is required to be finite. If $\rho \in \Sigma_\tau^*$, we write $s_0 \xrightarrow{\rho} s_n$ iff there are $s_1, \ldots, s_{n-1}$ such that for all $0 < i \leq n$, $(s_{i-1}, \rho_i, s_i) \in \Delta$. If there is an s_n s. t. $s_0 \xrightarrow{\rho} s_n$, we write $s_0 \xrightarrow{\rho}$. If $\rho \in \Sigma_\tau^\infty$, we write $s_0 \xrightarrow{\rho}$ iff $\exists : s_1, s_2, \ldots$ such that for all $i > 0$, $(s_{i-1}, \rho_i, s_i) \in \Delta$. If $\sigma \in (\Sigma^* \cup \Sigma^\infty)$, we write $s_0 \xRightarrow{\sigma} s_n$ ($s_0 \xRightarrow{\sigma}$) iff there is a $\rho \in (\Sigma_\tau^* \cup \Sigma_\tau^\infty)$ such that $s_0 \xrightarrow{\rho} s_n$ ($s_0 \xrightarrow{\rho}$) and $\sigma = vis(\rho)$. □

Definition 2.3 Let $L = (S, \Delta, s)$ be a labelled transition system.

- $\sigma \in \Sigma^*$ is a *trace* of L iff $s \xRightarrow{\sigma}$. $tr(L)$ is the set of traces of L.
- $\sigma \in \Sigma^\infty$ is an *infinite trace* of L, iff $s \xRightarrow{\sigma}$. $inftr(L)$ is the set of infinite traces of L.
- $\sigma \in \Sigma^*$ is a *divergence trace* of L, iff there is a $\rho \in \Sigma_\tau^\infty$ such that $s \xrightarrow{\rho}$ and $\sigma = vis(\rho)$. $divtr(L)$ is the set of divergence traces of L.
- $s' \in S$ is *stable*, iff not $s' \xrightarrow{\tau}$. Lts L is stable iff the initial state s is stable. We write $stable(L)$ if L is stable, and $\neg stable(L)$ if it is not.
- $(\sigma, A) \in \Sigma^* \times P(\Sigma)$, where $P(\Sigma)$ denotes the powerset of Σ, is a *failure* of L, iff there is an s' such that $s \xRightarrow{\sigma} s'$ and $s' \xRightarrow{a}$ for no $a \in A$,
- $(\sigma, A) \in \Sigma^* \times P(\Sigma)$ is a *stable failure* of L iff there is a stable s' such that $s \xRightarrow{\sigma} s'$ and $s' \xRightarrow{a}$ for no $a \in A$. $sfail(L)$ is the set of stable failures of L.
- $(\sigma, A) \in \Sigma^* \times P(\Sigma)$ is a *nondivergent failure* of L iff (σ, A) is a failure and σ is not a divergence trace. $ndfail(L)$ is the set of nondivergent failures of L.
- $\sigma \in \Sigma^*$ is a *deadlock trace* of L, iff (σ, Σ) is a stable failure of L. $dtr(L)$ is the set of deadlock traces of L.
- $\sigma \in \Sigma^*$ is a *nondivergent deadlock trace* of L, iff (σ, Σ) is a nondivergent failure of L. $nddtr(L)$ is the set of nondivergent deadlock traces of L. Note that $nddtr(L) = dtr(L) \setminus divtr(L)$. □

In addition to the preceding concepts we need some notation which does not ignore the τ transition labels.

Definition 2.4 Let $L = (S, \Delta, s)$ be a labelled transition system.

- $\rho \in \Sigma_\tau^*$ is a *path* of L iff $s \xrightarrow{\rho}$.
- $\rho \in \Sigma_\tau^\infty$ is an *infinite path* of L iff $s \xrightarrow{\rho}$. $infpath(L)$ is the set of infinite paths of L.
- $\rho \in \Sigma_\tau^*$ is a *deadlock path* of L iff there is a s' such that $s \xrightarrow{\rho} s'$ and for no ρ' $s' \xrightarrow{\rho'}$ holds. $dpath(L)$ is the set of deadlock paths of L. □

The following proposition lists some consequences of the definitions for later use.

Proposition 2.5 Let L be an lts.

- $tr(L) = divtr(L) \cup \{\sigma \mid (\sigma, \emptyset) \in sfail(L)\} = divtr(L) \cup \{\sigma \mid (\sigma, \emptyset) \in ndfail(L)\}$
- If $\rho \in dpath(L)$, then $vis(\rho) \in dtr(L)$.
- If $\rho \in infpath(L)$ and $vis(\rho) \in \Sigma^\infty$ then $vis(\rho) \in inftr(L)$.
- If $\rho \in infpath(L)$ and $vis(\rho) \in \Sigma^*$ then $vis(\rho) \in divtr(L)$.
- If $\rho \in dpath(L) \cup infpath(L)$, then $vis(\rho) \in nddtr(L) \cup divtr(L) \cup inftr(L)$
- If $\sigma \in dtr(L)$, there is a $\rho \in dpath(L)$ such that $vis(\rho) = \sigma$.
- If $\sigma \in divtr(L)$, there is a $\rho \in infpath(L)$ such that $vis(\rho) = \sigma$.
- If $\sigma \in inftr(L)$, there is a $\rho \in infpath(L)$ such that $vis(\rho) = \sigma$.
- If $\sigma \in nddtr(L) \cup divtr(L) \cup inftr(L)$, there is a $\rho \in dpath(L) \cup infpath(L)$ such that $vis(\rho) = \sigma$. □

On the basis of the definitions, the equivalence concepts can be easily defined.

Definition 2.6 Let L and L' be ltss. We say that L and L' are *CFFD* (NDFD) *equivalent* and write $L \overset{\text{cffd}}{\approx} L'$ ($L \overset{\text{ndfd}}{\approx} L'$) iff $\mathit{stable}(L) \Leftrightarrow \mathit{stable}(L')$, $\mathit{divtr}(L) = \mathit{divtr}(L')$, $\mathit{inftr}(L) = \mathit{inftr}(L')$, and $\mathit{sfail}(L) = \mathit{sfail}(L')$ ($\mathit{ndfail}(L) = \mathit{ndfail}(L')$). □

If the labelled transition systems examined are finite, the component *inftr* in the definition of CFFD-equivalence is superfluous. This corresponds to the original definition of CFFD-equivalence in [VT91], where only finite ltss were considered.

Proposition 2.7 Let L and L' be finite ltss. Then $L \overset{\text{cffd}}{\approx} L'$ ($L \overset{\text{ndfd}}{\approx} L'$) iff $\mathit{stable}(L) \Leftrightarrow \mathit{stable}(L')$, $\mathit{divtr}(L) = \mathit{divtr}(L')$ and $\mathit{sfail}(L) = \mathit{sfail}(L')$ ($\mathit{ndfail}(L) = \mathit{ndfail}(L')$).
Proof: By 2.5 $\mathit{divtr}(L) = \mathit{divtr}(L')$ and $\mathit{sfail}(L) = \mathit{sfail}(L')$ (or $\mathit{ndfail}(L) = \mathit{ndfail}(L')$) imply $\mathit{tr}(L) = \mathit{tr}(L')$.
If L is finite $\mathit{inftr}(L) = \{\omega \in \Sigma^\infty \mid \forall \sigma \prec \omega : \sigma \in \mathit{tr}(L)\}$ [VT92]. □

The following proposition is an immediate consequence of the definitions 2.3 and 2.6 and is essential for the preservation of linear temporal logic.

Proposition 2.8 If $L \overset{\text{cffd}}{\approx} L'$ ($L \overset{\text{ndfd}}{\approx} L'$), then $\mathit{inftr}(L) = \mathit{inftr}(L')$, $\mathit{divtr}(L) = \mathit{divtr}(L')$, and $\mathit{dtr}(L) = \mathit{dtr}(L')$ ($\mathit{nddtr}(L) = \mathit{nddtr}(L')$). □

Next we introduce some operators that can be used to combine labelled transition systems and state that CFFD and NDFD-equivalences are congruences with respect to these operators. The operators used are parallel composition $|[\ldots]|$, nondeterministic choice $[]$, hiding and renaming. The LOTOS operators >> and [> are not discussed, but the results presented here can be extended to them, too [VT92].

Definition 2.9 Let $L_1 = (S_1, s_1, \Delta_1)$ and $L_2 = (S_2, s_2, \Delta_2)$ be ltss, $G = \{g_1, \ldots, g_n\} \subset \Sigma$ and $H = \{h_1, \ldots, h_n\} \subset \Sigma$.
Then $L_1|[g_1, \ldots, g_n]|L_2$ is the lts $(S_1 \times S_2, (s_1, s_2), \Delta)$, where

- $((t,u), g_i, (t',u')) \in \Delta$, where $g_i \in G$, iff $(t, g_i, t') \in \Delta_1$ and $(u, g_i, u') \in \Delta_2$, and
- $((t,u), l, (t',u')) \in \Delta$, where $l \notin G$, iff either $(t, l, t') \in \Delta_1$ and $u = u'$ or $(u, l, u') \in \Delta_2$ and $t = t'$.

$L_1[]L_2$ is the lts $(\{s\} \times \{0\} \cup S_1 \times \{1\} \cup S_2 \times \{2\}, (s,0), \Delta)$, where

- $((t,i), l, (t',i)) \in \Delta$, where $i \in \{1,2\}$, iff $(t, l, t') \in \Delta_i$, and
- $((s,0), l, (t,i)) \in \Delta$, where $i \in \{1,2\}$, iff $(s_i, l, t) \in \Delta_i$.

hide $g_1, \ldots g_n$ **in** L_1 is the lts (S_1, s_1, Δ), where

- $(t, l, t') \in \Delta$, iff either $l \notin G$ and $(t, l, t') \in \Delta_1$ or $l = \tau$ and there is a $g_i \in G$ such that $(t, g_i, t') \in \Delta_1$.

$L_1[h_1/g_1, \ldots h_n/g_n]$ is the lts (S_1, s_1, Δ), where

- $(t, l, t') \in \Delta$, iff either $l \notin G$ and $(t, l, t') \in \Delta_1$ or $l = h_i$ and $(t, g_i, t') \in \Delta_1$. □

Definition 2.10 An equivalence $\approx$ between ltss is a *congruence* with respect to a syntactic operator f iff for every $L_1, \ldots, L_n$ and $L'_1, \ldots, L'_n$ such that $L_i \approx L'_i$ the following holds: $f(L_1, \ldots, L_n) \approx f(L'_1, \ldots, L'_n)$. □

Proposition 2.11 CFFD and NDFD equivalences are congruences with respect to all the operators defined in 2.9.
Proof: For the finite case CFFD see [VT91], for the general case [VT92]. □

3 Linear Temporal Logic

In this section we recall the definitions of linear temporal logic, by means of which properties of transition systems may be described, and discuss some aspects of the relation between process algebras and temporal logic.

Generally properties expressible using the temporal operator 'nexttime' $\bigcirc$, and in particular properties saying that something happens in fewer than n steps, are not preserved in concurrent combination of processes. Therefore, we follow [Lam83] in banishing it from the set of allowed operators. In other respects the definitions of our linear temporal logic LTL' are standard. The logic LTL^∞ consists of LTL' augmented with an additional operator $\overset{\infty}{\Diamond}$, *infinitely often in the proper future* [EC80].

Definition 3.1 The *alphabet of propositional linear temporal logic* consists of the countably infinite set AP of atomic propositions and of the symbols $(,), \neg, \vee, \mathcal{U}, \overset{\infty}{\Diamond}$. The *well-formed formulas* (wffs) of LTL' are defined by the rules 1, 2 and 4, and the wffs of LTL^∞ by the rules 1,2,3 and 4 below:

1 if $\phi \in AP$, then ϕ is a wff,
2 if ϕ_1 and ϕ_2 are wffs, then $(\neg\phi_1)$, $(\phi_1 \vee \phi_2)$ and $(\phi_1 \mathcal{U} \phi_2)$ are wffs,
3 if ϕ is a wff, $(\overset{\infty}{\Diamond}\phi)$ is a wff, and
4 there are no other wffs.

We use the abbreviations $\top \equiv_{df} (p \vee (\neg p))$ for some fixed $p \in AP$, $(\phi_1 \wedge \phi_2) \equiv_{df} (\neg((\neg\phi_1) \vee (\neg\phi_2)))$, $(\Diamond\phi) \equiv_{df} (\top\mathcal{U}\phi)$, $(\Box\phi) \equiv_{df} (\neg(\Diamond(\neg\phi)))$, and the ordinary precedence rules to reduce the number of parentheses. □

Definition 3.2 A *truth set* v is a set of atomic propositions, $v \subseteq AP$. A *truth set sequence* σ is a finite or infinite sequence of truth sets, $\sigma = (v_1, v_2, \ldots)$. The notation σ_n and $\sigma^{(n)}$ are as before. □

Definition 3.3 An LTL' (LTL^∞) -formula ϕ is true in a truth set sequence $\sigma = (v_1, v_2, \ldots)$, i.e. $\sigma \models \phi$, according to the following rules:

- If $\phi \in AP$, then $\sigma \models \phi$ iff $\phi \in v_1$.
- $\sigma \models \neg\phi$ iff not $\sigma \models \phi$.
- $\sigma \models \phi_1 \vee \phi_2$ iff $\sigma \models \phi_1$ or $\sigma \models \phi_2$.
- $\sigma \models \phi_1\mathcal{U}\phi_2$ iff $\exists : 0 \leq i < |\sigma|$, s. t. $\sigma^{(i)} \models \phi_2$ and for all $0 \leq j < i$, $\sigma^{(j)} \models \phi_1$.
- $\sigma \models \overset{\infty}{\Diamond}\phi$ iff there are infinitely many $i \geq 0$ such that $\sigma^{(i)} \models \phi$. □

If the nexttime-operator would be available, $\overset{\infty}{\Diamond}$ could be introduced as a derived operator $\overset{\infty}{\Diamond}\phi \equiv_{df} \Box\bigcirc\Diamond\phi$. Here $\overset{\infty}{\Diamond}$ is not definable in terms of the other operators. However, in all infinite truth set sequences $\overset{\infty}{\Diamond}\phi \Leftrightarrow \Box\Diamond\phi$. Therefore, what $\overset{\infty}{\Diamond}$ essentially brings in is the possibility of distinguishing a finite sequence from an infinite one, i.e. of distinguishing a deadlock from a divergence. The same expressive power could be obtained by the less general operator $\overset{\infty}{\Diamond}\top$, *the future is infinite*, as well.

The reflexivity of $\mathcal{U}$ and the lack of a nexttime-operator allow us to overlook truth sets in a truth set sequence if they do not differ from their predecessor.

Definition 3.4 Let $\sigma = (v_1, v_2, \ldots)$ be a truth set sequence. The *finitely reduced form of* σ, denoted by $fred(\sigma)$, is constructed by collapsing all finite continuous sequences $v_i, v_{i+1}, \ldots, v_n$ of identical elements $v_i = v_{i+1} = \ldots = v_n$ into one element v_i.

The *reduced form of* σ, denoted by $red(\sigma)$, is constructed by collapsing both finite and infinite continuous sequences $v_i, v_{i+1}, \ldots$ of identical elements $v_i = v_{i+1} = \ldots$ into one element v_i. □

Proposition 3.5 Let σ be a truth set sequence. If ϕ is an LTL^{∞}-formula, then $\sigma \models \phi$ iff $fred(\sigma') \models \phi$. If ϕ is an LTL'-formula, then $\sigma \models \phi$ iff $red(\sigma') \models \phi$.
Proof: Induction on the structure of ϕ. The result holds because LTL' is immune to both finite and infinite stuttering and LTL^{∞} to finite stuttering [Lam83]. □

Traditionally temporal logic expresses properties that are based on the truth-values of propositions in the states of a transition system. On the other hand, the process-algebraic equivalences and composition operators usually work purely on information that is based on transition labels. There seem to be several ways to bridge this gap. One is extending the notion of an lts to include state information, too, and modifying the equivalence concepts accordingly [Sti87]. Here we adopt another way of interpreting the transition labels as functional state transformers: an initial state description and a sequence of transformations induce a sequence of state descriptions on which temporal logic formulas may be interpreted.

The transformation function that corresponds to a transition label may be fixed in advance, as e.g. in [CLM89] where a transition l toggles a boolean state variable l between true and false. Here, as in [KV91], we leave the interpretation open and require only that the invisible τ-labels have to be mapped to a function that does not change state information in any way as far as atomic propositions in AP can see. As a result of not fixing the interpretation, we cannot talk of the truth of a formula only with respect to an lts, but also with respect to the interpretation mapping the transition labels to state transformers, and the initial state description.

The definition of a truth set modifier captures the intuitive idea of a state transformer. It expresses how a certain change affects a state as characterised by the atomic propositions true in that state. The temporal logical meaning of an lts is specified as a mapping assigning a truth set modifier to each of its transition labels.

Definition 3.6 A *truth set modifier sm* is a mapping $sm : P(AP) \to P(AP)$. The set of all truth set modifiers is denoted by TS. The identity truth set modifier I is the identity function. A *truth set modifier sequence* is a finite or infinite sequence of truth set modifiers. □

Definition 3.7 A *temporal semantics* for an lts L (a path ρ) is a mapping $f : \Sigma(L) \cup \{\tau\} \to TS$ ($f : \Sigma(\rho) \cup \{\tau\} \to TS$) such that $f(\tau) = I$. If $\rho = a_1a_2\ldots$ is a path of L, we write $f(\rho)$ for the sequence $(f(a_1), f(a_2), \ldots)$. □

Definition 3.8 The *truth set sequence induced by a truth set* $v \subseteq AP$ *and a truth set modifier sequence tsm*, denoted $tss(v, tsm)$, is a sequence of truth sets such that $tss(v,tsm)_1 = v$, $tss(v,tsm)_{i+1} = tsm_i(tss(v,tsm)_i)$, and if tsm is finite, then $|tss(v,tsm)| = |tsm| + 1$, otherwise $|tss(v,tsm)| = \infty$. □

Definition 3.9 Let $\rho \in \Sigma_\tau^* \cup \Sigma_\tau^\infty$, f a temporal semantics for ρ, v_0 a truth set and ϕ an LTL' (LTL^{∞})-formula. We say that ϕ *is true of* ρ *with respect to temporal semantics* f *and initial truth set* v_0, and write $\rho, f, v_0 \models \phi$ iff $tss(v_0, f(\rho)) \models \phi$. □

Due to the requirement that τ-transitions are not allowed to affect the atomic propositions in any way, the results of 3.5 carry easily over to ltss as well.

Proposition 3.10 Assume that ρ, f, v_0 are as in 3.9. If $\rho \in \Sigma^*$ or $vis(\rho) \in \Sigma^\infty$, then $\rho, f, v_0 \models \phi$ iff $vis(\rho), f, v_0 \models \phi$, for all LTL^{∞}-formulas ϕ.
If $\rho \in \Sigma^\infty$ and $vis(\rho) \in \Sigma^*$, then $\rho, f, v_0 \models \phi$ iff $vis(\rho) \cdot \tau^\infty, f, v_0 \models \phi$, for all LTL^{∞}-formulas ϕ.

Proof: By definition 3.7, removing τ-transitions from ρ can remove only identity truth set modifiers from $f(\rho)$. If $\rho \in \Sigma^*$ or $vis(\rho) \in \Sigma^\infty$, this means that $fred(tss(v_0, f(\rho))) = fred(tss(v_0, f(vis(\rho))))$. By 3.5, this means that $tss(v_0, f(\rho)) \models \phi$ iff $tss(v_0, f(vis(\rho))) \models \phi$.

If, on the other hand, $\rho \in \Sigma^\infty$ and $vis(\rho) \in \Sigma^*$, it means that $fred(v_0, tss(f(\rho))) = fred(tss(v_0, f(vis(\rho) \cdot \tau^\infty))$, and, consequently, that $tss(v_0, f(\rho)) \models \phi$ iff $tss(v_0, f(vis(\rho) \cdot \tau^\infty)) \models \phi$. In both cases the result follows. □

Proposition 3.11 Assume that ρ, f, v_0 are as in 3.9. Then $\rho, f, v_0 \models \phi$ iff $vis(\rho), f, v_0 \models \phi$, for all LTL'-formulas ϕ.
Proof: As in 3.10, except that there is no need to make difference between *vis* removing finitely or infinitely many consequent τ-transitions from ρ, and thus *red* may be used instead of *fred*. □

Usually linear temporal logic formulas are interpreted over the complete paths generated by a transition system. These correspond naturally to the infinite and deadlocking paths of an lts.

Definition 3.12 Let L be an lts, f a temporal semantics for L, v_0 a truth set, and ϕ an LTL' (LTL^∞) -formula. We say that ϕ is *true of L with respect to temporal semantics f and initial truth set v_0*, and write $L, f, v_0 \models \phi$ iff $\rho, f, v_0 \models \phi$ for all $\rho \in dpath(L) \cup infpath(L)$. □

Now a module of a system can be modelled by an lts and a temporal interpretation expressing the changes in the state information of that module caused by the transitions. These modules can then be combined to larger units by syntactic operators. We already know how an operator affects an lts. What needs to be defined is how an operator affects the temporal interpretation.

Discussing the parallel composition operator $|[\ldots]|$ first, the issue is deriving a temporal semantics f for $L = L_1|[\ldots]|L_2$ from the temporal semantics f_1 for L_1 and f_2 for L_2. Intuitively, the state information of L consists of the state information of both L_1 and L_2. The temporal semantics f_1 describes changes in the state information of L_1 and f_2 of L_2. If a transition of L corresponds to a transition taken by L_1 alone, the change in the state information of L is the same as in L_1. If a transition of L corresponds to a synchronised transition of both L_1 and L_2, the change in the state information of L consists of both the change to the state information of L_1 and of the change to the state information of L_2. Naturally this is not possible, if the temporal interpretations are mutually conflicting. Therefore the following definitions proceed by first stating when two temporal semantics can be combined, and by then defining the result of the operation.

Definition 3.13 A truth set modifier ts *affects* an atomic proposition a iff there is a $v \subseteq AP$ such that either $a \in v$ and $a \notin ts(v)$ or $a \notin v$ and $a \in ts(v)$. The set of all atomic propositions that ts affects is denoted by $aff(ts)$. We write $\overline{aff}(ts)$ for $AP \setminus aff(ts)$.

Truth set modifiers ts and ts' are *compatible* with each other iff for all atomic propositions $a \in aff(ts) \cap aff(ts')$ and for all $v \subseteq AP$, $a \in ts(v)$ iff $a \in ts'(v)$. If this is the case *the combination* of ts and ts', $ts \oplus ts'$, is the function $c : P(AP) \rightarrow P(AP)$ such that $c(v) = (ts(v) \cap aff(ts)) \cup (ts'(v) \cap aff(ts')) \cup (v \cap \overline{aff}(ts) \cap \overline{aff}(ts'))$ □

Definition 3.14 Temporal semantics f_1 and f_2 for ltss L_1 and L_2 are *compatible with respect to a synchronisation set* $G \subseteq \Sigma$ iff for all $g_i \in G$, $f_1(g_i)$ and $f_2(g_i)$ are compatible with each other and for all $l \in (\Sigma(L_1) \cap \Sigma(L_2)) \setminus G$, $f_1(l) = f_2(l)$.

If this is the case, the temporal semantics for $L_1|[G]|L_2$ is the function c such that $c(g_i) = f_1(g_i) \oplus f_2(g_i)$ for all $g_i \in G$, $c(l) = f_1(l)$ for all $l \in \Sigma(L_1) \setminus G$, and $c(l) = f_2(l)$ for all $l \in \Sigma(L_2) \setminus G$. □

In practice it is often the case that L_1 and L_2 are synchronised by all their common labels, i.e. that $(\Sigma(L_1) \cap \Sigma(L_2)) \setminus G = \emptyset$, and that the state information of L_1 is disjoint from that of L_2, i.e. that $\mathit{aff}(ts_1) \cap \mathit{aff}(ts_2) = \emptyset$ where $ts_1 = f_1(l_1)$ and $ts_2 = f_2(l_2)$, for all $l_1 \in \Sigma(L_1)$ and $l_2 \in \Sigma(L_2)$. In this case the temporal semantics for $L_1|[G]|L_2$ is easily seen to be well-defined.

Deriving the temporal semantics of an lts constructed by applying one of the other syntactic operators goes along the same lines. Please note in the following definition that the temporal semantics of the hide and rename constructs are well-defined only if the truth set modifier functions corresponding to transitions are not affected by applying the operator. This is fairly natural, as the operators should affect only the synchronisation capabilities of a module and not its internal state changes.

Definition 3.15 Let L_1 and L_2 be ltss and f_1 and f_2 temporal semantics for them.

If $L = L_1[]L_2$, temporal semantics f for L is well-defined iff $f_1(g) = f_2(g)$ for all $g \in \Sigma(L_1) \cap \Sigma(L_2)$, and f is defined by $f(g) = f_1(g)$ if $g \in \Sigma(L_1)$ and $f(g) = f_2(g)$ if $g \in \Sigma(L_2)$.

If $L =$ **hide** $g_1, \ldots, g_n$ **in** L_1, temporal semantics f for L is well-defined iff $f_1(g_i) = I$ for all $1 \leq i \leq n$, and f is defined by $f = f_1$.

If $L = L_1[h_1/g_1, \ldots, h_n/g_n]$, temporal semantics f for L is well-defined iff $f_1(g_i) = f_1(h_i)$ for all $1 \leq i \leq n$, and f is defined by $f = f_1$. □

4 Preservation of truth

Now that it is established how to interpret temporal logic formulas over labelled transition systems, we can formulate exactly what is required of an equivalence in order for it to be truth-preserving. Intuitively this means that no temporal logic formula should be able to distinguish an lts from an equivalent one. However, according to definition 3.12, the truth of a temporal logic formula in an lts depends on the temporal semantics and the initial truth set, as well. Remembering this, being truth-preserving is defined as meaning that no temporal logic formula should be able to make distinction between equivalent ltss *no matter what* the temporal semantics and the initial truth set are.

Definition 4.1 Let L and L' be ltss and ϕ an LTL' (LTL^∞) -formula. We say that L and L' *agree* on ϕ iff for every temporal semantics f and for every initial truth set v_0 it is the case that $L, f, v_0 \models \phi$ iff $L', f, v_0 \models \phi$. □

Definition 4.2 An equivalence $\approx$ between ltss is *LTL' (LTL^∞) -preserving* iff for any L, L' such that $L \approx L'$, L and L' agree on every LTL' (LTL^∞) formula ϕ. □

Next we point out some characteristics of an equivalence that guarantee its being truth-preserving, and notice that CFFD and NDFD-equivalences have these characteristics.

Proposition 4.3 Let L and L' be ltss such that $\mathit{inftr}(L) = \mathit{inftr}(L')$, $\mathit{divtr}(L) = \mathit{divtr}(L')$ and $\mathit{dtr}(L) = \mathit{dtr}(L')$. Then L and L' agree on every LTL^∞-formula ϕ.

Proof: Let ϕ be an LTL^∞-formula, and f, v_0 arbitrary temporal semantics and initial truth set, respectively. Now, $L, f, v_0 \models \phi$
iff $\rho, f, v_0 \models \phi$ for all $\rho \in dpath(L) \cup infpath(L)$ (by 3.12)
iff $vis(\rho), f, v_0 \models \phi$ for all $\rho \in dpath(L)$ and for all $\rho \in infpath(L)$ s. t. $vis(\rho) \in \Sigma^\infty$ and $vis(\rho) \cdot \tau^\infty, f, v_0 \models \phi$ for all $\rho \in infpath(L)$ s. t. $vis(\rho) \in \Sigma^*$ (by 3.10)
iff $\sigma, f, v_0 \models \phi$ for all $\sigma \in dtr(L)$ and for all $\sigma \in inftr(L)$ and $\sigma \cdot \tau^\infty, f, v_0 \models \phi$ for all $\sigma \in divtr(L)$ (by 2.5)
iff $\sigma, f, v_0 \models \phi$ for all $\sigma \in dtr(L')$ and for all $\sigma \in inftr(L')$ and $\sigma \cdot \tau^\infty, f, v_0 \models \phi$ for all $\sigma \in divtr(L')$ (by assumption)
iff $L', f, v_0 \models \phi$ (as above). The result follows by definition 4.1. □

Proposition 4.4 CFFD-equivalence is LTL^∞-preserving.
Proof: Immediate from 2.8 and 4.3. □

Proposition 4.5 Let L and L' be ltss such that $inftr(L) = inftr(L')$, $divtr(L) = divtr(L')$ and $nddtr(L) = nddtr(L')$. Then L and L' agree on every LTL'-formula ϕ.
Proof: Let ϕ be an LTL'-formula, and f, v_0 arbitrary temporal semantics and initial truth set, respectively. Now, $L, f, v_0 \models \phi$
iff $\rho, f, v_0 \models \phi$ for all $\rho \in dpath(L) \cup infpath(L)$ (by 3.12)
iff $\sigma, f, v_0 \models \phi$ for all $\sigma \in nddtr(L) \cup divtr(L) \cup inftr(L)$ (by 3.11 and 2.5)
iff $\sigma, f, v_0 \models \phi$ for all $\sigma \in nddtr(L') \cup divtr(L') \cup inftr(L')$ (by assumption)
iff $L', f, v_0 \models \phi$ (as above). The result follows by definition 4.1. □

Proposition 4.6 NDFD-equivalence is LTL'-preserving.
Proof: Immediate from 2.8 and 4.5. □

5 Minimality

In this section we show the minimality results, i.e. that CFFD (NDFD) equivalence is the weakest compositional equivalence preserving LTL^∞ (LTL'). This is done by showing that if any equivalence preserves LTL^∞ and is compositional, it must preserve all the properties that CFFD does, and alike for LTL' and NDFD. The proof proceeds as a series of lemmas, each lemma pointing out some particular characteristics that an equivalence must have in order to be congruent and truth-preserving. First some notation that will be needed in the proof.

Definition 5.1 If ϕ is a formula, $\neg^n \phi$ is ϕ if n is even and $\neg\phi$ if n is odd. □

Lemma 5.2 If $\approx$ is LTL^∞-preserving, then $L \approx L'$ implies $dtr(L) = dtr(L')$.
Proof: If $\approx$ is LTL^∞-preserving, then for all temporal semantics f, initial truth sets v_0 and LTL^∞-formulas ϕ the following holds: $L, f, v_0 \models \phi$ iff $L', f, v_0 \models \phi$.

In the proof we construct a particular temporal semantics f, initial truth set v_0, and a mapping ch from all strings of Σ^* to LTL^∞-formulas such that for any L and any $\sigma \in \Sigma^*$ the following holds: $L, f, v_0 \not\models \neg ch(\sigma)$ iff $\sigma \in dtr(L)$.

If $L \approx L'$, by the assumption $L, f, v_0 \models \neg ch(\sigma)$ iff $L', f, v_0 \models \neg ch(\sigma)$, and therefore $\sigma \in dtr(L)$ iff $\sigma \in dtr(L')$, which gives the required result $dtr(L) = dtr(L')$.

For the construction of f let cl be a symbol not in Σ. Without loss of generality we may assume that $AP = \{cl\} \cup \Sigma$. Denote by tsm_a the function $tsm_a : P(AP) \rightarrow P(AP)$ such that $tsm_a(A) = \{a\}$ if $cl \in A$ and $tsm_a(A) = \{a, cl\}$ if $cl \notin A$. Now the required f can be defined as $f : \Sigma_\tau \rightarrow TS$, $f(a) = tsm_a$ for all $a \in \Sigma$ and $f(\tau) = I$.

Intuitively the execution of a sets the atomic proposition a to true, negates the value of cl and sets all other propositions to false. cl may be thought of as a 'clock' which changes its value every time some visible event takes place.

The required v_0 is defined as $\{cl\}$.

Let $\sigma \in \Sigma^*$ and $|\sigma| = n$. The characteristic formula of σ, $ch(\sigma)$ is defined as follows:

$$cl\,\mathcal{U}(\sigma_1 \wedge ((\neg cl)\mathcal{U}(\sigma_2 \wedge (cl\,\mathcal{U} \ldots \mathcal{U}(\sigma_{n-1} \wedge ((\neg^{n-1} cl)\mathcal{U}(\sigma_n \wedge \Box(\neg^n cl) \wedge \neg \overset{\infty}{\Diamond}\top))) \ldots))))$$

The structure of $ch(\sigma)$ guarantees that $\rho, f, v_0 \models ch(\sigma)$ iff $vis(\rho) = \sigma$ and ρ is finite.

$L, f, v_0 \not\models \neg ch(\sigma)$ iff $\exists \rho \in infpath(L) \cup dpath(L) : \rho, f, v_0 \not\models \neg ch(\sigma)$, which means in turn that $\rho, f, v_0 \models ch(\sigma)$.

Now, $L, f, v_0 \not\models \neg ch(\sigma)$ iff $\exists \rho \in infpath(L) \cup dpath(L) : \rho, f, v_0 \models ch(\sigma)$ iff (since ρ must be finite) $\exists \rho \in dpath(L) : \rho, f, v_0 \models ch(\sigma)$ iff (by the above) $\exists \rho \in dpath(L) : vis(\rho) = \sigma$ iff (by 2.5) $\sigma \in dtr(L)$. Putting the stages together: $L, f, v_0 \not\models \neg ch(\sigma)$ iff $\sigma \in dtr(L)$. □

Lemma 5.3 If $\approx$ is LTL^∞-preserving, then $L \approx L'$ implies $divtr(L) = divtr(L')$.
Proof: As in 5.2, but this time $ch(\sigma)$ is defined as follows:

$$cl\,\mathcal{U}(\sigma_1 \wedge ((\neg cl)\mathcal{U}(\sigma_2 \wedge (cl\,\mathcal{U} \ldots \mathcal{U}(\sigma_{n-1} \wedge ((\neg^{n-1} cl)\mathcal{U}(\sigma_n \wedge \Box(\neg^n cl) \wedge \overset{\infty}{\Diamond}\top))) \ldots))))$$

The structure of $ch(\sigma)$ guarantees that $\rho, f, v_0 \models ch(\sigma)$ iff $vis(\rho) = \sigma$ and ρ is infinite. As in 5.2, $L, f, v_0 \not\models \neg ch(\sigma)$ iff $\exists \rho \in infpath(L) \cup dpath(L) : \rho, f, v_0 \models ch(\sigma)$ iff $\exists \rho \in infpath(L) : vis(\rho) = \sigma$ iff $\sigma \in divtr(L)$. □

Lemma 5.4 If $\approx$ is LTL'-preserving, then $L \approx L'$ implies $divtr(L) \cup dtr(L) = divtr(L') \cup dtr(L')$.
Proof: As in 5.2, but this time $ch(\sigma)$ is defined as follows:

$$cl\,\mathcal{U}(\sigma_1 \wedge ((\neg cl)\mathcal{U}(\sigma_2 \wedge (cl\,\mathcal{U} \ldots \mathcal{U}(\sigma_{n-1} \wedge ((\neg^{n-1} cl)\mathcal{U}(\sigma_n \wedge \Box(\neg^n cl)))) \ldots))))$$

The structure of $ch(\sigma)$ guarantees that $\rho, f, v_0 \models ch(\sigma)$ iff $vis(\rho) = \sigma$. As in 5.2, $L, f, v_0 \not\models \neg ch(\sigma)$ iff $\exists \rho \in infpath(L) \cup dpath(L) : \rho, f, v_0 \models ch(\sigma)$ iff $\exists \rho \in infpath(L) \cup dpath(L) : vis(\rho) = \sigma$ iff (by 2.5) $\exists \pi \in inftr(L) \cup divtr(L) \cup dtr(L) : \pi = \sigma$ iff (since σ is finite) $\exists \pi \in divtr(L) \cup dtr(L) : \pi = \sigma$ iff $\sigma \in divtr(L) \cup dtr(L)$. □

Lemma 5.5 If $L \approx L'$ implies $divtr(L) \cup dtr(L) = divtr(L') \cup dtr(L')$ and $\approx$ is a congruence with respect to parallel composition $|[\ldots]|$, then $L \approx L'$ implies $divtr(L) = divtr(L')$.
Proof: In the proof we construct for every $\sigma \in \Sigma^*$ of length n an lts L_σ such that for any L the following holds: $\sigma \in divtr(L)$ iff $\sigma \in divtr(L|[\sigma_1, \ldots, \sigma_n]|L_\sigma) \cup dtr(L|[\sigma_1, \ldots, \sigma_n]|L_\sigma)$.

The required result follows from this, as if $L \approx L'$
$\sigma \in divtr(L)$ iff $\sigma \in divtr(L|[\sigma_1, \ldots, \sigma_n]|L_\sigma) \cup dtr(L|[\sigma_1, \ldots, \sigma_n]|L_\sigma)$ iff
$\sigma \in divtr(L'|[\sigma_1, \ldots, \sigma_n]|L_\sigma) \cup dtr(L'|[\sigma_1, \ldots, \sigma_n]|L_\sigma)$ (assumption) iff $\sigma \in divtr(L')$.

Let $\sigma \in \Sigma^*$ be of length n. Choose some $a \in \Sigma$ such that $a \notin \{\sigma_1, \ldots, \sigma_n\}$. As Σ is infinite, such an a always exists. Now L_σ is defined as the following lts:

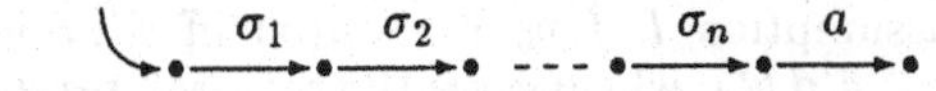

Let L be an arbitrary lts. We have to show that $\sigma \in divtr(L)$ iff
$\sigma \in divtr(L|[\sigma_1, \ldots, \sigma_n]|L_\sigma) \cup dtr(L|[\sigma_1, \ldots, \sigma_n]|L_\sigma)$.

If $\sigma \in divtr(L)$, then clearly $\sigma \in divtr(L|[\sigma_1, \ldots, \sigma_n]|L_\sigma)$.

If $\sigma \notin divtr(L)$, then $\sigma \notin divtr(L|[\sigma_1, \ldots, \sigma_n]|L_\sigma)$, since $\sigma \notin divtr(L_\sigma)$.

What is more, if $\sigma \notin divtr(L)$, then $\sigma \notin dtr(L|[\sigma_1, \ldots, \sigma_n]|L_\sigma)$, since $L|[\ldots]|L_\sigma$ can always execute a after having executed σ.
Therefore, $\sigma \in divtr(L)$ iff $\sigma \in divtr(L|[\sigma_1, \ldots, \sigma_n]|L_\sigma) \cup dtr(L|[\sigma_1, \ldots, \sigma_n]|L_\sigma)$. □

Lemma 5.6 If $L \approx L'$ implies $divtr(L) \cup dtr(L) = divtr(L') \cup dtr(L')$ and $\approx$ is a congruence with respect to parallel composition $|[\ldots]|$, then $L \approx L'$ implies $nddtr(L) = nddtr(L')$.
Proof: If $L \approx L'$, then by assumption $divtr(L) \cup dtr(L) = divtr(L') \cup dtr(L')$ and by 5.5 $divtr(L) = divtr(L')$. As $nddtr(L) = dtr(L) \setminus divtr(L) = (divtr(L) \cup dtr(L)) \setminus divtr(L)$, $nddtr(L) = nddtr(L')$ holds. □

Lemma 5.7 If $L \approx L'$ implies $dtr(L) = dtr(L')$, and $\approx$ is a congruence with respect to parallel composition $|[\ldots]|$, then $L \approx L'$ implies $sfail(L) = sfail(L')$.
Proof: Assume $L \approx L'$. In the proof we construct for every $(\sigma, A) \in \Sigma^* \times P(\Sigma)$ an lts $L_{(\sigma,A)}$ such that for both $L_x \in \{L, L'\}$ the following holds: $(\sigma, A) \in sfail(L_x)$ iff $\sigma \in dtr(L_x|[G]|L_{(\sigma,A)})$, where $G = \Sigma(L) \cup \Sigma(L')$. The required result follows as in 5.5.

Let $\sigma \in \Sigma^*$ be of length n, and $A \subset \Sigma$. Now $L_{(\sigma,A)}$ is defined as the following lts, where $\{a_1, \ldots, a_m\} = G \cap A$:

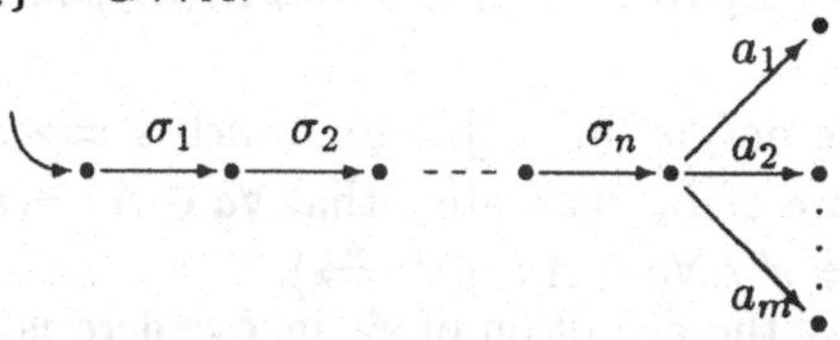

We have to show that $(\sigma, A) \in sfail(L_x)$ iff $\sigma \in dtr(L_x|[G]|L_{(\sigma,A)})$.

If $(\sigma, A) \in sfail(L_x)$, then $\sigma \in dtr(L_x|[G]|L_{(\sigma,A)})$, since corresponding to the path leading to the stable failure (σ, A) in L_x there is a path in $L_x|[G]|L_{(\sigma,A)}$ leading to a state where neither any $a \in A$ can be taken, since L_x does not allow it, nor any $a \in G \setminus A$, since $L_{(\sigma,A)}$ does not allow it.

If $\sigma \in dtr(L_x|[G]|L_{(\sigma,A)})$, then $(\sigma, A) \in sfail(L_x)$, since the path that leads to the deadlocking state in $L_x|[G]|L_{(\sigma,A)}$ corresponds to a path in L_x leading to a stable state where no $a \in A$ can be taken. □

Lemma 5.8 If $L \approx L'$ implies $nddtr(L) = nddtr(L')$, and $\approx$ is a congruence with respect to parallel composition $|[\ldots]|$, then $L \approx L'$ implies $ndfail(L) = ndfail(L')$.
Proof: As in 5.7. □

Lemma 5.9 If $L \approx L'$ implies $divtr(L) = divtr(L')$ and $sfail(L) = sfail(L')$, then $L \approx L'$ implies $ndfail(L) = ndfail(L')$.
Proof: Direct from definition 2.3. □

Lemma 5.10 If $L \approx L'$ implies $divtr(L) = divtr(L')$ and $ndfail(L) = ndfail(L')$, then $L \approx L'$ implies $tr(L) = tr(L')$
Proof: Immediate from 2.5. □

Lemma 5.11 If L is stable, then $\epsilon \notin divtr(L)$, and there is an $A \subset \Sigma$ such that for every $a \in A$, $a \notin tr(L)$, and $(\epsilon, A) \in ndfail(L)$ and for all $A' \subset \Sigma$: $(\epsilon, A') \in ndfail(L)$ implies $A' \subseteq A$.
Proof: $\epsilon \notin divtr(L)$, because otherwise $s \xrightarrow{\tau}$, which contradicts the stability of s.

If we define $A = \Sigma \setminus \{a \in \Sigma \mid s \xrightarrow{a}\}$ and L is stable, then $(\epsilon, A) \in ndfail(L)$ and for all $A' \subset \Sigma$, $(\epsilon, A') \in ndfail(L)$ implies $A' \subseteq A$. What is more, for all $a \in A$, not $s \xrightarrow{a}$, which means that $a \notin tr(L)$. □

Lemma 5.12 If $\approx$ is a congruence with respect to $[]$, and $L \approx L'$ implies $divtr(L) = divtr(L')$, $ndfail(L) = ndfail(L')$ and $tr(L) = tr(L')$, then $L \approx L'$ implies also $stable(L) \Leftrightarrow stable(L')$.

Proof: Let $L \approx L'$. If the conditions of 5.11 fail for L, they fail by assumption for L', too. Consequently, $\neg stable(L)$ and $\neg stable(L')$, i.e. $stable(L) \Leftrightarrow stable(L')$.

If the conditions of 5.11 hold, we construct an lts L_s such that for both $L_x \in \{L, L'\}$ the following holds: $(\epsilon, A) \in ndfail(L_x [] L_s)$ iff $\neg stable(L_x)$, where A is as in 5.11. The lts L_s is as follows:

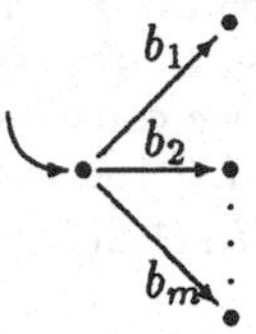

where $\{b_1, \ldots, b_m\} = \Sigma(L) \cup \Sigma(L')$.

Due to the structure of $L_x [] L_s$, $(\epsilon, A) \in ndfail(L_x [] L_s)$ iff $\exists s' : s \stackrel{\epsilon}{\Longrightarrow} s' \wedge \forall a \in A : \neg(s' \stackrel{a}{\Longrightarrow})$ in $L_x [] L_s$.

If $stable(L_x)$, then the only s' in $L_x [] L_s$ for which $s \stackrel{\epsilon}{\Longrightarrow} s'$ holds is the initial state s itself. The structure of L_s guarantees that $\forall a \in A : \neg(s \stackrel{a}{\Longrightarrow})$ does not hold. Consequently, $\neg \exists s' : s \stackrel{\epsilon}{\Longrightarrow} s' \wedge \forall a \in A : \neg(s' \stackrel{a}{\Longrightarrow})$.

If $\neg stable(L_x)$, then by the definition of A, in L_x there is a stable s' such that $s \stackrel{\epsilon}{\Longrightarrow} s' \wedge \forall a \in A : \neg(s' \stackrel{a}{\Longrightarrow})$. By the non-stability of L_x and s this means that $s' \neq s$. Consequently, also in $L_x [] L_s$ there is a stable s' such that $s \stackrel{\epsilon}{\Longrightarrow} s' \wedge \forall a \in A : \neg(s' \stackrel{a}{\Longrightarrow})$.

In both cases $(\epsilon, A) \in ndfail(L_x [] L_s)$ iff $\neg stable(L_x)$. As $ndfail(L [] L_s) = ndfail(L' [] L_s)$, this means that $stable(L)$ iff $stable(L')$. □

Now we can state the minimality results for finite ltss.

Proposition 5.13 If $\approx$ is a congruence with respect to $|[\ldots]|$ and $[]$ and is LTL^∞-preserving, then $L \approx L'$ implies $L \stackrel{\text{cffd}}{\approx} L'$ for all finite L and L'.

Proof: Immediate from 5.2, 5.3, 5.7, 5.9, 5.10, 5.12 and 2.7. □

Proposition 5.14 If $\approx$ is a congruence with respect to $|[\ldots]|$ and $[]$, and is LTL'-preserving, then $L \approx L'$ implies $L \stackrel{\text{ndfd}}{\approx} L'$ for all finite L and L'.

Proof: Immediate from 5.4, 5.5, 5.6, 5.8, 5.10, 5.12 and 2.7. □

Next these results are extended to deal with infinite ltss as well. However, the proof of the preservation of infinite traces requires that we are allowed to use arbitrary infinite ltss as given. Placing no restrictions in this sense is natural here, since we have not introduced any recursive or other operators by which infinite ltss could be constructed by finite means. Were it the case that some infinite ltss would be discarded as non-constructable, we could not be sure that the infinite lts L_σ required in the proof of 5.15 would be available. As the availability of a proof naturally depends on the particular recursion scheme used, we leave the issue open here.

Lemma 5.15 If $\approx$ is LTL'-preserving and a congruence with respect to parallel composition $|[\ldots]|$, then $L \approx L'$ implies $inftr(L) = inftr(L')$.

Proof: Let $L \approx L'$ and $\sigma \in \Sigma^\infty$. If the set $\{\sigma_1, \sigma_2, \ldots\}$ is infinite, $\sigma \notin inftr(L)$ and $\sigma \notin inftr(L')$ as the alphabet of any lts is required to be finite.

If $\{\sigma_1, \sigma_2, \ldots\}$ is finite, we construct an lts L_σ such that for both $L_x \in \{L, L'\}$ the following holds: $\sigma \in inftr(L_x)$ iff $inftr(L_x |[G]| L_\sigma) \neq \emptyset$, where $G = \Sigma(L) \cup \Sigma(L')$. L_σ is simply

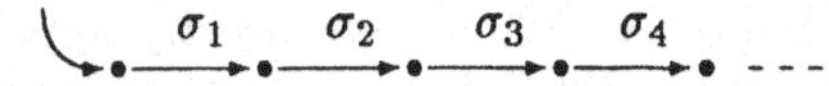

If we denote by ch_{fin} the formula $\Diamond(\Box cl \vee \Box\neg cl)$, then for any lts L: $L, f, v_0 \models ch_{fin}$ iff $inftr(L) = \emptyset$, where f and v_0 are as in 5.2. Therefore, $\sigma \in inftr(L_x)$ iff $L_x \|[G]\| L_\sigma, f, v_0 \not\models ch_{fin}$. Consequently, $inftr(L) = inftr(L')$. □

Finally we can state the general minimality results.

Proposition 5.16 If $\approx$ in a congruence with respect to $\|[\ldots]\|$ and $[]$ and is LTL^∞-preserving, then $L \approx L'$ implies $L \stackrel{cffd}{\approx} L'$.
Proof: Immediate from 5.2, 5.3, 5.7, 5.9, 5.10, 5.12 and 5.15. □

Proposition 5.17 If $\approx$ in a congruence with respect to $\|[\ldots]\|$ and $[]$, and is LTL'-preserving, then $L \approx L'$ implies $L \stackrel{ndfd}{\approx} L'$.
Proof: Immediate from 5.4, 5.5, 5.6, 5.8, 5.10, 5.12 and 5.15. □

It may be interesting to notice that all the proofs in this section could be carried through with the fixed temporal semantics of [CLM89], although the lack of an explicit 'clock' formula cl would require some changes to the definitions of characteristic formulas.

6 Reduction method

In this section we describe reduction methods preserving CFFD and NDFD-equivalences. In [VT91] a reduction algorithm for the CFFD-equivalence was presented. With minor modifications the algorithm can be used for NDFD-preserving reductions, as well. The method is related to the equivalence-checking methods presented in [BC89] and [CH90], and works as follows.

The lts to be reduced is transformed into an *acceptance graph (ag)*, which relies on the determinisation of the lts interpreted as a finite automaton. Accordingly, each state of the ag corresponds to a subset A of the states of the lts. The initial state of the ag corresponds to the states of the lts reachable by finite sequences of τ-transitions from the initial state of the lts. In the ag a state corresponding to A' is reachable from a state corresponding to A by an l-transition if for every $s' \in A'$ there is an $s \in A$ such that $s \xrightarrow{l} s'$, and if for every $s \in A$ and for every s' such that $s \xrightarrow{l} s'$, $s' \in A'$ holds. In addition to this, the states of the acceptance graph contain extra information specifying whether it is possible to diverge from the corresponding states of the lts, and what the stable failure sets of the corresponding states of the lts are.

The acceptance graph is minimised using the functional coarsest partition (fcp) algorithm in a way that guarantees that all the states of the original ag corresponding to a single state of the minimised ag have the same divergence trace and stable failure characteristics, as specified by the extra information attached to the states. Technically this is achieved by placing states having different divergence trace and stable failure characteristics into different elements of the initial partition which is then refined by the fcp-algorithm. The minimised acceptance graph is unwound into an lts using the information about the divergence traces and stable failures attached to the states of the acceptance graph.

Although the worst-case running-time of the algorithm is exponential in the size of the input, experiences from it [KLVC92] and from a related algorithm [CPS90] indicate that in practice this happens rarely.

Although the algorithm of [VT91] can be directly applied here as well, it has some drawbacks when the aim of the reduction is to verify temporal logic properties more efficiently. If we want to run a standard linear temporal logic model checker [LP85] on the model of a system, the states of the model must be labelled with the corresponding truth sets. For an arbitrary lts and temporal semantics this cannot always be done consistently when a state is reachable from the initial state by more than one route.

Assuming that an lts is generated by a program so that each state of the lts corresponds to a local state of the system, and each transition to an atomic statement, the states of the lts can be labelled consistently with truth sets. However, when the reduction algorithm is run on the lts, the reduced lts does not necessarily enjoy this property any more. Another algorithm [KV91] has to be applied in order to expand it into an equivalent lts that can be labelled consistently. As reducing an lts first and then partially expanding it again is neither very efficient nor elegant, we outline here a modification of the reduction algorithm guaranteeing that the reduced lts can be consistently labelled if the original lts can.

The modification is based on the fact that the truth set labelling of an lts may be consistently extended to the corresponding ag. Let V denote the mapping expressing the truth set labelling of the lts. The extension to ag is achieved by specifying: if a state of the ag corresponds to $A \subseteq S$, then $V(A) = V(s)$, where s is an arbitrary element of A. That $V(A)$ is well-defined is shown by induction on the transitions of the ag. If A corresponds to the initial state of the ag, A consists of the states of the lts reachable by τ-transitions from the initial state. The truth sets assigned to all these states must be the same, as τ-transitions cannot affect the truth sets. Therefore $V(A)$ is well-defined. If there is an l-transition from A to A' in the ag, then for all $s'_1, s'_2 \in A'$ there are $s_1, s_2 \in A$ so that $s_1 \xrightarrow{l} s'_1$ and $s_2 \xrightarrow{l} s'_2$. As $V(A)$ is well-defined, $V(s_1) = V(s_2)$. Consequently, $V(s'_1) = V(s'_2)$ and $V(A')$ is well-defined.

The only precaution that needs to be taken in order to keep the truth set labelling of the ag consistent in the minimisation step is seeing that states of the ag having different truth set labels are kept separate in the minimisation. This can be done in the same way as states having different divergence and stable failure characteristics are kept separate, i.e. by placing them in separate elements of the initial partition. In all other respects the algorithm of [VT91] can be used without modification.

7 Discussion

In this paper we have discussed equivalences facilitating the verification of nexttimeless linear temporal logic properties by using compositional reductions. The CFFD and NDFD-equivalences have a strong claim to being ideal in this respect: they are the weakest compositional equivalences preserving the linear temporal logics LTL^∞ and LTL', respectively. What is more, they have a reduction algorithm that works fast in practice [VT91]. Therefore, we believe that the CFFD and NDFD-equivalences have a lot to offer in the task of verifying realistic systems.

Acknowledgements: The authors would like to express their gratitude to Martti Tienari for his helpful comments and to the anonymous referee of [KV91] in CONCUR'91 for pointing out the importance of the $\bar{\otimes}$-operator. The work of A. Valmari was funded by the Technology Development Centre of Finland (Tekes).

References

[BB89] Bolognesi, T. & Brinksma, E.: Introduction to the ISO Specification Language LOTOS, in *The Formal Description Technique LOTOS*, North-Holland, 1989, pp. 23-73

[BC89] Bolognesi, T. & Caneve, M.: Equivalence Verification: Theory, Algorithms and a Tool, in *The Formal Description Technique LOTOS*, North-Holland, 1989, pp. 303-326

[BCG87] Browne, M. C. & Clarke, E. M. & Grümberg, O.: Characterizing Kripke Structures in Temporal Logic, in Ehrig, H. & Kowalski, R. & Levi, G. & Montanari, U. (eds.): *TAPSOFT '87, vol. I*, LNCS, vol. 249, Springer-Verlag, 1987, pp. 256-270

[CLM89] Clarke, E. M. & Long, D. & McMillan, K. L.: Compositional Model Checking, in *Proceedings of the Fourth IEEE Symposium on Logic in Computer Science*, 1989, pp. 353-362

[CH90] Cleaveland, R. & Hennessy, M.: Testing Equivalence as a Bisimulation Equivalence, in *Proceedings of the Workshop on Automatic Verification Methods for Finite State Systems*, LNCS, vol. 407, Springer-Verlag, 1990, pp. 11-23

[CPS90] Cleaveland, R. & Parrow, J. & Steffen, B.: The Concurrency Workbench, in *Proceedings of the Workshop on Automatic Verification Methods for Finite State Systems*, LNCS, vol. 407, Springer-Verlag, 1990, pp. 24-37

[EC80] Emerson, E. A. & Clarke, E. M.: Characterising Correctness Properties of Parallel Programs Using Fixpoints, in *Proceedings of the 7th ICALP*, LNCS, vol. 85, Springer-Verlag, 1980, pp. 169-181

[GS90] Graf, S. & Steffen, B.: Compositional Minimization of Finite-State Processes, in Kurshan, R. P. & Clarke, E. M. (eds.): *Proceedings of CAV'90*, LNCS, vol. 531, Springer-Verlag, 1990, pp. 186-196

[KV91] Kaivola, R. & Valmari, A.: Using Truth-Preserving Reductions to Improve the Clarity of Kripke-Models, in Baeten, J. C. M. & Groote, J. F. (eds.): *Proceedings of CONCUR'91*, LNCS, vol. 527, Springer-Verlag, 1991, pp. 361-375

[KLVC92] Kemppainen, J. & Levanto, M. & Valmari, A. & Clegg, M.: "ARA" Puts Advanced Reachability Analysis Techniques Together, in *Proceedings of the Fifth Nordic Workshop in Programming Environment Research*, Tampere University of Technology, Software Systems Laboratory Report 14, 1992

[Lam83] Lamport, L.: What Good is Temporal Logic?, in *Proceedings of the IFIP 9th World Computer Congress*, 1983, pp. 657-668

[LP85] Lichtenstein, O. & Pnueli, A.: Checking that Finite State Concurrent Programs Satisfy Their Linear Specification, in *Conference Record of the Twelfth Annual ACM Symposium on Principles of Programming Languages*, 1985, pp. 97-107

[Mil89] Milner, R.: *Communication and Concurrency*, Prentice Hall, 1989

[OH86] Olderog, E.-R. & Hoare, C. A. R.: Specification-Oriented Semantics for Communicating Processes, in *Acta Informatica*, vol. 23, 1986, pp. 9-66

[Sti87] Stirling, C.: *Comparing Linear and Branching Time Temporal Logics*, University of Edinburgh, LFCS Report Series ECS-LFCS-87-24, 1987

[Val90] Valmari, A.: Compositional State Space Generation, in *Proceedings of the 11th International Conference on Application and Theory of Petri Nets*, 1990, pp. 43-62, to appear also in *Advances in Petri Nets 92*, LNCS, Springer-Verlag, 1992

[VT91] Valmari, A. & Tienari, M.: An Improved Failures Equivalence for Finite-State Systems with a Reduction Algorithm, in *Protocol Specification, Testing and Verification XI*, North-Holland, 1991, pp. 3-18

[VT92] Valmari, A. & Tienari, M.: *Compositional Failure-based Semantic Models for Basic LOTOS*, A manuscript submitted for publication, 1992, 30 p.

[Wal87] Walker, D.: *Bisimulation Equivalence and Divergence in CCS*, University of Edinburgh, LFCS Report Series ECS-LFCS-87-29, 1987

Propositional Temporal Logics and Equivalences

Ursula Goltz[1], Ruurd Kuiper[2], Wojciech Penczek[3]

Abstract

We compare propositional temporal logics by comparing the equivalences that they induce on models. Linear time, branching time and partial order temporal logics are considered. The logics are interpreted on occurrence transition systems, generated by labelled prime event structures without autoconcurrency. The induced equivalences are also compared to directly defined equivalences, e.g., history preserving bisimulation, pomset bisimulation, pomset trace equivalence, and others. It is then shown which of the induced equivalences are and which are not preserved under action refinement.

Rather unexpectedly, the addition of the backward next step operator to the weakest logic considered yields a logic stronger than all others. It is shown that weak history preserving bisimulation can be obtained as the equivalence induced by a slightly constrained version of that logic.

1 Introduction

Currently a lot of formalisms to describe concurrent computations exist. Even only regarding logics still leaves a large set. Our aim is to bring some structure in this set.

Some comparisons between logics of course exist already. However, mostly these consider only logics interpreted over the same domain [EH86, Wo87, St87], whereas we consider logics which were originally interpreted over different domains. Also, these comparisons mainly address expressiveness, i.e., which sets of behaviours can be described. This is a rather strict notion, and logics quickly become incomparable ([La80]).

We wish to compare logics in the large. So we need, firstly, a common framework of interpretation and, secondly, a sufficiently relaxed measure for comparison.

Event structures ([NPW81, Wi88]) provide a very detailed model for concurrent computations. All features considered by many different logics are represented therein. Therefore we choose these as the natural candidate for a common framework of interpretation, satisfying our first desire.

Various equivalence notions have been defined on event structures to obtain more abstract system representations. These equivalences themselves can then be compared, thus providing a measure for the relative precision. This we use to provide an answer to our second desire, a measure for comparison. Each logic namely induces

---

[1]GMD, Bonn, Germany. Supported in part by Esprit Basic Research Action 3148 (DEMON)

[2]Eindhoven University of Technology, Department of Computer Science, The Netherlands. Supported by Esprit-BRA project 3096: Formal Methods and Tools for the Development of Distributed and Real-Time Systems (SPEC)

[3]Institute of Computer Science, Warsaw, Poland. Supported in part by the Dutch NFI/NWO project REX and also by The Wolfson Research Awards Scheme in The United Kingdom

an equivalence relation: the one that distinguishes just those event structures that can be distinguished by some formula. We then compare logics by comparing the induced equivalences. So the comparison is on precision rather than expressiveness.

For the interpretation of temporal logics we use transition systems derived from the event structures. The equivalence relations are redefined accordingly. We link event structures via configuration structures to *occurrence transition systems*. This idea has been used before, e.g., in [NRT91, Ro90]. These transition systems in turn are linked to the Kripke structures that serve as models for the logics. This idea is already present in [deNMV90], but motivated a little more extensively here. We, briefly, show how occurrence transition systems are obtained from the event structures because, as yet, we have no direct characterization of this class. As it turns out that none of the considered logics can handle autoconcurrency, we restrict the class of systems accordingly.

The following temporal logics are considered: the modal logic S4 extended with Next (S4N) [HC84], Linear Time Temporal Logic (LTTL) [MP91], Computation Tree Logic (CTL) [CES83], (Quantified) Interleaving Set Temporal Logic ((Q)ISTL) [KP87], and Concurrent Computation Tree Logic (CCTL) [Pe90], the latter three with their *- and/or concurrently fair versions. Furthermore, the extensions of these logics with added past modalities are considered. The selection is motivated in the paper by a table presenting various combinations of features of such logics.

A second aim was inspired by the fact that much research effort has been devoted already to devising useful equivalence notions directly. We compare induced and existing equivalences, to see whether or not the intuitions and motivations from the different fields lead to similar results.

The following equivalence notions have been selected: interleaving trace equivalence ($\sim_{i-t}$), forward bisimulation ($\sim_{f-b}$) [Pa81, Mi80], pomset trace equivalence ($\sim_{p-t}$), pomset bisimulation ($\sim_{p-b}$) [BC87], history preserving bisimulation ($\sim_{h-b}$) [TRH88], backward forward bisimulation ($\sim_{bf-b}$) [deNMV90]. Interleaving trace equivalence is the simplest interleaving equivalence where the branching structure (choices between alternative behaviours) is not taken into account. Pomset trace equivalence is the corresponding notion in causality based semantics, where causal dependencies in runs of systems are recorded. The selection of these equivalences will be, again, motivated later by a table of features. The decisions were influenced by intuitive expectations that the chosen equivalences would be close to induced ones. The combined result of these two efforts is a complete table, comparing all equivalences considered. Two interesting facts follow from this table about the extension of S4N with past modalities, Partial Order Logic (POL) [Si90]. Rather surprisingly, the expressively quite weak language POL induces a stronger equivalence than the expressively strong (though incomparable, expressively, to POL) concurrently fair version of CCTL*. Also, a slightly modified form of the simple logic POL induces $\sim_{h-b}$. The only other logic, to our knowledge, that provides this correspondence can be found in [deNF90].

Constructing the table of comparisons, we learned that there is a quite close connection between induced and existing equivalence relations. Where there was no direct match, our investigations gave enough insight to enable modifications to the logics, or, conversely, to the equivalence notions, to make them fit to one another. The main difference between equivalences induced by logics and those defined directly turned out to be that the former use completed pomsets or sequential traces whereas the latter use simply pomsets or sequential traces.

As a third, more or less independent aim, it is proven which of the equivalences are and which are not preserved under action refinement.

Here our results suggest that equivalences imposed by logics with backward modalities distinguishing branching points and concurrency are preserved under refinement of actions, whereas for similar logics without backward modalities, preservation under refinement can not be obtained.

The paper is organised as follows. Section 2 introduces the basic framework, event structures. It also contains the translation into configuration structures and the further translation of these into transition systems. In Section 3 the selected logics are presented, in Section 4 the chosen equivalence notions. In Section 5 the induced equivalences are derived. The comparison is made in Section 6. Preservation of the equivalences under action refinement is the subject of Section 7. Finally, some concluding remarks are made in Section 8. All the proofs can be found in the full version of this paper [GKP92].

2 Representations of Concurrent Systems

We are interested in comparing formalisms that describe concurrent systems performing actions from a given set *Act* of action names. These formalisms are interpreted over various different structures. As the first, basic, structure we have chosen event structures.

We use a subset of CCSP expressions to write down examples of concurrent systems in a concise and intuitive manner, assuming event structures as a semantics. An event structure semantics of CCSP can be found in, e.g., [Wi82, LG91].

2.1 Event Structures

Event structures represent a concurrent system by taking occurrences of actions as the starting point. Every occurrence of an action is modelled as a separate event; a label function indicates which action is represented. Two relations are provided that capture, respectively, the causality and conflict relationship between events. It turns out that for our purposes we only need a certain class of event structures; prime event structures with binary conflict. In the sequel we tacitly assume that event structures are taken from this class. Further details can be found in, e.g., [NPW81, Wi88].

Definition 2.1 *A* (prime labelled) event structure *over an alphabet Act is a 4-tuple* $\mathcal{E} = (E, <, \sharp, l)$, *where*

- E *is a countable set of* events,
- $< \subseteq E \times E$ *is an irreflexive partial order (the* causality relation*), satisfying the principle of* finite causes*:* $\forall e \in E : \{e' \in E \mid e' < e\}$ *is finite,*
- $\sharp \subseteq E \times E$ *is an irreflexive, symmetric relation (the* conflict relation*), satisfying the principle of* conflict heredity*:* $\forall d, e, f \in E : d < e$ *and* $d \sharp f$ *implies* $e \sharp f$, *and*

- $l : E \longrightarrow Act$ (labelling function).

The components of an event structure $\mathcal{E}$ are denoted by $E_{\mathcal{E}}$, $<_{\mathcal{E}}$, $\#_{\mathcal{E}}$, and $l_{\mathcal{E}}$. If clear from the context, the index $\mathcal{E}$ is omitted. As usual, we write $d \leq e$ for $d < e$ or $d = e$.

Throughout the paper, we assume a fixed set *Act* of *action names (labels)*. Causal independence (*concurrency*) of events is expressed by the derived relation $co \subseteq E \times E$: $d\ co\ e$ iff neither $d = e$ nor $d < e$ nor $e < d$ nor $d \# e$. For $x \subseteq E_{\mathcal{E}}$, the *restriction* of $\mathcal{E}$ to x is defined as $\mathcal{E}|x = (x, < \cap (x \times x), \# \cap (x \times x), l|x)$. To be able to abstract from names of events, we introduce a notion of isomorphism. Two event structures $\mathcal{E}$ and $\mathcal{F}$ are *isomorphic* ($\mathcal{E} \cong \mathcal{F}$) iff there exists a bijection between their sets of events preserving the relations $<$, $\#$, and labelling.

Example 2.1 *The CCSP program* $(a|b); (a+b)$ *is represented by the following event structure:* $\mathcal{E} = (E, <, \#, l)$, *where*

$$E = \{e_1, e_2, e_3, e_4\},\ < = \{(e_1, e_3), (e_1, e_4), (e_2, e_3), (e_2, e_4)\},$$

$$\# = \{(e_3, e_4), (e_4, e_3)\},\ \textit{and}\ l(e_1) = a,\ l(e_2) = b,\ l(e_3) = a,\ l(e_4) = b.$$

2.2 Configuration Structures

Configuration structures are a state oriented view of event structures - the first move towards structures that logics can be interpreted on. The notion of a state, called a *configuration*, that is used describes the situation at a certain moment during an execution. It is just the set of all events executed so far. The interpretation of $<$ as causal necessity means requiring that a configuration is *left-closed* with respect to $<$. Since $\#$ expresses mutual exclusion of events, a configuration must be *conflict-free*. It is assumed that in a finite period only finitely many actions are performed. We therefore consider only finite configurations.

Definition 2.2 *Let* $\mathcal{E}$ *be an event structure.*

- $x \subseteq E_{\mathcal{E}}$ *is* left-closed *in* $\mathcal{E}$ *if for all* $e \in x$, $d \in E_{\mathcal{E}}$: *if* $d < e$, *then* $d \in x$.
  $x \subseteq E_{\mathcal{E}}$ *is* conflict-free *if* $(x \times x) \cap \# = \emptyset$.
- *A finite, left-closed, conflict-free subset of* $E_{\mathcal{E}}$ *is called a* configuration *of* $\mathcal{E}$.
- *Let* $Conf(\mathcal{E})$ *denote the set of all configurations of* $\mathcal{E}$.
- *An isomorphism class of* $\mathcal{E}|x$, *for* $x \in Conf(\mathcal{E})$, *is called a* pomset *of* $\mathcal{E}$.

2.3 Occurrence Transition Systems

Transition systems are a way of looking at configuration structures not emphasizing the states but rather the actions that lead from one state to another, i.e., the structure. Winskel has shown that families of configurations may be used as an alternative representation of event structures [Wi88]. These families may be conceived of as special (possibly infinite) transition systems.

We now associate transition systems with event structures in a canonical way (see [NRT91, Ro90]), taking the configurations as nodes. The transition relation relates those configurations which differ by exactly one event and we label each transition by the corresponding action.

Definition 2.3 *The* occurrence transition system $OTS(\mathcal{E})$ *of an event structure* $\mathcal{E}$ *is given by the triple* $OTS(\mathcal{E}) = (Conf(\mathcal{E}), \rightarrow, q^0)$ *where*

- $\rightarrow \subseteq Conf(\mathcal{E}) \times Act \times Conf(\mathcal{E})$ *is defined by* $x \xrightarrow{a} y$ *iff* $y = x \uplus \{e\}$ *for some* $e \in E_{\mathcal{E}}$ *and* $l_{\mathcal{E}}(e) = a$,
- $q^0 = \emptyset$ *(*initial node*).*

We use occurrence transition systems generated by event structures for the representation of concurrent systems.

Example 2.2 *The occurrence transition system* $OTS(\mathcal{E})$*, generated by the event structure* $\mathcal{E}$ *of Example 2.1, is of the following form:* $OTS(\mathcal{E}) = (Q, \rightarrow, \emptyset)$.

- $Q = \{\emptyset, \{e_1\}, \{e_2\}, \{e_1, e_2\}, \{e_1, e_2, e_3\}, \{e_1, e_2, e_4\}\}$,
- $\rightarrow \; = \{(\emptyset, a, \{e_1\}), (\emptyset, b, \{e_2\}), (\{e_1\}, b, \{e_1, e_2\}), (\{e_2\}, a, \{e_1, e_2\}),$
  $(\{e_1, e_2\}, a, \{e_1, e_2, e_3\}), (\{e_1, e_2\}, b, \{e_1, e_2, e_4\})\}$.

Two occurrence transition systems *OTS* and *OTS'* are *isomorphic* (written, $OTS \cong OTS'$) iff there exists a bijection between their sets of nodes mapping initial node to initial node and respecting the labelled transition relation.

Let $x, y \in Q$ and $n \in \mathbb{N}$. We use the following notation:

- $x \rightarrow y$ if $x \xrightarrow{a} y$, for some $a \in Act$, $x \rightarrow^n y$ if there is a sequence of nodes $x_0, \ldots, x_n$ s.t. $x = x_0$, $y = x_n$, and $x_i \rightarrow x_{i+1}$, for $0 \leq i < n$.
- $x \rightarrow^* y$ if $x = y$ or $x \rightarrow^n y$, for some $n \in \mathbb{N}$.

Definition 2.4 *Let* $OTS = (Q, \rightarrow, q^0)$ *be an occurrence transition system and let* $Q' \subseteq Q$ *s.t.* Q' *has a smallest node, say* q'.

- *OTS is* directed *if for every* $x, y \in Q$ *there exists* $z \in Q$ *such that* $x \rightarrow^* z$ *and* $y \rightarrow^* z$.
- *The* restriction *of OTS to* Q' *is defined as* $TS|Q' = (Q', \rightarrow \cap (Q' \times Act \times Q'), q')$.
- Q' *is* directed *in OTS if* $OTS|Q'$ *is directed.*
- Q' *is* left-closed *in OTS if for all* $x \in Q'$, $y \in Q$: *if* $y \rightarrow^* x$, *then* $y \in Q'$.
- *A sequence of nodes and labels* $p = (x_0, a_0, x_1, a_1, \ldots)$ *is said to be a* path *in TS starting at* x_0 *if* $x_i \xrightarrow{a_i} x_{i+1}$ *for all* $i \geq 0$. *When convenient, we shall sometimes refer to a path p as a sequence of nodes* $p = (x_0, x_1, \ldots)$.
- $p|Act = a_0 a_1 \ldots$ *is the sequence of labels of p.*

- *OTS is* **without auto-concurrency** *if for all $x, y, z \in Q$: $y \xrightarrow{a} x$ and $z \xrightarrow{a} x$ implies $y = z$.*
- *OTS is* **finitly branching** *if for all $x \in Q$ the set $\{y \in Q \mid x \to y\}$ is finite.*

In what follows we consider occurrence transition systems with finite branching and without auto-concurrency. This is because the temporal logics are not able to characterize infinite branching and they cannot distinguish auto-concurrency.

To capture the causalities in executions of a system, we define the notion of a *(partial) run* of an occurrence transition system. We consider not only runs starting at the initial state, but also runs starting at arbitrary states.

Definition 2.5 *Let $OTS = (Q, \to, q^0)$ be an occurrence transition system, $x_0 \in Q$ and $p = (x_0, x_1, \ldots)$ be a path in OTS.*

- $\downarrow x_0 = \{x \in Q \mid x \to^* x_0\}$, $\uparrow x_0 = \{x \in Q \mid x_0 \to^* x\}$, $\downarrow p = \bigcup\{\downarrow x_i \mid i \geq 0\}$.
- *$R \subseteq Q$ is said to be a* **partial run** *of OTS starting at x_0 if R is a finite, left-closed, directed subset of the states of $OTS|\uparrow x_0$. By $R(x_0, y)$ we denote the partial run $\uparrow x_0 \cap \downarrow y$, for $y \in \uparrow x_0$. $PR_{x_0}(OTS)$ denotes the set of all partial runs of OTS starting at x_0. The set $PR_{q_0}(OTS)$ we simply denote as $PR(OTS)$. A partial run R induces the occurrence transition system $OTS|R$,*

  $R \subseteq Q$ is said to be a **run** *of OTS starting at x_0 if R is a maximal directed subset of the states of $OTS|\uparrow x_0$ (left-closedness follows from maximality). $R_{x_0}(OTS)$ denotes the set of all runs of OTS starting at x_0. The set $R_{q_0}(OTS)$ we simply denote as $R(OTS)$. A run R induces the occurrence transition system $OTS|R$.*
- *$o = (x_0, a_0, x_1, a_1, \ldots)$ is an* **observation** *of OTS starting at x_0 iff o is a maximal path in OTS s.t. there is a run R starting at x_0 in OTS satisfying $R = \{x \in Q \mid x_0 \to^* x \to^* x_i$, for some $i \geq 0\}$. We shall also say that o is an observation of R.*

The following intuitions may be helpful. A run represents all states that potentially occur during some execution of an OTS, they are pairwise comparable. An observation is a concurrently-fair (cF) linearization of a run.

3 Temporal Logics

In this section temporal logics are introduced. In order to define the logics in a uniform framework, we use more abstract structures as frames that subsume those originally used. To this end, we have to show how an occurrence transition system can be turned into a model for a modal logic.

Occurrence transition systems fit very well to our purpose in that no information is carried in the nodes, but all information is present in the structure. They are less nice in that the transitions are labelled, whereas temporal logics cannot use information from transition labels.

Therefore, the idea is to encode this information into the valuation function. This translation should be correct, i.e., yield a function. Also, no information should be lost, i.e., it should be possible to recover the original transition system. Finally, and perhaps most importantly in view of a comparison aim, no structural information should be transferred to the valuation function. This is done as follows, taking $PV = Act$. Note that we have chosen to use anchored models [MP88].

Definition 3.1 *The* model *corresponding to* $OTS = (Q, \longrightarrow', q^0)$ *is the 4-tuple* $M_{OTS} = (Q, \rightarrow, V, q^0)$, *where* $\rightarrow \subseteq Q \times Q$ *s.t.* $x \rightarrow y$ *iff* $x \overset{a}{\longrightarrow}' y$ *for some* $a \in Act$ *and* $V : Q \longrightarrow 2^{PV}$ *is a valuation function s.t.* $V(q^0) = \emptyset$ *and if* $x \overset{a}{\longrightarrow}' y$, *then* $V(y) = (V(x) - \{a\}) \cup (\{a\} - V(x))$.

There is a one-to-one correspondence between occurrence transition systems and the corressponding models, i.e, the transition labelling can be retrieved from models. Therefore, all notions provided in Definitions 2.4 and 2.5 can be extended to models.

Before introducing logics, we display a table, showing possibilities for defining logics with forward modalities. It appears that once the next operator is present in a logic, the decisive operators concerning induced equivalences are the universal and existential quantifiers. Furthermore, the equivalences depend on the precise objects these quantifiers range over. The objects considered in existing logics are successors, future states, runs, maximal paths and observations. In the table, the quantifiers ranging over successors and future states are mentioned explicitly in the case of S4N, where they are the only quantifiers, and left implicit in all other cases.

object / quantif.	successor fut. state	max. path	obser.	run: fut. state successor	run: max. path	run: obser.
universal	———	LTTL	$LTTL_{cF}$	??	??	$ISTL^{(*)}$
universal existential	S4N	$CTL^{(*)}$	$CTL^{(*)}_{cF}$	??	CCTL	$QISTL^{(*)}$ $CCTL^{(*)}_{cF}$

The question marks (??) indicate that for these slots we could not find existing logics. The absence of a star (*) indicates a limitation of the use of quantifiers, whereas cF means that path quantifiers range over observations.

The language of $ISTL^{(*)}$ contains universal and existential quantifiers ranging over observations, but only universal quantifiers ranging over runs.

Because of lack of space the definitions of the logics $LTTL_{cF}$, $CTL^{(*)}_{cF}$, $QISTL^{(*)}$, and $CCTL_{(cF)}$, which are variants of defined logics, are not given.

The most complicated logic $CCTL^{*}_{cF}$ is an extension of CCTL [Pe90].

The set of state, path, and run formulas is defined inductively:

S1. every member of PV is a state formula,

S2. if φ and ψ are state formulas, then so are $\neg\varphi$ and $\varphi \wedge \psi$,

S3. if φ is a path formula, then $E\varphi$ is a run formula,

S4. if φ is a run formula, then $\exists\varphi$ is a state formula,

P1. any state formula φ is also a path formula,

P2. if φ, ψ are path formulas, then so are $\varphi \wedge \psi$ and $\neg\varphi$,

P3. if φ, ψ are path formulas, then so are $X\varphi$ and $(\varphi U \psi)$.

P1'. any run formula φ is also a path formula,

R1. any state formula φ is also a run formula,

R2. if φ, ψ are run formulas, then so are $\varphi \wedge \psi$ and $\neg\varphi$.

Let $M_{OTS} = (Q, \rightarrow, V, q^0)$ be the model corresponding to an OTS, $R \in R_{x_0}(M_{OTS})$, for $x_0 \in Q$, and $o = (x_0, x_1, \ldots)$ be an observation. By o_i we denote the suffix $(x_i, x_{i+1}, \ldots)$ of o.

S1. $x \models p$ iff $p \in V(x)$, for $p \in PV$,

S2. $x \models \neg\varphi$ iff not $x \models \varphi$,

$x \models \varphi \wedge \psi$ iff $x \models \varphi$ and $x \models \psi$,

S3. $R \models E\varphi$ iff $o \models \varphi$ for some observation o of R,

S4. $x_0 \models \exists\varphi$ iff $R \models \varphi$ for some run $R \in R_{x_0}(M_{OTS})$,

P1. $o \models \varphi$ iff $x_0 \models \varphi$ for any formula φ,

P2. $o \models \varphi \wedge \psi$ iff $o \models \varphi$ and $o \models \psi$,

$o \models \neg\varphi$ iff not $o \models \varphi$,

P3. $o \models X\varphi$ iff $o_1 \models \varphi$,

$o \models (\varphi U \psi)$ iff $(\exists i \geq 0)\ o_i \models \psi$ and $(\forall j :\ 0 \leq j < i)\ o_j \models \varphi$,

P1'. $o \models \varphi$ iff $R \models \varphi$ for any run formula φ, where o is an observation of R,

R1. $R \models \varphi$ iff $x_0 \models \varphi$ for any state formula φ,

R2. $R \models \varphi \wedge \psi$ iff $R \models \varphi$ and $R \models \psi$,

$R \models \neg\varphi$ iff not $R \models \varphi$.

A $CCTL^*_{cF}$ formula φ is *valid* in a model M_{OTS} (written $M_{OTS} \models_{CCTL^*_{cF}} \varphi$) iff $M_{OTS}, q^0 \models \varphi$.

The syntax of ISTL* [KP87] is the same as that of CTL* [EH86], i.e, it is given by $S1 - S3$ and $P1 - P3$ (in $S3$ "run formula" is replaced by "state formula"). The semantics is the subset of the semantics of $CCTL^*_{cF}$ concerning ISTL* formulas, interpreted over models corresponding to runs of occurrence transition systems. $S3$ is now of the form: $x_0 \models E\varphi$ iff $o \models \varphi$ for some observation o starting at x_0.

An ISTL* formula φ is *valid* in a model M_{OTS} (written $M_{OTS} \models_{ISTL^*} \varphi$) iff $M_{OTS}|R, q^0 \models \varphi$, for all $R \in R(M_{OTS})$.

The semantics of CTL* differs from that of ISTL*. $S3$ is now of the form: $x_0 \models E\varphi$ iff $o \models \varphi$ for some maximal path o starting at x_0. CTL* formulas are interpreted over whole transition systems. Therefore, the validity in a model is defined like for $CCTL^*_{cF}$.

The language of ISTL [KP87] is the same as that of CTL [CES83] and it is a restriction of ISTL*. Therefore, only a single linear time operator (F, G, X, or U) can follow the path quantifier E.

The syntax of LTTL [MP91] is a restriction of that of CTL*, given by $S1 - S2$ and $P1 - P3$. LTTL formulas are interpreted over models corresponding to the maximal paths through transition systems in the present framework. Therefore, a LTTL formula φ is *valid* in a model M_{OTS} (written $M_{OTS} \models_{LTTL} \varphi$) iff $M_{OTS}|p \models \varphi$, for all maximal paths p in M_{OTS} starting at q^0.

S4N is the most basic logic that we consider. The language of S4N is an extension of that of S4 [HC84] by one modality: X corresponding to the forward next step relation ($\rightarrow$). S4N formulas are interpreted over models corresponding to transition systems in a standard way.

POL is the most basic logic referring to the past that we consider. The language of POL is an extension of that of S4N by two modalities: H corresponding to the relation $(\rightarrow^*)^{-1}$ and Y corresponding to the backward next step relation $\rightarrow^{-1}$. POL formulas are interpreted over models in a standard way.

4 Equivalence Notions for Transition Systems

We are interested in equivalence notions that are close to the ones induced by the logics. To construct a table, we use similar "objects" as before: paths, observations, paths & backward paths, ... As an analogon to the quantifiers, we consider using the objects in a linear or branching manner, i.e., comparing them only in the initial state or also in subsequent states.

approach \ object	(max.) path	obser.	pomset (part. run)	run	path/b-path	history
linear	$\sim_{(m)i-t}$	??	$\sim_{p-t}$	$\sim_{r-t}$ (*)	??	??
branching	$\sim_{f-b}$	??	$\sim_{p-b}$	$\sim_{r-b}$ (*)	$\sim_{bf-b}$	$\sim_{wh-b}$

(*) The equivalences in these slots have not been defined directly before. However, they correspond to equivalences induced by existing logics, namely $CCTL^*_{cF}$ and ISTL*. The remaining question marks (??) indicate that the equivalence has neither been defined directly before nor seems to correspond to an equivalence induced by any of the existing logics considered in the paper.

Now, we formally define equivalence notions for occurrence transition systems. We do this column by column according to the table. Let $OTS = (Q, \rightarrow, q^0)$, $OTS' = (Q', \rightarrow', q'^0)$ be two occurrence transition systems.

Definition 4.1 *(M)SeqTraces(OTS)* $= \{p|Act \mid p$ *is a (maximal, resp.) path in OTS starting at* $q^0\}$ *denotes the set of all sequences of labels of (maximal, resp.) paths of OTS.*

Two occurrence transition systems OTS and OTS′ are said to be (maximal) interleaving trace equivalent *(OTS* $\sim_{(m)i-t}$ *OTS′) iff* $(M)SeqTraces(OTS) = (M)SeqTraces(OTS')$.

Now, we give Park's and Milner's notion of forward bisimulation for occurrence transition systems.

Definition 4.2 *A relation* $Z \subseteq Q \times Q'$ *is an* f-bisimulation *between OTS and OTS′ iff* $(q^0, q'^0) \in Z$ *and if* $(x, x') \in Z$,

- *if* $x \xrightarrow{a} y$*, then there exists* y' *s.t.* $x' \xrightarrow{a}' y'$ *and* $(y, y') \in Z$*, and*
- *if* $x' \xrightarrow{a}' y'$*, then there exists* y *s.t.* $x \xrightarrow{a} y$ *and* $(y, y') \in Z$.

OTS and OTS′ are f-bisimilar *(OTS* $\sim_{f-b}$ *OTS′), if there exists an f-bisimulation between OTS and OTS′.*

Next, we define pomset trace equivalence and pomset bisimulation.

Definition 4.3 *OTS and OTS′ are* pomset trace equivalent *(OTS* $\sim_{p-t}$ *OTS′) iff*

- *for each partial run* $R \in PR(OTS)$*, there is a partial run* $R' \in PR(OTS')$ *s.t.* $OTS|R$ *and* $OTS'|R'$ *are isomorphic, and*
- *for each partial run* $R' \in PR(OTS')$*, there is a partial* $R \in PR(OTS)$ *s.t.* $OTS|R$ *and* $OTS'|R'$ *are isomorphic.*

A relation $Z \subseteq Q \times Q'$ *is called a* pomset bisimulation *for OTS and OTS′ iff* $(q^0, q'^0) \in Z$ *and if* $(x, x') \in Z$*, then*

- *for each partial run* $R(x, y)$ *in OTS, there is a partial run* $R(x', y')$ *in OTS′ s.t.* $OTS|R(x, y)$ *and* $OTS'|R(x', y')$ *are isomorphic and* $(y, y') \in Z$*, and*
- *for each partial run* $R(x', y')$ *in OTS′, there is a partial run* $R(x, y)$ *in OTS s.t.* $OTS|R(x, y)$ *and* $OTS'|R(x', y')$ *and* $(y, y') \in Z$.

OTS and OTS′ are pomset bisimulation equivalent *(OTS* $\sim_{p-b}$ *OTS′), if there exists a pomset bisimulation between OTS and OTS′.*

Then, we define the new notions of run trace equivalence and run bisimulation equivalence.

Definition 4.4 *OTS and OTS′ are* run trace equivalent *(OTS* $\sim_{r-t}$ *OTS′) iff*

- *for each run $R \in R(OTS)$, there is a run $R' \in R(OTS')$ s.t. $OTS|R$ and $OTS'|R'$ are isomorphic, and*

- *for each run $R' \in R(OTS')$, there is a run $R \in R(OTS)$ s.t. $OTS|R$ and $OTS'|R'$ are isomorphic.*

A relation $Z \subseteq Q \times Q'$ is a run bisimulation *for OTS and OTS' iff $(q^0, q'^0) \in Z$ and if $(x, x') \in Z$, then*

- *for each run $R \in R_x(OTS)$ starting at x, there is a run $R' \in R_{x'}(OTS')$ starting at x' s.t. Z restricted to $R \times R'$ is an isomorphism between $OTS|R$ and $OTS'|R'$, and*

- *for each run $R' \in R_{x'}(OTS')$ starting at x', there is a run $R \in R_x(OTS)$ starting at x s.t. Z restricted to $R \times R'$ is an isomorphism between $OTS|R$ and $OTS'|R'$.*

OTS and OTS' are run bisimulation equivalent *($OTS \sim_{r-b} OTS'$) iff there exists a run bisimulation for OTS and OTS'.*

Finally, we present backward-forward bisimulation and weak-history bisimulation.

Definition 4.5 *A relation $Z \subseteq Q \times Q'$ is a bf-bisimulation for OTS and OTS' iff Z is an f-bisimulation and if $(x, x') \in Z$,*

- *if $y \xrightarrow{a} x$, then there exists y' s.t. $y' \xrightarrow{a}{}' x'$ and $(y, y') \in Z$, and*

- *if $y' \xrightarrow{a}{}' x'$, then there exists y s.t. $y \xrightarrow{a} x$ and $(y, y') \in Z$.*

OTS and OTS' are bf-bisimilar *($OTS \sim_{bf-b} OTS'$), if there exists a bf-bisimulation for OTS and OTS'.*

A relation $Z \subseteq Q \times Q'$ is called a weak history preserving bisimulation *for OTS and OTS' iff Z is an f-bisimulation for OTS and OTS' and moreover, if $(x, x') \in Z$, then $OTS| \downarrow x$ and $OTS'| \downarrow x'$ are isomorphic.*

OTS and OTS' are weakly history preserving bisimilar *($OTS \sim_{wh-b} OTS'$) iff there exists a weak history preserving bisimulation for OTS and OTS'.*

For systems without auto-concurrency weak history preserving bisimulation coincides with history preserving bisimulation (see [vGG89]).

5 Modal Equivalences Imposed by Logics

In this section we investigate which equivalences are imposed on models corresponding to occurrence transition systems by the different logics considered in this paper. Each induced equivalence is shown to coincide with one defined in Section 4.

Let OTS and OTS' be two occurrence transition systems and M_{OTS} and $M_{OTS'}$ be the models corresponding to OTS and OTS'.

Definition 5.1 (modal equivalence) *The* modal equivalence *imposed by the logic* $L \in \{S4N, LTTL, CTL, CTL^*, POL, ISTL, ISTL^*, CCTL^*_{cF}\}$ *is defined as follows:* $M_{OTS} \equiv_L M_{OTS'}$ *iff* $(M_{OTS} \models_L \varphi \Leftrightarrow M_{OTS'} \models_L \varphi)$ *for each formula* φ *of the logic* L.

We start with standard results characterizing equivalences imposed on models by LTTL, S4N, CTL, and CTL*. It is known that the equivalence imposed by LTTL coincides with maximal interleaving trace equivalence, if the sets $MSeqTraces(OTS)$ and $MSeqTraces(OTS')$ are finite in size [St87].

The equivalences imposed by S4N, CTL, and CTL* coincide with forward bisimulation. A similar result for finite models can be found in [BCG88].

Theorem 5.1 *The following equivalences hold:*

- *OTS and OTS' are run trace equivalent iff* $M_{OTS} \equiv_{ISTL^{(*)}} M_{OTS'}$, *where the "only if" holds, provided* $R(OTS)$ *and* $R(OTS')$ *are finite.*
- *OTS and OTS' are run bisimulation equivalent iff* $M_{OTS} \equiv_{CCTL^*_{cF}} M_{OTS'}$, *where the "only if" holds, provided* $R(OTS)$ *and* $R(OTS')$ *are finite.*
- *OTS and OTS' are bf-bisimilar iff* $M_{OTS} \equiv_{POL} M_{OTS'}$.

6 Comparing Equivalences

The aim of this section is to compare equivalences, both those imposed on models by the temporal logics considered and those defined directly on occurrence transition systems. The results are presented as a complete picture at the end of this section. By complete we mean that for every two equivalences it is established whether they are comparable and if so, strict inclusion is shown.

Theorem 6.1 *The following strict inclusions hold:*

- $\cong \subset \sim_{bf-b}$, $\sim_{f-b} \subset \sim_{mi-t}$, $\sim_{p-b} \subset \sim_{p-t}$, $\sim_{r-b} \subset \sim_{r-t}$, $\sim_{mi-t} \subset \sim_{i-t}$,
- $\sim_{r-b} \subset \sim_{p-b}$, $\sim_{p-b} \subset \sim_{f-b}$, $\sim_{r-t} \subset \sim_{p-t}$, $\sim_{p-t} \subset \sim_{i-t}$, $\sim_{wh-b} \subset \sim_{p-b}$,
- $\sim_{r-t} \subset \sim_{mi-t}$, $\sim_{bf-b} \subset \sim_{wh-b}$, $\sim_{bf-b} \subset \sim_{r-b}$, $\sim_{wh-b} \subset \sim_{r-t}$.

Theorem 6.2 *The following equivalences are not comparable:*

- $\sim_{wh-b}$ *and* $\sim_{r-b}$; $\sim_{f-b}$ *and* $\sim_{p-t}$; $\sim_{p-b}$ *and* $\sim_{r-t}$,
- $\sim_{f-b}$ *and* $\sim_{r-t}$; $\sim_{p-t}$ *and* $\sim_{mi-t}$.

The following figure contains all the results relating equivalences.

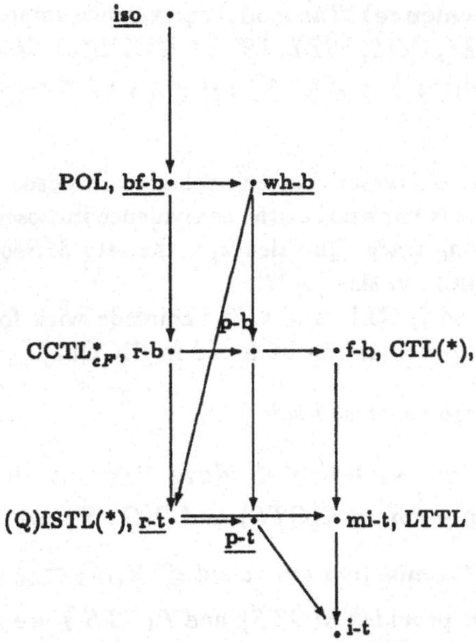

The equivalences preserved under action refinement are underlined (see Section 7).

6.1 Temporal Logics inducing $\sim_{wh-b}$, $\sim_{p-b}$, $\sim_{p-t}$, and $\sim_{i-t}$

In the former section we have shown that none of the existing temporal logics does correspond to weak history preserving bisimulation, pomset bisimulation, pomset trace equivalence, and interleaving trace equivalence.

Notice that if we restrict the language of POL s.t. no forward operator can be nested in a past operator, then it can be easily shown that the imposed equivalence coincides with $\sim_{wh-b}$.

If we change the definition of a model for ISTL(*) to be a partial run, then it can be easily shown that the imposed equivalence coincides with $\sim_{p-t}$.

If we change the definition of a model for LTTL to be a path, then it can be easily shown that the imposed equivalence coincides with $\sim_{i-t}$.

We conjecture that an appriopriate restriction of the language of CCTL$^*_{cF}$, where the quantification is changed to be over partial runs, imposes pomset bisimulation.

7 Refinement of actions

One of the most important features of equivalence notions is its preservation by refinement of actions. Since we have introduced new equivalence notions it is interesting to see whether or not they are preserved by refinement. We use the definition of refinement from [vGG89].

A refinement is a function r specifying, for each action a, an event structure $r(a)$, which is to be substituted for a.

Given an event structure $\mathcal{E}$ and a refinement function r, the refined event structure is constructed as follows. Each event e labelled by a is replaced by a disjoint copy, $\mathcal{E}_e$, of $r(a)$. The causality and conflict structure is inherited from $\mathcal{E}$: every event which was causally before e is causally before all events of $\mathcal{E}_e$, all events which causally followed e will causally follow all the events of $\mathcal{E}_e$, and all events in conflict with e are in conflict with all the events of $\mathcal{E}_e$.

Definition 7.1 *A refinement r is a function from a set of action names Act to the set of the event structures $\mathbb{E}$ labelled over Act that takes any action $a \in A$ into a finite, conflict-free, non-empty event structure $r(a) \in \mathbb{E}$. If $\mathcal{E}$ is an event structure and r is a refinement, then $r(\mathcal{E})$ is the event structure $(E_{r(\mathcal{E})}, <_{r(\mathcal{E})}, \sharp_{r(\mathcal{E})}, l_{r(\mathcal{E})})$, defined as follows:*

- $E_{r(\mathcal{E})} = \{(e, e') \mid e \in E_{\mathcal{E}}, e' \in E_{r(l_{\mathcal{E}}(e))}\}$,
- $(d, d') <_{r(\mathcal{E})} (e, e')$ *iff* $d <_{\mathcal{E}} e$ *or* $(d = e$ *and* $d' <_{r(l_{\mathcal{E}}(d))} e')$,
- $(d, d') \sharp_{r(\mathcal{E})} (e, e')$ *iff* $d \sharp_{\mathcal{E}} e$,
- $l_{r(\mathcal{E})}(e, e') = l_{r(l_{\mathcal{E}}(e))}(e')$.

Theorem 7.1 *The following conditions hold:*

i) $\sim_{bf-b}$ *is preserved under refinement of actions,*

ii) $\sim_{r-b}$ *is not preserved under refinement of actions,*

iii) $\sim_{r-t}$ *is preserved under refinement of actions.*

It is known that $\sim_{h-b}$ and $\sim_{p-t}$ are preserved under action refinement whereas $\sim_{p-b}$, $\sim_{f-b}$, and $\sim_{(m)i-t}$ are not preserved [vGG89].

8 Conclusions

The investigations carried out in this paper show that there is often a match between equivalences induced by temporal logics and those defined directly over structures without auto-concurrency. If not, adaptations are possible on either side to complete a pair. A particularly pleasant spin-off was finding simple logics that induce $\sim_{wh-b}$ and $\sim_{p-t}$. The structured approach to find definitions for logics as well as equivalences does not only order them, but also shows where new candidates in either family might be found. The fact that most obvious slots could be filled with existing notions leads credibility to the fact that the sets of logics and equivalences might be near completion. There were, however, some subtle differences in the definitions that might leave some options open for the addition of further notions. It would be very interesting to see which insights a similar investigation of event structure logics would give on these matters.

The fact that the addition of past modalities to a simple logic, like POL, gives rise to the strongest induced equivalence is quite surprising. One could derive from this, that those modalities could be used rather than the more complicated run quantifiers.

The relation with expressiveness should then also be investigated.

Another interesting issue is to look for extensions of temporal logics that would allow to distinguish auto-concurrency.

9 References

[BC87]: **Boudol, G., Castellani, I., On the Semantics of Concurrency: Partial Orders and Transition Systems, Proc. of TAPSOFT 87, LNCS 249, 123-137, 1987.**

[BCG88]: **Browne, M.C., Clarke, E.M., and Grumberg, O., Characterising Finite Kripke Structures in Propositional Temporal Logic, TCS 59 (1,2), 115 - 131, 1988.**

[CES83]: **Clarke, E.M., Emerson, E.A., Sistla, A.P., Automatic Verification of Finite State Concurrent Systems Using Temporal Logic Specifications: A Practical Approach, Proc. 10th Annual ACM Symp. on Principles of Programming Languages, Austin, 117-126, 1983.**

[EH86]: **Emerson, E.A., Halpern, J.Y., Sometimes and "Not Never" Revisited: On Branching versus Linear Time Temporal Logic, Journal of the ACM 33 (1), 151-178, 1986.**

[vGG89]: **Glabbeek, R. van. and Goltz, U.: Equivalence Notions for Concurrent Systems and Refinement of Actions, LNCS 379, pp. 237-248, 1989.**

[GKP92]: **Goltz, U., Kuiper, R., Penczek, W., Propositional Temporal Logics and Equivalences, deliverables of Esprit-BRA project 3096, 1992.**

[HC84]: **Hughes, G.E., Cresswell, M.J., A Companion to Modal Logic, London, Methuen, 1984.**

[KP87]: **Katz, S., Peled, D., Interleaving Set Temporal Logic, 6th ACM Symposium on Principles of Distributed Computing, Vancouver Canada, 178-190, 1987.**

[La80]: **Lamport, L., Sometime is sometimes not never, On the Temporal Logic of Programs, 7th ACM Symp. on Princ. of Programming Logic, 174-185, 1980.**

[LG91]: **Loogen, R. and Goltz, U., Modelling Nondeterministic Concurrent Processes with Event Structures, Fundamenta Informaticae 14, No. 1, 39-73, 1991.**

[MP88]: **Manna, Z., Pnueli, A., The Anchored Version of the Temporal Framework, LNCS 354, 1988.**

[MP91]: **Manna, Z., Pnueli, A., Linear Time Temporal Logic, Springer Verlag, 1991.**

[Mi80]: **Milner R., A Calculus for Communicating Systems, LNCS 92, 1980.**

[deNF90]: **deNicola, R., Ferrari, L., Observational Logics and Concurrency, Tech. Report, Dip. di Informatica, Univ. of Pisa, 1990.**

[deNV90]: **deNicola, R., Vaandrager, F., Three Logics for Branching Bisimulation, Proc. of LICS, 1990.**

[deNMV90]: **deNicola, R., Montanari, U., and Vaandrager, F., Back and Forth Bisimulations, Proc. of CONCUR'90, 1990.**

[NPW81]: **Nielsen, M., Plotkin, G., and Winskel, G., Petri Nets, Event Structures and Domains, Part I, TCS 13, Vol. 1, pp. 85 -109, 1981.**

[NRT91]: **Nielsen, M., Rosenberg, G., and Thiagarajan, Transition Systems, Event Structures and Unfoldings, in preparation, 1991.**

[Pa81]: **Park, D., Concurrency and automata on infinite sequences, 5th Conf. on Theoretical Comp. Sci., LNCS 104, 1981.**

[Pe90]: **Penczek, W., A Concurrent Branching Time Temporal Logic, Proceedings of the Workshop on Computer Science Logic, Kaiserslautern, LNCS 440, 337-354, 1990.**

[Ro90]: **Rozoy, B., On Distributed Languages and Models for Distributed Computation, Technical Report 563, L.R.I., 1990.**

[Si90]: **Sinachopoulos, A., Partial Order Logics for Elementary Net Systems: State- and Event - approches, Proc. of CONCUR'90, 1990.**

[St87]: **Stirling, C., Altrincham Workshop, LNCS 398, 1987.**

[TRH88]: **Trakhtenbrot, B.A., Rabinovich, A., and Hirschfeld J., Nets of Processes, TR 97/88, Tel Aviv Univ., 1988.**

[Wi82]: **Winskel, G., Event Structure Semantics for CCS and Related Languages, Proc. of ICALP, LNCS 224, 1982.**

[Wi88]: **Winskel, G., An Introduction to Event Structures, LNCS 354, pp. 364-387, 1988.**

[Wo87]: **Wolper, P., On the Relation of Programs and Computations to Models of Temporal Logic, Altrincham Workshop, LNCS 398, 75 -123, 1987.**

The Duality of Time and Information

Vaughan R. Pratt*

Stanford University

Abstract. The states of a computing system bear information and change time, while its events bear time and change information. We develop a primitive algebraic model of this duality of time and information for rigid local computation, or straightline code, in the absence of choice and concurrency, where time and information are linearly ordered. This shows the *duality* of computation to be more fundamental than the *logic* of computation for which choice is disjunction and concurrency conjunction.

To accommodate flexible distributed computing systems we then bring in choice and concurrency and pass to partially ordered time and information, the formal basis for this extension being Birkhoff-Stone dualtiy. A degree of freedom in how this is done permits a perfectly symmetric logic of computation amounting to Girard's full linear logic, which we view as the natural logic of computation when equal importance is attached to choice and concurrency.

We conclude with an assessment of the prospects for extending the duality to other organizations of time and information besides partial orders in order to accommodate real time, nonmonotonic logic, and automata that can forget, and speculate on the philosophical significance of the duality.

1 Introduction

The behavior of an automaton is to alternately wait in a *state* and perform a transition or *event*. We may think of the state as bearing information representing the "knowledge" of the automaton when in that state, and the event as modifying that information. At the same time we may think of the event as taking place at a moment in time, and the state as modifying or whiling away time.

Thus states *bear* information and *change* time, while events *bear* time and *change* information.

We often speak of events as "time-stamped;" we might similarly speak of states as "information-stamped."

One way of viewing this is to think of computational behavior as motion or flow in a space whose two dimensions are time and information, which we will conventionally plot horizontally and vertically respectively. Horizontal motion denotes a state, a quiescent period in which time passes while information is held constant. Vertical motion denotes an event, a transient phenomenon in which information changes but time does not. This accounts at least for rigid local behavior or nonbranching sequential computation; to extend it to flexible distributed behavior we might extend

* This work was supported by the National Science Foundation under grant number CCR-8814921 and a gift from Mitsubishi.

the linear geometry of each of time and space to spaces that can branch and increase in dimension, and to permit computation to be a broad flow like a river as opposed to motion of just a point. In this paper we confine attention to motion upwards (monotonically increasing information) and to the right (monotonically increasing time), leaving open the proper algebraic treatment of downward motion (forgetful automata and nonmonotonic logic), and raising as an interesting philosophical question what meaning might be attached to motion to the left, a feature of both Feynman diagrams and time machines.

Conventional automata are asymmetric with respect to time and information: states are vertices (points, 0-cells) while events are edges (line segments, 1-cells). Petri nets [Pet62, Rei85] on the other hand are symmetric: both states and events are vertices of a bipartite graph. The states, called places, are on one side, the transitions are on the other. The edges of a Petri net denote neither states nor events but rather connections between places and transitions. Edges from places to transitions specify static preconditions for a transition to *fire*, those from transitions to places specify the effect of the firing of a transition on the places. A Petri net may "be in" multiple places at the same time, represented by tokens placed on those places it is "in," the "token game." A conventional automaton can then be viewed as the special case of a Petri net whose every transition has both indegree and outdegree one, and corresponds to computation without concurrency. The dual notion in which every place has indegree and outdegree one is called an *occurrence net* or *causal net*, and corresponds to conflict-free computation, the notion of a deterministic computation or particular run of a net.

The token game can be formalized via a notion of global state as the multiset of marked places (two or more tokens may occupy the same place simultaneously). This however raises the "true concurrency" question of whether simultaneity of remote events is well-defined. One is therefore interested in models that are equally formal but manage somehow to sidestep this question. The occurrence subnets of an "unfolded" net, viewed as one run of the net, provide such a notion.

The algebra of Petri nets, under what we may view as their monotone combinators including (asynchronous) concurrency and choice, has been elegantly worked out, most notably by Winskel [Win86]. One might compare this "programmer's" algebra to the algebra of regular expressions [Kle56], whose operations are all monotone and analogous to program connectives.

Elsewhere [Pra90] we described a conservative extension of regular expressions to what we called "action logic," by adding two implications, *had-then* and *if-ever*. This made it into a logic by introducing nonmonotonicity, from which two negations, *never-before* and *never-again*, are derivable. The chief improvements over regular expressions are that the resulting equational theory is finitely axiomatizable (due to the nonmonotonicity permitting the expression of induction), and that it uniquely determines star in terms of union and concatenation, which the equational theory of regular expressions fails to do even in finite models (a four-element counterexample suffices).

Here we analogously extend the monotone algebra of Petri nets to a full logic, by adding implication to introduce nonmonotonicity as with action logic, this time with only one implication, and deriving a negation, namely the duality of schedules and automata, also describable as the duality of time and information. This extension is not strictly a conservative extension, the monotone operations being only loosely related to those of the extant algebras of Petri nets. The resulting logic while resembling action logic in some respects is much closer to Girard's full linear logic, in particular satisfying De Morgan's laws and thus having a Boolean-like symmetry of *and* and *or* absent from action logic.

In our system the Petri net formalism of a single bipartite graph is replaced by a dual pair of graphs, the schedule and its dual automaton. A *schedule* is a set of events distributed in time (temporal space). An *automaton* is a set of states distributed in information space. This reformulation is foreshadowed in Nielsen, Plotkin, and Winskel [NPW81], where Birkhoff-Stone duality makes its first albeit cryptic appearance in the theory of concurrent computation. The algebraic advantages of this passage from one bipartite graph to two dual graphs might loosely be compared to those of the passage from Aristotelian syllogistic to Boolean logic.

In practice events are distributed in space as well as time. Other than remarking that space seems to us to belong to the temporal side of the time-information duality, we shall make no attempt in this paper to incorporate space into our picture.

The information side of this duality ties in very satisfactorily with (Scott) domain theory, which emphasizes elements partially ordered by information content. The viewpoint of this paper adds the further interpretation that those elements are states, as opposed say to recursively defined functions, the motivating interpretation of domain theory.

It is possible, though not necessary, to make the duality of time and information perfectly symmetric, via our notion of event and state spaces. A considerable portion of domain theory (though not that involving stable functions) then can be reflected via this duality to apply not only to "information systems" but equally well to "temporal systems."

2 Rigid Local Computation and Chains

Straightline code contains neither branches nor concurrency but merely specifies a fixed sequence of operations. We think of a nonbranching program as rigid or inflexible, and one without concurrency as local or sequential.

Yet even in this computationally sterile setting one can not only find the duality of time and information but treat it algebraically. This simple setting has the virtue of revealing some of the basic principles of the duality without the distraction of such issues as choice and concurrency, respectively the disjunction and conjunction of computation. In this respect then the *duality* of time and information is more fundamental than its *logic*.

We now study certain categories of chains and their maps. The chains will be used to formalize both schedules and automata, with their elements corresponding respectively to events and states, and with the edges between consecutive elements corresponding respectively to states and events. Our goal here is to work out in detail the mathematics of this cross-connection, both as interesting mathematics in its own right and as a special case of a similar but richer cross-connection in the case of flexible distributed computation.

This paper is written for an audience interested in approaches to the formalization of concurrency, and assumes relatively little algebraic sophistication or familiarity with categories.. That we bring in category theory at all reflects the nature of duality as categorical rather than set-theoretic. The categories of finite chains that we begin with are simple enough that any mathematically mature reader should be able to follow every step and acquire the necessary category theory along the way.

A *chain* is a linearly ordered set $C = (X, \leq)$. For simplicity of exposition we shall restrict ourselves in this section to finite chains. The popularity of the term "fence-post error" suggests that we visualize a chain as made up of posts (the elements) and fences (consecutive pairs of elements).

A *map* $f : C \rightarrow D$ of chains is a monotone function, one such that $x \leq y$ implies $f(x) \leq f(y)$. We denote by **Fchn** the category[2] of all finite chains and their maps.

A variant on **Fchn** is **Fchn0**, the category of finite chains with bottom. The objects are those objects of **Fchn** that have a least element 0, and its maps are those maps of **Fchn** that preserve that least element, that is, $f(0) = 0$. The dual of this is **Fchn1**, the category of chains with top 1 and top-preserving maps, $f(1) = 1$. The intersection of these categories, **Fchn01**, consists of chains with top and bottom, with maps preserving both.

Let $\mathbf{m}$ denote the m-element chain $(\{0, 1, \ldots, m-1\}, \leq)$ standardly ordered by $\leq$. Of particular importance is $\mathbf{2}$, the two-element chain. For any chain C define the chain $\mathbf{2}^C$ to consist of all maps $f : C \rightarrow \mathbf{2}$, ordered pointwise, that is, $f \leq g$ just when $f(x) \leq g(x)$ for all x in C. It is readily seen that the maps $f : C \rightarrow \mathbf{2}$ are in bijective correspondence with the fences of C: each f corresponds to the fence whose posts are the greatest element mapped to 0 and the least element mapped to 1.

There are two issues here. First, the existence of the constantly 0 function $K0$ and its dual $K1$ demonstrates the need for two additional fences not meeting our description as consecutive pairs of posts, namely the fences lying outside the whole chain at either end. In order for this bijection to exist **Fchn** must admit both outlying fences, **Fchn0** must rule out the lower fence ($K1$ does not exist), dually **Fchn1** the upper fence ($K0$ is out), and **Fchn01** must rule out both.

Second, there is the question of the order of the fences (the alert reader will have noted the paradox of the largest function $K1$ corresponding to the bottom fence and vice versa). The larger functions are those with more 1's, but those correspond to the fences closer to the bottom. Hence although it is indeed a chain of fences of C, *the pointwise order of* $\mathbf{2}^C$ *is the reverse of the order in* C *of its posts.*

Let $C^\smile$ denote the order dual or *converse* of C, C turned upside down. Then either $\mathbf{2}^{C^\smile}$ or $(\mathbf{2}^C)^\smile$ denotes the fences of C in their natural order as defined by the order of their posts in C.

We now have three operations, $\mathbf{2}^C$, $C^\smile$, and their composition $\mathbf{2}^{C^\smile}$. We call these respectively the *dual*, *converse*, and *complement* of C, and denote them respectively $C^\perp$, $C^\smile$, and C^-.

Our next goal is to show that all three operations are involutions (self-inverses), and commute with each other. It is clear that converse is an involution, and we have

[2] A *category* is a reflexive multigraph, that is, a graph permitting multiple edges or *morphisms* from one vertex or *object* to another, including a distinguished self-loop at each vertex, along with an associative composition law for which the self-loops are identities. Any collection of sets and functions between them closed under ordinary function composition and containing the identity function for every set automatically forms a category. The reader should verify that our purported examples of categories are indeed so closed.

noted that it commutes with dual. The remaining commutativities then follow by expanding the definition of complement. We now show that dual is also an involution, from which it immediately follows that complement is an involution.

One thing that dualizing does is to turn fences into posts. The posts of C then can be viewed as turning into fences of $C^\perp$. Although the transformation takes place in one step we could imagine it happening continuously, with the fences of C shrinking to become posts of $C^\perp$, and the posts of C then stretching to fill in the resulting spaces, in parallel with rotating the whole chain so as to reverse the order. Complement also has this effect, but without the side effect of order reversal.

Given a chain C of **Fchn**, if we take $C^\perp$ to also belong to **Fchn** then C, $C^\perp$, $C^{\perp\perp}$, and so on will be progressively longer chains. But if we regard $C^\perp$ as belonging to **Fchn01**, with bottom $K0$ (the constantly 0 function) and top $K1$, then although $C^\perp$ will be one larger than C, $C^{\perp\perp}$ will shrink back down to the size of C again, and hence be isomorphic to C. If in addition, for any chain D in **Fchn01** we regard $D^\perp$ as belonging to **Fchn**, if C is in **Fchn** then so is $C^{\perp\perp}$. (We can rationalize this choice of category for $D^\perp$ with the observation that $D^\perp$ contains neither $K0$ nor $K1$ and so is a chain with neither a significant bottom nor top.) Moreover if D is in **Fchn01** so is $D^{\perp\perp}$. That is, dual switches back and forth between the two categories, and double dual is an isomorphism of **Fchn** and also of **Fchn01**. (Double dual of course reverses order twice.)

We thus have that converse is an isomorphism of **Fchn** with itself, as well as of **Fchn01** with itself, while dual is an isomorphism between **Fchn** and **Fchn01**. Converse is a true involution (composing it with itself yields the identity on **Fchn**, and similarly with **Fchn01**), while dual is an involution up to isomorphism, that is, $C^{\perp\perp}$ is isomorphic to C though not equal to it. We could turn isomorphism into equality by cutting **Fchn** and **Fchn01** down to what is called their *skeletons*, an arbitrarily chosen full subcategory consisting of one representative of each class of isomorphic objects, but the arbitrariness should serve to warn against such radical surgery.

We chose the terms "complement" and "converse" by analogy with the calculus of binary relations. Converse of course has its standard relational meaning with respect to the binary relation $\leq$ defining a chain. The converse $R^{-\smile}$ of the Boolean complement R^- of a binary relation R behaves analogously to dual, when **2** is replaced by the complement 0' of the identity relation 1', and exponentiation R^S is taken to mean the left residuation operation $R/S = (R^-;S^\smile)^-$ where $R;S$ is composition of binary relations (the right residual $S\backslash R$ will do just as well). Although no separate term for $R^{-\smile}$ has emerged in the relation calculus literature, as we have pointed out elsewhere [Pra92b] the significance of this operation as a form of negation was recognized by C.S. Peirce as long ago as 1882, in a Johns Hopkins circular *Remarks on [B.I. Gilman's "On Propositions and the Syllogism"]* [Klo86, p.345]. The same relationships obtain between relational converse, complement, and their composition thought of as dual, as with the operations of those names for chains. The two points of contact between relations and chains are via converse and dual, and the notion of complement for chains is then taken to be a derived notion by analogy with relations, namely as the composition of the other two operations. But whereas for relations complement is a Boolean operation, for chains it has no special Boolean character.

Now let us turn attention to **Fchn0** and **Fchn1**. Here dual performs the same exchange of posts and fences, but without growth or shrinkage of chains. For C in **Fchn0** the question arises as to whether $C^\perp$ should be considered as belonging to **Fchn0** or **Fchn1**. Following our earlier reasoning about constant functions, we observe that $C^\perp$ contains $K0$ but not $K1$, whence it belongs to **Fchn0**, $K0$ being the bottom of $C^\perp$. (Admittedly it is not clear why constancy is the appropriate criterion for this choice. However in the case of flexible distributed computation, where the appropriate generalizations of **Fchn0** and **Fchn1** are respectively state spaces and event spaces, this is the only possible choice.)

Converse on the other hand is a map from **Fchn0** to **Fchn1**, and its inverse (which we shall also call converse) going from **Fchn1** to **Fchn0**. Hence complement also runs between **Fchn1** and **Fchn0**.

So far we have only described the behavior of converse, dual, and complement on the objects of the categories. We shall now describe their behavior on the maps of the categories as well, making them into true functors between categories. In doing so we shall be obliged to distinguish between functors that preserve map direction and those that reverse it, called respectively covariant and contravariant functors. We shall follow the usual practice of dispensing altogether with the notion of a contravariant functor $F : A \rightarrow B$ from category A to category B by describing it instead as the covariant functor $F : A \rightarrow B^{\mathrm{op}}$, where B^{op}, the *opposite* of B, denotes B with its maps reversed.

For more insight into the significance of reversing maps, let us consider the maps of our four categories in more detail. In particular let us count the number of maps from a given chain $\mathbf{m}$ to $\mathbf{n}$. We call the set of such maps the *homset* from $\mathbf{m}$ to $\mathbf{n}$, notated $\mathrm{Hom}(\mathbf{m},\mathbf{n})$ or $A(\mathbf{m},\mathbf{n})$ if we wish to specify the category A to which the homset belongs.

We can easily enumerate such a homset using the following notation for a map $f : C \rightarrow D$. For each element d of D in order, list in order the elements of $f^{-1}(d)$ followed by d itself, creating a list of lists that we then "flatten" to a single list. Since f is monotone, the resulting list notating this function will be some merge of the elements of C with those of D. The last element of D must be the last element of this merge and hence is redundant, so we modify this notation to omit the last element of D. In **Fchn**, every merge of the elements of C with all but the last element of D arises in this way. There are $\binom{m+n}{m}$ merges of two lists having respectively m and n elements. Hence the number of monotone functions from the chain $\mathbf{m}$ to the chain $\mathbf{n}$ is $\binom{m+n-1}{m}$ (since we omitted the last element of $\mathbf{n}$).

As an example take the three maps from $\mathbf{2}$ to itself, namely $K0$, I (the identity), and $K1$. Write the elements of $\mathbf{2}$ as $a < b$ when $\mathbf{2}$ is the domain of the maps and as $0 < 1$ when the codomain (target). Then these three maps notated in full are respectively $ab01$, $a0b1$, and $0ab1$. When modified to omit the last element of the codomain these shorten to $ab0$, $a0b$, and $0ab$. These are all the possible merges of ab with 0.

In **Fchn0** the first element of the domain must occur first in our notation, whence we may omit it as well, leaving only $\binom{m+n-2}{m-1}$ maps from $\mathbf{m}$ to $\mathbf{n}$. But this of course equals $\binom{n+m-2}{n-1}$, whence in **Fchn0** the number of maps from $\mathbf{m}$ to $\mathbf{n}$ equals the number from $\mathbf{n}$ to $\mathbf{m}$. Put differently, **Fchn0** has the same number of maps from $\mathbf{m}$

to **n** as **Fchn0**^{op}. Hence as graphs (i.e. ignoring the composition law making a graph into a category), **Fchn0**^{op} is isomorphic to **Fchn0**. (Recall that we are allowing a graph to have a set of edges from one vertex to another.)

When A is isomorphic to B^{op} we say that A is *dual* to B. Thus we have shown that **Fchn0** is *self-dual* as a graph.

This proof is not constructive in that it does not specify any particular bijection of the set of edges from **m** to **n** with the equinumerous set from **n** to **m**. In the case of **Fchn0**, note that the first element of m and the last element of n is omitted, and dualizing reverses order. Thus an obvious choice of $f^\perp$ is just the function represented by writing our notation for f in reverse order. For example among the six functions from **3** to itself is $0bc1$ (mapping a to 0 and b and c to 1), whose dual is therefore $1cb0$ mapping 2 and 1 to c and 0 to a. (We adopt the convention, at least for **Fchn0**, of naming a function to **2** by the largest element it sends to 0. With this convention in **Fchn0** the dual of the chain 012 is written simply 210.)

In fact **Fchn0** is self-dual as a reflexive graph, noting that every chain has at least one map to itself, namely the identity map, and that the above bijection pairs up identities.

A similar duality holds between **Fchn** and **Fchn01**. The reader may verify that in **Fchn01** there are only $\binom{m+n-3}{m-2}$ maps from **m** to **n**. But the dual of **m** in **Fchn** grows to $\mathbf{m}+1$ in **Fchn01**, so the set of $\binom{m+n-1}{m}$ maps from **m** to **n** in **Fchn** should be compared with the set of $\binom{(n+1)+(m+1)-3}{n-1}$ maps from $\mathbf{n}+1$ to $\mathbf{m}+1$, and indeed these quantities are equal.

We now show that **Fchn0** is self-dual not only as a reflexive graph but as a category, by extending $C^\perp$ to a functor and then showing that it is still an isomorphism, but now between **Fchn0** and **Fchn0**^{op} rather than from **Fchn0** to itself (i.e. the object part of $C^\perp$ did not tell the whole story).

Given $f : C \to D$ in **Fchn0**, define 2^f, or $f^\perp$, to be the function which, given a function $g : D \to \mathbf{2}$, i.e. an element of $D^\perp$, yields the function $g \circ f : C \to \mathbf{2}$ (an element of $C^\perp$), monotone since composition preserves monotonicity. This then defines the dual of $f : C \to D$ to be a function $f^\perp : D^\perp \to C^\perp$. Since $C^\perp$ and $D^\perp$ are objects of **Fchn0** and $f^\perp$ is a monotone function between them, this makes $f^\perp$ a map of **Fchn0**, but running contravariantly to f.

We now have two competing very reasonable ways to create a bijection between **Fchn0(m, n)** and **Fchn0(n, m)**, notation reversal and dualization. Fortunately they are the same (exercise). The advantage of the latter description of this bijection is that it makes it easy to show that the bijection is functorial (is a homomorphism with respect to composition), as follows.

$$\begin{aligned}(g \circ f)^\perp(h) &= h \circ (g \circ f)\\ &= (h \circ g) \circ f\\ &= f^\perp(h \circ g)\\ &= f^\perp(g^\perp(h))\\ &= (f^\perp \circ g^\perp)(h)\end{aligned}$$

(The interchange of f and g is a side effect of $f^\perp$ being a duality as opposed to an isomorphism.)

We now similarly extend converse to a functor, in the process showing that **Fchn0** is isomorphic to **Fchn1** as a category. Define the converse of $f : C \to D$, namely $f^{\smile} : C^{\smile} \to D^{\smile}$, to be the same function as f on the underlying sets of C and D. Since the order of both domain and codomain have been reversed, $f^{\smile}$ remains monotone. And this description of converse makes it clear that $(g \circ f)^{\smile} = g^{\smile} \circ f^{\smile}$, and that the converse of the identity map is still the identity map. Hence converse is functorial.

Since complement is the composition of converse and dual, it follows that complement is also a functor.

We then have the following two diagrams showing all the above isomorphisms as functors between the indicated categories.

$$\begin{array}{ccc} \mathbf{Fchn} & \overset{\smile}{\leftrightarrow} & \mathbf{Fchn} \\ \perp\updownarrow & & \perp\updownarrow \\ \mathbf{Fchn01}^{op} & \overset{\smile}{\leftrightarrow} & \mathbf{Fchn01}^{op} \end{array} \qquad \begin{array}{ccc} \mathbf{Fchn0} & \overset{\smile}{\leftrightarrow} & \mathbf{Fchn1} \\ \perp\updownarrow & & \perp\updownarrow \\ \mathbf{Fchn0}^{op} & \overset{\smile}{\leftrightarrow} & \mathbf{Fchn1}^{op} \end{array}$$

In each diagram the complement functor, as the composition of converse with dual, or vice versa, runs along both diagonals. Technically speaking all arrows with distinct domain or codomain should be considered distinct, so in the case of the right hand diagram we really have a total of twelve functors consisting of four distinct clones of each of converse, dual, and complement. The left diagram has only six distinct functors since each appears twice.

Since each of **Fchn0** and **Fchn1** are self-dual, unlike either of **Fchn** and **Fchn01**, all four of **Fchn0**, **Fchn1**, **Fchn0**^{op}, and **Fchn1**^{op} appear in the one diagram, and hence all four are isomorphic to one another via the indicated isomorphisms. In the case of **Fchn** and **Fchn01** however, the only isomorphisms we have are between **Fchn** and **Fchn01**^{op}; there is an analogous but separate set of isomorphisms between **Fchn**^{op} and **Fchn01** which we did not bother to diagram. So counting only up to isomorphism, we have only three nonisomorphic categories of chains, **Fchn**, **Fchn01**, and **Fchn0**.

None of these categories have products or sums (coproducts), except for the empty sum or initial object (the object with exactly one map to every object including itself) and the empty product or final object (dually the object with exactly one map from every object including itself). In the case of **Fchn** the empty chain is initial and the unit chain is final. For **Fchn01** the two element chain is initial (both 0 and 1 must be sent to 0 and 1 in any map) and the one element chain is final (while any map to the unit chain is uniquely determined there are no maps from the one element chain to any chain of **Fchn01** except itself); in **Fchn01**^{op} these are reversed. The isomorphism between **Fchn** and **Fchn01**^{op} must therefore on these structural grounds pair the empty schedule with the unit automaton and the unit schedule with the two-element automaton (but we knew this already anyway).

In **Fchn0** and **Fchn1** the unit chain can be seen to be both initial and final, that is, a *zero* object.

3 Computational Interpretation of Chains

In any of the dualities we have seen, we want one chain to play the role of schedule and its complement the role of the corresponding automaton. There are four ways to do this depending on which of our four categories of chains we take as supplying the schedules.

When we choose **Fchn** for the schedules we are saying that a computation is a schedule S consisting of a finite sequence of 0 or more events. The corresponding complementary automaton S^- then consists of one or more states indicating how many events have been performed thus far. S^- has both a first and last state, which coincide just for the case when S is empty. In the first state nothing has been done, while in the last state everything is finished.

Now consider the choice of **Fchn01** for the schedules. What significance are we to attach to the initial and final events? Well, look at the complementary automaton to see what it permits. The first state records that the initial event has already been done, while the last state records that the final event still has not been done. We infer that the initial event is one that has always been done (was never not done), while the final event will never be done, the mathematical version of "over my dead body."

It is quite reasonable to have these "preordained events," since they can be used as single placeholders for the collection of all events of a schedule that we wish to consider as earlier than we care to think about, and dually those events we wish to imagine as happening later than we are interested in. This can be quite useful when using maps to massage or edit schedules. Each schedule consists of the pre-event, the active events, and the post-event. We think of a map as describing an editing operation in which the domain of the map is the schedule to be edited and the codomain the result of editing it. (Don't think of this as a function describing a generic editing operation applicable to many schedules, the map includes the particular schedule the operation was applied to. This is in general a good way to think about the role of maps when thinking of an object as an individual datum as opposed to a type of data.) Maps can add new active events, delete active events by moving them earlier to merge in with the pre-event or later to merge with the post-event, and combine active events. They can also combine the pre-event and post-event, but this yields inconsistency and prevents any further editing.

Now consider **Fchn1** for the schedules. Combining the reasoning for the two previous cases we infer that each schedule S has an impossible final event while its complementary automaton S^-, which must be in **Fchn0**^{op}, has an initial state in which nothing has yet been done, but no final state in which everything has been done. If we reverse this by taking **Fchn0** for the schedules then S has an always-done initial event while S^- has a final state in which everything has been done, but no initial state in which nothing has been done.

One can imagine uses for all four of these choices. If one's primary programming language is schedules rather than automata then **Fchn** is a natural choice in that always-done and never-done events might seem like redundant bells and whistles. As much could be said for automata, forcing the choice of **Fchn01**^{op} for schedules. But if one wants schedules and automata to be perfectly symmetric then one assigns **Fchn0** and **Fchn1**^{op} to them, one way round or the other.

4 Flexible Distributed Computation and Birkhoff-Stone Duality

For conventional automata the passage from linear to nonlinear computation is associated with the introduction only of branching as choice. For the notion of automaton that we shall treat in this paper the passage from linear computation as chains to nonlinear computation as posets is associated with the introduction of both branching and concurrency.

On the mathematical side, our categories of chains are not closed under three natural operations that we would like to bring in: sum, product, and exponentiation (though they are closed under equalizers and coequalizers). It is natural to think of the sum of two schedules, and dually the product of two automata, as their concurrent or conjunctive composition. The sum of automata (and hence by duality, but less obviously, the product of schedules) should represent their alternative or disjunctive composition. Exponentiation A^B is also useful, as the system resulting from A observing B. **Fchn** is not closed under the sum of two nonempty chains, the product of two chains with two or more elements, nor under C^D for C with at least three elements and D at least two. With small variations in these parameters the other categories of chains are similarly not closed.

The problem is the assumption of linearity of order, which so far has helped us by keeping the model very simple. Merely dropping it, thereby passing to partial orders, yields closure under sum and product, and in suitable analogues of **Fchn0** and **Fchn1** also exponentiation, which seems to thrive on symmetry.

Fchn most naturally turns into the category **Fpos** of finite posets (partially ordered sets), while its dual **Fchn01** turns into the category **FDL** of finite distributive lattices with top and bottom. The operations of converse, dual, and complement that we described for chains carries over without any essential changes, yielding diagrams of isomorphic categories exactly analogous to those for chains. Converse $A^{\smile}$ continues to be the result of reversing order, whether of posets or distributive lattices, and is obviously an isomorphism of **Fpos** with itself and of **FDL** with itself. Dual continues to be defined as before: if P is a finite poset, 2^P is a finite distributive lattice, and vice versa, making **Fpos** dual to **FDL** (isomorphic to $\mathbf{FDL}^{op}$) just as for **Fchn** and **Fchn01**, shown by Birkhoff in 1933 [Bir33]. Stone [Sto37] a little later found one extremal extension of this duality to infinite objects, much later characterized nicely by Priestley [Pri70] as the duality of partially ordered Stone spaces and distributive lattices; the other extremal extension is between posets and *profinite* distributive lattices, for whose definition, history, and many further extensions see Johnstone [Joh82, Ch.VII].

Complement continues to be $2^{A^{\smile}}$. All three of converse, dual, and complement continue to commute with each other and be involutions. And the same arguments showing that they are functorial continue to apply.

Fpos and **FDL** are each closed under finite sum (juxtaposition in the case of **Fpos**) and finite product (cartesian product in both cases), and the passage to infinite objects is then accompanied by an extension of sum and product to infinite arities referred to as closure under arbitrary sums and products. (An infinite sum of nonempty finite posets is necessarily infinite, and likewise for products of posets with at least two elements.) Although **Fpos** can be considered to be closed under exponentiation (since the set of monotone functions from poset P to poset Q forms a poset under pointwise order), we prefer not to since as we have seen we wish to view 2^P as a distributive lattice, just as we viewed 2^C as in **Fchn01** even though it was eligible for **Fchn**. And since any poset can arise as 2^L for some L in **DL**, **DL** is certainly not closed under exponentiation.

We now consider the corresponding generalization for **Fchn0** and **Fchn1**. It will turn out that in this symmetric situation infinite objects are handled with less fuss than in any previous situation, allowing us to dispense with the pedagogy of starting out with finite objects. And as we remarked earlier the symmetry also greatly helps exponentiation.

The most straightforward generalization of **Fchn0**, chains with the empty join (bottom), is to **CSLat**, posets with arbitrary joins, called join-complete semilattices. Their order dual, posets with arbitrary meets, called meet-complete semilattices, is isomorphic to itself, to its opposite, and to **CSLat**, whence if you are of the school of thought that routinely takes skeletal categories, i.e. recognizes only two groups of order four and one set of cardinality four, you would not give it a separate abbreviation (nor would you distinguish the categories **Fchn0** and **Fchn1**). For convenience in distinguishing the three isomorphisms of converse, dual, and complement we shall distinguish these anyway as $\mathbf{CSLat}_\vee$ and $\mathbf{CSLat}_\wedge$. With regard to the three isomorphisms these are the exact analogues of **Fchn0** and **Fchn1** respectively, the corresponding diagram being as follows.

$$\begin{array}{ccc}\mathbf{CSLat}_\vee & \breve{\leftrightarrow} & \mathbf{CSLat}_\wedge \\ \bot\updownarrow & & \bot\updownarrow \\ \mathbf{CSLat}_\vee{}^{op} & \breve{\leftrightarrow} & \mathbf{CSLat}_\wedge{}^{op}\end{array}$$

But as far as closure properties go it suffices to talk just about **CSLat**, being isomorphic to its four variants. **CSLat** is closed under arbitrary (including empty and infinite) sums and products (indeed under all limits and colimits), as well as under exponentiation.

However it turns out that sum and product are the same operation in **CSLat**. A logic using these connectives for disjunction and conjunction would be undesirably degenerate: what use would *or* be when it meant the same as *and*?

There is a very small variation on this generalization that has exactly the right effect. Instead of generalizing **Fchn0** to finite join semilattices, use posets that have bottom (as with **Fchn0**), but whose nonempty subsets have meets instead of joins (and with maps preserving bottom and nonempty meets). We have elsewhere [Pra91a, Pra92a] called such a structure a ***state space***, forming the category **St**. The dual notion is a poset with top and all nonempty joins, called an *event space*, forming the category **Ev**. The isomorphisms are thus:

$$\begin{array}{ccc}\mathbf{St} & \breve{\leftrightarrow} & \mathbf{Ev} \\ \bot\updownarrow & & \bot\updownarrow \\ \mathbf{St}^{op} & \breve{\leftrightarrow} & \mathbf{Ev}^{op}\end{array}$$

All the properties we have enumerated thus for complete semilattices also obtain for event and state spaces, except for the degeneracy of *or* and *and*. Converse reverses the order, while the dual of A is still 2^A and complement is their composition.

There is a simple alternative description of complement of an event space. Simply remove ∞ from the top and install q_0 at the bottom, leaving the other elements unchanged. Complement of chains can be seen to be a special case of this. The patient reader can confirm this by direct calculation with a few examples, for a proof see [Pra91a].

But this suggests that all the active (non-∞) events of an event space actually *become* the active (non-q_0) states of a state space. With the chains we were imagining that the fences between the elements (as posts) of a schedule were states, and that complementation turned those fences into posts and vice versa. In this new view we are leaving the whole of the active structure fixed and adjusting only the top and bottom. This new view appears to represent a slight phase shift. We do not have a good explanation of this shift, although it seems to us that there should be some way to explain the duality in the same fence-post terms that worked well for chains.

Now has this very small variation on **CSLat** broken the degeneracy sufficiently to be useful? Indeed it has, as we shall now argue briefly, more detailed discussion appears elsewhere [Pra91a, Pra92a].

In an event space, we take the final event, denoted ∞, to be the never-done event, as in **Fchn1**, and we take the join of a nonempty set Y of events, denoted $\bigvee Y$, to be the event expressing the completion of all events in the set. It is possible that $\bigvee Y = \infty$, in which case we say that Y is in *conflict*: it is not permitted for all events of Y to finish. This gives us a way of expressing conflict between events that a naive notion of schedule as a poset of events does not offer.

Dually in a state space, we take the initial state, denoted q_0, to be the state of ignorance ("original sin"), and we take the meet of a nonempty set Y of states, denoted $\bigwedge Y$, to be the last state at which every state of Y remains a possible future state. It is possible that $\bigwedge Y = q_0$, in which case we say that Y is in *dilemma*: knowing nothing, the automaton is nevertheless obliged to choose a proper subset of Y.

These notions of conflict and dilemma are internal to event and state spaces. External are the notions of sum, product, and exponential used to form larger spaces from smaller.

We can gain some insight into how these operations work by applying them to two-element event and state spaces, each of which represents one "active" event or state and one "dummy," either the final event or the initial state. Because they contain only one active event or state we think of them as the unit event space and unit state space respectively.

The sum of event spaces represents their asynchronous concurrent composition. The sum and product of two unit event spaces are given respectively by

$$\mathbf{I} + \mathbf{I} = \bigwedge \quad \text{and} \quad \mathbf{I} \times \mathbf{I} = \Diamond \, ,$$

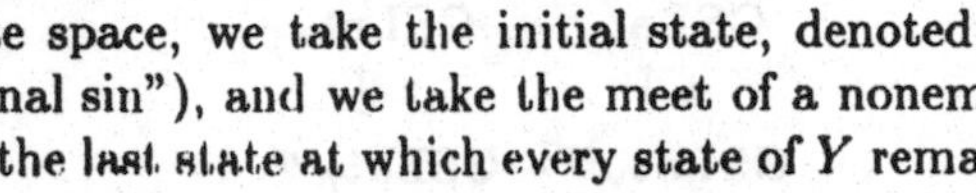

where time is assumed to flow from bottom to top. We see that $\bigwedge$ consists of two basic events, its leaves, and their join, the middle, expressing the concurrent execution of the basic events rather than their conflict because it is not equal to the final event ∞ at the top. But ◇ on the other hand has at bottom an event that we may regard as collecting the information needed to decide between the two basic events on either side. Since their join is ∞ those events are in conflict, the sense in which product is choice.

The same concepts appear in complementary form on the automaton side. Now the product and sum of two unit state spaces are given respectively by

$$ \mathbf{I} \times \mathbf{I} = \diamondsuit \quad \text{and} \quad \mathbf{I} + \mathbf{I} = \curlyvee . $$

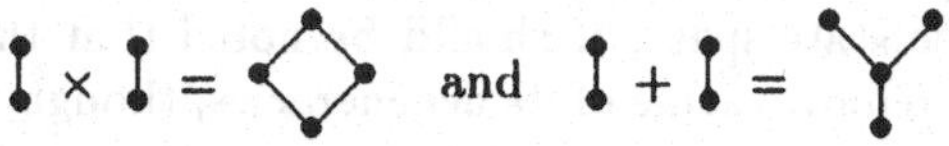

The product automaton can be thought of as the natural automaton accepting $ab + ba$, with the northwest axis corresponding to the a transition and the northeast the b.[3] Just as $\wedge$ contains no conflict, so does the choice automaton $\curlyvee$ contain no dilemma: it first performs a transition interpretable as the gathering of information for the choice, and then branches.

5 Linear Logic

We may find linear logic [Gir87] in event spaces in a natural way as follows. We start with two primitive operations: product $A \times B$, called *with* by Girard, defined as cartesian product of event spaces, and exponentiation A^B, called linear implication and written $A \multimap B$, defined as the event space of all event space maps from A to B.

We also have two primitive constants 1 and 2 denoting the event spaces of those cardinalities. The two so-called additive constants of linear logic, the respective units of the additive connectives, are both 1, while the two multiplicative constants are both 2. These are the only glaring degeneracies in this model of linear logic.

We then derive dual (called *perp* in linear logic) as $A^\perp = A \multimap 2$. Next we obtain sum $A + B$ as the De Morgan dual of product, $A + B = (A^\perp \times B^\perp)^\perp$. Sum and product are linear logic's "additive" connectives.

We next define tensor product $A \otimes B$ via $A \otimes B = (A \multimap B^\perp)^\perp$, and the tensor sum $A \oplus B$ (Girard notates this with an inverted ampersand) as the De Morgan dual of tensor product, $A \oplus B = (A^\perp \otimes B^\perp)^\perp$. These are the multiplicative connectives. The meaning of $A \otimes B$ is the flow of A through B, a symmetric relationship.

We now define a third primitive operation: $!A$ is the free event space on (generated by) the underlying poset of A, see [Pra91a, Pra92a] for details. Lastly we derive $?A$ as its dual, $?A = (!A^\perp)^\perp$, and an additive or intuitionistic implication $A \Rightarrow B$ defined by $A \Rightarrow B = !A \multimap B$.

---

[3] This seems like a violation of true concurrency; the solution we have proposed elsewhere [Pra91b] to this mismatch between truly concurrent schedules and seemingly falsely-concurrent automata is to regard the product automaton as a 2D surface rather than a hollow square along the lines of [Pap86] and [Shi85].

To remove these last degeneracies, take a larger model containing both event and state spaces mingled. Think of this as the product of **Ev** with the two-element category 2 with objects 0,1 and one nonidentity map from 0 to 1, and on which the operations of linear logic are all assigned their natural Boolean interpretation. For the event space A, interpret $(A, 1)$ as an event space and $(A, 0)$ as a state space. The operations are now all determined: dual for example becomes complement, while the product of an event space with a state space is a state space, which can be seen to be calculated by taking the converse of the event space to make it a state space and then multiplying by the state space. It should be noted that the same trick when applied to **CSLat** also removes some of its degeneracies, though not the degeneracy $!A = ?A$.

6 The Duality of Branching Time and True Concurrency

Branching time refers to the timing of decisions in semantic models: can all decisions be regarded as having been made at the beginning of time, or should the model record when (relative to other operations) a given decision was committed to? Prior to Milner's work on CCS this timing was not recognized as an essential feature of a semantic model, and continued not to be so recognized during the eighties, witness the number of papers in temporal logic that pitted linear time against branching.

True concurrency refers to whether there is any branching other than decision branching. That is, given the outcome of all decisions, is the resulting completely deterministic computation then a linear sequence of events? Or can two events occur in no particular order, suggesting that a deterministic computation might be a partially rather than linearly ordered set of events? Certainly there exist independent events, but their relative timing could be considered an either-or proposition—the events can happen in either order, as opposed to no well-defined order—with the (admittedly noncausal) choice of their relative order being lumped together with all the other choices. True concurrencists take the position that "either order" is different from "no order." Like branching time, true concurrency has its proponents and opponents.

We demonstrate here a connection between branching time and true concurrency that makes a much stronger connection between them than as mere coexisting imponderables of the eighties. We shall show that they are in fact dual phenomena structurally. That is, the dual of one in the sense of this paper *is* the other.

The following diagram depicts branching time on the left and true concurrency on the right. The upper row gives the state space perspective, the lower row the complementary event space account. In each of these four groups of two figures, the left figure is "before" or "bad" and the right "after" or "good."

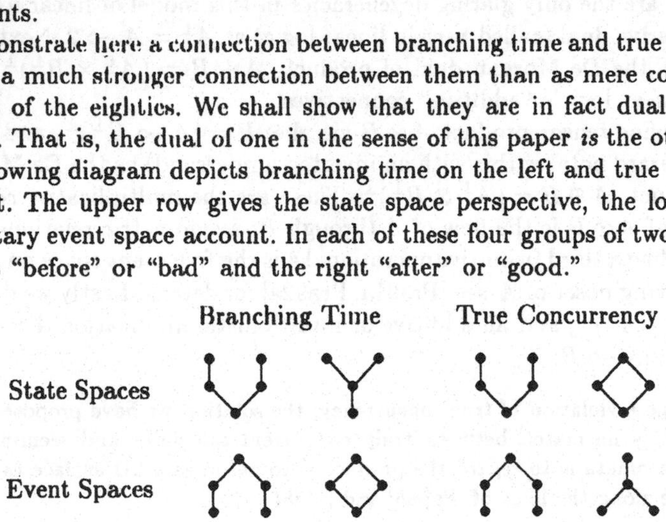

In the state space depiction of branching time (upper left), we give the conventional automata distinguishing $ab + ac$ from $a(b + c)$ (the initial state is at the bottom). The corresponding event spaces below are obtained as usual by deleting the initial state and installing the final event.

In the event space depiction of true concurrency (lower right), we have on the left two sequences of two events, corresponding to the choice ab or ba. On the right is the event space expressing the truly concurrent execution of a and b. The state spaces above are obtained as before.

The striking feature of this diagram is that the branching time contrast for state spaces is the order dual of the true concurrency contrast for event spaces, and vice versa (the other cross-connection), as promised.

We have shown only the structural similarity (no labels). The difference is that whereas with branching time the two sequences are ab and ac, with true concurrency they are ab and ba. The duality of branching time and true concurrency is therefore a structural one, and a distinction emerges when labels are introduced. Labels however are beyond the scope of this paper, which has focused exclusively on the underlying structure.

7 Future Work

The "automata" we have presented here are really "unfolded" automata, at least with respect to choice and iteration. Any choice leads to disjoint sets of states, the automaton accepting $a + b$ must be implemented with three states rather than two. (We think of the two-state version as one that, having made a choice of transitions, forgets the choice by coming back to a fixed state independent of that choice.) Furthermore iteration must be unwound, no state may be visited twice in a computation.

The question then naturally arises, can this duality be extended to handle automata that forget, and/or automata containing cycles?

One direction to pursue here is Pontryagin duality in locally compact Abelian groups. Consider the group G of complex numbers on the unit circle under multiplication. It is the dualizer (in the sense that the dual of H is G^H) for Pontryagin duality. Its own dual is the group of integers under addition, while the group of reals under addition is self-dual.

Groups rather than monoids make sense for automata that can forget because the inverse gives a means of taking things back, whether information or time. And of course groups are a natural setting for cyclic behavior.

There is also a connection with nonmonotonic logic here, whose essential characteristic is its ability to take back information that has accumulated in a theory. We view the problem of duality for forgetting automata and the problem of formalizing nonmonotonic logic as at least intimately related if not in fact the same problem.

Lastly we mention real time. We have only discussed ordered temporal and information spaces, whose metric is essentially two-valued: one event either does or does not precede another, and one state either does or does not contain less information than another. A natural extension of this notion that we have pursued elsewhere [CCMP91] is to richer measures of temporal distance between events such as causal time and real time. We have not explored the connection between these rich temporal structures with the duality of time and information, but it seems to us that such a connection should yield much additional insight into the nature of computation broadly construed to cover a wide range of such metric spaces of events and states.

8 Philosophical Significance

It seems that almost every proverb has its counterproverb. The sayings "Time is money" and "Look before you leap" express a tradeoff between response time and information gathered.

Theoretical computer science, at least at its most incestuous, i.e. not concerned for specific applications, can be broadly divided into those concerned about performance, the complexity theorists, backed by combinatorics, and those concerned about information without regard for time, once taken to be automata theory and formal languages but nowadays semantics and verification, backed by logic. This boundary of course is not hard and fast, and the advent of probabilistic computing has led to some rapprochement between these two foci. But it would seem that probabilistic computing gives up such a tiny amount of confidence in the answer that the rapprochement should be greatly improvable by giving up a much greater degree of confidence in return for yet more time.

In the silicon business, time to market is critical, but so is the correctness of large chips, calling for a similar juggling of priorities. This generalizes readily to almost any line of work: accuracy and response time are almost always both important.

A strikingly similar duality is found in quantum mechanics, where time is dual to energy, and space to momentum, to name two dualities. By setting $c = 1$ to eliminate the unit of time by equating nanoseconds to feet, the analogous complementarity of space and momentum becomes the same complementarity (but in 3D) by conferring its units, say cm and cm^{-1}, on time and energy respectively. By setting Planck's constant $\hbar = 1$ the now common unit of momentum and energy becomes simply cm^{-1}. These two simplifications are frequently adopted in quantum field theory.

We could just as well use take as units seconds and their inverse, sec^{-1}, a mere 3×10^{10} longer. This brings quantum mechanics within striking distance of computation: we are comparing QM's duality of time and energy with computation's duality of time and information. Now information is negative entropy, and, at least incrementally, entropy is proportional to energy, with temperature as the "constant" of proportionality, that is, $dQ = TdS$ where dQ is an increment of energy and dS an increment of entropy. But conventional computation is isothermal, with T held fixed at a few hundred degrees Kelvin, 300 for slow CMOS, closer to 350 for fast ECL, but in either case not varying significantly in the course of the computation.

This raises the question, is complementarity a feature of quantum mechanics because the universe is basically an information processor, with computation's duality of time and information showing up somehow as the duality of time and energy? We find this highly plausible. Another clue is Birkhoff and von Neumann's quantum logic, which resembles linear logic much more closely than say intuitionistic logic in its lack of distributivity of conjunction over disjunction yet satisfying double negation. We are presently working on finding more such clues and trying to fit them together into a comprehensive account of the relationship. One goal of this investigation is a less mystical explanation of quantum mechanics than the Copenhagen interpretation of Bohr, which on the one hand has had no really successful challengers in the past sixty years but on the other has left many physicists and philosophers very dissatisfied with the amount of disbelief that must be suspended.

References

[Bir33] G. Birkhoff. On the combination of subalgebras. *Proc. Cambridge Phil. Soc*, 29:441–464, 1933.

[CCMP91] R.T Casley, R.F. Crew, J. Meseguer, and V.R. Pratt. Temporal structures. *Math. Structures in Comp. Sci.*, 1(2):179–213, July 1991.

[Gir87] J.-Y. Girard. Linear logic. *Theoretical Computer Science*, 50:1–102, 1987.

[Joh82] P.T. Johnstone. *Stone Spaces*. Cambridge University Press, 1982.

[Kle56] S.C. Kleene. Representation of events in nerve nets and finite automata. In *Automata Studies*, pages 3–42. Princeton University Press, Princeton, NJ, 1956.

[Klo86] Christian Kloesel, editor. *Writings of Charles S. Peirce: A Chronological Edition*, volume 4, 1879-1884. Indiana University Press, Bloomington, IN, 1986.

[NPW81] M. Nielsen, G. Plotkin, and G. Winskel. Petri nets, event structures, and domains, part I. *Theoretical Computer Science*, 13, 1981.

[Pap86] C. Papadimitriou. *The Theory of Database Control*. Computer Science Press, 1986.

[Pet62] C.A. Petri. Fundamentals of a theory of asynchronous information flow. In *Proc. IFIP Congress 62*, pages 386–390, Munich, 1962. North-Holland, Amsterdam.

[Pra90] V.R. Pratt. Action logic and pure induction. In J. van Eijck, editor, *Logics in AI: European Workshop JELIA '90, LNCS 478*, pages 97–120, Amsterdam, NL, September 1990. Springer-Verlag.

[Pra91a] V.R. Pratt. Event spaces and their linear logic. In *Proc. Second International Conference on Algebraic Methodology and Software Technology*, Workshops in Computing, Iowa City, 1991. Springer-Verlag, to appear.

[Pra91b] V.R. Pratt. Modeling concurrency with geometry. In *Proc. 18th Ann. ACM Symposium on Principles of Programming Languages*, pages 311–322, January 1991.

[Pra92a] V.R. Pratt. Arithmetic + logic + geometry = concurrency. In *Proc. First Latin American Symposium on Theoretical Informatics, LNCS 583*, pages 430–447, São Paulo, Brazil, April 1992. Springer-Verlag.

[Pra92b] V.R. Pratt. Origins of the calculus of binary relations. In *Proc. 7th Annual IEEE Symp. on Logic in Computer Science*, Santa Cruz, CA, June 1992.

[Pri70] H.A. Priestley. Representation of distributive lattices. *Bull. London Math. Soc.*, 2:186–190, 1970.

[Rei85] W. Reisig. *Petri Nets: An Introduction*. Springer-Verlag, 1985.

[Shi85] M. Shields. Deterministic asynchronous automata. In E.J. Neuhold and G. Chroust, editors, *Formal Models in Programming*. Elsevier Science Publishers, B.V. (North Holland), 1985.

[Sto37] M. Stone. Topological representations of distributive lattices and brouwerian logics. *Časopis Pěst. Math.*, 67:1–25, 1937.

[Win86] G. Winskel. Event structures. In *Petri Nets: Applications and Relationships to Other Models of Concurrency, Advances in Petri Nets 1986, LNCS 255*, Bad-Honnef, September 1986. Springer-Verlag.

This article was processed using the LaTeX macro package with LLNCS style

Homology of Higher Dimensional Automata*

Eric Goubault[1] and Thomas P. Jensen[2]

[1] LIENS, Ecole Normale Supérieure, 45 rue d'Ulm, 75230 Paris Cedex 05, FRANCE, email:goubault@dmi.ens.fr
[2] Dept. of Computing, Imperial College, 180 Queen's Gate, London SW7 2BZ, U.K., email:tpj@doc.ic.ac.uk

Abstract. Higher dimensional automata can model concurrent computations. The topological structure of the higher dimensional automata determines certain properties of the concurrent computation. We introduce bicomplexes as an algebraic tool for describing these automata and develop a simple homology theory for higher dimensional automata. We then show how the homology of automata has applications in the study of branching-time equivalences of processes such as bisimulation.

1 Introduction

Geometry has been suggested as a tool for modeling concurrency using higher dimensional objects to describe the concurrent execution of processes. This contrasts with earlier models based on the interleaving of computation steps to capture all possible behaviours of a concurrent system. Interleaving models must necessarily commit themselves to a specific choice of *atomic action* which makes them unable to distinguish between the execution of two truly concurrent actions and two mutually exclusive actions as these are both modeled by their interleaving. In [9] and [1] Pratt and Glabbeek advocate a model of concurrency based on geometry and in particular on the notion of a higher–dimensional automaton (HDA). Higher–dimensional automata are generalisations of the usual non–deterministic finite automata as described in *e.g.* [2]. The basic idea is to use the higher dimensions to represent the concurrent execution of processes. Thus for two processes, a and b, we model the mutually exclusive execution of a and b by the automaton

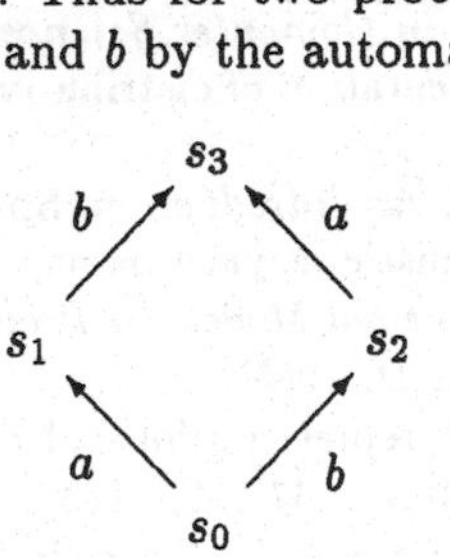

whereas their concurrent execution is modeled by including the two–dimensional surface delineated by the (one–dimensional) a– and b–transitions as a transition in the automaton. This is pictured as

* This work was partially supported by ESPRIT BRA 3074 SEMAGRAPH

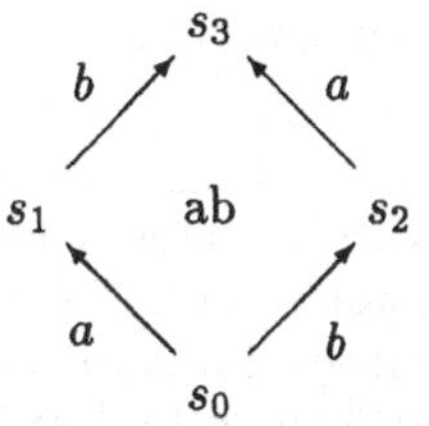

A computation is modeled by a path in this higher-dimensional automaton. Now several properties of computational relevance are determined by the topology of the HDA. *E.g.* a HDA is deterministic if for any two paths in the automaton one can be transformed into the other in a continuous fashion, *i.e.* non-determinism arises from ***holes*** in the automaton that prevent the transformation of one path into another. Furthermore certain differences in the topologies of two HDA imply that a computation is possible in one HDA but not the other, *i.e.* information about the topology of HDA can be used to answer questions about ***bisimulation*** between the HDA.

The field of algebraic topology offers several techniques for giving an algebraic description of topological properties of geometric objects. In this paper we develop a theory of ***homology*** of HDA. To each HDA we associate a sequence of groups that characterises the essential branchings and mergings in the HDA. These ***homology groups*** seem to be more amenable to automated computation than the fundamental groups associated with homotopy theory.

We introduce HDA in section 2. Section 3 defines the notion of bicomplex and show how HDA can be described by bicomplexes. In section 4 we give a translation of a CCS-like process language into bicomplexes. Section 5 develops the theory of homology of bicomplexes and show how our process language can be translated into homology groups. Finally section 6 shows how differences in the homology groups of two bicomplexes imply that the associated bicomplexes are not bisimular. Section 7 concludes.

Notation. We denote by Z_2 the group $(Z/2Z,+)$ (which is also a field with multiplication being multiplication modulo 2). For Q a set, we write $\overline{Q}$ for the free Z_2-vector space generated by Q (or Vect(Q)). For f a function from a set A to a set B, we define $\overline{f}$ from $\overline{A}$ to $\overline{B}$ as being the linear extension of f. We write $\otimes$ for the tensor product between two Z_2-vector spaces, $\oplus$ for the direct sum of two Z_2-vector spaces. The vector space generated by the cartesian product of two sets is the tensor product[3] of the vector spaces generated by each of these sets. We write $\{*\}$ for the trivial structure (e.g. group, or vector space), (x) for the Z_2-vector space generated by x, Ker f for the kernel of the function f, and Im f, for the image of the function f. Id is the identity function. Given V a Z_2-vector space, whose basis is $\{e_i$ / i=1...$\alpha\}$, we define the scalar product of two vectors x and y as being $\langle x,y\rangle = \Sigma_{i=1\ldots\alpha} x_i.y_i$ (with value in $\mathbf{Z}$), where $x = \Sigma_{i=1\ldots\alpha} x_i.e_i$ and $y = \Sigma_{i=1\ldots\alpha} y_i.e_i$.

[3] for a full definition see [4]

2 Higher–dimensional automata

We have already given one example of a higher–dimensional automaton, *viz.* the automaton in the introduction with the interior filled. In this section we define the concept of a higher–dimensional automaton (HDA) precisely and explain how this definition extends the usual definition of a finite automaton. Furthermore we define the notion of a *path* through a HDA, used to describe a concurrent computation.

The description of a finite automaton over an alphabet Σ consists of a set of states S together with a transition function $t : S \times \Sigma \rightarrow S$ such that $t(s, m)$ is the state reached when reading symbol m in state s. In addition, there is an initial state $s_0 \in S$ and a set of final states $F \subseteq S$. In this framework there is a clear distinction between states, where the automaton "rests" and transitions where the automaton is "in action". We call such an automaton a one–dimensional automaton, for reasons to become clear shortly.

This way of viewing an automaton is inadequate when the automaton is capable of performing several actions simultaneously. Such an automaton can be more or less active according to how many actions are being performed. We can picture such an automaton as a network of one–dimensional automata in which some automata rest and some are in action. A state of such a network is then a mixture of resting states and transitions/actions and the automaton changes from one state to another by initiating or terminating one or more actions. The number of actions is called the *dimension* of the state. We shall call such an automaton a higher–dimensional automaton.

The classical finite automaton can be described in this fashion. The new set of states consists of the old set of states, all elements of which have dimension zero, together with the states $(s, m, t(s, m))$ of dimension one, which represent the transitions. Note that the dimension of a state agrees with the standard way of drawing finite automata: A state is represented by a point, *i.e.* a zero–dimensional object, and a transition by a line, *i.e.* an object of dimension one.

The following is Glabbeek's definition of a HDA from [1]:

Definition 1. A higher–dimensional automaton is a tuple $(S, d, \sigma, \tau, s_0, F, \ell)$ where

- S is a set of states
- $d : S \rightarrow N$ is the dimension of a state
- $\sigma, \tau : S \times N \rightharpoonup S$ are partial functions. For $s \in S$ and $k < d(s)$, $\sigma(s, k)$ and $\tau(s, k)$ are the start state and the end state of the action in the k'th dimension. The functions σ, τ must satisfy the *cubical* laws (cf. [1]): For $i \leq j$

$$i) \quad d(\sigma(s,k)) = d(\tau(s,k)) = d(s) - 1$$

$$ii) \quad \sigma(\sigma(s,i),j) = \sigma(\sigma(s,j+1),i) \qquad iii) \quad \sigma(\tau(s,i),j) = \tau(\sigma(s,j+1),i)$$

$$iv) \quad \tau(\sigma(s,i),j) = \sigma(\tau(s,j+1),i) \qquad v) \quad \tau(\tau(s,i),j) = \tau(\tau(s,j+1),i)$$

- $s_0 \in S$ is the initial state and $F \subseteq S$ is the set of final states. They must satisfy $d(s_0) = d(f_i) = 0$ for all $f_i \in F$

- $\ell : S \rightarrow \Sigma$ is the labeling function, that assigns a label to every state of dimension one, *i.e.* $\ell(s)$ is defined if and only if $d(s) = 1$. Furthermore we require that

$$\ell(\sigma(s,i)) = \ell(\tau(s,i)) \qquad \text{for } i = 0, 1$$

Note Instead of specifying the dimension function $d : S \rightarrow N$ we shall sometimes present S as a family of sets $\{S_i\}_{i \in N}$ where S_i is the set of states of dimension i.
The first cubical law states that the dimension of state increases by one when an action is initiated and decreases by one when an action is terminated. Representing a state as a list of the actions that is being performed we can explain the other cubical laws. Let us as example take the law *iv)* and assume that $i < j$. The $j+1$'th element of the list representing s will be the j'th element in the list representing the state just before s where the i'th action has not been initiated, *i.e.* the state $\sigma(s,i)$. Hence the adjustment in the index. The rules *ii)*, *iii)*, *v)* can be explained in a similar way.

Example As an example we have the automaton from the introduction that performs the two actions a and b concurrently. It can be drawn as follows:

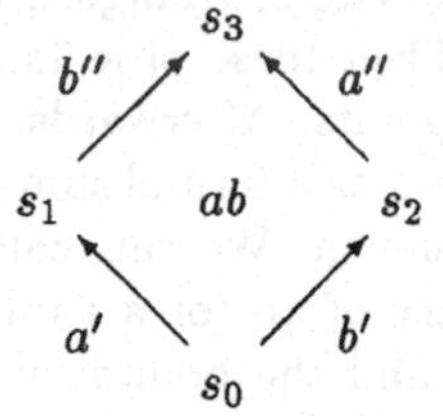

We name one–dimensional states (*i.e.* transitions) by their label and distinguish between different transitions with the same label using primes. In our example there are two transitions with the label a which are denoted by a' and a'' respectively[4]. Furthermore we shall write $\sigma(s,i)$ and $\tau(s,i)$ as $\sigma_i(s)$ and $\tau_i(s)$.
The labeling function ℓ extends to a function ℓ^* on higher–dimensional states by stipulating that $\ell^*(s)$ is the string of labels of the one–dimensional states constituting s (. is the string concatenation). Finally, ℓ^* of a sum of states is the set of ℓ^* of each of the summands.

Definition 2. A path in a higher-dimensional automaton $(S, d, \sigma, \tau, s_0, F, \ell)$ is a sequence p=$(p_i)_{0 \le i \le n}$ such that:

- $\forall i, p_i \in \bigcup_j S_j$
- $p_0 = s_0$
- $\forall i : 0 \le i \le n-1 : \sigma_j(p_{i+1}) = p_i$ or $\tau_k(p_i) = p_{i+1}$ for some $j, k < d(p_i)$

n+1 is the length of the path p. A word is w=$(\ell^*(p_i))_{0 \le i \le n}$. An acceptable path is a path such that $p_n \in F$. An acceptable word is a word whose associated path is acceptable. The language accepted by $(S, d, \sigma, \tau, s_0, F, \ell)$ is the set of all acceptable words.

[4] Thus the primes should not appear in the drawing. We have added them only for the sake of clarity

3 Complexes

In this section we introduce some concepts from algebraic topology that are useful for studying higher-dimensional automata. One of the fundamental problems studied in algebraic topology is how to characterise geometric objects that are equal modulo a continuous deformation. That is, two objects are considered equal if one can be deformed into the other by stretching, bending *etc* but not tearing nor piercing it. For connected objects, the essential problem here is to define rigorously the notion of a *hole* in an object and algebraic topology offers several possibilities. It was first done via the notion of fundamental group (a group of paths on a manifold[5]) in [8], and more generally (and later) by the homotopy groups. This notion is very close to the geometric intuition: A hole is something that prevents a path on a manifold passing on one side of the hole to be continuously deformed into another path passing on the other side of the hole. But it is difficult to compute these homotopy groups in general. To overcome this problem Poincaré introduced the notion of homology [6]. It defines holes in an even more algebraic and constructive way. First of all, a given manifold is represented in a discrete way by triangulating it. Thus the manifold is represented as a union of points, edges, triangles, tetrahedrons *etc.* These discrete sets are called simplicial sets. Then these objects are oriented geometrically; here it suffices to give an order on points. Afterwards, the notion of a ***boundary*** of an object of dimension n+1 is defined as a formal sum (keeping in mind the orientation problems) of objects of dimension n. We can call boundary everything which is the result of taking the boundary of an (or a sum of) object(s). The boundary of a line segment is its endpoints and the boundary of a triangle is the edge of the triangle. A *cycle* is a formal sum of objects whose boundary is null. All boundaries are cycles, but the converse is not true. It is false when there is a hole inside a cycle. A characterization of holes is therefore, objects which are cycles but not boundaries. More precisely an n-dimensional manifold can be represented as a collection of k-dimensional objects ($k = 0, \ldots, n$) and is fully described by listing for each object of dimension k its $(k-1)$-dimensional boundaries. Thus a manifold is given by the diagram

$$\cdots \quad Q_i \xrightarrow{\partial^i} Q_{i-1} \xrightarrow{\partial^{i-1}} \cdots \xrightarrow{\partial^1} Q_0$$

where Q_i is the collection of i-dimensional objects and ∂^i is the ***boundary*** map. Since the definition of boundaries and cycles involve formal sums of objects, it is more convenient to study this diagram where Q_i is replaced by $\overline{Q_i}$ and ∂^i is replaced by its linear extension. A boundary of an (i+1)-dimensional object can then be described as an object lying in the image of ∂^{i+1}. Similarly, as a cycle is an object with zero boundary, the cycles of dimension i are precisely the objects in the kernel of ∂^i. As all boundaries are cycles we have that $Im\partial^{i+1} \subseteq Ker\partial^i$, so $\overline{\partial}^i \circ \overline{\partial}^{i+1}$=0: this

[5] We shall not define an n dimensional manifold precisely, but just use it to denote a topological space that locally "looks like" a Euclidean space R^n

[6] Here, we just speak of simplicial homology

equation makes the diagram a *complex*. As both $Im\,\partial^{i+1}$ and $Ker\,\partial^i$ are subgroups of $\overline{Q_i}$ we can form the quotient

$$Ker\,\partial^i / Im\,\partial^{i+1}.$$

This is called the i–th ***homology group*** of the complex. As a first indication of the information present in the homology groups we note that non–zero homology in dimension i indicates an i–cycle that is not the boundary of an $i+1$–dimensional object, *i.e.* an i–dimensional hole in the manifold.

3.1 Transition systems and bicomplexes

In computer science, transition systems, which can also be seen as discrete formalizations of continuous processes, are central in the semantic definitions of programming languages. Paths are traces of execution, and one can imagine that a deformation of a path into another one is not possible because of branchings or mergings, equivalent of holes in the preceding case. In the sequential case, what we have just said is trivial. But if we generalize transition systems to represent concurrent processes, it is no more the case. A very nice geometrical way of doing it can be found in [1] and [9].Then, a semantic equivalence on these objects (in the manner of bisimulation) can be expressed in terms of deformation of paths, that is homotopy. It is then natural to consider an algebraic equivalent, much more computable, in the form of some kind of homology. A difficulty is to be able to take into account the (irreversible) flow of time *i.e.* to talk about the beginning and the end of a path. One can think of several ways to do it. One is to consider homology of monoids as in [3]. Another, as proposed in [9] is to use the formalism of n-categories. We suggest bicomplexes because they capture the difference between start–and end–states and fit well with the cubical laws while still being easy to understand and compute.

Definition 3. A bicomplex $(A_i, \delta_0^i, \delta_1^i)$ is a sequence of groups, A_i, together with two sequences of group homomorphisms δ_0^i, δ_1^i:

$$\cdots \quad A_3 \underset{\delta_0^3}{\overset{\delta_1^3}{\rightrightarrows}} A_2 \underset{\delta_0^2}{\overset{\delta_1^2}{\rightrightarrows}} A_1 \underset{\delta_0^1}{\overset{\delta_1^1}{\rightrightarrows}} A_0$$

such that

$$\delta_0^i \circ \delta_0^{i+1} = \delta_1^i \circ \delta_1^{i+1} = 0$$
$$\delta_0^i \circ \delta_1^{i+1} = \delta_1^i \circ \delta_0^{i+1}$$

Let S_i denote the states of dimension i. For each S_i we can construct the free Z_2–vector space generated by S_i, denoted by $\overline{S}_i$. The elements of $\overline{S}_i$ are formal sums

$$s_1 + \ldots + s_n \qquad s_i \in S_i, n \in N$$

where each $s_i \in S_i$ appears at most once. Addition of two elements is defined as usual by

$$\sum_{s\in S_i} n_j s + \sum_{s\in S_i} n_k s = \sum_{s\in S_i}(n_j + n_k)s$$

where all coefficients are in Z_2, *i.e.* an s appearing in both sums disappears. It is straightforward that each $\overline{S}_i$ forms an abelian group with the empty sum (all coefficients = 0) as neutral element.
The mappings $\sigma, \tau : S \times N \rightarrow S$ from a HDA induce a sequence of mappings

$$\sigma^i, \tau^i : S_{i+1} \rightarrow S_i$$

where σ^i and τ^i are the restrictions of σ and τ to states of dimension $i+1$. By linear extension we obtain two group homomorphisms $\delta^i_\sigma, \delta^i_\tau : \overline{S}_{i+1} \rightarrow \overline{S}_i$ by:

$\delta^i_\sigma(s_1+\ldots+s_n) = \delta^i_\sigma(s_1)+\ldots+\delta^i_\sigma(s_n)$ and $\delta^i_\tau(s_1+\ldots+s_n) = \delta^i_\tau(s_1)+\ldots+\delta^i_\tau(s_n)$.

The cubical laws can then be used to show the following theorem

Theorem 4. *Let $(S, d, \sigma, \tau, s_0, F, \ell)$ be a higher-dimensional automaton. Then*

$$\cdots \quad \overline{S_3} \underset{\delta^3_\tau}{\overset{\delta^3_\sigma}{\rightrightarrows}} \overline{S_2} \underset{\delta^2_\tau}{\overset{\delta^2_\sigma}{\rightrightarrows}} \overline{S_1} \underset{\delta^1_\tau}{\overset{\delta^1_\sigma}{\rightrightarrows}} \overline{S_0}$$

is a bicomplex.

Proof: Show $\delta_\sigma\delta_\sigma = 0$. Cubical law *ii)* implies that

$$\sum_{i=1}^{n}\left(\sum_{j=1}^{i-1}\sigma(\sigma(q,i),j) + \sum_{j=i}^{n-1}\sigma(\sigma(q,i),j)\right) = \sum_{i=1}^{n}\sum_{j=1}^{i-1}\sigma(\sigma(q,i),j) + \sum_{i=1}^{n}\sum_{j=i+1}^{n}\sigma(\sigma(q,j),i) =$$

$$\sum_{i=2}^{n}\sum_{j=1}^{i-1}\sigma(\sigma(q,i),j) + \sum_{j=1}^{n}\sum_{i=j+1}^{n}\sigma(\sigma(q,i),j) = \sum_{i=2}^{n}\sum_{j=1}^{i-1}\sigma(\sigma(q,i),j) + \sum_{i=2}^{n}\sum_{j=1}^{i-1}\sigma(\sigma(q,i),j) = 0.$$

The rest of the proof is similar. From now on, all HDA will be presented by bicomplexes (Q,∂_0,∂_1) and an initial state i, a set of final states F, and a labeling operator ℓ, written as a tuple (Q,∂_0,∂_1,i,F,ℓ).

4 Translation of process algebra into bicomplexes

We now develop a truly concurrent semantics of a process algebra by translating the terms of the algebra into bicomplexes. First a quick review of the CCS–like syntax: We assume given a set of actions $\Sigma \cup \{\tau\}$ such that when Σ contain an action ω it also contains its complementary action $\overline{\omega}$. The action τ is called the internal action.

Furthermore we have the idle process **nil**. Terms are then formed according to the following grammar:

$$t ::= \omega \mid \mathbf{nil} \mid (t_1 + t_2) \mid (t_1 \mid t_2) \mid (t_1 \ ; t_2)$$

where + is choice, | parallel composition and ; is sequential composition. For expository reasons we have divided the parallel operator in two operators: $t_1 \| t_2$, which is parallel composition without communication, and the general one |.

The translation is defined by structural induction over the terms. For each construct we specify the resulting bicomplex $(\overline{\{T_i\}}_{i\in N}, \partial_0, \partial_1, j, G, L)$. T_i are states of dimension i, *i.e.* generators of the i-th vector space and j and G are the initial state and the set of final states respectively. We introduce a special 0–dimensional state 1 (neutral for $\otimes$) to represent the idle action **nil**.

(case 1) $t = \omega \in \Sigma$. Then T_0={1, β}, T_1={γ}, j={1}, G={β}, $\partial_0(\gamma) = 1$, $\partial_1(\gamma) = \beta$, and L($\gamma$)=$\omega$, where β,γ are fresh state names.

(case 2) t=nil. Then T_0={1}, J={1}, G={1}.

(case 3) t=q+q'. We assume that the translation of q is $(Q, \partial_0, \partial_1, l, i, F)$, and of q' is $(Q', \partial_0, \partial_1, l', i', F')$. Then:
- $T_0 = (Q_0 \cup Q'_0)/\{i = i'\}$
- $\forall i \geq 1, T_i = Q_i \cup Q'_i$, *i.e.* $\overline{T_i} = \overline{Q_i} \oplus \overline{Q'_i}$
- j=i
- G=F∪F'
- $\partial_j[t] = \partial_j[q] \oplus \partial_j[q']$
- and for the labeling function:

$$\forall x \in T_1 : L(x) = \begin{cases} l(x) \text{ if x is in } Q_1 \\ l'(x) \text{ if x is in } Q'_1 \end{cases}$$

(case 4) t=q ; q'. We have the translation of q : $(Q, \partial_0, \partial_1, l, i, F)$, and of q' : $(Q', \partial_0, \partial_1, l', i', F')$. Then, if F={$f_1$, ..., f_m}

$$\forall i \geq 0, T_i = Q_i \ \cup \ f_1 \times Q'_i \ \cup \ ... \ \ f_m \times Q'_i$$

We have also,
- j=i
- $G = f_1 \times F' \ \cup \ ... \ \cup \ f_m \times F'$
- $\partial_j[T] = \partial_j[Q] \oplus Id \otimes \partial_j[Q']$
- for x in T_1, L(x)=l(x) if x is in Q_1, otherwise x=(f_i,y) with y element of Q'_1 and L(x)=l'(y).

(case 5) t=(q || q'). By considering the valid transitions for t, we see that we need that the paths of t be isomorphic to the cartesian product of the paths of q and q'. Thus it is natural to form the product of the cubical sets underlying the corresponding automata:

$$\forall k \geq 0, T_k = \bigcup_{i=0...k} Q_i \times Q'_{k-i}.$$

And for the boundary operators:

$$\partial_j^k[T] = \bigoplus_{i=0\ldots k} (\partial_j^i[Q] \otimes Id \oplus Id \otimes \partial_j^{k-i}[Q'])$$

Moreover, j=i⊗i', G=F×F'. Finally, the labeling operator is given by:

$$L^*_{|\oplus_{i,j}\overline{Q}_i \otimes \overline{Q}_j} = \Sigma_{i,j} l^*_{|\overline{Q}_i} . l^*_{|\overline{Q}_j}$$

(case 6) t=(q | q'). There is here a possibility of communication between q and q', represented by equations on tensor products between states which can interact. We define the new bilinear product, quotient of the tensor product by these equations, as follows:

- x $\in Q_1$, y $\in Q'_1$ and l(x)=$\overline{l}(y)$, then x$\otimes_c$y=s, where s is in T_1 such that $L(s) = \tau$, $\partial_0(s) = \partial_0(x) \otimes \partial_0(y)$, $\partial_1(s) = \partial_1(x) \otimes \partial_1(y)$.
- otherwise x$\otimes_c$y=x$\otimes$y (no interaction).

The symbol τ is a distinguished element in Σ, used (see for instance [7]) to represent the internal synchronisation action.

Now we define T as being given by almost the same equations as in (case 5).

- $\overline{T}_0 = \overline{Q}_0 \otimes \overline{Q}'_0$
- $\overline{T}_1 = \overline{Q}_0 \otimes \overline{Q}'_1 \oplus \overline{Q}_1 \otimes \overline{Q}'_0 \oplus$ Vect($\{x \otimes_c y / x \in Q_1, y \in Q'_1$ and $l(x) = \overline{l(y)}\}$)
- $\overline{T}_2 = \overline{Q}_0 \otimes \overline{Q}'_2 \oplus \overline{Q}_2 \otimes \overline{Q}'_0 \oplus$ Vect($\{x \otimes y / x \in Q_1, y \in Q'_1$ and $l(x) \neq \overline{l(y)}\}$)
- $\forall k \geq 3, T_k = \bigcup_{i=0\ldots k} Q_i \times Q'_{k-i}$.

And for the boundary operators, for $q \in Q_i$ and $q' \in Q'_{k-i}$:

$$\partial_j^k[T](q \otimes q') = \partial_j^i[Q](q) \otimes q' \oplus q \otimes \partial_j^{k-i}[Q'](q')$$

and

$$\partial_j^{k-1}[T](q \otimes_c q') = \partial_j^i[Q](q) \otimes \partial_j^{k-i}[Q'](q'), \text{ if } k = 2, i = 1, l(q) = \overline{l'(q')}$$

Moreover, j=i⊗i', G=F×F'. Finally, the labeling operator is given (on 1-states) by:

$$L(x \otimes_c y) = \tau \quad \text{if } x \in Q_1, y \in Q'_1 \text{ and } l(x) = \overline{l'(y')}$$

otherwise

$$L(x \otimes y) = \begin{cases} l(x) & \text{if } x \in Q_1, y \in Q'_0 \\ l'(y) & \text{if } x \in Q_0, y \in Q'_1 \end{cases}$$

For a term t, we denote by $[\![t]\!]$ the result of its translation into a higher-dimensional automaton, as described above. We then have a correctness result about the translation, which states that the paths of $[\![t]\!]$ describe the fully concurrent execution of t (as can be inferred from a presentation of CCS via rules of transition, in [7]). We prefer in this article to give a few examples instead of a fully abstract treatment of that property.

Examples

(1) We use the inductive construction above to translate the CCS-term (a|b) (which is equal to (a||b)). We have:

$$1 \xrightarrow{a} \alpha \qquad\qquad 1 \xrightarrow{b} \beta$$

to represent $[\![a]\!]$ and $[\![b]\!]$ respectively. We now form the tensor product:

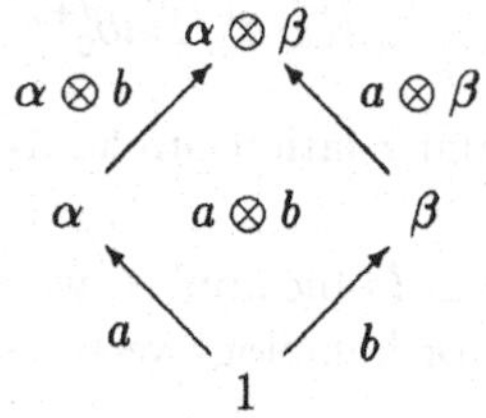

The paths of $([\![a||b]\!])$ are:

$1.(1, a, \alpha, \alpha \otimes b, \alpha \otimes \beta)$	$4.(1, b, \beta, a \otimes \beta, \alpha \otimes \beta)$
$2.(1, a, a \otimes b, \alpha \otimes b, \alpha \otimes \beta)$	$5.(1, b, a \otimes b, \alpha \otimes b, \alpha \otimes \beta)$
$3.(1, a, a \otimes b, a \otimes \beta, \alpha \otimes \beta)$	$6.(1, b, a \otimes b, a \otimes \beta, \alpha \otimes \beta)$

Paths 3 and 5 are not considered in [1]. We can nevertheless give an intuitive meaning to each of them. Associate action a to processor p_1 and action b to processor p_2. Then for path 3 we have:
1: p_1 and p_2 are idle
a: p_1 is computing a while p_2 is idle
a⊗b: p_1 continues to compute a while p_2 is computing b
a⊗β: p_1 still computes a while p_2 is idle (it has finished its computation)
$\alpha \otimes \beta$: p_1 and p_2 are idle (they have both finished their computations)

(2) We now compute $[\![(a+b)||c]\!]$.

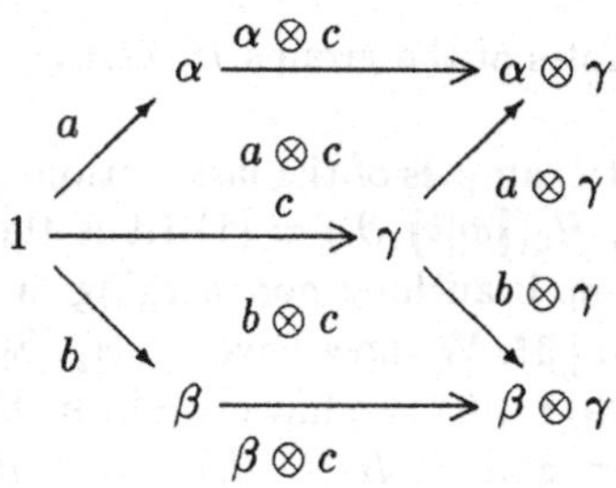

5 Homology of bicomplexes

Let $(Q,\partial_0,\partial_1)$ be a bicomplex arising from a HDA. Then if two transitions of dimension one a, b have a common start point, *i.e.* $\partial_0(a) = \partial_0(b)$, then the sum a+b will belong to the kernel of ∂_0 since we work modulo 2:

$$\partial_0(a+b) = \partial_0(a) + \partial_0(b) = 0$$

i.e. $a + b \in Ker\partial_0$. So a+b is a potential branching. However it is not a non-deterministic choice if a and b are boundaries of a higher-dimensional transition, *i.e.* if there exists a 2-dimensional transition A such that $\partial_0(A) = a+b$. These branchings could have been defined in a more standard way, but this is particularly amenable to generalisation for higher-dimensional transitions. Intuitively, a non-deterministic choice of dimension i is an element of the kernel of ∂_0^i modulo the boundaries of i+1-dimensional objects, *i.e.* is an element of the i-th homology group:

$$Ker\partial_0^i / Im\partial_0^{i+1}$$

This is for branching. A similar relationship holds for mergings and the maps ∂_1^i.

Definition 5. For (Q,∂_0,∂_1) a Z_2-bicomplex, we define two sequences of homology (see for instance [5]) groups (or homology vector spaces):

$$H_i(Q,\partial_0) = Ker\partial_0^i / Im\partial_0^{i+1} \qquad H_i(Q,\partial_1) = Ker\partial_1^i / Im\partial_1^{i+1}$$

An element of $Ker\partial_j^i$ is an i-cycle, and an element of $Im\partial_j^{i+1}$ is an i-boundary. An element of $H_i(Q,\partial_0)$ is called a branching of dimension i. An element of $H_i(Q,\partial_1)$ is called a merging of dimension i. An element x satisfying $\langle x, H_i(Q,\partial_0)\rangle \neq 0$ is called a branching choice of dimension i. We write $H_*(T,\partial_j)$ for $\bigoplus_{k\geq 0} H_k(T,\partial_j)$.

We begin by giving some intuition about these homology groups. We consider first of all the homology groups of dimension 0.

Lemma 6. Consider a finite automaton given by (Q,∂_0,∂_1,i,F,ℓ), with no cycle

- *i=$H_0(Q,\partial_1)$ implies that all states of Q are reachable*
- *F=$H_0(Q,\partial_0)$ implies that all states of Q are co-reachable*

Now we give a few examples of the groups $H_i(Q,\partial_j)$.

(1) We consider the first examples of the last section. Then, $H_0([\![a||b]\!],\partial_0) = (\alpha\otimes\beta)$ (it is the end state), $H_0([\![a||b]\!],\partial_1) = (1)$ (it is the initial state), and the other H_i are null (there is no branching nor merging, it is deterministic).

(2) We study now Q=$[\![a \mid \overline{a}]\!]$. We now have, $H_0(Q,\partial_0) = (\alpha\otimes\beta)$, $H_0(Q,\partial_1) = (1)$, $H_1(Q,\partial_0) = (a+\tau)\oplus(\tau+\overline{a})$ (we have two branchings of dimension one, that is a choice between a, τ, and $\overline{a}$), $H_1(Q,\partial_1) = (\alpha\otimes\beta+\tau)\oplus(\tau+\alpha\otimes\overline{a})$ (we have two mergings of dimension 1), and the other H_i are null.

5.1 Some results from homology theory

We now list some results concerning the calculation of homology groups of direct sums and tensor products of complexes. These results are needed when we later are to interpret operators of our process algebra as operators on homology groups.

We first look at the sum of two manifolds. If $(\overline{Q}_*, \partial[Q])$ and $(\overline{Q}'_*, \partial[Q'])$ are the complexes corresponding to two manifolds, then the one arising from their disjoint union $(\overline{T}_i, \partial[T])$, is given by:

$$\forall i, \overline{T}_i = \overline{Q}_i \oplus \overline{Q'}_i$$

and,

$$\forall i, \partial^i[T] = \partial^i[Q] \oplus \partial^i[Q']$$

Now for bicomplexes, the analogous is fairly obvious: let $(\overline{Q}_*, \partial_0[Q], \partial_1[Q])$ and $(\overline{Q'}_*, \partial_0[Q'], \partial_1[Q'])$ be two bicomplexes. Then we form $(\overline{T}_*, \partial_0[T], \partial_1[T])$ with

$$\forall i, \overline{T}_i = \overline{Q}_i \oplus \overline{Q'}_i,\ \partial_0^i[T] = \partial_0^i[Q] \oplus \partial_0^i[Q'],\ \partial_1^i[T] = \partial_1^i[Q] \oplus \partial_1^i[Q']$$

We can compute the homology groups of T given those of Q and Q' as follows:

Lemma 7. *The homology groups of T, disjoint union of Q and Q' are:*

$$\forall i, j, H_i(T, \partial_j) = H_i(Q, \partial_j) \oplus H_i(Q', \partial_j).$$

More generally, we have a relation between the homologies of bicomplexes Q, Q' and $Q \cap Q'$, using the so-called Mayer-Vietoris sequence (see for instance [6]). In particular, it permits us to adapt the preceding lemma to a case where we sum two bicomplexes, and identify their initial states (used in next section).
Now for products: One can compute the homology of the cartesian product of two manifolds, given the homology of both of them. If the cartesian product of two (concrete) manifolds M_1 and M_2 is based on the set of pairs of points of M_1 and M_2 respectively, we have to describe the discrete equivalent of such an operation on the simplicial sets and complexes arising from a triangulation of them. Let Q_i and Q'_i be the simplicial sets of objects of dimension i associated with the manifolds M_1 and M_2. Then the simplicial sets T_i associated with the product of M_1 and M_2 are:

$$\forall k \geq 0, T_k = \bigcup_{i=0\ldots k} Q_i \times Q'_{k-i}.$$

Therefore, the generated simplicial complex is:

$$\forall k \geq 0, \overline{T}_k = \bigoplus_{i=0\ldots k} \overline{Q}_i \otimes \overline{Q}'_{k-i}.$$

And the boundary operator is :

$$\partial^k[T] = \bigoplus_{i=0\ldots k} (\partial^i[Q] \otimes Id \oplus Id \otimes \partial^{k-i}[Q']).$$

This construction is the tensor product of the two complexes associated with M_1 and M_2 (see [6]). Now the homology groups are given by the Künneth formula:

Lemma 8.

$$H_k(T, \partial) = \bigoplus_{i=0\ldots k} (H_{k-i}(Q, \partial) \otimes H_i(Q', \partial))$$

The reader can verify that the definition of tensor product we have given is also correct for bicomplexes.

5.2 Translation of process algebra into homology groups

In this section we demonstrate that the operators of our process algebra can be interpreted as operators on homology groups such that the interpretation of a term gives the homology groups of the bicomplex corresponding to the term.

(case 1) $t = \omega \in \Sigma$. We have $H_i = \{*\}$ for all i≥1. $H_0([\![t]\!], \partial_0) = (\alpha)$, and $H_0([\![t]\!], \partial_1) = (1)$.

(case 2) t=**nil**. Here we also have $H_i = \{*\}$ for all i≥1. $H_0([\![t]\!], \partial_j) = (1)$.

(case 3) t=q+q'. We interpret choice as a connected sum of the complexes of q and q'. The homology groups are:

- $H_1(T, \partial_j) = H_1(Q, \partial_j) \oplus H_1(Q', \partial_j) \oplus N$ where, N=({0,X+X'},+) with i=$\partial_0(X)$, i'=$\partial_0(X')$, $X \in Q$, $X' \in Q'$
- $\forall i \neq 1, H_i(T, \partial_j) = H_i(Q, \partial_j) \oplus H_i(Q', \partial_j)$.

(case 4) t=q ; q'. The homology groups are then:

$$H_0(T, \partial_0) = \overline{F} \otimes H_0(Q', \partial_0) \quad H_0(T, \partial_1) = (1)$$

$$\forall i \geq 1, H_i(T, \partial_j) \oplus \overline{F} \otimes H_i(Q', \partial_j).$$

(case 5) t=(q || q'). The parallel composition without communication is interpreted as the tensor product of complexes. For the homology groups, we then have (Künneth formula):

$$H_0(T, \partial_j) = H_0(Q, \partial_j) \otimes H_0(Q', \partial_j)$$

$$H_k(T, \partial_j) = \bigoplus_{i=0\ldots k} (H_{k-i}(Q, \partial_j) \otimes H_i(Q', \partial_j))$$

(case 6) t=(q | q'). There is here a possibility of communication between p and q, represented by equations on tensor products between states which can interact. For the homology groups, we have:

$$H_1(T, \partial_j) = H_1(Q, \partial_j) \otimes H_0(Q', \partial_j) \oplus H_0(T, \partial_j) \otimes H_1(Q', \partial_j) \oplus I(Q, Q'),$$

where I(Q,Q') is the interaction term:

$$I(Q, Q') = Vect\{x \otimes_c (y + \partial_j(y)), (x + \partial_j(x)) \otimes_c y \;/\; x \in Q_1, y \in Q'_1, l(x) = \overline{l(y)}\}$$

and for $k > 1$, $H_k(T, \partial_j)$ is given by the Künneth formula from above.

Remark In what we have presented here, we have always $F = H_0(T, \partial_0)$. One can show (using the Mayer-Vietoris sequence) that as soon as we have the restriction operator \, we can have elements of $H_0(T, \partial_0)$ not in F. They are called deadlocks. One can use such a definition, and tools from homological algebra, to study deadlocks of processes, or also failure pairs if one wishes to study failure equivalence. In the next section we concentrate on bisimulation.

6 Bisimulation equivalence

Glabbeek defined the notion of bisimulation of HDA. In this section we demonstrate how the homology groups of HDA can be used to show that two HDA are not bisimulation equivalent.

Definition 9. S is a bisimulation between (Q,∂_0, ∂_1,l,I,F) and (Q',∂_0, ∂_1,l',I',F') if:

- S is a relation between $\cup_i Q_i$ and $\cup_j Q'_j$
- all $s \in I$ are related to an $s' \in I'$ and vice-versa
- (s,s')∈S ⇒ (s∈F ⇔ s'∈F')
- (s,s')∈S ⇒ (∀ q a path for Q such that ∃i, q_i=s, ∃q' a path for Q' such that ∃j, q'_j=s' and (q_{i+1}, q'_{j+1}) ∈S, $l^*(q_{i+1}) = l^*(q'_{j+1})$)
- (s,s')∈S ⇒ (∀ q' a path for Q' such that ∃j, q'_j=s', ∃q a path for Q such that ∃i, q_i=s and (q_{i+1}, q'_{j+1}) ∈S, $l^*(q_{i+1}) = l^*(q'_{j+1})$)

(Q,∂_0, ∂_1,l,I,F) and (Q',∂_0, ∂_1,l',I',F') are bisimulation equivalent if and only if there exists a bisimulation between them.

This notion of bisimulation equivalence "naturally" generalizes the usual notion (as found in [7]) of observational equivalence, or bisimulation equivalence on one dimensional automata. In our setting, the description of bisimulation equivalence is more complex than in the sequential case. Nevertheless, we can show that it is still a branching (in our sense) time equivalence, that is, it locally preserves some geometric shapes.

We introduce now local invariants of bisimulation equivalence. Let $\equiv$ be the smallest congruence on Λ containing the relation G, defined by: $uGv \Leftrightarrow u = l^*(x)$, $v = l^*(y)$ and $\exists z \in Im\partial_0$ such that $\langle x, z\rangle$=1, $\langle y, z\rangle$=1 Then l^* induces a map $[l^*] : H_i(Q, \partial_0) \rightarrow \Lambda/\equiv$. Let S(x), for $x \in \oplus_i H_i(Q, \partial_0)$ be $[l^*](x)$ if card $l^*(x) \geq 2$, 0 otherwise. Under the hypothesis that the number of states at a finite distance (equal to the length of the minimal path to reach them) of a given state is finite, we have the following result,

Proposition 10. *If (Q,∂_0, ∂_1,l,I,F) and (Q',∂_0, ∂_1,l',I',F') are bisimulation equivalent then (∀i, $S(H_i(Q, \partial_0)) = S(H_i(Q', \partial_0))$ (as a subset of $\Lambda/\equiv = \Lambda/\equiv'$) provided that all (i-1)-states (for i≥2) of elements of H_i are in the ∂_0- boundary of i-states which have all different labels in $\Lambda/\equiv$).*

This states that the branchings whose branches have distinct labels are preserved (and possibly duplicated) by bisimulation equivalence.

Examples:

(1) Let q=a.(b+c) and q'=a.b+a.c. We compute their homology groups: $H_1([\![q]\!], \partial_0) = H_0([\![a]\!], \partial_0) \otimes H_1([\![b+c]\!], \partial_0) = H_0([\![a]\!], \partial_0) \otimes (H_1([\![b]\!], \partial_0) \oplus H_1([\![c]\!], \partial_0) \oplus (b+c)) = (\alpha \otimes b + \alpha \otimes c)$, and, $H_1([\![q']\!], \partial_0) = (a + a')$ (where a' is a distinct copy of a). Thus S($H_1([\![q]\!], \partial_0)$)={b+c}, and S($H_1([\![q']\!], \partial_0)$)={0}. Thus, by proposition 10 $[\![q]\!]$ and $[\![q']\!]$ are not bisimulation equivalent.

(2) Let q=(a+b)||(c+d) and q'=a||c+b||c+a||d+b||d. We compute their homology groups: $H_1([\![q]\!],\partial_0) = H_1([\![a+b]\!],\partial_0)\otimes H_0([\![c+d]\!],\partial_0)\oplus H_0([\![a+b]\!],\partial_0)\otimes H_1([\![c+d]\!],\partial_0) = (a\otimes\gamma+b\otimes\gamma)\oplus(a\otimes\delta+b\otimes\delta)\oplus(\alpha\otimes c+\alpha\otimes d)\oplus(\beta\otimes c+\beta\otimes d)$. And $H_2([\![q]\!],\partial_0) = H_1([\![a+b]\!],\partial_0)\otimes H_1([\![c+d]\!],\partial_0) = (a\otimes c+b\otimes c+a\otimes d+b\otimes d)$. Now, $H_1([\![q']\!],\partial_0) = H_1([\![a||c]\!],\partial_0)\oplus H_1([\![b||c]\!],\partial_0)\oplus H_1([\![a||d]\!],\partial_0)\oplus H_1([\![b||d]\!],\partial_0)\oplus(a+b)\oplus(b+c)\oplus(c+d)$. And $H_2([\![q']\!],\partial_0)=\{*\}$. In particular, S($H_2([\![q]\!],\partial_0)$)={a.c+b.c+a.d+b.d}, and S($H_2([\![q']\!],\partial_0)$={0}. Thus q and q' are not bisimulation equivalent.

7 Conclusion

In this paper we have presented a technique for modeling concurrency centered around the description of higher dimensional automata by bicomplexes. From this we derived a notion of homology of HDA and demonstrated the pertinence of homology to concepts like non–determinism and bisimulation. The notion of HDA is taken directly from [1] and although the original definition is intuitively clear we find that the bicomplexes provide a more streamlined presentation of HDA for some purposes. The idea of applying algebraic topology to the study of concurrency was largely inspired by Pratt's article [9] where he introduced the notion of monoidal homotopy in HDA to model "true non–determinism". With this paper we hope to have provided some initial evidence of the usefulness of the related notion of homology. Further work on the subject will consider the restriction and recursion operators of CCS. Restriction is modeled by projection on vector spaces and recursion by limits in a suitable category of bicomplexes. Finally invariance under refinement of action should be related to the independence of choice of triangulation for calculating homology groups.

Acknowledgement. Thanks are due to Vaughan Pratt and Samson Abramsky for encouraging this work, to Chris Hankin and Patrick Cousot for support, and to the referees for helpful comments. Diagrams were drawn using Paul Taylor's tex macros.

References

1. Rob van Glabbeek. Bisimulation semantics for higher dimensional automata. Technical report, Stanford University, 1991.
2. J.E. Hopcroft and J.D. Ullman. *Introduction to Automata Theory, Languages and Computation*. Addison–Wesley, 1979.
3. Yves Lafont and Alain Prouté. Church-Rosser property and homology of monoids. Technical report, Ecole Normale Supérieure, 1990.
4. Serge Lang. *Algebra*. Addison Wesley, second edition, 1984.
5. Saunders Mac Lane. *Categories for the working mathematician*. Springer-Verlag, 1971.
6. William S. Massey. Homology and cohomology theory. In *Monographs and Textbooks in Pure and Applied Mathematics*, number 46. Marcel DEKKER, INC., 1978.
7. Robin Milner. *Communication and Concurrency*. Prentice Hall, 1989.
8. Henri Poincaré. De analysis situ. *Journal de l'Ecole Polytechnique*, 1895.
9. Vaughan Pratt. Modeling concurrency with geometry. In *Proc. 18th ACM Symposium on Principles of Programming Languages*. ACM Press, 1991.

Posets for Configurations!

Arend Rensink

University of Twente
P.O.Box 217, 7500 AE Enschede, the Netherlands
email: rensink@cs.utwente.nl

Abstract. We define *families of posets*, ordered by prefixes, as the counterpart of the usual families of configurations ordered by subsets. On these objects we define two types of morphism: *event* and *order* morphisms, resulting in categories **FPos** and $\mathbf{FPos}^{\sqsubseteq}$. We then show the following:

- Families of posets, in contrast to families of configurations, are always prime algebraic; in fact the category $\mathbf{FPos}^{\sqsubseteq}$ is equivalent to the category of prime algebraic domains;
- On the level of events, **FPos** is equivalent to the category of prime algebraic domains with an additional relation encoding *event identity*.
- The (abstract) event identity relation can be used to characterize concrete relations between events such as binary conflict and causal flow;
- One can characterize a wide range of event-based models existing in the literature as families of posets satisfying certain specific structural conditions formulated in terms of event identity.

1 Introduction

This paper focusses on *partial order*, in contrast to *interleaving*, models of behaviour. The basic building blocks of behaviour are *actions* as usual; in the partial order approach however it has been found convenient to introduce the notion of *events* to distinguish the different occurrences of actions. The information that is added to the models by the introduction of this new notion is abstracted away again by taking isomorphic variants in which only the *structure of the state space* is preserved, containing only aspects of the behaviour that are independent of the event identities.

With the introduction of events, there are (at least) two levels on which we can be interested in behaviour models: the more abstract *order level*, in which only the order structure of the state space is considered relevant, or the more concrete *event level*, on which the identities of events are also taken into account. On the order level, the question what events *are* is a non-issue, and it is therefore all right to take a very rigid attitude towards them. One often used assumption is that events are different if they have different histories (with respect to the relevant partial ordering). The resulting models, called *prime event structures*, are in a sense the simplest event-based models around, and on the order level they are quite sufficient.

Having introduced the level of events however, it turns out that there is something to be gained from studying such models on this level, that is *before* abstracting away from event identities. There are operations, notably that of *product* (corresponding to synchronization in process algebraic terms) which are awkward to define over prime event structures —a claim to the contrary by Vaandrager [21] notwithstanding. Some other treatments of this problem can be found in e.g. [7, 11]. It is mainly this problem which has inspired a host of other event-based models in which such constructions as product are simpler. Notable examples are *stable event structures* [26, 27], *flow event structures* [1, 3, 4], and *bundle event structures* [10]. Most of these models have in common that on the order level they have precisely the expressive power of prime event structures (more precisely, they are categorically equivalent), wherease on the event level they are more expressive. Unlike prime event structures however *these other models generally lack a precise, intuitive notion of what events are*, and consequently they do not have characterizations on the level of events.[1]

All the event-based models mentioned above have in common that they are compared by regarding the *families of configurations* that they yield. A configuration, here, is a set of events that have occurred during a (partial) execution according to the model. In this paper we generalize the notion of configuration: we will use not sets but *posets* of events, in which the partial ordering relation over the events is explicitly represented. There is a well-known notion of *poset prefix* which is a partial ordering over posets; our models will be certain sets of posets ordered by the prefix relation. We call our models *families of posets* to express the obvious connection to families of configurations. They are on the order level still equivalent to prime event structures. However, we can characterize them on the event level by another structural relation besides the prefix ordering, called *top equivalence*, such that on this level families of posets have a richer, yet well-defined structure. Better yet, it turns out that we can also show precisely under which additional conditions (phrased in terms of prefix and top equivalence) the families of posets are constrained to be in one-to-one correspondence to families of configurations of, say, stable event structures. We give such characterizations for a range of event-based models from the literature.

The main thesis of this paper is that expressed in the form of a slogan by the title. We think that using posets instead of sets of events as configurations —i.e. using families of posets rather than families of configurations— has several theoretical and practical advantages. On the theoretical side, the order structure is better behaved, and we have an abstract notion of event which we can play around with to characterize existing models. On the practical side, the model is more expressive so that constructions which are awkward in families of configurations, such as sequential composition, are easy to define.[2] Of course, there are cases in which families of configurations suffice and are perhaps conceptually easier than families of posets. On the whole however, we find that there is much to be gained and little to be lost by using posets for configurations.

2 Preliminaries

Partially ordered sets play a central role in this paper. In order to deal conveniently with them we introduce some notation and conventions. We assume a universe of events **Evt** and we consider the *finite posets over* **Evt**, i.e., those structures $p =$

[1] An exception is a characterization through so-called *prime intervals* [8, 26, 27].

[2] Unfortunately, limits on the available space prevent us from discussing model constructions in this paper; cf. [17]. For the same reason all proofs have had to go; cf. [18].

$\langle E_p, \leq_p \rangle$ such that $E_p \subseteq_{\text{fin}} \mathbf{Evt}$ is a finite set of events and $\leq_p \subseteq E_p \times E_p$ is a partial ordering relation. We use **Pos** to denote the universe of finite posets, P, Q to denote sets of posets, p, q to denote single posets and $a, \ldots, e$ to denote events. An event e will sometimes also be used to stand for the poset $\langle \{e\}, \varnothing \rangle$.

A poset is called *triangular* if it has a greatest element. If P is a set of posets, then P^Δ denotes the triangular posets in P. If p is a triangular poset then the the greatest element of p is denoted $p\Delta$ (hence $\forall e \in E_p.\, e \leq_p p\Delta$); it is called the *top event* of p. The *context of* $p\Delta$ *in* p, denoted $\lhd p$, is defined as p without its top event. Two posets p and q are called *top equivalent*, denoted $p \triangleq q$, if they are triangular with the same top event, i.e. $p\Delta = q\Delta$. Note that $\triangleq$ is an *partial* equivalence relation over **Pos**, that is, an equivalence relation over a (proper) subset of **Pos**, viz. $\mathbf{Pos}^\Delta$.

$$p \subseteqq q :\Leftrightarrow (E_p = E_q) \wedge (\leq_p \subseteq \leq_q) \tag{1}$$

$$p \preccurlyeq q :\Leftrightarrow (E_p \subseteq E_q) \wedge (\leq_p = \leq_q \cap (E_q \times E_p)) \ . \tag{2}$$

If $p \subseteqq q$ then p *is augmented by* q or q *is an augment of* p; if $p \preccurlyeq q$ then p is a *prefix* or *initial segment* of q. The condition on $\leq_p$ and $\leq_q$ can be interpreted as follows: if $a, b \in E_q$ are such that $a \leq_q b$, then $b \in E_p$ implies $a \in E_p$ and $a \leq_p b$. A set $P \subseteq \mathbf{Pos}$ is said to be *prefix closed* if it is left-closed w.r.t. $\preccurlyeq$; hence $\forall p \in \mathbf{Pos}, q \in P.\, p \preccurlyeq q \Longrightarrow p \in P$. $\curlywedge P$ denotes the infimum w.r.t. $\preccurlyeq$ of a set P of posets and $\curlyvee P$ the supremum, if they exist.

$$p \frown q :\Leftrightarrow \leq_p \cap (E_p \times E_q) = \leq_q \cap (E_q \times E_p) \ . \tag{3}$$

If $p \frown q$ we say that p and q are called *compatible*. A set of posets P is *pairwise compatible*, denoted $P \frown$, if $p \frown q$ for all pairs $p, q \in P$. Comparing this property to the prefix relation, we see that $p \frown q$ implies that the intersection of p and q is a prefix of both, and also $p \preccurlyeq q \iff (p \frown q) \wedge (E_p \subseteq E_q)$. Compatibility is a prerequisite for some of the poset operations defined below.

$$p \restriction E := \langle E_p \cap E, \leq_p \cap (E \times E) \rangle \tag{4}$$

$$p; q := \langle E_p \cup E_q, \leq_p \cup (E_p \times E_q) \cup \leq_q \rangle \tag{5}$$

$$p \cup q := \langle E_p \cup E_q, \leq_p \cup \leq_q \rangle \tag{6}$$

$$p \cap q := \langle E_p \cap E_q, \leq_p \cap \leq_q \rangle \ . \tag{7}$$

$p \restriction E$ denotes the restriction of p to the event set E; it is well enough known. $p; q$ denotes the sequential composition of p and q; it is defined only if E_p and E_q are disjoint. We sometimes use expressions of the form $p; e$, where e is interpreted as a poset. This yields a triangular poset; see above. The union $p \cup q$, also used in the form $\bigcup P$, is a useful construction only if it is a poset such that $\forall p \in P.\, p \preccurlyeq \bigcup P$; in other words, if $\bigcup P$ is an upper bound for P w.r.t. prefix ordering. Similarly for the intersection $p \cap q$ or $\bigcap P$. As an example, consider $p_1 = \boxed{a \to b}$, $p_2 = \boxed{a \to c}$ and $p_3 = \boxed{b \to c}$: then $p_1 \frown p_2$ and $p_1 \cup p_2 =$ 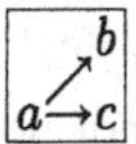

which is an upper bound for p_1 and p_2; but for instance $p_2 \not\frown p_3$, and consequently $p_2 \cup p_3 =$ 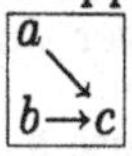

is not an upper bound for p_2 or p_3.

Lemma 1. *If P is a set of posets, then the following three statements are equivalent:*
1. P has an upper bound;
2. $P\frown$;
3. $\bigcup P = \curlyvee P$.
Moreover, if P is nonempty then any of these statements implies $\bigcap P = \curlywedge P$.

Hereafter we will use the notation $\curlyvee$ and $\curlywedge$ for the poset operations $\cup$ and $\cap$.

3 Families of Posets

This section introduces and motivates the class of models central to this paper: families of posets. As the name suggest —certainly in combination with the title of this paper— there is a strong relation with families of configurations, the traditional semantic models for event structures.

Definition 2. A family of posets is a nonempty prefix closed set of posets. The class of families of posets is denoted **FPos**, and ranged over by $\mathcal{P}$, $\mathcal{Q}$.

A finite subset $P \subseteq_{\text{fin}} \mathcal{P}$ is called *consistent in $\mathcal{P}$*, denoted $P\uparrow_{\mathcal{P}}$, if $P\frown$ and $\curlyvee P \in \mathcal{P}$. An arbitrary subset $Q \subseteq \mathcal{P}$ is called *finitely consistent in $\mathcal{P}$*, or simply *consistent*, if $P\uparrow_{\mathcal{P}}$ for all finite subsets $P \subseteq_{\text{fin}} Q$. In addition we introduce $p \uparrow_{\mathcal{P}} q$ as shorthand notation for the binary case, and we usually drop the index $\mathcal{P}$ when the family of posets is clear from the context. Consistency implies compatibility but not vice versa, as the following example shows.

Example 1. Let $\mathcal{P}$ be a family of posets containing $p_1 = \boxed{a\to b}$, $p_2 = \boxed{c\to b}$ and $p_3 = \boxed{a\to c}$. Clearly $p_1 \not\frown p_2$; this implies $p_1 \not\uparrow p_2$, hence p_1 and p_2 are inconsistent. On the other hand let $q = \boxed{\begin{smallmatrix} & b \\ \nearrow & \\ a\to & c\end{smallmatrix}}$; then $p_1 \frown p_3$ and $p_1 \curlyvee p_3 = q$. Now if $q \notin \mathcal{P}$ then $p_1 \not\uparrow p_3$, and in executing p_1 a system makes an irrevocable choice, rejecting p_3. If however $q \in \mathcal{P}$ then there is no real choice involved in doing p_1: the system may still proceed to q.

There are two viewpoints from which families of posets may be motivated: posets as executions or posets as states. Representing system executions is an obvious application of posets. Examples of this principle are pomset traces [13], pomset transitions [2], and also Mazurkiewicz traces [12]. Note however that in the cited papers the partial orders are applied on the level of *actions*: events are either not present or abstracted away from by isomorphism. From this viewpoint, a family of posets $\mathcal{P}$ contains all the (partial) executions of a given system. If $p \preccurlyeq q$ for two posets $p, q \in \mathcal{P}$, this means that the system may execute q by first doing p and then what remains of q.

Example 2. If $p = \boxed{a\to b}$ and $q = \boxed{\begin{smallmatrix} & b \\ \nearrow & \\ a\to & c\end{smallmatrix}}$ then a system may proceed from p to q by executing c. Note however that c does not depend on all events in p, but just on a.

The dual viewpoint, posets as states, resembles the traditional interpretation of configurations as states. The posets, resp. the configurations represent the *previous* behaviour, i.e. the work that went into arriving at a state. The question then is how much to record of the previous behaviour. Families of configurations and families of posets give different answers: using posets one may record more about the previous behaviour than using configurations –besides the event identities, the *dependencies* of events are also recorded. This implies that further behaviour may also depend on this information; see Ex. 3 below. The prefix relation becomes a state transition relation, whereas consistency of a set of posets (states) implies that there is a next state reachable from all of them. It is proved in Sect. 4 that as state spaces, families of posets have a very clear algebraic structure.

Example 3. Consider a family of posets containing $\boxed{a \rightarrow b}$ and $\boxed{\begin{array}{c} a \\ b \rightarrow c \end{array}}$. When the system described by this family is in the state $\boxed{a}$ then there are two possible contexts in which b may occur, viz. the empty poset and $\boxed{a}$, yielding respectively states $\boxed{\begin{array}{c} a \\ b \end{array}}$ from which the system may execute c, and $\boxed{a \rightarrow b}$ from which it may not.

We construct two categories with objects from **FPos** and different morphisms as arrows. First we define some auxiliary terms.

Definition 3. If $\langle D_i, \sqsubseteq_i \rangle$ are partial orders for $i = 1, 2$ with consistency predicates $\uparrow_i$, suprema $\sqcup_i$ and infima $\sqcap_i$, then a function $\phi: D_1 \rightarrow D_2$ is called

additive if $\quad \forall x, y \in D_1 . x \uparrow_1 y \Longrightarrow (\phi(x) \uparrow_2 \phi(y)) \wedge (\phi(x \sqcup_1 y) = \phi(x) \sqcup_2 \phi(y))$

stable if $\quad \forall x, y \in D_1 . x \uparrow_1 y \Longrightarrow \phi(x \sqcap_1 y) = \phi(x) \sqcap_2 \phi(y)$

image closed if $\forall x \in D_2, y \in D_1 . x \sqsubseteq_2 \phi(y) \Longrightarrow \exists z \in D_1 . (z \sqsubseteq_1 y) \wedge (\phi(z) = x)$.

The conditions of additivity and stability are inspired by [26, Definition 2.1.6]. Note that additivity implies monotonicity. Image closedness ensures that every left-closed part of D_1 is mapped onto a left-closed part of D_2.

Definition 4. Let $\mathcal{P}, \mathcal{Q}$ be families of posets.

1. An *order morphism from $\mathcal{P}$ to $\mathcal{Q}$* is a function $g: \mathcal{P} \rightarrow \mathcal{Q}$ which is additive, stable and image closed with respect to prefix.
2. An *event morphism from $\mathcal{P}$ to $\mathcal{Q}$* is a partial function $f: E_{\mathcal{P}} \rightarrow E_{\mathcal{Q}}$ which is injective on E_p for all $p \in \mathcal{P}$, such that $f^*(p) \in \mathcal{Q}$ for all $p \in \mathcal{P}$ (where f^* is the pointwise extension of f to posets).

Proposition 5. *Let* $\mathbf{FPos}^{\sqsubseteq}$ (**FPos**) *have families of posets as objects and order morphisms (event morphisms) as arrows; let arrow composition be defined by function composition. Then* $\mathbf{FPos}^{\sqsubseteq}$ (**FPos**) *is a category with the obvious choice of identity arrows.*

An order morphism $g: \mathcal{P} \rightarrow \mathcal{Q}$ models how $\mathcal{P}$ can simulate an initial part of the behaviour of $\mathcal{Q}$. Order isomorphism (also discussed in [8] and as *equivalence* in [1]) can be characterized in terms of the preservation of prefix structure (the symbol $\cong$ is used to denote structural isomorphism):

Proposition 6. *If $\mathcal{P}$ and $\mathcal{Q}$ are families of posets, then $\mathcal{P}$ and $\mathcal{Q}$ are order isomorphic — i.e. isomorphic in* $\mathbf{FPos}^{\sqsubseteq}$ *— if and only if* $\langle \mathcal{P}, \preccurlyeq \rangle \cong \langle \mathcal{Q}, \preccurlyeq \rangle$.

Example 4. Let $\mathcal{P}, \mathcal{Q}$ be families of posets such that $\mathcal{P}$ has maximal elements $\boxed{\begin{smallmatrix}a\to c\\ b\end{smallmatrix}}$ and $\boxed{b\to c}$, and $\mathcal{Q}$ has maximal elements $\boxed{a\to d}$ and $\boxed{\begin{smallmatrix}a\\ b\to c\end{smallmatrix}}$. Then $\mathcal{P}$ and $\mathcal{Q}$ are order isomorphic; for instance the following mappings induce an order isomorhpism g:

$$\boxed{a} \mapsto \boxed{b},\ \boxed{b} \mapsto \boxed{a},\ \boxed{a\to c} \mapsto \boxed{b\to c},\ \boxed{b\to c} \mapsto \boxed{a\to d}\ .$$

Event morphisms are inspired by the family of configuration morphisms defined in e.g. [24, 26]. If $f: \mathcal{P} \to \mathcal{Q}$ is an event morphism then $\mathcal{P}$ can also simulate an initial part of the behaviour described by $\mathcal{Q}$, but such that different events in a given run of $\mathcal{P}$ correspond to different events in $\mathcal{Q}$ and vice versa. $\mathcal{P}$ may be able to do events that fall outside the scope of $\mathcal{Q}$, since f is a partial function. This ensures that products in **FPos** can be used to model partial synchronization, as in the standard theory of categories of models for concurrency; cf. [25]. Event isomorphism (called *isomorphism* in [8]) can also be characterized in terms of the preservation of structure.

Theorem 7. *If $\mathcal{P}, \mathcal{Q}$ are two families of posets, then $\mathcal{P}$ and $\mathcal{Q}$ are event isomorphic —i.e. isomorphic in* **FPos**— *if and only if $\langle \mathcal{P}, \preccurlyeq, \triangleq \rangle \cong \langle \mathcal{Q}, \preccurlyeq, \triangleq \rangle$.*

To prove this, we establish a connection between order and event morphisms, stating that an event morphism is essentially an order morphism that also preserves top equivalence.

Theorem 8. *Let $\mathcal{P}$ and $\mathcal{Q}$ be families of posets.*

1. *If $f: E_{\mathcal{P}} \to E_{\mathcal{Q}}$ is an event morphism, then f^* is an order morphism.*
2. *If $g: \mathcal{P} \to \mathcal{Q}$ is an order morhpism, then the following conditions are equivalent:*
   (a) There exists an event morphism $f: E_{\mathcal{P}} \to E_{\mathcal{Q}}$ such that $f^* = g$;
   (b) $p \triangleq q \Longrightarrow g(\lhd p) = g(p) \vee g(p) \triangleq g(q)$ for all $p, q \in P^{\Delta}$.
   If these conditions hold then the event morphism f is uniquely determined.

Example 5. In Ex. 4 above the order morphism g does not preserve top equivalence: for instance, $\boxed{a\to c} \triangleq \boxed{b\to c}$ but $g(\boxed{a\to c}) \neq g(\lhd\boxed{a\to c})$ and $g(\boxed{a\to c}) = \boxed{b\to c} \not\triangleq \boxed{a\to d} = g(\boxed{b\to c})$. The function $f = \{(a, b), (b, a), (c, c), (d, c)\}$ is an event morphism from $\mathcal{Q}$ to $\mathcal{P}$ but not an isomorphism: it does not even have an inverse. In fact $\mathcal{P}$ and $\mathcal{Q}$ are not event isomorphic: the number of events in $E_{\mathcal{P}}$ and $E_{\mathcal{Q}}$ differ, for one thing.

4 Event Bases

It has been shown before that some event-based models interpreted as state spaces yield finitary prime algebraic domains. This is in particular true for the configurations of stable event structures and prime event structures: see e.g. Winskel [26]. In fact every such domain corresponds to the state space of some prime event structure. For more general classes of models however, e.g. the (general) event structures also to be found in [26], the corresponding configurations interpreted as states do not form prime algebraic domains; for the geometrical automata described by Gunawardena [9] they do not even form domains. Using posets instead of configurations as states however, we get prime algebraic domains even in those cases. Our results are actually formulated in terms of *compact* and *distributive bases*. We do some order theoretic juggling first, defining the models and connecting them to prime algebraic domains. We assume standard order-theoretic terms such as 'domain' and 'compact element' to be familiar; see e.g. [5]. Since the discussion below is the only place where these notions appear, and the remainder of this section relies only on Def. 9, the uninitiated reader can proceed regardless.

Definition 9. A *compact basis* is a partially ordered structure $\langle X, \sqsubseteq \rangle$ such that the following holds:

1. X is nonempty;
2. every nonemtpy subset of X has a meet;
3. no infinite subset of X has a join.

A *distributive basis* is a compact basis such that

4. $x \sqcup \bigsqcap Y = \bigsqcap_{y \in Y}(x \sqcup y)$ whenever $Y \subseteq X$ and $x \sqcup y$ is defined for all $y \in Y$.

The class of distributive bases is denoted **DBas**.

Compact bases can be turned into finitary domains in a process similar to *ideal completion*. The situation is the same as with the configuration structures in Van Glabbeek and Goltz [23] and the finite configurations used in Boudol [1], which are the compact elements of the families of configurations used in Winskel [26]. We use compact bases instead of domains for simila reasons: our models contain only finite elements, and the inclusion of the infinite "ideals" is unnecessary and complicating.

$$X^{pr} := \{x \in X \mid \forall Y \subseteq X.\, x \sqsubseteq \bigsqcap Y \Longrightarrow \exists y \in Y.\, x \sqsubseteq y\} \tag{8}$$

defines the set of *primes* of a partially ordered structure $\mathcal{X} = \langle X, \sqsubseteq \rangle$. $\mathcal{X}$ is called *prime algebraic* if it satisfies

$$\forall x \in X.\, x = \bigsqcup \{y \in X^{pr} \mid y \sqsubseteq x\}\ . \tag{9}$$

Lemma 10. *A partially ordered structure $\mathcal{X} = \langle X, \sqsubseteq \rangle$ is a compact basis if and only if X consists of the compact elements of a finitary domain $\mathcal{Y} = \langle Y, \sqsubseteq \rangle$. $\mathcal{Y}$ is uniquely determined by this up to isomorphism. $\mathcal{X}$ is distributive if and only if $\mathcal{X}$ is prime algebraic if and only if $\mathcal{Y}$ is prime algebraic.*

Lemma 11. *If $\langle X, \sqsubseteq \rangle$ is a prime algebraic partially ordered structure then*

$$x \in X^{pr} \iff \exists y \in X.\, y \sqsubset x \wedge \forall z \in X.\, z \sqsubset x \Longrightarrow z \sqsubseteq y\ .$$

This y is uniquely determined; it is called the direct predecessor *of x, denoted $pred(x)$.*[3]

Definition 12. An *event basis* is a structure $\langle X, \sqsubseteq, \simeq \rangle$ such that $\langle X, \sqsubseteq \rangle$ is a distributive basis and $\simeq\, \subseteq X^{pr} \times X^{pr}$ is an equivalence relation called *event identity*, satisfying:[4]

$$\begin{aligned} &\textit{uniqueness:} && \forall x, y.\, x \simeq y \wedge x \uparrow y \Longrightarrow x = y \\ &\textit{determinism:} && \forall x, y.\, x \simeq y \wedge pred(x) = pred(y) \Longrightarrow x = y\ . \end{aligned}$$

The class of all event bases is denoted **EBas**.

[3] It follows that direct predecessors are defined only for primes.

[4] $x \uparrow y$ denotes that x and y are *consistent*, i.e. have a common upper bound in X.

Intuitively, when distributive bases (or prime algebraic domains) appear as the state spaces of event-based models, the primes correspond to *events* occurring in a given *context*, determined by their predecessors. In the case of prime event structures for instance, each prime corresponds *uniquely* to an event, whereas in the domains obtained from stable event structures the mapping from primes to events is generally not injective. Event bases generalize this principle by modelling events as equivalence classes of primes, i.e. primes are equivalent if and only if they correspond to the same event. The conditions on the event identity can be motivated as follows.

uniqueness If two states are consistent then they have a common successor state, which forms the combination of the two. Hence if two different event identical primes are consistent, then in the common successor state the same event has occurred twice, which contradicts the intuition behind the event concept.

determinism There does not exist more information about an execution than the events that have occurred and the contexts in which they have occurred, i.e. the previous events on which they depend. It follows that the events and their contexts completely determine a state.

Definition 13.

1. Let $\mathcal{X}_i = \langle X_i, \sqsubseteq_i \rangle$ be distributive bases for $i = 1, 2$. A function $h: X_1 \to X_2$ is an *order morphism* if h is additive, stable and image closed.
2. Let $\mathcal{X}_i = \langle X_i, \sqsubseteq_i, \simeq_i \rangle$ be event bases for $i = 1, 2$. A function $h: X_1 \to X_2$ is an *event morphism* from $\mathcal{X}_1$ to $\mathcal{X}_2$ if it is an order morphism from $\langle X_1, \sqsubseteq_1 \rangle$ to $\langle X_2, \sqsubseteq_2 \rangle$ such that
$$\forall x, y \in X_1^{pr}. \, x \simeq_1 y \Longrightarrow h(x) \simeq_2 h(y) \vee h(x) = h(pred(x)) \ .$$

Proposition 14. *Let* **DBas** (**EBas**) *have distributive bases (event bases) as objects and order morphisms (event morphisms) as arrows, and let arrow composition be defined by function composition. Then* **DBas** (**EBas**) *is a category with the obvious choice of identity arrows.*

It is obvious from the definitions that **DBas** and **EBas** are closely related. In fact, there exists an adjunction between **DBas** and **EBas**: the right adjoint "forgets" the event identity structure and the left adjoint effectively creates an event for each prime.

Theorem 15. *Let* U *be a mapping defined over the objects and arrows of* **EBas** *such that* $\mathrm{U}: \langle X, \sqsubseteq, \simeq \rangle \mapsto \langle X, \sqsubseteq \rangle$ *and* $\mathrm{U}: h \mapsto h$. *Let* F *be a mapping defined over the objects and arrows of* **DBas** *such that* $\mathrm{F}: \langle X, \sqsubseteq \rangle \mapsto \langle X, \sqsubseteq, \{ (x, x) \mid x \in X^{pr} \} \rangle$ *and* $\mathrm{F}: h \mapsto h$. *Then* U *and* F *are functors such that* $\mathrm{F} \dashv \mathrm{U}$.

The following theorem, establishing the connection between these new categories and families of posets, is one of the main results of this paper.

Theorem 16.

1. *Let* $\mathrm{B}^{\sqsubseteq}$ *be a mapping defined over the objects and arrows of* $\mathbf{FPos}^{\sqsubseteq}$ *such that* $\mathrm{B}^{\sqsubseteq}: \mathcal{P} \mapsto \langle \mathcal{P}, \preccurlyeq \rangle$ *and* $\mathrm{B}^{\sqsubseteq}: g \mapsto g$; *then* $\mathrm{B}^{\sqsubseteq}$ *is a functor from* $\mathbf{FPos}^{\sqsubseteq}$ *to* **DBas** *which induces an equivalence.*

2. *Let* B *be a mapping defined over the objects and arrows of* **FPos** *such that* $B: \mathcal{P} \mapsto \langle \mathcal{P}, \preccurlyeq, \triangleq \rangle$ *and* $B: f \mapsto f^*$; *then* B *is a functor from* **FPos** *to* **EBas** *which induces an equivalence.*

The proof consists of constructing left and right adjoints for $B^{\sqsubseteq}$ and B. These are slightly more complicated because we have to construct posets from the elements of the event basis. The main intuition is that events are $\simeq$-classes of primes: if $\mathcal{X} = \langle X, \sqsubseteq, \simeq \rangle$ is an event basis then the events are given by

$$E_{\mathcal{X}} := X^{pr}/\simeq \ .$$

For an arbitrary $x \in X$ the *poset of* x is the structure $p_x = \langle E_x, \leq_x \rangle$ defined by

$$E_x := \{ a \in E_{\mathcal{X}} \mid \exists y \in a.\, y \sqsubseteq x \}$$
$$a \leq_x b :\Leftrightarrow \exists y \in a, z \in b.\, y \sqsubseteq z \sqsubseteq x \ .$$

One can prove $x \sqsubseteq y \iff p_x \preccurlyeq p_y$ and $x \in X^{pr} \iff p_x \in \mathbf{Pos}^{\Delta}$ for all $x, y \in X$. We can then construct adjoints for B and $B^{\sqsubseteq}$.

Theorem 17.
1. *Let* $P^{\sqsubseteq}$ *be a mapping defined on the objects and arrows of* **DBas** *such that if* $\mathcal{X}$ *is a distributive basis and* $F\mathcal{X} = \langle X, \sqsubseteq, \simeq \rangle$ *then* $P^{\sqsubseteq}: \mathcal{X} \mapsto \{ p_x \mid x \in X \}$, *and if* $\mathcal{X}_i = \langle X_i, \sqsubseteq_i \rangle$ *are arbitrary event bases for* $i = 1, 2$ *and* $h: \mathcal{X}_1 \to \mathcal{X}_2$ *is an order morphism then* $(P^{\sqsubseteq}h): p_x \mapsto p_{h(x)}$. *Then* $P^{\sqsubseteq}$ *is a functor from* **DBas** *to* $\mathbf{FPos}^{\sqsubseteq}$ *which is left and right adjoint to* $B^{\sqsubseteq}$.
2. *Let* P *be a mapping defined over the objects and arrows of* **EBas** *such that if* $\mathcal{X} = \langle X, \sqsubseteq, \simeq \rangle$ *is an event basis then* $P: \mathcal{X} \mapsto \{ p_x \mid x \in X \}$, *and if* $\mathcal{X}_i = \langle X_i, \sqsubseteq_i, \simeq_i \rangle$ *are arbitrary event bases for* $i = 1, 2$ *and* $h: \mathcal{X}_1 \to \mathcal{X}_2$ *is an event morphism between then* $P: h \mapsto f$, *where* $f: E_{\mathcal{X}_1} \to E_{\mathcal{X}_2}$ *is defined by*
$$f: e \mapsto [h(x)]_{\simeq_2} \text{ if } x \in X_1^{pr} \text{ and } e = [x]_{\simeq_1} \text{ and } h(x) \neq h(pred(x)) \ .$$
Then P *is a functor from* **EBas** *to* **FPos** *which is left and right adjoint to* B.

Through the equivalence between $\mathbf{FPos}^{\sqsubseteq}$ and **DBas** and between **FPos** and **EBas** it follows that the adjunction proved in Th. 15 also holds between $\mathbf{FPos}^{\sqsubseteq}$ and **FPos**. The following corollary of Th. 16 is a direct generalization of [26, Theorem 1.2.8].

Corollary 18.
1. *If* $\mathcal{P}$ *is a family of posets then* $\langle \mathcal{P}, \preccurlyeq, \triangleq \rangle$ *is an event basis;*
2. *If* $\mathcal{X}$ *is an event basis then* $\mathcal{X} \cong \langle \mathcal{P}, \preccurlyeq, \triangleq \rangle$ *for some family of posets* $\mathcal{P}$.

5 Families of Configurations

Those readers who have been exposed to event structures will have recognized the similarity between those and families of posets. To describe the semantics of event structures one uses a "lower level model," so-called *families of configurations*. Much of the mathematical treatment as well as the underlying intuitions of families of posets and families of configurations are very much alike.

Definition 19 [22, Definition IV.3.1]. A *family of configurations* is a set $\mathcal{C}$ of finite sets of events satisfying

1. $\varnothing \in \mathcal{C}$;
2. $\forall F, G, H \in \mathcal{C}.\, F \cup G \subseteq H \Longrightarrow F \cup G \in \mathcal{C}$;
3. $\forall F \in \mathcal{C}.\, \forall a, b \in F.\, a \neq b \Longrightarrow \exists G \in \mathcal{C}.\, G \subseteq F \wedge (a \in G \iff b \notin G)$.

The set of all families of configurations is denoted **FCnf**.

We will use the letters F, G, H to denote configurations and $\mathcal{C}$ to denote families of configurations. Note that configurations in our definition are finite sets; in this we follow [22] but differ from [26]. The infinite configurations can always be reconstructed from the finite ones, in the manner described in Sect. 4. We define morphisms over **FCnf**, which by now should look familiar from those over families of posets (Def. 4) and event bases (Def. 13).

Definition 20. Let $\mathcal{C}_i$ are families of configurations and $E_i = \bigcup \mathcal{C}_i$ for $i = 1, 2$.

1. An *order morphism from* $\mathcal{C}_1$ *to* $\mathcal{C}_2$ is a function $g: \mathcal{C}_1 \to \mathcal{C}_2$ which is additive, stable and image closed w.r.t. the subset ordering.
2. An *event morphism from* $\mathcal{C}_1$ *to* $\mathcal{C}_2$ is a partial function $f: E_1 \to E_2$ which is injective on F for all $F \in \mathcal{C}$, such that $f^*(F) \in \mathcal{C}_2$ for all $F \in \mathcal{C}_1$ (where f^* is the pointwise extension of f to sets of events).

The definition of event morphisms was taken from [26]. Again, this provides us with two categories based on the same objects: one in which only the structure imposed by the subset relation is considered to be relevant and one in which the event identities are also respected.

Proposition 21. *Let* $\mathbf{FCnf}^{\sqsubseteq}$ (**FCnf**) *have families of configurations as objects and order morphisms (event morphisms) as arrows, and let arrow composition be defined by function composition. Then* $\mathbf{FCnf}^{\sqsubseteq}$ (**FCnf**) *is a category with the obvious choice of identity arrows.*

We aim to establish a connection between the categories of families of posets and families of configurations. Intuitively, the elements of a family of posets are just configurations with extra information in the form of ordering relations; hence a functor should essentially forget those ordering relations. The order and event morphisms over families of posets can be lifted to families of configurations without much change.

Definition 22.

1. Let $\mathrm{C}^{\sqsubseteq}$ be a mapping defined over the objects and arrows of $\mathbf{FPos}^{\sqsubseteq}$ such that $\mathrm{C}^{\sqsubseteq}: \mathcal{P} \mapsto \{E_p \mid p \in \mathcal{P}\}$ and $(\mathrm{C}^{\sqsubseteq} g): E_p \mapsto E_{g(p)}$.
2. Let C be a mapping defined over the objects and arrows of **FPos** such that $\mathrm{C}: \mathcal{P} \mapsto \{E_p \mid p \in \mathcal{P}\}$ and $\mathrm{C}: f \mapsto f$.

Unfortunately neither mapping is a functor. For instance, the following example shows a family of posets which does not yield a family of configurations under C.

Example 6. Let $\mathcal{P}$ be the family of posets with $\boxed{a \to b \to c}$ and $\boxed{c \to b \to a}$ as maximal elements. Now $\mathrm{C}\mathcal{P}$ is not a family of configurations: in particular, $\{a\}, \{c\}, \{a, b, c\} \in \mathrm{C}\mathcal{P}$ but $\{a, c\} \notin \mathrm{C}\mathcal{P}$, hence Rule 2 of Def. 19 is not satisfied.

Another, quite different problem is that even if $C\mathcal{P}$ is a family of configurations, then its order structure w.r.t. $\subseteq$ may differ from the structure of $\mathcal{P}$ under $\preccurlyeq$. This is follows from the fact that families of posets are order equivalent to distributive bases, as proved in Sect. 4, whereas it is well-known that (general) families of configurations are not distributive. The following example shows that there are morphisms in $\mathbf{FPos}^{\sqsubseteq}$ whose image under $\mathrm{C}^{\sqsubseteq}$ is not a morphism in $\mathbf{FCnf}^{\sqsubseteq}$.

Example 7. Let $\mathcal{P}$ be the family of posets with maximal elements $\boxed{\begin{array}{l} a\rightarrow c \\ b \end{array}}$ and $\boxed{\begin{array}{l} a \\ b\rightarrow d \end{array}}$, and let $\mathcal{Q}$ have maximal elements $\boxed{\begin{array}{l} a\rightarrow c \\ b \end{array}}$ and $\boxed{\begin{array}{l} a \\ b\rightarrow c \end{array}}$. We get the following:

$$
\mathrm{C}\mathcal{P} = \begin{array}{ccccccc} & \nearrow & \{a\} & \rightarrow & \{a,c\} & \rightarrow & \{a,b,c\} \\ \varnothing & & & \searrow & \{a,b\} & \nearrow \searrow & \\ & \searrow & \{b\} & \rightarrow & \{b,d\} & \rightarrow & \{a,b,d\} \end{array} \quad ; \quad \mathrm{C}\mathcal{Q} = \begin{array}{ccccccc} & \nearrow & \{a\} & \rightarrow & \{a,c\} & \searrow & \\ \varnothing & & & \searrow \nearrow & \{a,b\} & \rightarrow & \{a,b,c\} \\ & \searrow & \{b\} & \rightarrow & \{b,c\} & \nearrow & \end{array} .
$$

$f = \{(a,a),(b,b),(c,c)\}$ is an event morphism from $\mathcal{P}$ to $\mathcal{Q}$ and from $\mathrm{C}\mathcal{P}$ to $\mathrm{C}\mathcal{Q}$, and f^* is an order morphism; however $\mathrm{C}^{\sqsubseteq} f^*$ is not an order morphism from $\mathrm{C}\mathcal{P}$ to $\mathrm{C}\mathcal{Q}$ (it is not image closed).

This suggests that we should forget about comparing on the order level, and concentrate on the event level. Hence we will only compare **FPos** and **FCnf**. We have C of Def. 22 as a "proto-functor," for we are pretty much convinced that is the right choice; now we construct a subcategory of **FPos** over which C is actually a functor.

Property Sim **(Simulativity).** $\mathcal{P}$ is said to be *simulative* if for all $p,q \in \mathcal{P}$

$$
(\exists q' \in \mathcal{P}.\, p \preccurlyeq q' \wedge E_{q'} = E_q) \iff E_p \subseteq E_q
$$

$$
(\exists q' \in \mathcal{P}.\, E_p \preccurlyeq E_{q'} \wedge E_q \preccurlyeq E_{q'}) \implies p \not\curlywedge q \vee p \uparrow q \; .
$$

The full subcategory of **FPos** consisting of the simulative families of posets is denoted **FPos**[Sim]. This method of denoting subcategories is also used for other properties defined below. Whenever we state something like 'C is a functor from **FPos**[Sim] to something else' then we mean the appropriate restriction of C.

Lemma 23. C *is a full and faithful functor from* **FPos**[Sim] *to* **FCnf**.

C is actually part of an adjunction between **FPos**[Sim] and **FCnf**. To define the reverse functor we have to extract the ordering relations of the posets (in the prospective family of posets) from the $\subseteq$-structure of the family of configurations: they are not explicitly available in the family of configurations that we start with. This is rather more awkward than just throwing away the orderings, which was all we had to do in C. Our first solution is to generate *all* posets that comply with the $\subseteq$-structure, i.e. whose left-closed subsets are configurations.

Theorem 24. *Let* R *be a mapping defined over the objects and arrows of* **FCnf** *such that* $\mathrm{R}: \mathcal{C} \mapsto \{p \in \mathbf{Pos} \mid \forall q \preccurlyeq p.\, E_q \in \mathcal{C}\}$ *and* $\mathrm{R}: f \mapsto f$. *Then* R *is a functor from* **FCnf** *to* **FPos**[Sim] *such that* $\mathrm{C} \dashv \mathrm{R}$.

It follows that every family of configurations can be obtained from a simulative family of posets. The following example however shows that in general this family of posets is not uniquely determined even up to event isomorphism; hence $\mathrm{C} \dashv \mathrm{R}$ is not an equivalence.

Example 8. Let $\mathcal{P}, \mathcal{Q}$ be families of posets such that $\mathcal{P}$ has maximal elements $\boxed{a \to b}$ and $\boxed{\begin{smallmatrix} a \\ b \end{smallmatrix}}$ and $\mathcal{Q}$ has maximal elements $\boxed{b \to a}$ and $\boxed{\begin{smallmatrix} a \\ b \end{smallmatrix}}$; then $\mathcal{P}$ and $\mathcal{Q}$ are simulative (Sim) and $\mathrm{C}\mathcal{P} = \mathrm{C}\mathcal{Q}$. However, $\mathcal{P}$ and $\mathcal{Q}$ are not event isomorphic.

Nevertheless we can still strengthen the result. For a given configuration structure $\mathcal{C}$ the set $\{\mathcal{P} \in \mathbf{FPos}[\mathsf{Sim}] \mid \mathrm{C}\mathcal{P} = \mathcal{C}\}$ is a boolean lattice w.r.t. $\subseteq$ such that if $\mathrm{C}\mathcal{P} = \mathrm{C}\mathcal{Q} = \mathcal{C}$ then the posets in $\mathcal{Q} \setminus \mathcal{P}$ are *augments* (1) of posets in $\mathcal{P}$ and vice versa. It can be argued that augments do not add information to the model, and hence one can freely choose to include all or none of them. The first choice is the one incorporated in R above and yields *augment saturated* families of posets, which form a *reflective* subcategory of **FPos**[Sim]; the second choice yields *augment free* families of posets, which form a *coreflective* subcategory of **FPos**[Sim]. Of the boolean lattices mentioned above, the augment saturated families are the top elements and the augment free families are the bottom elements.

Definition 25.

1. Let AS be a mapping defined over the objects and arrows of **FPos** such that $\mathrm{AS}: \mathcal{P} \mapsto \{p \in \mathbf{Pos} \mid \exists q \in \mathcal{P}.\, q \subseteqq p\}$ and $\mathrm{AS}: f \mapsto f$. For every family of posets $\mathcal{P}$, let $\eta_{\mathcal{P}}: \mathcal{P} \to \mathrm{AS}\,\mathcal{P}$ be the event morphism defined by the identity over $E_{\mathcal{P}}$.
2. Let AF be a mapping defined over the objects and arrows of **FPos** such that $\mathrm{AF}: \mathcal{P} \mapsto \{p \in \mathcal{P} \mid \forall q \in \mathcal{P}.\, q \subseteqq p \Longrightarrow q = p\}$ and $\mathrm{AF}: f \mapsto f$. For every family of posets $\mathcal{P}$, let $\theta_{\mathcal{P}}: \mathcal{P} \to \mathrm{AF}\,\mathcal{P}$ be the morphism defined by the identity over $E_{\mathcal{P}}$.

Property ASat & AFr **(Augment saturation and freedom).** $\mathcal{P}$ is called

augment saturated if $\mathrm{AS}\,\mathcal{P} = \mathcal{P}$
augment free if $\mathrm{AF}\,\mathcal{P} = \mathcal{P}$

Theorem 26.

1. **FPos**[Sim|ASat] *is a reflective subcategory of* **FPos**[Sim] *with reflector* AS *and unit* η*;* R *is an isomorphism from* **FCnf** *to* **FPos**[Sim|ASat] *with inverse* C.
2. **FPos**[Sim|AFr] *is a coreflective subcategory of* **FPos**[Sim] *with coreflector* AF *and counit* θ*;* AF $\cdot$ R *is an isomorphism from* **FCnf** *to* **FPos**[Sim|AFr] *with inverse* C.

We prefer augment free families to augment saturated ones because they are more parsimonious and provide a better generalization of previous results concerning stable event structures in [26, Proposition 1.2.6].

Up to now we have dealt with families of configurations in their most general form. In practice however mostly subclasses of families of configurations are used. The reason is that by imposing some well-chosen restrictions on the families of configurations to be used, they become much more tractable. In fact, this tractability appears in our setting in the fact that we can once more say something about the models on the order level.

Definition 27. A family of configurations $\mathcal{C}$ is called

stable if $\forall F, G \in \mathcal{C}.\, F \cup G \in \mathcal{C} \Longrightarrow F \cap G \in \mathcal{C}$
prime if $\forall F, G \in \mathcal{C}.\, F \cap G \in \mathcal{C}$

The full subcategories of stable and prime configuration structures are denoted $\mathbf{FCnf}_{\mathrm{S}}$ and $\mathbf{FCnf}^{\sqsubseteq}_{\mathrm{S}}$, resp. $\mathbf{FCnf}_{\mathrm{P}}$ and $\mathbf{FCnf}^{\sqsubseteq}_{\mathrm{P}}$.

For instance, the family of configurations $C\mathcal{Q}$ in Ex. 7 is not stable: in particular, if $F = \{a, c\}$ and $G = \{b, c\}$ then $F \cup G \in C$ but $F \cap G = \{c\} \notin C$. Stability exactly ensures distributivity, with the result that stable configuration strauctures are prime algebraic and can be linked to $\mathbf{FPos}^{\sqsubseteq}$, i.e. on the order level. Prime configuration structures, in their equivalent guise of prime event structures, are probably the best known event-based model. They have a very restricted notion of events: essentially each event in a prime event structure has a uniquely determined causal history, with the effect that there is a one-to-one correspondence between the events and the triangular posets. As promised, we can once more say something about order structure of the respective subcategories.

Property Stb & Prm **(Stability and primeness).** A family of posets $\mathcal{P}$ is called

stable if $\forall p, q \in \mathcal{P}.\, p \triangleq q \wedge \lhd p \uparrow q \Longrightarrow p = q$

prime if $\forall p, q \in \mathcal{P}.\, p \triangleq q \Longrightarrow p = q$.

Theorem 28.

1. C *is an isomorphism from* **FPos**[Sim] *to* $\mathbf{FCnf}_\mathrm{S}$*;*
2. $\mathrm{C}^{\sqsubseteq}$ *is an isomorphism from* $\mathbf{FPos}^{\sqsubseteq}$[Sim] *to* $\mathbf{FCnf}^{\sqsubseteq}_\mathrm{S}$*;*
3. C *and* $\mathrm{C}^{\sqsubseteq}$ *coincide on* **FPos**[Prm]*, and are isomorphic to* $\mathbf{FCnf}_\mathrm{P}$.

6 Event Relations

We have seen that in families of posets, the concept of "event" is encoded by the abstract, structural notion of top equivalence. Interestingly, we can formulate all kinds of predicates over top equivalence that correspond to intuitive properties of events. The families of posets that satisfy such predicates then incorporate the corresponding intuition about events. For instance, *uniqueness* and *determinism* in Def. 12 are two such predicates. Hence we can use top equivalence as an instrument to formalize ideas about the nature of events. This principle is demonstrated below through the relations *binary conflict* and *causal flow*.

In event based models it makes sense to talk about *conflicting events*. These are pairs of events that never under any circumstance occur in the same run.[5] In a given model we can simply find those events for which this is the case; using the top equivalence relation we can define event conflict # as the *largest* relation such that

$$\forall p, q \in \mathcal{P}^{\Delta}.\, p\Delta \#_{\mathcal{P}} q\Delta \Longrightarrow p \not\uparrow_{\mathcal{P}} q \,. \tag{10}$$

Things become interesting when we try to *use event conflict to explain inconsistency*, allowing two states to be inconsistent only if they contain conflicting events.[6]

Property Cfl **(Conflict reflection).** $\mathcal{P}$ is said to *reflect conflict* if

$$\forall p, q \in \mathcal{P}^{\Delta}.\, (\lhd p \uparrow q) \wedge (p \uparrow \lhd q) \Longrightarrow (p \uparrow q) \vee (p \triangleq q) \vee (p\Delta \# q\Delta) \,.$$

[5] It is also possible to model *larger* conflicting sets; we are however exclusively concerned with *binary* conflict here.

[6] This actually conforms more to the *configuration* than to the *poset* view of behaviour, since the *dependencies* between events are not seen as valid explanations for inconsistency.

If p and q are triangular posets satisfying the left hand of the implication then their top events are the only hope of explaining why they are inconsistent. Actually we have not done exactly as we said, because we also *incompatibility* of p and q as an explanation of inconsistency ($p \triangleq q$ implies either $p \uparrow q$ or $p \not\nearrow q$). A stronger property ensues if we rule out that kind of explanation.

Property SCfl **(Strong conflict reflection).** $\mathcal{P}$ *strongly reflects conflict* if

$$\forall p, q \in \mathcal{P}^{\Delta} .\, (\lhd p \uparrow q) \wedge (p \uparrow \lhd q) \Longrightarrow (p \uparrow q) \vee (p\Delta \# q\Delta) \ .$$

So far we have only discussed *pairs* of runs; implicitly we have assumed that if every pair of a set of runs is consistent then the run is consistent. This is explicitly formulated in the following property.

Property Coh **(Coherence).** $\mathcal{P}$ is called *coherent* if

$$\forall P \subseteq_{\mathrm{fin}} \mathcal{P} .\, (\forall p, q \in P .\, p \uparrow q) \Longrightarrow P\uparrow \ .$$

The properties we have now reasoned out actually characterize certain classes of families of configurations. Unfortunately there is no space to present the details of the characterizations; cf. the full paper [18]. $\mathbf{FCnf}_{\#}$, $\mathbf{FCnf}_{\mathrm{S}\#}$ and $\mathbf{FCnf}_{\mathrm{P}\#}$ denote resp. the general, stable and prime subcategories of **FCnf** with binary conflict. The relevant event structures and their configurations are defined e.g. in [27].

Theorem 29.

1. C *is an isomorphism from* **FPos**[Sim|AFr|Cfl|Coh] *to* $\mathbf{FCnf}_{\#}$.
2. C *is an isomorphism from* **FPos**[SCfl|Coh] *to* $\mathbf{FCnf}_{\mathrm{S}\#}$.
3. C *is an isomorphism from* **FPos**[P|Coh] *to* $\mathbf{FCnf}_{\mathrm{P}\#}$.

These characterizations are nontrivial; for instance, they improve on an incorrect result in [27, Theorem 3.7], which states that coherence is enough to guarantee representability through binary conflict.

Example 9. Let $\mathcal{P}$ is the family of posets with maximal elements $p = \boxed{a \rightarrow c}$, $\boxed{\begin{matrix} a \\ b \end{matrix}}$ and $p' = \boxed{b \rightarrow c}$. $\mathcal{P}$ is coherent but does not respect conflict: if $q = \boxed{b}$ then $p \not\triangleq q$, $\lhd p \uparrow q$ and $q \uparrow \lhd q$, but $p \not\uparrow q$ and $\neg(c \# b)$. However C$\mathcal{P}$ cannot be obtained from an event structure with binary conflict.

Where binary conflict attempts to explain inconsistency of states in terms of a relation over the events, causal flow attempts to explain the partial ordering within states in the same fashion. This is based on the intuition that there is a fixed cause-and-effect relation between the events in a behaviour model. Such a relation can be given the form of a binary *flow relation* $\rightarrow$, as the largest relation satisfying

$$\forall p, q \in \mathcal{P}^{\Delta} .\, p\Delta \rightarrow q\Delta \wedge p \uparrow \lhd q \Longrightarrow p \prec q \ . \tag{11}$$

The transitive closure of the flow relation within a run should coincide with the ordering relation of the run. Again, we draw up a characterization of those families of posets which are compatible with such an intuition.

Property Flw **(Causal flow reflection).** $\mathcal{P}$ is said to *reflect causal flow* if

$$\forall p, q \in \mathcal{P}^{\Delta} .\, p \preccurlyeq q \Longrightarrow (\exists q' \in \mathcal{P}^{\Delta} .\, p \prec q' \prec q) \vee p\Delta \rightarrow q\Delta \ .$$

Causal flow is for instance reflected in *flow event structure configurations*; cf. [3, 4].

Theorem 30. AF *is an injective, full and faithful functor from* $\mathbf{FCnf}_{\mathrm{F}}$, *the subcategory of flow configurations, to* $\mathbf{FPos}$[SCfl|Flw], *with left inverse* C.

This theorem is unfortunately not a full characterization: there are families of posets which strongly reflect conflict and also reflect causal flow but for which there is no corresponding flow event structure. Although we have tried hard we have not been able to improve on it; it has proven very hard to formulate a general criterion to determine which families of posets "misbehave" in this manner.

7 Conclusions and Further Work

We have presented a dual view on event-based models through *families of posets* and *event bases*. The emphasis has been on families of posets, which are very closely related to families of configurations, but offer the following advantages:

- Families of posets are inherently prime algebraic, even those that are (event) equivalent to non-prime algebraic families of configurations;
- In families of posets we can characterize event isomorphism as a structure-preserving notion (it preserves top equivalence)·instead of the internal notion that it generally is in families of configurations;
- Families of posets are more expressive than families of configurations, and this expressivity is very useful in some model constructions;
- Using families of posets we can characterize in a rather nice way many versions of families of configurations.

The *event basis* view provides an interesting insight in the nature of events: families of posets are exactly those event bases in which events are *unique* and in a sense *deterministic.* This reflects the usual intuition in which uniqueness and determinism are inherent properties of events —but for instance not the Petri Net intuition, where events are thought of as repeating entities. The restrictions under which families of posets are reduced to other event-based models can be seen as stronger conditions on the nature of events. Hence, in a sense, different event-based models actually embody different ideas about the nature of events. The following figure represents the hierarchy of event-based models, ordered according to their expressiveness on the level of events. It includes results not presented in this paper due to space limitations, viz. about the class of *selection structures* **Sel** [16] and the class of *geometric automata* **GA**) due to Gunawardena [9]. Like general families of configurations, geometric automata do not form prime algebraic domains, but all the same they are (event) isomorphic to a subcategory of **FPos**.

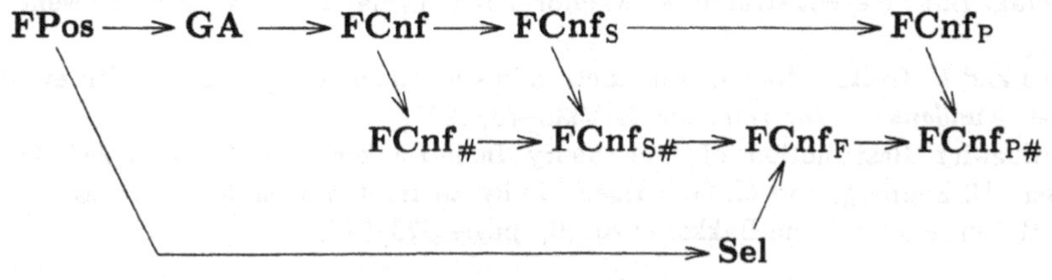

In terms of event bases, we may raise a question that does not let itself be formulated in families of posets: what happens if we *weaken* the properties of uniqueness or determinism introduced in Def. 12? For one thing, the resulting model would no longer be expressible as a family of posets. Still, such models would be interesting if only by the virtue of fitting in the framework of event bases. We conjecture that it is possible to obtain for instance *Mazurkiewicz traces* [12] by requiring a stronger form of determinism and an appropriate form of *preservation of independence* —where independence would be a property over the "events" $E_{\mathcal{X}}$ of an event base $\mathcal{X}$ similar to conflict and causal flow— but *no uniqueness*. It would be especially interesting if in this way one could also find a link to *transition system-like* models, i.e. in which the state graph may contain cycles. Some existing work suggests that this might be the case; cf. [14, 15, 19, 28]. This is a subject for further study.

Acknowledgements. This work was developed at the Tele-Informtics and Open Systems Group at the University of Twente, under the supervision of Ed Brinksma. The ideas presented have been sharpened in many discussions with Rom Langerak. Thanks are also due to Ed Brinksma and Ulla Goltz for encouragement and helpful suggestions.

References

1. G. Boudol. Flow event structures and flow nets. In I. Guessarian, editor, *Semantics of Systems of Concurrent Processes*, volume 469 of *Lecture Notes in Computer Science*, pages 62–95. Springer-Verlag, 1990.
2. G. Boudol and I. Castellani. Concurrency and atomicity. *Theoretical Comput. Sci.*, 59:25–84, 1988.
3. G. Boudol and I. Castellani. Permutations of transitions: An event structure semantics for CCS and SCCS. In de Bakker et al. [6], pages 411–427.
4. G. Boudol and I. Castellani. Flow models of distributed computations: Event structures and nets. Rapports de Recherche 1482, INRIA, July 1991.
5. B. Davey and H. A. Priestley. *Introduction to Lattices and Order*. Cambridge Mathematical Textbooks. Cambridge University Press, 1990.
6. J. W. de Bakker, W.-P. de Roever, and G. Rozenberg, editors. *Linear Time, Branching Time and Partial Order in Logics and Models for Concurrency*, volume 354 of *Lecture Notes in Computer Science*. Springer-Verlag, 1989.
7. P. Degano, R. De Nicola, and U. Montanari. Partial orderings descriptions and observations of nondeterministic concurrent processes. In de Bakker et al. [6], pages 438–466.
8. M. Droste. Event structures and domains. *Theoretical Comput. Sci.*, 68:37–47, 1989.
9. J. Gunawardena. Geometric logic, causality and event structures. In J. C. M. Baeten and J. F. Groote, editors, *Concur '91*, volume 527 of *Lecture Notes in Computer Science*, pages 266–280. Springer-Verlag, 1991.
10. R. Langerak. Bundle event structures. Memoranda informatica, University of Twente, 1992.
11. R. Loogen and U. Goltz. Modelling nondeterministic concurrent processes with event structures. *Fundamenta Informaticae*, XIV:39–73, 1991.
12. A. Mazurkiewicz. Basic notions of trace theory. In de Bakker et al. [6], pages 285–363.
13. M. Nielsen, U. Engberg, and K. G. Larsen. Fully abstract models for a process language with refinement. In de Bakker et al. [6], pages 523–549.

14. M. Nielsen, G. Rozenberg, and P. S. Thiagarajan. Elementary transition systems. In [20], 1991.
15. M. Nielsen, G. Rozenberg, and P. S. Thiagarajan. Transition systems, event structures and unfoldings. In [20], 1991.
16. A. Rensink. Selection structures. Memoranda Informatica 90–71, University of Twente, Dec. 1990.
17. A. Rensink. Poset semantics, bisimulation. Memoranda informatica, University of Twente, 1992.
18. A. Rensink. Posets for configurations! Memoranda informatica, University of Twente, 1992.
19. E. W. Stark. Connections between a concrete and an abstract model of concurrent systems. In M. Main, A. Melton, M. Mislove, and D. Schmidt, editors, *Mathematical Foundations of Programming Semantics*, volume 442 of *Lecture Notes in Computer Science*, pages 53–79. Springer-Verlag, 1990.
20. P. S. Thiagarajan. Some models and logics for concurrency. Course Notes of the Advanced School on Concurrency, Gargnano del Garda, Italy, Sept. 1991.
21. F. W. Vaandrager. A simple definition for parallel composition of prime event structures. Report ACMCS-R8903, Centre for Mathematics and Computer Science, 1989.
22. R. van Glabbeek. *Comparative Concurrency Semantics and Refinement of Actions.* PhD thesis, Free University of Amsterdam, 1990.
23. R. van Glabbeek and U. Goltz. Refinement of actions in causality based models. In J. W. de Bakker, W. P. de Roever, and G. Rozenberg, editors, *Stepwise Refinement of Distributed Systems — Models, Formalisms, Correctness*, volume 430 of *Lecture Notes in Computer Science*, pages 267–300. Springer-Verlag, 1990.
24. G. Winskel. Event structure semantics for CCS and related languages. In M. Nielsen and E. M. Schmidt, editors, *Automata, Languages and Programming*, volume 140 of *Lecture Notes in Computer Science*, pages 561–576. Springer-Verlag, 1982.
25. G. Winskel. Categories of models for concurrency. In S. D. Brookes, A. W. Roscoe, and G. Winskel, editors, *Seminar on Concurrency*, volume 197 of *Lecture Notes in Computer Science*, pages 246–267. Springer-Verlag, 1985.
26. G. Winskel. Event structures. In W. Brauer, W. Reisig, and G. Rozenberg, editors, *Petri Nets: Applications and Relationships to Other Models of Concurrency*, volume 255 of *Lecture Notes in Computer Science*, pages 325–392. Springer-Verlag, 1987.
27. G. Winskel. An introduction to event structures. In de Bakker et al. [6], pages 364–397.
28. G. Winskel. Categories of models for concurrency. Course Notes of the Advanced School on Concurrency, Gargnano del Garda, Italy, Sept. 1991.

This article was processed using the LaTeX macro package with LLNCS style

On the Semantics of Petri Nets

José Meseguer† *Ugo Montanari*‡

Vladimiro Sassone‡

† SRI International, Menlo Park, California 94025

‡ Dipartimento di Informatica – Università di Pisa, Italy

Abstract

Petri Place/Transition (PT) nets are one of the most widely used models of concurrency. However, they still lack, in our view, a satisfactory semantics: on the one hand the "token game" is too intensional, even in its more abstract interpretations in term of nonsequential processes and monoidal categories; on the other hand, Winskel's basic unfolding construction, which provides a coreflection between nets and finitary prime algebraic domains, works only for safe nets.

In this paper we extend Winskel's result to PT nets. We start with a rather general category PTNets *of PT nets, we introduce a category* DecOcc *of decorated (nondeterministic) occurrence nets and we define adjunctions between* PTNets *and* DecOcc *and between* DecOcc *and* Occ, *the category of occurrence nets. The role of* DecOcc *is to provide natural unfoldings for PT nets, i.e. acyclic safe nets where a notion of family is used for relating multiple instances of the same place.*

The unfolding functor from PTNets *to* Occ *reduces to Winskel's when restricted to safe nets, while the standard coreflection between* Occ *and* Dom, *the category of finitary prime algebraic domains, when composed with the unfolding functor above, determines a chain of adjunctions between* PTNets *and* Dom.

Introduction

Petri nets, introduced by C.A. Petri in [Pet62] (see also [Pet73,Rei85]), are a widely used model of concurrency. This model is attractive from a theoretical point of view because of its simplicity and because of its intrinsically concurrent nature, and has often been used as a semantic basis on which to interpret concurrent languages (see for example [Win82,Old87,vGV87,DDM88]).

For *Place/Transition (PT) nets*, having a satisfactory semantics—one that does justice to their truly concurrent nature, yet is abstract enough—remains in our view an unresolved problem. Certainly, many different semantics have been proposed in the literature; we briefly discuss some of them below.

At the most basic operational level we have of course the "*token game*". To account for computations involving many different transitions and for the *causal connections* between transition events, various notions of *process* have been proposed [Pet77,GR83,BD87], but process models do not provide a satisfactory semantic denotation for a net as a whole. In fact, they specify only the meaning of single, deterministic computations, while the accurate description of the fine interplay between concurrency and nondeterminism is one of the most valuable features of nets.

Research partially supported by ESPRIT Basic Research Action CEDISYS. The first and the third author have been supported by the US Office of Naval Research Contract N00014-88-C-0618.

Other semantic investigations have capitalized on the *algebraic structure* of PT nets, first noticed by Reisig [Rei85] and later exploited by Winskel to identify a sensible notion of *morphism* between nets [Win84,Win87]. More recently, a different interpretation of the algebraic structure of PT nets in terms of monoidal categories has been proposed in a paper by two of the authors [MM90].

One particular advantage of the algebraic approaches based on category theory is that they provide useful net combinators, associated to standard categorical constructions such as product and coproduct, which can be used to give a simple account of corresponding compositional operations at the level of a concurrent programming language, such as various forms of *parallel* and *non-deterministic* composition [Win87,MM90].

A unification of the process-oriented and algebraic views has recently been proposed by two of the present authors in joint work with P. Degano [DMM89], by showing that the *commutative processes* [BD87] of a net N are isomorphic to the arrows of a symmetric monoidal category $T[N]$. Moreover, they introduced the *concatenable processes* of N—a slight variation of Goltz-Reisig processes [GR83] on which sequential composition is defined—and structured them as the arrows of the symmetric monoidal category $\mathcal{P}[N]$. That would individuate in the category of the symmetric monoidal categories a semantic domain for PT nets. However, in spite of accounting for algebraic and process aspects in a simple unified way, this semantics is still too concrete, and a more abstract semantics—one allowing greater semantic identifications between nets—would be clearly preferable.

A very attractive formulation for the semantics that we seek would be an *adjoint functor* assigning an abstract denotation to each PT net and preserving certain compositional properties in the assignment. This is exactly what Winskel has done for the subcategory of safe nets [Win86]. In that work—which builds on the previous work [NPW81]—the denotation of a safe net is a *Scott domain* [Sco70], and Winskel shows that there exists a coreflection—a particularly nice form of adjunction—between the category <u>Dom</u> of (coherent) *finitary prime algebraic domains* and the category <u>Safe</u> of *safe Petri nets*. This construction is completely satisfactory: from the intuitive point of view it gives the "truly concurrent" semantics of safe nets in the most universally accepted type of model, while from the formal point of view the existence of an adjunction guarantees its "naturality". Winskel's coreflection factorizes through the chain of coreflections

$$\underline{\mathbf{Safe}} \;\underset{\hookleftarrow}{\xrightarrow{\;\mathcal{U}_w[\_]\;}}\; \underline{\mathbf{Occ}} \;\underset{\xleftarrow[\;\mathcal{N}[\_]\;]{}}{\xrightarrow{\;\mathcal{E}[\_]\;}}\; \underline{\mathbf{PES}} \;\underset{\xleftarrow[\;\mathcal{P}r[\_]\;]{}}{\xrightarrow{\;\mathcal{L}[\_]\;}}\; \underline{\mathbf{Dom}}$$

where <u>**PES**</u> is the category of *prime event structures* (with binary conflict relation), which is equivalent to <u>**Dom**</u>, <u>**Occ**</u> is the category of *occurrence nets* [Win86], and $\hookleftarrow$ is the inclusion functor.

Recently, various attempts have been made to extend this chain or, more generally, to identify a suitable semantic domain for PT nets. Among them, we recall [Pra91], where, in order to obtain a model "mathematically more attractive than Petri nets", a *geometric* model of concurrency based on n-categories as models of higher dimensional automata is introduced, but the modelling power obtained is not greater than that of ordinary PT nets; [HKT92], in which the authors give semantics to PT nets in terms of generalized *trace languages* and discuss how using their work it could perhaps be possible to obtain a concept of unfolding for PT nets; and [Eng91], where the unfolding of Petri nets is given in term of a *branching process*. However, the nets considered in [Eng91] are not really PT nets because their transitions are restricted to have pre- and post-sets where all places have no multiplicities.

The present work extends Winskel's approach from safe nets to the category of PT nets. We define the *unfoldings* of PT nets and relate them by an *adjunction* to occurrence nets and therefore—exploiting the already existing adjunctions—to prime event structures and finitary prime algebraic domains. The adjunctions so obtained are extensions of the correspondent Winskel's coreflections.

The category **PTNets** that we consider is quite general. Objects are PT nets in which markings may be infinite and transitions are allowed to have infinite pre- and post-sets, but, as usual, with finite multiplicities. The only technical restriction we impose, with respect to the natural extension to nets with infinite markings of the general formulation in [MM90], is the usual condition that transitions must have non-empty pre-sets. Actually, the objects of **PTNets** strictly include those of the categories considered in [Win86,Win87]. Although a technical restriction applies to the morphisms—they are required to map places belonging to the initial marking or to the post-set of the same transition to disjoint multisets—they are still quite general. In particular, the category **PTNets** has *initial* and *terminal* objects, and has *products* and *coproducts* which model, respectively, the operations of parallel and non-deterministic composition of nets as in [Win87] and [MM90]. It is worth remarking that, while coproducts do *not* exist in the categories of generally marked, non-safe PT nets considered in the above cited works, they do in **PTNets**. However, due to the lack of space, such a result cannot be given here. It will be presented in a forthcoming full version of this work.

Concerning the organization of the paper, in Section 1 we introduce a new kind of nets, the *decorated occurrence nets*, which naturally represent the unfoldings of PT nets and can account for the multiplicities of places in transitions. They are occurrence nets in which places belonging to the post-set of the same transition are partitioned into *families*. Families are used to relate places corresponding in the unfolding to multiple instances of the same place in the original net. When all the families of a decorated occurrence net have cardinality one, we have (a net isomorphic to) an ordinary occurrence net. Therefore, **Occ** is (isomorphic to) a full subcategory of **DecOcc**, the category of decorated occurrence nets.

Then, we show an adjunction $\langle (\_)^+, \mathcal{U}[\_] \rangle : \mathbf{DecOcc} \rightharpoonup \mathbf{PTNets}$ whose right adjoint $\mathcal{U}[\_]$ gives the unfoldings of PT nets. This adjunction restricts to Winskel's coreflection from **Occ** to **Safe** as illustrated by the following commutative diagrams:

$$
\begin{array}{ccc}
\underline{\mathbf{PTNets}} & \xrightarrow{\mathcal{U}[\_]} & \underline{\mathbf{DecOcc}} \\
\uparrow & & \uparrow \\
\underline{\mathbf{Safe}} & \xrightarrow{\mathcal{U}_\omega[\_]} & \underline{\mathbf{Occ}}
\end{array}
\qquad\qquad
\begin{array}{ccc}
\underline{\mathbf{PTNets}} & \xleftarrow{(\_)^+} & \underline{\mathbf{DecOcc}} \\
\uparrow & & \uparrow \\
\underline{\mathbf{Safe}} & \hookleftarrow & \underline{\mathbf{Occ}}
\end{array}
$$

i.e. the left and the right adjoint, when restricted respectively to **Safe** and **Occ**, coincide with the correspondent adjoints of Winskel's coreflection.

In Section 2, we relate decorated occurrence nets to occurrence nets by showing an adjunction $\langle \mathcal{D}[\_], \mathcal{F}[\_] \rangle : \mathbf{Occ} \rightharpoonup \mathbf{DecOcc}$, where $\mathcal{F}[\_]$ is the *forgetful* functor which forgets about families. Moreover, the diagram

$$
\begin{array}{ccc}
\underline{\mathbf{PTNets}} & \xrightarrow{\mathcal{U}[\_]} & \underline{\mathbf{DecOcc}} \\
\uparrow & & \downarrow{\scriptstyle \mathcal{F}[\_]} \\
\underline{\mathbf{Safe}} & \xrightarrow{\mathcal{U}_\omega[\_]} & \underline{\mathbf{Occ}}
\end{array}
\tag{1}
$$

commutes.

Therefore, we get the desired adjunction between **Dom** and **PTNets** as the composition of the chain of adjunctions

$$\begin{array}{ccccccc} \underline{\mathbf{PTNets}} & \underset{(\_)^+}{\overset{\mathcal{U}[\_]}{\rightleftarrows}} & \underline{\mathbf{DecOcc}} & & & & \\ & & \mathcal{D}[\_]\uparrow\downarrow\mathcal{F}[\_] & & & & \\ & & \underline{\mathbf{Occ}} & \underset{\mathcal{N}[\_]}{\overset{\mathcal{E}[\_]}{\rightleftarrows}} & \underline{\mathbf{PES}} & \underset{\mathcal{P}r[\_]}{\overset{\mathcal{L}[\_]}{\rightleftarrows}} & \underline{\mathbf{Dom}} \end{array}$$

It follows from the commutative diagram (1) that, when **PTNets** is restricted to **Safe**, all the right adjoints in the above chain coincide with the corresponding functors defined by Winskel. In this sense, this work generalizes the work of Winskel and gives an abstract, truly concurrent semantics for PT nets. Moreover, the existence of left adjoints guarantees the "naturality" of this generalization.

1 PT Net Unfoldings

In this section we define the categories **PTNets** of *Place/Transition (PT) nets* and **DecOcc** of *decorated occurrence nets*. We define the *unfolding* of a PT net as a decorated occurrence net and show that it is a functor from **PTNets** to **DecOcc** which has a left adjoint.

A *pointed set* is a pair (S, s) where S is a set and $s \in S$ is a chosen element of S: the pointed element. Morphisms of pointed sets are functions that preserve the pointed elements. Therefore, pointed set morphisms provide a convenient way to treat partial functions between sets as total functions.

Given a set S, we denote by $S^{\mathcal{M}}$ the set of *multisets* of S, i.e. the set of all functions from S to the set of natural numbers ω, and by $S^{\mathcal{M}_\infty}$ the set of *multisets* with (possibly) *infinite multiplicities*, i.e. the functions from S to $\omega_\infty = \omega \cup \{\infty\}$. For $\mu \in S^{\mathcal{M}_\infty}$, we write $[\![\mu]\!]$ to denote the subset of S consisting of those elements s such that $\mu(s) > 0$.

A multiset $\mu \in S^{\mathcal{M}_\infty}$ can be represented as a formal sum $\bigoplus_{s \in S} \mu(s) \cdot s$. Given an arbitrary index set I and $\{\eta_i \in \omega_\infty \mid i \in I\}$, we define $\Sigma_{i \in I}\eta_i$ to be the usual sum in ω if only finitely many η_i are non-zero and ∞ otherwise. Then, we can give meaning to linear combinations of multisets, i.e. multisets of multisets, by defining

$$\bigoplus_{\mu \in S^{\mathcal{M}_\infty}} \eta_\mu \cdot \mu = \bigoplus_{\mu \in S^{\mathcal{M}_\infty}} \eta_\mu \cdot \left(\bigoplus_{s \in S} \mu(s) \cdot s\right) = \bigoplus_{s \in S}\left(\Sigma_{\mu \in S^{\mathcal{M}_\infty}} \eta_\mu \mu(s)\right) \cdot s.$$

A $(\_)^{\mathcal{M}_\infty}$-homomorphism from $S_0^{\mathcal{M}_\infty}$ to $S_1^{\mathcal{M}_\infty}$ is a function $g : S_0^{\mathcal{M}_\infty} \to S_1^{\mathcal{M}_\infty}$ such that

$$g(\mu) = \bigoplus_{s \in S_0} \mu(s) \cdot g(1 \cdot s),$$

where $1 \cdot s$ is the formal sum corresponding to the function which yields 1 on s and zero otherwise. Actually, it is worth noticing, that $(\_)^{\mathcal{M}_\infty}$ can be seen as an endofunctor on **Set**, the category of sets. As such, it defines a *commutative monad* [MM90] which sends S to $S^{\mathcal{M}_\infty}$, whose multiplication is the operation of linear combination of multisets and whose unit maps $s \in S$ to $1 \cdot s$. In these terms, $S^{\mathcal{M}_\infty}$ is a $(\_)^{\mathcal{M}_\infty}$-algebra and a $(\_)^{\mathcal{M}_\infty}$-homomorphism is a homomorphism between $(\_)^{\mathcal{M}_\infty}$-algebras.

We will regard $S^{\mathcal{M}}$ also as a pointed set whose pointed element is the empty multiset, i.e. the function which always yields zero, that, in the following, we denote by 0. In the paper, we will often denote a multiset $\mu \in S^{\mathcal{M}_\infty}$ by $\bigoplus_{i\in I} \eta_i s_i$ where $\{s_i \mid i \in I\} = [\![\mu]\!]$ and $\eta_i = \mu(s_i)$, i.e. as a sum whose summands are all nonzero. In case of multisets in $S^{\mathcal{M}}$, instead of η_i, we will use $n_i, m_i, \ldots$, the standard variables for natural numbers. Moreover, given $S' \subseteq S$, we will write $\bigoplus S'$ for $\bigoplus_{s\in S'} 1 \cdot s$.

Definition 1.1 *(PT Nets)*
A PT net *is a structure* $N = \left(\partial^0_N, \partial^1_N : (T_N, 0) \to S^{\mathcal{M}}_N, u^I_N\right)$ *where* S_N *is a set of* places; T_N *is a pointed set of* transitions; $\partial^0_N, \partial^1_N$ *are pointed set morphisms; and* $u^I_N \in S^{\mathcal{M}}_N$ *is the* initial marking. *Moreover, we assume the standard constraint that* $\partial^0_N(t) = 0$ *if and only if* $t = 0$.
A morphism *of PT nets from* N_0 *to* N_1 *consists of a pair of functions* $\langle f, g\rangle$ *such that:*

i. $f : T_{N_0} \to T_{N_1}$ *is a pointed set morphism;*

ii. $g : S^{\mathcal{M}_\infty}_{N_0} \to S^{\mathcal{M}_\infty}_{N_1}$ *is a* $(\_)^{\mathcal{M}_\infty}$*-homomorphism;*

iii. $\partial^0_{N_1} \circ f = g \circ \partial^0_{N_0}$ *and* $\partial^1_{N_1} \circ f = g \circ \partial^1_{N_0}$, *i.e.* $\langle f, g\rangle$ *respects source and target;*

iv. $g(u^I_{N_0}) = u^I_{N_1}$, *i.e.* $\langle f, g\rangle$ *respects the initial marking;*

v. $\forall b \in [\![u^I_{N_1}]\!]$, $\exists! a \in [\![u^I_{N_0}]\!]$ *such that* $b \in [\![g(a)]\!]$
$\forall b \in [\![\partial^1_{N_1}(f(t))]\!]$, $\exists! a \in [\![\partial^1_{N_0}(t)]\!]$ *such that* $b \in [\![g(a)]\!]$.

This, with the obvious componentwise composition of morphisms, gives the category **PTNets**.□

A PT net is thus a graph whose arcs are the transitions and whose nodes are the multisets on the set of places, i.e. *markings* of the net. As usual, transitions are restricted to have pre- and post-sets, i.e. sources and targets, in which each place has only finitely many tokens, i.e. finite multiplicity. The same is required for the initial marking. To be consistent with the use of zero transitions as a way to treat partial mappings, they are required to have empty pre- and post-sets. Moreover, they are the only transitions which can have empty pre-sets. To simplify notation, we assume the standard constraint that $T_N \cap S_N = \emptyset$—which of course can always be achieved by an appropriate renaming.

Morphisms of PT nets are graph morphisms in the precise sense of respecting source and target of transitions, i.e. they make the diagram

$$
\begin{array}{ccccc}
T_{N_0} & \overset{\partial^0_{N_0}}{\underset{\partial^1_{N_0}}{\rightrightarrows}} & S^{\mathcal{M}}_{N_0} & \hookrightarrow & S^{\mathcal{M}_\infty}_{N_0} \\
f \big\downarrow & & & & \big\downarrow g \\
T_{N_1} & \overset{\partial^0_{N_1}}{\underset{\partial^1_{N_1}}{\rightrightarrows}} & S^{\mathcal{M}}_{N_1} & \hookrightarrow & S^{\mathcal{M}_\infty}_{N_1}
\end{array}
$$

commute. Moreover they map initial markings to initial markings. To simplify notation, we will sometimes use a single letter to denote a morphism $\langle f, g\rangle$. In these cases, the type of the argument will identify which component we are referring to.

A $(\_)^{\mathcal{M}_\infty}$-homomophism $g : S^{\mathcal{M}_\infty}_{N_0} \to S^{\mathcal{M}_\infty}_{N_1}$, which constitutes the place component of a morphism $\langle f, g\rangle$, is completely defined by its behaviour on S_{N_0}, the generators of $S^{\mathcal{M}_\infty}_{N_0}$. Therefore, we will often define morphisms between nets giving their transition components and a map

$g : S_{N_0} \to S_{N_1}^{\mathcal{M}_\infty}$ for their place components: it is implicit that the latter have to be thought of as lifted to the correspondent $(\_)^{\mathcal{M}_\infty}$-homomorphisms.

The last condition in the definition means that morphisms are not allowed to map two different places in the initial marking or in the post-set of some transition to two multisets having a place in common. This is pictorially described in the figure below, where dashed arrows represent the forbidden morphisms. We use the standard graphical representation of nets in which circles are places, boxes are transitions, the initial marking is given by the number of "tokens" in the places, and sources and targets are directed arcs whose weights represent multiplicities. Unitary weights are omitted.

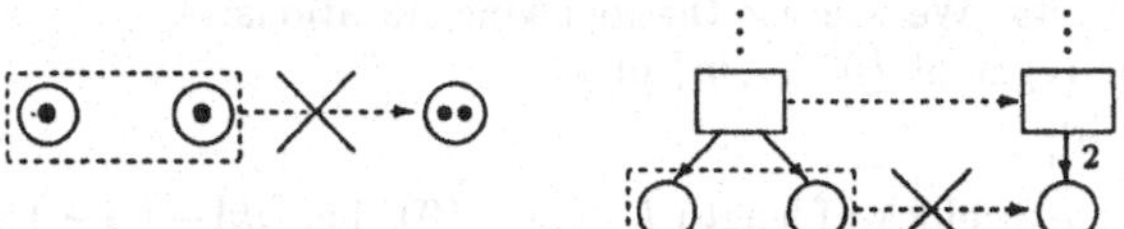

Such a condition will play an important role while establishing the adjunction between PTNets and DecOcc. In fact, it is crucial for showing the *universality* of the *counit* of the adjunction.

Now, we recall the definition of a well-known class of nets: *safe nets*.

Definition 1.2 *(Safe Nets)*
A PT Net N is safe *if and only if*

$$\forall t \in T_N,\ \bigoplus[\![\partial^i_N(t)]\!] = \partial^i_N(t),\ \textit{for } i = 0,1 \textit{ and } \forall v \in \mathcal{R}[N],\ \bigoplus[\![v]\!] = v,$$

where $\mathcal{R}[N]$ is the set of reachable markings of N (see, for instance, [Rei85]).
This defines the category Safe *as a full subcategory of* PTNets. □

Observe that $\bigoplus[\![v]\!] = v$ is a compact way of saying that each $s \in S$ has multiplicity at most one in v. Therefore this definition is exactly the classical definition of safe nets.

Another important class of nets is that of *occurrence nets*. In the following, we will use ${}^\bullet a$ to mean the *pre-set* of a, that is ${}^\bullet a = \{t \in T_N \mid a \in [\![\partial^1_N(t)]\!]\}$. Symmetrically, the *post-set* of a is indicated as $a^\bullet = \{t \in T_N \mid a \in [\![\partial^0_N(t)]\!]\}$. These notations are extended in the obvious way to the case of sets of places.

Definition 1.3 *(Occurrence Nets)*
A (non-deterministic) occurrence net *is a safe net Θ such that*

i. $a \in [\![u^I_\Theta]\!]$ *if and only if* ${}^\bullet a = \emptyset$;

ii. $\forall a \in S_\Theta,\ |{}^\bullet a| \leq 1$, *where* $|\_|$ *gives the cardinality of sets;*

iii. $\prec$ *is irreflexive, where* $\prec$ *is the transitive closure of the relation*

$$\prec^1 = \{(a,t) \mid a \in S_\Theta, t \in T_\Theta, t \in a^\bullet\} \cup \{(t,a) \mid a \in S_\Theta, t \in T_\Theta, t \in {}^\bullet a\};$$

moreover, $\forall t \in T_\Theta,\ \{t' \in T_\Theta \mid t' \prec t\}$ *is finite;*

iv. the binary "conflict" relation $\#$ *on* $T_\Theta \cup S_\Theta$ *is irreflexive, where*

$$\forall t_1, t_2 \in T_\Theta,\ t_1 \#_m t_2 \Leftrightarrow [\![\partial^0_\Theta(t_1)]\!] \cap [\![\partial^0_\Theta(t_2)]\!] \neq \emptyset \textit{ and } t_1 \neq t_2,$$

$$\forall x, y \in T_\Theta \cup S_\Theta,\ x \# y \Leftrightarrow \exists t_1, t_2 \in T_\Theta :\ t_1 \#_m t_2 \textit{ and } t_1 \preceq x \textit{ and } t_2 \preceq y,$$

where $\preceq$ *is the reflexive closure of* $\prec$.

This defines the category Occ *as a full subcategory of* Safe □

It is easy to see that Winskel's categories of safe nets, called **Net**, and of occurrence nets, here called $\underline{\mathbf{Occ}}_{\mathbf{W}}$, are *full* subcategories of **Safe** and **Occ**. In fact, the objects of **Net** ($\underline{\mathbf{Occ}}_{\mathbf{W}}$) are the objects of **Safe** (**Occ**) with sets of places, initial markings and post-sets which are non-empty, and without *isolated places*—places belonging neither to the initial marking nor to the pre- or post-set of any transition—while the morphisms between any pair of nets in **Net** ($\underline{\mathbf{Occ}}_{\mathbf{W}}$) coincide with the morphisms between the same pair of nets in **Safe** (**Occ**). However, since all the results in [Win86] easily extend to **Safe** and **Occ**, in the following we will ignore any difference between **Safe** and **Net** and between **Occ** and $\underline{\mathbf{Occ}}_{\mathbf{W}}$.

We now introduce the category of *decorated occurrence nets*, a type of occurrence nets in which places are grouped into families. They allow a convenient treatment of multiplicity issues in the unfolding of PT nets. We will use the following notations:

- $[n,m]$ for the segment $\{n,\ldots,m\}$ of ω;
- $[n]$ for $[1,n]$;
- $[k]_i$ for the i-th block of length k of $\omega - \{0\}$, i.e. $[ik] - [(i-1)k]$.

Definition 1.4 *(Block Functions)*
We call a function $f : [n] \to [m]$ *a* block function *if and only if* $n = km$ *and* $f([k]_i) = \{i\}$, *for* $i = 1,\ldots,m$. □

The place component g of a PT net morphism $\langle f,g\rangle : N_0 \to N_1$ can be thought of as a *multirelation* (with possibly infinite multiplicities) between S_{N_0} and S_{N_1}, namely the multirelation g such that $a\, g\, \eta b$ if and only if $g(a)(b) = \eta$. Indeed, this is a (generalization of a) widely used formalization of net morphisms due to Winskel [Win84,Win87]. In the case of morphisms between occurrence nets, we have that g is a *relation* and that the inverse relation g^{op}, defined by $bg^{op}a$ if and only if agb, restricts to (total) functions $g^{op}_{\emptyset} : [u^I_{N_1}] \to [u^I_{N_0}]$ and $g^{op}_{\{t\}} : [\partial^1_{N_1}(f(t))] \to [\partial^1_{N_0}(t)]$ for each $t \in T_{N_0}$. We will use these functions in the next definition.

Definition 1.5 *(Decorated Occurrence Nets)*
A decorated occurrence net *is an occurrence net* Θ *such that:*

i. S_Θ *is of the form* $\bigcup_{a\in A_\Theta}\{a\} \times [n_a]$, *where the set* $\{a\} \times [n_a]$ *is called the family of* a. *We will use* a^F *to denote the family of* a *regarded as a multiset;*

ii. $\forall a \in A_\Theta$, $\forall x,y \in \{a\} \times [n_a]$, ${}^\bullet x = {}^\bullet y$.

A morphism *of decorated occurrence nets* $\langle f,g\rangle : \Theta_0 \to \Theta_1$ *is a morphism of occurrence nets which respects families, i.e. for each* $[a^F] \subseteq S_{\Theta_0}$, *given* $x = {}^\bullet[a^F]$*—which is a singleton set or the empty set by ii above and the definition of occurrence nets—we have:*

i. $g(a^F) = \bigoplus_{i\in I_a} b_i^F$, *for some index set* I_a;

ii. $\pi_a \circ g_i^{op} \circ in_{b_i}$ *is a block function, where*
π_a *is the projection of* $\{a\} \times [n_a]$ *to* $[n_a]$,
in_a *is the bijection from* $[n_a]$ *to* $\{a\} \times [n_a]$, *and*
$g_i^{op} : \{b_i\} \times [n_{b_i}] \to \{a\} \times [n_a]$ *is* g_x^{op} *restricted to* $\{b_i\} \times [n_{b_i}]$.

This defines the category **DecOcc**. □

A family is thus a collection of finitely many places with the same pre-set, and a decorated occurrence net is an occurrence net where each place belongs to exactly one family. Families, and therefore decorated occurrence nets, are capable of describing *relationships* between places by grouping them together. We will use families to relate places which are *instances* of the same place obtained in a process of unfolding. Therefore, morphisms treat families in a special way: they map families to families (condition *i*) and they do that in a unique pre-determined way (condition *ii*). This is because what we want to describe is that a^F is mapped to b^F. Hence, since the way to map a family to another family is fixed by definition, in the following we will often define morphisms by just saying what families are sent to what families.

Observe that the full subcategory of <u>**DecOcc**</u> consisting of all nets Θ such that $S_\Theta = \bigcup_{a\in A_\Theta}\{a\}\times[1]$ is (isomorphic to) <u>**Occ**</u>. Observe also that, since the initial marking consists exactly of the elements with empty pre-set and, by point *ii* in Definition 1.5, elements of a family have the same pre-set, for a decorated occurrence net u_Θ^I is of the form $\bigoplus_{i\in I} a_i^F$.

We have seen that for occurrence nets and decorated occurrence nets simple concepts of causal dependence ($\prec$) and conflict ($\#$) can be defined. The orthogonal concept is that of concurrency.

Definition 1.6 *(Concurrent Elements)*
Given a (decorated) occurrence net Θ (which defines $\prec$, $\preceq$ and $\#$), we can define

- *For $x, y \in T_\Theta \cup S_\Theta$, x co y iff $\neg(x \prec y$ or $y \prec x$ or $x\#y)$.*
- *For $X \subseteq T_\Theta \cup S_\Theta$, $\mathrm{Co}(X)$ iff $(\forall x, y \in X,\ x$ co $y)$ and $|\{t\in T_\Theta \mid \exists x \in X,\ t \preceq x\}| \in \omega$.* □

As a first step in relating the categories <u>**DecOcc**</u> and <u>**PTNets**</u>, we define a functor from decorated occurrence nets to PT nets.

Proposition 1.7 *($(\_)^+$: from <u>**DecOcc**</u> to <u>**PTNets**</u>)*
Given the decorated occurrence net $\Theta = \left(\partial_\Theta^0, \partial_\Theta^1 : (T_\Theta, 0) \to (\bigcup_{a\in A_\Theta}\{a\}\times[n_a])^{\mathcal{M}}, u_\Theta^I\right)$ let $(\_)^+$ denote the $(\_)^{\mathcal{M}_\infty}$-homomorphism from $S_\Theta^{\mathcal{M}_\infty}$ to $A_\Theta^{\mathcal{M}_\infty}$ such that $(a,j)^+ = a$.
Then, we define Θ^+ to be the net $\left((\_)^+\circ\partial_\Theta^0,\ (\_)^+\circ\partial_\Theta^1 : (T_\Theta, 0) \to A_\Theta^{\mathcal{M}}, (u_\Theta^I)^+\right)$.
Given a morphism $\langle f, g\rangle : \Theta_0 \to \Theta_1$, let $\langle f, g\rangle^+ : \Theta_0^+ \to \Theta_1^+$ be $\langle f, (\_)^+\circ g \circ in\rangle$ where $in : A_{\Theta_0}^{\mathcal{M}_\infty} \to S_{\Theta_0}^{\mathcal{M}_\infty}$ is the $(\_)^{\mathcal{M}_\infty}$-homomorphism such that $in(a) = (a,1)$.
*Then, $(\_)^+ :$ <u>**DecOcc**</u> $\to$ <u>**Occ**</u> is a functor.* □

The following example shows the result of applying $(\_)^+$ to a decorated occurrence net. In all the pictures to follow, a family is represented by drawing its elements from left to right in accordance with its ordering, and enclosing them into an oval. Families of cardinality one are not explicitly indicated.

Example 1.8

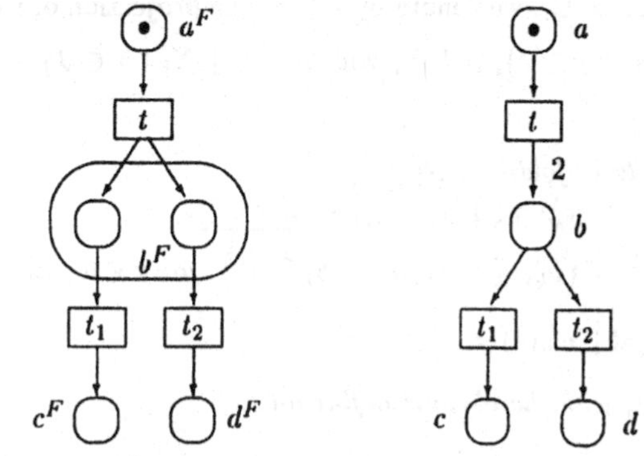

A decorated occurrence net Θ and the net Θ^+

Nets obtained via $(\_)^+$ from decorated occurrence nets have a structure very similar to that of occurrence nets. In particular they have places whose pre-sets contain at most one transition and the places in the initial marking are exactly those with empty pre-set. Moreover, the causal dependence relation $\prec$, defined as in the case of occurrence nets, is irreflexive. Observe that, if Θ is (isomorphic to) an occurrence net, then Θ^+ is an occurrence net isomorphic to Θ. We will denote by **DecOcc⁺** the full subcategory of **PTNets** consisting of (nets isomorphic to) nets of the form Θ^+.

Let $\mathcal{B}$ range over **Occ**, **DecOcc** and **DecOcc⁺**. For any net in $\mathcal{B}$, we can define the concept of *depth* of an element of the net, thanks to their nice tree-like structure.

Definition 1.9 *(Depth)*
Let Θ be a net in $\mathcal{B}$. The depth *of an element in $T_\Theta \cup S_\Theta$ is inductively defined by:*

- $\text{depth}(b) = 0$ *if* $b \in [u^I_\Theta]$;
- $\text{depth}(t) = \max\{\text{depth}(b) \mid b \prec t\} + 1$;
- $\text{depth}(b) = \text{depth}(t)$ *if* $\{t\} = {}^\bullet b$. □

Given a net Θ in $\mathcal{B}$ its *subnet* of depth n is the net $\Theta^{(n)}$ consisting of the elements of Θ whose depth is not greater than n. Clearly, for each $n \leq m$ there is a morphism $in_{n,m} : \Theta^{(n)} \to \Theta^{(m)}$ whose components are both set inclusions. In the following we will call such net morphisms simply *inclusions* and we will denote the inclusion of $\Theta^{(n)}$ in $\Theta^{(n+1)}$ by in_n.

Now, consider the category $\underline{\omega} = \{0 \to 1 \to 2 \cdots\}$ and the class $\mathcal{D}$ of diagrams $D : \underline{\omega} \to \mathcal{B}$ such that $D(n \to n+1) : D(n) \to D(n+1)$ is an inclusion. For such a class we have the following results. The reader is referred to [ML71, III.3] for the definition of the categorical concepts involved.

Proposition 1.10 (Colim(D) *exists and Θ is the colimit of its subnets)*

i. For any $D \in \mathcal{D}$, the colimit of D in $\mathcal{B}$ exists.

ii. Given a net Θ in $\mathcal{B}$, let $D_\Theta : \underline{\omega} \to \mathcal{B}$ be the diagram such that $D_\Theta(n) = \Theta^{(n)}$ and $D_\Theta(n \to n+1) = in_n$. Then $\Theta = \text{Colim}(D_\Theta)$. □

Next, we define a functor from **PTNets** to **DecOcc** which will be the right adjoint to $(\_)^+$. We start by giving the object component of such a functor. To this aim, given a net N, we define a family of decorated occurrence nets, one for each $n \in \omega$, where the n-th net approximates the unfolding of N up to depth n, i.e. it reflects the behaviour of the original net up to step sequences of length at most n. Clearly, the unfolding of N will be defined to be the colimit of an appropriate ω-shaped diagram built on the approximant nets. We will use the following notation: given $s \in X_1 \times \ldots \times X_n$, we denote by $s \downarrow X_i$ the projection of s on the X_i component. Moreover, given $S = \bigcup\{s_j \mid j \in J\}$, $S \downarrow X_i$ will be $\{s_j \downarrow X_i \mid j \in J\}$ and $S \overset{\oplus}{\downarrow} X_i$ will denote $\bigoplus_{j \in J}(s_j \downarrow X_i)$.

Definition 1.11 *(PT Nets Unfoldings: $\mathcal{U}[\_]^{(n)}$)*
Let $N = \left(\partial^0_N, \partial^1_N : (T_N, 0) \to S^{\mathcal{M}}_N, u^I_N\right)$ be a net in **PTNets**.
We define the nets $\mathcal{U}[N]^{(k)} = \left(\partial^0_k, \partial^1_k : (T_k, 0) \to S^{\mathcal{M}}_k, u^I_k\right)$, for $k \in \omega$, where:

- $S_0 = \bigcup\left\{\{(\emptyset, b)\} \times [n] \mid u^I_N(b) = n\right\}$;
- $T_0 = \{0\}$, *and the* ∂^i_0 *with the obvious definitions;*
- $u^I_0 = \bigoplus S_0$;

for $k > 0$,

- $T_k = T_{k-1} \cup \Big\{ (B,t) \;\Big|\; B \subseteq S_{k-1},\ \mathrm{Co}(B),\ t \in T_N,\ B \overset{\oplus}{\downarrow} S_N = \partial^0_N(t) \Big\}$;
- $S_k = S_{k-1} \cup \Big(\bigcup \Big\{ \big\{ (\{t_0\}, b) \big\} \times [n] \;\Big|\; t_0 \in T_k,\ b \in S_N,\ \partial^1_N(t_0 \downarrow T)(b) = n \Big\} \Big)$;
- $\partial^0_k(B,t) = \bigoplus B$, *and* $\partial^1_k(B,t) = \bigoplus \Big\{ \big((\{(B,t)\}, b), i \big) \in S_k \Big\}$;
- $u^I_k = \bigoplus \Big\{ \big((\emptyset, b), i \big) \in S_k \Big\} = \bigoplus S_0 = u^I_0$. □

Therefore, informally speaking, the net $\mathcal{U}[N]^{(0)}$ is obtained by exploding in families the initial marking of N, and $\mathcal{U}[N]^{(n+1)}$ is obtained, inductively, by generating a new transition for each possible subset of concurrent places of $\mathcal{U}[N]^{(n)}$ whose corresponding multiset of places of N constitutes the source of some transition t of N; the target of t is also exploded in families which are added to $\mathcal{U}[N]^{(n+1)}$. As a consequence, the transitions of the n-th approximant net are instances of transitions of N, in the precise sense that each of them corresponds to a unique occurrence of a transition of N in one of its step sequences of length at most n.

Definition 1.12 *(PT Net Unfoldings: $\mathcal{U}[\_]$)*
*We define $\mathcal{U}[N]$ to be the colimit of the diagram $D : \underline{\omega} \rightarrow$ **DecOcc** such that $D(n) = \mathcal{U}[N]^{(n)}$ and $D(n \rightarrow n+1) = in_n$. Since for all $n \in \omega$, $\mathcal{U}[N]^{(n)}$ is a decorated occurrence net of depth n and moreover for any $n \in \omega$ there is an inclusion $in_n : \mathcal{U}[N]^{(n)} \rightarrow \mathcal{U}[N]^{(n+1)}$, then D belongs to $\mathcal{D}$ and so, by Proposition 1.10 (i), the colimit exists and is a decorated occurrence net.* □

Example 1.13

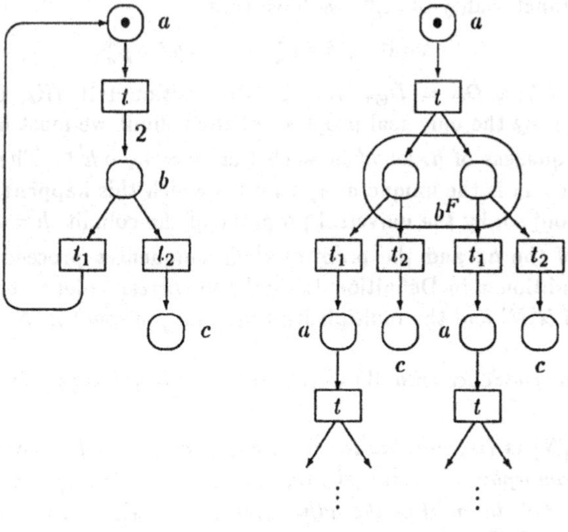

A PT Net N and (part of) its unfolding $\mathcal{U}[N]$

The correspondence between elements of the unfolding and elements of the original net is formalized by the folding morphism, which will also be the counit of the adjunction.

Proposition 1.14 *(Folding Morphism)*
Consider the map $\epsilon_N = \langle f_\epsilon, g_\epsilon \rangle : \mathcal{U}[N]^+ \to N$ *defined by*

- $f_\epsilon(B,t) = t$ *and* $f_\epsilon(0) = 0$;
- $g_\epsilon(\bigoplus_i (x_i, y_i)) = \bigoplus_i y_i$.

Then, ϵ_N *is a morphism in* **PTNets**, *called the* folding *of* $\mathcal{U}[N]$ *into* N. □

Finally, we are ready to prove that $\mathcal{U}[\_]$ is right adjoint to $(\_)^+$.

Theorem 1.15 *(*$(\_)^+ \dashv \mathcal{U}[\_]$*)*
The pair $\langle (\_)^+, \mathcal{U}[\_] \rangle$: **DecOcc** $\rightharpoonup$ **PTNets** *constitutes an adjunction.*

Proof. Let N be a PT Net and $\mathcal{U}[N]$ its unfolding. By [ML71, Theorem 2, pg. 81], it is enough to show that the folding $\epsilon_N : \mathcal{U}[N]^+ \to N$ is universal from $(\_)^+$ to N, i.e. for any decorated occurrence net Θ and any morphism $k : \Theta^+ \to N$ in **PTNets**, there exists a unique $h : \Theta \to \mathcal{U}[N]$ in **DecOcc** such that $k = \epsilon_N \circ h^+$.

$$\begin{array}{ccccccc} N & & \mathcal{U}[N] & & \mathcal{U}[N]^+ & \xrightarrow{\epsilon_N} & N \\ \forall k \uparrow & & \exists! h \uparrow & \text{s.t.} & h^+ \uparrow & \nearrow_{k} & \text{commutes.} \\ \Theta^+ & & \Theta & & \Theta^+ & & \end{array}$$

Consider the diagram in **DecOcc** given by $D_\Theta(n) = \Theta^{(n)}$, the subnet of Θ of depth n and $D_\Theta(n \to n+1) = in_n : \Theta^{(n)} \to \Theta^{(n+1)}$. We define a sequence of morphisms of nets $h_n : \Theta^{(n)} \to \mathcal{U}[N]$, such that for each n, $h_n = h_{n+1} \circ in_n$.
Since by Proposition 1.10 *(ii)* $\Theta = \text{Colim}(D_\Theta)$, there is a unique $h : \Theta \to \mathcal{U}[N]$ such that $h \circ \mu_n = h_n$ for each n. At the same time, we show that

$$\forall n \in \omega,\; k \circ \mu_n^+ = \epsilon_N \circ h_n^+ \qquad (1)$$

and that the h_n are the unique sequence of morphisms $h_n : \Theta^{(n)} \to \mathcal{U}[N]$ such that (1) holds. Now, by functoriality of $(\_)^+$, we have that

$$\forall n \in \omega,\; k \circ \mu_n^+ = \epsilon_N \circ h^+ \circ \mu_n^+.$$

Therefore, since $(\_)^+ \circ D_\Theta = D_{\Theta^+}$ and, by Proposition 1.10 *(ii)*, $\Theta^+ = \text{Colim}(D_\Theta^+) = \text{Colim}((\_)^+ \circ D_\Theta)$, by the universal property of the colimit we must have $k = \epsilon_N \circ h^+$.
To show the uniqueness of h, let h' be such that $k = \epsilon_N \circ h'^+$. Then we have $k \circ \mu_n^+ = \epsilon_N \circ h'^+ \circ \mu_n^+$. But h_n is the unique morphism for which this happens. Therefore, for each n, $h_n = h' \circ \mu_n$ and so, by the universal property of the colimit, $h = h'$.
The definition of the h_n and the proof of their uniqueness proceed by induction on n, exploiting k, condition v in Definition 1.1 and the correspondence between the structure of the families of $\mathcal{U}[N]$ and the multiplicities originally present in N. □

Theorem 1.16 *(Correspondence with Winskel's Safe Net Unfoldings [Win86])*
Let N *be a safe net.*
Then, its unfolding $\mathcal{U}[N]$ *is (isomorphic to) an occurrence net and, therefore,* $\mathcal{U}[N]^+ \cong \mathcal{U}[N]$. *Moreover,* $\mathcal{U}[N]$ *is (isomorphic to) Winskel's unfolding of* N. *Finally, whenever* N *is (isomorphic to) an occurrence net, the unit of the adjunction* $(\_)^+ \dashv \mathcal{U}[\_]$, $\eta_N : N \to \mathcal{U}[N^+] \cong \mathcal{U}[N]$, *is an isomorphism.*
Therefore, the adjunction $\langle (\_)^+, \mathcal{U}[\_] \rangle$: **DecOcc** $\rightharpoonup$ **PTNets** *restricts to Winskel's coreflection* $\langle (\_)^+_{\textbf{Occ}}, \mathcal{U}[\_]_{\textbf{Safe}} \rangle$: **Occ** $\rightharpoonup$ **Safe**. □

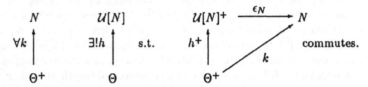

2 PT Nets, Event Structures and Domains

In this section, we show an adjunction between occurrence nets and decorated occurrence nets. Composing this adjunction with that given in Section 1, we obtain an adjunction between Occ and PTNets. Moreover, exploiting Winskel's coreflections in [Win86], we obtain adjunctions between PES and PTNets and between Dom and PTNets, as explained in the Introduction.

We first define a functor from decorated occurrence nets to occurrence nets. It is simply the *forgetful* functor which forgets about the structure of families.

Definition 2.1 *($\mathcal{F}[\_]$: from DecOcc to Occ)*
Given a decorated occurrence net Θ define $\mathcal{F}[\Theta]$ to be the occurrence net Θ. Furthermore, given $\langle f,g\rangle : \Theta_0 \rightarrow \Theta_1$, define $\mathcal{F}[\langle f,g\rangle]$ to be $\langle f,g\rangle$. □

In order to define a left adjoint for $\mathcal{F}[\_]$, we need to identify, for any occurrence net Θ, a decorated occurrence net which is, informally speaking, a "saturated" version of Θ, in the sense that it can match in a *unique* way the structure of the families of any decorated occurrence net built on (subnets of) Θ. Because of the uniqueness requirement, saturating occurrence nets is a delicate matter: we need to identify a suitable set of families which can "factorize" uniquely all the others. To this aim are devoted the following definition and lemma, where the relation $\mapsto$ is introduced to capture the behaviour of decorated occurrence net morphisms on families and *prime strings* are meant to represent—in a sense that will be clear later—exactly the families which we must add to Θ in order to saturate it.

In the following, given a string s on an alphabet Σ, as usual we denote the i-th element of s by s_i and its length by $|s|$. Moreover, σ^n, for $\sigma \in \Sigma$ and $n \in \omega$, will denote the string consisting of the symbol σ repeated n times.

Definition 2.2 *(Prime Strings)*
Let Σ be an alphabet, i.e. a set of symbols. Define the binary relation $\mapsto$ on Σ^+, the language of non-empty strings on Σ, by

$$\sigma_1^{n_1}\ldots\sigma_k^{n_k} \mapsto \sigma_1^{m_1}\ldots\sigma_k^{m_k} \Leftrightarrow \sigma_i \neq \sigma_{i+1} \text{ and } \exists q \in \omega \text{ s.t. } qn_i = m_i,\ i = 1,\ldots,k.$$

Define the language of prime strings *on Σ to be*

$$\Sigma^P = \Sigma^+ - \left\{\sigma_1^{n_1}\sigma_2^{n_2}\ldots\sigma_k^{n_k} \mid \sigma_i \in \Sigma,\ \sigma_i \neq \sigma_{i+1},\ \gcd(n_1,\ldots,n_k) > 1\right\},$$

where gcd *is the greatest common divisor.* □

Lemma 2.3 *(Prime Strings are primes)*
Given $s' \in \Sigma^+$ there exists a unique $s \in \Sigma^P$ s.t. $s \mapsto s'$. □

We start relating strings and nets by looking at sets of places as alphabets and by looking at families as strings on such alphabets.

Given a (decorated) occurrence net Θ and a transition $t \in T_\Theta$, we denote by $\Sigma_{\{t\}}$ the alphabet $[\partial^1_\Theta(t)]$. By analogy, since the places in the initial marking are in the post-set of no transition, $\Sigma_\emptyset$ will consist of the places $[u^I_\Theta]$; following the analogy, in the rest of the section u^I_Θ will also be denoted by $\partial^1_\Theta(\emptyset)$.

Since a family b^F of a decorated occurrence net Θ is nothing but an ordered subset of the initial marking or of the post-set of a transition, it corresponds naturally to a string in Σ^+_x where

$x = {}^\bullet[\![b^F]\!]$, namely, the string of length $|[\![b^F]\!]|$ whose i-th element is (b,i). We will write $\hat{b}^F$ to indicate such a string.

Now, we can define the saturated net corresponding to an occurrence net Θ. It is the net $\mathcal{D}[\Theta]$ whose transitions are the transitions of Θ, and whose families are the prime strings on the alphabets defined by Θ. It is immediate to see that this construction is well-defined, i.e. that $\mathcal{D}[\Theta]$ is a decorated occurrence net.

Definition 2.4 *($\mathcal{D}[\_]$: from Occc to DecOcc)*
Let Θ be a net in Occ.
We define the decorated occurrence net $\mathcal{D}[\Theta] = \left(\partial^0_{\mathcal{D}[\Theta]}, \partial^1_{\mathcal{D}[\Theta]} : (T_\Theta, 0) \to S^{\mathcal{M}}_{\mathcal{D}[\Theta]}, u^I_{\mathcal{D}[\Theta]}\right)$, *where*

- $S_{\mathcal{D}[\Theta]} = \bigcup\left\{\{s\} \times \left[|\, s\,|\right] \mid s \in \Sigma^P_x \text{ and } (x = \{t\} \subseteq T_\Theta \text{ or } x = \emptyset)\right\}$;
- $\partial^0_{\mathcal{D}[\Theta]}(t) = \bigoplus\left\{(s,i) \in S_{\mathcal{D}[\Theta]} \mid s_i \in [\![\partial^0_\Theta(t)]\!]\right\}$;
- $\partial^1_{\mathcal{D}[\Theta]}(t) = \bigoplus\left\{(s,i) \in S_{\mathcal{D}[\Theta]} \mid s_i \in [\![\partial^1_\Theta(t)]\!]\right\} = \bigoplus\left\{s^F \mid s \in \Sigma^P_{\{t\}}\right\}$;
- $u^I_{\mathcal{D}[\Theta]} = \bigoplus\left\{s^F \mid s \in \Sigma^P_\emptyset\right\}$. □

Example 2.5

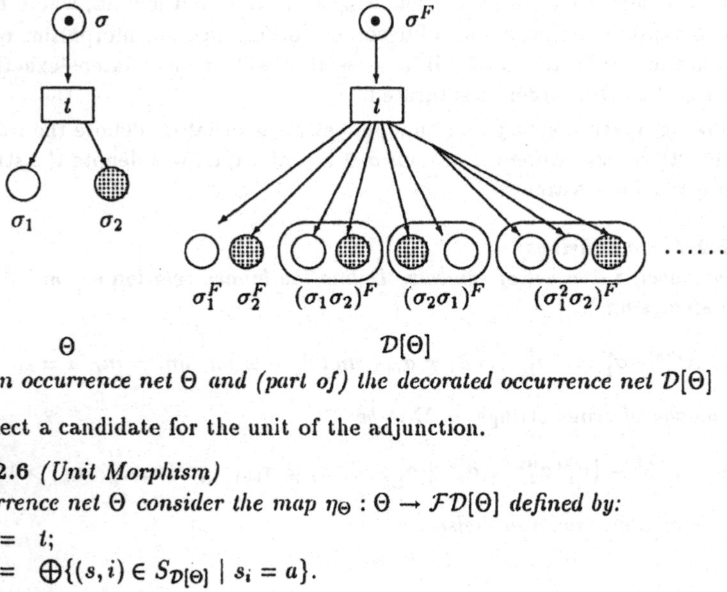

An occurrence net Θ and (part of) the decorated occurrence net $\mathcal{D}[\Theta]$

We now select a candidate for the unit of the adjunction.

Proposition 2.6 *(Unit Morphism)*
Given an occurrence net Θ consider the map $\eta_\Theta : \Theta \to \mathcal{FD}[\Theta]$ defined by:

$$\eta_\Theta(t) = t;$$
$$\eta_\Theta(a) = \bigoplus\{(s,i) \in S_{\mathcal{D}[\Theta]} \mid s_i = a\}.$$

Then η_Θ is a morphism in Occ. □

In order to illustrate the above definition, considering again the net Θ of Example 2.5. For such a net we have that

$$\eta_\Theta(\sigma_1) = (\sigma_1,1) \oplus (\sigma_1\sigma_2,1) \oplus (\sigma_2\sigma_1,2) \oplus (\sigma_1^2\sigma_2,1) \oplus (\sigma_1^2\sigma_2,2) \oplus \ldots;$$
$$\eta_\Theta(\sigma_2) = (\sigma_2,1) \oplus (\sigma_1\sigma_2,2) \oplus (\sigma_2\sigma_1,1) \oplus (\sigma_1^2\sigma_2,2) \oplus \ldots.$$

Before showing that η_Θ is universal, we need to develop further the relation between nets and strings. Since a morphism maps post-sets to post-sets, it naturally induces a (contravariant) mapping between the languages associated to transitions related by the morphism. To simplify notation, in the rest of this section, for k a morphism of nets, $k(\{t\})$ and $k(\emptyset)$, denote, respectively, $\{k(t)\}$ and $\emptyset$. Moreover, $\partial^1_\Theta(\{t\})$ denotes $\partial^1_\Theta(t)$.

Definition 2.7 *(S_k^x: from $\Sigma^+_{k(x)}$ to Σ^+_x)*
Let Θ_0 and Θ_1 be (decorated) occurrence nets, let $k = \langle f, g\rangle : \Theta_0 \to \Theta_1$ be a morphism and let $x = \{t\} \subseteq T_{\Theta_0}$ or $x = \emptyset$ and y be such that $f(x) = y$. Then k induces a unique semigroup homomorphism S_k^x from Σ^+_y to Σ^+_x defined on the generators $b \in [\![\partial^1_{\Theta_1}(y)]\!]$ by

$$S_k^x(b) = a \in [\![\partial^1_{\Theta_0}(x)]\!] \text{ s.t. } g(a) = b.$$

From the properties of safe net morphisms, it is easy to see that S_k^x is well-defined, i.e. there exists one and only one $a \in [\![\partial^1_{\Theta_0}(x)]\!]$ such that $g(a) = b$. □

To clarify the relation between $\mapsto$ and decorated occurrence net morphisms, observe that, in the condition of the previous definition, if Θ is a decorated occurrence net and k is a decorated occurrence net morphism, then $\hat{a}^F \mapsto S_k^x(\hat{b}^F)$ if and only if $[\![b^F]\!] \subseteq [\![k(a^F)]\!]$.

Example 2.8
Consider the following figure, where the morphism $\langle f, g\rangle$ is such that $g(\sigma_1) = s_2 \oplus s_3$ and $g(\sigma_2) = s_1$.

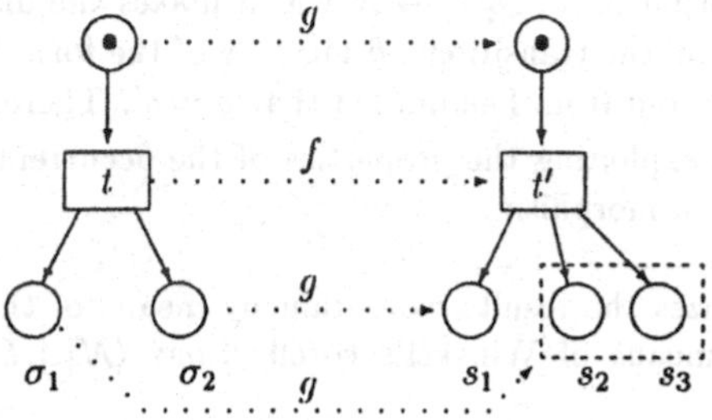

Then, for instance, we have that $S^{\{t\}}_{\langle f,g\rangle}(s_1 s_2 s_3 s_2 s_1) = \sigma_2 \sigma_1^3 \sigma_2$. □

Finally, we show that $\mathcal{D}[\_]$ extends to a functor which is left adjoint to $\mathcal{F}[\_]$.

Theorem 2.9 *($\mathcal{D}[\_] \dashv \mathcal{F}[\_]$)*
The pair $\langle \mathcal{D}, \mathcal{F}\rangle$: Occ $\rightharpoonup$ DecOcc constitutes an adjunction.

Proof. Let Θ be an occurrence net. By [ML71, Theorem 2, pg. 81] it is enough to show that the morphism $\eta_\Theta : \Theta \to \mathcal{F}\mathcal{D}[\Theta]$ is universal from Θ to $\mathcal{F}$, i.e. for any decorated occurrence net Θ' and any $k : \Theta \to \mathcal{F}[\Theta']$ in Occ, there exists a unique $\langle f, g\rangle : \mathcal{D}[\Theta] \to \Theta'$ in DecOcc such that $k = \mathcal{F}[\langle f, g\rangle] \circ \eta_\Theta$.

$$\begin{array}{ccc} \Theta & & \mathcal{D}[\Theta] \\ \forall k \downarrow & \exists! \langle f,g\rangle & \downarrow \\ \mathcal{F}[\Theta'] & & \Theta' \end{array} \quad \text{s.t.} \quad \begin{array}{ccc} \Theta & \xrightarrow{\eta_\Theta} & \mathcal{F}\mathcal{D}[\Theta] \\ & {\scriptstyle k} \searrow & \downarrow \mathcal{F}[\langle f,g\rangle] \\ & & \mathcal{F}[\Theta'] \end{array} \quad \text{commutes.}$$

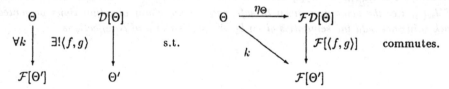

Given Θ' and k, we define $\langle f, g\rangle : \mathcal{D}[\Theta] \to \Theta'$ as follows:

$f(t) = k(t)$
$[\![b^F]\!] \subseteq [\![g(s^F)]\!] \Leftrightarrow s \mapsto S_k^x(\hat{b}^F)$, where $x = {}^\bullet[\![s^F]\!]$ and $k(x) = {}^\bullet[\![b^F]\!]$

First remark that $\langle f, g\rangle$ is well-defined: if $s = \sigma_1^{n_1} \ldots \sigma_r^{n_r} \mapsto S_k^x(\hat{b}^F)$ then there is one and only one way to have $[\![b^F]\!] \subseteq [\![g(s^F)]\!]$, namely

$$g(s, i) = \bigoplus \{b\} \times [q]_i,$$

where q is the unique integer such that $\sigma_1^{qn_1} \ldots \sigma_r^{qn_r} = S_k^x(\hat{b}^F)$.

Let $x = \{t_0\}$ or $x = \emptyset$. Observe that $\forall a \in [\![\partial^1_{\Theta'}(x)]\!]$

$$\forall (b,j) \in [\![k(a)]\!]\ \exists!(s,i) \text{ such that } (s,i) \in [\![\partial^1_{\mathcal{D}[\Theta]}(x)]\!] \text{ and } (b,j) \in [\![g(s,i)]\!]. \qquad (1)$$

Moreover, (s,i) is the unique place in $\mathcal{D}[\Theta]$ such that $s_i = a$ and $(b,j) \in [\![g(s,i)]\!]$. (2)
Now, if $(b,j) \in [\![g(s,i)]\!]$ then $s \mapsto \mathcal{S}^x_k(\hat{b}^F)$ and therefore, by definition of $\mapsto$, we have $\mathcal{S}^x_k(\hat{b}^F)_{(i-1)q+1}, \ldots, \mathcal{S}^x_k(\hat{b}^F)_{iq} = s_i$. Thus, by definition of $\mathcal{S}^x_k$, $\{b\} \times [q]_i \subseteq [\![k(s_i)]\!]$. So we have $\bigcup\{[\![g(s,i)]\!] \mid s_i = a\} = [\![k(a)]\!]$. Obviously, all the $[\![g(s,i)]\!]$ are disjoint and $\bigoplus[\![g(s,i)]\!] = g(s,i)$, since the families are disjoint. Therefore,

$$\bigoplus\{g(s,i) \mid s_i = a\} = k(a). \qquad (3)$$

It is now easy to see that the diagram commutes. For transitions this is clear. Concerning places, we have:

$$\mathcal{F}[\langle f,g\rangle] \circ \eta_\Theta(a) = \bigoplus g\Big(\{s_i \mid s_i = a\}\Big) = k(a).$$

Now, consider any morphism $h : \mathcal{D}[\Theta] \to N$ which makes the diagram commute. Because of the definition of η_Θ on the transitions, h must be of the form $\langle f, g'\rangle$. It follows from the definitions of g and $\mathcal{S}^x_k$ and from Lemma 2.3 that $g = g'$. Therefore, $h = \langle f, g\rangle$.
From (1), (2) and (3), exploiting the properties of the occurrence net morphism h, it can be shown that $\langle f, g\rangle$ is a morphism. □

The next corollary summarizes the results we obtain by means of the adjunction $\langle \mathcal{D}[\_], \mathcal{F}[\_]\rangle$: **Occ** $\rightharpoonup$ **DecOcc** and by means of Winskel's coreflections $\langle \mathcal{N}[\_], \mathcal{E}[\_]\rangle$: **PES** $\rightharpoonup$ **Occ** and $\langle \mathcal{P}r[\_], \mathcal{L}[\_]\rangle$: **Dom** $\rightharpoonup$ **PES**.

Corollary 2.10 *(Extensions of Winskel's coreflections [Win86])*
The following are adjunctions whose right adjoints relate PT nets to, respectively, occurrence nets, prime event structures and prime algebraic domains.

- $\langle (\_)^+ \mathcal{D}[\_], \mathcal{FU}[\_]\rangle$: **Occ** $\rightharpoonup$ **PTNets**;
- $\langle (\_)^+ \mathcal{DN}[\_], \mathcal{EFU}[\_]\rangle$: **PES** $\rightharpoonup$ **PTNets**;
- $\langle (\_)^+ \mathcal{DNP}r[\_], \mathcal{LEFU}[\_]\rangle$: **Dom** $\rightharpoonup$ **PTNets**.

Moreover, $\mathcal{FU}[\_]_{\mathbf{Safe}} = \mathcal{U}_w[\_]$ *and, therefore,* $\mathcal{EFU}[\_]_{\mathbf{Safe}} = \mathcal{EU}_w[\_]$ *and* $\mathcal{LEFU}[\_]_{\mathbf{Safe}} = \mathcal{LEU}_w[\_]$, *i.e. the semantics given to safe nets by the chain of adjunctions presented in this work coincides with the semantics given by Winskel's chain of coreflections.* □

ACKNOWLEDGMENTS. We thank Narciso Martí-Oliet for his careful reading of the manuscript.

References

[BD87] E. BEST, AND R. DEVILLERS. Sequential and Concurrent Behaviour in Petri Net Theory. *Theoretical Computer Science*, n. 55, pp. 87–135, 1987.

[DDM88] P. DEGANO, R. DE NICOLA, AND U. MONTANARI. A Distributed Operational Semantics for CCS based on Condition/Event Systems. *Acta Informatica*, n. 26, pp. 59–91, 1988.

[DMM89] P. DEGANO, J. MESEGUER, AND U. MONTANARI. Axiomatizing Net Computations and Processes. In proceedings of *4th LICS Symposium*, pp. 175–185, IEEE, 1989.

[Eng91] J. ENGELFRIET. Branching Processes of Petri Nets. *Acta Informatica*, n. 28, pp. 575–591, 1991.

[GR83] U. GOLTZ, AND W. REISIG. The Non-Sequential Behaviour of Petri Nets. *Information and Computation*, n. 57, pp. 125–147, 1983.

[HKT92] P.W. HOOGERS, H.C.M. KLEIJN, AND P.S. THIAGARAJAN. A Trace Semantics for Petri Nets. To appear in proceedings of *ICALP '92*, 1992.

[ML71] S. MACLANE. *Categories for the Working Mathematician*. GTM, Springer-Verlag, 1971.

[MM90] J. MESEGUER, AND U. MONTANARI. Petri Nets are Monoids. *Information and Computation*, n. 88, pp. 105–155, 1990.

[NPW81] M. NIELSEN, G. PLOTKIN, AND G. WINSKEL. Petri Nets, Event Structures and Domains, Part 1. *Theoretical Computer Science*, n. 13, pp. 85–108, 1981.

[Old87] E.R. OLDEROG. A Petri Net Semantics for CCSP. In *Advances in Petri Nets*, LNCS, n. 255, pp. 196–223, Springer-Verlag, 1987.

[Pet62] C.A. PETRI. *Kommunikation mit Automaten*. PhD thesis, Institut für Instrumentelle Mathematik, Bonn, FRG, 1962.

[Pet73] C.A. PETRI. Concepts of Net Theory. In proceedings of *MFCS '73*, pp. 137–146, Mathematics Institute of the Slovak Academy of Science, 1973.

[Pet77] C.A. PETRI. *Non-Sequential Processes*. Interner Bericht ISF-77-5, Gesellschaft für Mathematik and Datenverarbeitung, Bonn, FRG, 1977.

[Pra91] V. PRATT. Modeling Concurrency with Geometry. In proceedings of *POPL '91*, pp. 311-322, ACM, 1991.

[Rei85] W. REISIG. *Petri Nets*. Springer-Verlag, 1985.

[Sco70] D. SCOTT. Outline of a Mathematical Theory of Computation. In proceedings of *4th Annual Princeton Conference on Information Science and Systems*, pp. 169–176, 1970.

[vGV87] R. VAN GLABBEEK, AND F. VAANDRAGER. Petri Net Model for Algebraic Theories of Concurrency. In proceedings of *PARLE*, LNCS, n. 259, pp. 224–242, Springer-Verlag, 1987.

[Win82] G. WINSKEL. Event Structure Semantics for CCS and related languages. In proceedings of the *9th ICALP*, LNCS, n. 140, pp. 561–576, Springer-Verlag, 1982.

[Win84] G. WINSKEL. A New Definition of Morphism on Petri Nets. In proceedings of *STACS '84*, LNCS, n. 166, pp. 140–150, Springer-Verlag, 1984.

[Win86] G. WINSKEL. Event Structures. In proceedings of *Advanced Course on Petri Nets*, LNCS n. 255, pp. 325–392, Springer-Verlag, 1987.

[Win87] G. WINSKEL. Petri Nets, Algebras, Morphisms and Compositionality. *Information and Computation*, n. 72, pp. 197–238, 1987.

Structural Operational Specifications and Trace Automata

Eric Badouel and Philippe Darondeau

Irisa, Campus de Beaulieu , F-35042 Rennes Cedex, France
E-mail : ebadouel@irisa.fr - darondeau@irisa.fr

Abstract. Structural Operational Specifications (SOS) are supplied with concurrent models based on permutations of proved transitions. Those models take the form of trace automata which are deterministic automata equipped with an explicit relation of independence on actions. In order to characterize the finite trace automata which may be realized in SOS-algebras, we introduce a new kind of nets which encode exactly the concurrent behaviour of systems specified in SOS and we establish a correspondence between nets and the so-called 'separated' trace automata which may be realized in SOS.

1 Introduction

We show that Structural Operational Specifications (SOS) induce models which exhibit the concurrent behaviour of programs, extending Boudol and Castellani's construction for finite CCS [BC88]. Their approach, based on permutations of proved transitions, is to decide, for each pair of co-initial transitions A and B, whether A survives B and symmetrically. Transitions A and B are said to be independent when they both survive the other, in which case they may be amalgamated to a concurrent transition A.(B/A)= B.(A/B), where B/A is the residual of B after A. Since we intend to cover a variety of specifications, the definition of independence should conform to the obligation to deal uniformly with all operators in the signature, forbidding us to distinguish so-called parallel operators. It appears that no satisfactory rule of independence can be stated unless every operator is replaced by exactly one operator in every transition, maintaining the structure of terms. We therefore concentrate on a restriction of De Simone's format [deS84], called *basic SOS*, which ensure that property, and still covers all existing process algebras. Operational specifications given in basic SOS may equivalently be translated to *SOS automata* whose states are term constructors and whose transitions are labelled by proof constructors encoding the inductive rules of the specification. Transition systems induced from process terms are called *process automata.* Each process automaton is a synchronized product (*à la* Arnold-Nivat) of copies of the SOS automaton indexed by the (invariant) tree domain of the process term. In the SOS automaton, pairs of independent moves are exhibited by diamonds; and in the process automaton, two transitions are independent if and only if they project to locally independent moves at each occurrence in the tree domain.

Both SOS automata and process automata happen to be instances of trace automata introduced by Stark (see [Sta89b] and [Sta89c]). Those are ordinary deterministic automata equipped with an explicit relation of independence on actions.

Trace automata provide a concrete representation for a subset of Stark's concurrent transition systems [Sta89a] in which transitions are classes for the equivalence by permutations. Boudol and Castellani's model for finite CCS falls in that case, and would lead to infinite trace automata if rational process terms were taken into account. Although most statements expressed in this paper do not depend on the finiteness of terms or automata, the main question we address is the realization of finite trace automata in SOS. The answer is non trivial, and we shall elaborate the example of a finite trace automaton which has no realization in SOS. In order to solve the above question, we introduce a new kind of (SOS) nets which encode exactly the concurrent behaviour of SOS processes, and we construct an adjunction between the so-called saturated nets and separated trace automata, preserving that behaviour. For that purpose, we follow the idea of Nielsen, Rozenberg, and Thiagarajan [NRT90] to form nets places from regions in automata, defined as set of states uniformly entered or exited by all transitions labelled by the same action. The concept of region was introduced originally to relate elementary transition systems and elementary nets. An adapted version was used by Winskel [Win91] to support an adjunction between Bednarczyk's asynchronous transition systems [Bed88] and a variant of elementary nets. In both kind of nets, transitions normally wait for emptiness of their output places. On the contrary, regions associated with process automata correspond to places that may be emptied and refilled by their input and output transitions. We notice that the axiom of uniform commutation for synchronous transition systems (condition 4.1.1.(ii) in [Bed88]) is not satisfied by process automata, which do not either fit in with elementary transition systems due to self looping states and multiple transitions between pair of states. Now, not every trace automaton corresponds to a net: This holds only for those trace automata which satisfy an axiom of separation ensuring that they have enough regions to separate a state disabling an action from the set of states where it is allowed. Similarly, not every net corresponds to a trace automaton: this holds only for those nets which enjoy a property of saturation dual to the property of separation. We show that every net is equivalent to a saturated net, and the finite trace automata realized by SOS processes are exactly those which satisfy the axiom of separation.

Boudol and Castellani have established in [BC90] the consistency of the model based on proved transitions with two different models for CCS, constructed in the respective settings of flow event structures and flow nets. A similar fact holds in general for arbitrary process algebras defined in basic SOS: every specification in basic SOS may be given three equivalent models in the respective frameworks of trace automata, (SOS) nets, and event structures with a binary conflict. The justification for that claim is twofold. On the one hand, Stark's correspondence between concurrent transition systems and event structures with arbitrary conflict restricts to a correspondence between trace automata and event structure with binary conflict, preserving domain of configurations (this observation will not be elaborated in the present paper). On the other hand, the adjunction between separated trace automata and saturated nets preserves concurrent behaviour, and we obtain by the way compositional net models for structural operational specifications.

The body of the paper is organized as follows. Section 2 introduces basic SOS, and constructs concurrent models for that format in the setting of trace automata. Section 3 introduces nets and constructs the adjunction between separated trace automata and saturated nets. Relying on that adjunction, we show that the finite trace automata realized in SOS are exactly the separated ones. For lack of space no proof is provided, the interested reader is referred to the full version of the paper, [BD92].

2 Diamonds in SOS

The objective of this section is to construct inductively from structural operational specifications, see [Plo81], process automata where diamonds reflects concurrency. In a first attempt, let us identify process graphs with pointed transition graphs whose states are the finite or infinite terms over some signature Σ, and whose labelled transitions are proved from some set R of rules which conform to the de Simone format [deS84]:

$$\frac{u_{i_1} \xrightarrow{a_1} v_{i_1}, \ldots, u_{i_k} \xrightarrow{a_k} v_{i_k}}{f(u_1, \ldots, u_n) \xrightarrow{a} C[v_1, \ldots, v_n]}$$

where all variables $\{u_1, \ldots, u_n, v_{i_1}, \ldots, v_{i_k}\}$ are different, and $v_j = u_j$ for $j \notin \{i_1, \ldots, i_k\}$, while $f \in \Sigma_n$ is an n-ary operator symbol, and $C[v_1, \ldots, v_n]$ is a Σ-expression where each variable v_j occurs at most once. A diamond is then any subgraph of the form:

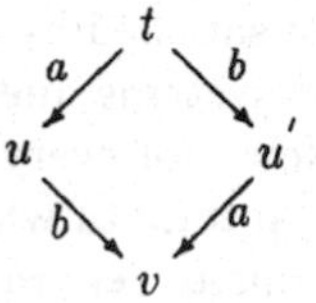

where opposite transitions bear identical labels and $a \neq b$. Thus, both terms $(a.b.nil + b.a.nil)$ and $a.nil \| b.nil$ originate diamonds in CCS, see [Mil80], even though the first diamond does not reflect concurrency. In order to correct that fault, one may consider a different graph whose arcs are labelled by proofs of transitions. However, labelling arcs by plain proofs is too strong: two transitions with the same proof must be identical and the second diamond of the example therefore disappears, even though it reflects concurrency. Arcs of the proved transition graph must therefore be labelled by *schematic proofs* abstracting from the sources of labelled transitions. Let us give a definition before resuming the discussion. We recall that a Σ-tree is a partial mapping $T : (\mathbb{N}_+)^* \to \Sigma$ whose domain $Dom(T)$ is a prefix closed set of paths (encoded by sequences of positive integers) including the empty path ϵ, and such that $T(s) \in \Sigma_k \Rightarrow (s.i \in Dom(T)$ **iff** $i \in \{1, \ldots, k\})$.

Definition 2.1 (Schematic proofs) *Given a set R of SOS rules for Σ-terms, we shall also let R denote the signature formed by extracting from each rule a corresponding operator $\rho = <\{i_1, \ldots, i_k\},\ <a_1, \ldots, a_k, a>,\ C(1, \ldots, n)>$ with arity k. The action a is called the action of ρ. denoted as $a = act(\rho)$. A schematic proof is then a finite R-tree* A *, satisfying for every $s \in Dom($* A *$)$ the following condition of local correctness:*

$\mathrm{A}(n) = <\{i_1, \ldots, i_k\}, <a_1, \ldots, a_k, a>, C(1, \ldots, n)> \Rightarrow [(\forall j \in \{1, \ldots, k\} \quad act(\mathrm{A}(n.j)) = a_j)]$

By forgetting the sources of all moves, proofs for transitions $t \xrightarrow{a} u$ are projected to schematic proof trees A where a is the action of the root A(ϵ). Now, the proved transition graph $\Gamma(\Sigma, R)$ is the graph formed from the resulting arcs $t \xrightarrow{A} u$.

Let us examine on examples whether diamonds in the proved transition graph are representative of concurrency. Suppose R is Milner's set of rules for CCS. Then $(a.nil \| b.nil)$ is the origin of a diamond in $\Gamma(\Sigma, R)$ whereas $(a.b.nil + b.a.nil)$ is not. Unfortunately, no diamond originates from $((a.nil \| b.nil) + c.nil)$, because the schematic proof $\quad$ A $=<\{1\}, <a,a>, 1> (<\{1\}, <a,a>, 1\|2> (<\{\}, <a>, 1>))$

while it applies to $((a.nil \| b.nil) + c.nil)$ and then produces the proved transition $((a.nil \| b.nil) + c.nil) \xrightarrow{A} nil \| b.nil$, does not apply to $(a.nil \| nil)$. In that case, the apparent cause of failure is that operator + vanishes when acted upon: CCS-transitions do not strictly maintain the structure of terms. A different cause of failure may be observed for $\Sigma = \{f, nil\}$ when R is the set of rules:

$$\vdash f(u) \xrightarrow{a} f(f(u)) \qquad\qquad u \xrightarrow{a} v \vdash f(u) \xrightarrow{a} f(v)$$

$R = \{\varphi, \psi\}$ with $\varphi =<\{\}, <a>, f(f(1))>$ and $\psi =<\{1\}, <a,a>, f(1)>$, then $f(f(nil)) \xrightarrow{\varphi} f(f(f(nil))) \xrightarrow{\psi(\varphi)} f(f(f(f(nil))))$ and $f(f(nil)) \xrightarrow{\psi(\varphi)} f(f(f(nil))) \xrightarrow{\varphi} f(f(f(f(nil))))$. Unfortunately, this diamond does not represent concurrency: the computation proved by the sequence $\varphi.\psi(\varphi)$ is inherently sequential, while the computations that duplicate the two occurrences of f in parallel are proved by $\psi(\varphi).\varphi$ and $\varphi.\psi(\psi(\varphi))$ which form no diamond! The apparent cause of failure is again that transitions do not strictly maintain the structure of terms, since precisely the operator f may duplicate itself. In order to avoid disagreements between the intuitive notion of concurrency on the one hand and the notion of concurrency based on diamonds on the other hand, we should impose further restrictions on De Simone's format. We call *basic SOS* the restriction defined by forcing rules ρ to the basic format:

$$\frac{u_{i_1} \xrightarrow{a_1} v_{i_1}, \ldots, u_{i_k} \xrightarrow{a_k} v_{i_k}}{f(u_1, \ldots, u_n) \xrightarrow{a} g(v_1, \ldots, v_n)}$$

So, in basic SOS the structure of terms is strictly maintained by transitions: terms t and u have identical domains whenever $t \xrightarrow{a} u$. The usual definition of CCS in De Simone's format does not fit in basic SOS, but CCS may nevertheless be redefined in basic SOS. The price to pay is the introduction of three auxiliary operators (id, π_1 and π_2 such that e.g. $\alpha.nil + nil \xrightarrow{\alpha} \pi_1(id(nil), nil)$). The alternative definition of CCS may be found in table 2.

An operational specification in basic SOS may be equivalently represented by a graph $\gamma(\Sigma, R)$ on Σ, called an SOS-automaton, in which each SOS rule

$$\frac{u_{i_1} \xrightarrow{a_1} v_{i_1}, \ldots, u_{i_k} \xrightarrow{a_k} v_{i_k}}{f(u_1, \ldots, u_n) \xrightarrow{a} g(v_1, \ldots, v_n)}$$

is encoded by an arc $f \xrightarrow{\rho} g$ labelled by $\rho = <\{i_1, \ldots, i_k\}, <a_1, \ldots, a_k, a>, g>$. Thus, in an SOS automaton $\gamma(\Sigma, R)$, states $f \in \Sigma$ are operators on process terms and transition labels $\rho \in R$ are operators on proof terms. SOS automata proceed from the same inspiration as the (more general) context-systems of Larsen and Xinxin [LX90]. It is worth noting that an SOS automaton associated with SOS rules generated

nil : no rule

$a.-$: $a.t \xrightarrow{a} id(t)$

$$id : \frac{t \xrightarrow{\mu} t'}{id(t) \xrightarrow{\mu} id(t')}$$

$$+ : \frac{t \xrightarrow{\mu} t'}{t + u \xrightarrow{\mu} \pi_1(t', u)} \qquad \frac{u \xrightarrow{\mu} u'}{t + u \xrightarrow{\mu} \pi_2(t, u')}$$

$$\pi_i : \frac{t \xrightarrow{\mu} t'}{\pi_1(t, u) \xrightarrow{\mu} \pi_1(t', u)} \qquad \frac{u \xrightarrow{\mu} u'}{\pi_2(t, u) \xrightarrow{\mu} \pi_2(t, u')}$$

$$\| : \frac{t \xrightarrow{\mu} t'}{t\|u \xrightarrow{\mu} t'\|u} \qquad \frac{u \xrightarrow{\mu} u'}{t\|u \xrightarrow{\mu} t\|u'} \qquad \frac{t \xrightarrow{a} t', u \xrightarrow{\overline{a}} u'}{t\|u \xrightarrow{\tau} t'\|u'}$$

$$\backslash_a : \frac{t \xrightarrow{\mu} t'}{t\backslash_a \xrightarrow{\mu} t'\backslash_a} \; \mu \notin \{a, \overline{a}\}$$

$$[\Phi] : \frac{t \xrightarrow{\mu} t'}{t[\Phi] \xrightarrow{\Phi(\mu)} t'[\Phi]}$$

Table 1. alternative SOS rules for CCS

from a finite set of schemes of rules, has a finite presentation. For instance the SOS automaton For CCS has its transitions labelled by the instances of the following schemes of proof constructors each of which corresponds to one scheme of rules of table 1:

$\alpha * a =< \emptyset, \{a\}, id >$, $id * \mu =< \{1\}, < \mu, \mu >, id >$,
$\pi_1 * \mu =< \{1\}, < \mu, \mu >, \pi_1 >$, $\pi_2 * \mu =< \{2\}, < \mu, \mu >, \pi_2 >$,
$\|_1 * \mu =< \{1\}, < \mu, \mu >, \| >$, $\|_2 * \mu =< \{2\}, < \mu, \mu >, \| >$, $\sigma * a =< \{1, 2\}, < a, \overline{a}, \tau >, \| >$,
$\backslash_a * \mu =< \{1\}, < \mu, \mu\backslash_a >, \backslash_a >$, and $\Phi * \mu =< \{1\}, < b, \Phi(\mu) >, \Phi >$,

where a and μ are generic actions (respectively visible and arbitrary), $\backslash_a$ is the operator of restriction given by $\mu\backslash_a = \mu$ if $\mu \notin \{a, \overline{a}\}$ and undefined elsewhere, and $\Phi : \Lambda \to \Lambda$ is a total renaming function.

An important property of the SOS-automaton is to generate the proved transition graph by means of synchronized products. This fact is made wholly clear in the restating of the definition of proved transition graph given below. First, we notice that the SOS automaton $\gamma(\Sigma, R)$ is deterministic since $f \xrightarrow{\rho} g$ implies that ρ is of the form $< \{i_1, \ldots, i_k\}, < a_1, \ldots, a_k, a >, g >$. This seems an appropriate place for fixing some notations: if $T \subset Q \times \Lambda \times Q$ is a deterministic transition system we use notation $x \xrightarrow{a}$ as an abbreviation for $\exists y \;.\; (x \xrightarrow{a} y)$, and we denote $x.a$ the unique y such that $(x \xrightarrow{a} y)$ when it exists. And we inductively define $q \xrightarrow{u}$ and $q.u$ for $q \in Q$ and $u \in A^*$ by: $q \xrightarrow{\epsilon}$ always holds with $q.\epsilon = q$ and $q \xrightarrow{a.u}$ **iff** $q \xrightarrow{a}$ & $q.a \xrightarrow{u}$ with $q.(a.u) = (q.a).u$.

Definition 2.2 (Proved transition graph) *Given a signature Σ and a set R of rules in basic SOS, the proved transition graph $\Gamma(\Sigma, R)$ is the deterministic transition system on set of (finite or infinite) Σ-terms $t \xrightarrow{\mathrm{A}} u$ labelled by schematic proofs such that, if we let φ_A denote the injective mapping $\varphi_A : Dom(A) \rightarrow 2^{(\mathbb{N}_+)^*}$ given by $\varphi_A(\epsilon) = \epsilon$ and $\varphi_A(n.j) = \varphi_A(n).i_j$, then a schematic proof* A *is enabled in t ($t \xrightarrow{\mathrm{A}}$) if, and only if*

1. A *matches the domain of t, i.e. $Im(\varphi_A) \subset Dom(t)$, and*
2. *the proof is locally enabled, i.e. $\forall s \in Dom(A) \quad t(\varphi_A(s)) \xrightarrow{A(s)}$ in $\gamma(\Sigma, R)$.*

And then, the unique u such that $t \xrightarrow{\mathrm{A}} u$ is given by

1. $\forall s \in Dom(A) \quad t(\varphi_A(s)) \xrightarrow{A(s)} u(\varphi_A(s))$ *in $\gamma(\Sigma, R)$, and*
2. $\forall s \notin Im(\varphi_A) \quad t(s) = u(s)$.

We notice that the part of $\Gamma(\Sigma, R)$ accessible from a Σ-term t corresponds to synchronized product of automata *à la* Arnold-Nivat (see [Niv79] and [AN82]). More precisely we have as many copies of the SOS-automaton $\gamma(\Sigma, R)$ as there are occurrences $s \in Dom(t)$ in the domain of t, synchronized by the set of vectors $V_A \in (R \cup \{\epsilon\})^{Dom(t)}$ defined from the schematic proofs A enabled in t by: $V_A(s) =$ A $(\varphi_A^{-1}(s))$ if $s \in Im(\varphi_A)$ and $V_A(s) = \epsilon$ otherwise. The specific contribution of SOS is the inductive characterization of the synchronization vectors via the structure of logical proofs. For instance, the proof term shown on the left of Fig. 1, encoding the proof

$$\cfrac{\cfrac{a.t \xrightarrow{a} id(t) \qquad \overline{a}.u \xrightarrow{\overline{a}} id(u)}{a.t\|\overline{a}.u \xrightarrow{\tau} id(t)\|id(u)}}{(a.t\|\overline{a}.u) + v \xrightarrow{\tau} \pi_1(id(t)\|id(u), v)}$$

may be seen as a synchronization vector with four non-empty components. It indicates that the four copies of the SOS-automaton dedicated to the circled nodes should synchronized their respective local moves, indicated by the dashed arrows, while the other copies standing at uncircled nodes stay inactive. The result of the synchronized move is the state vector represented by the rightmost term.

We will now concentrate our attention on concurrency and analyze in terms of synchronized products the circumstances under which diamonds appear in the proved transition graph.

Definition 2.3 (Diamond) *A diamond $\diamond(q, a, b)$ in a deterministic transition system $T \subset Q \times A \times Q$ consists of a state $q \in Q$ and a pair of distinct actions $a, b \in A$ such that $q \xrightarrow{ab}$, $q \xrightarrow{ba}$ and $q.ab = q.ba$.*

Such a diamond represents two actions a and b performed in parallel. Since the subgraph of $\Gamma(\Sigma, R)$ accessible from t coincides with an accessible subgraph of the synchronized product $\gamma(\Sigma, R)^{Dom(t)}$, there is a diamond $\diamond(t,$ A , B $)$ in $\Gamma(\Sigma, R)$ if and only if, A $\neq$ B and for every $s \in Dom(t)$, $V_A(s) = \epsilon$ or $V_B(s) = \epsilon$ or there is a *local diamond* $\diamond(t(s), V_A(s), V_B(s))$ in $\gamma(\Sigma, R)$ i.e. a subgraph

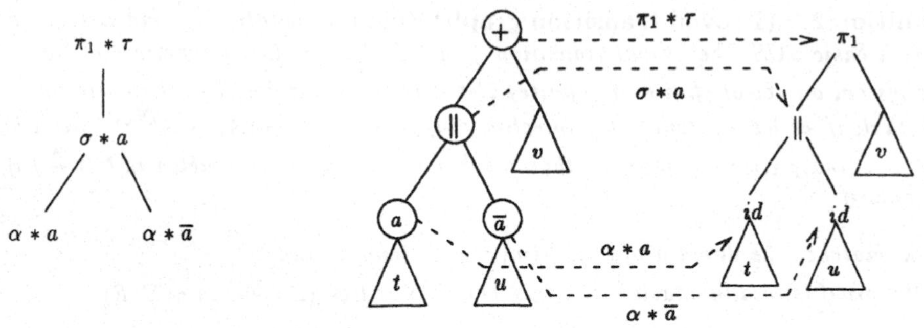

Fig. 1. a proved transition and its schematic proof

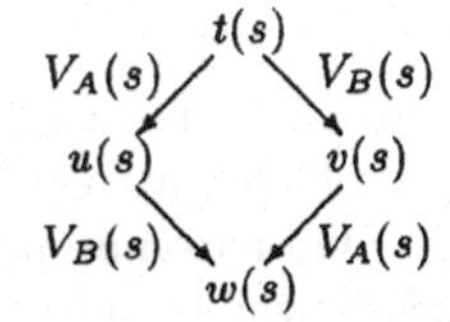

with possibly $V_A(s) = V_B(s)$. A typical diamond in $\Gamma(\Sigma, R)$ is shown in Fig. 2. By the way, since $f \xrightarrow{\alpha} g$ implies $\alpha = <\{i_1, \ldots, i_k\}, <a_1, \ldots, a_k, a>, g>$, it may be observed that a local diamond is any subgraph of $\gamma(\Sigma, R)$ of the particular form:

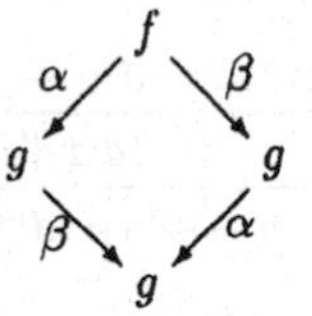

with possibly $\alpha = \beta$. Thus, two different transitions $t \xrightarrow{\text{A}} u$ and $t \xrightarrow{\text{B}} v$ in the proved transition graph are concurrent if and only if they can be executed jointly in one "unit of logical time", i.e. with at most one (synchronized) change of local state for all copies of the SOS-automaton dedicated to occurrences $s \in Dom(\text{A}) \cap Dom(\text{B})$. Notice that two concurrent transitions $t \xrightarrow{\text{A}} u$ and $t \xrightarrow{\text{B}} v$ need not restrict to concurrent transitions $t[s] \xrightarrow{\text{A}[s]} u[s]$ and $t[s] \xrightarrow{\text{B}[s]} v[s]$, (where $s \in Dom(t)$ and $t[s]$ and $\text{A}[s]$ stand respectively for the subtrees of t and A at the occurrence s) even though $s \in Dom(\text{A}) \cap Dom(\text{B})$, because $\text{A}[s]$ and $\text{B}[s]$ may be equal for some occurrence s: for synchronized systems two occurrences of the same event in a sequence $t \xrightarrow{\text{A}} t' \xrightarrow{\text{A}} t"$ cannot be amalgamated into a concurrent transition, nevertheless two concurrent events may coincide on some component of the synchronized system. And this phenomenon call for the possibility of "self-independent" events in the SOS automaton, where events describe only partially the activity of the system. Let us state that an event a in a trace automaton is *auto-concurrent* whenever every transition bearing that label leads to the same state: $q_1 \xrightarrow{a} q_1'$ and $q_2 \xrightarrow{a} q_2'$ entail $q_1' = q_2'$. Consequently, any chain $q_1 \xrightarrow{a} q_1' \xrightarrow{a} q_1' \ldots \xrightarrow{a} q_1'$ labelled by an auto-concurrent event a involves at

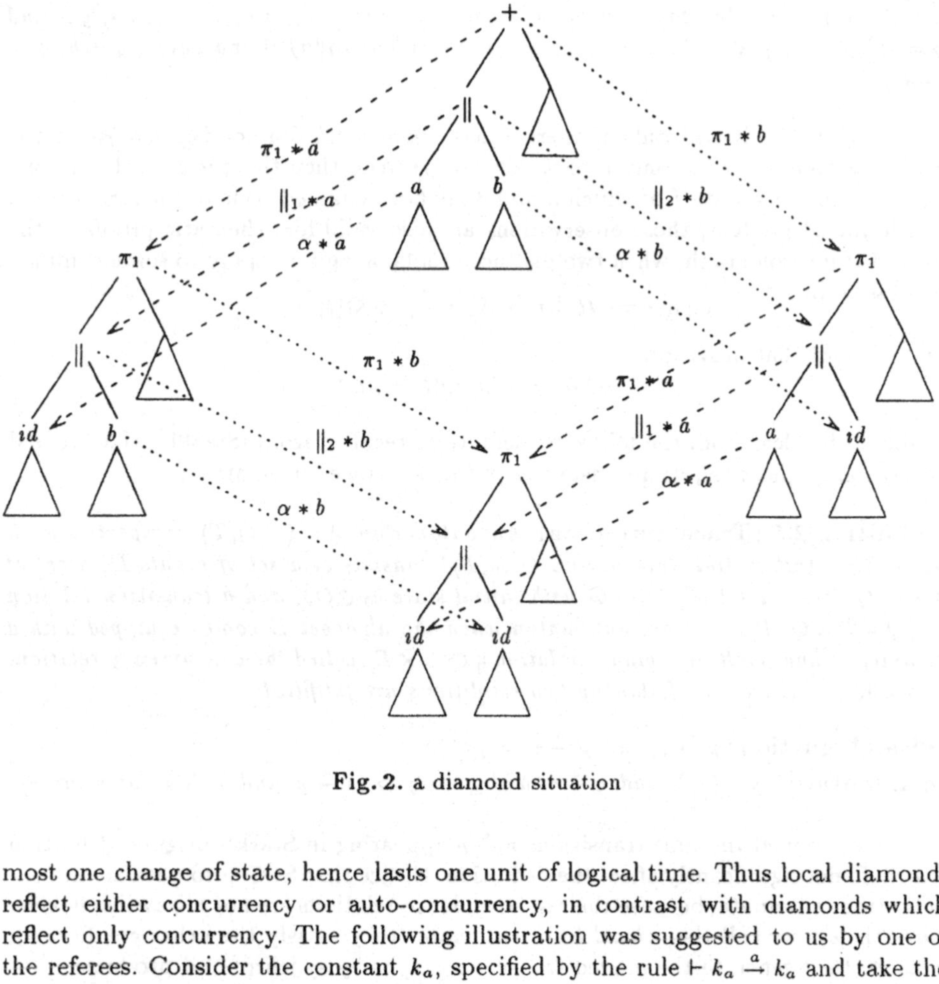

Fig. 2. a diamond situation

most one change of state, hence lasts one unit of logical time. Thus local diamonds reflect either concurrency or auto-concurrency, in contrast with diamonds which reflect only concurrency. The following illustration was suggested to us by one of the referees. Consider the constant k_a, specified by the rule $\vdash k_a \xrightarrow{a} k_a$ and take the function $f(x, y, z)$ specified by the rules

$$\frac{x \xrightarrow{a} x' \quad y \xrightarrow{b} y'}{f(x, y, z) \xrightarrow{b} f(x', y', z)} \, r_1 \qquad\qquad \frac{x \xrightarrow{a} x' \quad z \xrightarrow{c} z'}{f(x, y, z) \xrightarrow{c} f(x', y, z')} \, r_2$$

then the two transitions obtained by applying respectively r_1 and r_2 to $f(k_a, k_b, k_c)$ are concurrent. Now if we replace the constant k_a by the "two-state clock", specified by $\vdash k_a \xrightarrow{a} k'_a$ and $\vdash k'_a \xrightarrow{a} k_a$ the corresponding transitions are no longer concurrent. A one-state clock can be *read* in parallel by an arbitrary number of agents in one "unit" of time, whereas a two-state clock beat time and forces the actions b and c to be performed sequentially (in either order).

Now, we can define formally the independence of schematic proofs.

Definition 2.4 (Independence of schematic proof) *Two schematic proofs* A *and* B *are independent* (A $\|$ B) *if and only if* A $\neq$ B *and* $\forall s \in Dom($ A $) \cap Dom($ B $)$,

$A(s) \| B(s)$, *where two proof constructors* $\alpha = <\{i_1, \ldots, i_k\}, <a_1, \ldots, a_k, a>, g>$ *and* $\beta = <\{j_1, \ldots, j_l\}, <b_1, \ldots, b_l, b>, h>$ *are independent* $(\alpha \| \beta)$ *if and only if* $g = h$, $g \xrightarrow{\alpha}$ *and* $g \xrightarrow{\beta}$.

Notice that $\alpha \| \beta$ if and only if there exists some local diamond $\diamond(f, \alpha, \beta)$ in the SOS automaton (for the only-if part take $f = g$). And then there is a local diamond $\diamond(f, \alpha, \beta)$ in every state f at which α and β are both enabled. Due to the structure of synchronized product, those observations are also valid for schematic proofs in the proved transition graph: when two distinct proofs A and B apply to some common process term,

$$A \| B \iff \forall t \ [(t \xrightarrow{A} \ \& \ t \xrightarrow{B}) \Rightarrow \Diamond(t, A, B)]$$

and without that provision,

$$A \| B \iff \exists t \ \Diamond(t, A, B)$$

It should be clear from the following definition, recalled from [Sta89b], that proved transition graphs $\Gamma(\Sigma, R)$ are instances of Stark's trace automata.

Definition 2.5 (Trace automata) *An automaton* $A = (E, Q, T)$ *(respectively an automaton with initial state* $A = (E, Q, q_0, T)$*) consists of a set of events* E, *a set of states* Q *(resp. a set of states* Q *with initial state* $q_0 \in Q$*), and a transition relation* $T \subset Q \times E \times Q$. *It is a trace automaton when the alphabet* E *comes equipped with a symmetric and irreflexive binary relation* $\| \subset E \times E$, *called the concurrency relation, and when moreover the following two conditions are fullfilled.*

disambiguation : $p \xrightarrow{a} q$ *and* $p \xrightarrow{a} r \Rightarrow q = r$
commutativity : $(a \| b$ *and* $q \xrightarrow{a} r$ *and* $q \xrightarrow{b} s) \Rightarrow (s \xrightarrow{a} p$ *and* $r \xrightarrow{b} p$ *for some* $p)$

We have dropped the unit transitions $p \xrightarrow{\epsilon} p$ appearing in Stark's original definition which do not significantly affect the definition. Notice that SOS automata are not exactly trace automata, because their independence relations are not always irreflexive. Nevertheless we call them *local trace automata* (they satisfy the other conditions), because they generate the proved transition graphs by synchronized products.

Definition 2.6 (Local trace automata) *A local trace automaton* $A = (E, \|, Q, T)$ *(respectively a local trace automaton with initial state* $A = (E, \|, Q, q_0, T)$*) is an automaton (resp. an automaton with initial state) equipped with a symmetric relation* $\| \subset E \times E$ *fulfilling the property of disambiguation and commutativity.*

3 Trace Automata and Nets

A question which arises naturally is to find an axiomatic characterization for the class of (finite) process automata. In order to solve that question, we establish in the present section a connection between trace automata and a new type of nets introduced for that purpose.

For any relation $\nabla \subset A \times B$ we shall denote $()^{\nabla} : A \rightarrow 2^B$ and ${}^{\nabla}() : B \rightarrow 2^A$ the mappings given by:

$$a \nabla b \text{ iff } b \in a^{\nabla} \text{ iff } a \in {}^{\nabla} b$$

We say that a family of subsets $X_i \subset A$ is a *partitioning* of A whenever they are pairwise disjoint: $X_i \cap X_j = \emptyset$ for $i \neq j$, and cover the whole of A: $\bigcup_i X_i = A$. It is like a partition except that some of the components of a partitioning may be empty. Notice that a family $\nabla_i \subset A \times B$ of binary relations is a partitioning of $A \times B$ **iff** for all $a \in A$, the family a^{∇_i} is a partitioning of B **iff** for all $b \in B$ the family ${}^{\nabla_i}b$ is a partitioning of A.

Definition 3.1 (Nets) *A net* $\mathcal{N} = (P, A, \overset{0}{\leftarrow}, \overset{1}{\leftarrow}, \overset{0}{\rightarrow}, \overset{1}{\rightarrow}, \perp, M_0)$ *consists of a set of places* P*, a set of actions (or events)* A*, flow relations* $\overset{0}{\leftarrow}, \overset{1}{\leftarrow}, \overset{0}{\rightarrow}, \overset{1}{\rightarrow}, \perp$ *which give a partitioning of* $A \times P$*, and an initial marking* $M_0 \subset P$.

If $M \subset P$ is a marking with $x \in M$ (respectively $x \notin M$) we say that the place x is *full* (resp. *empty*) for the marking M. x is said to be an *input place* for a if $a \leftarrow x$ where $\leftarrow = \overset{0}{\leftarrow} \cup \overset{1}{\leftarrow}$ and x is said to be an *output place* for a if $a \rightarrow x$ where $\rightarrow = \overset{0}{\rightarrow} \cup \overset{1}{\rightarrow}$. The behaviour of the net is the following: an action a is *enabled* at a marking M (in notation $M \overset{a}{\rightarrow}$) if every input place of a is full for the marking M, i.e. the input places are the *pre-conditions* for an action to be *fired*. When an action a is fired, the places x such that $a\mathbf{0}x$ (where $\mathbf{0} = \overset{0}{\leftarrow} \cup \overset{0}{\rightarrow}$) are emptied if they were not already empty, and the places x such that $a\mathbf{1}x$ (where $\mathbf{1}\overset{1}{\leftarrow} \cup \overset{1}{\rightarrow}$) are filled if they were not already full. Therefore, if $\mathcal{M} = 2^P$ is the set of *markings*, the set of transitions $T \subset \mathcal{M} \times A \times \mathcal{M}$ for the net $\mathcal{N}$ is given by:

$$M \overset{a}{\rightarrow} M' \text{ iff } a^{\leftarrow} \subset M \text{ and } M' = (M \setminus a^{\mathbf{0}}) \cup a^{\mathbf{1}}$$

To sum up, there are two types of input places for an action a: the first type corresponds to pre-conditions that must hold for a to execute but which are unaffected by this action (we say that the action a *tests* the condition x when $a \overset{1}{\leftarrow} x$), whereas pre-conditions of the second type do no longer hold after a has executed (we say that the action a *consumes* the resource x when $a \overset{0}{\leftarrow} x$). As for the output places of an action a, they do not condition the firing of a but on the contrary, they are set unconditionally to a value at each execution of a. The remaining case when $a \perp x$ (read a and x are *orthogonal*) corresponds to conditions which are neither tested nor affected by the action a. A net may be depicted as shown in Fig. 3 where an arc is drawn from a place x (depicted by a circle) to an action a (depicted by a box) when x is an input place for a, and conversely an arc is drawn from an action a to every of its ouput places. An arc between a place and an action is labelled according to the value of the place after the execution of the action (empty: 0 or full: 1). A marking M may then be represented by setting one token in each of its places. In Fig. 3 we have also represented the transition system associated to the net with initial marking as indicated. In this example, the actions a and b consume their respective resources x and y and each of them turns on the condition z, the action c tests the condition z and thus awaits for a or b (or both) to be performed, and then consumes its resource t and turns on the condition u. Intuitively the actions a and b may be performed in parallel, and when for instance a has been fired, the actions b and c may also be performed in parallel. This is reflected by the presence of *diamond situations* $\Diamond(M, a, b)$ in the transition system, which are defined by a marking M and two actions a and b such that $M \overset{ab}{\rightarrow}$, $M \overset{ba}{\rightarrow}$ and $M.ab = M.ba$. A diamond situation occurs when two independent actions are both enabled at a given marking where:

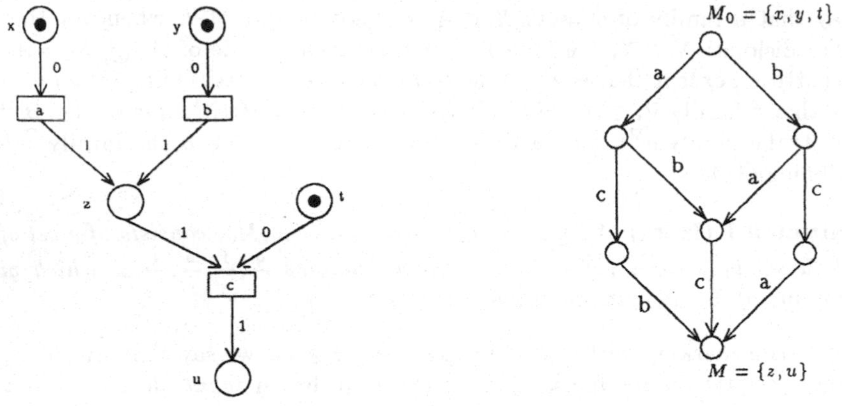

Fig. 3. a net and its transition system

Definition 3.2 (Independence in nets) *Two* distinct *actions a and b in a net* $\mathcal{N}$ *are said to be* independent *(in notation* $a \| b$*) when* (i) *there exists an accessible marking M at which they are both enabled,* (ii) $a^{\overset{0}{\leftarrow}} \subset b^{\perp}$ & $a^{0} \cap b^{1} = \emptyset$*, and symmetrically* (iii) $b^{\overset{0}{\leftarrow}} \subset a^{\perp}$ & $b^{0} \cap a^{1} = \emptyset$.

Two actions are structurally independent (conditions *(ii)* and *(iii)*) when no resource consumed by one action is either tested or possibly modified by the other one, and they do not deliver incompatible values to shared places. The condition *(i)* is not structural but it ensures that all the information about independence is *visible* on the transition system in the sense that two actions a and b are independent if and only if they occur in a diamond situation. The following definition and proposition state this property more precisely.

Definition 3.3 (Canonical trace automata) *A trace automaton* $(A, \|, Q, q_0, T)$ *is said to be canonical when* (i) *every state is accessible from the initial state,* (ii) *every action is enabled in at least one state,* (iii) *each pair of independent actions is enabled in at least one state (giving rise to at least one diamond situation),* and (iv) *independence reflects diamond situations: two* distinct *actions a and b are independent whenever* $q \xrightarrow{ab}$, $q \xrightarrow{ba}$ *and* $q.ab = q.ba$ *for some state q.*

Proposition 3.4 *Let* $T_a \subset \mathcal{M}_a \times A \times \mathcal{M}_a$ *be the restriction of the transition system associated with a net* $\mathcal{N}$ *to its accessible markings* $\mathcal{M}_a \subset 2^P$*, then* $(A, \|, \mathcal{M}_a, M_0, T_a)$ *is a canonical trace automaton.*

Trace automata may be derived from nets as the induced transitions systems. Conversely we can associate to a trace automaton a net whose places are *regions* in the following sense.

Definition 3.5 (Regions) *Let* $(A, \|, Q, q_0, T)$ *be a trace automaton, we define the following (disjoint) flow relations on* $A \times 2^Q$*:*

$a \overset{0}{\leftarrow} x$ **iff** $q \overset{a}{\rightarrow} q' \Rightarrow (q \in x \;\;\&\;\; q' \notin x)$
$a \overset{0}{\rightarrow} x$ **iff** $(q \overset{a}{\rightarrow} q' \Rightarrow q' \notin x) \;\&\; (\exists q \overset{a}{\rightarrow} q' \;\, with \;\, q \notin x)$
$a \overset{1}{\leftarrow} x$ **iff** $q \overset{a}{\rightarrow} q' \Rightarrow (q \in x \;\;\&\;\; q' \in x)$
$a \overset{1}{\rightarrow} x$ **iff** $(q \overset{a}{\rightarrow} q' \Rightarrow q' \in x) \;\&\; (\exists q \overset{a}{\rightarrow} q' \;\, with \;\, q \notin x)$
$a \bot x$ **iff** $[q \overset{a}{\rightarrow} q' \Rightarrow (q \in x \; iff \; q' \in x)] \;\&\; (\exists q \overset{a}{\rightarrow} q' \; with \; q \notin x) \;\&\; (\exists q \overset{a}{\rightarrow} q' \; with \; q \in x)$

A set of states $x \subset Q$ is said to be a region *(or place) if $\{\overset{0}{\leftarrow} x, \overset{1}{\leftarrow} x, \overset{0}{\rightarrow} x, \overset{1}{\rightarrow} x, {}^{\bot}x\}$ is a partitioning of the set A of actions. The above defined relations then restrict to homonymic flow relations on $A \times P$ where P is the set of regions. The net derived from the trace automaton is then defined as $(P, A, \overset{0}{\leftarrow}, \overset{1}{\leftarrow}, \overset{0}{\rightarrow}, \overset{1}{\rightarrow}, \bot, M_0)$ with $M_0 = \{x \in P / \; q_0 \in x\}$ as initial state.*

Like we did for nets we also define on $A \times 2^Q$ the derived flow relations $\rightarrow = \overset{0}{\rightarrow} \cup \overset{1}{\rightarrow}$, $\leftarrow = \overset{0}{\leftarrow} \cup \overset{1}{\leftarrow}$, $\mathbf{0} = \overset{0}{\leftarrow} \cup \overset{0}{\rightarrow}$, and $\mathbf{1} = \overset{1}{\leftarrow} \cup \overset{1}{\rightarrow}$. The last two have a clear characterization:

$$a\mathbf{0}x \text{ iff } q \overset{a}{\rightarrow} q' \Rightarrow q' \notin x$$
$$a\mathbf{1}x \text{ iff } q \overset{a}{\rightarrow} q' \Rightarrow q' \in x$$

In order to obtain a simple characterization of regions, let us finally state:

$$a \dagger x \text{ iff } q \overset{a}{\rightarrow} q' \Rightarrow (q \in x \;\text{ iff }\; q' \in x)$$

then x is a region if and only if $A = {}^{\mathbf{0}}x \cup {}^{\mathbf{1}}x \cup {}^{\dagger}x$. Thus $x \subset Q$ is a region if and only if none of two *forbidden cases* depicted in Fig. 4 occurs for x, more precisely x is not

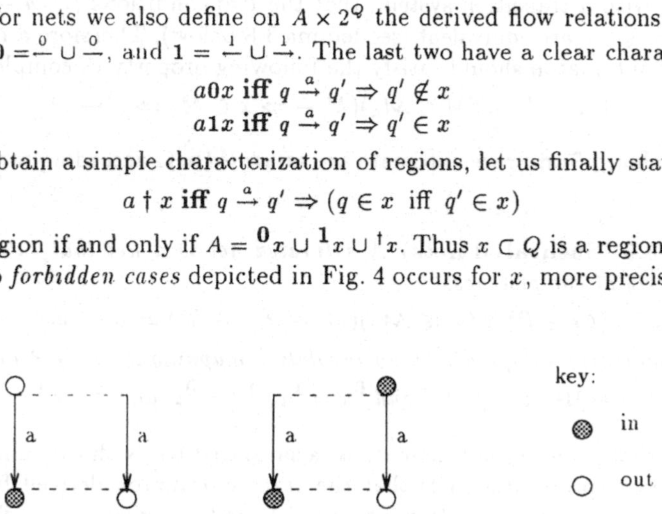

Fig. 4. forbidden cases for a region

a region if and only if

$$\exists a \in A \;\; \exists q_1 \overset{a}{\rightarrow} q_1' \in T \;\; \exists q_2 \overset{a}{\rightarrow} q_2' \in T \;\; \text{ with } \;\; q_1' \in x \;\&\; q_2' \notin x \;\&\; (q_1 \notin x \text{ or } q_2 \in x)$$

Therefore, if $x \subset Q$ is a region, its complement $\overline{x} = Q \setminus x$ is also a region, with ${}^{\mathbf{0}}\overline{x} = {}^{\mathbf{1}}x$, ${}^{\mathbf{1}}\overline{x} = {}^{\mathbf{0}}x$, and ${}^{\bot}\overline{x} = {}^{\bot}x$.

Since all information about independence is reflected in the structure of its transition system, the concurrent behaviour of a *canonical* trace automaton may be identified with its class for the rooted bisimulation equivalence. The concurrent behaviour of a net is then the behaviour of its associated canonical trace automaton (Prop. 3.4). We know how to associate a net with a canonical trace automaton and *vice versa*; now it is not true that the objects set in correspondence always have the same behaviour. For instance if we consider the transition system depicted in Fig. 3 and identify the states M_0 and M we obtain a cyclic trace automaton with only the trivial regions. That case of collapsing is due to *contact situations* (there exist a state q, an action a and transitions $q_1 \xrightarrow{a} q$ and $q \xrightarrow{a} q_2$) which can easily be eliminated by a finite unfolding operation which produces an equivalent trace automaton. But there exists more involved cases, as we shall see, which cannot be dealt with so easily. We are therefore looking for additional conditions to impose on nets and trace automata, such that translations in both directions between nets and automata preserve behaviours. On the one hand, since the complement of a region is a region, we know that a net which is produced from a trace automaton comes up equipped with a duality operation $\overline{()} : P \to P$ on places, which is an involutive mapping such that $(\forall x \in P)\ x \in M_0 \Leftrightarrow \overline{x} \notin M_0$, and ${}^{\mathbf{0}}\overline{x} = {}^{\mathbf{1}}x$, ${}^{\mathbf{1}}\overline{x} = {}^{\mathbf{0}}x$, and ${}^{\perp}\overline{x} = {}^{\perp}x$. Notice that by the very definition of the transition system associated with a net, the above property of the initial marking is shared by all accessible markings: $(\forall M \in \mathcal{M}_a)$ $x \in M \Leftrightarrow \overline{x} \notin M$. On the other hand, there follows from the definition of a region x associated with a transition system, that the two conditions *(i):* $a \leftarrow x$ and, *(ii):* $\forall q \in Q\ q \xrightarrow{a} \Rightarrow q \in x$ are equivalent (see lemma 3.8 below). Therefore a net produced from a trace automaton should satisfy the following property of completeness:

$$(\forall x \in P)\ [(\forall M \in \mathcal{M}_a)(M \xrightarrow{a} \Rightarrow x \in M) \Rightarrow a \leftarrow x]$$

The nets produced from trace automata are *saturated* nets according to the following definition.

Definition 3.6 (Saturated nets) *A saturated net is a net that satisfies the following property of completeness:*

$$(\forall x \in P)\ [(\forall M \in \mathcal{M}_a)(M \xrightarrow{a} \Rightarrow x \in M) \Rightarrow a \leftarrow x]$$

and that is moreover equipped with an involutive mapping $\overline{()} : P \to P$ on places such that $(\forall x \in P)\ (x \in M_0 \Leftrightarrow \overline{x} \notin M_0)$ and ${}^{\mathbf{0}}\overline{x} = {}^{\mathbf{1}}x$, ${}^{\mathbf{1}}\overline{x} = {}^{\mathbf{0}}x$ and ${}^{\perp}\overline{x} = {}^{\perp}x$.

We shall see that for every net there exists a saturated net with the same behaviour. We already saw (proposition 3.4) that the trace automaton derived from a net is *canonical*. Now, whenever an action a cannot be fired at a given (accessible) marking in the net, a token should be missing in some input place of a, and this dynamic property of nets entails the following property of separation for the trace automaton derived from the net.

Definition 3.7 (Separated trace automata) *A Separated trace automaton is a canonical trace automaton (definition 3.3) which satisfies the following property of separation:*

$$q \overset{a}{\not\rightarrow} \;\Rightarrow\; \exists x \in P\ s.t.\ q \notin x\ \&\ x \in a^{\leftarrow}$$

Lemma 3.8 *Let x be a region, then conditions* (i) *$a \leftarrow x$ and* (ii) *$\forall q \in Q \quad q \xrightarrow{a} \Rightarrow q \in x$ are equivalent.*

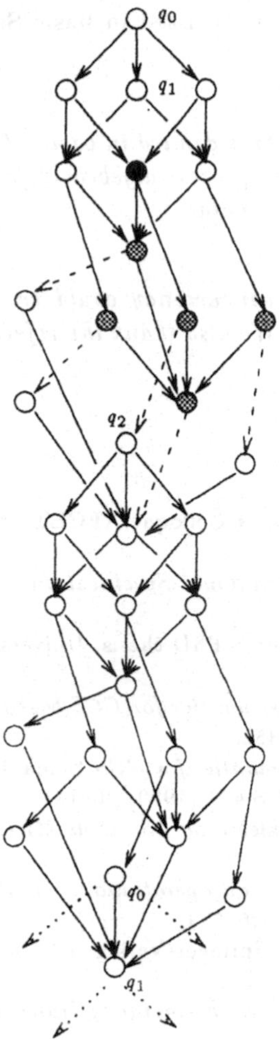

In view of the above lemma, the separation property means that for every action a and state q at which a is not enabled, we can find a region which separate q from a in the sense that it contains every state at which a is enabled but not q. Figure 5 displays the simplest example we were able to produce of a contact-free ($q \xrightarrow{a} \Rightarrow not(\xrightarrow{a} q)$) but not separated trace automaton. The vertical arrows represent one action, say a, and the dashed arrows represent another action, say b. Action b is enabled in the grey states but not in the black one. Let us prove by contradiction, that an input region of b must contain the black state. Assume $b \overset{0}{\leftarrow} x$ and x does not contain the black state. Then all grey states belong to x and necessarily $a1x$, but this is impossible since there exists one state which is reached by both a and b. Assume $b \overset{1}{\leftarrow} x$ and x does not contain the black state. Then all sources and targets of dashed arrows and all targets of vertical arrows are in x, Using the second excluded pattern shown in Fig. 4, it is easy to propagate downwards the property $q \in x$ from state q_2 to all the other states q such that $q \xrightarrow{a}$, including the state q_0 and the black state. Due to the existence of one action (namely a) which may never be taken but remains always live, that case of non-separation cannot be escaped by any sort of limited unfolding.

Fig. 5. a non separated automaton

With adequate notions of morphisms for separated trace automata and saturated nets, the above correspondences between nets and trace automata restrict to an adjunction such that the objects set in correspondence have the same behaviour.

Theorem 3.9 *There exists an adjunction between the categories of separated trace automata and saturated nets, such that both left and right adjoint functors preserve behaviours.*

The following corollary emphasizes the fact (already mentioned) that all classes of behaviourally equivalent nets are represented in that correspondance.

Corollary 3.10 (Saturation of nets) *For every net there exists a saturated net with the same behaviour.*

We are now able to establish that the finite trace automata realized in basic SOS are exactly the separated ones.

Proposition 3.11 (Representation results)
(i) All finite nets are realized by finite terms in process algebras defined in basic SOS, and (ii)The set of trace automata realized by finite terms in process algebras defined in basic SOS is the set of all finite and separated trace automata.

Acknowledgement: *The idea according to which true concurrency could be extracted from SOS was suggested to us by Frits Vaandrager. We also thank the referees for their valuable comments.*

References

[AN82] ARNOLD, A., and NIVAT, M., *Comportements de processus.* Colloque AFCET « Les Mathématiques de l'Informatique », (1982) 35–68.

[BD92] BADOUEL, E., and DARONDEAU, PH., *Structural Operational Specifications and Trace Automata,* INRIA research report no 1631 (1992).

[Bed88] BEDNARCZYK, M.A., *Categories of asynchronous systems.* PhD thesis, University of Sussex, report no.1/88 (1988).

[BC88] BOUDOL, G., and CASTELLANI, I., *A non-interleaving semantics for CCS based on proved transitions.* Fundamenta Informaticae XI (1988) 433–452.

[BC90] BOUDOL, G., and CASTELLANI, I., *Three Equivalent Semantics for CCS.* Semantics of Systems of Concurrent Processes, I. Guessarian (Ed.) LNCS 469 (1990) 96–141.

[deS84] DE SIMONE, R., *Calculabilité et expressivité dans l'algèbre de processus MEIJE.* Thèse de 3ème Cycle, Université de Paris VII (1984).

[LX90] LARSEN, K., and XINXIN, L., *Compositionality through an operational semantics of contexts.* Proc. 17th ICALP (Warwick), LNCS 443 (1990) 526–539.

[Mil80] MILNER, R., *A calculus of communicating systems.* Springer-Verlag LNCS 92 (1980).

[NRT90] NIELSEN, M., ROZENBERG, G., AND THIAGARAJAN, P.S., *Elementary Transition Systems.* DAIMI PB-310 Aarhus (1990).

[Niv79] NIVAT, M., *Sur la synchronisation des processus.* Revue technique Thomson-CSF, 11, (1979) 899–919.

[Plo81] PLOTKIN, G., *A structural approach to operational semantics.* DAIMI-FN-19 Aarhus (1981).

[Sta89a] STARK, E.W., *Concurrent transition systems.* Theoretical Computer Science 64 (1989) 221–269.

[Sta89b] STARK, E.W., *Connections between a concrete and an abstract model of concurrent systems.* 5th Mathematical Foundations of programming semantics (1989) 53–79.

[Sta89c] STARK, E.W., *Compositional Relational Semantics for Indeterminate Dataflow Networks.* Summer Conference on Category Theory and Computer Science, LNCS 389 (1989) 52–74.

[Win91] WINSKEL, G., *Categories of Models for Concurrency.* Advanced School on the Algebraic, Logical, and Categorical Foundations of Concurrency, Gargnano del Garda (1991).

At-Most-Once Message Delivery

A Case Study in Algorithm Verification

Butler Lampson
Digital Equipment Corp.

Nancy Lynch
MIT

Jørgen Søgaard-Andersen,
MIT and
Department of Computer Science
Technical University of Denmark

1 Introduction

The ***at-most-once message delivery problem*** involves delivering a sequence of messages submitted by a user at one location to another user at another location. If no failures occur, all messages should be delivered in the order in which they are submitted, each exactly once. If failures (in particular, node crashes or timing anomalies) occur, some messages might be lost, but the remaining messages should not be reordered or duplicated.

This talk examines two of the best-known algorithms for solving this problem: the *clock-based* protocol of [3] and the *five-packet interchange* protocol of [2]. It is shown that both of these protocols can be understood as implementations of a common (untimed) protocol that we call the *generic* protocol. It is also shown that the generic protocol meets the problem specification.

The development is carried out in the context of (timed and untimed) automata [7, 8] and [6], using simulation techniques [7]. It exercises many aspects of the relevant theory, including timed and untimed automata, refinement mappings, forward and backward simulations, history and prophecy variables. The theory provides insight into the algorithms, and vice versa.

In this short paper, we simply give formal descriptions of the problem specification and of the two algorithms, leaving detailed discussion of the proof for the talk and for a later paper.

2 The Specification S

The transitions of the specification we use for the at-most-once message delivery problem are given below. Formally, the object denoted by the specification is an I/O automaton [5, 6]. The notation used is somewhat standard for describing I/O automata (see, for example, [4]). The user interface is a set of *external* (input and output) actions. Even though we in S have a *central*, i.e., not *distributed*, view of the system, the external actions can be logically partitioned into actions on the "sender" side (*send_msg*, *ack*, $crash_s$, and $recover_s$) and actions on the "receiver" side (*receive_msg*, $crash_r$, and $recover_r$). Furthermore, there is an internal action *lose*. All these actions then manipulate shared data structures like, e.g., *queue*.

```
send_msg(m)
  Effect:
    if rec_s = false then
      append m to queue
      status := ?

receive_msg(m)
  Precondition:
    rec_r = false
    m is first on queue
  Effect:
    remove first element of queue
    if queue is empty and
      status = ? then
      status := true

ack(b)
  Precondition:
    rec_s = false
    status = b ∈ {true, false}
  Effect:
    none

crash_s
  Effect:
    rec_s := true

crash_r
  Effect:
    rec_r := true

lose
  Precondition:
    rec_s = true or rec_r = true
  Effect:
    delete arbitrary elements
      of queue
    if last element of queue
      is deleted then
      status := false
    else optionally
      status := false

recover_s
  Precondition:
    rec_s = true
  Effect:
    rec_s := false

recover_r
  Precondition:
    rec_r = true
  Effect:
    rec_r := false
```

We specify fairness by partitioning the actions that the protocol controls (output and internal action) in *fairness classes.* In the execution of the protocol it must not be the case that actions from a fairness class are continuously enabled without actions from that class being executed infinitely often.

For the specification S we use the following five classes:

1. *ack* actions
2. *receive_msg* actions
3. *recover$_s$*
4. *recover$_r$*
5. *lose*

3 The Clock-Based Protocol C

Code for the clock-based protocol of [3] is given below. Since at this level of abstraction we have a distributed view of the system, the code is partitioned into code for the sender and code for the receiver part of the protocol. Formally, the sender and receiver protocols are *timed automata* in the style of [8].

In C, the sender protocol associates a *time* value with each message it wishes to deliver. The *time* values are obtained from a local clock. The receiver protocol uses the associated time value to decide whether or not to accept a received message – as a rough strategy, it will accept a message provided the associated time is greater than the time of the last message that was accepted. However, the receiver protocol cannot always remember the time of the last accepted message: it might forget this information because of a crash, or simply because a long time has elapsed since the last message was accepted and it is no longer efficient to remember it. Thus, the receiver protocol uses safe time estimates determined from its own local clock to decide when to accept a message.

Correctness of this protocol requires that the two local clocks be synchronized to real time, to within a tolerance ϵ, when crashes do not occur. It also requires reliability bounds and upper time bounds on the low-level channels connecting the sender and receiver protocols.

Sender

$send\_msg(m)$
  Effect:
    if $mode_s \neq$ rec then
      append m to buf_s

$choose\_id(m, t)$
  Precondition:
    $mode_s =$ acked,
    m is first on buf_s,
    $time_s = t$,
    $t > last_s$
  Effect:
    $mode_s :=$ send
    remove first element of buf_s
    $current\text{-}msg_s := m$
    $last_s := t$

$send\_pkt_{sr}(m, t)$
  Precondition:
    $mode_s =$ send,
    $current\text{-}msg_s = m$,
    $last_s = t$
  Effect:
    none

$receive\_pkt_{rs}(t, b)$
  Effect:
    if $mode_s =$ send and
      $last_s = t$ then
      $mode_s :=$ acked
      $current\text{-}ack_s := b$
      $current\text{-}msg_s :=$ nil

$ack(b)$
  Precondition:
    $mode_s =$ acked
    buf_s is empty
    $current\text{-}ack_s = b$
  Effect:
    none

$crash_s$
  Effect:
    $mode_s :=$ rec

$recover_s$
  Precondition:
    $mode_s =$ rec
  Effect:
    $mode_s :=$ acked
    $last_s := time_s$
    empty buf_s
    $current\text{-}msg_s :=$ nil
    $current\text{-}ack_s := false$

$tick_s(t)$
  Effect:
    $time_s := t$

We only need one class of locally controlled actions for the sender protocol:

1. *choose_id*, $send\_pkt_{sr}$, *ack*, and $recover_s$ actions

We put an upper time bound of l on all the classes, meaning that if actions from a class get enabled, then an action from that class must be executed within time l unless the actions are disabled in the meantime.

Receiver

$receive\_pkt_{sr}(m, t)$
  Effect:
    if $mode_r \neq$ rec then
      if $lower_r < t \leq upper_r$ then
        $mode_r :=$ rcvd
        add m to $buffer_r$
        $last_r := t$
        $lower_r := t$
      else if $last_r < t \leq lower_r$ then
        add t to $nack\text{-}buffer_r$
      else if $mode_r =$ idle and
          $t = last_r$ then
        $mode_r =$ ack

receive_msg(m)
  Precondition:
    $mode_r =$ rcvd,
    m is first on buf_r
  Effect:
    remove first element of buf_r
    if buf_r is empty then
      $mode_r :=$ ack

$send\_pkt_{rs}(t, true)$
  Precondition:
    $mode_r =$ ack,
    $last_r = t$
  Effect:
    $mode_r :=$ idle

$send\_pkt_{rs}(t, false)$
  Precondition:
    $mode_r \neq$ rec
    t is first on $nack\text{-}buf_r$
  Effect:
    remove first element of $nack\text{-}buf_r$

$crash_r$
  Effect:
    $mode_r :=$ rec

$recover_r$
  Precondition:
    $mode_r =$ rec,
    $upper_r + 2\epsilon < time_r$
  Effect:
    $mode_r :=$ idle
    $last_r := 0$
    empty buf_r
    $lower_r := upper_r$
    $upper_r := time_r + \beta$
    empty $nack\text{-}buf_r$

increase-lower(t)
  Precondition:
    $mode_r \neq$ rec,
    $lower_r \leq t < time_r - \rho$
  Effect:
    $lower_r := t$

increase-upper(t)
  Precondition:
    $mode_r \neq$ rec,
    $upper_r \leq t = time_r + \beta$
  Effect:
    $upper_r := t$

$tick_r(t)$
  Effect:
    $time_r := t$

For the receiver protocol we use the following classes of locally controlled actions:

1. *receive_msg*, $send\_pkt_{rs}(, true)$, and $recover_r$ actions
2. $send\_pkt_{rs}(, false)$ actions
3. *increase-lower* actions
4. *increase-upper* actions

4 The Five-Packet Protocol *5P*

Code for the five-packet handshake protocol of [2] is given below. As for C, the code is partitioned into code for the sender protocol and code for the receiver protocol. For the *5P* protocol we assume that the sender and receiver protocols communicate via channels that may lose or dublicate packets, the latter only a finite number of times for each packet instance. In order to prove liveness properties of the *5P* protocol, we furthermore assume that if the same packet is sent an infinite number of times, then it will also be received an infinite number of times.

In this protocol, for each message that the sender protocol wishes to deliver, there is an initial exchange of packets between the sender and receiver protocols to establish a commonly-agreed-upon message identifier. The sender protocol then associates this identifier with the message. The receiver protocol uses the associated identifier to decide whether or not to accept a received message – it will accept a message provided the associated identifier is current. Additional packets are required in order to tell the receiver protocol when it can throw away a current identifier.

4.1 Sender

send_msg(m)
 Effect:
  if $mode_s \neq$ rec then
   append m to buf_s

choose_jd(jd)
 Precondition:
  $mode_s =$ acked,
  m first on buf_s,
  $jd \notin jd\text{-}used_s$
 Effect:
  $mode_s :=$ needid
  $jd_s := jd$
  add jd to $jd\text{-}used_s$
  remove first element of buf_s
  $current\text{-}msg_s := m$

send_pkt$_{sr}$(needid, nil, jd)
 Precondition:
  $mode_s =$ needid, $jd = jd_s$
 Effect:
  none

receive_pkt$_{rs}$(accept, jd, id)
 Effect:
  if $mode_s \neq$ rec then
   if $mode_s =$ needid and
    $jd = jd_s$ then
    $mode_s :=$ send
    $id_s := id$
    add id to the end of $used_s$
   else if $id \neq id_s$ then
    add id to the end
     of $acked\text{-}buf_s$

send_pkt$_{sr}$(send, id, m)
 Precondition:
  $mode_s =$ send,
  $id = id_s$,
  $m = current\text{-}msg_s$
 Effect:
  none

receive_pkt$_{rs}$(ack, id, b)
 Effect:
  if $mode_s \neq$ rec then
   if $mode_s =$ send and
    $id = id_s$ then
    $mode_s :=$ acked
    $current\text{-}ack_s := b$
    $jd_s :=$ nil
    $id_s :=$ nil
    $current\text{-}msg_s :=$ nil
   if $b = true$ then
    add id to $acked\text{-}buf_s$

send_pkt$_{sr}$(acked, id, nil)
 Precondition:
  id is first on $acked\text{-}buf_s$
 Effect:
  remove first element of $acked\text{-}buf$

ack(b)
 Precondition:
  $mode_s =$ acked, buf_s is empty,
  $b = current\text{-}ack_s$
 Effect:
  none

crash$_s$
 Effect:
  $mode_s :=$ rec

recover$_s$
 Precondition:
  $mode_s =$ rec
 Effect:
  $mode_s :=$ acked
  $jd_s :=$ nil
  $id_s :=$ nil
  empty buf_s
  $current\text{-}msg_s :=$ nil
  $current\text{-}ack_s := false$
  empty $acked\text{-}buf_s$

grow-jd-used$_s$
 Precondition:
  none
 Effect:
  add some *JD*s to $jd\text{-}used_s$

We define the following fairness classes of the locally controlled actions of the sender protocol:

1. *ack*, *choose_jd*(*jd*), *send_pkt*$_{sr}$(**needid**, ,), *send_pkt*$_{sr}$(**send**, ,), and *recover*$_s$ actions
2. *send_pkt*$_{sr}$(**acked**, ,) actions
3. *grow-jd-used*$_s$

4.2 Receiver

receive_pkt$_{sr}$(needid, nil, *jd*)
  Effect:
    if *mode*$_s$ = **idle** then
      mode$_r$:= **accept**
      choose an *id* not in *issued*$_r$
      jd$_r$:= *jd*
      id$_r$:= *id*
      add *id* to *issued*$_r$

send_pkt$_{rs}$(accept, *jd*, *id*)
  Precondition:
    mode$_r$ = accept,
    jd = *jd*$_r$,
    id = *id*$_r$
  Effect:
    none

receive_pkt$_{sr}$(send, *id*, *m*)
  Effect:
    if *mode*$_r$ ≠ **rec** then
      if *mode*$_r$ = accept and
        id = *id*$_r$ then
        mode$_r$:= **rcvd**
        append *m* to *buf*$_r$
        last$_r$:= *id*
      else if *id* ≠ *last*$_r$ then
        append *id* to *nack-buf*$_r$

receive_msg(*m*)
  Precondition:
    mode$_r$ = **rcvd**, *m* first on *buf*$_r$
  Effect:
    remove the first element of *buf*$_r$
    if *buf*$_r$ is empty then
      mode$_r$:= **ack**

send_pkt$_{rs}$(ack, *id*, *true*)
  Precondition:
    mode$_r$ = ack, *id* = *last*$_r$
  Effect:
    none

send_pkt$_{rs}$(ack, *id*, *false*)
  Precondition:
    mode$_r$ ≠ **rec**,
    id is first on *nack-buf*$_r$
  Effect:
    remove first element of *nack-buf*$_r$

receive_pkt$_{sr}$(acked, *id*, nil)
  Effect:
    if (*mode*$_r$ = accept and
        id = *id*$_r$) or
      (*mode*$_r$ = ack and
        id = *last*$_r$) then
      mode$_r$:= **idle**
      jd$_r$:= **nil**
      id$_r$:= **nil**
      last$_r$:= **nil**

crash$_r$
  Effect:
    mode$_r$:= **rec**

recover$_r$
  Precondition:
    mode$_r$ = **rec**
  Effect:
    mode$_r$:= **idle**
    jd$_r$:= **nil**
    id$_r$:= **nil**
    last$_r$:= **nil**
    empty *buf*$_r$
    empty *nack-buf*$_r$

grow-issued$_r$
  Precondition:
    none
  Effect:
    add some *ID*s to *issued*$_r$

We define the following three fairness classes of the locally controlled actions of the receiver protocol:

1. *receive_msg*, $recover_r$, $send\_pkt_{rs}$(**accept**, ,), and $send\_pkt_{rs}$(**ack**, , *true*) actions
2. $send\_pkt_{rs}$(**ack**, , *false*) actions
3. $grow\text{-}issued_r$

5 Discussion

Both protocols share a common high-level description: both involve association of identifiers with messages, and acceptance of messages by the receiver based on recognition of "good" identifiers. Both also involve very similar strategies for acknowledgement of messages. It is thus desirable to base correctness proofs on this common structure.

We define a high-level (untimed) *generic* protocol G, which represents the common structure, and show that both C and *5P* implement G. We also show that the generic protocol meets the problem specification S. The proof that G satisfies S uses a ***backward simulation*** [7] (or ***prophecy variables*** [1]). The proof that *5P* implements G uses a ***forward simulation*** [7] (or ***history variables*** [9]). The proof that C implements G uses a ***timed forward simulation*** [7].

References

1. M. Abadi and L. Lamport. The existence of refinement mappings. In *Proceedings of the 3rd Annual Symposium on Logic in Computer*, pages 165–175, Edinburgh, Scotland, July 1988.
2. D. Belsnes. Single message communication. *IEEE Transactions on Communications*, Com-24(2), February 1976.
3. B. Liskov, L. Shrira, and J. Wroclawski. Efficient at-most-once messages based on synchronized clocks. Technical Report MIT/LCS/TR-476, Laboratory for Computer Science, Massachusetts Institute of Technology, April 1990.
4. N. Lynch and I. Saias. Distributed Algorithms. Fall 1990 Lecture Notes for 6.852. MIT/LCS/RSS 16, Massachusetts Institute of Technology, February 1992.
5. N. Lynch and M. Tuttle. Hierarchical correctness proofs for distributed algorithms. Techical Report MIT/LCS/TR-387, Laboratory for Computer Science, Massachusetts Institute Technology, Cambridge, MA, 02139, April 1987.
6. N. Lynch and M. Tuttle. An introduction to Input/Output automata. *CWI-Quarterly*, 2(3):219–246, September 1989.
7. N. Lynch and F. Vaandrager. Forward and backward simulations for timing-based systems. In *Proceedings of REX Workshop "Real-Time: Theory in Practice"*, Mook, The Netherland, 1992. Springer-Verlag, LNCS 600.
8. F. Modugno, M. Merritt, and M. Tuttle. Time constrained automata. In *CONCUR'91 Proceedings Workshop on Theories of Concurrency: Unification and Extension*, Amsterdam, August 1991.
9. S. Owicki and D. Gries. An axiomatic proof technique for parallel programs i. *Acta Informatica*, 6(4):319–340, 1976.

This article was processed using the LaTeX macro package with LLNCS style

Games I/O Automata Play*

(Extended Abstract)

Nick Reingold[1] and Da-Wei Wang[2] and Lenore D. Zuck[1]

[1] Department of Computer Science
Yale University
[2] Department of Computer Science
University of Delaware

Abstract. We introduce a *game* approach for specifying reactive systems. In particular, we define a simple two-player game between *System* and *Environment*, and consider the outcomes of such a game as a specification of a reactive system. We introduce six classes of game languages. We then show that the class of languages generated by I/O automata equals one of our game classes. An immediate corollary to the proof is that the fairness condition of I/O automata, which is defined as an extrinsic property by Lynch and Tuttle, can be incorporated as an intrinsic part of the automata. We also show closure properties of the six game classes. For example, we show that the class of languages defined by I/O automata is closed under union and hiding but not under intersection or complementation. The closure results are obtained by reasoning directly about games, thus demonstrating the advantage of the game-based approach.

1 Introduction

Designing correct reactive systems has been a central research topic for several years. While there is some agreement as to the importance of being able to specify reactive systems, there is not much agreement on the best way to go about it. Indeed, many different models for specification of reactive systems have been studied, for example, I/O automata [14, 16], statecharts [9], knowledge-based protocols [7], and functional approach [2, 4, 10, 12, 18], and many have been applied successfully to problems of distributed computing (e.g., [6, 8, 9, 13]). A natural question is how these models compare to one another. An obvious way to compare various formalisms is in terms of their expressive power. Unfortunately, to date, little research has been done in comparing the expressive power of models for reactive systems.

Unlike "normal" sequential systems, whose specifications are easily expressed as a relation between the input domain and the output domain, reactive systems are specified by their interaction with the environment. For example, consider a simple first-in-first-out buffer system buf over a domain D. The set of *actions* of buf is $\{read(d) : d \in D\} \cup \{write(d) : d \in D\}$, where $read(d)$ denotes that the value d is

* This work was supported in part by the National Science Foundation under grants CCR-8910289, IRI-9015570, and CCR-8958528, by the Air Force under grant AFOSR-890382(G), and by an IBM graduate fellowship.

read by the buffer, and *write*(d) denotes that the value d is written by the buffer. A specification of buf can be given by the set of first-in-first-out sequences over D.

Generally, a reactive system (buf) is associated with two types of actions, the *input* actions (e.g., *read*(d)), which are imposed on the system by the environment, and *output* actions (e.g., *write*(d)) which are imposed on the environment by the system. We adopt the trace semantics and consider the specification of a reactive system to consist of the set of sequences over the input and output actions that are allowed to occur when the system interacts with its environment.

We consider another formalism for specifying reactive systems which we believe is quite natural and of interest in its own right. Our formalism involves viewing a reactive system as a game between the system and the environment. Our games are essentially the same as the games used by Abadi and Lamport [1] and Moschovakis [19, 20]. One advantage of this formalism is its simplicity. Another advantage is that it gives us tools to characterize the expressive power of I/O automata [14, 16] and allows us to give simple proofs of various closure and non-closure properties for systems characterized by I/O automata.

The games we consider are two-player games between *System* and *Environment*. Starting with an empty sequence, the players take alternate moves and add elements to the sequence; the Environment adds elements from Σ_E^*, and the System adds elements from $\Sigma_S \cup \{\lambda\}$. A pair of strategies, one for System and one for Environment, defines an infinite sequence of alternating elements from Σ_E^* and $\Sigma_S \cup \{\lambda\}$. By concatenating the elements of this sequence we obtain its *behavior*, a sequence over the actions in $\Sigma_E \cup \Sigma_S$. Since System models the reactive system whose strategy is assumed to be known, and since Environment models the environment whose strategy is unknown, we are interested in the set of behaviors obtained when System plays its strategy against any Environment strategy. Hence, a strategy for System defines a set of sequences over the actions in Σ. We say that a set of sequences L over Σ is *deterministic game realizable*, denoted by game(L), if and only if L is definable by some system strategy. We denote the class of sets L which are deterministic game realizable by game.

We extend our notion of games to nondeterministic strategies for System and to unions of deterministic strategies for System in the obvious way. The resulting sets of sequences over Σ are termed *nondeterministic game realizable* and *union game realizable* respectively, and the resulting classes of sets of sequences are denoted Ngame and Ugame respectively. We also consider restricting strategies for System to ones which depend only on the observable behavior of the play. This leads us to three additional classes of games, game_b, Ngame_b, and Ugame_b, which are the behavior counterparts of the classes described above. These *behavior games* were also studied by Broy et al. [4].

The *I/O automaton* model [14, 16], which offers an elegant and appealing approach to reasoning about asynchronous concurrent computation, was used for specification, verification, and upper and lower bound proofs in many papers, e.g., [3, 5, 11, 14, 15, 21, 23]. I/O automata are state machines that have three types of *actions*: *input actions*, which are generated by the environment and imposed on the I/O automaton, *output actions*, which are generated by the I/O automaton and imposed on the environment, and *internal actions*. Each action is a (possibly nondeterministic) state transformer. Each I/O automaton has an associated (weak) fairness condition. An *execution* of an I/O automaton is a possibly infinite sequence of alternating states and actions. The *behavior* of an execution is its restriction on the observable (input and output) actions. The specification of an I/O automaton is the set of fair behaviors it generates. Given a set of sequences L over the input and output actions, we say that L can be *generated by an automaton*, denoted by $\mathsf{Ioa}(L)$, if there exists an I/O automaton, whose set of fair behaviors is exactly L. We denote by Ioa the class of sets that can be generated by some I/O automaton.

We first show that Ioa, the class of languages (sets of sequences over Σ) generated by I/O automata, where the input actions are Σ_E and the output actions are Σ_S, equals Ugame. Since our notion of game has no explicit fairness condition, an immediate corollary to the proof is that the fairness condition of I/O automata, which is defined as an extrinsic property in [16], can be incorporated as an intrinsic part of the automaton.

We then compare the expressive power of the six game classes. We show that no inclusions hold among them, apart from the obvious inclusions. This demonstrates the direct correlation between the power given to the System, and the expressive power of the resulting reactive system.

Lastly, we consider closure properties for the classes we define. For example, we show that Ioa is closed under union and hiding but not under intersection or complementation. The proofs for closure under union and hiding are obtained by reasoning directly about games, thus demonstrating the advantage of the game-based approach.

2 Games, I/O Automata, and Languages

We describe reactive systems by sequences of *actions*. Actions can be internal actions of the system, which have no effect on the environment, or external actions, which describe interactions with the environment. In describing these interactions, it is helpful to classify the actions involving the system as "input" and "output" actions. Input actions originate in the environment and are imposed on the system, while output actions are generated by the system and imposed by it on the environment.

In this section, we define two models of reactive systems. The first is our *game model*, and the second is the *I/O automaton model* [14, 16]. Both these models are described as interactions between a system and an environment. Let Σ_E and Σ_S be disjoint alphabets, and let $\Sigma = \Sigma_E \cup \Sigma_S$. Σ_E denotes the set of the environment's input actions, and Σ_S denotes the set of the system's output actions.

2.1 Games

Informally, our games are two-player games between System and Environment, where, starting with an empty "play", the players take turns making moves by appending strings to the "play". System can append a single element of Σ_S, and Environment can append any finite (possibly empty) string of elements in Σ_E. The players alternate turns, with Environment making the first move.

Formally, a *move* is an element of $\Sigma_E^* \cup \Sigma_S$. A move is a *move for Environment* if it is in Σ_E^*, and a *move for System* if it is in $\Sigma_S \cup \{\lambda\}$, where λ denotes the empty string. Note that λ can be a move for either player.

A *partial play* is either the empty sequence, $\langle\rangle$, or a sequence of moves which alternates between moves for System and moves for Environment. We do not concatenate together the elements of the sequence. Intuitively, the reason for avoiding concatenation of moves is that information could be lost, since each player is allowed to take λ-moves. We will discuss this further in Section 4. Since λ can be a move for either player, a partial play should also specify the player of the first move. We will usually ignore this, as the identity of the first player will be clear from context.

For any finite partial play $\alpha = \langle a_0, a_1, \ldots, a_i \rangle$, and any finite or infinite partial play $\beta = \langle b_0, b_1, \ldots \rangle$ we define the *concatenation* of α and β, denoted $\alpha \cdot \beta$, by

$$\alpha \cdot \beta = \begin{cases} \langle a_0, \ldots, a_i, b_0, \ldots \rangle & a_i \text{ and } b_0 \text{ are moves for different players} \\ \langle a_0, \ldots, a_i, \lambda, b_0, \ldots \rangle & \text{otherwise.} \end{cases}$$

Let G_{odd} be the set of odd length partial plays which start with a move for Environment, let G_{even} be the set of even length partial plays which start with a move for Environment, let $G_f = G_{\mathrm{even}} \cup G_{\mathrm{odd}}$, let G_ω be the set of all infinite partial plays which start with a move for Environment, and let $G = G_f \cup G_\omega$. The elements of G_ω are called *plays*.

Players usually make moves according to their "strategies," which are functions from partial plays to moves. A *deterministic System strategy* is a function S mapping elements of G_{odd} to $\Sigma_S \cup \{\lambda\}$. A *deterministic Environment strategy* is a function E mapping elements of G_{even} to Σ_E^*. For a System strategy S and Environment strategy E, we define $play(S,E)$ to be the play $\langle a_0, b_0, a_1, \ldots \rangle$, where $a_0 = E(\langle\rangle)$, $b_0 = S(\langle a_0 \rangle)$, $a_1 = E(\langle a_0, b_0 \rangle)$, etc.

Let $\mathcal{S}$ denote the set of deterministic System strategies and $\mathcal{E}$ denote the set of deterministic Environment strategies. For a deterministic System strategy S, let $play(S,\mathcal{E})$ denote the set $\cup_{E \in \mathcal{E}} play(S,E)$, i.e., $play(S,\mathcal{E})$ is the set of plays of System playing S and Environment playing any deterministic strategy. For a set $\mathcal{S}' \subseteq \mathcal{S}$ of deterministic System strategies, let $play(\mathcal{S}',\mathcal{E})$ denote the set $\cup_{S \in \mathcal{S}'} play(S,\mathcal{E})$.

A *nondeterministic System strategy* is a function $\hat{S}$ mapping elements of G_{odd} to $2^{\Sigma_S \cup \{\lambda\}}$. For a nondeterministic System strategy $\hat{S}$, we define $D(\hat{S})$ to be the set of all deterministic System strategies S such that for all $\alpha \in G_{\mathrm{odd}}$, $S(\alpha) \in \hat{S}(\alpha)$. For an Environment strategy E, we define $play(\hat{S},E)$ to be $play(D(\hat{S}),E)$. That is, each $S \in D(\hat{S})$ corresponds to "fixing" choices that $\hat{S}$ might make. We say that a play α is *consistent* with a nondeterministic System strategy $\hat{S}$ if $\alpha \in play(\hat{S},\mathcal{E})$.

For a partial play α define the *behavior* of α, written $beh(\alpha)$, as the sequence over Σ which is obtained by concatenating together all the elements of α. *Behavior games* are the games resulting when both System and Environment have to base their moves according to the behavior of partial plays rather than the plays themselves. These games are considered in [4].

For behavior games, the notions of moves, plays, and partial plays are defined as before. A *deterministic System behavior strategy* is a function S mapping Σ^* to $\Sigma_S \cup \{\lambda\}$. A *deterministic Environment behavior strategy* is a function E mapping Σ^* to Σ_E^*. For a System behavior strategy S and Environment behavior strategy E, we define $play(S,E)$ to be the play $\langle a_0, b_0, a_1, \ldots \rangle$, where $a_0 = E(\langle\rangle)$, $b_0 = S(a_0)$, $a_1 = E(a_0 b_0)$, etc. We also define a *nondeterministic System behavior strategy* and plays of such behavior strategies in the obvious way.

Let $\mathcal{S}_b$ denote the set of deterministic System behavior strategies and $\mathcal{E}_b$ denote the set of deterministic Environment behavior strategies. For a deterministic System behavior strategy S, let $play(S, \mathcal{E}_b)$ denote the set $\cup_{E \in \mathcal{E}_b} play(S, E)$, i.e., $play(S, \mathcal{E}_b)$ is the set of plays of System playing S and Environment playing any deterministic behavior strategy. For a set $\mathcal{S}' \subseteq \mathcal{S}_b$ of deterministic System behavior strategies, let $play(\mathcal{S}', \mathcal{E}_b)$ denote the set $\cup_{S \in \mathcal{S}'} play(S, \mathcal{E}_b)$.

2.2 I/O Automata

The I/O automaton model was first defined in [16]. See [14, 16] for a complete description of the model. Here, we provide a brief summary of those parts of the model used in this paper.

I/O automata are state machines whose state to state transitions are caused by actions. Actions can be internal actions, which have no effect on the environment, or Σ actions, which describe interactions with the environment. Formally, an *I/O automaton* A (which we often call simply an *automaton*) is described by:

1. Three mutually disjoint sets of actions: $in(A) = \Sigma_E$, $out(A) = \Sigma_S$, and $int(A)$. We denote $acts(A) = \Sigma \cup int(A)$, and $loc(A) = int(A) \cup \Sigma_S$, i.e., $acts(A)$ is the set of A's actions, and $loc(A)$ is the set of A's locally controlled actions.
2. A set $states(A)$ of A's *states* and a set $initial(A) \subseteq states(A)$ of A's *initial states*.
3. A transition relation, $trans(A) \subseteq states(A) \times acts(A) \times states(A)$, that is *input enabled*, i.e., for every input action π and state s, there exists some state s' such that $(s, \pi, s') \in trans(A)$. (In general, an action π is *enabled* from a state s if for some s', $(s, \pi, s') \in trans(A)$. We denote by $enabled(s)$ the set of actions that are enabled from state s.)
4. A fairness condition, $fair(A)$, described as a partition on A's local actions with countably many equivalence classes. If $fair(A) = \{loc(A)\}$ then we say that A has the *trivial fairness partition*.

A *partial execution* η of an automaton A is a (possibly infinite) sequence of the form:

$$s_0 \pi_1 s_1 \pi_2 \ldots,$$

where for every $i \geq 0$, $s_i \in states(A)$, $\pi_{i+1} \in acts(A)$, and $(s_i, \pi_{i+1}, s_{i+1})$ is a transition of A. If η is finite then it terminates in a state. We refer to s_0 as the *initial state* of η.

A partial execution is called an *execution* if its initial state is an initial state of A. A finite execution is *fair* if no local action is enabled from its last state. An infinite execution η is *fair* if for every set of local actions $\Pi \in fair(A)$, either actions from Π are taken infinitely many times in η (i.e., for infinitely many i, $\pi_i \in \Pi$), or actions from Π are not enabled infinitely many times in η (i.e., for infinitely many i, $enabled(s_i) \cap \Pi = \emptyset$).

Let B be some set and let B' be some subset of B. For every sequence α over B, we denote by $\alpha | B'$ the *restriction* of α on B', that is, the sequence obtained from α when all non-B' elements are deleted. Similarly, for a set of sequences A over B, $A | B'$ is defined in the obvious way.

The *behavior* of a partial execution η of A (and more generally, of any sequence of actions and states of A), $beh(\eta)$, is defined to be the sequence $\alpha|\Sigma$. A *fair behavior* of an automaton A is any sequence of the form $beh(\eta)$, where η is a fair execution of A. Note that our definition of behavior differs from the one in [16] where a behavior is defined as the restriction of execution to the actions and not only to the observable actions.

2.3 Languages

We now define the language (over Σ) described by games and by I/O automata. Denote $\Sigma^\infty = \Sigma^* \cup \Sigma^\omega$. For a language $L \subseteq \Sigma^\infty$, let $pref(L)$ denote the set of prefixes for sequences in L, i.e.,

$$pref(L) = \{\sigma' \in \Sigma^* : \exists \tau \in \Sigma^\infty \quad \sigma'\tau \in L\}.$$

Languages Defined by Games. Recall that for a partial play α, $beh(\alpha)$ is the sequence over Σ which is obtained by concatenating together all the elements of α. For a set of plays, $\mathcal{A}$, we define $beh(\mathcal{A})$ to be $\bigcup_{\alpha \in \mathcal{A}} beh(\alpha)$. Define the *linearization* of α, written $lin(\alpha)$, as the sequence of actions contained within α where λ-moves of Environment are replaced by a new symbol, λ_E. Note that $lin(\alpha)$ is a sequence over $\Sigma \cup \{\lambda, \lambda_E\}$.

Let $L \subseteq \Sigma^\infty$. We say that L is *game realizable*, or simply $\mathsf{game}(L)$, if L is the set of behaviors of play between some System strategy S and every Environment strategy E, i.e.,

$$\mathsf{game}(L) \quad \text{iff} \quad beh(play(S, \mathcal{E})) = L \quad \text{for some } S \in \mathcal{S}.$$

We say that L is *union-game realizable*, or simply $\mathsf{Ugame}(L)$, if there exists a set of System strategies $\mathcal{S}' \subseteq \mathcal{S}$ such that $L = beh(play(\mathcal{S}', \mathcal{E}))$.

We denote by Ugame (resp. game) the set of languages L over Σ such that $\mathsf{Ugame}(L)$ (resp. $\mathsf{game}(L)$).

For a language $L \subseteq \Sigma^\infty$, we say that L is *nondeterministically game realizable*, written $\mathsf{Ngame}(L)$, if there is a nondeterministic System strategy $\hat{S}$ such that $L = beh(play(\hat{S}, \mathcal{E}))$. Since we always take the union over all possible environment strategies, we need not explicitly consider nondeterministic Environment strategies.

Let $L \subseteq \Sigma^\infty$. We say that L is *behavior game realizable*, or simply $\mathsf{game}_\mathsf{b}(L)$, if L is the set of behaviors of plays in a behavior game between some System behavior strategy S and every Environment behavior strategy E, i.e.,

$$\mathsf{game}_\mathsf{b}(L) \quad \text{iff} \quad beh(play(S, \mathcal{E}_\mathsf{b})) = L \quad \text{for some } S \in \mathcal{S}_\mathsf{b}.$$

We say that L is *union behavior game realizable*, or simply $\mathsf{Ugame}_\mathsf{b}(L)$, if there exists a set of System behavior strategies $\mathcal{S}' \subseteq \mathcal{S}_\mathsf{b}$ such that $L = beh(play(\mathcal{S}', \mathcal{E}_\mathsf{b}))$.

We say that L is *nondeterministically behavior game realizable* if there is a nondeterministic System behavior strategy S such that $L = beh(play(S, \mathcal{E}_\mathsf{b}))$.

The six game classes are summarized in Table 1.

Table 1. Game classes

	Behavior Game	General Game
single deterministic strategy	$\mathsf{game_b}$	game
single nondeterministic strategy	$\mathsf{Ngame_b}$	Ngame
union of deterministic strategies	$\mathsf{Ugame_b}$	Ugame

Languages Defined by I/O Automata. Let A be an automaton. The *language generated by* A, or simply $L(A)$, is the set of A's fair behaviors. A language $L \subseteq \Sigma^\infty$ is *automaton realizable*, or simply $\mathsf{loa}(L)$, if $L = L(A)$ for some automaton A. We denote by loa the set $\{L \subseteq \Sigma^\infty : \mathsf{loa}(L)\}$.

A special class of automaton realizable languages is the class of languages realizable by automata with the trivial fairness partition. If L is realizable by such an automaton, we denote it by $\mathsf{loat}(L)$. Similarly, we denote by loat the class of all languages L such that $\mathsf{loat}(L)$.

Examples of Game Languages. Below are two examples of non-trivial game languages.

Example 1:
Assume $\Sigma_S = \{0,1\}$, and let h be the homomorphism on Σ defined by:

$$h(a) = \begin{cases} \lambda & a \in \Sigma_E \\ a & a \in \Sigma_S. \end{cases}$$

For any language L, $h^{-1}(L)$ is all strings obtained by arbitrarily inserting elements of Σ_E into sequences from L.

Consider the nondeterministic strategy $\hat{S}$ of System such that for every $\alpha \in G_{\text{even}}$,

- if $beh(\alpha)$ is empty then $\hat{S}(\alpha) = \{\lambda, 0, 1\}$.
- if $beh(\alpha)$ has some 1's, then $\hat{S}(\alpha) = \{1\}$.
- in all other cases, let n_α be the number of λ-moves for Environment in α before the first non-λ move (of either Environment or System). If the number of 0-moves in α is less than n_α, then $\hat{S}(\alpha) = \{0\}$, else $\hat{S}(\alpha) = \{1\}$.

Clearly, $L_1 = beh(play(\hat{S}, \mathcal{E})) = h^{-1}(\{0^n 1^\omega : n \geq 0\} \cup \{\lambda\})$. Hence L_1 is in Ngame.

Example 2:
Let Σ_S and h be as in Example 1. Let $L_2 = h^{-1}(\{0^n 1^\omega : n \geq 0\})$, where h is the homomorphism defined above. Since every prefix of 0^ω can be extended to be in L_2 but 0^ω is not in L_2, then, as we will show in the full version of the paper, $L_2 \notin \mathsf{Ngame}$. However, L_2 is in $\mathsf{Ugame_b}$: For every n, let S_n be the strategy whose first n moves are 0 and whose subsequent moves are all 1's. Then $L_2 = beh(\bigcup_n (play(S_n, \mathcal{E}_b)))$ thus $L_2 \in \mathsf{Ugame_b}$.

Closure Properties of Language Classes. Let $\mathcal{L}$ be some class of languages over an alphabet $\Sigma = \Sigma_E \cup \Sigma_S$. We define the following closure properties of $\mathcal{L}$:

union: $\mathcal{L}$ is closed under *finite union* (resp. *countable union*, *arbitrary union*) if the union of any finite (resp. countable, arbitrary) collection of sets in $\mathcal{L}$ is in $\mathcal{L}$.

intersection: $\mathcal{L}$ is closed under *intersection* if the intersection of any finite collection of sets in $\mathcal{L}$ is in $\mathcal{L}$.

complementation: $\mathcal{L}$ is closed under *complementation* if the complement of every set in $\mathcal{L}$ (i.e., $\Sigma^\infty - L$) is also in $\mathcal{L}$.

hiding: $\mathcal{L}$ is closed under *hiding* if for every L in $\mathcal{L}$ and $a \in \Sigma$, $L \setminus a = L|(\Sigma - \{a\})$ is in $\mathcal{L}$.

renaming: $\mathcal{L}$ is closed under *renaming* if for every $\Sigma_1, \Sigma_2 \subseteq \Sigma$, every bijection $g: \Sigma_1 \rightarrow \Sigma_2$, and every set $L \subseteq \Sigma_1^\infty$ in $\mathcal{L}$, $g(L)$ is in $\mathcal{L}$ set ($g(L)$ is obtained by applying g to every sequence in L, and applying g to a sequence means applying it to each element of the sequences).

parallel composition: $\mathcal{L}$ is closed under *parallel composition* if for every $\Sigma_1, \Sigma_2 \subseteq \Sigma$ and every $L_1 \subseteq \Sigma_1^\infty$ and $L_2 \subseteq \Sigma_2^\infty$ in $\mathcal{L}$,

$$L_1 \parallel L_2 = \{\sigma \in \Sigma^\infty : \sigma|\Sigma_1 \in L_1 \text{ and } \sigma|\Sigma_2 \in L_2\}$$

is also in $\mathcal{L}$.

3 Games and I/O Automata

We establish the equivalence of Ioa, Ioat, and Ugame. The following theorem establishes that Ugame $\subseteq$ Ioat.

Theorem 1. *For every language* $L \subseteq \Sigma^\infty$, *if* Ugame($L$) *then* Ioat($L$).

Outline of Proof:

Assume that Ugame(L). Let $S' \subseteq \mathcal{S}$ be such that $L = beh(play(S', \mathcal{E}))$. Let A be an automaton with the trivial fairness partition such that $int(A) = \{\lambda\}$, $states(A) = S' \times G_f$, $initial(A) = S' \times \{\langle\rangle\}$, and $trans(A)$ consists of

- $((S, \sigma), a, (S, \sigma \cdot \langle a \rangle))$, if either $a \in \Sigma_E$ or $S(\sigma) = a$,
- $((S, \sigma), a, (S, \sigma \cdot \langle \lambda, a \rangle))$, if $a = S(\sigma \cdot \langle \lambda \rangle)$.

In the full version of the paper we show that $L = L(A)$. □

The following theorem establishes that Ioa $\subseteq$ Ugame.

Theorem 2. *For every language* $L \subseteq \Sigma^\infty$, *if* Ioa($L$) *then* Ugame($L$).

Outline of Proof:

Let A be an I/O automaton which generates L and assume that $fair(A) = \{\Pi_1, \ldots\}$. Let $dove = dove(0), dove(1), \ldots$ be an infinite sequence over the Π_i's such that every Π_i in $fair(A)$ appears in $dove$ infinitely many times.

In the full version of the paper we show that for every fair execution η of A whose initial state is s_0, there exists a deterministic strategy S_η for System such that (1) $beh(play(S_\eta, \mathcal{E})) \subseteq L$ and (2) there exists some $E_\eta \in \mathcal{E}$ such that $play(\eta) = play(S_\eta, E_\eta)$.

Intuitively, S_η follows η when possible. If Environment makes a move which makes it impossible to follow η then S_η uses *dove* to ensure that the play corresponds to some fair execution of A. □

Corollary 3. *For every language* $L \subseteq \Sigma^\infty$,

$$\mathsf{loa}(L) \iff \mathsf{Ugame}(L) \iff \mathsf{loat}(L).$$

4 Inclusion Relations Among Game Classes

It is easy to see that each general game class contains the corresponding behavior game class since the general strategy could simply be a behavior strategy. It is also easy to see that for either general or behavior games, a single deterministic strategy can be considered a nondeterministic strategy, and that a nondeterministic strategy can be considered a union of deterministic strategies. In the full version of the paper we show that all above inclusions are strict, and no other inclusion relations hold other than those implied by transitivity. This is summarized in (1).

$$\begin{array}{ccccc} \mathsf{game_b} & \subset & \mathsf{Ngame_b} & \subset & \mathsf{Ugame_b} \\ \cap & & \cap & & \cap \\ \mathsf{game} & \subset & \mathsf{Ngame} & \subset & \mathsf{Ugame} \end{array} \qquad (1)$$

Here we show that some of the subset relations in (1) are strict. In particular, we show that both **game** classes are strict subsets of their **Ngame** classes and that **game** is not a subset of $\mathsf{Ugame_b}$. From the latter claim it follows that each of the behavior game classes is a strict subset of its non-behavior counterpart.

Lemma 4. *There is a language which is in* Ngame *but not in* game, *and there is a language which is in* $\mathsf{Ngame_b}$ *but not in* $\mathsf{game_b}$.

Proof. Suppose that $\{0,1\} \subseteq \Sigma_s$.

For the first part of the claim, let $L_3 = beh(play(\hat{S}, \mathcal{E}))$, where $\hat{S}$ is the nondeterministic System strategy such that $\hat{S}(\alpha) = \{0, 1, \lambda\}$ for every $\alpha \in G_{\mathrm{odd}}$. Obviously, $L_3 \in$ **Ngame**. To see that $L_3 \notin$ **game**, note that from the definition of deterministic System strategies it immediately follows that for every language $L \in$ **game**, for every $a_1, a_2 \in \Sigma_s$, both a_1 and a_2 are in $pref(L)$ if and only if $a_1 = a_2$.

For the second part of the claim, let $L_4 = beh(play(\hat{S}, \mathcal{E}_b))$, where $\hat{S}$ is the nondeterministic System behavior strategy such that $\hat{S}(\langle\lambda\rangle) = \{0, 1\}$. Obviously, L_4 is in $\mathsf{Ngame_b}$. To see that $L_4 \notin \mathsf{game_b}$, observe that for every language L in $\mathsf{game_b}$ and for every $a_1, a_2, b \in \Sigma_s$, if both βa_1 and βa_2 are in $pref(L)$ for some $\beta \in \Sigma^*$, then $a_1 = a_2$. Since both 0 and 1 are in $pref(beh(play(\hat{S}, \mathcal{E}_b)))$, $L_4 \notin \mathsf{game_b}$. □

Lemma 5. *There is a language which is in* **game** *but not in* $\mathsf{Ugame_b}$.

Proof. We first show that if $L \subseteq \Sigma^\infty$ is in $\mathsf{Ugame_b}$, $i \in \Sigma_E$, and $i^\omega \in L$, then $i^n \in L$ for infinitely many n's: Since $i^\omega \in L$ there exists a System behavior strategy S and an Environment strategy E such that $play(S, E) = \langle i^{m_1}, \lambda, i^{m_2}, \lambda, \ldots \rangle$. For every $n \geq 1$ let E_n be the environment strategy which plays i^{m_k} on its kth move for $k \leq n$, and plays λ thereafter. It therefore follows that $beh(play(S, E_n)) = i^{(\sum_{k=1}^{n} m_k)}$. Since $\sum_k m_k = \infty$ and each m_k is finite, $i^n \in L$ for infinitely many n's:

Next, assume $0 \in \Sigma_S$, and let $L_5 = beh(play(S, \mathcal{E}))$ where S is the deterministic System strategy which plays 0 after every λ move of Environment, and λ after every non-λ move of Environment. Obviously, no play of S and $E \in \mathcal{E}$ has two consecutive λ moves, hence there are no finite sequences in L_5, and, in particular, $i^n \notin L_5$ for every n. However, let E be the Environment strategy which always plays i. It is easy to see that $beh(play(S, E)) = i^\omega$. From the observation above in now follows that L_5 is not in $\mathsf{Ugame_b}$. □

5 Closure Properties

Table 2. Closure Properties of Game Classes

	union	hiding	intersection/complementation / parallel composition	renaming
$\mathsf{game_b}$/game	no	no	no	yes
$\mathsf{Ngame_b}$/Ngame	finite	no	no	yes
$\mathsf{Ugame_b}$	arbitrary	no	no	yes
Ugame	arbitrary	yes	no	yes

In this section we discuss the closure properties of the six game classes, game, $\mathsf{game_b}$ Ngame, $\mathsf{Ngame_b}$ Ugame, and $\mathsf{Ugame_b}$. Our results are summarized in Table 2.

We first consider renaming.

Theorem 6. *All six game classes are closed under renaming.*

Proof. We prove the claim only for the case of game; the other cases are similar. Let $L = beh(play(S, \mathcal{E}))$ for some $S \in \mathcal{S}$, and assume $L \subseteq \Sigma_1^\infty$. Let $\Sigma_1 \subseteq \Sigma_2$, and let g be a bijection from Σ_1 to Σ_2. It suffices to show that $g(L)$ is in game. Define a strategy $S' \in \mathcal{S}$ as follows:

$$S'(\eta) = g(S(g(\eta))) \qquad \text{for every } \eta \in G_{\text{even}}.$$

It is easy to see that $g(L) = beh(play(S', \mathcal{E}))$. □

Next we consider closure under union. Here, the "amount" of closure depends on whether System is allowed to play a deterministic strategy, a nondeterministic strategy, or a union of strategies, but does not depend on whether the strategies are required to be behavior strategies.

The following theorem follows immediately from the definition of union classes.

Theorem 7. Ugame *and* $\mathsf{Ugame_b}$*, and are closed under arbitrary unions. Note that it then follows from Theorem 3 that* loa *is also closed under arbitrary union.*

Theorem 8. Ngame, $\mathsf{Ngame_b}$ *are closed under finite unions but are not closed under countable unions.*

Proof. We prove the claim only for the case of **Ngame**. The case of $\mathsf{Ngame_b}$ is similar and left to the reader.

Let $\hat{S}_1, \hat{S}_2$ be two nondeterministic System strategies. Consider the System strategy $\hat{S}$ defined as follows. For every $\alpha \in G_{\text{odd}}$,

$$\hat{S}(\alpha) = \begin{cases} \hat{S}_1(\alpha) \cup \hat{S}_2(\alpha) & \alpha \text{ is consistent with both } \hat{S}_1 \text{ and } \hat{S}_2 \\ \hat{S}_1(\alpha) & \alpha \text{ is consistent with } \hat{S}_1 \text{ but not with } \hat{S}_2 \\ \hat{S}_2(\alpha) & \alpha \text{ is consistent with } \hat{S}_2 \text{ but not with } \hat{S}_1 \\ \{\lambda\} & \alpha \text{ is not consistent with either } \hat{S}_1 \text{ or } \hat{S}_2. \end{cases}$$

It is easy to see that every behavior that is realizable by $\hat{S}_1$ or $\hat{S}_2$ is also realizable by $\hat{S}$. To prove that every behavior realizable by $\hat{S}$ is also realizable by $\hat{S}_1$ or $\hat{S}_2$, observe that each move performed by $\hat{S}$ can only make the play inconsistent with at most one strategy. If a play is inconsistent with, say $\hat{S}_1$, then the following moves performed by $\hat{S}$ are the same as those performed by $\hat{S}_2$. Therefore every play generated by $\hat{S}$ is consistent with either $\hat{S}_1$ or $\hat{S}_2$. Consequently, **Ngame** is closed under finite unions.

To see that **Ngame** is not closed under countable unions, consider the language L_2 of Example 2 (Section 2), which, as we show in the final version of the paper, is not in **Ngame**. Yet, it is easy to see that L_2 is a countable union of languages in **game**, thus L_2 is a countable union of languages in **Ngame**. Consequently, **Ngame** is not closed under countable unions. □

Theorem 9. game *and* $\mathsf{game_b}$ *are not closed under finite union.*

Proof. Assume $\Sigma_S = \{0,1\}$. For every $a \in \{0,1\}$, let S_a be the behavior strategy for System that always plays a, and let L_a be $beh(play(S_a, \mathcal{E}_b)) = (\Sigma_E^* a)^\omega$. From the observation in the proof of Lemma 4 it follows that $L_0 \cup L_1$ is not in **game**. □

We now show that no game class is closed under intersection, complementation, or parallel composition. That **loa** is not closed under intersection or complementation was previously shown in [22].

Theorem 10. *No game class is closed under intersection, complementation, or parallel composition.*

Proof. To show that none of the game classes is closed under intersection, note that for L_0 and L_1 as in the proof of Theorem 9, $L_0 \cap L_1 = \emptyset$ and $\emptyset$ is not in any game class.

By De Morgan's laws, **Ugame**, **Ngame**, $\mathsf{Ugame_b}$, and $\mathsf{Ngame_b}$ are not closed under complementation. It is easy to see that **game**, $\mathsf{game_b}$ are not closed under complementation.

As for parallel composition, note that if $L_1, L_2 \subseteq \Sigma'$, then $L_1 \parallel L_2 = L_1 \cap L_2$. Therefore, since no game class is closed under intersection, none is closed under parallel composition. □

We next consider closure under hiding. While it is easy to show that each of the game classes (and therefore loa) is closed under hiding of output actions, no game class except Ugame (and hence loa) is closed under hiding of input actions. Hence, no game class except Ugame is closed under hiding.

Theorem 11. Ugame *is closed under hiding.*

Proof. It suffices to show that if L be in game and a be in Σ, then $L \setminus a$ is in Ugame. Let $S \in \mathcal{S}$ be such that $L = beh(play(S, \mathcal{E}))$.

If $a \in \Sigma_S$, then we construct a system strategy $S' \in \mathcal{S}$ that replaces any a-move of S by a λ-move such that $beh(play(S', \mathcal{E})) = L \setminus a$. For every partial play η over $\Sigma - \{a\}$, we define ext(η) to be the unique prefix of $play(S, E)$ which is obtained from η by replacing some λ-moves of system by a. If no such prefix exists, then we ext(η) is undefined.

For every $\eta \in G_{\mathrm{odd}}$, define

$$S'(\alpha) = \begin{cases} S(\mathrm{ext}(\eta))\ \mathrm{ext}(\eta)) \text{ is defined and not equal } b \\ \lambda \qquad\qquad\quad \text{otherwise} \end{cases}$$

We leave it to the reader to verify that $L \setminus a = beh(play(S', \mathcal{E})$.

If $a \in \Sigma_E$, then for every $E \in \mathcal{E}$, let $\sigma(E)$ denote the sequence $beh(play(S, E))$, so that L is the union of $\sigma(E)$ over all $E \in \mathcal{E}$. For every Environment strategy $E \in \mathcal{E}$, we construct a system strategy $S(E) \in \mathcal{S}$ such that $beh(play(S(E), \mathcal{E})$ includes $\sigma(E) \setminus a$ and is a subset of $L \setminus a$. Hence, if we let S' be the union (over all $E \in \mathcal{E}$) of $S(E)$, then $beh(play(S', \mathcal{E})$ is exactly $L \setminus a$. It remains to show how to construct $S(E)$ for a given Environment strategy E.

For every partial play η over $\Sigma - \{a\}$, we say that η is *extendible* if there exists a prefix η' of game(S, E), such that η' is obtained from η by replacing some λ-moves of Environment with elements of a^+. If such an η' exists, then we say that η' is an *extension* of η. Note that if η is extendible then it has a unique extension.

We next define a function g from partial plays over $\Sigma - \{a\}$ to partial plays over Σ. For every extendible η, $g(\eta)$ is to be the extension of η. For an η which is not extendible, let η' be the longest extendible prefix of η and assume that $\eta = \eta' \cdot \eta''$, and let $g(\eta)$ be $g(\eta') \cdot \eta''$. For every $\eta \in G_{\mathrm{odd}}$, define $S(E)(\eta) = S(g(\eta))$.

We leave it to the reader to check that $beh(play(S(E), \mathcal{E})$ includes $\sigma(E) \setminus a$ and that it is a subset of $L \setminus a$. □

To contrast the Lemma 11 with the other game classes, we have:

Theorem 12. *None of* $\mathsf{Ugame_b}$, Ngame, game, $\mathsf{Ngame_b}$ *and* $\mathsf{game_b}$ *is closed under hiding of input actions.*

Proof. Assume $\Sigma_E = \{a, b\}$ and $\Sigma_S = \{0, 1\}$. Consider the deterministic behavior strategy S of System such that for every $\alpha \in G_{\mathrm{even}}$,

$$S(\alpha) = \begin{cases} 0 \text{ number of 0's in } \alpha < \text{number of } B\text{'s in the first element of } \alpha \\ 1 \text{ otherwise} \end{cases}$$

Let $L = beh(play(S, \mathcal{E}))$. It is easy to see that $L \setminus a$ is L_2 of Example 2 (Section 2). Hence, $L \setminus a$ is not in Ngame. Consequently, none of Ngame, game, $\mathsf{Ngame_b}$, and $\mathsf{game_b}$ is closed under hiding.

To show that $\mathsf{Ugame_b}$ is not closed under hiding, consider the behavior system strategy S defined by:

$$S(\alpha) = \begin{cases} \lambda \text{ last play of Environment is } ab \\ 0 \text{ otherwise} \end{cases}$$

for every $\alpha \in G_{\text{even}}$.

It is easy to see that $a^\omega \in L \setminus b$. However, $a^n \notin L \setminus a$ for every $n > 0$. It therefore follows that $L \setminus a$ is not in $\mathsf{Ugame_b}$. □

6 Comparison with Broy, et al.

Broy, Dederichs, Dendorfer, and Rainer [4] considered various formalisms for describing reactive systems. In this subsection we compare our results to those of [4]. The classes defined there are *input enabled and input free.* Formally, A language $L \subseteq \Sigma^\infty$ is *input enabled and input free* if for every $\alpha \in pref(L)$:

1. for every $i \in \Sigma_E$, $\alpha i \in pref(L)$, and
2. there exists some $\beta \in \Sigma_S^\infty$ such that $\alpha\beta \in L$.

(1) means that every prefix of a string in L can be followed by any element in Σ_E to yield a prefix of a string in L, and (2) means that every prefix of a string in L can be extended by a string over Σ_S to yield a string in L.

The class of input enabled and input free languages is termed IEF in [22], and *Local-SL* in [4].

Broy et al. define a class of languages which are generated by automata with a *strong fairness* condition. More specifically, the authors define I/O automata similar to those we define in Section 2, with the following differences:

1. the fairness partition is finite.
2. an infinite execution η is fair if actions from each fairness class that is enabled infinitely many times are taken infinitely many times. Note that we only require that, in fair executions, only actions from fairness classes that are eventually permanently enabled are infinitely many times taken.

The class of languages that are generated by automata with strong fairness is denoted *Automatic* in [4]. The class of languages that are generated by the standard automata is denoted *Automatic-WF.* Broy et al. also consider certain games, which are exactly our behavior games. They use *Strategic*, *Strategic-ND*, and *Fully-realizable* to denote our $\mathsf{game_b}$, $\mathsf{Ngame_b}$, and $\mathsf{Ugame_b}$ respectively. They claim that

$$\mathsf{game_b} \subset \mathsf{Ngame_b} \subset \mathsf{Ugame_b} \subset \textit{Local-SL}$$

(where $\subset$ stands for proper inclusion), and conjecture that $\mathsf{Ugame_b} = \mathsf{loa}$.

However, from Corollary 3 and (1) it immediately follows that $\mathsf{Ugame_b}$ is a proper subclass of loa, refuting the conjecture. Abadi had previously observed that the conjecture was false.

As to *Local-SL*, we show in [22] that loa is a strict subclass of it, which coincides with the results of [4].

7 Conclusion and Future Work

We presented games between a powerful Environment and a System, where the players use strategies to obtain infinite sequences of symbols over a given alphabet. The games define finite- and infinite-string languages over Σ in a natural way. We considered different restrictions on the class of strategies allowed, therefore obtaining several classes of languages.

Our main result is the equivalence between the class of languages generated by I/O automata and one of our game classes. We also include comparisons between the different game classes, and a study of their closure properties. Finally, we compare our results to those in [4].

We believe that the game approach is a powerful tool for analyzing the expressive power of reactive systems. In fact, this work began when the authors tried to investigate the closure properties of loa languages, and failed to do so using the I/O automaton model. The same results were extremely easy to obtain by reasoning about Ugame. It would be interesting to apply the game approach to other reactive models.

Closed sets have generated much interest lately, partly because they describe *safety* properties of systems [17]. Generally speaking, a language L over Σ is *closed* if for every $\sigma \in \Sigma^\omega$, $\sigma \in L$ if and only if every prefix of σ is in $pref(L)$. We currently do not know the exact relation between our six game classes and their closed counterparts. We do know, however, that the closed loa languages are in Ngame but not necessarily in game, and that they do not include all the $\mathsf{game_b}$ languages.

Another interesting topic is the notion of strong fairness and whether it can be incorporated as (a version of) a game. While it's obvious how to incorporate strong fairness in the I/O automaton model, there seems no natural way of doing that in the game model. If, indeed, there is no way, then it might imply that strong fairness is a state-dependent notion. Games have a much different notion of states, and it's possible that no game-like model can capture the notion of state needed to define strong fairness.

References

1. M. Abadi and L. Lamport. Composing specifications. Research Report 66, DEC SRC, October 1990.
2. J.A. Bergstra and J.W. Klop. Process theory based on bisimulation semantics. In *LNCS 345*, 1989.
3. B. Bloom. Constructing two-writer atomic registers. In *Proc. 6th ACM Symp. on Principles of Distributed Computing*, pages 249–259, 1987.

4. M. Broy, F. Dederichs, C. Dendorfer, and R. Weber. Characterizing the behavior of reactive systems by trace sets (extended abstract). In *Third Workshop on Concurrency and Compositionality,* March 1991.
5. A. Fekete, N. Lynch, M. Merritt, and W. Weihl. Commutativity-based locking for nested transactions. *Journal of Computer and System Science.* to appear.
6. A. Fekete, N. Lynch, and L. Shrira. *A Modular Proof of Correctness for a Network Synchronizer,* volume 312 of *Lecture Notes in Computer Science,* pages 219–256. Springer-Verlag, Amsterdam, Netherlands, July 1987.
7. J. Y. Halpern and R. Fagin. Modeling knowledge and action in distributed systems. Technical Report, IBM, RJ6303, 1988.
8. J. Y. Halpern and L. D. Zuck. A little knowledge goes a long way: Simple knowledge-based derivations and correctness proofs for a family of protocols. In *Proc. 6th ACM Symp. on Principles of Distributed Computing,* pages 269–280, 1987. To appear in JACM.
9. D. Harel. Statecharts: A visual formalism for complex systems. *Science of Computer Programming,* 8(33):231–278, 6 1987.
10. M. Hennessy. *Algebraic Theory of Process.* MIT Press, 1988.
11. M. Herlihy. Impossibility and universality results for wait-free synchronization. In *Proc. 7th ACM Symp. on Principles of Distributed Computing,* pages 276–290, 1988.
12. C.A.R. Hoare. *Communicating Sequential Process.* Prentice-Hall, 1985.
13. L. Lamport. Specifying concurrent program modules. *ACM Trans. on Programming Languages and Systems,* 5(2):190–222, 4 1983.
14. N. Lynch and M. Tuttle. An introduction to input/output automata. *CWI Quarterly,* 2(3):219–246, September 1989.
15. N. A. Lynch, Y. Mansour, and A. Fekete. Data link layer: Two impossibility results. In *Proc. 7th ACM Symp. on Principles of Distributed Computing,* pages 149–170, August 1988.
16. N. A. Lynch and M. R. Tuttle. Hierarchical correctness proofs for distributed algorithms. In *Proc. 6th ACM Symp. on Principles of Distributed Computing,* pages 137–151, August 1987.
17. Z. Manna and A. Pnueli. The anchored version of temporal logic. In *LNCS 354,* pages 201–284, 1989.
18. R. Milner. *Concurrency and Communication.* Prentice-Hall, 1989.
19. Y. N. Moschovakis. A game-theoretic modeling of concurrency. In *Proc. of the 4th annual Symp. on Logic in Computer Science,* pages 154-163. IEEE Computer Society Press, 1989.
20. Y. N. Moschovakis. Computable processes. In *Proc. 18th ACM Symp. on Principles of Programming Languages,* pages 72–80. IEEE Computer Society Press, 1990.
21. D.-W. Wang and L. D. Zuck. Real-time sequence transmission problem. In *Proc. 10th ACM Symp. on Principles of Distributed Computing,* August 1991.
22. D.-W. Wang, L. D. Zuck, and Nick Reingold. The power of I/O automata. Unpublished manuscript, December 1990.
23. J. Welch, L. Lamport, and N. Lynch. A lattice-structured proof of a minimum spanning tree algorithm. In *Proc. 7th ACM Symp. on Principles of Distributed Computing,* pages 28-43, August 1988.

This article was processed using the LaTeX macro package with LLNCS style

Minimization of Timed Transition Systems

R. Alur
AT&T Bell Laboratories
Murray Hill, New Jersey

C. Courcoubetis
University of Crete
Heraklion, Greece

N. Halbwachs
IMAG Institute
Grenoble, France

D. Dill, H. Wong-Toi
Stanford University
Stanford, California

1 Introduction

Model checking is a powerful technique for the automatic verification of finite-state systems [10, 13, 8]. A model-checking algorithm determines whether a finite-state system, represented by its state-transition graph, satisfies its specification given as a temporal logic formula. For *speed independent* or *delay insensitive* systems, the correctness can be proved by abstracting away real-time retaining only the sequencing of state-transitions. For such systems, model checking has a long history spanning over ten years, and has been shown to be useful in validating protocols and circuits [7]. Only recently there have been attempts to extend these techniques to verification of timing properties that explicitly depend upon the actual magnitudes of the delays [15, 4, 2, 1, 18, 3]. Because of the practical need for some support for developing reliable *real-time* systems, the interest in studying these techniques further is considerable. The initial theoretical results indicate that the addition of timing constraints makes the model-checking problem harder: in addition to the state-explosion problem inherent in qualitative model checking, now we also have to deal with the blow-up caused by the magnitudes of the delay bounds. Clearly, to make the proposed algorithms applicable to substantial examples there is a need to develop heuristics. In this paper, we show how to apply state-minimization techniques to verification algorithms for real-time systems.

We use *timed automata* as a representation of real-time systems [12, 2]. A timed automaton provides a way of annotating a state-transition graph of the system with timing constraints. It operates with a finite-state control and a finite number of fictitious time-measuring elements called *clocks*. Various problems have been studied in the framework of timed automata [2, 1, 3, 9, 19]. Before we can say how we improve the existing algorithms, let us recall how these algorithms work. First notice that a *state* of a timed automaton needs to record the location of the control and the (real) values for all its clocks, and thus, a timed automaton has infinitely many states. The algorithms for timed automata rely on partitioning the uncountable state space into finitely many *regions* and constructing a quotient called the *region graph*. States in the same region are in some sense equivalent, and the region graph is adequate for solving many problems. For instance, it can be used for testing emptiness of a timed automaton [2], real-time model-checking [1], testing bisimulation equivalence [9], finding bounds on the delays [11], and controller synthesis [20]. The main hurdle in implementing such algorithms using the region graph is that it's too big – it is exponential in the number of clocks and in the length of timing constraints. Recently, to overcome this problem Henzinger et al. have shown how to compute certain timing properties of timed automata symbolically [16]. We propose another approach, namely, of applying a state-minimization algorithm while constructing the region graph to reduce its size.

The objective of the minimization algorithm is to construct a minimal reachable region graph from a timed automaton. Note that we want to construct such a minimal graph without constructing the full region graph first. Recently, algorithms have been proposed for performing simultaneously the reachability analysis and minimization from an implicitly defined transition system [5, 6, 17]. First we show how these algorithms can be adapted to our needs to construct the minimal region graph. Next we extend these methods to propose an algorithm for the problem of deciding whether a timed automaton meets a specification in TCTL — a real-time extension of the branching-time logic CTL. The minimal region graph, in itself, is not adequate for checking TCTL properties. Firstly, it does not incorporate the "non-Zeno" assumption about real-time behaviors which requires that time progresses without any bound along an infinite sequence of transitions. Secondly, the minimization algorithm concerns only with reachability, and not with "timed" reachability (e.g. to check a temporal property of the form "within time 3" we need to check whether a sequence of transitions is possible within the specified bound 3). We show how to refine the minimal region graph to incorporate these requirements, and this leads to an algorithm for model checking. A nice feature of the algorithm is that it splits the minimal graph only as much as needed depending on the TCTL-formula to be checked. We remind the reader that model-checking for TCTL has been shown to be computationally hard, namely, PSPACE-complete [1]. However, examples indicate that the minimized region graph is much smaller than the worst-case exponential bound, and consequently, our methods should result in a big saving.

The rest of the paper is organized as follows. Section 2 reviews the definitions of timed automata and region graphs. In Section 3 we review the minimization algorithm, and in the following section we show how to construct the minimal region graph using it. Section 5 gives examples illustrating the construction of the minimal region graph. In the final section we consider extensions needed to do model checking for TCTL.

2 Timed automata and region graphs

In this section we recall the definition of *timed automata* and the principles of their analysis by means of finite *region graphs* [12, 2, 1].

2.1 Timed Automata

Timed automata have been proposed to model finite-state real-time systems. Each automaton has a finite set of *locations* and a finite set of *clocks* which are real-valued variables. All clocks proceed at the same rate and measure the amount of time that has elapsed since they were started (or reset). Each transition of the system might reset some of the clocks, and has an associated enabling condition which is a constraint on the values

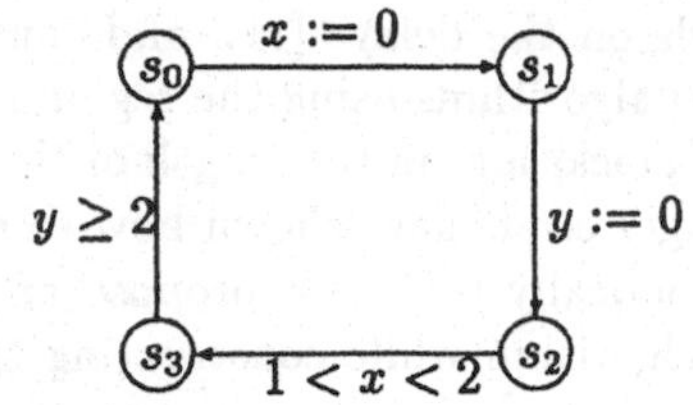

Figure 1: An example of a timed automaton

of the clocks. A transition can be taken only if the current clock values satisfy its enabling condition.

For example, the automaton of Figure 1 represents a system with four locations and two clocks x, and y. The clock x gets initialized on the transition from s_0 to s_1. At any instant, the value of x equals the time elapsed since the last time this transition was taken. The enabling condition associated with the s_2 to s_3 transition expresses the following timing constraint: the delay between the transition from s_0 to s_1 and the transition from s_2 to s_3 has lower bound 1 and upper bound 2. Similarly, the clock y constrains the transition from s_3 to s_0 to occur at least two units later than the transition from s_1 to s_2. Thus to express a bound on the delay between two transitions, we reset a clock with the first transition, and associate an enabling condition with the other transition.

For each transition, the enabling condition is required to be a convex polyhedron of $\mathbb{R}^n$ ($\mathbb{R}$ denotes the set of nonnegative reals, and n is the number of clocks in the system), consisting of all the solutions of a system of linear inequalities of the form

- $x \leq k$, $x < k$, $x \geq k$, $x > k$, where x is a clock and k is an integer, or
- $x - y \leq k$, $x - y < k$, where x and y are clocks and k is an integer.

In this paper, such a polyhedron will be called a (time) *zone*. Let $\mathcal{Z}(n)$ (or simply $\mathcal{Z}$) be the set of zones of $\mathbb{R}^n$. We consider also a set of *reset actions* $\mathcal{A}(n)$ (or simply $\mathcal{A}$), which are functions from $\mathbb{R}^n$ to $\mathbb{R}^n$. For each $a \in \mathcal{A}$, there is a set of indexes $I_a \subseteq \{1 \ldots n\}$ such that

$$\forall \vec{x} \in \mathbb{R}^n, \forall i = 1, \ldots, n, \quad a(\vec{x})[i] = \begin{cases} 0 & \text{if } i \in I_a \\ \vec{x}[i] & \text{otherwise} \end{cases}$$

A timed automaton G is a tuple (S, C, s_{init}, T) where

1. S is a finite set of locations,
2. $C = \{x_1, \ldots, x_n\}$ is a set of clocks,
3. $s_{init} \in S$ is an initial location,
4. $T \subseteq S \times \mathcal{Z}(n) \times \mathcal{A}(n) \times S$ is a transition relation. A transition (s, z, a, s') in T will be denoted by $s \xrightarrow{z,a} s'$.

The automaton G starts with the control at the location s_{init} with all its clocks initialized to 0. The values of all the clocks increase uniformly with time. At any point in time, the automaton can make a transition, if the current values of the clocks belong to the associated zone. The transitions are instantaneous. With each transition $s \xrightarrow{z,a} s'$, the clocks in I_a get reset to 0 and start counting time with respect to that transition. At any instant, the state of the system can be fully described by specifying the current location and the values of all its clocks. So, a *state* of the system is a pair $\langle s, \vec{x} \rangle$, where $s \in S$ and $\vec{x} \in \mathbb{R}^n$.

Now we can define a timed consecution relation on the states of a timed automaton. For $\delta \in \mathbb{R}$, a state $\langle s', \vec{x'} \rangle$ is said to be δ-successor of another state $\langle s, \vec{x} \rangle$, written $\langle s, \vec{x} \rangle \stackrel{\delta}{\Rightarrow} \langle s', \vec{x'} \rangle$, iff either

- $\delta = 0$ and there is a transition $s \xrightarrow{z,a} s' \in T$ such that $\vec{x} \in z$ and $\vec{x'}$ equals $a(\vec{x})$, or
- $\delta > 0$ and $s' = s$ and $\vec{x'} = \vec{x} + \vec{\delta}$ (where $\vec{\delta}$ denotes the n-tuple $[\delta, \delta, \ldots] \in \mathbb{R}^n$).

A state $\langle s', \vec{x'} \rangle$ is said to be a successor of another state $\langle s, \vec{x} \rangle$, written $\langle s, \vec{x} \rangle \Rightarrow \langle s', \vec{x'} \rangle$, iff there exists a $\delta \in \mathbb{R}$ such that $\langle s, \vec{x} \rangle \stackrel{\delta}{\Rightarrow} \langle s', \vec{x'} \rangle$.

The behavior of a timed automaton can now be formally defined using the consecution relation $\Rightarrow$. A run of the automaton started in a state $\langle s, \vec{x} \rangle$ is obtained by iterating the relation $\Rightarrow$. Formally, a *run* r is an infinite sequence of locations $s_i \in S$, clock vectors $\vec{x_i} \in \mathbb{R}^n$, and time values $\delta_i \in \mathbb{R}$ of the form

$$\langle s_0, \vec{x_0} \rangle \stackrel{\delta_0}{\Rightarrow} \langle s_1, \vec{x_1} \rangle \stackrel{\delta_1}{\Rightarrow} \langle s_2, \vec{x_2} \rangle \stackrel{\delta_2}{\Rightarrow} \cdots \stackrel{\delta_{n-1}}{\Rightarrow} \langle s_n, \vec{x_n} \rangle \stackrel{\delta_n}{\Rightarrow} \cdots$$

Note that the above definition allows more than one transitions to occur at the same time. This means that, the time of the clocks is stopped, and the system can perform instantaneously several transitions which are enabled one after the other; each transition is enabled by the clock values which the previous one produced. Such an assumption allows the modeling of simultaneous actions of different components by interleaving; however, it is not essential for our algorithms.

The run r is called *progressive* iff the sequence of sums $\Sigma_{i=0}^{k} \delta_i$ is unbounded. This requirement corresponds to the "non-Zeno" constraint which rules out the runs in which an infinite number of transitions occur in a bounded interval of time. Thus the actual behavior of a real-time system gives rise only to progressive runs, and hence, while checking temporal properties of a timed automaton, we will restrict attention only to the progressive runs.

2.2 Region graphs

The key to solving verification problems for timed automata is construction of a finite region graph [1]. This solution constructs a specific region graph, we generalize this notion here.

A *region* $F \subseteq S \times \mathbb{R}^n$ is a set of states. Typically F will be of the form $\{\langle s, \vec{x} \rangle \mid \vec{x} \in Z\}$, denoted by $\langle s, Z \rangle$, for a zone Z. For a state $\langle s, \vec{x} \rangle$ in a region F and a region F', the consecution relation $\langle s, \vec{x} \rangle \Rightarrow F'$ holds iff one of the following two conditions are met:

- *Elapse of time:* Starting from the state $\langle s,\vec{x}\rangle$, as time elapses, the state enters the region F' while staying in the region $F \cup F'$ along the way. That is, for some $\delta > 0$, $\langle s,\vec{x}+\vec{\delta}\rangle \in F'$, and the set of states $\{\langle s,\vec{x}+\vec{\delta'}\rangle \mid 0 \le \delta' \le \delta\}$ is entirely included in the region $F \cup F'$.

- *Eventual explicit transition:* Starting from the state $\langle s,\vec{x}\rangle$, the state stays in the region F as time elapses, and then enters F' because of an explicit transition. That is, for some $\delta \ge 0$ and some $\langle s',\vec{x'}\rangle \in F'$, the set $\{\langle s,\vec{x}+\vec{\delta'}\rangle \mid 0 \le \delta' \le \delta\}$ is entirely included in F, and $\langle s,\vec{x}+\vec{\delta}\rangle \stackrel{0}{\Rightarrow} \langle s',\vec{x'}\rangle$.

A partition R of the state space $S \times \mathbb{R}^n$ into regions is said to be *stable* iff

1. *R is stable with respect to the elapsing of time*: For every $\langle s,\vec{x}\rangle$ in F, if $\langle s,\vec{x}\rangle$ can lead to a region $F' \in R$ by letting the time elapse, then every other state $\langle s',\vec{x'}\rangle$ in F can also lead to F' by letting the time elapse.

2. *R is stable with respect to explicit transitions*: For every $\langle s,\vec{x}\rangle$ in F, if $\langle s,\vec{x}\rangle$ can lead to a region $F' \in R$ by eventually enabling an explicit transition, then every other state $\langle s',\vec{x'}\rangle$ in F can also lead to F' by eventually enabling an explicit transition (not necessarily the same transition as $\langle s,\vec{x}\rangle$).

Intuitively, stability of R means that all states in a region are equivalent with respect to the reachability analysis: if for some state $\langle s,\vec{x}\rangle \in F$, there is a state $\langle s',\vec{x'}\rangle \in F'$ such that $\langle s,\vec{x}\rangle \Rightarrow^* \langle s',\vec{x'}\rangle$, then for every state $\langle u,\vec{y}\rangle \in F$ there is a state $\langle u',\vec{y'}\rangle \in F'$ such that $\langle u,\vec{y}\rangle \Rightarrow^* \langle u',\vec{y'}\rangle$. Also our definitions ensure that the paths leading $\langle s,\vec{x}\rangle$ and $\langle u,\vec{y}\rangle$ to F' visit the same sequence of regions of R along the way. Thus, the reachability questions about the states of a timed automaton can be reduced to reachability questions about the regions of a stable partition. In general, given an initial partition of the state space, we will be interested in constructing a partition that is stable and refines the initial partition (a partition R refines another partition R' if every region F of R is entirely contained in some region F' of R'). This motivates the following definition.

A *region graph* corresponding to a timed automaton G and an initial partition R_0 of the state space of G, is a graph $RG(G,R_0) = (R,E)$ such that

1. R is a stable partition of $S \times \mathbb{R}^n$,
2. R refines the initial partition R_0, and
3. there is an edge from F to F' in E iff $\langle s,\vec{x}\rangle \Rightarrow F'$ for some state $\langle s,\vec{x}\rangle$ in F.

Clearly, we can define a region graph in which every region contains a single state. But this is not useful, because a timed automaton has infinitely many states. The following proposition, which is the main result of [2], states that it can always be folded into a finite region graph:

Proposition : For any timed automaton G and the initial partition $R_0 = \{\langle s,\mathbb{R}^n\rangle \mid s \in S\}$, there exists a *finite* region graph $RG(G,R_0)$. □

The proof of this proposition is based on the existence of the *detailed region graph* *DRG*(G) (the initial partition is assumed to contain a region $\langle s,\mathbb{R}^n\rangle$ for every location

s). The constructive proof defines an equivalence relation $\cong$ on $\mathbb{R}^n$. Let c be the largest constant used in defining a zone Z used in an enabling condition of G. Then, for $\vec{x}$ and $\vec{y}$ in $\mathbb{R}^n$, define $\vec{x} \cong \vec{y}$ iff for *every* zone $Z \in \mathcal{Z}$ that is defined using integer constants not greater than c, $\vec{x} \in Z$ iff $\vec{y} \in Z$. This equivalence relation has the following properties:

- The quotient $[\mathbb{R}^n/\cong]$ is finite.
- $S \times [\mathbb{R}^n/\cong]$ is a stable partition of $S \times \mathbb{R}^n$.

Any region graph is adequate for doing a finite reachability analysis, however, as we will see later, it is not fine enough to do TCTL model-checking. On the other hand, the detailed region graph is adequate to solve the model-checking problem. The only stumbling block is its size: the number of regions of $DRG(G)$ is $O(n!|S|c^n)$.

So, the problems of interest, which will be addressed in the remainder of the paper, are

- Is it possible to symbolically build a region graph smaller than the detailed region graph?
- Is it possible to use such a reduced region graph to perform full TCTL model-checking?

3 Minimization Algorithm

Bouajjani et al [6] (see also [5]) describe a general algorithm to directly generate a minimal state graph from an implicit description (e.g., a program). Let us briefly recall this algorithm, before adapting it to the generation of region graphs.

We start from a transition system $\mathcal{S} = (S, s_0, \rightarrow)$, where S is the set of states, $s_0 \in S$ is the initial state, and $\rightarrow \subseteq S \times S$ is the transition relation. A state s is said to be accessible from s_0 if and only if $s_0 \rightarrow^* s$, where $\rightarrow^*$ denotes the reflexive-transitive closure of $\rightarrow$. For a state s and a set $X \subseteq S$, we will use the notation $s \Rightarrow X$ to denote $s \rightarrow s'$ for some $s' \in X$. Let ρ be a partition of S. A class $X \in \rho$ is said to be *stable* with respect to ρ if and only if

$$\forall Y \in \rho . [(\exists x \in X, x \Rightarrow Y) \text{ implies } (\forall x \in X, x \Rightarrow Y)].$$

A partition ρ is a *bisimulation* if and only if every class of ρ is stable with respect to ρ.

The *reduction* of $\mathcal{S}$ according to a partition ρ is the transition system $\mathcal{S}|\rho$ given by $(Acc(\rho), [s_0]_\rho, \rightarrow_\rho)$, where

- $Acc(\rho)$ is the set of classes of ρ which contain at least one state accessible from s_0;
- $[s_0]_\rho$ denotes the class of ρ which contains s_0;
- $X \rightarrow_\rho Y$ iff $x \Rightarrow Y$ for some $x \in X$.

Given a transition system $\mathcal{S}$ and an initial partition ρ_0, the algorithm described in [6] explicitly builds the transition system $\mathcal{S}|\overline{\rho}$, where $\overline{\rho}$ is the coarsest bisimulation compatible with ρ_0 (that is, every class of ρ_0 is a union of classes of $\overline{\rho}$). The termination of the algorithm requires that the bisimulation $\overline{\rho}$ must have a finite number of classes. The algorithm is given below, with the following notations:

- The function *split* "splits" a class X of a partition ρ into a minimal set of subclasses which are all stable with respect to ρ;
- For a class X of ρ, $post_\rho(X)$ denotes the set of classes of ρ which contain at least one state directly accessible from a state of X: $post_\rho(X) = \{Y \mid \exists x \in X, x \Rightarrow Y\}$.
- Conversely, $pre_\rho(X)$ denotes the set of classes of ρ which contain at least one state from which a state of X is directly accessible: $pre_\rho(X) = \{Y \mid \exists y \in Y, y \Rightarrow X\}$.

In the following algorithm, ρ is the current partition, α is the set of classes of ρ which have been found accessible from (the class of) the initial state, and σ is the set of classes of ρ which have been found stable with respect to ρ.

Minimization Algorithm:

```
ρ = ρ0; α = {[s0]ρ}; σ = ∅;
while α ≠ σ do
   choose X in α \ σ;
   let α' = split(X, ρ);
   if α' = {X} then
      σ := σ ∪ {X};  α := α ∪ postρ(X);
   else
      α := α \ {X};
      if ∃Y ∈ α' such that s0 ∈ Y then α := α ∪ {Y};
      σ := σ \ preρ(X);
      ρ := (ρ \ {X}) ∪ α';
   fi
od
```

4 Constructing the minimal region graph

Given a timed automaton $G = (S, C, s_{init}, T)$, we can use the algorithm of Section 3 to generate a minimal region graph. Recall that the automaton G can be viewed as a transition system over $S \times \mathbb{R}^n$ with the initial state $\langle s_{init}, \vec{0}\rangle$ and the transition relation $\Rightarrow$ (which is the union of $\overset{\delta}{\Rightarrow}$, $\delta \geq 0$). For simplicity of implementation, we require every region F to be of the form $\langle s, Z\rangle$ for a zone Z. We start with some definitions.

The *set of time predecessors* of a zone Z is

$$Z_{\swarrow} = \{\vec{y} \mid \exists \vec{x} \in Z,\ \exists \delta \in \mathbb{R},\ \vec{y} = \vec{x} + \vec{\delta}\}.$$

For zones Z and Z', $Z \setminus Z'$ is some set of disjoint zones such that the set $\{Z'\} \cup Z \setminus Z'$ forms a partition of Z, and

$$Z \sqcup Z' = \{Z \cap Z'\} \ \cup\ (Z \setminus Z') \ \cup\ (Z' \setminus Z).$$

We generalize this operator to accept any finite number of arguments: For any finite set $\{Z_1, \ldots, Z_k\}$ of zones, $\bigsqcup_{i=1}^k Z_i$ is a partition of $\bigcup_{i=1}^k Z_i$ into a set $\{Z'_1, \ldots, Z'_p\}$ of disjoint zones, such that for each $i = 1 \ldots k$, $j = 1 \ldots p$, either $Z'_j \subseteq Z_i$ or $Z'_j \cap Z_i = \emptyset$. The operator $\sqcup$ extends over regions also: $\langle s, Z\rangle \sqcup \langle s, Z'\rangle = \{\langle s, Z''\rangle \mid Z'' \in Z \sqcup Z'\}$.

In order to adapt the algorithm of Section 3 to generate a minimal region graph, we could define the "precondition" function: *pre*(F) is the set of states $\langle s', \vec{x}'\rangle$ which may lead to some $\langle s, \vec{x}\rangle \in F$ either by letting the time elapse (if $s = s'$), or by an explicit transition. For a region $F = \langle s, Z\rangle$ this definition translates to:

$$pre(\langle s, Z\rangle) = \langle s, Z_{\swarrow}\rangle \cup \bigcup_{s' \xrightarrow{z,a} s} \langle s', (a^{-1}(Z) \cap z)_{\swarrow}\rangle.$$

However, such a formalization doesn't take into account the fact that one cannot reach $\langle s, Z\rangle$ from $\langle s, Z'\rangle$ without going through some zone Z'' "separating" Z' from Z. For instance, one cannot reach $\langle s, \{x \geq 2\}\rangle$ from $\langle s, \{x < 1\}\rangle$ without going through $\langle s, \{1 \leq x < 2\}\rangle$ (Recall the definition of $\langle s, \vec{x}\rangle \Rightarrow F'$ for stability of regions from Section 2). In fact, we cannot formalize the right abstraction of "time elapsing", by means of a single precondition function. Instead of looking for such a precondition, we will make precise in what case a region may directly lead to another region (following [16]), and use this notion to define the function for splitting a region into stable regions.

Let $Z \Uparrow Z'$ denote the set of $\vec{x} \in Z$ for which there exists $\delta \in \mathbb{R}$ such that $\vec{x} + \vec{\delta} \in Z'$ and $\vec{x} + \vec{\delta'} \in Z \cup Z'$ for all $0 \leq \delta' \leq \delta$. It is easy to show that $Z \Uparrow Z'$ is a zone.

Now the stability of a region can be expressed as follows. A region $\langle s, Z\rangle$ is stable with respect to another region $\langle s', Z'\rangle$ if and only if

- if $s = s'$ then $Z \Uparrow Z' \in \{Z, \emptyset\}$, and
- for every transition $s \xrightarrow{z,a} s'$ [1],
  - either $a(Z \cap z) \cap Z' = \emptyset$ (this includes the case where $Z \cap z = \emptyset$),
  - or $a(Z \cap z) \subseteq Z'$ and $Z \Uparrow (Z \cap z)$ equals Z.

From this definition, we derive the function *split*: For any locations s, s' ($s \neq s'$), for any zones Z, Z',

$$split(\langle s, Z\rangle, \langle s', Z'\rangle) = \langle s, Z\rangle \sqcup \bigsqcup_{s \xrightarrow{z,a} s'} \langle s, Z \Uparrow (Z \cap z \cap a^{-1}(Z'))\rangle$$

$$split(\langle s, Z\rangle, \langle s, Z'\rangle) = \langle s, Z\rangle \sqcup \langle s, Z \Uparrow Z'\rangle \sqcup \bigsqcup_{s \xrightarrow{z,a} s} \langle s, Z \Uparrow (Z \cap z \cap a^{-1}(Z'))\rangle$$

Now all the definitions needed for applying the algorithm can be given. Let ρ be any partition of the states into regions, and let $\langle s, Z\rangle$ be a region. Then,

$$pre_{\rho}(\langle s, Z\rangle) = \{\langle s, Z'\rangle \in \rho \mid Z' \Uparrow Z \neq \emptyset\} \cup \bigcup_{s' \xrightarrow{z,a} s} \{\langle s', Z'\rangle \in \rho \mid a(Z' \cap z) \cap Z \neq \emptyset\},$$

[1] and this includes the case where $s = s'$ and there is a looping transition on s.

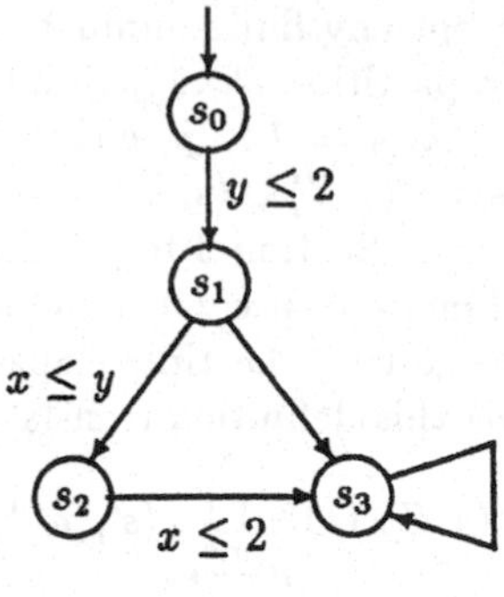

Figure 2: The timed automaton of Example 1

$$post_\rho(\langle s, Z\rangle) = \{\langle s, Z'\rangle \in \rho \mid Z \Uparrow Z' \neq \emptyset\} \cup \bigcup_{s \xrightarrow{z,a} s'} \{\langle s', Z'\rangle \in \rho \mid a(Z \cap z) \cap Z' \neq \emptyset\},$$

$$split(\langle s, Z\rangle, \rho) = \bigsqcup_{\langle s', Z'\rangle \in \rho} split(\langle s, Z\rangle, \langle s', Z'\rangle).$$

To implement the algorithm, we simply need efficient ways for representing zones and computing simple operations on them such as $Z \sqcup Z'$, $Z \Uparrow Z'$, $a(Z)$, and $a^{-1}(Z)$.

5 Examples

We will demonstrate the effectiveness of minimization procedure on simple examples.

5.1 Example 1

We consider first the very simple timed automaton shown on Fig. 2.

We start with an initial partition which only distinguishes regions according to their node component: $\rho = \rho_0 = \{C_0, C_1, C_2, C_3\}$, with $C_i = \langle s_i, \mathbb{R}^2\rangle$ for $i = 0, 1, 2, 3$. Since the initial state belongs to C_0, we have $\alpha = \{C_0\}, \sigma = \emptyset$.

So, we consider first $X = C_0$. Obviously $split(C_0, C_2) = split(C_0, C_3) = \{C_0\}$, since there is no transition from s_0 to s_2 or s_3. So, $split(C_0, \rho) = split(C_0, C_1) = \{C_{00}, C_{01}\}$, with

$$C_{00} = \langle s_0, \{y \leq 2\}\rangle \quad C_{01} = \langle s_0, \{y > 2\}\rangle$$

The initial state $\langle s_0, \{x = y = 0\}\rangle$ belongs to C_{00}, so α is updated to $\{C_{00}\}$. Considering $X = C_{00}$, we find it stable with respect to $\rho = \{C_{00}, C_{01}, C_1, C_2, C_3\}$, since all of its elements can lead to C_{01} and to C_1. So, we get $\alpha = \{C_{00}, C_{01}, C_1\}$ and $\sigma = \{C_{00}\}$.

The region $X = C_{01}$ is stable with respect to ρ, and it doesn't lead to any other region. Considering $X = C_1$, we find

$$split(C_1, C_{00}) = split(C_1, C_{01}) = split(C_1, C_3) = \{C_1\}$$

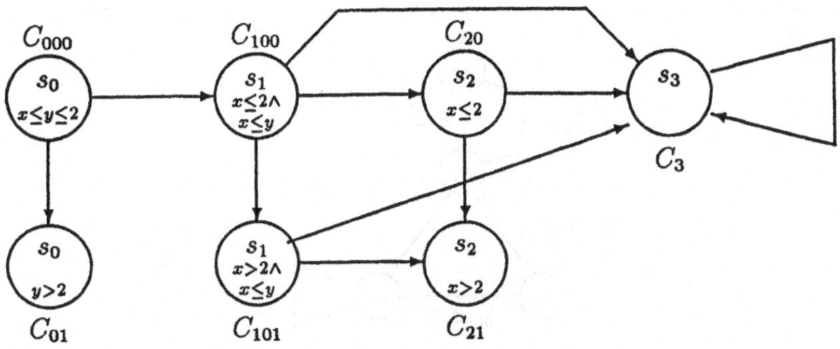

Figure 3: The minimal region graph of Example 1

so

$$split(C_1, \rho) = split(C_1, C_2) = \{C_{10}, C_{11}\}$$

with

$$C_{10} = \langle s_1, \{x \leq y\} \rangle \quad C_{11} = \langle s_1, \{x > y\} \rangle$$

Splitting C_1 questions about the stability of C_{00}, which is removed from σ, and considered again.

Now, we have $\rho = \{C_{00}, C_{01}, C_{10}, C_{11}, C_2, C_3\}$, $\alpha = \{C_{00}, C_{01}\}$, $\sigma = \{C_{01}\}$ and $X = C_{00}$. We get $split(C_{00}, \rho) = split(C_{00}, C_{10}) = split(C_{00}, C_{11}) = \{C_{000}, C_{001}\}$, with

$$C_{000} = \langle s_0, \{x \leq y \leq 2\} \rangle \quad C_{001} = \langle s_0, \{y \leq 2 \wedge y < x\} \rangle$$

The initial state belongs to C_{000} which is stable, and can lead either to C_{01} or to C_{10}. We get $\rho = \{C_{000}, C_{001}, C_{01}, C_{10}, C_{11}, C_2, C_3\}$, $\alpha = \{C_{000}, C_{01}, C_{10}\}$, $\sigma = \{C_{000}, C_{01}\}$.

$X = C_{10}$ is found stable with respect to ρ, leading to C_2 and C_3, which become both accessible. $X = C_2$ is split into

$$C_{20} = \langle s_2, \{x \leq 2\} \rangle \quad C_{21} = \langle s_2, \{x > 2\} \rangle$$

so C_{10} must be considered again. It is split into

$$C_{100} = \langle s_1, \{x \leq y \wedge x \leq 2\} \rangle \quad C_{101} = \langle s_1, \{2 < x \leq y\} \rangle$$

C_{000} is also considered again, it is found stable and leads to C_{100}.

We have $\rho = \{C_{000}, C_{001}, C_{100}, C_{101}, C_{20}, C_{21}, C_3\}$, $\alpha = \{C_{000}, C_{01}, C_{100}\}$ and $\sigma = \{C_{000}, C_{01}\}$. Now, C_{100} is found stable, leading to C_{101}, C_{20} and C_3. C_{101} is stable and leads to C_{21} and to C_3. C_{20} is stable, and leads to C_{21} and C_3. C_{21} and C_3 are stable. The resulting graph is shown on Fig. 3. Notice that the detailed region graph of this example has 160 regions, 24 of which are accessible.

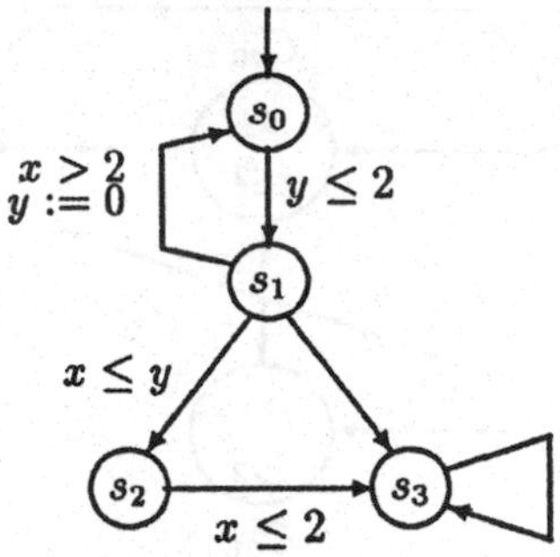

Figure 4: The timed automaton of Example 2

5.2 Example 2

Let us slightly complexify our example as shown by Fig. 4. The first steps of the algorithm are similar, but now C_{101} can lead to C_{001} which becomes accessible. C_{001} is found stable, leading to C_{11}. C_{11} is stable and leads to C_3. Our reduced graph has 9 accessible regions, instead of 40 in the detailed graph.

6 Model checking for TCTL

In this section we show how to use the algorithm for constructing the minimal region graph to check properties specified in the branching-time logic TCTL.

6.1 The logic TCTL

Let us briefly review the logic TCTL of [1]. It is a real-time extension of the branching-time logic CTL of [14]. The syntax of TCTL allows putting subscripts on the temporal operators of CTL to restrict their scope in time. Thus one can write $\exists\Diamond_{<3}\, p$ meaning "along some run within time 3." Formally, let AP be a set of atomic propositions, then the formulas ϕ of TCTL are defined inductively as:

$$\phi \; := \; p \mid \neg\phi \mid \phi_1 \wedge \phi_2 \mid \exists\, \phi_1 \mathcal{U}_{\sim c} \phi_2 \mid \forall\, \phi_1 \mathcal{U}_{\sim c} \phi_2,$$

where p is in AP and c is an integer and $\sim$ stands for one of the binary relations $<, >, =, \leq, \geq$.

Informally, $\exists\, \phi_1 \mathcal{U}_{<c} \phi_2$ means that for some run, there exists an initial prefix of time length less than c such that ϕ_2 holds at the last state of the prefix, and ϕ_1 holds at all its intermediate states. Similarly, $\forall\, \phi_1 \mathcal{U}_{<c} \phi_2$ means that for every run, there is an initial prefix with the above property. Formally, the semantics of TCTL is defined with respect to *continuous computation trees*, but for our purposes it suffices to interpret TCTL formulas over timed automata. To interpret TCTL formulas over a timed automaton, first we need to know which atomic propositions are true in every location of the automaton. A *labeled timed automaton* is a pair (G, μ), where G is a timed automaton and μ is a labeling function from the locations of G to 2^{AP}.

Given a labeled timed automaton (G, μ), we define the satisfaction relation $\langle s, \vec{x}\rangle \models \phi$ inductively as follows:

$\langle s, \vec{x}\rangle \models p$ iff $p \in \mu(s)$.
$\langle s, \vec{x}\rangle \models \neg\phi$ iff $\langle s, \vec{x}\rangle \not\models \phi$.
$\langle s, \vec{x}\rangle \models \phi_1 \wedge \phi_2$ iff both $\langle s, \vec{x}\rangle \models \phi_1$ and $\langle s, \vec{x}\rangle \models \phi_2$.
$\langle s, \vec{x}\rangle \models \exists \phi_1 \mathcal{U}_{\sim c}\phi_2$ iff for some progressive run r of G starting at $\langle s, \vec{x}\rangle$, $r \models \phi_1 \mathcal{U}_{\sim c}\phi_2$.
$\langle s, \vec{x}\rangle \models \forall \phi_1 \mathcal{U}_{\sim c}\phi_2$ iff for every progressive run r of G starting at $\langle s, \vec{x}\rangle$, $r \models \phi_1 \mathcal{U}_{\sim c}\phi_2$.

For a run $r = \langle s_0, \vec{x_0}\rangle \stackrel{\delta_0}{\Rightarrow} \langle s_1, \vec{x_1}\rangle \stackrel{\delta_1}{\Rightarrow} \cdots$, the relation $r \models \phi_1 \mathcal{U}_{\sim c}\phi_2$ holds iff there exists k and $\delta \leq \delta_k$ such that (1) $(\delta + \Sigma_{i<k}\delta_i) \sim c$, and (2) $\langle s_k, \vec{x_k}\rangle \models \phi_2$, and (3) for all $0 \leq i < k$, for all $0 \leq \delta' < \delta_i$, $\langle s_i, \vec{x_i} + \vec{\delta'}\rangle \models \phi_1$, and (4) for all $0 \leq \delta' < \delta$, $\langle s_k, \vec{x_k} + \vec{\delta'}\rangle \models \phi_1$.

A labeled timed automaton (G, μ) satisfies a TCTL-formula ϕ iff $\langle s_{init}, \vec{0}\rangle \models \phi$. The model-checking problem for TCTL is to decide if (G, μ) satisfies ϕ. The problem is known to be PSPACE-complete [1].

6.2 Model checking algorithm

We sketch how to adapt the minimization algorithm to do model-checking for TCTL. Let (G, μ) be the labeled timed automaton with state space $S \times \mathbb{R}^n$. For a TCTL-formula ϕ, let F_ϕ be the set of states $\langle s, \vec{x}\rangle$ such that $\langle s, \vec{x}\rangle \models \phi$. The detailed region graph of [1] is adequate for TCTL model-checking: for any TCTL-formula ϕ, the set F_ϕ is a union of regions of the detailed region graph. Now our objective is to construct the set F_ϕ through only a "minimal" splitting. In our analysis, the set F_ϕ will always be a union of regions of the form $\langle s, Z\rangle$ for $Z \in \mathcal{Z}$.

The construction of F_ϕ is defined inductively on the structure of ϕ. The cases when ϕ is an atomic proposition, or is a boolean combination are simple:

$$\begin{aligned} F_p &= \cup_{p \in \mu(s)} \langle s, \mathbb{R}^n\rangle \\ F_{\neg\phi} &= (S \times \mathbb{R}^n) \setminus F_\phi \\ F_{\phi_1 \wedge \phi_2} &= F_{\phi_1} \cap F_{\phi_2} \end{aligned}$$

The interesting case is when ϕ is a "timed until" formula. For simplicity of presentation, we only consider the case when ϕ is of the form $\exists\Diamond_{\sim c}\psi$ (that is, $\exists\, true\, \mathcal{U}_{\sim c}\psi$) or $\forall\Diamond_{\sim c}\psi$; the changes necessary to handle the "until" formulas should be obvious.

First consider an unbounded temporal formula $\phi = \exists\Diamond\psi$ (that is, $\exists\Diamond_{\geq 0}\psi$). Suppose we have constructed the set F_ψ. Let R_ψ be the partition of the states of G into two regions: F_ψ and $F_{\neg\psi}$. Now we run the minimization algorithm of Section 4 to construct a region graph $RG(G, R_\psi) = (R, E)$. Since R refines R_ψ, for any region F of R, either ψ holds at all states in F or $\neg\psi$ holds in all states in F. Suppose we want to determine the truth of the formula ϕ at the state $\langle s, \vec{x}\rangle$ in the region F of R. From the semantics of TCTL, it follows that ϕ holds at $\langle s, \vec{x}\rangle$ iff some state in F_ψ appears on a ***progressive*** run of G starting at $\langle s, \vec{x}\rangle$. Since every finite run can be extended to obtain a progressive infinite run, ϕ holds at $\langle s, \vec{x}\rangle$ iff some state in F_ψ is reachable from $\langle s, \vec{x}\rangle$. This holds precisely when a region $F' \in R$ such that $F' \subseteq F_\psi$ is reachable from F in the region graph. Thus, the desired set F_ϕ is a union of regions F for which some $F' \subseteq F_\psi$ is reachable from F in (R, E).

6.2.1 Progressiveness

Now consider the formula $\phi = \forall\Diamond\psi$. Suppose we construct the region graph $RG(G, R_\psi) = (R, E)$ as before. Now, ϕ holds at $\langle s, \vec{x}\rangle \in F$ iff some state in F_ψ appears on every *progressive* run of G starting at $\langle s, \vec{x}\rangle$. However, this is not equivalent to saying that every infinite path in the region graph, starting at F, contains some region $F' \subseteq F_\psi$. To determine the truth of ϕ we need to account for the progressiveness assumption while constructing the region graph.

From the results in [1] it follows that the progressiveness assumption can be modeled as *fairness constraints* on the detailed region graph which require that a path of $DRG(G)$ infinitely often visits certain sets of regions. In particular, these constraints require that for every clock i, the constraint $\vec{x}[i] = 0$ or $\vec{x}[i] > c_i$ holds at infinitely many regions along the path (here, c_i is the largest constant in a constraint involving x in the enabling conditions of G). We can use this fact to handle progressiveness in our reduced region graphs. For each clock i, let R_i be the partition of the states of G into three regions:

$$R_i^0 = \{\langle s, \vec{x}\rangle \mid \vec{x}[i] = 0\},\ R_i^{>c_i}\{\langle s, \vec{x}\rangle \mid \vec{x}[i] > c_i\},\ R_i^{\leq c_i}\{\langle s, \vec{x}\rangle \mid 0 < \vec{x}[i] \leq c_i\}.$$

Now, as the initial partition we choose the coarsest partition R'_ψ that refines R_ψ and also refines R_i for each clock i. The next step is to construct a region graph $RG(G, R'_\psi) = (R, E)$. An infinite path in this region graph is called *progressive* iff for every $i = 1 \ldots n$:

- it contains an infinite number of regions $F \subseteq R_i^0 \cup R_i^{>c_i}$.
- it contains an infinite number of regions $F \subseteq R_i^{\leq c_i} \cup R_i^{>c_i}$.

The set F_ϕ is now the union of regions F in the region graph such that every progressive path starting at F contains a region $F' \subseteq F_\psi$.

6.2.2 Timed Reachability

Now consider a formula $\phi = \exists\Diamond_{<3}\psi$. To compute whether ϕ holds at a state, we need to determine whether some state in F_ψ can be reached within 3 time units. The region graph constructed for the case $\phi = \exists\Diamond\psi$ has information only about reachability, but not about "timed" reachability. The timed reachability analysis can be performed by introducing an auxiliary clock x_0. The new state space is $S \times \mathbb{R}^{n+1}$, and the timed consecution relation $\overset{\delta}{\Rightarrow}$ on this new space is defined as before; the transitions corresponding to the elapse of time increment the value of x_0 along with the other clocks, and the transitions corresponding to the change of location do not depend upon the value of x_0 and leave x_0 unchanged. For $\vec{x} \in \mathbb{R}^n$ and $t \in \mathbb{R}$, let $[t]\vec{x}$ denote the $(n+1)$-vector that assigns t to the clock x_0 and agrees with $\vec{x}$ on the values of the remaining n clocks. Conversely, for $\vec{x} \in \mathbb{R}^{n+1}$, let $\vec{x}^n$ denote the n-vector obtained by discarding the value of the clock x_0.

To compute the value of ϕ at $\langle s, \vec{x}\rangle$ we consider the paths starting at $\langle s, [0]\vec{x}\rangle$. The value of x_0 is 0 at the beginning of the path and at later points its value reflects the elapsed time. The formula ϕ holds at $\langle s, \vec{x}\rangle$ iff there is a state $\langle u, \vec{y}\rangle$ reachable from $\langle s, [0]\vec{x}\rangle$ (in the extended state space) such that $\vec{y}[0] < 3$ and $\langle s, \vec{y}^n \rangle \in F_\psi$. To test this condition, we construct a region graph for the extended state space $S \times \mathbb{R}^{n+1}$. The initial partition needs to distinguish between the cases $x_0 = 0$, $0 < x_0 < 3$, and $x_0 \geq 3$ and also on the basis of the truth of ψ.

Let R_ψ be the partition of $S \times \mathbb{R}^{n+1}$ into two regions: $F'_\psi = \{\langle s, \vec{x} \rangle \mid \langle s, \vec{x}^n \rangle \in F_\psi\}$ and its complement. Let R_0 be the partition of the state space into three regions:

$$R_0^0 = \{\langle s, \vec{x} \rangle \mid \vec{x}[0] = 0\},\ R_0^{<3}\{\langle s, \vec{x} \rangle \mid 0 < \vec{x}[0] < 3\},\ R_0^{\geq 3}\{\langle s, \vec{x} \rangle \mid \vec{x}[0] \geq 3\}.$$

As the initial partition R'_ψ we choose the coarsest partition that refines both R_ψ and R_0 above, and build the region graph $RG(G, R'_\psi) = (R, E)$. Now, the truth of ϕ can be evaluated by a simple reachability analysis on this region graph. The set $F'_\phi \subseteq S \times \mathbb{R}^{n+1}$ is union of the regions $F \subseteq R_0^0$ of R for which there is a region $F' \subseteq F'_\psi \cap (R_0^0 \cup R_0^{<3})$ reachable from F. The set $F_\phi \subseteq S \times \mathbb{R}^n$ is simply the projection of F'_ϕ: $\{\langle s, \vec{x}^n \rangle \mid \langle s, \vec{x} \rangle \in F'_\phi\}$.

For a formula $\phi = \exists\Diamond_{\sim c}\psi$, the algorithm is the same; the initial partition now distinguishes between the cases $\vec{x}[0] = 0$ and $0 < \vec{x}[0] \sim c$ and $0 < \vec{x}[0] \not\sim c$. The analysis for $\phi = \forall\Diamond_{\sim c}\psi$ is similar; the initial partition now needs to account for the progressiveness assumption also (as in the case of $\forall\Diamond\psi$).

References

[1] R. Alur, C. Courcoubetis, and D. Dill. Model-checking for real-time systems. In *Proceedings of the Fifth IEEE Symposium on Logic in Computer Science*, pages 414–425, 1990.

[2] R. Alur and D. Dill. Automata for modeling real-time systems. In *Automata, Languages and Programming: Proceedings of the 17th ICALP*, Lecture Notes in Computer Science 443, pages 322–335. Springer-Verlag, 1990.

[3] R. Alur, T. Feder, and T. Henzinger. The benefits of relaxing punctuality. In *Proceedings of the Tenth ACM Symposium on Principles of Distributed Computing*, pages 139–152, 1991.

[4] R. Alur and T. Henzinger. A really temporal logic. In *Proceedings of the 30th IEEE Symposium on Foundations of Computer Science*, pages 164–169, 1989. Journal version to appear in the Journal of the ACM.

[5] A. Bouajjani, J. Fernandez, and N. Halbwachs. Minimal model generation. In *Proceedings of the Second Workshop on Computer-Aided Verification, Rutgers University*, 1990.

[6] A. Bouajjani, J. Fernandez, N. Halbwachs, P. Raymond, and C. Ratel. Minimal state graph generation. *Science of Computer Programming*, 1992. To appear.

[7] M. Browne, E. Clarke, D. Dill, and B. Mishra. Automatic verification of sequential circuits using temporal logic. *IEEE Transactions on Computers*, C-35(12):1035–1044, 1986.

[8] J. Burch, E. Clarke, D. Dill, L. Hwang, and K. L. McMillan. Symbolic model checking: 10^{20} states and beyond. In *Proceedings of the Fifth IEEE Symposium on Logic in Computer Science*, pages 428–439, 1990.

[9] K. Cerans. Decidability of bisimulation equivalence for parallel timer processes. In *Proceedings of Chalmers Workshop on Concurrency, Goteborg*, 1991.

[10] E. Clarke, E. A. Emerson, and A. P. Sistla. Automatic verification of finite-state concurrent systems using temporal-logic specifications. *ACM Transactions on Programming Languages and Systems*, 8(2):244–263, 1986.

[11] C. Courcoubetis and M. Yannakakis. Minimum and maximum delay problems in real-time systems. In *Proceedings of the Third Workshop on Computer-Aided Verification, Aalborg University, Denmark*, 1991.

[12] D. Dill. Timing assumptions and verification of finite-state concurrent systems. In J. Sifakis, editor, *Automatic Verification Methods for Finite State Systems*, Lecture Notes in Computer Science 407. Springer–Verlag, 1989.

[13] E. Emerson and C. Lei. Modalities for model-checking: Branching time logic strikes back. In *Proceedings of the 12th ACM Symposium on Principles of Programming Languages*, pages 84–96, 1985.

[14] E. A. Emerson and E. M. Clarke. Using branching-time temporal logic to synthesize synchronization skeletons. *Science of Computer Programming*, 2:241–266, 1982.

[15] E. A. Emerson, A. Mok, A. P. Sistla, and J. Srinivasan. Quantitative temporal reasoning. Presented at the First Workshop on Computer-aided Verification, Grenoble, France, 1989.

[16] T. Henzinger, X. Nicollin, J. Sifakis, and S. Yovine. Symbolic model-checking for real-time systems. In *Proceedings of the Seventh IEEE Symposium on Logic in Computer Science*, 1992. To appear.

[17] D. Lee and M. Yannakakis. Online minimization of transition systems. In *Proceedings of ACM Symposium of Theory of Computing*, 1992. To appear.

[18] H. Lewis. A logic of concrete time intervals. In *Proceedings of the Fifth IEEE Symposium on Logic in Computer Science*, pages 380–389, 1990.

[19] X. Nicollin, J. Sifakis, and S. Yovine. From ATP to timed graphs and hybrid systems. In *Proccedings of REX workshop "Real-time: theory in practice"*, Lecture Notes in Computer Science 600. Springer-Verlag, 1991.

[20] H. Wong-Toi and G. Hoffmann. The control of dense real-time discrete event systems. In *Proceedings of the 30th IEEE Conference on Decision and Control*, pages 1527–1528, 1991.

Using CSP to Verify a Timed Protocol over a Fair Medium

Jim Davies and Steve Schneider

Programming Research Group, Oxford University
Oxford OX1 3QD, UK

Abstract. Standard timed models of CSP are based upon finite observations, and are thus unsuitable for the analysis of fairness conditions. The addition of infinite observations to the standard timed failures model permits an adequate treatment of fairness in a timed context. The resulting model admits a complete proof system for admissible specifications, and supports a theory of timed refinement for untimed programs. This is demonstrated with a study of a familiar protocol—the alternating bit protocol—communicating over an unreliable but fair medium.

1 Introduction

There are several denotational models for the language of CSP, supporting the analysis of safety and liveness constraints in both timed and untimed contexts. The standard models of [Ree88] and [Dav91] are based upon finite observations of program execution, and are unsuitable for the analysis of fairness properties. In this paper, we introduce a model based upon infinite observations, and show how it may be applied to the verification of a communication protocol over a fair medium.

The paper begins with a review of CSP, and an introduction to the new denotational model. We show how this model may be used to capture safety, liveness and fairness constraints, and describe a complete, compositional proof system for behavioural specifications. We then introduce a theory of refinement between the new timed model and the untimed traces model of [Ree88]. If a result is independent of timing considerations, and we are able to produce a proof in the traces model, we may conclude that an equivalent result holds in the timed model.

To demonstrate the application of the model to protocol verification, we consider an alternating bit protocol—a sending program and a receiving program—communicating across unreliable media. Each medium may delay a message for an arbitrary period before delivery, or lose an arbitrary number of messages, but it will always eventually pass each message. A formal analysis of a protocol over such a medium requires a semantic model with an adequate treatment of unbounded nondeterminism.

Having formalised the protocol requirements, we use the language of CSP to propose implementations for the sending and receiving programs. The compositional proof system, and the theory of refinement, are used to demonstrate that these implementations combine to provide the specified service. The paper ends with a brief comparison between this work and similar approaches to protocol verification.

2 The Infinite Timed Model

The language used in this paper is an extension of Hoare's Communicating Sequential Processes. Several denotational models have been formulated for the untimed language, the most important being the traces model of [Hoa85], and the failures-divergences model of [BrR85]. The language was extended to include timing information in [ReR86], and provided with a uniform hierarchy of semantic models [Ree88].

The untimed traces model M_T is sufficient for the analysis of untimed safety conditions: constraints that proscribe certain events or sequences of events in a program execution. To reason about real-time systems, we may employ the timed failures model TM_F, introduced in [Ree88]; this model supports reasoning about timed liveness conditions: requirements upon the availability of certain synchronisations or communications.

In these models, each program is identified with a set of observations—the models are named according to the type of observations made. In the existing timed models, every observation of a program must be completed in a finite time. The infinite timed model extends TM_F to include infinite observations; these describe complete executions, and permit a satisfactory treatment of fairness.

2.1 Language

The language of Timed CSP is generated by the following grammar rule:

$$P ::= STOP \mid SKIP \mid WAIT\ t \mid a \rightarrow P \mid P\,;P \mid P \,\Box\, P \mid P \sqcap P \mid$$
$$a : A \rightarrow P_a \mid P \overset{t}{\rhd} P \mid \|_{A_i} P_i \mid P \,|||\, P \mid P \setminus A \mid f(P) \mid \mu X \bullet F(X)$$

In this rule, event a is drawn from the set of all synchronisations Σ, event set A ranges over the set of subsets of Σ, and t is a non-negative real number.

STOP is a program which will never engage in external communication; it is a broken program. *SKIP* is a program which does nothing except terminate, and is ready to terminate immediately. *WAIT t* is a delayed form of *SKIP*; it does nothing, but is ready to terminate successfully after time t.

The prefix program $a \rightarrow P$ is initially prepared to engage in synchronisation a; if this event occurs, it immediately begins to behave as P. The following abbreviation will prove useful:

$$a \overset{t}{\longrightarrow} P = a \rightarrow WAIT\ t\,; P$$

The program $a \overset{t}{\longrightarrow} P$ will delay for t time units after the first occurrence of a, before behaving as P. We consider events to be instantaneous; if the duration of an action is of interest, then that action may be modelled by considering the beginning and the end of the action to be separate events.

In the sequential composition $P\,;Q$, control is passed from P to Q if and when program P performs the termination event $\checkmark$. $P \,\Box\, Q$ is an external choice between programs P and Q. If the environment is prepared to cooperate with P but not Q, then the choice is resolved in favour of P, and vice versa. $P \sqcap Q$ is a internal choice between P and Q; the outcome of this choice is nondeterministic.

In the timeout program $P \overset{t}{\triangleright} Q$ control is transferred from P to Q at time t if no communications have occurred. The program $a : A \rightarrow P_a$ offers an external choice of initial event a, drawn from a set A, which may be infinite. If channel c carries values of type T, then the program $c?x : T \rightarrow P_x$ is prepared to accept any value v of type T on channel c, and then behave accordingly. We use the expression $c!v$ to denote the output of value v on channel c.

The parallel combination of a set of programs is parameterised by a corresponding set of interfaces: for each program P, we provide an interface set A_P. In the network $\|_{A_P} P$ each event a requires the participation of every subprogram P for which event a appears in interface set A_P; a simple form of network is the binary parallel combination $P\ {}_A\|_B\ Q$. In an interleaving parallel combination $P \,|||\, Q$ both subprograms evolve concurrently without interacting.

The program $P \backslash A$ behaves as P, except that events from set A are concealed from the environment of the program. Hidden events no longer require the cooperation of the environment, and so occur as soon as P is ready to perform them. The relabelled program $f(P)$ has a similar control structure to P, with observable events renamed according to function f.

The recursive program $\mu X \circ F(X)$ behaves as $F(X)$, with each instance of variable X representing a recursive invocation; this program satisfies the equation $P = F(P)$. More generally, programs may be defined by sets of mutually recursive equations; these programs have a well-defined semantics if the equation set is *guarded* for the list of recursive variables—if there is a non-zero lower bound upon the time between successive recursive calls.

2.2 A Semantic Model

In the infinite timed model TM_I, programs are sets of pairs, each consisting of a timed trace—a record of events seen to be performed—and a timed refusal—a record of events seen to be refused during the same execution. Every event is a synchronous communication, drawn from the universal set of events Σ. Events are labelled with the times at which they are observed; our domain of time values is the non-negative real numbers $[0, \infty)$ and the set of all timed events is given by $TE = [0, \infty) \times \Sigma$. We use $TE^{\mathfrak{i}}$ to denote the set of all timed traces: possibly-infinite sequences of timed events. If a timed trace u has infinite length, it must take an infinite amount of time to observe a performance of u: the set of times appearing in the trace is unbounded.

Timed refusals are sets of timed events, with the structure of a step function upon our time domain: events are refused over intervals. We use $IRSET$ to denote the set of all timed refusals. The presence of a timed event (t, a) in a timed refusal $\aleph$ corresponds to the observation that event a is refused at time t. An observation in the timed infinite model is a pair $(u, \aleph)$, where u is a timed trace, and $\aleph$ a timed refusal:

$$TM_I \subseteq \mathbb{P}(TE^{\mathfrak{i}} \times IRSET)$$

To support reasoning about the model, we define a number of healthiness conditions, or axioms. For example, we insist that the empty observation $(\langle\rangle, \{\})$ should be a possible observation of every program. These conditions are presented in [Sch92], and will not be reproduced here; they are not required for applications of the model.

We define a semantic function $\mathcal{F}_I$ from the language of Timed CSP to the semantic model TM_I. For example, $STOP$ is given a semantics by

$$\mathcal{F}_I[\![STOP]\!] = \{(u, \aleph) \mid u = \langle\rangle\}$$

The only timed trace corresponding to this program is the empty trace $\langle\rangle$, and any set of timed events may be refused. For reasons of space, we omit the definition of the semantic function $\mathcal{F}_I$, and the fixed point theory underlying the model: these will be found in [Sch92].

2.3 Specifications

The infinite timed model may be used to reason about the timed properties of CSP programs. The semantic function $\mathcal{F}_I$ associates each program P with the set of observations that may be made of an execution of P. We formalise requirements upon P as predicates upon the semantic set $\mathcal{F}_I[\![P]\!]$.

A behavioural specification is a predicate upon an arbitrary observation, in the current semantic model. A program P satisfies a behavioural specification $S(u, \aleph)$ in the infinite timed model precisely when the predicate $S(u, \aleph)$ is true for every observation $(u, \aleph)$ of P. We employ a familiar notation for satisfaction

$$P \textbf{ sat } S(u, \aleph) \text{ in } TM_I = \forall (u, \aleph) \in \mathcal{F}_I[\![P]\!] \bullet S(u, \aleph)$$

If the identity of the model is obvious from the context, we omit the qualification.

To facilitate reasoning about observations, we define a number of projection functions and relations upon timed traces, timed refusals, and timed observations. We require three subsequence relations. We write $u \leqslant v$ if u is a prefix of v, $u \preceq v$ if u is a subsequence of v, and $u \underline{\text{in}}\, v$ if u is a contiguous subsequence of v. If u is an infinite trace, then $u \leqslant v \Leftrightarrow u = v$. We use $\langle\rangle$ to denote the empty trace, and $u \frown v$ to denote the catenation of two traces u and v, and $\#u$ to denote the length of trace u.

The functions ***head*** and ***tail*** are defined on traces with at least one element, returning the first pair in the trace, and the remainder of the trace, respectively. The expressions 'times(u)' and $\sigma(u)$ denote the set of times appearing in u, and the set of events appearing in u, respectively; similar functions are defined upon timed refusal sets.

The expressions $\mathrm{m}(\aleph)$ and $\overline{\mathrm{m}}(\aleph)$ denote the measure of refusal set $\aleph$, and the measure of the complement of $\aleph$:

$$\mathrm{m}(\aleph) = \mathrm{M}(\mathrm{times}(\aleph))$$
$$\overline{\mathrm{m}}(\aleph) = \mathrm{M}([0, \infty) - \mathrm{times}(\aleph))$$

where $\mathrm{M}(T)$ denotes the measure of set T: the integral of the characteristic function χ_T.

The projections ***begin*** and ***end*** return the times of occurrence of the earliest and latest events in a trace. We take the minimum of the empty set of times to be ∞; this is the value of $begin(\langle\rangle)$. If u is an infinite trace, then $end(u) = \infty$. We define before ($\upharpoonleft$) and after ($\upharpoonright$) operators on timed refusals, returning the sets of timed events refused before or after a specified time:

$$\aleph \upharpoonleft t = \aleph \cap [0,t) \times \Sigma$$
$$\aleph \upharpoonright t = \aleph \cap [t,\infty) \times \Sigma$$

We define a distributed addition function $+$ upon timed traces, timed refusals, and timed observations, shifting each recorded time by a constant amount. For this function to return valid traces and refusals, we require that the resulting time values are all non-negative.

The function 'strip' removes the timing information from a timed trace: the result is an untimed trace. The projection function '$\upharpoonright$' returns the part of an observation containing events from the specified set. The count function '$\downarrow$', defined upon timed and untimed traces, returns the number of occurrences of events from a given set.

2.4 Examples

Suppose that a program P accepts inputs on a channel *in* and delivers outputs on a channel *out*. If we wish to guarantee that the number of outputs provided by P never exceeds the number of inputs, we should prove that

$$P \textbf{ sat } u \downarrow OUT \leqslant u \downarrow IN$$

where IN and OUT are the sets of possible communications on channels *in* and *out*, respectively. This is a safety specification upon program P.

The inclusion of infinite observations in our semantic model means that we can express fairness conditions. The following specification states that program P cannot perform an infinite sequence of inputs without performing an infinite sequence of outputs

$$P \textbf{ sat } u \downarrow IN = \infty \Rightarrow u \downarrow OUT = \infty$$

If a trace of program P contains an infinite number of input events, then it must also contain an infinite number of outputs.

As an example of a liveness condition, consider the requirement that program P be ready to output a value one time unit after the arrival of the same value on channel *in*. This is captured by the following specification:

$$P \textbf{ sat } \langle(t, in.v)\rangle \underline{\text{in}}\ u \Rightarrow \exists\, t' > t \bullet \langle(t', out.v)\rangle \underline{\text{in}}\ u \\ \vee \\ out.v \notin \sigma(\aleph \upharpoonright t + 1)$$

If an input $in.v$ is observed at time t, then the output $out.v$ must either be observed at some time $t' > t$, or made available from time $t + 1$ onwards.

The inclusion of infinite observations in our semantic model allows us to express eventual liveness requirements: conditions which place no upper bound on the time by which a program must be ready to communicate:

$$P \textbf{ sat } \langle(t, in.v)\rangle \underline{\text{in}}\ u \Rightarrow \exists\, t' > t \bullet \langle(t', out.v)\rangle \underline{\text{in}}\ u \\ \vee \\ \exists\, t'' > t \bullet out.v \notin \sigma(\aleph \upharpoonright t'')$$

In this case, if the program has not yet performed the output event, there must be some time t'' at which the event will become available.

The above specifications require that an output event is made available until it occurs: once an offer of output has been made, the program may not retract until the output event has been observed. Another form of liveness condition may be expressed using the measure of the times at which an event may be refused:

$$P \textbf{ sat } \langle(t, in.v)\rangle \;\underline{\text{in}}\; u \Rightarrow \exists\, t' > t \bullet \langle(t', out.v)\rangle \;\underline{\text{in}}\; u \quad \vee \quad \mathrm{m}(\aleph \downharpoonright out.v) < \infty \qquad (1)$$

If P has not yet performed the output $out.v$, then the set of times at which P may refuse to perform the event has only finite measure. If a program is prepared to wait for a value on channel out, then the event $out.v$ will occur eventually.

We may use the complementary function $\overline{\mathrm{m}}$ to define a weaker form of liveness condition. The specification

$$P \textbf{ sat } \langle(t, in.v)\rangle \;\underline{\text{in}}\; u \Rightarrow \exists\, t' > t \bullet \langle(t', out.v)\rangle \;\underline{\text{in}}\; u \quad \vee \quad \overline{\mathrm{m}}(\aleph \downharpoonright out.v) = \infty \qquad (2)$$

states that the set of times at which output is available has infinite measure. Consider a program which is prepared to output only during intervals of the form $[2n, 2n+1)$ for natural numbers n: $[0,1)$, $[2,3)$, $[4,5)$, etc. Such a program would satisfy a liveness specification of the form *(2)*, but not one of the form *(1)*.

3 Proof and Satisfaction

The semantic equations for the infinite timed model may be used to derive a number of inference rules. These rules relate the properties of programs to the properties of their syntactic subcomponents, and form the basis of a compositional proof system for behavioural specifications. The following inference rules are representative.

Sequential Programs. The broken program *STOP* is unable to engage in external communication:

$$\frac{}{STOP \textbf{ sat } u = \langle\rangle}$$

Any trace of this program must be equal to the empty trace, but we can infer nothing about a typical refusal set.

The event prefix operator is associated with the following rule:

$$\frac{P \textbf{ sat } S(u, \aleph)}{\begin{array}{l} a \rightarrow P \textbf{ sat } u = \langle\rangle \wedge a \notin \sigma(\aleph) \\ \qquad\qquad \vee \\ \qquad\qquad u = \langle(t, a)\rangle \frown u' \wedge a \notin \sigma(\aleph \upharpoonright t) \wedge S((u', \aleph) - t) \end{array}}$$

If the trace is empty, then a may not be refused, and is therefore absent from the refusal set $\aleph$. Otherwise, the first event must be a. If a occurs at time t, we know that a may not be refused before this time; the subsequent behaviour is due to program P, and must satisfy specification S.

In the timeout program $P \overset{t}{\rhd} Q$, control is transferred to Q unless P performs an external action before time t.

$$\frac{\begin{array}{l} P \textbf{ sat } S(u,\aleph) \\ Q \textbf{ sat } T(u,\aleph) \end{array}}{\begin{array}{l} P \overset{t}{\rhd} Q \textbf{ sat } \mathit{begin}(u) \leqslant t \wedge S(u,\aleph) \\ \qquad\qquad\qquad \vee \\ \qquad\qquad\qquad \mathit{begin}(u) \geqslant t \wedge S(\langle\rangle, \aleph \upharpoonright t) \wedge T((u,\aleph) - t) \end{array}}$$

Parallel Programs. Suppose that $(u,\aleph)$ is an observation of the network $\|_{A_i} P_i$. For every index i, the restriction of trace u to the set A_i is the trace of events performed by the corresponding program P_i. Further, the timed trace u contains only events drawn from set $\bigcup_i A_i$.

Suppose that $(u_i, \aleph_i)$ is an observation of component P_i which corresponds to the observation $(u,\aleph)$. If a program can refuse the events in $\aleph_i$ while performing u, it could refuse any subset of $\aleph_i$ instead. We may therefore choose these observations such that $\sigma(\aleph_i) \subseteq A_i$. Finally, any event from set A_i will require the cooperation of P_i, so we may deduce that $\aleph_i \subseteq \aleph \downharpoonright A_i$.

$$\frac{\forall i \in I \bullet P_i \textbf{ sat } S_i(u,\aleph)}{\begin{array}{l} \|_{A_i} P_i \textbf{ sat } \forall i \in I \bullet \exists u_i, \aleph_i \bullet S_i(u_i,\aleph_i) \wedge \\ \qquad\qquad\qquad u_i = u \downharpoonright A_i \wedge u \downharpoonright \bigcup_i A_i = u \wedge \\ \qquad\qquad\qquad \aleph_i \subseteq \aleph \downharpoonright A_i \wedge \aleph \downharpoonright \bigcup_i A_i = \bigcup(\aleph_i) \end{array}}$$

Abstraction. To reason about the observations of $P \setminus A$, we identify the observations of P in which every event from A occurs as soon as possible. These are precisely the observations of P in which events from A may be continuously refused; if we can show that these observations satisfy a behavioural specification S' such that

$$S'(u, \aleph \cup [0,\infty) \times A) \Rightarrow S(u \setminus A, \aleph)$$

where $u \setminus A = u \downharpoonright (\Sigma - A)$, then we may conclude that $P \setminus A$ satisfies S. The resulting inference rule for the hiding operator is

$$\frac{\begin{array}{l} P \textbf{ sat } [0,\infty) \times A \subseteq \aleph \Rightarrow S'(u,\aleph) \\ S'(u, \aleph \cup [0,\infty) \times A) \Rightarrow S(u \setminus A, \aleph) \end{array}}{P \setminus A \textbf{ sat } S(u,\aleph)}$$

The second antecedent states that if S' holds for observation $(u,\aleph)$ when events from A are continuously refused, then S should hold of the same observation when events from A are removed from the trace. If we are to find a suitable specification S' for program P, the external specification S must not depend upon the occurrence of events from A.

Recursive Programs. We may define recursive programs using the μ-notation of [Hoa85]. The equality

$$P = \mu X \bullet F(X)$$

defines a program P that behaves as program $F(X)$, with each occurrence of X denoting a recursive invocation of P. The definition of programs by mutually-recursive sets of equations is discussed in [Dav91], together with a theory of recursion induction for behavioural specifications. In this paper, we exhibit only the following inference rule:

$$\frac{X \textbf{ sat } S(u,\aleph) \;\Rightarrow\; F(X) \textbf{ sat } S(u,\aleph)}{\mu X \bullet F(X) \textbf{ sat } S(u,\aleph)} \quad [\, S \text{ admissible} \,]$$

For this rule to be applicable, we must ensure that predicate S is admissible, and that the recursive program $\mu X \bullet F(X)$ has a valid semantics.

A predicate S on timed observations is admissible if

$$\neg\, S(u,\aleph) \Rightarrow \exists (s,\aleph') \bullet \neg\, S(s,\aleph') \;\wedge\; \exists\, t \bullet (s,\aleph') = (u,\aleph) \upharpoonright t$$

If S fails to hold for an infinite observation $(u,\aleph)$, then S must also fail for some finite approximation $(s,\aleph')$ of $(u,\aleph)$. In this definition, $\mathit{end}(u,\aleph)$ denotes the supremum of the time values recorded in observation $(u,\aleph)$, and $(u,\aleph) \upharpoonright t$ denotes the restriction of the observation to time values less than t.

To show that a recursive program has a valid semantics, we must eliminate the possibility of an infinitely number of recursive invocations in a finite time. If we consider the program to be defined by a recursive equation, it is sufficient to show that the body of the definition is *guarded* for the recursive variable. A program $F(X)$ is guarded for variable X if every free occurrence of X is preceded by a non-zero time delay.

The proof system is complete for admissible behavioural specifications of guarded programs: if such a specification is true of a program P, then we can establish this fact by reasoning within our proof system. This result follows easily from a similar result for the finite timed failures model; the proof is omitted for lack of space.

4 Timed Refinement

The denotational semantic models of CSP form a hierarchy, ordered by the information content of the semantic sets. To establish that a program P satisfies a specification of the form $S \wedge T$ we begin by choosing the simplest model M in which both S and T may be expressed. If one of these conditions—say S—can be expressed as a behavioural specification S' in a smaller model M', without loss of information, then we may be able to establish S by reasoning within M'. If the two specifications are equivalent, and we can prove that P satisfies S' in M', then we may deduce that P satisfies S in the larger model M.

For example, suppose that we wish to establish that a program P meets a conjunction of safety and liveness properties. Suppose also that the safety condition depends only upon the order of occurrence of certain events, while the liveness condition is dependent upon the fairness properties of components of P. In this case, we may be able to establish safety by reasoning within the untimed trace model, but a proof of liveness will require the timed infinite model.

We formalise the specification of P using the larger of the two models

$$SAFE(u, \aleph) \wedge LIVE(u, \aleph)$$

and restrict the domain of predicate $SAFE$ to finite timed traces. For this to be valid, we must check that the predicate $SAFE$ is admissible. A predicate $S(u)$ on timed traces is admissible if and only if $(\forall s < u \bullet S(s)) \Rightarrow S(u)$: if $S(u)$ fails to hold, then $S(s)$ must fail for some finite prefix s of u.

We must also find an untimed trace specification $SAFE'(tr)$ which is equivalent to $SAFE$, but is expressed in terms of the untimed traces model:

$$\forall s \in TT \bullet SAFE(s) \Leftrightarrow SAFE'(\text{strip } s)$$

Our proof obligation is then reduced to showing that

$$P \textbf{ sat } SAFE'(tr) \wedge P \textbf{ sat } LIVE(u, \aleph)$$

This reduction is justified by a refinement proof rule from M_T to TM_I:

$$\frac{\begin{array}{l} P \textbf{ sat } S'(tr) \text{ in } M_T \\ \forall s \in TT \bullet S'(\text{strip } s) \Leftrightarrow S(s) \end{array}}{P \textbf{ sat } S(u) \text{ in } TM_I} \quad [\, S \text{ admissible} \,]$$

This refinement rule will be required for the protocol example in section 5.

These rules follow from a more general theory of refinement, linking all of the semantic models in our hierarchy. If models M and M' are linked by a projection mapping π, then we may define a refinement relation $\sqsubseteq_\pi$ on programs, such that

$$\frac{P' \textbf{ sat } S' \text{ in } M'}{P \textbf{ sat } S \text{ in } M} \quad [\, P' \sqsubseteq_\pi P \,]$$

where S' is an equivalent form of specification S, expressed in terms of model M'.

5 A Simple Protocol

We consider the simplest protocol sufficient for communication over an unreliable but fair medium: the *alternating bit protocol*. This protocol consists of a sending program and a receiving program, linked by two media which may lose messages, but never corrupt them. Each medium may delay a message for an arbitrary period before delivery, or lose an arbitrary number of messages, but it will always eventually pass each message.

Informal Description. Each message accepted for transmission is tagged with a bit value; successive messages are tagged with *0* and *1* alternately, beginning with value *0*. After each transmission, the sending program waits for confirmation that the message was received. If no confirmation is forthcoming within a certain time interval, the sending program retransmits.

The receiving program examines the bit value of any message which arrives. This should be the complement of the bit value found on the last message to pass through successfully (or *0*, in the initial state). If this is not the case, then the transmitter must have retransmitted a message which was successfully received. A confirmation is sent, but the message is discarded. If the bit value is satisfactory, a confirmation is sent and the message is output.

Formal Description. Our formal description of the protocol comprises a sending program S and a receiving program R, communicating across unreliable media $M1$ and $M2$. The programs and media are linked by simple channels lm, rm, lc and rc: the first two channels are for messages, the others for confirmations. Messages to and from the protocol are carried by channels ***in*** and ***out***. We assume a type $MESSAGE$ of messages carried by the protocol, and define a type $BITS$ with precisely two elements: *0* and *1*. For each channel ***chan***, we let $CHAN$ be the set of possible communications involving ***chan***: e.g.

$$LM = \{lm.b.m \mid m \in MESSAGE \; ; b \in BIT\}$$

The projection functions 'messages' and 'bits' are defined upon timed and untimed traces, and return the sequence of messages and the sequence of bit values passed, respectively.

The Protocol. The protocol should satisfy the following safety specification:

$$\begin{aligned} SAFE(u) = {} & \text{messages}(u \upharpoonright OUT) \leqslant \text{messages}(u \upharpoonright IN) \\ & \wedge \\ & u \downarrow IN \leqslant u \downarrow OUT + 1 \end{aligned}$$

This states that the sequence of messages delivered on channel ***out*** is a prefix of the sequence of messages presented on channel ***in***, and that no more than one message may be held within the network.

In any incomplete execution, the protocol will eventually be ready to communicate on either the input or the output channel:

$$\begin{aligned} LIVE(u, \aleph) = \#u < \infty \Rightarrow {} & \exists\, m \bullet \text{m}(\aleph \upharpoonright out.m) < \infty \\ & \vee \\ & \forall\, m \bullet \text{m}(\aleph \upharpoonright in.m) < \infty \end{aligned}$$

In combination with the safety specification, this liveness condition is enough to guarantee that: if a message is being held, the protocol will eventually be ready to output the current message on channel ***out***; if no message is being held, the protocol will eventually be ready to accept any message on channel ***in***.

Component Specifications. The essential properties of the sending program are as follows: (*S1*) if a message is transmitted with a bit value b, then b should be the complement of the last bit value received as a confirmation (except in the initial case); (*S2*) if a message is transmitted with a bit value, and another is transmitted with the complementary value, then there must be an input between the two transmissions; (*S3*) the program should not retransmit without a confirmation unless it has been waiting for at least T time units; (*S4*) the program should be live on some combination of the input and transmission channels.

$$S1(u) = \langle rc.b, lm.c.m\rangle \ \underline{\text{in}}\ \text{strip}(u \upharpoonright LM \cup RC) \Rightarrow b = \overline{c}$$

$$S2(u) = \langle lm.b.m\rangle^\frown s^\frown\langle lm.\overline{b}.n\rangle \ \underline{\text{in}}\ \text{strip}\ u \Rightarrow s \upharpoonright IN \neq \langle\rangle$$

$$S3(u, \aleph) = T \times \text{timeouts}(u) \leqslant \overline{\text{m}}(\aleph \upharpoonright rc)$$

$$S4(u, \aleph) = \forall m \bullet in.m \notin \sigma(\aleph \uparrow end(u) + 1) \vee \exists b, m \bullet lm.b.m \notin \sigma(\aleph \uparrow end(u) + T)$$

The number of timeouts in a trace u is the difference between the number of transmissions and the number of confirmations, less one for the initial transmission: that is, $u \downarrow LM - u \downarrow RC - 1$.

The essential properties of the receiving program are: (*R1*) the program should send precisely one confirmation for every message received; (*R2*) the sequence of bits transmitted on the confirmation channel should be a prefix of the sequence of bits received on the message channel; (*R3*) the program should become ready to communicate on at least one of its three channels, within one time unit of the last event.

$$R1(u) = 0 \leqslant u \downarrow RM - u \downarrow LC \leqslant 1$$

$$R2(u) = \text{bits}(u \upharpoonright LC) \leqslant \text{bits}(u \upharpoonright RM)$$

$$R3(u, \aleph) = \forall m, b \bullet rm.b.m \notin \sigma(\aleph \uparrow end(u) + 1) \vee \exists b \bullet lc.b \notin \sigma(\aleph \uparrow end(u) + 1) \vee \exists m \bullet out.m \notin \sigma(\aleph \uparrow end(u) + 1)$$

The essential properties of the media are: (*M1*) the sequence of messages output is a subsequence of the sequence of messages input—messages are not reordered or invented; (*M2*) each medium should be live on its input channel: if only finitely many messages have been accepted, then it will become ready to accept another; (*M3*) each medium should be fair: if an infinite number of inputs are observed, then an infinite number of outputs should be possible.

In the case of the medium linking message channels lm and rm, these properties correspond to the following specifications:

$$M1(u) = \text{messages}(u \upharpoonright RM) \preceq \text{messages}(u \upharpoonright LM)$$

$$M2(u, \aleph) = \#(u \upharpoonright LM) < \infty \Rightarrow \text{m}(\aleph \upharpoonright LM) < \infty$$

$$M3(u, \aleph) = \#(u \upharpoonright LEFT) = \infty \Rightarrow \#(u \upharpoonright RM) = \infty \vee \exists m \bullet \text{m}(\aleph \upharpoonright right.m) < \infty$$

If an infinitely many inputs have been accepted, then either infinitely many outputs have occurred or some output is offered infinitely often—more precisely, there is a message m such that the set of times at which event $right.m$ may be refused has only finite measure.

Implementation. We may use the program algebra of Timed CSP to suggest an implementation of the alternating bit protocol. This algorithmic description complements the property-oriented description provided by the formal specification above.

The sending program accepts messages on channel *in*, augmenting each with a bit value of *0* or *1* before transmitting them on channel *lm*. Once a message *m* has been transmitted with bit value *b*, the sending process waits for the corresponding confirmation on channel *rc*. If no confirmation arrives within time *T* of the transmission, the program retransmits the current message.

$$\begin{aligned} S &= in?m \xrightarrow{1} lm!1.m \xrightarrow{1} S_{1.m} \\ S_{b.m} &= rc?c \xrightarrow{1} \text{ if } b = c \text{ then } in?n \xrightarrow{1} lm!\overline{b}.n \xrightarrow{1} S_{\overline{b}.m} \\ &\qquad\qquad\qquad\qquad\quad \text{else } lm!b.m \xrightarrow{1} S_{b.m} \\ &\overset{T}{\rhd} lm!b.m \xrightarrow{1} S_{b.m} \end{aligned}$$

If a confirmation *c* arrives which does not match the bit value *b* of the last message transmitted, then the sending program behaves as if no confirmation had arrived—the current message is transmitted again.

There are several time delays in our model of the sending program. In a sequential implementation, there will be unavoidable delays following each transmission, confirmation, or input. These delays should be negligable. In our description, we have set each delay to one time unit, content in the knowledge that we could replace each with a time variable, or a nondeterministic delay, should it prove useful.

Initially, the receiver expects a message on *rm*, with a bit value of *0*. If a message arrives with bit value *c*, and this matches the value *b* expected, the message is output on channel *out*. If not, then the message is discarded. In either case, the bit value received is transmitted as a confirmation on channel *lc*.

$$\begin{aligned} R &= R_0 \\ R_b &= rm?c.m \xrightarrow{1} \text{ if } b = c \text{ then } out!m \xrightarrow{1} lc!c \xrightarrow{1} R_{\overline{c}} \\ &\qquad\qquad\qquad\qquad\quad \text{else } lc!c \xrightarrow{1} R_{\overline{c}} \end{aligned}$$

After sending an acknowledgement, the receiver waits for another message, expecting a bit value which complements the last one received.

We do not provide implementations for the media—our information about these entities is limited to the knowledge that they are lossy but fair, and this has already been expressed in the formal specification above. Our protocol may be represented by the following Timed CSP program:

$$PROTOCOL = (\|_{A_P} P) \setminus COMMS$$

This is a network of four programs, in which each program *P* cooperates on all events from interface set A_P. The communications from set *COMMS* are hidden from the environment: these are the events on the internal channels:

$$COMMS = LM \cup RM \cup LC \cup RC$$

The identity of program *P* ranges over the set $\{S, R, M1, M2\}$, and the corresponding interfaces are given by

$$\begin{aligned} A_S &= IN \cup LM \cup RC & A_{M1} &= LM \cup RM \\ A_R &= OUT \cup RM \cup LC & A_{M2} &= LC \cup RC \end{aligned}$$

This completes our Timed CSP implementation of the alternating bit protocol. We are now obliged to demonstrate that it satisfies the correctness criterion formulated above: we must verify our proposed implementation.

Verification. The safety specification is independent of timing information, requiring only that the sequence of messages output should be a prefix of the sequence of messages input, and that these sequences should differ by at most one message. We may verify that our proposed implementation meets this specification by reasoning within the untimed trace model, and applying the refinement rule given in section 4.

To establish liveness, we must reason within TM_I. To show that $LIVE$ holds once the events from $COMMS$ have been concealed, we must prove that the network satisfies a similar liveness condition:

$$\begin{aligned} LIVE'(u, \aleph) = \#(u \setminus COMMS) < \infty \Rightarrow & \exists\, m \bullet out.m \notin \aleph \restriction t \\ & \vee \\ & \forall\, m \bullet in.m \notin \aleph \restriction t \end{aligned}$$

This is a stronger specification, the antecedent requires only that the number of inputs and outputs is finite. Using the infinite timed model, we are able to prove that the implementation is satisfactory without demonstrating the absence of divergent behaviour—the possibility of an infinite sequence of consecutive internal events is no cause for alarm.

Having applied the inference rule for abstraction, we may apply the inference rule for networked parallel combination to the new proof obligation. We have then to show that the sending program and the receiving program meet the corresponding specifications. In each case, we have only to demonstrate that a simple sequential process satisfies a conjunction of safety and liveness conditions. The proof may be completed by applying the inference rules for recursion, timeout, and prefixing in the appropriate models.

6 Discussion

We have shown how the infinite timed model for CSP may be applied to the specification and verification of a protocol over a fair medium. We demonstrated that the alternating bit protocol is sufficient for reliable communication over a medium which may lose messages, but is guaranteed to pass any message which is repeated infinitely often. Our choice of such a familiar protocol should facilitate comparisons with other formal methods of specification.

The correctness criterion for the protocol consisted of a safety condition and a liveness condition, both expressed as behavioural specifications in the infinite timed model. Although the proposed implementation involved timing constraints, we were able to establish the safety condition using the untimed trace model of CSP. Our theory of timed refinement allows us to conclude that the proof remains valid when timing information is added to the program description.

We introduced a compositional proof system for the infinite timed model: a set of inference rules relating properties of programs to properties of their syntactic subcomponents. This proof system was used to verify that the proposed implementation satisfied the liveness condition for the protocol. This verification would have been impossible in a model based upon finite observations, as there is no time by which delivery of a message can be guaranteed.

A variety of other approaches have been developed for the analysis of timed concurrent systems. Real-time logics [AlH91, JaM86] have been formulated, and may be used in conjunction with timed automata [ACD90], and hence with process algebra [NSY91]. In these approaches, verification is often based upon model-checking, and little emphasis is placed upon compositionality. Closer in spirit to timed CSP are the process algebras of TCCS [MoT90], IPA [Mur90], APA [Jef92], ATP [Nic90], TPL [HeR91], and RTPA [BaB91]. However, the approach to program verification is different from that taken in this paper. In general, programs are used as specifications, and an adequate treatment of fairness requires the inclusion of fair constructs within the abstract program language.

A characteristic of the use of denotational models is the separation of programs and specifications. A specification language—such as a temporal logic—can be given an interpretation in the denotational domain, and used to capture program requirements in a property-oriented fashion. An important feature of the CSP models is the provision of synchronous communication: every event on a channel requires the participation of both source and destination. A liveness condition which insisted upon output would place a requirement upon the user of the program, as well as upon the program itself.

Accordingly, a liveness condition upon a Timed CSP program takes the form of a disjunction: a communication must be performed, or it must be offered to the other programs involved. In the case of the medium studied in this paper, either infinitely many outputs must be performed, or there must be an output ready, waiting for the cooperation of the user program connected to the output channel.

Although [OrF91] presents a timed semantics for a language similar to CSP, offers made during executions are not recorded, and events are not instantaneous: a design decision which reflects a different computational model, and a level of abstraction more appropriate to "true" concurrency. An approach which invites detailed comparison is that of [Hoo91], in which an occam-like language is given a timed ready-set semantics and a compositional proof system for a metric temporal logic.

In [Hoo91], observations consist of finite and infinite executions, and a specification on a program is a property which must hold for every possible observation. A fairness constraint upon a medium similar to the one employed in this paper may be expressed in the associated temporal logic as follows:

$$\Box \neg done \land \Box \Diamond comm(in) \rightarrow \Box(\Diamond(wait(out!)\ \mathcal{U}\ comm(out))$$

As in Timed CSP, the compositional nature of the semantics means that a reasonable specification cannot insist that output is performed infinitely often, since the medium has no control over the context in which it will be placed.

The inclusion of time permits a finer treatment of fairness even where explicit timing constraints do not form part of the specification. Using the untimed infinite traces model of [Ros88], a much stronger restriction must be placed upon the medium studied in this paper: any trace containing an infinite number of inputs must also contain an infinite number of outputs. This requires the medium to refuse further input at some point, if it will be unable to output messages indefinitely; this is not necessary in a timed context.

Future research will include a more detailed study of fairness for Timed CSP, extending the treatment of fairness—for a sublanguage of untimed CSP—given in [Fra86]. A temporal logic for Timed CSP has already been defined [Jac90], and will assist in the analysis of the various forms of fairness for real-time systems.

References

[ACD90] R. Alur, C. Courcoubetis and D. Dill, *Model checking for real time systems*, Proceedings of the 5th Logics in Computer Science, 1990

[AlH91] R. Alur and T. Henzinger, *Logics and models of real-time: a survey*, Proceedings of REX '91, to appear in Springer LNCS

[BaB91] J. C. M. Baeten and J. A. Bergstra, *Real time process algebra*, Formal Aspects of Computing, Volume 3, Number 2, 1991

[BrR85] S. D. Brookes and A. W. Roscoe, *An improved failures model for communicating sequential processes*, Proceedings of the Pittsburgh Seminar on Concurrency, Springer LNCS 197, 1985

[Dav91] J. Davies, *Specification and proof in real-time systems*, Programming Research Group Monograph PRG–93, Oxford University, 1991

[Fra86] N. Francez, *Fairness*, Springer-Verlag 1986

[HeR91] M. Hennessy and T. Regan, *A process algebra for timed systems*, Report 5–91, School of Cognitive and Computing Sciences, University of Sussex 1991

[Hoa85] C. A. R. Hoare, *Communicating Sequential Processes*, Prentice-Hall 1985

[Hoo91] J. Hooman, *Specification and compositional verification of real-time systems*, Ph.D thesis, University of Eindhoven, 1991

[Jac90] D. M. Jackson, *Specifying timed communicating sequential processes using temporal logic*, PRG Report TR–5–90, Oxford University 1990

[JaM86] F. Jahanian and A.K. Mok, *Safety analysis of timing properties in real-time systems*, IEEE Transactions on Software Engineering, SE–12, 1986

[Jef92] A. S. Jeffrey, *Observation spaces and timed processes*, Oxford University D.Phil thesis, 1992

[MoT90] F. Moller and C. Tofts, *A temporal calculus of communicating systems*, Proceedings of CONCUR 90, Springer LNCS 458, 1990

[Mur90] D. V. J. Murphy, *Time, causality and concurrency*, Surrey University Ph.D thesis, 1990

[Nic90] X. Nicollin, J.-L. Richier, J. Sifakis and J. Voiron, *ATP: an algebra for timed processes*, Proceedings of the IFIP Conference on Programming Concepts and Methods, 1990

[NSY91] X. Nicollin, J. Sifakis and S. Yovine, *From ATP to timed graphs and hybrid systems*, Proceedings of REX '91, to appear in Springer LNCS

[OrF91] Y. Ortega-Mallen and D. de Frutos-Escrig, *A complete proof system for timed observations*, Proceedings of TAPSOFT 91, Springer LNCS 493, 1991

[Ree88] G. M. Reed, *A uniform mathematical theory for real-time distributed computing*, Oxford University D.Phil thesis, 1988

[ReR86] G. M. Reed and A. W. Roscoe, *A timed model for communicating sequential processes*, Proceedings of ICALP'86, Springer LNCS 226, 1986

[Ros88] A. W. Roscoe, *Unbounded nondeterminism in CSP*, Programming Research Group Technical Monograph PRG–67, Oxford University, 1988

[Sch92] S. Schneider, *Unbounded nondeterminism in Timed CSP*, to appear

Timed Ethernet: Real-Time Formal Specification of Ethernet*

Henri B. Weinberg and Lenore D. Zuck

Department of Computer Science
Yale University

Abstract. The goal of this paper is to show how formal specification can be applied to a full-fledged, real-world protocol while maintaining, or even enhancing, readability. The system we formally specify is Ethernet as it appears in IEEE 802.3. We focus on the specification of the Medium Access Control (MAC) layer—the part of the Data Link Layer that implements a 1-persistent CSMA/CD protocol—and its interfaces with adjacent layers. The specification method is based on one of Henzinger's real-time models. We believe that the readability of our specification is due to the graphical presentation using transition graphs of real-time programs.

1 Introduction

1.1 Formal Specification

In this paper we show how formal specification can be applied to a full-fledged, real-world protocol.

Reasoning about complex systems, especially those that involve interaction among several concurrent components, is notoriously difficult. We follow [12] and use the term "reactive systems" when referring to discrete systems that maintain an ongoing interaction with their environment. Subtle bugs are often found in seemingly correct specifications of reactive systems (see, for example, [9, 30]). Consequently, over the last several years we have witnessed the evolution of a formal methodology for the specification, verification, and development of reactive systems. Temporal logic [25, 26], the state machine approach [3, 4, 20, 32, 17], Floyd-Hoare-style methods [10], model checking [5, 27], and interval logic [29] have all been advocated, and indeed, have been used successfully.

Yet, these formal specification models seem to be unsatisfactory to people who are dealing with real systems. For one, most of the models have focused on qualitative aspects of reactive systems. In particular, they haven't focused on timing constraints of reactive systems. For another, researchers dealing with development of real systems claim that the models are too theoretical, that the formal specification is unreadable and therefore the advantages it offers, namely, the ability to prove the correctness of the system, is irrelevant. They also claim that these models are

* This work was supported in part by the National Science Foundation under grant CCR-8910289. Note: Current address of first co-author is Bellcore, Morristown NJ.

only good for sufficiently small and simple systems, which real systems are obviously not.

To answer the first criticism, there has been a considerable effort in incorporating timing constraints in models of reactive systems. Examples are timed automata [7, 16], timed transition systems [24, 15, 14], timed I/O automata [21], and timed process algebra [28, 23]. See §1.4.1 of [1] for a survey.

In this paper we attempt to deal with the second point. In particular, we show that a real system can indeed be specified in a way which is both formal and clear. The real system we chose is Ethernet. The formal model we chose is a variant of one of Henzinger's real time models. We believe that our formal specification is much clearer and readable, as well as more succinct, than the informal specification of Ethernet in IEEE 802.3 [31], especially the specification of the MAC layer.

This work is a continuation of a work with Michael Fischer ([8]), which attempted to formally specify 1-persistent CSMA/CD protocols.

1.2 Ethernet

Most communication networks are organized as a multi-layer architecture, where, conceptually, each layer communicates with its peer layers on other machines. In reality, all layers but the bottom-most communicate only with adjacent layers on the same machine. Inter-layer communication is through protocols which are often referred to as "interfaces".

One of the most commonly used network models is the Open System Interconnection (OSI) model, which is based on a proposal by the International Standards Organization (ISO) and is often referred to as the *ISO OSI* model. See [6], for details on the model, and Tanenbaum's review of this and other network models [33]. The OSI model has seven layers; the bottom two layers are the *Physical Layer* (PL) and the *Data Link Layer*. The Physical layer deals with actual transmission of bits over the communication channel. The Data Link layer gets frames from the higher (network) layer and transmits them to the Physical layer. It also receives bit streams from the Physical layer and derives the sent frames for them. Since the Physical layer is prone to errors, it is the task of the Data Link layer to overcome them. In some models the Data Link layer is divided into two sublayers: the Logical Link Control (LLC) sublayer and the Medium Access Control (MAC) sublayer. The LLC sublayer provides mechanisms for detecting frame deletions, reorderings, and duplications, while the lower MAC sublayer provides mechanisms for corruption detection and frame collision avoidance.

IEEE has produced several standards for Local Area Networks in a family of standards referred to as IEEE 802. These standards have been adopted by ANSI, NBS, and ISO. IEEE 802.1 describes the set of standards, IEEE 802.2 describes the LLC sublayer, IEEE 802.3–802.5 describe different MAC and Physical layers. IEEE 802.3 describes *Ethernet*, IEEE 802.4 describes the *token bus*, and IEEE 802.5 describes the *token ring*. In this paper we focus on Ethernet—the protocols described by IEEE 802.3.

1.3 Overview

Section 2 gives a review of the part of Henzinger's model [14] which we use for the specification of Ethernet. We start by describing *real-time transition system*—basically a generator of *timed behavior*—and then we describe *real-time programs* and how several concurrent real-time programs uniquely define a real-time transition system.

Section 3 describes the architecture of an Ethernet system and the basic ideas of its operation.

Section 4 gives a general overview of our formal model of Ethernet, which follows IEEE 802.3. This model uses shared variables and CSP-like interprocess communication (for message passing as well as for the external procedure calls of IEEE 802.3). We then describe each of the processes participating in the system as a real-time program, expressed as a graph.

Section 5 shows how some properties of Ethernet can be formally proved from the specification of Section 4.

We conclude in Section 6 with a discussion of why we think our specification is clearer than the one in IEEE 802.3 and where and why we have to diverge from IEEE 802.3. We also discuss future work.

2 Real Time

2.1 Real-time Transition Systems

The specification model we use is a variant of Henzinger's *time transition system* (cf. [14]). We describe our model and refer the reader to [14] for a detailed discussion.

We evaluate our time over the set $\mathbf{R}^*$ of nonnegative reals. Timed state sequences are infinite sequences of states each with an associated time. The associated times are monotonically increasing and non-converging. Formally, let Σ be some set of *states*. A pair $\rho = (\sigma, \mathsf{T})$, where σ is an infinite sequence over Σ and T is an infinite monotonic sequence over $\mathbf{R}^*$ is called a *timed state sequence* if for every $r \in \mathbf{R}^*$, there exists some i such that $\mathsf{T}_i > r$. (Unless otherwise stated, we index sequences starting from zero.) A timed state sequence represents a sequence of snapshots taken from a real time system. Each snapshot is identified by the time it is taken and the state observed.

A timed state sequence (σ, T), is a *behavior* if each two consecutive snapshots have either the same time or the same state but not both.

A *transition system* is a tuple $\mathcal{S} = \langle \Sigma, \Theta, \mathcal{T} \rangle$ where Σ is a set of *states*, $\Theta \subseteq \Sigma$ is a set of *initial states*, and $\mathcal{T}$ is a set of *transitions*. Each transition $\tau \in \mathcal{T}$ is a function from Σ to 2^{Σ}. The set $\mathcal{T}$ includes τ_I, the *idle* transition, i.e., $\tau(s) = \{s\}$ for every $s \in \Sigma$. A transition $\tau \in \mathcal{T}$ is said to be *enabled* from a state $s \in \Sigma$ if $\tau(s) \neq \emptyset$.

A finite sequence $\sigma = s_0, s_1, \ldots$ over Σ is a computation of $\mathcal{S}$ if $s_0 \in \Theta$ and for every $i \geq 0$, $s_{i+1} \in \tau(s_i)$ for some $\tau \in \mathcal{T}$ (we then say that τ is *taken* at position i). The case of the idle transition being taken is called a *stuttering step*.

A *timed transition system* is a tuple $\mathcal{S} = \langle \Sigma, \Theta, \mathcal{T}, l, u \rangle$ where $\mathcal{S}^- = \langle \Sigma, \Theta, \mathcal{T} \rangle$ is a transition system, $l: \mathcal{T} \to \mathbf{R}^*$ and $u: \mathcal{T} \to \mathbf{R}^* \cup \{\infty\}$. We require that for every $\tau \in \mathcal{T}$, $l(\tau) \leq u(\tau)$, and that $l(\tau_I) = 0$ and $u(\tau_I) = \infty$.

An infinite behavior (σ, T) is a computation of the timed transition system $\mathcal{S}$ if σ is a computation of the untimed transition system $\mathcal{S}^-$ and the following holds:

Lower bound For every transition τ and every $i \geq 0$, if τ is taken at position i, then for every $j \leq i$ such that $\mathsf{T}_j > \mathsf{T}_i - l(\tau)$, τ is enabled on σ_j. In other words, if τ is taken, it must have been enabled for at least $l(\tau)$ time units beforehand.

Upper bound For every $i \geq 0$, for every transition τ which is enabled from σ_i, either for some $j > i$ such that $\mathsf{T}_j < \mathsf{T}_i + u(\tau)$, τ is not enabled from σ_j, or for some $j > i$ such that $\mathsf{T}_j \leq \mathsf{T}_i + u(\tau)$, τ is taken from σ_j. In other words, once τ is enabled, it is either taken or disabled within $u(\tau)$ time units.

2.2 Real-time programs

The real-time concurrent program we describe here corresponds to Henzinger's message passing system. We refer the reader to [14] for more details.

A sequential program is often described by the variables it uses, their domains and initial values, and the transition graph of the program. See [19] for complete definitions and examples. The program is often denoted by $\{\theta\}P$, where θ is the *data precondition* of the program and P is the transition graph. A transition graph consists of nodes denoting program locations (one of them is designated as the initial location), and edges of the form $c \rightarrow [\bar{x} := \bar{e}]$, where c is a boolean expression (often called *guard*), $\bar{x}$ is a vector of variables, and $\bar{e}$ is an (equally typed) vector of expressions. An example of an edge label is:

$$x \geq 0 \rightarrow [(x, y) := (x + y, 0)].$$

The intended operation of an edge as above connecting node ℓ and node ℓ' is that, once control is in ℓ, if c is true then control can pass to ℓ' and the current values of $\bar{e}$ are assigned to the variables $\bar{x}$.

A concurrent program of m processes is similarly described; the set of variables is the set of shared as well as local variables, and each process is described by a transition graph (where each transition is restricted to refer only to the variables accessible to the process). To account for (synchronous) message passing, we allow for guarded send and receive commands; the send operation, $\alpha!e$, outputs the value of the expression e on channel α, and the receive operation, $\alpha?x$, reads an input value from the channel α and stores it in the variable x. A send operation and a receive operation *match* if they belong to different processes and refer to the same channel.

A concurrent real-time program is a concurrent program in which each transition τ is assigned a time interval $[l, u]$. For any two *matching* communications instructions with intervals $[l, u]$ and $[l', u']$, we require that $\max(l, l') \leq \min(u, u')$. The intended operational meaning of synchronous message passing is that if

1. o_s and o_r are matching operations whose time intervals are $[l_s, u_s]$ and $[l_r, u_r]$ respectively, and
2. for $\max(l_s, l_r)$ time units the control of the two processes has been in the originating locations with the guards continuously enabled

then the two processes *may* proceed to the target location. If, however, we replace "$\max(l_s, l_r)$"in 2 by "$\min(u_s, u_r)$", then the processes *must* proceed to the target location.

A concurrent real-time program

$$P = \{\theta\}[P_1 \parallel \ldots \parallel P_m]$$

can be associated with a timed transition system $\mathcal{S}_P = \langle \Sigma, \Theta, \mathcal{T}, l, u \rangle$ as follows. Σ contains all interpretations of the set of data and control variables; the control variable of a process that has not started is its entry location[2]. Θ is the set of states where θ is true and the control of each process is its entry location. $\mathcal{T}$ contains, in addition to the idle transition, a transition τ_E for every transition in an individual process program, and a transition $\tau_{E_s E_r}$ for every matching pair of communication instructions E_s and E_r:

If E is a transition in P_i, connecting ℓ_i to ℓ_i' and labeled by $c \to \bar{x} := \bar{e}$, then $\sigma' \in \tau_E(\sigma)$ iff the following all hold: (1) the location of P_i in σ is ℓ_i and σ' is ℓ_i', (2) c is true in σ, and the value of $\bar{x}$ in σ' is $\bar{e}$, (3) the values of the other variables in σ' is exactly as in σ, and (4) $l(\tau_E)$ is the minimal delay labelling E and $u(\tau_E)$ is the maximal delay labelling E.

If E_s and E_r are matching edges in P_s and P_r respectively, where E_s connects ℓ_s to ℓ_s' and is labeled by $c_s \to \alpha!e$, and E_r connects ℓ_r to ℓ_r' and is labeled by $c_r \to \alpha?x$, then σ' is in $\tau_{E_s E_r}(\sigma)$ iff the following all hold: (1) for every $i \in \{s, r\}$, the location of P_i is ℓ_i in σ and ℓ_i' in σ', (2) both c_s and c_r are true in σ, and the value of x in σ' is e, (3) the values of the other variables in σ' is exactly as in σ, (4) if E_s and E_r are labelled by the intervals $[l_s, u_r]$ and $[l_r, u_r]$ respectively, then $l(\tau_{E_s E_r}) = \max(l_s, l_r)$ and $u(\tau_{E_s E_r}) = \min(u_s, u_r)$.

3 General Description of Ethernet

An Ethernet system consists of N nodes connected through a single cable. By "specifying" Ethernet we mean specifying the MAC layer by describing it and its interfaces with the adjacent LLC layer and PL.

The LLC sublayer sends and receives *raw frames* (unencapsulated frames) to and from the MAC sublayer. The LLC is informed whether raw frames are successfully sent or not. We do not elaborate on the LLC sublayer since it is not a part of the IEEE 802.3.

The MAC sublayer, whose task it is to send and receive (encapsulated) frames to and from the PL, is basically a 1-persistent Carrier Sense Multiple Access with Collision Detect (CSMA/CD) protocol. In 1-persistent CSMA protocols, the transmitter of a frame listens to the channel to check whether somebody else is transmitting (this is the "carrier sense" part). If the channel is busy, the transmitter waits until it is idle. When the channel is idle, the transmitter transmits the frame (this is the "1-persistent" part) . If a collision occurs, the transmitter waits a random number of time units before it attempts to retransmit the frame. In Ethernet, the wait time is determined by a *binary-exponential backoff* algorithm (cf. [22, 2, 11, 13]). CSMA

[2] This assumption corresponds to systems that start synchronously in [14]

with Collision Detection (CSMA/CD) is a CSMA protocol where a transmission of a frame is aborted as soon as a collision is detected. In fact in Ethernet, once a collision is detected, the transmitter transmits a special *jam sequence* to ensure that all other participants are alerted to the collision.

On the sending side, the MAC sublayer receives raw frames from the LLC layer, encapsulates them with some other information and error detection codes, and attempts to send them. The LLC layer is notified if the frame is eventually sent successfully or if the MAC layer gave up after too many attempts. On the receiving side, the MAC layer receives a stream of bits from the PL and attempts to assemble them into whole frames to send up to the LLC layer. It discards the frame fragments that result from collision and frames that have been corrupted.

The PL provides bit-wise send and receive facility between the MAC layer and the communication medium. It also provides two flags to the MAC layer that signal the status of the medium. *Carrier Sense* indicates whether or not there is a transmission on the medium, and *Collision Detect* signals whether more than one node is transmitting concurrently (as far as the node can tell). *Collision Detect* is reliable only when the node is transmitting.

Each Ethernet node connects to the cable at a different location. There are several parameters that are associated with each Ethernet system:

Bit rate is the number of bits per time sent. A typical value of bit rate is 10 Megabits per second. Its inverse is **bit time**, and is the lengh of time that it takes to transmit a single bit. Since the bit rate (time) defines a bijection between time and length (in bits), we interchange "time" and "length" in the following.

Signal propagation time between two nodes is the time that it takes a signal originating at one node to reach the other node.

Round-trip propagation time is an upper bound on collision detection in the system and can be shown to be twice the propagation time between the two most distant nodes.

Jam sequence is the special sequence that the transmitter sends when it detects a collision.

Slot time is a upper bound on a length of a frame fragment resulting from a collision. We can show that the slot time must be greater than the round-trip propagation time plus the length of the jam sequence. The slot time is used as a parameter in the exponential backoff algorithm.

4 Formal Specification of Ethernet

In the specification, Ethernet consist of several components that communicate through message passing and shared variables. Some of the message passing we use corresponds to external procedure calls of IEEE 802.3; a process (procedure) is "invoked" by a message, and the "invoked" process sends a message to the invoking process to signal termination (and an optional return value).

We assume that raw frames, which are sent to and from the LLC sublayer, are over $\Sigma_2 = \{0, 1\}$, and all the other frames (sent among the MAC processes and from them to the lower layers) are over $\Sigma_3 = \{0, 1, \#\}$, where # is a special symbol. In the sequel, a "bit" means a Σ_3 element.

Figure 2 describes a global view of node n in the system. Each of the node's processes is represented by a graph node (subscripted by n), which we shall call a *vertex* to distinguish it from Ethernet nodes. Message passing between processes is represented by arrows leading from the vertex representing the sending process, to the vertex representing the receiving process, and labelled by the message type. These edges are the "channels" of interprocess communication. For example, LLC_n sends FT_n (Frame Transmitter) a raw frame, and FT_n sends LLC_n either an "okay" message (when the frame is successfully delivered) or a "fail"message (when the transmission of the frame fails). Dashed arrows represent shared variables; they are labelled by the name of the variable, and are directed from the writer to the reader(s). For example, the shared variable $\textsc{cs}_n$ (Carrier Sense) is written by CSG_n (Carrier Sense Generator) and read by DG_n (Deference Generator) and FR_n (Frame Receiver).

We next describe each of the processes. For each process, we describe its local variables, shared variables and the initial values of the variables it writes to. The process itself is described by a transition graph. In the descriptions of the processes we adopted the following conventions:

• We omit subscripts of nodes and messages when possible.

• We omit conditions and assignments when trivial, hence, true $\rightarrow [\bar{x} := \bar{e}]$ is abbreviated to $[\bar{x} := \bar{e}]$, and $c \rightarrow []$ is abbreviated to c.

• Every process description has an initial node, pointed to by the entry transition. Variables whose inital value is significant are noted as if set on the entry transition.

• While there is no need to have more than one channel connecting two processes—each message on the channel can be tagged with its type—we choose to allocate a separate channel to each message type that is sent between the processes. For example, LLC can send FT a "raw frame", and FT can send LLC a "fail" or an "ok" message. Each of these is represented as if a different channel carries it. Thus, the channel names are rawframe, fail, and ok respectively. Note that some channels (e.g. fail and ok) carry messages without content and some (e.g. rawframe) carry messages with a value.

• We assume that, unless otherwise stated, the time delay on edges is some $[l, u]$ that depends on the actual implementation. Both l and u are usually considerably smaller than the explicit timing delays, since they correspond to fast internal steps and are therefore negligible with respect to the external delays.

• The functions (e.g., **rand** which determined the wait time in the the exponential backoff algorithm) used by Ethernet are sequential and do not involve communication. We therefore chose to omit their description. See the appendix for a complete list of the functions used.

• To ease readability, we prefix each send (receive) instruction in the transition graphs with the name of the receiving (sending) process. E.g., "LLC: rawframe?r" in FT's transition graph indicates that raw frame r is received from the LLC.

• The messages, shared variables, parameters, and functions used by the processes are summarized in the appendix. We denote process names by upper case SANS SERIF, messages by lower case sans serif, shared vairables by SMALL CAPS, parameters and functions by **bold face**, and local variables by *italics*.

• For lack of space we defer a complete program for the medium to the final version of this paper. Informally, each process n has a vector $\textsc{bits}_n[1..N]$ of elements over

$\Sigma_3 \times \{0,1\}$. The first element of BITS$_n[m]$ is the last bit received from node m. The second element is a parity bit, which is complemented each time a new bit is received. The value of BITS$_n[m]$ is updated by $\mathsf{CHL}_{m \to n}$, which acts as a buffer from m to n with a fixed transmission time—$\delta_{\mathbf{mn}}$. When node m sends bits, PLS_m sends each bit to $\mathsf{CHL}_{m \to i}$ for every i. When $\mathsf{CHL}_{m \to n}$ receives this bit, it updates BITS$_n[m]$ appropriately (after $\delta_{\mathbf{nm}}$ time units).

4.1 The MAC Layer

The MAC layer is divided into four subprocesses: Frame Transmitter (FT), Deference Generator (DG), Bit Transmitter (BT), and Frame Receiver (FR).

Frame Transmitter (FT): A description of FT is given by Figure 3. Its local variables are: r—raw frame received from LLC, f—raw frame after being encoded by **transEncap**, *att*—number of failures to transmit current frame, t—number of time-cycles (in units of **ct**) to wait for next attempt.

Deference Generator (DG): When DG notices that CS is true (i.e., there is a passing frame on the channel), it sets DEF to true. When it notices there are no passing frames on the channel, it sets DEF to false after waiting (at least $\mathbf{intSP}^-$ and at most $\mathbf{intSP}^+$ time units). It then waits until WAITING is false, to ensure that FT notices that DEF is false. DG is described by Figure 4. It has no local variables.

Bit Transmitter (BT): Upon receiving a frame from FT, BT encapsulates it (using **phyEncap**), and sends it, bit by bit, to the Sender (PLS). The synchronization of the transmission is through message passing: after BT sends a bit, it waits until the Sender signals it is ready for the next one. If, during the transmission of a frame, BT receives a jam message, it replaces the transmitted frame with the jam sequence. When it completes transmitting, either the frame or the jam sequence, it sends a message to FT to indicate it. BT is described by figure 5. Its local variables are: f—frame as received from FT, p—f after being encapsulated, and x—current position in frame.

Frame Receiver (FR): When CS is true, FRreceives bits from PLR, which it accumulates until Carrier Sense is false. FR then checks whether the accumulated bits constitute of a legal frame. If so, FR sends the frame to LLC. Otherwise, FR discards it and awaits the next frame. FR is described by Figure 6. Its local variables are: p—incoming frame, as received by PLR, r—p after decapsulation, as sent to LLC, and x—incoming bit.

4.2 The Physical Layer

The Physical Layer is divided into five subprocesses: Sender (PLS), clock (Clock), Receiver (PLR), Carrier Sense Generator (CSG), and Collision Detect Generator (CDG). Due to space limitations, we postpone the prose description of the Physical Layer processes for the full version of the paper. They are described in Fig. 7 through Fig. 10, respectively.

5 Some Properties of Ethernet

Let $\mathcal{E}$ denote the concurrent real time program

$$\{\theta\}[\text{medium} \parallel \|_{i\in N}(\mathsf{FT}_i \parallel \mathsf{DG}_i \parallel \mathsf{BT}_i \parallel \mathsf{FR}_i \parallel \mathsf{Clock}_i \parallel \mathsf{PLS}_i \parallel \mathsf{PLR}_i \parallel \mathsf{CSG}_i \parallel \mathsf{CDG}_i)].$$

where θ, the data precondition of $\mathcal{E}$, is the set of states where all variables are assigned their initial values and the control of all the processes is the entry location.

In the full version of the paper we prove several properties that hold over all executions of $\mathcal{E}$. Here we prove that, for a given Ethernet node, whenever WAITING is true and DEF is true, if CS is false for sufficiently long then within a bounded number of time units WAITING is set to false. We prove this property through a series of invariance properties, all stated in *Metric Temporal Logic* (MTL) and proved in proof system of [15].

To simplify notation, denote the state formula "FT is in location i" by ℓ_i, and the state formula "DG is in location i" by m_i. Let τ_3 and τ_4 denote the transitions in DG leading from location 3 to location 4 and the transition in DG leading from location 4 to location 5. Similarly, let τ_4' and τ_5' be the transitions in FT leading from locations 4 and 5 respectively. Let t_3, t_4, t_4', and t_5' be the upper time bound of τ_3, τ_4, τ_4', and τ_5' respectively. Our main claim can be now stated as the following invariance property in MTL:

$$(\text{WAITING} \land \text{DEF} \land \Box_{\geq t_3}\neg\text{CS}) \rightarrow \Diamond_{\leq t_3+t_4+t_4'+t_5'}\neg\text{WAITING} \tag{1}$$

where $\Box_{\geq c}$ means "continuously for at least c time units" and $\Diamond_{\leq c}$ means "eventually within c time units". Hence, (1) states that for every timed state execution of the timed transition system defined by $\mathcal{E}$, it is always the case that if both WAITING and DEF are true and CS is false continuously for t_3 time units, then within $t_3+t_4+t_4'+t_5'$ time units WAITING becomes false.

We first state two simple invariance properties of $\mathcal{E}$ which can be proven by any proof system for "timeless" temporal logic properties (see, e.g., [18]).

$$\text{DEF} \equiv \ell_3 \lor \ell_4 \qquad \text{and} \qquad \text{WAITING} \equiv m_4 \lor m_5 \tag{2}$$

The other invariance properties we establish require the $\Diamond$-ss and $\Diamond$-trans proof rules of [15]. A somewhat simplified version of these rules is given is Figure 1.

$$\begin{array}{ll} \Diamond\text{-ss} & p \rightarrow (p_1 \lor p_2) \\ & p_1 \rightarrow enabled(\tau) \\ & \{p_1\}\mathcal{T}\{p_1 \lor p_2\} \\ & \{p_1\}\tau\{p_2\} \\ \hline & p \rightarrow \Diamond_{\leq u_\tau} p_2 \end{array} \qquad\qquad \begin{array}{ll} \Diamond\text{-trans} & p \rightarrow \Diamond_{\leq u_1} p_1 \\ & p_1 \rightarrow \Diamond_{\leq u_2} p_2 \\ \hline & p \rightarrow \Diamond_{\leq u_1+u_2} p_2 \end{array}$$

Fig. 1. Proof Rules

$\Diamond$-ss states that for any state formulae p, p_1 and p_2 and any transition τ whose upper bound is u_τ, if (a) $p \rightarrow (p_1 \lor p_2)$, (b) τ is enabled (its guard is true) in p_1-states, (c) any transition that is enabled on p_1-states leads to p_1- or p_2-states, and

(d) τ leads from p_1-states to p_2-states, then whenever the system is in a p-state it reaches a p_2-state within u_τ time units. $\diamond$-trans states transitivity among successive applications of $\diamond$-ss.

The following invariance properties all follow from $\diamond$-ss:

$$\ell_3 \rightarrow \diamond_{\leq t_3}(\ell_4 \vee \text{CS}) \quad (3) \qquad (\ell_5 \wedge m_4) \rightarrow \diamond_{\leq t'_4} m_5 \quad (5)$$

$$\ell_4 \rightarrow \diamond_{\leq t_4} \ell_5 \quad (4) \qquad m_5 \rightarrow \diamond_{\leq t'_5}(\neg\text{WAITING}) \quad (6)$$

(3) follows from $\diamond$-ss when setting $p_1 = \neg\text{CS}$ and $\tau = \tau_3$. (4), (5), and (6) follows from $\diamond$-ss when $p_1 = p$ and τ is τ_4, τ'_4, and τ'_5 respectively.

Since $\diamond$ distributes over disjunctions and since $\square_{\geq c}\varphi \rightarrow \neg(\diamond_{\leq c}\neg\varphi)$, it follows from (3) that

$$(\ell_3 \wedge \square_{\geq t_3}\neg\text{CS}) \rightarrow \diamond_{\leq t_3} \ell_4 \quad (7)$$

is an invariance property of $\mathcal{E}$.

Using $\diamond$-trans and recalling (2), it follows that the following are also invariance properties of $\mathcal{E}$:

$$(\text{DEF} \wedge \square_{\geq t_3}\neg\text{CS}) \rightarrow \diamond_{\leq t_3+t_4} \ell_5 \quad (8)$$

$$(\text{WAITING} \wedge \ell_4) \rightarrow \diamond_{\leq t'_4+t'_5} \neg\text{WAITING} \quad (9)$$

(1) now follows from $\diamond$-trans on (8) and (9) noting that $\ell_5 \rightarrow (\text{WAITING} \vee \neg\text{WAITING})$ and $\diamond_{\leq c_1}\varphi \rightarrow \diamond_{\leq c_2}\varphi$ for every formula φ and $c_2 \geq c_1$.

6 Discussion and Future Work

The paper presents a formal specification of the Ethernet. While there are no yardsticks to the success of a specification, it is obvious that a successful specification of a complex system be clear, readable, lend itself to formal correctness proofs, and isolate the different portions of a system and define their interfaces. We believe our specification has all those properties.

The only other specification of Ethernet that we are aware of is the one in IEEE 802.3. The specification there is divided into several parts. For example, §4 describes the MAC layer using "Concurrent Pascal", §6 describes the interface between the MAC and the Physical layers, §7 contains the Physical Layer Specification using state diagrams. We found parts of the specification hard to understand. Some of the difficulties seem to stem from the lack of well defined syntax and semantics of the "Concurrent Pascal" which is often used. Others of inconsistencies in the name space and assumptions. Yet others of incompleteness of the specification. Below are three examples of difficulties of the latter types.

• In §4.3.3 it says " the collision detect signal is generated only during transmission and is not true at any other time; in particular, it cannot be used during frame reception to detect collisions between overlapping transmissions from two or more other stations." In §6.3.2.2, "collision detect" is referred to as "PLS_SIGNAL.indicate", and there is no such restriction as to the when the signal is set. (Note that our model clearly specifies that CD is read by FT only).

• In §4.3.3 frames are assumed to be over Σ_2 and an additional shared variable, "transmitting", is used to indicate that FT is transmitting. In §6.2.3, §6.3.1.1.2, and §7.2.2.1.1 frames are assumed to be over Σ_3, and there is no variable "transmitting".
• It seems that there is some implicit assumption that bits get to FR at some fixed rate. We explicitly guarantee this in the specification of PLS. The most explicit mention of this in IEEE 802.3 is in §4.3.3, where it says "The ReceiveBit operation is synchronous. Its duration is the entire reception of a single bit". Yet, the "reception of a single bit" is not a well-defined quantity during a collision. AS the specification of the PL points out (§7.2.1.2.1), "If the *signal_quality_error* [collision detect] message is being sent from the MAU [our medium], the input waveform is unpredictable... NOTE: This signal is not necessarily retimed."

Evidently, the "right" way to resolve the inconsistencies was not always clear. We chose what was simpler and fitted our model best. Consequently, we sometimes diverge from parts of IEEE 802.3.

Our main conclusion is that formal specification of complex systems is feasible and advantageous and that, just like in many other areas of specification, a sound formal model together with some visual tools can be combined to yield a coherent specification of real-time systems.

The variant of Henzinger's model which we used in the work might not be the only real-time model in which Ethernet can be specified. Yet, it seems to have offered us ease of expression which we failed to find in some other models we had tried earlier. We are currently working on checking whether other real-time models simplify the exposition.

Another direction we wish to explore is the visual aspect of the specification. The tools used here are rather simple. It would be interesting to see how the specification can be embedded in a more sophisticated visualization tool.

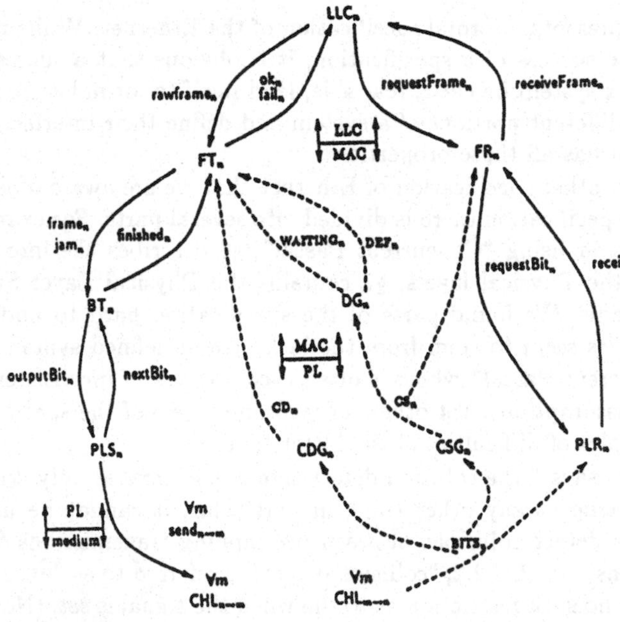

Fig. 2. The processes of Node *n*

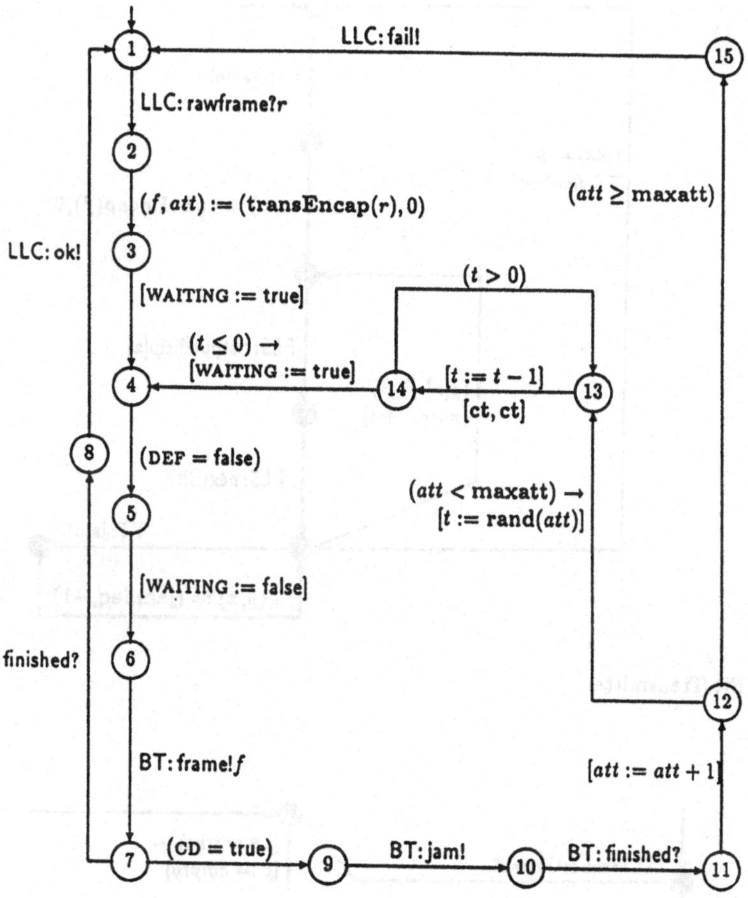

Fig. 3. Frame Transmitter

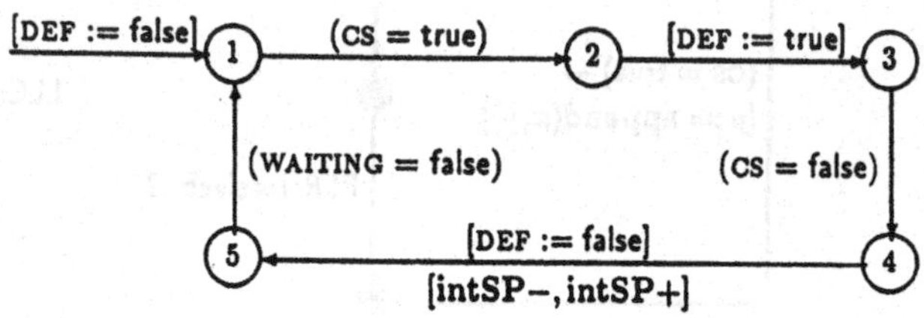

Fig. 4. Deference Generator

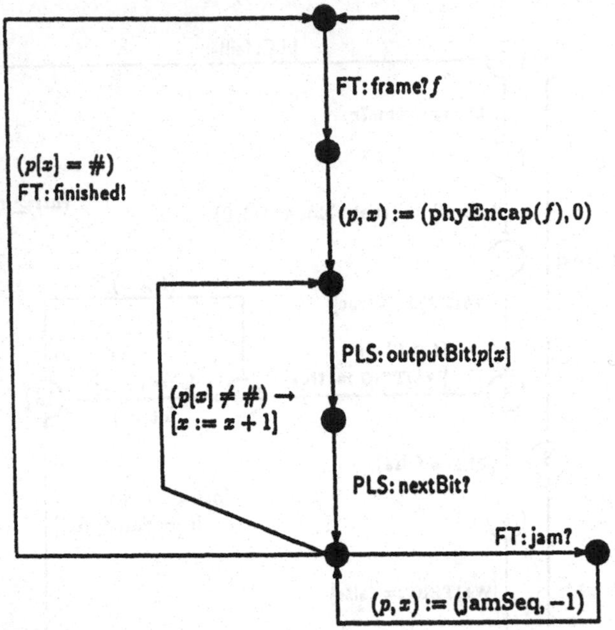

Fig. 5. Bit Transmitter

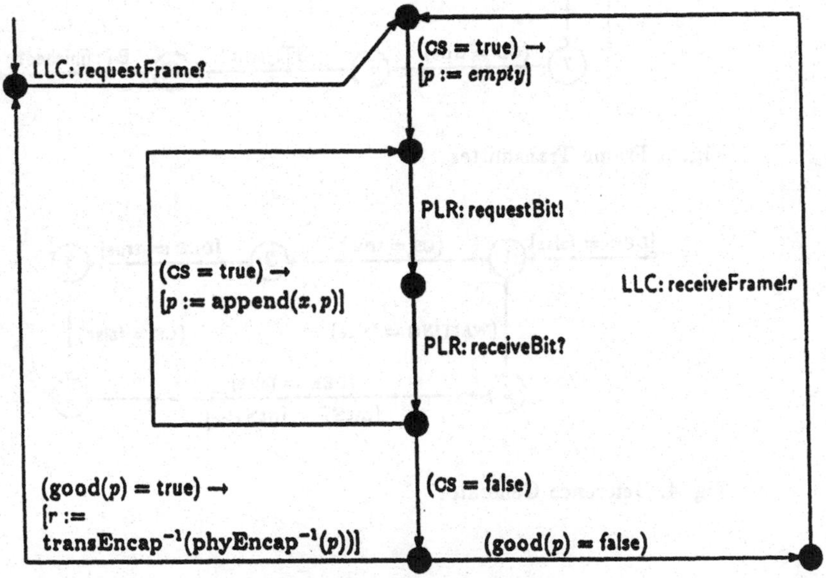

Fig. 6. Frame Receiver

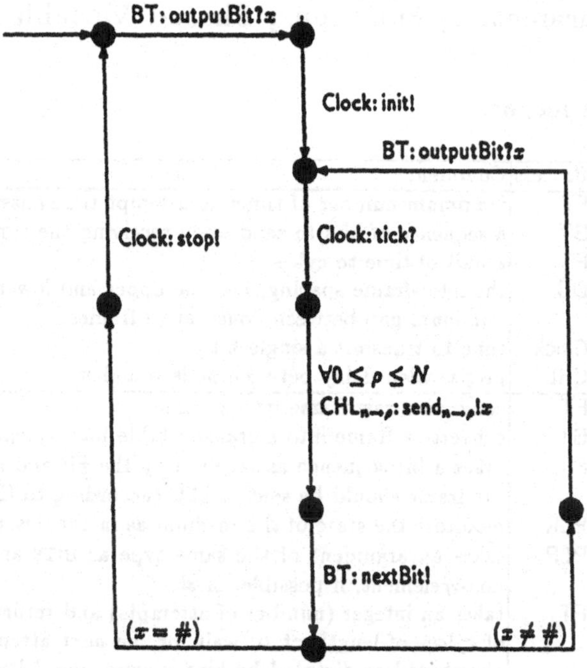

Fig. 7. Sender

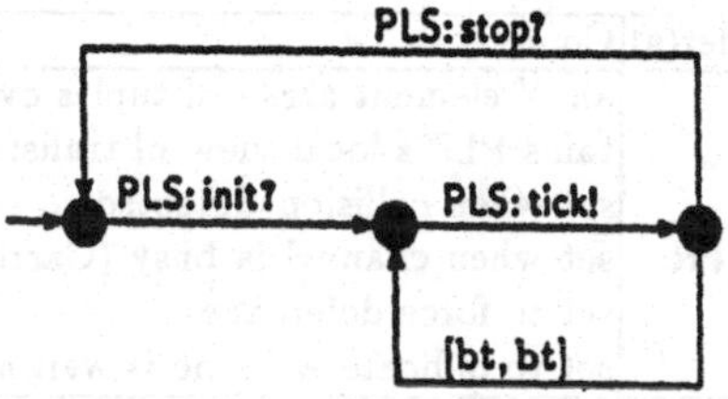

Fig. 8. Clock

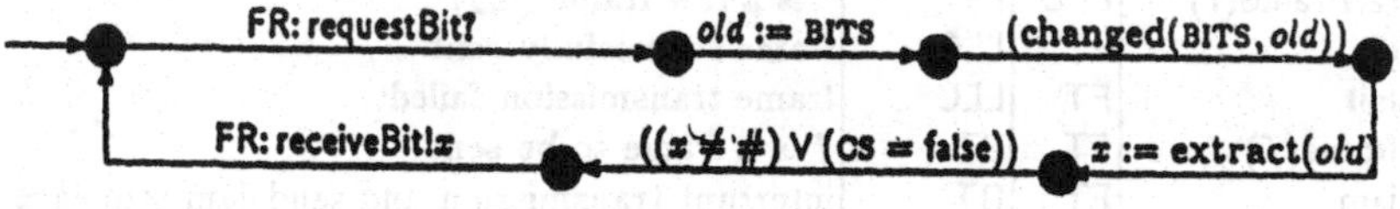

Fig. 9. Receiver

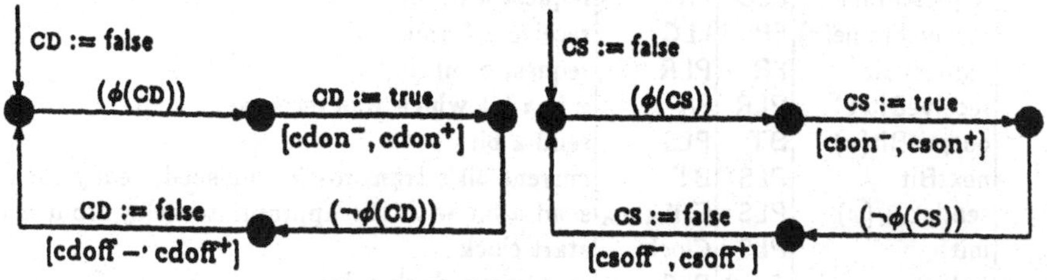

Fig. 10. Collision Detect and Carrier Sense Generators

Appendix: Parameters, Functions, Shared Variables, Channels

Parameters and Functions

Name	Process	Comment
maxatt	FT	maximum number of times to attempt transmission of a frame
jamSeq	BT	a sequence of bits to send upon receiving the jam signal
ct	FT	a unit of time to cylce
$[\mathbf{intSP}^-, \mathbf{intSP}^+]$	DG	the inter-frame spacing, i.e., the upper and lower bound of the minimum gap between consecutive frames.
bt	Clock	time to transmit a single bit
δnm	CHL	propagation delay between node n and m
transEncap	FT	converts a raw frame into a frame
phyEncap	BT	converts a frame into a transmittable bit sequence
good	FR	takes a bit sequence as received by the FR and returns true iff the frame should be sent to LLC (according to IEEE 802.3)
changed	PLR	monitors the state of the medium as in the description of PLR
extract	PLR	takes an argument of the same type as BITS and returns any non-#element, if possible, or #.
rand	FT	takes an integer (number of attempts) and returns the number of cylces of length ct to wait for the next attempt, randomly distributed as dictated by binary exponential backoff

Shared Variables

Name	Writer(s)	Reader(s)	Comment
BITS	CHL	PLR	an N element array of tuples over $\{0,1\} \times \Sigma_3$, which contains PLR's local view of transmissions.
CD	CDG	FT	set when collision detected
CS	CSG	DG, FR	set when channel is busy (Carrier Sense)
DEF	DG	FT	set to force deference
WAITING	FT	DG	set to indicate a frame is waiting

Channels

Name	From	To	Comment
rawframe(r)	LLC	FT	r is a raw frame (Σ_2)
ok	FT	LLC	frame successfully sent
fail	FT	LLC	frame transmission failed
frame(f)	FT	BT	f is a frame to be sent
jam	FT	BT	interrupt transmission and send jam sequence
finished	BT	FT	frame/jam transmission finished
requestFrame	LLC	FR	request a frame
receiveFrame(r)	FR	LLC	receive a frame
requestBit	FR	PLR	request a bit
receiveBit(x)	PLR	FR	x is a bit which was received
outputBit(x)	BT	PLS	send a bit
nextBit	PLS	BT	current bit's transmission finished, ready for next bit
$\mathsf{send}_{n \to m}(x)$	PLS	$\mathsf{CHL}_{n \to m}$	send a bit with the appropriate delay from n to m
init	PLS	Clock	start clock
tick	Clock	PLS	sent every clock cylce
stop	PLS	Clock	stop clock

References

1. R. Alur. *Techniques for Automatic Verification of Real-Time Systems.* PhD thesis, Stanford University, August 1991.
2. D. Bertsekas and R. Gallager. *Data Networks.* Prentice-Hall, 1987.
3. G. V. Bochmann and J. Gecsei. A unified method for the specification and verification of protocols. In B. Gilchrist, editor, *Information Processing 77*, pages 229–234, Amsterdam, 1977. North-Holland Publishing Co.
4. G. V. Bochmann and C. A. Sunshine. Formal methods in communication protocol design. *IEEE Transactions on Communications*, COM-28:624–631, 1980.
5. E. M. Clarke, E. A. Emerson, and A. P. Sistla. Automatic verification of finite-state concurrent systems using temporal logic specifications. *ACM Trans. on Programming Languages and Systems*, 8(2):244–263, 1986. An early version appeared in *Proceedings of the 10th ACM Symposium on Principles of Programming Languages*, 1983.
6. J. D. Day and H. Zimmerman. The OSI reference model. In *Proc. of IEEE*, volume 71, pages 1334–1340, December 1983.
7. D. Dill. Timing assumption and verification of finite-state concurrent system. In J. Sifakis, editor, *Automatic Verification MEthods for Finite State Systems, LNCS 407.* Springer-Verlag, 1989.
8. Michael Fischer and Lenore Zuck, August 1991. Preliminary work on spcification of CSMA/CD protocols.
9. V. D. Gligor and S. H. Shattuck. On deadlock detection in distributed systems. *IEEE Transactions on Software Engineering*, SE-6(5):435–440, 1980.
10. B. T. Hailpern and S. S. Owicki. Modular verification of communication protocols. *IEEE Transactions on Communications*, COM-31(1):56–68, 1983.
11. J. L. Hammonds and P. J. P. O'Reilly. *Performance Analysis of Local Computer Networks.* Addison-Wesley, 1986.
12. D. Harel. Statecharts: A visual formalism for complex systesm. *Sci. COmp. Prog.*, 8:231–274, 1987.
13. J. F. Hayes. *Modeling and Analysis of Computer Communication Networks.* Plenum Press, 1984.
14. T. A. Henzinger. *The Temporal Specification and Verification of Real-Time Systems.* PhD thesis, Stanford University, August 1991.
15. T. A. Henzinger, Z. Manna, and A. Pnueli. Temporal proof methodologies for real-time systems. In *Proc. of 18th POPL*, pages 353–366, 1991.
16. H. R. Lewis. Finiste-state analysis of asynchrocous circuits with bounded temporal uncertainty. Technical Report TR-15-89, Haravard University, 1989.
17. N. A. Lynch and M. R. Tuttle. Hierarchical correctness proofs for distributed algorithms. In *Proc. 6th ACM Symp. on Principles of Distributed Computing*, pages 137–151, August 1987.
18. Z. Manna and A. Pnueli. The anchor version of the temporal framework. In J. W. de Bakker, W.-P. de Roever, and G. Rozenberg, editors, *Linear Time, Branching time, and Parital Order Models and Logics for Concurrency*, pages 201–284. Springer Verlag, LNCS 354, 1989.
19. Z. Manna and A. Pnueli. *The Temporal Logic of Reactive and Concurrent Systems (Specification).* Springer-Verlag, 1992.
20. P. M. Merlin. A methodology for the design and implementation of communication protocols. *IEEE Transactions on Communications*, COM-24(4):614–621, 1976.
21. M. Merritt, F. Modugno, and M. Tuttle. Time constrained automata. Manuscript, August 1990.
22. R. M. Metcalfe and D. R. Boggs. Ethernet: Distributed packet swithching for local computer networks. *Communications of the ACM*, 19:395–404, July 76.
23. F. Moller and C. Tofts. A temporal calculus of communicating processes. In J. C. M. Baeton and J. W. Klop, editors, *CONCUR 90, LNCS 458*, pages 401–415. Springer-Verlag, 1990.
24. J. Orsroff. *Temporal Logic of Real Time Systems.* Research Studies Press, 1990.
25. S. Owicki and L. Lamport. Proving liveness properties of concurrent programs. *ACM Trans. on Programming Languages and Systems*, 4(3):455–495, 1982.
26. A. Pnueli. The temporal logic of programs. In *Proc. 18th IEEE Symp. on Foundations of Computer Science*, pages 46–57, 1977.
27. J. P. Queille and J. Sifakis. Specification and verification of concurrent systems in CESAR. In *Proc. 5th Int'l Symp. on Programming*, 1981.
28. G. M. Reed and A. W. Roscoe. A timed model for communicating sequential processes. *Theoretical Computer Science*, 58:249–26, 1988.
29. R. L. Schwartz and P. M. Melliar-Smith. From state machines to temporal logic: Specification methods for protocol standards. *IEEE Transactions on Communications*, 1982.
30. S. R. Soloway and P. A. Humblet. On distributed network protocols for changing topologies. Technical Report LIDS-P-1564, MIT, 1986.
31. ANSI/IEEE std. *Information Processing Systems– Local Area Networks– Part 3: Carrier sense multiple access with collision detection (CSMA/CD) access method and physical layer specificaitons.* The IEEE, Inc., NY, October 1991.
32. C. A. Sunshine. Formal techniques for protocol specification and verification. *IEEE Computer*, 12:20–27, 1979.
33. A. Tanenbaum. *Computer Networks.* Prentice Hall, 2nd edition, 1989.

Implementing LOTOS Specifications by Communicating State Machines

Günter Karjoth*

IBM Research Division, Zurich Research Laboratory,
8803 Rüschlikon, Switzerland
Email gka@zurich.ibm.com

Abstract. This paper presents algorithms to translate a LOTOS specification into a network of extended finite state machines, a representation which is more tractable for simulation, verification by model checking, and code generation purposes. Objectives are efficient executability and the coverage of a wide range of LOTOS specifications.

1 Introduction

The specification language LOTOS was standardized by ISO [1] to be applied specifically to the formal description of ISO Open Systems Interconnection (OSI) standards. However, the language has gained wider acceptance owing not least to its elegance, which is due to its process algebra foundation. Several approaches have been developed to translate a LOTOS specification into a representation which is more tractable for simulation, verification by model checking, and code generation purposes. The first LOTOS compilers [2, 3, 4] map the LOTOS processes on C functions which are executed as coroutines in a specialized run-time environment. Accordingly, the structure of the generated code closely resembles the architecture of the LOTOS specification.

In [5, 6, 7], a finite representation of nondeterministic and concurrent processes without value passing expressed in some process algebra, e.g. CCS, is given by a systematic construction of the corresponding automata. In LOTOS, the process algebra has been supplemented with an abstract data type formalism to describe data values and their transfer between processes. The above-mentioned construction techniques can be extended to process algebras with value passing, as shown in [8]. There the target is an automaton combined with concepts from imperative languages – essentially data variables and assignment statements – and is often referred to as an extended finite state machine. Terms of the process algebra correspond to control states, and data parameter passing corresponds to assignment statements.

The restriction to regular processes imposes a severe limitation on the proposed method [8]. Dubuis [9] has therefore extended the target to encompass a set of state machines and ports. Ports are addressed by state machines when they wish to synchronize. They implement the multiway synchronization capability of LOTOS

* This work was partially supported by the European Communities under RACE project no. 1046, Specification and Programming Environment for Communication Software (SPECS). It represents the views of the author.

processes. State machines can execute special operations to create new state machines and to modify the characteristics of ports. Valenzano et al [10] also use a set of cooperating state machines. Additionally, they introduce a stack to serialize the execution of state machines. Their state machines, however, cannot be parameterized and therefore cannot be used as recursive procedures.

In [11], we have developed the XFSM model, a network of communicating state machines, to overcome our single state machine approach. In the way processes are composed and communicate, XFSM closely follows the LOTOS model. State machines possess specialized transitions whose execution affects the configuration of the network; state machines and/or gates can be added or removed. In the XFSM model, state machines may serve as processes or as procedures. Procedures enable the translation of 'context-free' LOTOS processes where enabling and recursion are intertwined.

In this paper, we define a translation from LOTOS to XFSM. If recursion is intertwined with parallel composition, we replace all occurrences of such dynamic process creation by a *fork* action, thus achieving a regular process structure (control only). The special action **fork**(Q, S, H) creates the process Q and the gates defined in H. The new process Q is interconnected with the current process by the gates in S:

$$\mathbf{fork}(Q,S,H); P \equiv \mathbf{i}; \mathbf{hide}\ H\ \mathbf{in}\ (Q\ |[S]|\ P)$$

Note that dynamic process creation always starts with an internal action **i**. This assumption leaves it to the creating process to decide when to fork a new process. Next, we translate each regular process into a single state machine. Section 3 defines the accepted subset of LOTOS, followed by a preprocessing of the LOTOS text which replaces all occurrences of dynamic process creation by a fork action. Section 4 gives the definition of the syntax-driven state machine construction. Section 5 presents two examples.

2 Communicating State Machines

In this section, we informally explain the semantics of communicating state machines, a subset of the XFSM model as formally defined in [11].

2.1 Data

The set of *terms* is constructed from functions and variables, and ranged over by t. By T we denote the set of ground terms, by $T(V)$ the set of terms with variables of $V = (V_s)_{s\in S}$, and by $Var(t)$ the set of variables occurring in a term t. We do not deal with the evaluation of terms of the presumed underlying language. However, we make the assumption that evaluation is purely applicative.

We fix the set of *predicates*, T_{Bool}, to contain the constant *true*. Let p range over predicates. Further, we introduce *statements* which successively update variables by evaluating terms and assigning the result to the variables. The set C of statements is defined as

$$\begin{array}{lll} C ::= & nop & \textit{(skip)} \\ & \mid v_1,\ldots,v_n{:=}t_1,\ldots,t_n & \textit{(assignment)} \\ & \mid C;C & \textit{(sequential composition)} \end{array}$$

2.2 Actions

Let G be a set of *gates* over which state machines can communicate. Elements of G are denoted by g. Each gate has assigned a fixed signature, of which its arguments must be: $G \rightarrow S_1 \times \ldots \times S_n, \quad n \geq 0$. Let S and H be lists of gate names from G. The distinguished symbols τ, δ, $fork$, and $call$ shall not be members of G. A *relabeling function* f is a function from G to G.

We distinguish three kinds of actions: internal actions, communication actions, and commands. An *internal action* is denoted by τ. A *communication action* consists of a gate name, a list of experiment offers, and a selection predicate. The experiment offers may be of the kind $!t$, where t is a value expression denoting the communication of a fixed value, or they have the form $?v$, where v is a typed value variable denoting that any value of $type(v)$ may be received and assigned to v. The selection predicate puts a restriction on the values communicated. Although selection predicates are optional, we shall assume without loss of generality that they are part of each communication action. If a communication action has no selection predicate, we assume the default form $true$, which is always satisfied. With g being a gate name, $d_1, \ldots, d_n$ experiment offers, and p a predicate let $Act = \{gd_1 \ldots d_n[p] \mid g \in G, t \in T(V), v \in V, p \in T_{Bool}(V)\}$ be the set of communication actions. There is also the special communication action $\delta!t_1 \ldots !t_n$ to signal successful termination.

Commands have a special effect on the configuration of the network. They add state machines or execute coroutines by pushing a state machine's current state on the stack:

- $fork(N, S, H, f)$ - parallel task creation,
- $call(X[g_1, \ldots, g_m]((t_1, \ldots, t_n)(v_1, \ldots, v_k)))$ - procedure call,

where N is a task, f is a relabeling, S and H are sets of gates, X is a task identifier, $g_1, \ldots, g_m$ are gates, $t_1, \ldots, t_n$ are data terms, and $v_1, \ldots, v_k$ are data variables.

The letter a ranges over actions. For each action a the function $gate(a)$ yields the gate name in a or τ:

$$gate(a) = \begin{cases} g \text{ if } a = g\ d_1 \ldots d_n[p]; \\ \delta \text{ if } a = \delta\ d_1 \ldots d_n; \\ \tau \text{ otherwise.} \end{cases}$$

2.3 State Machines

A state machine can be viewed as a collection of *rules* $R = \{r_1, \ldots, r_n\}$ on some set of *state variables* $V = \{v_1, \ldots, v_n\}$. Each variable is associated with some domain of values. The set of possible assignments of values to the state variables constitutes the set of *data states*. Each rule r_i is of the form $\langle j, p \rangle \xrightarrow{a} \langle j', c \rangle$. It defines that from control state j the state machine can perform an action a and thereby enter control state j' provided the enabling predicate p is true under the current assignment of the state variables. Furthermore, statement c is executed and may change the values of some state variables. Thereby, a rule defines a class of transitions. In addition, a state machine has an *initialization statement* c_0 that assigns initial values to the variables in V, and an *initial control state* j_0.

Following Milner [5], a state may be labeled by zero or more (process) identifiers; a process identifier X indicates states at which the behavior of the state machine may be extended by substitution of another state machine for X. However, our extensions also carry information on the update of data variables and a predicate making the continuation conditional. Let $\mathcal{X}$ be the set of process identifiers. A *state machine* is a 6-tuple $M = \langle J, E, R, V, j_0, c_0 \rangle$ where

- J is a finite set, the *control states* of M,
- $E \subseteq J \times \mathcal{X} \times T(V)^* \times T_{Bool}(V)$ is a finite set, the *extensions* of M,
- R is a finite set, the *rules* of M,
- V is a finite set, the *variables* of M,
- j_0 is the *initial control state* ($j_0 \in J$) and
- c_0 is the *initialization statement* ($c_0 \in C^{init}$) of the variables of M.

$\mathcal{M}$ denotes the set of machines and M, N range over $\mathcal{M}$. Let $action(r)$ refer to the action of rule r. We shall frequently write:

$$\begin{aligned} R(j) &= \{\langle a, j', p, c\rangle \mid \langle a, j, j', p, c\rangle \in R\}, &&\text{the rules of } j;\\ E(j) &= \{\langle X, c, p\rangle \mid \langle j, X, c, p\rangle \in E\}, &&\text{the extensions of } j;\\ E(j, X) &= \{\langle c, p\rangle \mid \langle j, X, c, p\rangle \in E\}, &&\text{the extensions of } j \text{ w.r.t } X. \end{aligned}$$

2.4 A Network of State Machines

Let π be a stack of state machine activations. The empty stack is denoted by ϵ; $\pi :: \langle\langle v_1, \ldots, v_k\rangle : M\rangle$ pushes state machine M together with its return parameters $v_1, \ldots, v_k$ on top of stack π. The set $\mathcal{N}$ of network expressions is defined as

$$\begin{aligned} \mathcal{N} ::= \;& M_\pi^f && \textit{sequential task}\\ &\mid X[g_1, \ldots, g_m](t_1, \ldots, t_n) && \textit{task instantiation}\\ &\mid \mathcal{N}[f] && \textit{relabeling}\\ &\mid \mathcal{N}\backslash H && \textit{hiding}\\ &\mid \mathcal{N} |[S]| \mathcal{N} && \textit{parallel task} \end{aligned}$$

Parentheses may be used for grouping, but otherwise \ has the weakest precedence. The association of binary network operators is to the right.

An XFSM specification is a pair $\langle N, E\rangle$ with $E = \{E_1, \ldots, E_n\}$ a set of task declarations. Each declaration E_i is of the form $X_i[g_1, \ldots, g_m](v_1, \ldots, v_n) := N_i$ where X_i is a task identifier, the formal parameters g_j are distinct gate names and v_j are distinct variables, and N_i is a network expression.

A primitive task consists of a state machine together with a stack π and a relabeling f. The task can perform the same communication actions (possibly renamed), internal actions, and commands as the state machine. A primitive task may only terminate if the state machine terminates successfully and the stack is empty. The *call* command closely matches the effect of calling and returning from procedures within a single processor. The successor state and the return variables are stored on the stack. On successful termination of the invocation of a state machine, the task will continue with the behavior on top of the stack. The exit values will be assigned to the return parameters of the calling process.

A state machine can create new tasks through execution of the *fork* command. The command $fork(N, S, H, f)$ creates task N. Furthermore, new gates are created as contained in H. Access to these gates is defined by a relabeling through f for the creating task M. Tasks N and $M'[f]$ synchronize via the gates in S. Gates in H are not accessible (and not visible) from the outside. Synchronization between parallel tasks is as in LOTOS.

3 Preprocessing

In the translation, we accept all well-formed LOTOS specifications as defined in the standard [1] but with two major exceptions. First, we exclude any kind of value generation such as generalized choice over data (**choice** $v : S$ [] $[f(v)] \rightarrow a!v; B$), and hiding of actions with input offers (**hide** a **in** $a?v : S; B$, **exit**(**any** S) $\gg$ **accept** $v : S$ **in** B) from translation unless it can be resolved during parallel process merging. Second, a process may not recursively instantiate itself on the left-hand side of a [> operator, nor may the left-hand side create dynamically parallel processes.

For the purpose of this paper, we also exclude the intertwining of renaming and hiding with recursion. (Example: $P[a, b] := a; P[b, a]$.) We refer to the work of Taubner [7] how to extend our translation algorithm. We also assume that each gate has a fixed signature such that the action is uniquely identified by its gate name. If there are actions with the same gate but different number of offers and/or sorts, gate splitting may be used to satisfy this constraint. Finally, we assume that the derived operators within LOTOS, |||, ||, **let**, **choice**, and **par**, are replaced as defined in Annex B of the LOTOS standard [1].

3.1 Flattening

As the first step, the canonical LOTOS specification has to be obtained by performing the mapping '$\sharp.\sharp$' as defined in [1], paragraph 7.3.4. Let $\langle AS, BS\rangle$ be the flattened LOTOS text produced. The behavioral specification part BS is a pair $\langle X_0, Xdefs\rangle$ with $Xdefs = \{Xdef_0, \ldots, Xdef_n\}$ a set of process definitions. Each process definition $Xdef$ is of the form $X[g_1, \ldots, g_m](v_1, \ldots, v_n) := B$ where X is a process name and B is a behavior expression. Let B_X be the body of process X if $X[g_1, \ldots, g_m](v_1, \ldots, v_n) := B_X \in Xdefs$. All identifiers are global which implies that each process declaration is uniquely identified by its process identifier.

3.2 Identification of non-regular processes

The following algorithm determines all processes which dynamically create new processes. Function *dyn* returns a set of process identifiers. Parameter f keeps all processes (*fathers*) which occurred before the last parallel composition operator and parameter s records all process identifiers (*sons*) which occurred after that parallel composition operator. Parameter f indicates whether a parallel composition operator was found between process definition and instantiation. The function *dyn* is inductively defined as follows:

- $dyn(\Box, f, s) = \emptyset$ for $\Box$ ranging over **stop** and **exit**.
- $dyn(B_1 \Box B_2, f, s) = dyn(B_1, f, s) \cup dyn(B_2, f, s)$ for $\Box$ ranging over $[]$, $\gg$, $[>$.
- $dyn(\Box B, f, s) = dyn(B, f, s)$ for $\Box$ ranging over a, **hide** S **in** , and $[p] \rightarrow$.
- $dyn(B_1 \,|[S]|\, B_2, f, s) = dyn(B_1, f \cup s, \emptyset) \cup dyn(B_2, f \cup s, \emptyset)$.
- $dyn(X[g_1, \ldots, g_m](t_1, \ldots, t_n), f, s) = \begin{cases} \emptyset & \text{if } X \in s; \\ \{X\} & \text{if } X \in f; \\ dyn(B_X, f, \{X\} \cup s) & \text{otherwise.} \end{cases}$

Application of $dyn(X_0, \emptyset, \emptyset)$ determines all processes in the LOTOS specification whose instantiations must be expressed by the fork command. This does not imply that all the other processes can be expanded into one state machine. Conditions that allow processes to be merged safely are a subject of further investigation. For the moment, we either replace each parallel composition by a fork command, or we use annotations to distinguish between static and dynamic process creation.

Another source of non-regularity is the intertwining of enabling and recursion. The next algorithm is similar to the previous one. Application of $cf(X_0, \emptyset, \emptyset)$ determines all processes in the LOTOS specification whose instantiations must be expressed by the call command. The function cf is inductively defined as follows:

- $cf(\Box, f, s) = \emptyset$ for $\Box$ ranging over **stop** and **exit**.
- $cf(B_1 \Box B_2, f, s) = cf(B_1, f, s) \cup cf(B_2, f, s)$ for $\Box$ ranging over $[]$, $|[S]|$, $[>$.
- $cf(\Box B, f, s) = cf(B, f, s)$ for $\Box$ ranging over a, **hide** S **in** , and $[p] \rightarrow$.
- $cf(B_1 \gg B_2, f, s) = cf(B_1, f \cup s, \emptyset) \cup cf(B_2, f, s)$.
- $cf(X[g_1, \ldots, g_m](t_1, \ldots, t_n), f, s) = \begin{cases} \emptyset & \text{if } X \in s; \\ \{X\} & \text{if } X \in f; \\ cf(B_X, f, \{X\} \cup s) & \text{otherwise.} \end{cases}$

3.3 Introduction of fork commands

If necessary by using new equations, the flattened LOTOS specification has to be transformed such that process instantiations are the only operands of parallel composition. Let $\tilde{X} = X[\tilde{g_1}, \ldots, \tilde{g_m}](t_1, \ldots, t_n)$ be the operand of a parallel composition operator and let $X[g_1, \ldots, g_m](v_1, \ldots, v_n) \in Xdefs$. If H is a set of newly created gates, then $f_i = \{\tilde{g_j}/g_j \mid \tilde{g_j} \in H\}$ defines a relinking of the gates of the father process from its environment to its son. Let $\bar{f_i}$ be the inverse renaming. For each dynamic process X, its body has to be replaced by $\Gamma(X[g_1, \ldots, g_m](v_1, \ldots, v_n), X, \emptyset, \emptyset\})$.

- $\Gamma(\Box, pid, H, ids) = \Box$ for $\Box$ ranging over **stop** and **exit**.
- $\Gamma(a; B, pid, H, ids) = \begin{cases} \mathbf{i}; (B, pid, H, ids) \text{ if } gate(a) \in H; \\ a; (B, pid, H, ids) \text{ otherwise.} \end{cases}$
- $\Gamma(B_1 \Box B_2, pid, H, ids) = \Gamma(B_1, pid, H, ids) \Box \Gamma(B_2, pid, H, ids)$ for $\Box$ ranging over $[]$, $\gg$, and $[>$.
- $\Gamma([p] \rightarrow B, pid, H, ids) = [p] \rightarrow \Gamma(B, pid, H, ids)$.

- $\Gamma(\textbf{hide } S \textbf{ in } B, pid, H, ids) = \Gamma(B, pid, H \cup S, ids)$.
- $\Gamma(\tilde{X}_1 \,|[S]|\, \tilde{X}_2, pid, H, ids)$

$$= \begin{cases} \textbf{fork}(\tilde{X}_2, S, H, f_1); \tilde{X}_1[\bar{f}_1] & \text{if } pid = \tilde{X}_1; \\ \textbf{fork}(\tilde{X}_1, S, H, f_2); \tilde{X}_2[\bar{f}_2] & \text{if } pid = \tilde{X}_2; \\ \tilde{X}_1 \,|[S]|\, \tilde{X}_2 & \text{otherwise.} \end{cases}$$

- $\Gamma(X[g_1, \ldots, g_m](t_1, \ldots, t_n), pid, H, ids)$

$$= \begin{cases} \left.\begin{cases} X[g_1, \ldots, g_m](t_1, \ldots, t_n) & \text{if } H = \emptyset; \\ \textbf{hide } H \textbf{ in } X[g_1, \ldots, g_m](t_1, \ldots, t_n) & \text{otherwise.} \end{cases}\right\} & \text{if } X \in ids; \\ \Gamma(\hat{B_X}, pid, H, \{X\} \cup ids) & \text{otherwise.} \end{cases}$$

where $\hat{B_X} = B_{X[\hat{g_1}, \ldots, \hat{g_m}](v_1, \ldots, v_n)}[g_1/\hat{g_1}, \ldots, g_m/\hat{g_m}, t_1/v_1, \ldots, t_n/v_n]$

In the rule for parallel composition, the relinking of gates has to be expressed by the function f_i due to the dynamic nature of the implicit relabeling operator of LOTOS. The inverse function $\bar{f}_i$ reverses the renaming within the creating process.

4 State Machine Construction

In this section, we present the algorithm to translate the flattened LOTOS process definitions into state machines. Let $X_{Dyn} \supseteq dyn(X_0)$ define all processes that shall be created *dynamically* and let $X_{cf} = cf(X_0)$ define all processes that shall be used as *procedures* within BS. In particular, the set X_{Dyn} is chosen such that $root(X_0)$ is defined:

$$root(B) = \begin{cases} X & \text{if } X \in (X_{Dyn} \cup X_{cf}); \\ root(B') \backslash H & \text{if } B = \textbf{hide } H \textbf{ in } B'; \\ root(B_1) \,|[S]|\, root(B_2) & \text{if } B = B_1 \,|[S]|\, B_2; \\ undefined & \text{otherwise.} \end{cases}$$

Here $transl(B, P, Q)$ denotes the translation of the behavior expression B. Parameter P keeps track of encountered process instantiations and Q defines all 'procedures' that shall be excluded from the current translation. The complete translation of a flattened LOTOS text is defined as:

$$\begin{aligned} & translate(\langle AS, BS \rangle) = \\ & \quad \langle root(X_0), \{X[g_1, \ldots, g_m](v_1, \ldots, v_n) := \\ & \qquad transl(X[g_1, \ldots, g_m](v_1, \ldots, v_n), \emptyset, X_{cf} - \{X\}) \mid X \in (X_{Dyn} \cup X_{cf})\}\rangle \end{aligned}$$

In the following translation rules, assume that $transl(B', P, Q)$ yields a state machine $M' = \langle J', E', V', R', j_0', c_0' \rangle$. B'' is analogous. Let j_0 and j_1 be control states neither in J' nor J'', and assume J' and J'' to be disjoint. By the substitution below, the effect of a statement $\tilde{c}$ and a predicate $\tilde{p}$ on the subsequent behavior will be considered.

$$\Psi_{\tilde{c}}^{\tilde{p}}(\langle a, j, p, c \rangle) = \langle a[\Omega_{\tilde{c}}], j, \tilde{p} \wedge p[\Omega_{\tilde{c}}], \tilde{c}; c \rangle$$

Thereby, the substitution Ω performs the update of the statement $\tilde{c}$: $\Omega_{nop} = \imath$; $\Omega_{v_1, \ldots, v_n := t_1, \ldots, t_n} = t_1, \ldots, t_n / v_1, \ldots, v_n$; $\Omega_{c_1; c_2} = \Omega_{c_1} \circ \Omega_{c_2}$. As usual, we denote the identity function by $\imath$.

Atomic expressions

The LOTOS atomic expression **stop** denotes the process that is unable to proceed as there is no transition rule for **stop**.

$$transl(\textbf{stop}, P, Q) = \langle \{j_0\}, \emptyset, \emptyset, \emptyset, j_0, nop\rangle$$

The LOTOS expression **exit**$(t_1, .., t_n)$ denotes a process that terminates by passing values denoted by $t_1, .., t_n$ to the environment through the special gate δ. For $1 \leq i \leq n$, t_i is either a term or the expression **any** s with s a sort in AS. In the latter case a nondeterministic choice of values of sort s is offered to the environment upon termination. We assume that there are enough free variables $\tilde{v_1}, \ldots, \tilde{v_k}$ such that

$$\begin{aligned} & transl(\textbf{ exit}(d_1, \ldots, d_k), P, Q) \\ & = \langle \{j_0, j_1\}, \emptyset, \{v \mid \tilde{d_i} = ?v\}, \{\langle \delta\tilde{d_1} \ldots \tilde{d_k}, j_0, j_1, true, nop\rangle\}, j_0, nop\rangle \end{aligned}$$

where $\tilde{d_i} = \begin{cases} ?\tilde{v_i} : s \text{ if } d_i = \textbf{any s}; \\ !d_i \quad \text{otherwise.} \end{cases}$

Action-Prefix

The LOTOS expression $a; B$ denotes that the process may execute the a action and behaves like process B afterwards.

$$transl(a; B', P, Q) = \langle J' \cup \{j_0\}, E', V' \cup Var(a), R' \cup \{\langle a, j_0, j_0', true, c_0'\rangle\}, j_0, nop\rangle$$

Choice

The LOTOS choice expression $B_1 \, [] \, B_2$ means that the process may behave like B_1 or B_2 where the choice is determined by the first action.

$$transl(B' \, [] \, B'', P, Q) = \langle J, E, V' \cup V'', R, j_0, c_0'; c_0''\rangle$$

where

$$J = (J' - \{j_0'\}) \cup (J'' - \{j_0''\}) \cup \{j_0\}$$

$$E(j) = \begin{cases} E'(j_0') \cup E''(j_0'') & \text{if } j = j_0; \\ E'(j) & \text{if } j \in (J' - \{j_0'\}); \\ E''(j) & \text{if } j \in (J'' - \{j_0''\}). \end{cases}$$

$$R(j) = \begin{cases} R'(j_0') \cup R''(j_0'') & \text{if } j = j_0; \\ R'(j) & \text{if } j \in (J' - \{j_0'\}); \\ R''(j) & \text{if } j \in (J'' - \{j_0''\}). \end{cases}$$

The sum of two processes is formed by adding a new control state j which has the rules and extensions of both M' and M''.

Parallel expression

The LOTOS parallel composition operator $|[g_1, \ldots, g_n]|$ forces communication of actions on gates $g_1, \ldots, g_n$ to take place and synchronous termination on δ.

Action matching. In LOTOS, synchronization is defined on events. But at translation time we do not know which actual events will occur. Therefore we introduce the concept of action matching which assures that only "compatible" actions will be joined. We say that two actions match, $ad_1 \dots d_n[sp_a] \equiv bd'_1 \dots d'_m[sp_b]$, whenever

- the name a is equal to the name b,
- $m = n$,
- $\left.\begin{cases} t_i = t'_i & \text{if } d_i = !t_i \text{ and } d'_i = !t'_i; \\ type(d_i) = type(d'_i) & \text{otherwise.} \end{cases}\right\}$ for $0 \leq i \leq n$.

Let $v^j_1, \dots, v^j_k$ be the variables of the input offers of an action, then v^m_n denotes the n-th variable from the m-th offer.

The result of joining two actions, $gd_1 \dots d_n[sp_a] \otimes gd'_1 \dots d'_m[sp_b]$, is the action $gd''_1 \dots d''_n[sp]$ where

$$d''_i = \begin{cases} !t & \text{if } d_i = !t; \\ !t' & \text{if } d'_i = !t'; \\ ?v'_i & \text{if } d'_i = ?v'_i \wedge v'_i \in sp_b \wedge v'_i \notin sp_a; \\ ?v_i & \text{otherwise.} \end{cases}$$

$$sp = sp_a \wedge sp_b[\sigma(v^1_i)/v^1_i, \dots, \sigma(v^k_j)/v^k_j].$$

In selection predicate sp_b, all variables $v^1_i, \dots, v^k_j$ have to be replaced by the term of the corresponding offer in the left action

$$\sigma(v^j_i) = \begin{cases} t \text{ if } d_i = !t; \\ v \text{ if } d_i = ?v. \end{cases}$$

Chart merging. The parallel composition of two processes yields a state machine having the Cartesian product of the control states of the state machines of the subprocesses as its control state space. The new rule set is composed such that synchronization determines the reachable control states.

Because of preprocessing, we get B' and B'' with $E' = E'' = \emptyset$. State machines have a flat name space. In case of variable name clashes, the data variables in M' and M'' will be renamed to be mutually disjoint.

$$transl(B' \,|[S]|\, B'') = \langle J' \times J'', \emptyset, V' \cup V'', R, \langle j'_0, j''_0 \rangle, c'_0; c''_0 \rangle$$

where

$$\begin{aligned} R(\langle j', j'' \rangle) = {} & \{\langle a, \langle to', j'' \rangle, p, c \rangle \mid \langle a, to', p, c \rangle \in R'(j') \text{ and } gate(a) \notin S\} \\ & \cup \\ & \{\langle b, \langle j', to'' \rangle, p, c \rangle \mid \langle b, to'', p, c \rangle \in R''(j'') \text{ and } gate(b) \notin S\} \\ & \cup \\ & \{r_1 \otimes r_2 \mid r_1 \in R'(j'),\ r_2 \in R''(j'') \text{ and } gate(action(r_1)) \in (S \cup \{\delta\})\}. \end{aligned}$$

When two actions r_1 and r_2 match, then the joined action is determined by the following function:

$$r_1 \otimes r_2 = \langle a \oplus b, \langle j', j'' \rangle, \langle to', to'' \rangle, p' \wedge p'' \wedge P(a,b), F(a,b); c'; c'' \rangle \text{ if } a \equiv b.$$

For each pair of output offers, a predicate will be included in the provided clause to check at run time whether the values will be equal.

$$P(gd_1 \dots d_n[sp_a], gd'_1 \dots d'_m[sp_b]) = \bigwedge_{1 \le i \le n} Pi(d_i, d'_i)$$

$$Pi(d, d') = \begin{cases} t = t' \text{ if } d = !t \text{ and } d' = !t'; \\ true \;\; \text{otherwise.} \end{cases}$$

Similarly, for each pair of input offers an assignment statement will be introduced into the statement clause which also stores the received value in the second variable.

$$F(gd_1 \dots d_n[sp_a], gd'_1 \dots d'_m[sp_b]) = Fi(d_1, d'_1); \dots; Fi(d_n, d'_n)$$

$$Fi(d, d') = \begin{cases} nop \;\; \text{if } d = !t \text{ and } d' = !t'; \\ v' := t \text{ if } d = !t \text{ and } d' = ?t'; \\ v := t' \text{ if } d = ?v \text{ and } d' = !t'; \\ v' := v \text{ otherwise.} \end{cases}$$

Abstraction

Hiding of gates transforms some observable actions of a process into unobservable ones. These actions are thus made unavailable for synchronization with other processes. By restriction, B' is regular, and we may assume $E' = \emptyset$.

$$transl(\textbf{hide } S \textbf{ in } B', P, Q) = \langle J', \emptyset, V', R, j'_0, c'_0 \rangle$$

where

$$R = \{r \mid r \in R' \wedge gate(action(r)) \notin S\}$$
$$\cup \{r[\tau/action(r)] \mid r \in R' \wedge gate(action(r)) \in S\}$$

The translation substitutes an internal action for every action with a gate contained in the set S.

Guarded expression

The guarded command only makes the behavior possible if its predicate holds, otherwise the whole expression is equivalent to **stop**.

$$transl([p] \rightarrow B', P, Q) = \langle J', E, V', R, j'_0, c'_0 \rangle$$

where

$$E(j) = \begin{cases} \{\langle X, c, p \wedge p' \rangle \mid \langle X, c, p' \rangle \in E'(j)\} \text{ if } j = j_0; \\ E'(j) \qquad \text{otherwise.} \end{cases}$$

$$R(j) = \begin{cases} \{\langle a, j, p \wedge p', c \rangle \mid \langle a, j, p', c \rangle \in R'(j)\} \text{ if } j = j'_0; \\ R'(j) \qquad \text{otherwise.} \end{cases}$$

In all rules of the initial control state j'_0 the provided clause has to be extended with the condition of the guard. Also the extensions of j_0 have to be adopted.

Enabling expression

The LOTOS enabling expression can be looked upon as a sequential composition with value passing, or as a restricted parallel composition in which only the termination action of the enabling process and a special first accepting action of the enabled process communicate.

$$\begin{aligned} &\mathit{transl}(B' \gg \textbf{accept}\ v_1 : s_1, \ldots, v_k : s_k\ \textbf{in}\ B'', P, Q) \\ &= \langle J' \cup J'', E'', V' \cup V'' \cup \{v_1, \ldots, v_k\}, R, j_0', c_0' \rangle \end{aligned}$$

with $R = R'' \cup \{r \mid r \in R' \text{ and } \mathit{gate}(\mathit{action}(r)) \neq \delta\}$
$\cup \{\langle \tau, j, j_0'', p, v_1, \ldots, v_k{:=}t_1, \ldots, t_k; c_0'' \rangle \mid \langle \delta!t_1 \ldots !t_k, j, j', p, \mathit{nop} \rangle \in R'\}$
$\cup \{\langle j_0'', \mathit{call}(X[g_1, \ldots, g_m]((\mathbf{t})(v_1, \ldots, v_k))), p, c_0'' \rangle \mid \langle j, X, \mathbf{v}{:=}\mathbf{t}, p \rangle \in E'\}$

The sequential composition of two processes is formed by connecting all end nodes of the first process with the root of the second process through internal transitions. These internal transitions perform either the value passing or push the current state on the stack.

We can distinguish between two cases. If the first process is finite, i.e. self-contained, value-passing is performed by assigning the exit values to the variables in the accept statement. If not, the continuing process (stored in $E(j)$) has to be called. To receive the return values, the variables of the accept statement have to be included (*call by reference*) in the parameter list of the process.

Disabling expression

The disable expression B' [> B'' denotes the nondeterministic choice of behavior B'' in every state of B'. By restriction, B' is regular and we may assume $E' = \emptyset$.

$$\mathit{transl}(B' \,[\!>\, B'', P, Q) = \langle J' \cup J'', E, V' \cup V'', R, j_0', c_0' \rangle$$

where

$$J_E' = \{j \mid \langle \delta d_1, \ldots, d_n, j', j, p, c \rangle \in R'\}$$

$$E(j) = \begin{cases} E''(j_0'') & \text{if } j = j_0'; \\ E''(j) & \text{if } j \in J''. \end{cases}$$

$$R(j) = \begin{cases} R'(j) \cup \Psi_{c_0''}(R''(j_0'')) & \text{if } j \in (J' - J_E'). \\ R''(j) & \text{if } j \in J''. \end{cases}$$

Disruption of the first process is achieved by attaching copies of the initial rules $R''(j_0'')$ to all states of the first process. Also, the extensions of the root of the second process have to be attached to all nodes (without outgoing δ-rules) of the first process.

Process Instantiation

Let $X[h_1,\ldots,h_m](v_1,\ldots,v_n) := B' \in Xdefs$ be a process definition. The interpretation of a process instantiation $X[g_1,..,g_m](t_1,..,t_n)$ is the right-hand side B' of the process definition with all actions on formal gates $h_1,..,h_m$ renamed to $g_1,..,g_m$ and actual parameters $t_1,..,t_n$ substituted for the formal variables $v_1,..,v_n$ respectively.

$$transl(X[g_1,\ldots,g_m](t_1,\ldots,t_n),P,Q) = \begin{cases} \langle\{j_0\},\tilde{E},\emptyset,\emptyset,j_0,nop\rangle & \text{if } X\in P;\\ \langle\{j_0,j_1\},\emptyset,\{v_1,\ldots,v_k\},\tilde{R},j_0,nop\rangle & \text{if } X\in Q\wedge Func(X)=\{s_1,\ldots,s_k\};\\ \langle J,E,V',R[g_1/h_1,\ldots,g_m/h_m],j_0',c_0\rangle & \text{otherwise.}\end{cases}$$

In the construction for the (recursive) instantiation of processes we have to distinguish between three cases:

Process Call. In the extension, all information is stored to record that at this node the state machine can behave like process X where the actual parameters are $t_1,\ldots,t_n$ and the actual gate names are $g_1,\ldots,g_m$.

$$\tilde{E} = \{\langle j_0, X, v_1,\ldots,v_n{:=}t_1,\ldots,t_n\rangle, true\}$$

Procedure Call. If process X is used as a procedure within the LOTOS specification, i.e. $X \in cf(BS)$ and $Func(X) = s_1,\ldots,s_k$ are the sorts of the return values, then a state machine will be constructed whose only rule performs a call command for process X.

$$\tilde{R} = \{\langle j_0, j_1, call(X[g_1,\ldots,g_m]((t_1,\ldots,t_n)(v_1,\ldots,v_k))), true, nop\rangle\}$$

The names of the variables $v_1,\ldots,v_k$ shall not occur in BS and each variable v_i should be of sort s_i.

Process Abstraction. Finally, when we have constructed the state machine for the body B' of process X, we must then replace each extension X in $transl(B',P\cup\{X\},Q)$ by the rules and extensions (except for X) of j_0'.

$$c_0 = \langle v_1,\ldots,v_n{:=}t_1,\ldots,t_n; c_0'\rangle.$$

$$J = J' - \{j \mid R(j)=\emptyset \text{ and } |E(j,X)|=1\}$$

and for all $j \in J$:

$$E(j) = \begin{cases} (E'(j)\cup E'(j_0')) - \{\langle X,c,p\rangle\} & \text{if } \langle j,X,c,p\rangle \in E';\\ E'(j) & \text{otherwise.}\end{cases}$$

$$R(j) = \begin{cases} \{\langle a, j_0', p, c; \tilde{c}; c_0'\rangle \mid \langle a,j',p,c\rangle \in R'(j)\} & \text{if } E'(j') = \{\langle X,\tilde{c},true\rangle\} \wedge R'(j')=\emptyset;\\ \tilde{R}(j) & \text{otherwise.}\end{cases}$$

The first clause deals with simple recursion. All rules leading to leaves labeled by only one X extension are redirected to the root of the graph. Note that this redirection is safe because of the introduction of a new root in the construction of $B_1 \,[]\, B_2$. The assignment statements of the extension $\tilde{c}$ as well as the initialization statement c'_0 are appended to the statement element of the redirected rule.

$$\tilde{R}(j) = \begin{cases} R'(j) \cup \Psi^{\tilde{p}}_{\tilde{c};c'_0}(R'(j'_0)) & \text{if } \langle j, X, \tilde{c}, \tilde{p}\rangle \in E' \wedge R'(j'_0) \neq \emptyset; \\ \langle \tau, j', \tilde{p}, \tilde{c}\rangle & \text{if } \langle j, X, \tilde{c}, \tilde{p}\rangle \in E' \wedge R'(j'_0) = \emptyset; \\ R'(j) & \text{otherwise.} \end{cases}$$

For all other cases, new transitions and extensions are introduced. For each X extension, a copy of $R'(j'_0)$ and $E'(j'_0) - \{\langle X, c, p\rangle\}$ originates from the labeled node. Application of function Ψ computes the effects of the assignment $\tilde{c}$. The third clause models divergence and ensures guardedness.

5 Examples

Recursive process calls. Process X in the following LOTOS specification calls itself recursively. At the first occurrence, process instantiation is in conflict with action b. But at the second occurrence, it is simple tail recursion.

```
process X [ a,b,c] : noexit :=
   (a ; (X [] b ; stop))  []  (c ; X)
endproc (* X *)
```

In the following picture, we see the last step in the generation of the graph of the state machine. Resolving recursion, the right X extension leads to a redirection of the c transition to the root (tail recursion). For the other extension, a copy of the former two root transitions originating from the node of the extension is introduced.

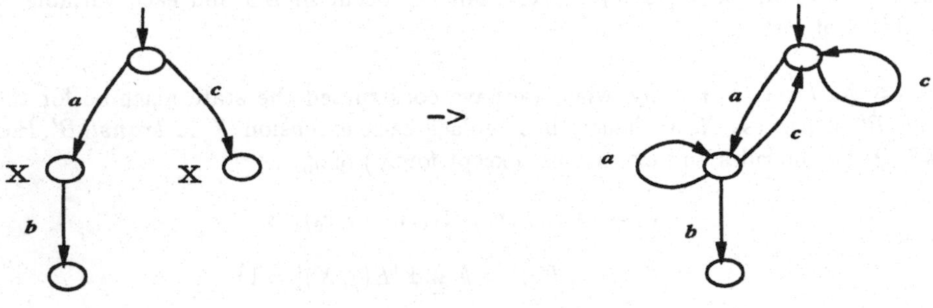

Recursive procedure calls. The LOTOS process *Factorial* acts as a recursive procedure. The value of $v!$ may be obtained by calling and returning from process *Factorial*, copying the value of the parameter and the result.

```
process Factorial( n: Nat) : exit ( Nat) :=
   [ n == 0] -> exit( 1)
   []
   [ n != 0] -> ( Factorial( n - 1) >> accept v: Nat in exit( n * v) )
endproc (* Factorial *)
```

The translation constructs a state machine with two states and three transitions. The rules of the state machine are given in tabular form. Instead of an action we may denote by $\Rightarrow t_1, \ldots, t_n$ the passing of return values upon successful termination of a process. We omit conditions and statements if equal with *true* or *nop*.

Process Factorial[ ](n : Nat)
Variables v : Nat;
Rules:

from	condition	action	to
1	n==0	$\Rightarrow 1$	
	n!=0	call(Factorial((n-1)(v)))	2
2		$\Rightarrow n * v$	

6 Conclusion

We have presented a method to translate LOTOS processes into a network of communicating state machines. Algorithms are defined for identifying non-regular processes, replacing parallel process compositions with dynamic processes by a special fork action, and translating LOTOS processes into communicating state machines. The algorithms have been implemented in the LOTOS compiler of the LOEWE workbench [13].

The presented translation algorithm seems to generate the smallest state machine (in terms of a minimum of control states) for all LOTOS operators except parallel composition, hiding and choice. Analysis methods shall be applied to remove at least the τ-transitions where possible. For the choice operator, isomorphic subtrees within a behavior expression lead to redundant control states. Enhancing our translation algorithm with common subexpression recognition would further compress the generated state machines as shown in [14].

From the state machine representation, the LOEWE compiler generates code in the imperative programming language C which can be executed in a specific runtime environment [4] that provides the LOTOS communication facilities. Further, the XFSM simulator can be used to produce a labeled transition graph of a finite state specification to enable the application of a number of effective analyzing methods, e.g. model checking [15], bisimulation [16]. The XFSM simulator of the LOEWE workbench [13] is capable of generating state graphs of the order of one million states.

Likewise, the CÆSAR system of Garavel and Sifakis [17] translates LOTOS specifications into extended Petri nets for which a powerful simulator exists. Their compilation technique imposes static control constraints which are a sufficient condition for generating finite graphs. The accepted subset of LOTOS is smaller. In particular, no dynamic process creation or recursive procedure calls are allowed. Process enabling is expanded into parallel composition and hiding because there is no concept of procedures in Petri nets. However, the use of ϵ-transitions in the Petri nets allows behavior to be simulated where parallel composition is combined with choice, e.g. "$B_1 \,[]\, (B_2 ||| B_3)$", without computing the product automaton corresponding to "$B_2 ||| B_3$". In our approach, this would correspond to "$B_1 \,[]\, \mathrm{i}; (B_2 ||| B_3)$".

Acknowledgments

My sincere thanks go to the members of the RACE/SPECS project, in particular to Carl Binding, Michel Dauphin, Jan Friso Groote and Heinrich Nirschl. Liba Svobodova's thoughtful comments on the manuscript improved its readability.

References

1. ISO. LOTOS — a formal description technique based on the temporal ordering of observational behaviour. IS 8807, 1988.
2. J.A. Mañas and T. de Miguel. From LOTOS to C. In K. Turner, editor, *Formal Description Techniques I*, pages 79–84. North-Holland, 1988.
3. S. Nomura, T. Hasegawa, and T. Takizuka. A LOTOS compiler and process synchronization manager. In *Protocol Specification, Testing, and Verification X*, pages 165–184. North-Holland, 1990.
4. C. Binding. Executing LOTOS behavior expressions. Research Report RZ 2118, IBM Research Division, 04/25/91.
5. R. Milner. A complete inference system for a class of regular behaviours. *Journal of Computer and System Sciences*, 28:439–466, 1984.
6. J.A. Bergstra and J.W. Klop. A complete inference system for regular processes with silent moves. In F.R. Drake and J.K. Truss, editors, *Logic Colloquium*. North-Holland, 1986.
7. D. Taubner. *Finite Representations of CCS and TCSP Programs by Automata and Petri Nets*, volume 369 of *Lecture Notes in Computer Science*. Springer Verlag, 1989.
8. G. Karjoth. Implementing process algebra specifications by state machines. In *Protocol Specification, Testing and Verification VIII*, pages 47–60. North-Holland, 1988.
9. E. Dubuis. An algorithm for translating LOTOS behavior expressions into automata and ports. In *Formal Description Techniques, II*, pages 163–177. North-Holland, 1989.
10. A. Valenzano, R. Sisto, and L. Ciminiera. Modeling the execution of LOTOS specifications by cooperating extended finite state machines. In *IEEE Symp. on Parallel and Distributed Processing*, Dallas, Tx, 9–13 December 1990.
11. G. Karjoth. XFSM: A formal model of communicating state machines for implementation specifications. Research Report RZ 2209, IBM Research Division, 09/12/1991.
12. ISO. Guidelines for the Application of ESTELLE, LOTOS and SDL. Technical Report 10167, September 1991.
13. G. Karjoth, C. Binding, and J. Gustafsson. LOEWE: A LOTOS engineering workbench. Research Report RZ 2143, IBM Research Division, 06/17/91. A revised version will appear in *Computer Networks and ISDN Systems*, special issue on "Tools for FDTs".
14. G. Karjoth. Generating Transition Graphs from LOTOS Specifications, Research Report RZ 2312, IBM Research Division, 05/18/1992.
15. E.M. Clarke, E.A. Emerson, and A.P. Sistla. Automatic verification of finite state concurrent systems using temporal logic specifications. *ACM Transactions on Programming Languages and Systems*, 8:244–263, April 1986.
16. J.F. Groote and F. Vaandrager. An efficient algorithm for branching bisimulation and stuttering equivalence. In M.S. Paterson, editor, *ICALP 90*, volume 443 of *Lecture Notes in Computer Science*, pages 626–638. Springer Verlag, 1990.
17. H. Garavel and J. Sifakis. Compilation and verification of LOTOS specifications. In *Protocol Specification, Testing, and Verification X*. pages 359–376, North-Holland, 1990.

Discrete Time Process Algebra

J.C.M. Baeten
Department of Computing Science, Eindhoven University of Technology, P.O.Box 513, 5600 MB Eindhoven, The Netherlands

J.A. Bergstra
Programming Research Group, University of Amsterdam, Kruislaan 403, 1098 SJ Amsterdam, The Netherlands
and
Department of Philosophy, Utrecht University, Heidelberglaan 8, 3584 CS Utrecht, The Netherlands

The axiom system ACP of [BEK84] is extended to ACP_{dt}, which involves discrete time delay, and then to ACP_{dt} + ATP, an axiomatisation that adds key features of ATP [NIS90] to ACP. We give an interpretation of all discrete time constructs in the real time theory $ACP\rho\sqrt{I}$.

Note: This work was done in the context of ESPRIT Basic Research Action 7166, CONCUR2. The second author received partial support from RACE project 1046, SPECS. This document does not necessarily reflect the views of the SPECS consortium.

1. INTRODUCTION

Process algebra in the form of ACP [BEK84, BAW90], CCS [MIL89] or CSP [BRHR84] describes the main features of imperative concurrent programming without any explicit mention of time. Implicitly, time is present in the interpretation of sequential composition: in $p \cdot q$ (ACP notation) the process p must be executed before q. It may be felt as a weakness of the mentioned process algebra frameworks that time features implicitly only. Indeed a quantative view on the relation between process execution and progress of time is definitely absent in these calculi. This seems worse than it actually is. Nothing prevents one to introduce timer processes that represent clocks, to have processes interacting with these clocks and to have synchronisations between the clocks in ACP, CCS or CSP.

Of course timing aspects can be described more uniformly if process algebras are given that provide standardised features to incorporate a quantative view on time. Obvious as this may seem, the more obvious observation is that in a few years so many different formats for timed process algebras have been introduced that this uniformity seems to be further away than before. There are several reasons for the emerging spectrum of timed process algebras. We explain this by considering various options for adding time. The most obvious option is to represent time by means of non-negative reals, and to have time stamps on actions. This is done in [BAB91a] for ACP, in [RER88] for CSP and in [MOT90] for CCS. The timed versions of ACP, CSP and CCS in these papers differ concerning the degree to which time stamping is explicit in the notational format of the proposed process algebras. This in turn influences the form of equations, axiomatisations and the appearance of examples.

Another option is to divide time in slices indexed by natural numbers, to have an implicit or explicit time stamping mechanism that provides each action with the time slice in which it occurs and to have a time order within each slice only. This has been worked out in ATP [NIS90], a process algebra that adds time slicing to a version of ACP based on action prefixing rather than sequential composition (this version was called aprACP in [BAB91b]). Further, [GRO90a] has extended ACP with time slices and [MOT89] have added these features to CCS. We propose to use the phrase *real time process algebra* for algebras that involve explicit or implicit time stamping with real numbers (or any dense subfield of the reals) and *discrete time process algebra* if an enumeration of time slices is used.

The objective of this paper is to extend ACP to a discrete time process algebra. This is done by adding a new process constructor σ_d: P → P. The notation σ has been taken from [HER90], the subscript d draws attention to the discrete time setting. $\sigma_d(x)$ delays the process x till the next time slice. In addition to σ_d four auxiliary functions and an auxiliary sort are used. The booleans serve as auxiliary sort and the conditional operator $\_ \triangleleft \_ \triangleright \_ : P \times B \times P \to P$ is used. Further, an operator $1 \gg_d \_ : P \to P$ (unit time shift) reduces a process x to those options left if it idles one step (and to δ if it cannot idle). Then, the operator $\_ \gg_d 1 : P \to P$ restricts a process x to those options that do not involve initial idling. Finally, $D: P \to B$ determines whether or not a process can idle at all.

For practical application, the primitives of ACP_{dt} are too low level. So we have added the key operators of ATP [NSVR90, NIS90] to ACP_{dt}, thus obtaining ACP_{dt} + ATP. These features include: unit delay $\lfloor x \rfloor(y)$ and maximal progress composition $\oplus$. It turns out that these operators can be eliminated in the presence of those of ACP_{dt}. (It should be noted that ATP's $\oplus$ is already present as + in TCCS.) As ACP_{dt}, ATP and TCCS each use strong bisimulation equivalence to obtain a semantic domain, they may be considered as different (and intertranslatable) sets of generators for the same semantic world of processes. Some less important modifications with respect to ATP in the form of [NIS90] have been made. ACP_{dt} + ATP contains time stops (as does [MOT89]), a feature that is rejected on philosophical grounds in [NIS90]. In addition, the cancel mechanism has been omitted because that is covered automatically due to the presence of proper termination in ACP_{dt}.

Both the selection of ACP_{dt} + ATP and the casting of ATP in an ACP framework require some motivation. We have the following remarks on this.

1. ATP contains a combination of features (time outs, time posteriority, mode transfers) which turn it into a nontrivial extension of the untimed (symbolic) case. The features of ATP turn out to be of significant expressive power in comparison to [GRO90a], [GM91]. Moreover we found these features quite pleasant when dealing with examples.
2. The claim of [NIS90] that progress of time by itself should not be allowed to introduce non-determinism is convincing in our view. If time is represented by an action σ then $\sigma\cdot(x + y) = \sigma\cdot x + \sigma\cdot y$ is an appropriate axiom called time factorisation in [NIS90] ([GRO90a], [GM91] do not satisfy time factorisation).
3. ATP does not involve an action that represents the flow of time. This is a conceptual advantage as intuitively, flow of time is not an action, and the progress of an independent clock is not an observable system action.
4. Building ATP on top of ACP has several advantages. First of all, both termination and general sequential composition (which are in ACP but not in ATP) combine very well with ATP's key features. We hope that this is sufficiently illustrated by the protocol verifications in section 5. Secondly, the meta theory of ACP + ATP can be organised just as the meta theory of ACP (see e.g. [BAW90]) which allows to simply copy most of the results and proof techniques.

We design our discrete time process algebra in an incremental way, through stages BPA_{dt}, $BPA_{\delta dt}$, $PA_{\delta dt}$, ACP_{dt}, extensions of ACP_{dt}. This resembles the segmentation of ACP that can be found in [BAW90]. At each step we will motivate our choice and compare it with the sources of discrete time process algebra that we have used: TCCS [MOT89], ACP_t [GRO90a], TPL [HER90], ATP [NIS90], CCSS [GM91]. Some of the axioms we use already appear in [BAB91b].

2. REAL TIME PROCESS ALGEBRA WITH INITIAL ABSTRACTION OPERATOR

This paper intends to present discrete time process algebra in a clear connection with real time process algebra. To this end, we will first outline the axiom system $ACP\rho\surd I$. It describes the real time process algebra of [BAB91a] while integrating absolute and relative time notation using the initial abstraction operator $\surd$ that was introduced in [BAB92]. For $ACP\rho\surd I$, a bisimulation model M_A^* is outlined (here, A denotes the underlying set of atomic actions). This model is introduced in a sketchy way. Its relevance lies in the clear operational meaning it provides for processes. It serves as a conceptual standard model.

Having available M_A^*, the various operators and constants of discrete time process algebra are defined in it. Then, as a subject, discrete time process algebra reduces to an investigation of certain reducts and subalgebras of M_A^* as well as their axiomatic description.

We start with a quick overview of the real time process algebra introduced in [BAB91a]. We follow the presentation of [BAB92]. We also use the initial abstraction operator introduced in [BAB92], which allows to present the absolute time and relative time versions of the theory of [BAB91a] in a unified framework. We give an operational semantics in the style of [BAB91a].

2.1 ATOMIC ACTIONS

We start from a set A of (symbolic) atomic actions. The set A is a parameter of the theory. Further, we have a special constant δ, denoting inaction, the absence of any execution. In particular, δ cannot terminate. We put $A_\delta = A \cup \{\delta\}$. The set of atomic actions with time parameter, AT is

$$AT = \{a(t) \mid a \in A_\delta, t \in \mathbb{R}_{\geq 0}\}.$$

In [BAB91a], we considered the use of locations, in order to express that actions can occur at the same time but at different locations. This led to the introduction of *multi-actions*. In [BAB92], we extended this concept by looking at actions parametrised by a point in real space. As we will not need multi-actions in discrete time process algebra (since within a time slice, $a \| b = a \cdot b + b \cdot a + a \mid b$, i.e. interleaving works), we will not consider multi-actions in this paper. The theory can be set up for multi-actions, just like we did in [BAB92], without any problems.

2.2 BASIC PROCESS ALGEBRA

Process algebra (see [BEK84, BAW90]) starts from a given *action alphabet*, here AT. Elements of AT are constants of the sort P of *processes*. The process a(t) can let time progress until t, will then execute action a at time t, and then terminate successfully. The process δ(t) can also let time progress until t, but then nothing more is possible (in particular, time cannot progress anymore). Therefore, the δ actions are often called *time stops*.

Real Time Basic Process Algebra with Deadlock (BPAρδ) has two binary operators $+, \cdot : P \times P \to P$; + stands for alternative composition and · for sequential composition. Moreover, there is the additional operator $\gg : \mathbb{R}_{\geq 0} \times P \to P$, the *initialisation operator*. This operator was called the time shift

$X + Y = Y + X$	A1	$a \mid b = b \mid a$	C1
$(X + Y) + Z = X + (Y + Z)$	A2	$a \mid (b \mid c) = (a \mid b) \mid c$	C2
$X + X = X$	A3	$\delta \mid a = \delta$	C3
$(X + Y) \cdot Z = X \cdot Z + Y \cdot Z$	A4	$t \neq s \Rightarrow a(t) \mid b(s) = \delta(\min(t,s))$	ATC1
$(X \cdot Y) \cdot Z = X \cdot (Y \cdot Z)$	A5	$a(t) \mid b(t) = (a \mid b)(t)$	ATC2
$X + \delta = X$	A6		
$\delta \cdot X = \delta$	A7	$X \| Y = X \lfloor\!\lfloor Y + Y \lfloor\!\lfloor X + X \mid Y$	CM1
		$X = r \gg Z \Rightarrow a(t) \lfloor\!\lfloor X = (a(t) \gg U(X)) \cdot X$	ATCM2'
$a(0) = \delta(0) = \delta$	ATA1	$Y = r \gg Z \Rightarrow a(t) \cdot X \lfloor\!\lfloor Y = (a(t) \gg U(Y)) \cdot (X \| Y)$	
$\delta(t) \cdot X = \delta(t)$	ATA2		ATCM3'
$t < r \Rightarrow \delta(t) + \delta(r) = \delta(r)$	ATA3	$(X + Y) \lfloor\!\lfloor Z = X \lfloor\!\lfloor Z + Y \lfloor\!\lfloor Z$	CM4
$a(t) + \delta(t) = a(t)$	ATA4	$(a(t) \cdot X) \mid b(r) = (a(t) \mid b(r)) \cdot X$	CM5'
$a(t) \cdot X = a(t) \cdot (t \gg X)$	ATA5	$a(t) \mid (b(r) \cdot X) = (a(t) \mid b(r)) \cdot X$	CM6'
		$(a(t) \cdot X) \mid (b(r) \cdot Y) = (a(t) \mid b(r)) \cdot (X \| Y)$	CM7'
$t < r \Rightarrow t \gg a(r) = a(r)$	ATB1	$(X + Y) \mid Z = X \mid Z + Y \mid Z$	CM8
$t \geq r \Rightarrow t \gg a(r) = \delta(t)$	ATB2	$X \mid (Y + Z) = X \mid Y + X \mid Z$	CM9
$t \gg (X + Y) = (t \gg X) + (t \gg Y)$	ATB3		
$t \gg (X \cdot Y) = (t \gg X) \cdot Y$	ATB4	$\partial_H(a) = a$ if $a \notin H$	D1
$r \geq t \Rightarrow a(r) \gg t = \delta(t)$	ATB5	$\partial_H(a) = \delta$ if $a \in H$	D2
$r < t \Rightarrow a(r) \gg t = a(r)$	ATB6	$\partial_H(a(t)) = (\partial_H(a))(t)$	ATD
$(X + Y) \gg t = (X \gg t) + (Y \gg t)$	ATB7	$\partial_H(X + Y) = \partial_H(X) + \partial_H(Y)$	D3
$(X \cdot Y) \gg t = (X \gg t) \cdot Y$	ATB8	$\partial_H(X \cdot Y) = \partial_H(X) \cdot \partial_H(Y)$	D4
$U(a(t)) = t$	ATU1		
$U(X + Y) = \max\{U(X), U(Y)\}$	ATU3		
$U(X \cdot Y) = U(X)$	ATU4		

TABLE 1. ACPρ.

operator in [BAB91a]. $t \gg X$ denotes the process X starting at time t. This means that all actions that have to be performed at or before time t are turned into deadlocks because their execution has been delayed too long. BPAρδ has the axioms A1-7 and ATA1-5, ATB1-4 from table 1 ($a \in A_\delta$).

The letter A in the names of the axioms in table 5 refers to *absolute time* (versions with relative time will be considered in the following). Next we consider parallel composition. Notice that we do not have a counterpart of PA ([BEK84]) in real time process algebra, because merge without communication is not an option here. Indeed, even if a and b do not communicate, the term $a(2) \| b(2)$ cannot be evaluated as $a(2)\cdot b(2) + b(2)\cdot a(2)$. Thus, we consider merge with communication straight away.

2.3 ALGEBRA OF COMMUNICATING PROCESSES

In order to formulate communication between processes, we assume we have given a *communication function* | on A_δ. This function $| : A_\delta \times A_\delta \to A_\delta$ is also a parameter of the theory, and should be commutative, associative and have δ as a neutral element (see axioms C1,2,3). An axiomatization of parallel composition with communication uses the left merge operator $\lfloor\!\lfloor$, the communication merge operator |, and the encapsulation operator ∂_H of [BEK84]. Moreover, two extra auxiliary operators introduced in [BAB91a] are needed: the ultimate delay operator and the bounded initialization operator.

The *ultimate delay* operator U takes a process expression X, and returns an element of $\mathbb{R}_{\geq 0}$. The intended meaning is that X can idle before U(X), but X can never reach time U(X) or a later time by just idling. The *bounded initialization* operator (or *time out* operator) is also denoted by $\gg$, and is the counterpart of the operator with the same name that we saw in the axiomatization of BPAρδ. With $X \gg t$ we denote the process X with its behaviour restricted to the extent that its first action must be performed at a time before $t \in \mathbb{R}_{\geq 0}$. The axioms of ACPρ are in table 1. There, $H \subseteq A$, $a,b,c \in A_\delta$.

Note that in [BAB91a], axioms ATCM2, ATCM3 appear that coincide with the axioms ATCM2', ATCM3' above except for the absence of the conditions. Because in the setting of ACPρ there, we have absolute time notation only, each process X satisfies $X = 0 \gg X$. We have that in the context of ACPρ, the new axioms are slightly weaker than the old, unprimed ones, but are in fact equivalent on finite process expressions. We have to add the conditions in order to deal with the following extension to processes that can also involve relative time notation. Notice that by substitution of $0 \gg Z$ for X resp. Y both conditional equations are turned into ordinary equations.

2.4 RELATIVE TIME NOTATION

The following is based on [BAB92]. In some cases, relative time notation is profitable. In [BAB91a], we have introduced a relative time notation as follows: a[t] denotes action a t time units after the previous atomic action or, if such action doesn't exist, after system initialisation. In this paper, we will deal with relative time in a different way. The basis for this approach to relative time notation is the *initial abstraction operator*. Let for $t \in \mathbb{R}_{\geq 0}$, F(t) be a process in P, then $\surd t.F(t)$ denotes a process that, when started at time r, proceeds as F(r). Semantically, $\surd t.F(t)$ is just a function f from $\mathbb{R}_{\geq 0}$ into P that satisfies $f(t) = t \gg f(t)$ for all $t \geq 0$. We make things more precise in the following.

2.5 DEFINITION

Let P be the sort of processes. Put $P^* = \{f : f: \mathbb{R}_{\geq 0} \to P \text{ such that } f(t) = t \gg f(t)\}$.

We introduce the initial abstraction operator $\surd: T \times P^* \to P^*$, where T is a set of variables ranging over $\mathbb{R}_{\geq 0}$ called *time variables*. In the expression $\surd t.F$, free occurrences of t in F become bound.

2.6 AXIOMS

All axioms of real time process algebra with integration presented before remain valid on the extended domain. On top of the old axioms, we have the extra axioms in table 2, governing the use of initial abstraction. We intend to axiomatise in such a way that all finite process expressions can be written in a form $\surd t.X$, with X not containing initial abstraction. In all axioms in table 2, it is assumed that t is not free in G. Axiom IA0 is the definition of the square bracket notation, axiom IA1 gives α-conversion, axiom IA2 achieves β-conversion, i.e. function application. IA4 gives the embedding of absolute time notation processes in the extended domain. IA5 is an extensionality axiom. We call the theory with axioms in tables 1-2 ACPρ√.

$a[r] = \surd t.\ a(t + r)$	IA0	$(\surd t.F) + G = \surd t.(F + t \gg G)$	IA6
$\surd s.G = \surd t.G[t/s]$	IA1	$(\surd t.F)\cdot G = \surd t.(F\cdot G)$	IA7
$s \gg \surd t.F = s \gg F[s/t]$	IA2	$(\surd t.F) ⌊⌊ G = \surd t.(F ⌊⌊ t \gg G)$	IA8
$\surd t.\ \surd r.\ F = \surd t.\ F[t/r]$	IA3	$G ⌊⌊ (\surd t.F) = \surd t.(t \gg G ⌊⌊ F)$	IA9
$G = \surd t.G$	IA4	$(\surd t.F) \mid G = \surd t.(F \mid t \gg G)$	IA10
$\forall t{\geq}0\ t \gg X = t \gg Y \Rightarrow X = Y$	IA5	$G \mid (\surd t.F) = \surd t.(t \gg G \mid F)$	IA11
		$\partial_H(\surd t.F) = \surd t.\partial_H(F)$	IA12
		$(\surd t.F) \gg r = \surd t.(F \gg r)$	IA13
		$U(F) = U(0 \gg F)$	IA14

TABLE 2. Axioms for initial abstraction.

2.7 EXAMPLES

1. $a(3)\cdot(\surd t.\surd r.a(t + r + 1)) = a(3)\cdot(3 \gg \surd t.\surd r.a(t + r + 1)) = a(3)\cdot(3 \gg \surd r.a(r + 4)) =$ $= a(3)\cdot(3 \gg a(7)) = a(3)\cdot a(7)$.
2. $a(t)\cdot b[r] = a(t)\cdot(\surd s.b(r + s)) = a(t)\cdot(t \gg \surd s.b(r + s)) = a(t)\cdot(t \gg b(r + t)) = a(t)\cdot b(r + t)$.
3. $a(t)\cdot\partial_H(b[r]\cdot X \parallel Y) = a(t)\cdot\partial_H(b(t{+}r)\cdot X \parallel Y)$.
4. For all closed terms $s \gg s \gg X = s \gg X$. From this follows: $\surd t.\ F = \surd t.\ (t \gg F)$.

2.8 RIGIDITY

We call a process X *rigid* iff $X = 0 \gg X$. The rigid processes in P^* correspond to the objects of P, now embedded into P^* by $p \mapsto \lambda t.\ t \gg p$.
On rigid processes, the axioms ATCM2', ATCM3' reduce to their unprimed versions in ACPρ.

2.9 BOOLEANS

A useful feature in the language is the addition of *conditionals* or *guards*. We follow the presentation of [BAB91c]. We use the auxiliary sort BOOL with constants true, false and add the ternary operator
$.\lhd.\rhd.: P^* \times BOOL \times P^* \rightarrow BOOL$.
Here, $x \lhd b \rhd y$ should be read as: if b then x else y. On occasion, we also this operator without the third component, the *guarded command* defined by $b :\rightarrow x = x \lhd b \rhd \delta$.

Axioms for these operators are straightforward (table 3). In the case of expressions involving the initial abstraction operator, we will also encounter Boolean expressions parametrised by time variables, so we turn the sort BOOL into a function domain: $BOOL^* = \{f : f: \mathbb{R}_{\geq 0} \rightarrow BOOL\}$, and we have the abstraction operator $\surd: T \times BOOL^* \rightarrow BOOL^*$. Boolean functions as $\wedge,\vee,\neg$ are defined on $BOOL^*$ pointwise.

The signature has sort $BOOL^*$, constants true, false and functions $\neg: BOOL^* \rightarrow BOOL^*$, $\wedge,\vee$: $BOOL^* \times BOOL^* \rightarrow BOOL^*$, $<,>,\leq,\geq,=: \mathbb{R}_{\geq 0} \times \mathbb{R}_{\geq 0} \rightarrow BOOL^*$, $\gg: \mathbb{R}_{\geq 0} \times BOOL^* \rightarrow BOOL^*$.

$\neg\text{true} = \text{false}$	$\neg\text{false} = \text{true}$
$b \wedge \text{false} = \text{false}$	$b \vee \text{false} = b$
$b \wedge \text{true} = b$	$b \vee \text{true} = \text{true}$
$b \wedge c = c \wedge b$	$b \vee c = c \vee b$
$t \gg \text{true} = \text{true}$	$t \gg \text{false} = \text{false}$
$t \gg \surd r.\, b = t \gg b[t/r]$	
$\neg(\surd t.\, b) = \surd t.\, \neg b$	$(\surd t.\, b) \wedge c = \surd t.\, (b \wedge (t \gg c))$
$(\surd t.\, b) \vee c = \surd t.\, (b \vee (t \gg c))$	
$x \lhd \text{true} \rhd y = x$	$x \lhd \text{false} \rhd y = y$
$b :\rightarrow x = x \lhd b \rhd \delta$	$x \lhd b \rhd y = (b :\rightarrow x) + (\neg b :\rightarrow y)$
$(b \vee c) :\rightarrow x = (b :\rightarrow x) + (c :\rightarrow x)$	
$b :\rightarrow (c :\rightarrow x) = (b \wedge c) :\rightarrow x$	
$x \lhd \surd t.\, b \rhd y = \surd t.\, (t \gg x \lhd b \rhd t \gg y)$	
$t \gg (x \lhd b \rhd y) = t \gg x \lhd t \gg b \rhd t \gg y$	

TABLE 3. Axioms for conditionals.

2.10 EXAMPLE

With the help of Boolean expressions, we can find normal forms for all closed process expressions of the form $\surd t.P$, with P using atomic actions, +, $\cdot$ and conditionals only. Thus, the operators $\|, \lfloor\!\lfloor, |$ can be eliminated. We just give two examples to illustrate this:

1. $a(3) = \surd t.\ (t < 3) :\rightarrow a(3)$.
2. $a(3) \| b[2] = \surd t.\ (a(3) \| b(t+2)) = \surd t.\ t \gg (a(3) \| b(t+2)) =$

$= \surd t.\ ((t < 1) :\rightarrow (b(t+2) \cdot a(3)\ +\ (t = 1) :\rightarrow (a \mid b)(3)\ +\ (t > 1) :\rightarrow a(3) \cdot b(t+2))$

2.11 INTEGRATION

An extension of ACP$\rho\surd$ (called ACP$\rho\surd$I) that is very useful in applications is the extension with the integral operator, denoting a choice over a continuum of alternatives. That is, if V is a subset of $\mathbb{R}_{\geq 0}$, and v is a variable over $\mathbb{R}_{\geq 0}$, then $\int_{v \in V} P$ denotes the alternative composition of alternatives $P[t/v]$ for $t \in V$ (expression P with nonnegative real t substituted for variable v). In this expression, a free occurrence of v in P becomes bound. Alpha conversion must be allowed to avoid name clashes. For more information, we refer the reader to [BAB91a] and [KLU91].

Of course, we cannot give a complete axiomatisation of general integration because the general theory is undecidable, but the axioms in [BAB91a] are very useful in derivations (for a complete axiomatisation of a subtheory, see [KLU91]).

In closed process expressions with a restricted form of integration, we can eliminate the operators $\|$, $\lfloor\!\lfloor$, $|$, ∂_H, $\gg$, $\gg$. [FOK92] explains in detail how such an elimination theorem can be obtained formally. They use the special case of prefixed integration. Generalisation of their results to a setting with initial abstraction seems unproblematic. This leads to normal forms of the form $\surd t.P$ with P using $a(t)$, $\delta(t)$, +, $\cdot$, conditionals and prefixed integration.

2.12 SEMANTICS

Suppose we have an unspecified set of *states* S of cardinality $\leq 2^{\aleph_0}$ (when we use structured operational semantics further on, S will consist of closed process expressions plus the special symbol

$\surd$). S contains a special state $\surd$ (the terminated state). A *real time transition system* over a set of actions A (not including δ) has domain $S \times \mathbb{R}_{\geq 0}$, has two relations on this domain:

$\quad$ idle $\subseteq S \times \mathbb{R}_{\geq 0} \times S \times \mathbb{R}_{\geq 0}$, notation $\langle s, t\rangle \rightarrow \langle s', t'\rangle$

$\quad$ step $\subseteq S \times \mathbb{R}_{\geq 0} \times A \times \mathbb{R}_{\geq 0} \times S \times \mathbb{R}_{\geq 0}$, notation $\langle s, t\rangle \xrightarrow{a(r)} \langle s', t'\rangle$,

and satisfies the following requirements:

0. there is a *root* $\langle s, 0\rangle$ for some $s \in S$; no transition starts from $\langle s', 0\rangle$ when $s' \neq s$
1. if $\langle s, t\rangle \rightarrow \langle s', t'\rangle$, then $t < t'$
2. if $\langle s, t\rangle \xrightarrow{a(r)} \langle s', t'\rangle$, then $t < r = t'$
3. if $\langle s, t\rangle \rightarrow \langle s', t'\rangle$ and $t < t^* < t'$, then there is $s^* \in S$ with $\langle s, t\rangle \rightarrow \langle s^*, t^*\rangle \rightarrow \langle s', t'\rangle$
4. if $\langle s, t\rangle \xrightarrow{a(r)} \langle s', t'\rangle$, and $t < t^* < r$, then there is $s^* \in S$ with $\langle s, t\rangle \rightarrow \langle s^*, t^*\rangle \xrightarrow{a(r)} \langle s', t'\rangle$
5. if $\langle s, t\rangle \rightarrow \langle s^*, t^*\rangle \rightarrow \langle s', t'\rangle$, then $\langle s, t\rangle \rightarrow \langle s', t'\rangle$
6. if $\langle s, t\rangle \rightarrow \langle s^*, t^*\rangle \xrightarrow{a(r)} \langle s', t'\rangle$, then $\langle s, t\rangle \xrightarrow{a(r)} \langle s', t'\rangle$
7. no transition starts from a state $\langle \surd, t\rangle$.

Let us call the set of such transition systems RTTS.

In order to deal with initial abstraction, we have to extend this set to the function space $RTTS^* = \{f : f: \mathbb{R}_{\geq 0} \rightarrow RTTS\}$. We can call elements of $RTTS^*$ *type 1 transition systems.*

2.13 STRUCTURED OPERATIONAL SEMANTICS

Now we interpret the constants and operators of ACP$\rho\surd$I in the semantic domain of type 1 transition systems. We define the mapping $I^*: P^* \rightarrow RTTS^*$ by means of a mapping $I: P \rightarrow RTTS$. We have the identity $\qquad I^*(F) = \lambda t.\ I(t \gg F) \qquad$ (t not free in F),
so it remains to define I. The simplest way to define I is by means of structured operational semantics. As set of states we can take the set of closed process expressions plus $\surd$, the root of $I(x)$ is $\langle x, 0\rangle$ and the transitions are exactly those that can be derived on the basis of the rules in the following table 4. On the basis of these rules, the operators can also be directly defined on real time transition systems in a straightforward way. We have $t,s,r \in \mathbb{R}_{\geq 0}$, $x,x',y,y' \in S - \{\surd\}$, $a,b,c \in A$.

$t<r \Rightarrow \langle a(r), t\rangle \xrightarrow{a(r)} \langle \surd, r\rangle$		
$t<s<r \Rightarrow \langle a(r), t\rangle \rightarrow \langle a(r), s\rangle$		$t<s<r \Rightarrow \langle \delta(r), t\rangle \rightarrow \langle \delta(r), s\rangle$
$\dfrac{\langle x,t\rangle \xrightarrow{a(r)} \langle x',r\rangle}{\langle x+y,t\rangle \xrightarrow{a(r)} \langle x',r\rangle,\ \langle y+x,t\rangle \xrightarrow{a(r)} \langle x',r\rangle}$		$\dfrac{\langle x,t\rangle \xrightarrow{a(r)} \langle \surd,r\rangle}{\langle x+y,t\rangle \xrightarrow{a(r)} \langle \surd,r\rangle,\ \langle y+x,t\rangle \xrightarrow{a(r)} \langle \surd,r\rangle}$
$\dfrac{\langle x,t\rangle \rightarrow \langle x,r\rangle}{\langle x+y,t\rangle \rightarrow \langle x+y,r\rangle,\ \langle y+x,t\rangle \rightarrow \langle y+x,r\rangle}$		
$\dfrac{\langle x,t\rangle \xrightarrow{a(r)} \langle x',r\rangle}{\langle x\cdot y,t\rangle \xrightarrow{a(r)} \langle x'\cdot y,r\rangle}$	$\dfrac{\langle x,t\rangle \rightarrow \langle x,r\rangle}{\langle x\cdot y,t\rangle \rightarrow \langle x\cdot y,r\rangle}$	$\dfrac{\langle x,t\rangle \xrightarrow{a(r)} \langle \surd,r\rangle}{\langle x\cdot y,t\rangle \xrightarrow{a(r)} \langle y,r\rangle}$
$t<r<s \Rightarrow \langle s \gg x, t\rangle \rightarrow \langle s \gg x, r\rangle$		
$\dfrac{\langle x,t\rangle \xrightarrow{a(r)} \langle x',r\rangle,\ r>s}{\langle s \gg x,t\rangle \xrightarrow{a(r)} \langle x',r\rangle}$	$\dfrac{\langle x,t\rangle \rightarrow \langle x,r\rangle}{\langle s \gg x,t\rangle \rightarrow \langle s \gg x,r\rangle}$	$\dfrac{\langle x,t\rangle \xrightarrow{a(r)} \langle \surd,r\rangle,\ r>s}{\langle s \gg x,t\rangle \xrightarrow{a(r)} \langle \surd,r\rangle}$
$\dfrac{\langle x,t\rangle \rightarrow \langle x,r\rangle,\ r<s}{\langle x \gg s,t\rangle \rightarrow \langle x \gg s,r\rangle}$	$\dfrac{\langle x,t\rangle \xrightarrow{a(r)} \langle x',r\rangle,\ r<s}{\langle x \gg s,t\rangle \xrightarrow{a(r)} \langle x',r\rangle}$	$\dfrac{\langle x,t\rangle \xrightarrow{a(r)} \langle \surd,r\rangle,\ r<s}{\langle x \gg s,t\rangle \xrightarrow{a(r)} \langle \surd,r\rangle}$
$\dfrac{\langle x,t\rangle \xrightarrow{a(r)} \langle x',r\rangle,\ a \notin H}{\langle \partial_H(x),t\rangle \xrightarrow{a(r)} \langle \partial_H(x'),r\rangle}$	$\dfrac{\langle x,t\rangle \rightarrow \langle x,r\rangle}{\langle \partial_H(x),t\rangle \rightarrow \langle \partial_H(x),r\rangle}$	$\dfrac{\langle x,t\rangle \xrightarrow{a(r)} \langle \surd,r\rangle,\ a \notin H}{\langle \partial_H(x),t\rangle \xrightarrow{a(r)} \langle \surd,r\rangle}$

$$\frac{\langle x,t\rangle \xrightarrow{a(r)} \langle x',r\rangle,\ \langle y,t\rangle \to \langle y,r\rangle}{\langle x\|y,t\rangle \xrightarrow{a(r)} \langle x'\|y,r\rangle,\ \langle y\|x,t\rangle \xrightarrow{a(r)} \langle y\|x',r\rangle,\ \langle x \mathbin{|\!\!\lfloor} y,t\rangle \xrightarrow{a(r)} \langle x'\|y,r\rangle}$$

$$\frac{\langle x,t\rangle \xrightarrow{a(r)} \langle \surd,r\rangle,\ \langle y,t\rangle \to \langle y,r\rangle}{\langle x\|y,t\rangle \xrightarrow{a(r)} \langle y,r\rangle,\ \langle y\|x,t\rangle \xrightarrow{a(r)} \langle y,r\rangle,\ \langle x \mathbin{|\!\!\lfloor} y,t\rangle \xrightarrow{a(r)} \langle y,r\rangle}$$

$$\frac{\langle x,t\rangle \to \langle x,r\rangle,\ \langle y,t\rangle \to \langle y,r\rangle}{\langle x\|y,t\rangle \to \langle x\|y,r\rangle,\ \langle x \mathbin{|\!\!\lfloor} y,t\rangle \to \langle x \mathbin{|\!\!\lfloor} y,r\rangle}$$

$$\frac{\langle x,t\rangle \xrightarrow{a(r)} \langle x',r\rangle,\ \langle y,t\rangle \xrightarrow{b(r)} \langle y',r\rangle,\ a\,|\,b=c\neq\delta}{\langle x\|y,t\rangle \xrightarrow{c(r)} \langle x'\|y',r\rangle,\ \langle x\,|\,y,t\rangle \xrightarrow{c(r)} \langle x'\|y',r\rangle}$$

$$\frac{\langle x,t\rangle \xrightarrow{a(r)} \langle x',r\rangle,\ \langle y,t\rangle \xrightarrow{b(r)} \langle \surd,r\rangle,\ a\,|\,b=c\neq\delta}{\langle x\|y,t\rangle \xrightarrow{c(r)} \langle x',r\rangle,\ \langle y\|x,t\rangle \xrightarrow{c(r)} \langle x',r\rangle,\ \langle x\,|\,y,t\rangle \xrightarrow{c(r)} \langle x',r\rangle,\ \langle y\,|\,x,t\rangle \xrightarrow{c(r)} \langle x',r\rangle}$$

$$\frac{\langle x,t\rangle \xrightarrow{a(r)} \langle \surd,r\rangle,\ \langle y,t\rangle \xrightarrow{b(r)} \langle \surd,r\rangle,\ a\,|\,b=c\neq\delta}{\langle x\|y,t\rangle \xrightarrow{c(r)} \langle \surd,r\rangle,\ \langle x\,|\,y,t\rangle \xrightarrow{c(r)} \langle \surd,r\rangle}$$

$$\frac{\langle x,t\rangle \xrightarrow{a(r)} \langle x',r\rangle}{\langle x \triangleleft \mathrm{true} \triangleright y,t\rangle \xrightarrow{a(r)} \langle x',r\rangle,\ \langle y \triangleleft \mathrm{false} \triangleright x,t\rangle \xrightarrow{a(r)} \langle x',r\rangle}$$

$$\frac{\langle x,t\rangle \xrightarrow{a(r)} \langle \surd,r\rangle}{\langle x \triangleleft \mathrm{true} \triangleright y,t\rangle \xrightarrow{a(r)} \langle \surd,r\rangle,\ \langle y \triangleleft \mathrm{false} \triangleright x,t\rangle \xrightarrow{a(r)} \langle \surd,r\rangle}$$

$$\frac{\langle x,t\rangle \to \langle x,r\rangle}{\langle x \triangleleft \mathrm{true} \triangleright y,t\rangle \to \langle x \triangleleft \mathrm{true} \triangleright y,r\rangle,\ \langle y \triangleleft \mathrm{false} \triangleright x,t\rangle \to \langle y \triangleleft \mathrm{false} \triangleright x,r\rangle}$$

$$\frac{\langle x(u),t\rangle \to \langle x(u),r\rangle,\ u\in V}{\langle \int_{v\in V} x(v),t\rangle \to \langle \int_{v\in V} x(v),r\rangle} \qquad \frac{\langle x(u),t\rangle \xrightarrow{a(r)} \langle x',r\rangle\ \ u\in V}{\langle \int_{v\in V} x(v),t\rangle \xrightarrow{a(r)} \langle x',r\rangle} \qquad \frac{\langle x(u),t\rangle \xrightarrow{a(r)} \langle \surd,r\rangle\ \ u\in V}{\langle \int_{v\in V} x(v),t\rangle \xrightarrow{a(r)} \langle \surd,r\rangle}$$

TABLE 4. Structured operational semantics.

All that remains is to define the ultimate delay operator U on RTTS*. If $f: \mathbb{R}_{\geq 0} \to$ RTTS, then $U(f) = U(f(0))$, and if $R \in$ RTTS, then $U(R) = \sup\{t \in \mathbb{R}_{\geq 0} : \text{there is } s \in S \text{ with root}(R) \to \langle s, t\rangle\}$.

2.14 BISIMULATIONS

A *bisimulation* on RTTS is a binary relation R on $S \times \mathbb{R}_{\geq 0}$ such that ($a \in A$):

i. whenever $R(\langle s, t\rangle, \langle s', t'\rangle)$ then $t = t'$
ii. for each p and q with R(p, q): if there is a step a(t) possible from p to p', then there is $q' \in S \times \mathbb{R}_{\geq 0}$ such that R(p', q') and there is a step a(t) possible from q to q'.
iii. for each p and q with R(p, q): if there is a step a(t) possible from q to q', then there is $p' \in S \times \mathbb{R}_{\geq 0}$ such that R(p', q') and there is a step a(t) possible from p to p'.

iv. for each p and q with $R(p, q)$: a termination step $a(t)$ to $\langle\surd,t\rangle$ is possible from p iff it is possible from q.
v. for each p and q with $R(p, q)$: if there is an idle step possible from p to p', then there is $q' \in S \times \mathbb{R}_{\geq 0}$ such that $R(p', q')$ and there is an idle step possible from q to q'.
vi. for each p and q with $R(p, q)$: if there is an idle step possible from q to q', then there is $p' \in S \times \mathbb{R}_{\geq 0}$ such that $R(p', q')$ and there is an idle step possible from p to p'.

We say expressions p and q are *bisimilar*, denoted $p \leftrightarrow q$, if there exists a bisimulation with $R(p,q)$. We can show that bisimulation is a congruence relation on real time transition systems, and that real time transition systems modulo bisimulation determine a model for ACPρI. This model also contains solutions of recursive process definitions, but that is an issue that is outside the scope of the present paper. We obtain a model of ACPρI that we will call M_A.

This definition of bisimulation naturally extends to a definition on type 1 real time transition systems: two elements $f,g \in RTTS^*$ are bisimilar if for all $t \geq 0$, $f(t)$ and $g(t)$ are bisimilar. This gives us a model M_A^* of type 1 real time transition systems modulo bisimulation.

2.15 STANDARD INITIALISATION AXIOMS

We present some axioms that are useful in the calculations to come, that hold in the model we described, and that can be derived from the theory for all closed process expressions. Thus, these axioms have the same status as the so-called *Standard Concurrency* axioms of [BEK84].

$t \leq r \Rightarrow t \gg (x \gg r) = (t \gg x) \gg r$	$t \geq r \Rightarrow t \gg (x \gg r) = \delta(t)$
$t \leq r \Rightarrow t \gg (r \gg x) = r \gg x$	$t \geq r \Rightarrow (t \gg x) \gg r = \delta(r)$

TABLE 5. Standard initialisation axioms.

3. THE SYNTAX OF DISCRETE TIME PROCESS ALGEBRA

3.1 SIGNATURE

The signature of discrete time process algebra provides constants and operators parametrised by natural numbers rather than real numbers. We add the following ingredients to real time process algebra.

- constants

$a(n)$	a in the n-th time slice ($a \in A$, $n \in \mathbb{N}$)
$\underline{a}$	a in the current time slice (= $a[1]$)
$\overline{a}[n]$	a in the (n-1)-st time slice after the current one ($n \in \mathbb{N}$)
$\delta(n)$	time stop at the end of the n-th time slice ($n \in \mathbb{N}$)
δ	δ at the end of the current time slice (= $\delta[1]$)
$\bar{\delta}[n]$	δ at the end of the (n-1)-st time slice after the current one ($n \in \mathbb{N}$)
$\underline{a}$	delayable a ([BAB91a])
$\underline{\delta}$	livelock ([BAB91a])
T	discrete time step ([GRO90a])

- functions

$\gg_d 1: P^* \to P^*$	time out in the current time slice
$1 \gg_d: P^* \to P^*$	initialisation in the following time slice
$\sigma_d: P^* \to P^*$	discrete time unit delay
$\sigma^t: P^* \to P^*$	real time delay by time $t \in \mathbb{R}_{\geq 0}$ ([MOT89])
$D: P^* \to BOOL^*$	delayability
$\surd_d: N \times P^* \to P^*$	discrete time initialisation ([HER90])
$\oplus: P^* \times P^* \to P^*$	strong choice
$\lfloor . \rfloor(.): P^* \times P^* \to P^*$	unit delay

$\lfloor . \rfloor^n(.) : P^* \times P^* \to P^*$ start delay within n ($n \in \mathbb{N}$)
$\lfloor . \rfloor^\omega : P^* \to P^*$ unbounded start delay
$\lceil . \rceil^n(.) : P^* \times P^* \to P^*$ execution delay.

The last five operators are taken from ATP [NIS90]. We will call these the Nicollin-Sifakis functions for discrete time process algebra.

3.2 DEFINITIONS

Now we give the definitions of the new signature ingredients in table 6, except for the execution delay $\lceil . \rceil^n(.)$. For $t \in \mathbb{R}$, $\lfloor t \rfloor$ is the floor of t, the largest integer less than or equal to t. We will also on occasion need the largest integer less than t, and define $\lfloor t \rfloor' = \max\{0, \max\{n \in \mathbb{Z} : n < t\}\}$.

$$\underline{a} = \int_{t>0} a(t) \qquad \underline{\delta} = \int_{t>0} \delta(t)$$

$$a(0) = \delta \qquad a(n+1) = n \gg (\underline{a} \gg (n+1)) \qquad \underline{\delta}(n) = \delta(n)$$

$$a[0] = \delta \qquad \underline{a} = a[1] \qquad a[1] = \surd t.\ \underline{a} \gg \lfloor t+1 \rfloor \qquad a[n+2] = \sigma_d(a[n+1])$$

$$\delta[0] = \delta \qquad \underline{\delta} = \delta[1] \qquad \delta[1] = \surd t.\ \delta(\lfloor t+1 \rfloor) \qquad \delta[n+2] = \sigma_d(\delta[n+1])$$

$$T = \surd r.\ t(\lfloor r+1 \rfloor)$$

$$x \gg_d 1 = \surd t.\ (\lfloor t \rfloor \gg x) \gg \lfloor t+1 \rfloor \qquad 1 \gg_d x = \surd t.\ \lfloor t+1 \rfloor' \gg (\lfloor t \rfloor' \gg x)$$

$$\sigma_d(x) = \surd t.\ \lfloor t+1 \rfloor \gg x \qquad \sigma^r(x) = \surd t.\ (t+r) \gg x$$

$$D(x) = \lambda t.\ U(\lfloor t \rfloor \gg x) \geq \lfloor t+1 \rfloor \qquad \surd_d n.\ X = \surd t.\ X(\lfloor t \rfloor / n)$$

$$x \oplus y = \surd t.\ \Big(\lfloor t \rfloor \gg x + y \gg \lfloor t+1 \rfloor + \sum_{i=0}^{\infty} \lfloor t+i+1 \rfloor \gg (\lfloor t \rfloor \gg x + y) \gg \lfloor t+i+2 \rfloor$$

$$\lhd\ D(\lfloor t \rfloor \gg x)(\lfloor t+i \rfloor) \wedge D(\lfloor t \rfloor \gg y)(\lfloor t+i \rfloor) \rhd \underline{\delta} \Big)$$

$$\lfloor x \rfloor(y) = x \gg_d 1 + \sigma_d(y)$$

$$\lfloor x \rfloor^0(y) = y \qquad \lfloor x \rfloor^{n+1}(y) = \lfloor x \rfloor(\lfloor x \rfloor^n(y))$$

$$\lfloor x \rfloor^\omega = \surd t.\ \sum_{i=0}^{\infty} \lfloor t+i \rfloor \gg x \gg \lfloor t+i+1 \rfloor$$

TABLE 6. Discrete time.

3.3 COMMENTS

We will make several comments concerning the definitions in table 6.

1. The definition of T uses a special atomic action t, that satisfies $t \mid t = t$ and $t \mid a = \delta$ for $a \neq t$. T is Groote's time step of [GRO90a]. Thus, Groote's t becomes a 'grote t', avoiding confusion for Dutchmen.

2. We follow [MOT89] by taking $\underline{\delta}$ (0 in [MOT89]) to be a time stop. $\underline{\delta}$ is like δ in [GRO90a] and like nil in [GM91]. We notice that TPL has no constant that introduces a time stop, but $\underline{\delta}$ seems to be equivalent to an infinite τ-loop in TPL [HER90]. Although time stops do not occur in actual computing, we see no objection to their presence in the calculus. So we so not follow ATP's design rationale, which excludes time stops on a priori grounds. $\underline{a}$ is like atomic actions in ATP.

3. We have a (unit) delay operator σ_d rather than an action representing the progress of time. The name σ for the delay operator has been taken from TPL [HER90] where σ is used to name the action that

represents delay. TCCS [MOT89] has a unary operator for delay just like we have. ATP [NIS90] also introduces delay via an operator, be it a binary one ($\lfloor . \rfloor(.)$). $\underline{a}$ models the delayable atoms of TPL.
4. Our + has the interpretation of weak choice in TCCS [MOT89].
5. The notation for the delayability predicate has been taken from TCCS (where it denotes a slightly different operator, however).

3.4 EXECUTION DELAY

In order to give the definition of the execution delay operator $\lceil . \rceil^n(.)$, we first need to define an extra real time operator $\lceil . \rceil^t_\rho(.)$. We want to stick to the ATP notation for the execution delay operator, and so we use a distinguishing subscript ρ for the real time variant. We have the axioms in table 7.

$\lceil a(r) \rceil^t_\rho(x) = a(r) \lhd r < t \rhd t \gg (0 \gg x)$	$\lceil a(r) \cdot x \rceil^t_\rho(y) = a(r) \cdot \lceil r \gg x \rceil^t_\rho(y) \lhd r < t \rhd t \gg (0 \gg y)$
$\lceil x + y \rceil^t_\rho(z) = \lceil x \rceil^t_\rho(z) + \lceil y \rceil^t_\rho(z)$	$\lceil \surd s.\ x \rceil^t_\rho(y) = \lceil 0 \gg x[0/s] \rceil^t_\rho(y)$
$\lceil x \rceil^0(y) = y$	$\lceil x \rceil^{n+1}(y) = \surd r.\ \lceil r \gg x \rceil^{r+n+1}_\rho(\lfloor r+n+1 \rfloor \gg y)$

TABLE 7. Execution delay.

3.5 PROPERTIES

1. $n \gg a(m) = a(m) \lhd n < m \rhd \delta(n)$.
2. $\sigma_d(x) + \sigma_d(y) = \sigma_d(x + y)$
3. $\sigma_d(x \cdot y) = \sigma_d(x) \cdot y$.
4. $\lfloor x \rfloor^\omega = \lfloor x \rfloor(\lfloor x \rfloor^\omega)$
5. $x \oplus y = x + y \gg_d 1 + \sigma_d(1 \gg_d x \oplus 1 \gg_d y) \lhd D(x) \wedge D(y) \rhd \delta$.
6. $\sigma_d(x) \gg_d 1 = \delta$.
7. $\lceil \underline{a} \rceil^{n+1}(x) = \underline{a}$.
8. $\lceil \underline{a} \cdot x \rceil^{n+1}(y) = \underline{a} \cdot \lceil x \rceil^{n+1}(y)$.
9. $\lceil x + y \rceil^{n+1}(z) = \lceil x \rceil^{n+1}(z) + \lceil y \rceil^{n+1}(z)$.
10. $\lceil \sigma_d(x) \rceil^{n+1}(y) = \sigma_d(\lceil x \rceil^n(y))$.
11. $\lceil \lfloor x \rfloor(y) \rceil^{n+1}(z) = \lfloor \lceil x \rceil^{n+1}(z) \rfloor(\lceil y \rceil^n(z))$.

3.6 LIVENESS

We have that the rigid processes in M^*_A correspond to the real time processes involving absolute time notation only. We can also define a notion that gives the processes that can be specified using relative time notation only by means of the following definition.

We define the predicate $\mathcal{L}: M^*_A \to BOOL^*$ as follows: $\mathcal{L}(X) = \lambda t.\ t \gg X \neq \delta(t)$.

We call a process X *live* if $\mathcal{L}(X) = \lambda t.$ true. We obtain the following properties:

1. $\mathcal{L}(a[0]) = \mathcal{L}(\delta[0]) = \mathcal{L}(a(0)) = \mathcal{L}(\delta(0)) = \lambda t.$ false
2. $\mathcal{L}(a[n+1]) = \mathcal{L}(\delta[n+1]) = \lambda t.$ true
3. $\mathcal{L}(a(n+1)) = \mathcal{L}(\delta(n+1)) = \lambda t.\ t < n+1$
4. $\mathcal{L}(a(r)) = \mathcal{L}(\delta(r)) = \lambda t.\ t < r$
5. $\mathcal{L}(x + y) = \lambda t.\ \mathcal{L}(x)(t) \vee \mathcal{L}(y)(t)x$
6. $\mathcal{L}(x \cdot y) = \mathcal{L}(x)$
7. $\mathcal{L}(x \parallel y) = \mathcal{L}(x \mathbin{⫫} y) = \mathcal{L}(x \mid y) = \lambda t.\ \mathcal{L}(x)(t) \wedge \mathcal{L}(y)(t)$
8. $\mathcal{L}(\int_{v \in V} X) = \lambda t.\ \bigvee_{v \in V} \mathcal{L}(X)(t)$.

In the previous sections, we have sketched a theory ACPρ√I and a model M^*_A that contain both real time and discrete time processes. Now we will concentrate on the reduct of the theory and the subalgebra of the model that contain only discrete time processes. For this reason, we introduce the discretisation operator $\mathcal{D}$.

$\mathcal{D}(a(t)) = a(\lfloor t+1 \rfloor)$	
$\mathcal{D}(x + y) = \mathcal{D}(x) + \mathcal{D}(y)$	$\mathcal{D}(x \cdot y) = \mathcal{D}(x) \cdot \mathcal{D}(y)$
$\mathcal{D}(\int_{v \in V} X) = \int_{v \in V} \mathcal{D}(X)$	$\mathcal{D}(\surd t.\ X) = \surd t.\ \mathcal{D}(X)$

TABLE 8. Discretisation.

3.7 DEFINITION

First of all, we define discretisation on P^*. We have the axioms in table 8.

3.8 PROPERTIES

1. $\mathcal{D}$ does not distribute over $\|$, $\lfloor\!\lfloor$, $|$.
2. $\mathcal{D}(\partial_H(x)) = \partial_H(\mathcal{D}(x))$
3. $\lfloor U(x) \rfloor \leq U(\mathcal{D}(x)) \leq \lfloor U(x) + 1 \rfloor$. Here, both extremes are possible.

3.9 DEFINITION

We say that a process x is *discretised*, $\mathcal{DIS}(x)$, if there is a process $y \in P^*$ such that $\mathcal{D}(y) = x$. Also, we define this concept on the model M_A^*. Let $f \in M_A^*$. Then f is discretised, $f \in \mathcal{DIS}M_A^*$, if:

i. for all $t \geq 0$, all actions of $f(t)$ have a timestamp not in $\mathbb{N}$.

ii. for all $t \geq 0$ we have the following: let σ be a sequence of actions in $f(t)$ of the form $a_1(r_1)\, a_2(r_2) \ldots a_n(r_n)$ with $r_1 < r_2 < \ldots < r_n < r$. Let in $f(t)$ the sequence of actions σ from the root be followed by the step $p \xrightarrow{a(r)} q$. Then:

 a. for each r' with $r < r' < \lfloor r+1 \rfloor$ there is a q' with $p \xrightarrow{a(r')} q'$ and $q' \underline{\leftrightarrow}\, r' \gg q$.

 b. for each r' with $\max(r_n, \lfloor r \rfloor) < r' < r$ there is a q' with $p \xrightarrow{a(r')} q'$ and $q \underline{\leftrightarrow}\, r \gg q'$.

3.10 EXAMPLES

1. $\mathcal{DIS}(a(n))$, $\mathcal{DIS}(a[n])$, $\mathcal{DIS}(\delta(n))$, $\mathcal{DIS}(\delta[n])$, $\neg\mathcal{DIS}(a(t))$.
2. $\mathcal{DIS}(x) \Rightarrow \mathcal{DIS}(\partial_H(x)), \mathcal{DIS}(\sigma_d(x)), \mathcal{DIS}(1 \gg_d x), \mathcal{DIS}(x \gg_d 1)$.
3. $\mathcal{DIS}(x) \wedge \mathcal{DIS}(y) \Rightarrow \mathcal{DIS}(x+y), \mathcal{DIS}(x \cdot y), \mathcal{DIS}(x \| y), \mathcal{DIS}(x \lfloor\!\lfloor y), \mathcal{DIS}(x \mid y)$.
4. for all $v \in V$ $\mathcal{DIS}(X) \Rightarrow \mathcal{DIS}(\int_{v \in V} X)$.
5. $\mathcal{DIS}(x) \wedge \neg D(x) \Rightarrow x = x \gg_d 1$.
6. $\mathcal{DIS}(x) \wedge D(x) \Rightarrow x = x \gg_d 1 + \sigma_d(1 \gg_d x)$.
7. $\mathcal{DIS}(x) \Rightarrow \mathcal{L}(x)$
8. $\mathcal{DIS}(x) \Rightarrow \underline{a} \lfloor\!\lfloor x = \underline{\ddot{a}} \cdot x$
9. $\mathcal{DIS}(x) \wedge \mathcal{DIS}(y) \Rightarrow \underline{a} \cdot x \lfloor\!\lfloor y = \underline{a} \cdot (x \| y)$
10. $\mathcal{DIS}(x) \wedge \overline{\mathcal{DIS}}(y) \wedge \bar{D}(y) \Rightarrow \sigma_d(x) \lfloor\!\lfloor y = \sigma_d(x \lfloor\!\lfloor 1 \gg_d y)$.

The remarkable point is that $\mathcal{L}(\underset{\sim}{\delta})$. Thus, the discretisation of δ is live. The discretisation of δ is certainly not equal to $\delta = \delta(0)$.

3.11 EQUIVALENT MODEL

We can define a model that is conceptually much simpler than $\mathcal{DIS}M_A^*$, but contains an isomorphic copy of it. This is the model $G/\underline{\leftrightarrow}$ that we will use in the following chapter.

We define a set of process graphs as in [BAW90] with labels from $A \cup \{\sigma\}$ satisfying two extra conditions:

i. every node has *at most one* outgoing σ-labeled edge;

ii. a σ-labeled edge may not lead to a termination node.

Let G be the set of such process graphs with cardinality $\leq 2^{\aleph_0}$. To state this precisely, an element of G is a quadruple $\langle N, E, r, T \rangle$ where N is the set of *nodes*, $E \subseteq N \times A \cup \{\sigma\} \times N$ is the set of *edges*, $r \in N$ is the *root node*, and $T \subseteq N$ is the set of *termination nodes*. We will always have that a termination node has no outgoing edges. A node without outgoing edges that is not a termination node is called a *deadlock node*.

We define a mapping from $\mathcal{DIS}M_A^*$ to $G/\underline{\leftrightarrow}$ by defining a mapping ϕ from real time transition systems to process graphs. Let R be a real time transition system. For simplicity, we assume that R is actually a tree, so each node has at most one incoming transition. If $\langle r, 0 \rangle$ is the root of R ($r \in S$), we take r as the root of $\phi(R)$. The set of states of $\phi(R)$ is S, the set of states of R. Suppose $\langle r, 0 \rangle \xrightarrow{a(t)} \langle s, t \rangle$ is a transition in R from the root. In $\phi(R)$, we take from the root then $\lfloor t \rfloor$ σ-transitions followed by an a-transition to state s. In case of several transitions in R from the root, we share σ-transitions in $\phi(R)$ as much as possible. We proceed likewise for the other states: if $\langle s, u \rangle \xrightarrow{a(t)} \langle s', t \rangle$ is a transition in R, we take from s in $\phi(R)$ $\lfloor t \rfloor - \lfloor u \rfloor$ σ-transitions followed by an a-transition to s'. This describes *f*. We then obtain a mapping from $\mathcal{DIS}M_A^*$ to $G/\underline{\leftrightarrow}$ by mapping a representative of $f \in \mathcal{DIS}M_A^*$ to $R(f(0))/\underline{\leftrightarrow}$.

The inverse mapping can be defined along the same lines. We identify the domain of G with some subset of S, the fixed domain of states that was mentioned in 2.12. If r is the root of $g \in G$, and $r \xrightarrow{\sigma} \dots \xrightarrow{\sigma} \xrightarrow{a}$ s a sequence of transitions with n σ-steps, add transitions $\langle r, 0\rangle \xrightarrow{a(t)} \langle s, t\rangle$ for each t between n and n+1. Next, if node s of g corresponds to node $\langle s, u\rangle$, and $s \xrightarrow{a} s'$ is a transition in g, add transitions $\langle s, u\rangle \xrightarrow{a(t)} \langle s', t\rangle$ for each t between u and $\lfloor u+1 \rfloor$. On the other hand, if $s \xrightarrow{\sigma} \dots \xrightarrow{\sigma} \xrightarrow{a} s'$ is a sequence of transitions in g with n σ-steps, $n \geq 1$, add transitions $\langle s, u\rangle \xrightarrow{a(t)} \langle s', t\rangle$ for each t between $\lfloor u+n \rfloor$ and $\lfloor u+n+1 \rfloor$.
We leave the verification that this indeed defines an isomorphism to the reader.

4. DISCRETE TIME PROCESS ALGEBRA

Now we will incrementally define axiomatisations for discrete time subalgebras of our theory.

4.1 BPA_{dt}

We start from:

- a unary operator for (unit) delay like in TCCS [MOT89];
- time factorisation as in ATP [NIS90], TCCS [MOT89], TPCCS [HAN91] and TPL [HER90];
- the interpretation of + as the weak choice of TCCS;
- a significant difference with TPL [HER90] because visible actions cannot idle (here, we follow TCCS, ACP_t and ATP).

The signature of BPA_{dt} has constants a ($a \in A$), functions $+, \cdot : P \times P \to P$ and the discrete time unit delay σ_d: $P \to P$. The axioms of BPA_{dt} are A1-5 and DT1,2 in table 9. Axioms DT1,2 appear in [BAB91b]. They can be derived from the theory in the previous section.

4.2 INACTION

$BPA_{\delta dt}$ is obtained by introducing δ as a constant representing inaction (time stop at the end of the current time slice). The axioms for δ are standard (table 9).

Next, we add parallel composition. We start by describing a system with merge without communication, a so-called free merge.

4.3 $PA_{\delta dt}$

We start by adding to the signature of $BPA_{\delta dt}$ the extra sort BOOL with constants and functions as 2.9. We add the initialisation in the following time slice, $1 \gg_d .: P \to P$, and the delayability predicate D: $P \to$ BOOL. Axioms for this additional syntax are in table 3 (those without the √ operator). They already appear in [BAB91b]. Then, the merge with its auxiliary operator left merge is introduced. Axioms for this theory are UTB1-4, DEL1-4, CM1 without third summand and CM2-4 in table 9. Axiom DTM1 is from [BAB91b].

Our definition of merge is similar to the definitions in TCCS, ATP, ACP_t and TPL (the correspondence with the TPL definition holds in the absence of communication; in the presence of communication TPL's merge will give priority to internal communications). So it follows that without communication the relation between merge and discrete time is not controversial. We depart from ATP by allowing time stops. Indeed, in our setting $\sigma_d(a \cdot b) \| \delta$ is unable to perform a σ-step. This is a matter of taste. The presence of time stops seems not to spoil the algebra at all.

4.4 COMMUNICATION

If we add communication, we have a given *communication function* $| : A_\delta \times A_\delta \to A_\delta$ as an extra parameter of the theory. This communication function is required to be commutative, associative and has δ as a neutral element (axioms C1,2,3 of table 1 in 2.3). Furthermore, in order to axiomatise the merge, we have an additional auxiliary operator | (communication merge). The DTM axioms are from

[BAB91b]. Encapsulation ∂_H: $P \rightarrow P$ for each $H \subseteq A$ has the same equations as in ACP plus the additional axiom DTM7.

At this point, some further motivation is needed. The merge of ACP_{dt} works just as the merge of TCCS [MOT89] (taking weak choice of TCCS for +). It is also equivalent to the merge of ATP. It differs from merge in [GRO90a], set up in ACP_t. In fact, ACP_t contains axioms $\sigma \mathbin{\|\!\_} x = \delta$ and $\sigma x \mathbin{\|\!\_} y = \delta$. Our main objection against these axioms (which occur in ATP as well) is that they render it impossible to injectively embed ACP_t into $ACP\rho$ of [BAB91a].

It follows that the only disagreement in the literature is about the proper semantics of left merge and communication merge. Therefore, our choice for the merge operator itself seems reasonably well motivated.

$x + y = y + x$	A1	$\underline{a} \mid \underline{b} = \underline{c}$	if $a \mid b = c$	C4
$(x + y) + z = x + (y + z)$	A2			
$x + x = x$	A3	$x \| y = x \mathbin{\|\!\_} y + y \mathbin{\|\!\_} x + x \mid y$		CM1
$(x + y) \cdot z = x \cdot z + y \cdot z$	A4	$\underline{a} \mathbin{\|\!\_} x = \underline{a} \cdot x$		CM2
$(x \cdot y) \cdot z = x \cdot (y \cdot z)$	A5	$(\underline{a} \cdot x) \mathbin{\|\!\_} y = \underline{a} \cdot (x \| y)$		CM3
$x + \underline{\delta} = x$	A6	$(x + y) \mathbin{\|\!\_} z = x \mathbin{\|\!\_} z + y \mathbin{\|\!\_} z$		CM4
$\underline{\delta} \cdot x = \underline{\delta}$	A7	$\sigma_d(x) \mathbin{\|\!\_} y = \sigma_d(x \mathbin{\|\!\_} (1 \gg_d y)) \lhd D(y) \rhd \underline{\delta}$		DTM1
		$(\underline{a} \cdot x) \mid \underline{b} = (\underline{a} \mid \underline{b}) \cdot x$		CM5
$\sigma_d(x) + \sigma_d(y) = \sigma_d(x + y)$	DT1	$\underline{a} \mid (\underline{b} \cdot x) = (\underline{a} \mid \underline{b}) \cdot x$		CM6
$\sigma_d(x) \cdot y = \sigma_d(x \cdot y)$	DT2	$(\underline{a} \cdot x) \mid (\underline{b} \cdot y) = (\underline{a} \mid \underline{b}) \cdot (x \| y)$		CM7
		$(x+y) \mid z = x \mid z + y \mid z$		CM8
$1 \gg_d \underline{a} = \underline{\delta}$	UTB1	$x \mid (y+z) = x \mid y + x \mid z$		CM9
$1 \gg_d (x + y) = 1 \gg_d x + 1 \gg_d y$	UTB2	$\underline{a} \mid \sigma_d(x) = \underline{\delta}$		DTM2
$1 \gg_d (x \cdot y) = (1 \gg_d x) \cdot y$	UTB3	$\sigma_d(x) \mid \underline{a} = \underline{\delta}$		DTM3
$1 \gg_d \sigma_d(x) = x$	UTB4	$(\underline{a} \cdot x) \mid \sigma_d(y) = \underline{\delta}$		DTM4
		$\sigma_d(x) \mid (\underline{a} \cdot y) = \underline{\delta}$		DTM5
$D(\underline{a}) = \text{false}$	DEL1	$\sigma_d(x) \mid \sigma_d(y) = \sigma_d(x \mid y)$		DTM6
$D(x + y) = D(x) \vee D(y)$	DEL2			
$D(x \cdot y) = D(x)$	DEL3	$\partial_H(\underline{a}) = \underline{a}$	if $a \notin H$	D1
$D(\sigma_d(x)) = \text{true}$	DEL4	$\partial_H(\underline{a}) = \underline{\delta}$	if $a \in H$	D2
		$\partial_H(x + y) = \partial_H(x) + \partial_H(y)$		D3
		$\partial_H(x \cdot y) = \partial_H(x) \cdot \partial_H(y)$		D4
		$\partial_H(\sigma_d(x)) = \sigma_d(\partial_H(x))$		DTM7

TABLE 9. ACP_{dt}.

4.6 STRUCTURED OPERATIONAL SEMANTICS

The semantic domain for BPA_{dt} consists of process graphs as defined in 3.11. We will map every closed term to a graph in $\mathbb{G}$ by using a structured operational semantics. We can give a transition system specification by means of the rules in table 10 ($a \in A$). The operational meaning of $\underline{\delta}$ is a process that allows no step whatsoever.

Now we can show that these rules give rise to a unique transition relation on closed terms. This can be done with the theory of [GRO90b].

Bisimulation is defined as usual, so a symmetric binary relation R on process terms is a bisimulation iff the following transfer conditions hold:

i. if R(p,q) and $p \xrightarrow{u} p'$ ($u \in A \cup \{\sigma\}$), then there is q' such that $q \xrightarrow{u} q'$ and R(s',t');

ii. if R(p,q), then $p \xrightarrow{a} \surd$ ($a \in A$) iff $q \xrightarrow{a} \surd$.

Two terms p,q are bisimilar, $p \underline{\leftrightarrow} q$, if there exists a bisimulation relating them. We have that the set of process terms modulo bisimulation is a model for ACP_{dt}, and moreover, the axiomatisation of ACP_{dt} is sound and complete for the set of transition graphs modulo bisimulation.

$$\underline{a} \xrightarrow{a} \surd \qquad\qquad \sigma_d(x) \xrightarrow{\sigma} x$$

$$\frac{x \xrightarrow{a} x'}{x \cdot y \xrightarrow{a} x' \cdot y} \qquad \frac{x \xrightarrow{a} \surd}{x \cdot y \xrightarrow{a} y} \qquad \frac{x \xrightarrow{\sigma} x'}{x \cdot y \xrightarrow{\sigma} x' \cdot y}$$

$$\frac{x \xrightarrow{a} x'}{x+y \xrightarrow{a} x', \; y+x \xrightarrow{a} x'} \qquad \frac{x \xrightarrow{a} \surd}{x+y \xrightarrow{a} \surd, \; y+x \xrightarrow{a} \surd}$$

$$\frac{x \xrightarrow{\sigma} x', \; y \xrightarrow{\sigma} y'}{x+y \xrightarrow{\sigma} x'+y'} \qquad \frac{x \xrightarrow{\sigma} x', \; y \xrightarrow{\sigma} \!\!\!\!/}{x+y \xrightarrow{\sigma} x', \; y+x \xrightarrow{\sigma} x'}$$

$$\frac{x \xrightarrow{a} x'}{x \| y \xrightarrow{a} x' \| y, \; y \| x \xrightarrow{a} y \| x', \; x \mathbin{\underline{\|}} y \xrightarrow{a} x' \| y} \qquad \frac{x \xrightarrow{a} \surd}{x \| y \xrightarrow{a} y, \; y \| x \xrightarrow{a} y, \; x \mathbin{\underline{\|}} y \xrightarrow{a} y}$$

$$\frac{x \xrightarrow{\sigma} x', \; y \xrightarrow{\sigma} y'}{x \| y \xrightarrow{\sigma} x' \| y', \; x \mathbin{\underline{\|}} y \xrightarrow{\sigma} x' \mathbin{\underline{\|}} y', \; x \mid y \xrightarrow{\sigma} x' \mid y'} \qquad \frac{x \xrightarrow{a} x', \; y \xrightarrow{b} y', \; a \mid b=c}{x \| y \xrightarrow{c} x' \| y', \; x \mid y \xrightarrow{c} x' \| y'}$$

$$\frac{x \xrightarrow{a} x', \; y \xrightarrow{b} \surd, \; a \mid b=c}{x \| y \xrightarrow{c} x', \; x \mid y \xrightarrow{c} x', \; y \| x \xrightarrow{c} x', \; y \mid x \xrightarrow{c} x'} \qquad \frac{x \xrightarrow{a} \surd, \; y \xrightarrow{b} \surd, \; a \mid b=c}{x \| y \xrightarrow{c} \surd, \; x \mid y \xrightarrow{c} \surd}$$

$$\frac{x \xrightarrow{\sigma} y, \; y \xrightarrow{a} z}{1 \gg_d x \xrightarrow{a} z} \qquad \frac{x \xrightarrow{\sigma} y, \; y \xrightarrow{a} \surd}{1 \gg_d x \xrightarrow{a} \surd} \qquad \frac{x \xrightarrow{\sigma} y, \; y \xrightarrow{\sigma} z}{1 \gg_d x \xrightarrow{\sigma} z}$$

$$\frac{x \xrightarrow{a} x', \; a \notin H}{\partial_H(x) \xrightarrow{a} \partial_H(x')} \qquad \frac{x \xrightarrow{a} \surd, \; a \notin H}{\partial_H(x) \xrightarrow{a} \surd} \qquad \frac{x \xrightarrow{\sigma} x'}{\partial_H(x) \xrightarrow{\sigma} \partial_H(x')}$$

TABLE 10. Operational semantics of ACP_{dt}.

4.7 SYNTAX

The syntax of ACP_{dt} is not tailored to practical application. The operators are too low level. We will extend ACP_{dt} with ATP operators in order to extend the expressive power of the language. In addition, we need an auxiliary operator $\_ \gg_d 1$. We have functions

$$\gg_d 1 : P \to P \qquad\qquad \text{time out in the current time slice}$$

$\oplus: P \times P \to P$	strong choice
$\lfloor . \rfloor(.): P \times P \to P$	unit delay
$\lfloor . \rfloor^n(.): P \times P \to P$	start delay within n ($n \in \mathbb{N}$)
$\lfloor . \rfloor^\omega: P \to P$	unbounded start delay
$\lceil . \rceil^n(.): P \times P \to P$	execution delay.

The last five functions are the Nicollin-Sifakis functions.

4.8 AXIOMS

Axioms for these auxiliary operators are in table 11. Axioms UTB5-8 and ATP1,2 are taken from [BAB91b]. For ATP1,3,6 we refer to table 7, the other axioms can be shown to hold in the theory of section 3.

$\underline{a} \gg_d 1 = \underline{a}$	UTB5	$(x \cdot y) \gg_d 1 = (x \gg_d 1) \cdot y$	UTB7
$(x+y) \gg_d 1 = x \gg_d 1 + y \gg_d 1$	UTB6	$\sigma_d(x) \gg_d 1 = \underline{\delta}$	UTB8
$\lfloor x \rfloor(y) = x \gg_d 1 + \sigma_d(y)$			ATP1
$x \oplus y = x+y \gg_d 1 + \sigma_d(1 \gg x \oplus 1 \gg y) \lhd D(x) \wedge D(y) \rhd \underline{\delta}$			ATP2
$\lfloor x \rfloor^0(y) = y$	ATP3	$\lfloor x \rfloor^{n+1}(y) = \lfloor x \rfloor(\lfloor x \rfloor^n(y))$	ATP4
$\lfloor x \rfloor^\omega = \lfloor x \rfloor(\lfloor x \rfloor^\omega)$	ATP5	$\lceil x \rceil^0(y) = y$	ATP6
$\lceil \underline{a} \rceil^{n+1}(x) = \underline{a}$	ATP7	$\lceil x + y \rceil^{n+1}(z) = \lceil x \rceil^{n+1}(z) + \lceil y \rceil^{n+1}(z)$	ATP9
$\lceil \underline{a} \cdot x \rceil^{n+1}(y) = \underline{a} \cdot \lceil x \rceil^{n+1}(y)$	ATP8	$\lceil \sigma_d(x) \rceil^{n+1}(y) = \sigma_d(\lceil x \rceil^n(y))$	ATP10

TABLE 11. Additional axioms for Nicollin-Sifakis functions.

4.9 OPERATIONAL SEMANTICS

For the new operators, we have the following operational rules (table 12).

4.10. ABSTRACTION

An advantage of concentrating on the subtheory ACP_{dt} + ATP is that it is easier to add the silent step τ. In real time, this is still under investigation (see [KLU92]), but on our discrete time subtheory, it is relatively straightforward. We will not interpret the constant τ in real time. Clearly, that is an option, but it requires real time abstraction to be worked out. (Besides [KLU92], other approaches exist.)

We add a constant $\underline{\tau}$ ($\tau \notin A$), and for each $I \subseteq A$ an operator τ_I: $P \to P$. Further, we have an operator τ_σ that substitutes $\underline{\tau}$ for time steps. All axioms in tables xxx are now valid for $a \in A \cup \{\delta, \tau\}$. Additional axioms are shown in table 13 (where also $a \in A \cup \{\delta, \tau\}$). Here, we follow the presentation of the silent step according to *branching bisimulation semantics* (see [BAW90]). Alternatively, it is possible to present an axiomatisation of the silent step in *weak bisimulation semantics* (see [MIL89]).

4.11SEMANTICS

The operational semantics is obtained by using the rules of tables xxx also in case $a = \tau$, but dividing out a different congruence relation. Let E be the set of process terms enriched with the extra element $\surd$. Write $x \Rightarrow y$ for elements of E if either $x=y$ or there is path from x to y (generated by the operational rules) that only contains τ-labels.

A *branching bisimulation* is a symmetric binary relation R on E with the following transfer property:

if $R(x,y)$ and $x \xrightarrow{a} x'$ ($a \in A \cup \{\tau\}$), then either $a=\tau$ and $R(x',y)$,or there are y^*, y' in E such that $y \Rightarrow y^* \xrightarrow{a} y'$ and $R(x,y^*)$, $R(x',y')$.

$\dfrac{x+y \xrightarrow{a} z}{x\oplus y \xrightarrow{a} z}$	$\dfrac{x+y \xrightarrow{a} \surd}{x\oplus y \xrightarrow{a} \surd}$	$\dfrac{x \xrightarrow{\sigma} x',\ y \xrightarrow{\sigma} y'}{x\oplus y \xrightarrow{\sigma} x'\oplus y'}$
$\lfloor x\rfloor(y) \xrightarrow{\sigma} y$	$\dfrac{x \xrightarrow{a} x'}{\lfloor x\rfloor(y) \xrightarrow{a} x'}$	$\dfrac{x \xrightarrow{a} \surd}{\lfloor x\rfloor(y) \xrightarrow{a} \surd}$
$\dfrac{y \xrightarrow{a} y'}{\lfloor x\rfloor^0(y) \xrightarrow{a} y'}$	$\dfrac{y \xrightarrow{a} \surd}{\lfloor x\rfloor^0(y) \xrightarrow{a} \surd}$	$\dfrac{y \xrightarrow{\sigma} y'}{\lfloor x\rfloor^0(y) \xrightarrow{\sigma} y'}$
$\lfloor x\rfloor^{n+1}(y) \xrightarrow{\sigma} \lfloor x\rfloor^n(y)$	$\dfrac{x \xrightarrow{a} x'}{\lfloor x\rfloor^{n+1}(y) \xrightarrow{a} x'}$	$\dfrac{x \xrightarrow{a} \surd}{\lfloor x\rfloor^{n+1}(y) \xrightarrow{a} \surd}$
$\lfloor x\rfloor^\omega \xrightarrow{\sigma} \lfloor x\rfloor^\omega$	$\dfrac{x \xrightarrow{a} x'}{\lfloor x\rfloor^\omega \xrightarrow{a} x'}$	$\dfrac{x \xrightarrow{a} \surd}{\lfloor x\rfloor^\omega \xrightarrow{a} \surd}$
$\dfrac{y \xrightarrow{a} y'}{\lceil x\rceil^0(y) \xrightarrow{a} y'}$	$\dfrac{y \xrightarrow{a} \surd}{\lceil x\rceil^0(y) \xrightarrow{a} \surd}$	$\dfrac{y \xrightarrow{\sigma} y'}{\lceil x\rceil^0(y) \xrightarrow{\sigma} y'}$
$\dfrac{x \xrightarrow{a} x'}{\lceil x\rceil^{n+1}(y) \xrightarrow{a} \lceil x'\rceil^{n+1}(y)}$	$\dfrac{x \xrightarrow{a} \surd}{\lceil x\rceil^{n+1}(y) \xrightarrow{a} \surd}$	$\dfrac{x \xrightarrow{\sigma} x'}{\lceil x\rceil^{n+1}(y) \xrightarrow{\sigma} \lceil x'\rceil^n(y)}$

TABLE 12. Operational semantics of Nicollin-Sifakis functions.

$x\underline{\tau} = x$	B1	$\underline{\tau} \mid \underline{a} = \underline{\delta}$		C4
$x(\underline{\tau}(y+z)+y) = x(y+z)$	B2			
		$\tau_I(\underline{a}) = \underline{a}$	if $a \notin I$	TI1
$\tau_\sigma(\underline{a}) = \underline{a}$	TS1	$\tau_I(\underline{a}) = \underline{\tau}$	if $a \in I$	TI2
$\tau_\sigma(x+y) = \tau_\sigma(x) + \tau_\sigma(y)$ if $D(x) \vee D(y) = \text{false}$	TS2	$\tau_I(x+y) = \tau_I(x) + \tau_I(y)$		TI3
$\tau_\sigma(x \cdot y) = \tau_\sigma(x) \cdot \tau_\sigma(y)$	TS3	$\tau_I(x \cdot y) = \tau_I(x) \cdot \tau_I(y)$		TI4
$\tau_\sigma(\sigma_d(x)) = \underline{\tau} \cdot \tau_\sigma(x)$	TS4	$\tau_I(\sigma_d(x)) = \sigma_d(\tau_I(x))$		TI8

TABLE 13. Axioms of ACP^{τ}_{dt}.

We say that two process terms are *branching bisimilar*, $x \leftrightarrow_b y$, if there is a branching bisimulation R on E with R(x,y).
Now let R be a branching bisimulation on E. We say that x,y *satisfy the root condition* if we have:

if R(x,y) and $x \xrightarrow{a} x'$ $(a \in A \cup \{\tau\})$, then there is y' in E such that $y \xrightarrow{a} y'$ and R(x',y').

Then, we say that two process terms are *rooted branching bisimilar*, $x \leftrightarrow_{rb} y$, if there exists a branching bisimulation R such that:

i. x,y satisfy the root condition;
ii. whenever R(x',y') and there is a path from x to x' with the last transition a σ-transition, then x',y' satisfy the root condition.

4.12 FAIR ABSTRACTION

Besides KFAR and CFAR (see [BAW90]), there is a rule for time:

$\dfrac{x = \sigma_d(x) + y}{\tau\cdot\tau_\sigma(x) = \tau\cdot\tau_\sigma(y)}$	σFAR

TABLE 14. Time abstraction.

Using Nicollin-Sifakis functions, the fair abstraction principle can be formulated as follows:

$$\frac{x = \lfloor y \rfloor(x)}{\tau\cdot\tau_\sigma(x) = \tau\cdot\tau_\sigma(y)}, \text{ or alternatively: } \qquad \tau\cdot\tau_\sigma(\lfloor x \rfloor^\omega) = \tau\cdot\tau_\sigma(x) \qquad \sigma\text{FAR}.$$

5. EXAMPLE.

We describe the alternating bit protocol with time out of [NIS90]. Put $B = \{0,1\}$.

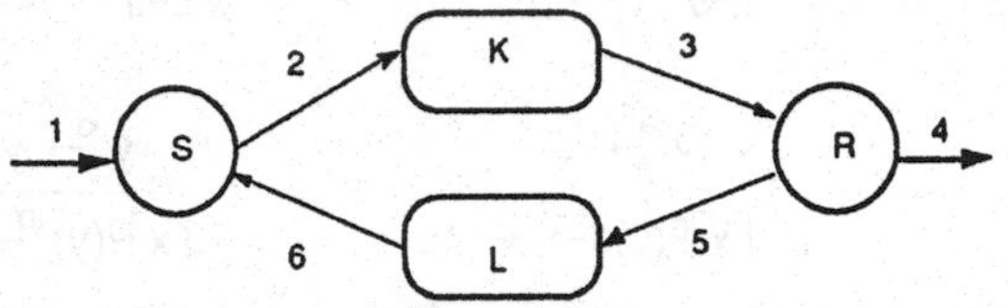

FIGURE 1.

Sender: $\quad S = S0\cdot S1\cdot S$

$$Sb = \sum_{d\in D} \lfloor r1(d) \rfloor^\omega \cdot Sb_d \qquad Sb_d = s2(db)\cdot\lfloor r6(b) + r6(1-b)\cdot Sb_d \rfloor(Sb_d) \qquad (b=0,1,\ d\in D).$$

Unreliable data transmission channel: $K = \sum_{f\in D\times B} \lfloor r2(f) \rfloor^\omega\cdot(i\cdot K + i\cdot s3(f)\cdot K)$.

Unreliable acknowledgement transmission channel: $L = \sum_{b\in B} \lfloor r5(b) \rfloor^\omega\cdot(i\cdot L + i\cdot s6(b)\cdot L)$.

Receiver: $\quad R = R1\cdot R0\cdot R$

$$Rb = \sum_{d\in D} \lfloor r3(db) \rfloor^\omega\cdot s5(b)\cdot Rb + \sum_{d\in D} \lfloor r3(d(1-b)) \rfloor^\omega\cdot s4(d)\cdot s5(1-b).$$

Then the protocol is described by: $\quad ABP = \tau_\sigma\circ\tau_I\circ\partial_H(S \,\|\, K \,\|\, L \,\|\, R)$, where
$H = \{ri(x), si(x) \mid i\in\{2,3,5,6\}, x\in\{0,1\}\cup D\times\{0,1\}\}$, $I = \{ci(x) \mid i\in\{2,3,5,6\}, x\in\{0,1\}\cup D\times\{0,1\}\}\cup\{i\}$.
We can derive that the process $X = \partial_H(S \,\|\, K \,\|\, L \,\|\, R)$ satisfies the following recursive specification:

$$X = \sum_{d\in D} \lfloor r1(d) \rfloor^\omega\cdot X^1_d \qquad Y = \sum_{d\in D} \lfloor r1(d) \rfloor^\omega\cdot Y^1_d$$

$$X^1_d = c2(d0)\cdot\left(i\cdot\sigma_d(X^1_d) + i\cdot c3(d0)\cdot s4(d)\cdot X^2_d\right) \qquad Y^1_d = c2(d1)\cdot\left(i\cdot\sigma_d(Y^1_d) + i\cdot c3(d1)\cdot s4(d)\cdot Y^2_d\right)$$

$$X^2_d = c5(0)\cdot(i\cdot\sigma_d(X^3_d) + i\cdot c6(0)\cdot Y) \qquad Y^2_d = c5(1)\cdot(i\cdot\sigma_d(Y^3_d) + i\cdot c6(1)\cdot X)$$

$$X^3_d = c2(d0)\cdot(i\cdot\sigma_d(X^3_d) + i\cdot c3(d0)\cdot X^2_d) \qquad Y^3_d = c2(d1)\cdot(i\cdot\sigma_d(Y^3_d) + i\cdot c3(d1)\cdot Y^2_d).$$

Due to the fact that a complete cycle takes place in one time slice, the sender never receives a wrong bit, and so the specification could be simplified. We see that this protocol is more a PAR protocol (Positive Acknowledgement with Retransmission) than an ABP protocol.

After abstraction from internal actions, and using RSP, we obtain the following specification:

$$\tau_I(X) = \sum_{d\in D} \lfloor r1(d) \rfloor^{\omega}\cdot Z^1_d \qquad Z^1_d = \tau\cdot\sigma_d(Z^1_d) + \tau\cdot s4(d)\cdot Z^2_d$$

$$Z^2_d = \tau\cdot\sigma_d(Z^3_d) + \tau\cdot\tau_I(X) \qquad Z^3_d = \tau\cdot\sigma_d(Z^3_d) + \tau\cdot Z^2_d.$$

Abstraction from time and using fair abstraction gives the desired result:

$$ABP = \tau_\sigma\circ\tau_I(X) = \tau\cdot\sum_{d\in D} r1(d)\cdot s4(d)\cdot ABP.$$

Notice that we need a generalisation of the fair abstraction rule in order to obtain this result.

6. CONCLUDING REMARKS

We have proposed a theory on discrete time process algebra that extends ACP with the features of ATP in a setting consistent with that of ACPρ. Many options for further work remain. We mention some:

i. All extensions of ACP that have been developed can be integrated in the discrete time setting, such as state operators, process creation, signals, priorities, asynchronous communication (see [BAW90] for these ACP extensions).

ii. The proof theory of ACP_{dt} with initialisation abstraction requires further work. Interesting are principles dealing with infinite behaviour as RDP, RSP, AIP. In particular, the special case of regular processes needs attention because regularity requires a new definition in the presence of discrete time.

iii. The analysis of ACP_{dt} and its extensions as a term rewriting system may be worth attention.

iv. Like in the case of ACP, many more models than the bisimulation model are possible. Investigation of failures, ready, trace models as well as projective limit constructions will be worth while.

v. A theory that explains how discrete time process algebra can be used to develop simulation techniques for real time processes would be very interesting. We think that along this line the best perspective for application of discrete time process algebra can be found.

REFERENCES

[BAB91a] J.C.M. BAETEN & J.A. BERGSTRA, *Real time process algebra*, Formal Aspects of Computing 3 (2), 1991, pp. 142-188.

[BAB91b] J.C.M. BAETEN & J.A. BERGSTRA, *A survey of axiom systems for process algebras*, report P9111, Programming Research Group, University of Amsterdam 1991.

[BAB91c] J.C.M. BAETEN & J.A. BERGSTRA, *Process algebra with signals and conditions*, report CS-R9103, CWI, Amsterdam 1991. To appear in Proc. NATO Summer School, Marktoberdorf 1990 (M. Broy, ed.), Springer Verlag.

[BAB92] J.C.M. BAETEN & J.A. BERGSTRA, *Real space process algebra*, report CSN 92/03, Dept. of Comp. Sci., Eindhoven University of Technology 1992.

[BAW90] J.C.M. BAETEN & W.P. WEIJLAND, *Process algebra*, Cambridge Tracts in Theor. Comp. Sci. 18, Cambridge University Press 1990.

[BEK84] J.A. BERGSTRA & J.W. KLOP, *Process algebra for synchronous communication*, Inf. & Control 60, 1984, pp. 109-137.

[BRHR84] S.D. BROOKES, C.A.R. HOARE & A.W. ROSCOE, *A theory of communicating sequential processes*, JACM 31 (3), 1984, pp. 560-599.

[GRO90a] J.F. GROOTE, *Specification and verification of real time systems in ACP*, in: Proc. 10th Symp. on Protocol Specification, Testing and Verification, Ottawa (L. Logrippo, R.L. Probert & H. Ural, eds.), North-Holland, Amsterdam 1990, pp. 261-274.

[GRO90b] J.F. GROOTE, *Transition system specifications with negative premises*, in: Proc. CONCUR'90, Amsterdam (J.C.M. Baeten & J.W. Klop, eds.), LNCS 458, Springer 1990, pp. 332-341.

[GM91] D.P. GRUSKA & A. MAGGIOLO-SCHETTINI, *A timed process description language based on CCS*, report TR-9/91, Universitá di Pisa 1991.

[HAN91] H. HANSSON, *Time and probabilities in formal design of distributed systems*, Ph.D. thesis, report DoCS 91/27, Uppsala University, 1991.

[HER90] M. HENNESSY & T. REGAN, *A temporal process algebra*, report 2/90, University of Sussex 1990.

[KLU91] A.S. KLUSENER, *Completeness in real time process algebra*, report CS-R9106, CWI Amsterdam 1991. Extended abstract in Proc. CONCUR'91, Amsterdam (J.C.M. Baeten & J.F. Groote, eds.), Springer LNCS 527, 1991, pp. 376-392.

[KLU92] A.S. KLUSENER, *Abstraction in real time process algebra*, to appear in Proc. REX Workshop on Real Time: Theory in Practice (J.W. de Bakker, C. Huizing, W.P. de Roever & G. Rozenberg, eds.), Mook 1991, Springer LNCS.

[MIL89] R. MILNER, *Communication and concurrency*, Prentice-Hall 1989.

[MOT89] F. MOLLER & C. TOFTS, *A temporal calculus of communicating systems*, report LFCS-89-104, University of Edinburgh 1989.

[MOT90] F. MOLLER & C. TOFTS, *A temporal calculus of communicating systems*, in: Proc. CONCUR'90, Amsterdam (J.C.M. Baeten & J.W. Klop, eds.), LNCS 458, Springer 1990, pp. 401-415.

[NIS90] X. NICOLLIN & J. SIFAKIS, *The algebra of timed processes ATP: theory and application*, report RT-C26, IMAG, Grenoble 1990.

[NSVR90] X. NICOLLIN, J.-L. RICHIER, J. SIFAKIS & J. VOIRON, *ATP: an algebra for timed processes*, in Proc. IFIP TC2 Conf. on Progr. Concepts & Methods, Sea of Gallilee, Israel 1990.

[RER88] G.M. REED & A.W. ROSCOE, *A timed model for communicating sequential processes*, TCS 58, 1988, pp. 249-261.

The Silent Step in Time *

A.S. Klusener

Dept. of Softw. Techn., CWI, P.O. Box 4079, 1009 AB Amsterdam, The Netherlands
e-mail stevenk@cwi.nl

Abstract. In untimed process algebras such as CCS and ACP the silent step enables abstraction from internal actions. Several formalizations of abstraction have been introduced, such as observational congruence, delay bisimulation and branching bisimulation. However, in real time process algebras, such as ACPρ, the silent step has not yet received much attention, so far. Therefore, we formalize these semantics regarding the silent step into real time process algebra. We study the characterizing laws, which correspond closely to the untimed laws, and we investigate which of the semantics is appropriate in the context of ACPρ.

1985 Mathematics Subject Classification: 68Q60.
1982 CR Categories: D.3.1, F.3.1, J.7.
Key Words & Phrases: Real Time Process Algebra, ACP, Abstraction

1 Introduction

Over the years several semantics of the silent step have been introduced. The silent step, normally denoted by the symbol τ, is due to Milner. It enables abstraction from internal activity. As abstraction mechanism it is for proving that an implementation meets its specification. However, there are several options for defining an equivalence on processes involving silent steps. Milner and Henessy have defined *observation equivalence* ([HM80]), later Milner defined a slightly stricter equivalence to which we refer as *delay bisimulation* ([Mil81]). Since observation equivalence may be confused with observational equivalence, which coincides with strong bisimulation for finitely branching processes, we refer to observation equivalence by *weak bisimulation.* Van Glabbeek & Weijland have argued in ([GW91]) that neither of these equivalences respect the branching structures of the processes fully and therefore they introduced *branching bisimulation.* The picture is completed by η*-bisimulation* of Baeten & Van Glabbeek ([BG87]). In this paper we will not mention η-bisimulation any further.

Recently, Baeten & Bergstra introduced ACPρ (ACP with real time) in [BB91]. To each occurrence of an atomic action a time stamp (from $[0,\infty]$) is assigned, denoting the time at which the action is being executed. This leads to two languages, one in which the time stamp is interpreted *absolute*, i.e. from the starting time zero, and one in which it is interpreted *relative*, i.e. with respect to the previous action. They have also introduced the powerful mechanism of integration, which can be considered as an alternative composition over a continuum of alternatives. We refer to ACPρ with integration by ACPρI. Using integration it can be expressed that

* This is an extended abstract. The full version will appear as a technical CWI report with the same title.

an action must be executed somewhere within a certain interval. Moreover, time dependencies between consecutive actions can be expressed. For example

$$\int_{v\in\langle 0,1]} a(v) \quad \text{and} \quad \int_{v\in\langle 0,1]}(a(v)\cdot \int_{w\in\langle v+1,v+3\rangle} a'(w)).$$

In this paper we denote a left and/or right open interval with respectively $\langle$ and $\rangle$. Note that this latter term is equivalent to the following term in relative time:

$$\int_{v\in\langle 0,1]}(a[v]\cdot \int_{w\in\langle 1,3\rangle} a'[w])$$

In fact, every term in relative time can be translated easily to a term in absolute time. The converse is not trivial, but it is implied by the fact that every term in absolute time (without recursion) can be equated to a *basic* term [Klu91a]. Each basic term can be translated directly to a term in relative time. In this paper we will consider absolute time only.

In [Klu91a] the restricted syntax of *prefixed integration* without recursion is studied, for this restriction it is proven that the theory of ACPρI characterizes strong bisimulation completely. Finally, Fokkink has proven in [Fok91] that equality in ACPρI is decidable for prefixed integration without recursion.

In the next section we start with a brief presentation of (untimed) process algebra and recall three bisimulations which involve the silent step. These bisimulations are given in detail together with the rootedness condition on these bisimulations for being a congruence. We restrict ourselves to Basic Process Algebra (BPA$_{\tau\delta}$) without recursion.

Then we present the syntax of real time Basic Process Algebra (BPA$\rho_{\tau\delta}$) in Section 3. We discuss strong timed bisimulation and its axiomatization. After having presented the silent step in untimed process algebra and strong bisimulation in the timed case we can adapt the definition of untimed branching bisimulation by adding time in Section 4. We see that the silent step may correspond with idling and that we need a rootedness condition in order to have a congruence. We obtain one new law which characterizes completely those identities of rooted branching bisimulation over BPAρ which are not strong bisimilar.

We do similarly for weak bisimulation in Section 5. Again we have to apply a rootedness condition and for rooted timed weak bisimulation we obtain a law which is very similar to Milner's third τ-law

$$a\cdot(\tau\cdot X+Y)=a\cdot(\tau\cdot X+Y)+a\cdot X$$

Unfortunately this equivalence is not a congruence anymore after adding the parallel merge and encapsulation, which is essential for an equational theory like ACPρ.

In Section 6 we add prefixed integration. We have to adapt the real time law which we have found in the section without integration. Furthermore, we have a law which does correspond closely to Van Glabbeek & Weijland's branching bisimulation law.

In the original paper of Baeten & Bergstra ([BB91]) the authors suggest to interpret the τ transitions in the operational semantics as idling transitions. This idea is investigated in [Klu91b] for the case without integration, resulting in a timed τ-equivalence. In Section 7 this equivalence is defined in a different style by introducing timed delay bisimulation. We obtain a timed version of Milner's second τ-law

$$\tau\cdot X=\tau\cdot X+X.$$

We do not discuss recursion. This may seem a very severe restriction, but it is our idea that the calculus without recursion has to be understood sufficiently before adding recursion.

2 The Silent Step without Time

2.1 Basic Process Algebra and Strong Bisimulation

In this section we introduce some main notions of process algebra. Our point of departure is Basic Process Algebra with τ and δ, abbreviated by $\mathrm{BPA}_{\tau\delta}$. The symbol τ denotes the silent step and the constant δ denotes inaction, comparable with the *NIL* and the 0 of Milner ([Mil80],[Mil89]) and Hoare's *STOP* ([Hoa85]). This section is based on [BW90], while the part on bisimulations-with-silent-step is based as well on [GW91].

We have an alphabet A, not containing τ and δ. $A \cup \{\tau, \delta\}$ is abbreviated by $A_{\tau\delta}$. The set of terms over $\mathrm{BPA}_{\tau\delta}$ is denoted by $\mathcal{T}_u$ (the subscript u refers to *untimed*) and it has typical elements p, p_i.

$$p ::= a \mid p_1 + p_2 \mid p_1 \cdot p_2 \quad (a \in A_{\tau\delta})$$

Here, $p_1 + p_2$ is the alternative composition of p_1, p_2 while $p_1 \cdot p_2$ is the sequential composition of p_1, p_2. An element of $\mathcal{T}_u$ is referred to as a term or a process term.

The semantics of a process term p is a labeled transition system with p as root. The states of such a transition system are taken from $\mathcal{T}_u \cup \{\surd\}$. The symbol $\surd$ denotes termination. These transition systems are defined as the least relations satisfying the action rules of Table 1. This style of giving operational semantics is due to Plotkin ([Plo81]).

Table 1. Action Rules for $\mathcal{T}_u$ $(a \in A_\tau)$

$$a \xrightarrow{a} \surd \qquad \frac{p \xrightarrow{a} p'}{p \cdot q \xrightarrow{a} p' \cdot q} \quad \frac{p \xrightarrow{a} \surd}{p \cdot q \xrightarrow{a} q}$$

$$\frac{p \xrightarrow{a} p'}{p + q \xrightarrow{a} p'} \quad \frac{p \xrightarrow{a} p'}{q + p \xrightarrow{a} p'} \qquad \frac{p \xrightarrow{a} \surd}{p + q \xrightarrow{a} \surd} \quad \frac{p \xrightarrow{a} \surd}{q + p \xrightarrow{a} \surd}$$

We may identify process terms of which the transition systems represent the same behavior. However, there are several options of formalizing the idea of "having the same behavior", especially when dealing with the silent step.

The first equivalence we give is strong bisimulation equivalence, denoted by $\leftrightarrow_s$, it does not yet regard the special role of the silent action. $p \leftrightarrow_s q$ means roughly that every transition of p can be mimicked by q such that the resulting pair is strong bisimilar again and vice versa. First we need the definition of a strong bisimulation. In the sequel $p\mathcal{R}q$ abbreviates that the binary relation $\mathcal{R}$ contains the pair (p, q).

Definition 1. $\mathcal{R} \subset \mathcal{T}_u \times \mathcal{T}_u$ is a *strong bisimulation* if whenever $p\mathcal{R}q$ then

1. $p \xrightarrow{a} p'$ implies $\exists q'$ such that $q \xrightarrow{a} q'$ and $p'\mathcal{R}q'$.
2. $p \xrightarrow{a} \surd$ implies $q \xrightarrow{a} \surd$.
3. Respectively (1) and (2) with the role of p and q interchanged.

Strong bisimulation equivalence is now defined by

Definition 2. $p \leftrightarrow_s q$ iff there is a strong bisimulation $\mathcal{R}$ relating p and q.

This equivalence is known to be a congruence. Hence, the transition system model modulo strong bisimulation is a well-defined algebra; in fact it can be completely axiomatized by the axiom system of Table 2.

Table 2. BPA$_\delta$ = BPA + A6 + A7

A1 $X + Y$ $= Y + X$	A4 $(X + Y) \cdot Z = X \cdot Z + Y \cdot Z$
A2 $(X + Y) + Z = X + (Y + Z)$	A5 $(X \cdot Y) \cdot Z = X \cdot (Y \cdot Z)$
A3 $X + X$ $= X$	
BPA = A1 − A5	
A6 $X + \delta$ $= X$	A7 $\delta \cdot X$ $= \delta$

2.2 Semantics for the Silent Step

We consider three different bisimulation equivalences (and their rooted verions)
which regard the silent step: branching bis.([GW91]), delay bis. [Mil83] and weak
bis. ([Mil80]). $\quad \underline{\leftrightarrow}_b \,(\underline{\leftrightarrow}_{rb}) \quad \subset \quad \underline{\leftrightarrow}_d \,(\underline{\leftrightarrow}_{rd}) \quad \subset \quad \underline{\leftrightarrow}_w \,(\underline{\leftrightarrow}_{rw})$
Each of these bisimulation equivalences allow that an a-transition on one side may
be mimicked by a a-transition possibly preceded or followed by silent steps on the
other side. This is shown by the following picture.

Fig. 1. Three bisimulations with τ

$$\quad b \qquad\qquad d \qquad\qquad w$$

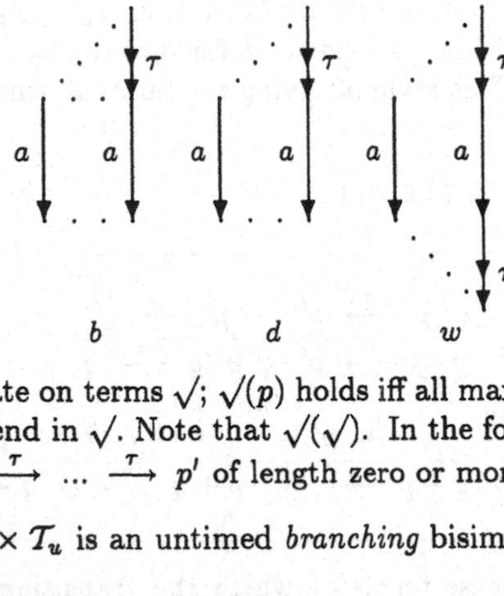

We have one predicate on terms $\sqrt{}$; $\sqrt{}(p)$ holds iff all maximal paths starting in p
consist of τ's only and end in $\sqrt{}$. Note that $\sqrt{}(\sqrt{})$. In the following $p \Longrightarrow p'$ denotes
that there is a path $p \xrightarrow{\tau} \ldots \xrightarrow{\tau} p'$ of length zero or more.

Definition 3. $\mathcal{R} \subset \mathcal{T}_u \times \mathcal{T}_u$ is an untimed *branching* bisimulation if whenever $p\mathcal{R}q$
then
1. If $p \xrightarrow{a} p'$ then either $a = \tau$ and $p'\mathcal{R}q$
   or $\exists z, q'$ such that $q \Longrightarrow z \xrightarrow{a} q'$, $p\mathcal{R}z$ and $p'\mathcal{R}q'$.
2. If $p \xrightarrow{a} \sqrt{}$ then $\exists z, z'$ such that $q \Longrightarrow z \xrightarrow{a} z'$ with $\sqrt{}(z')$ and $p\mathcal{R}z$.
3. Respectively (1) and (2) with the role of p and q interchanged.

Definition 4. $\mathcal{R} \subset \mathcal{T}_u \times \mathcal{T}_u$ is an untimed *delay* bisimulation if whenever $p\mathcal{R}q$ then
1. If $p \xrightarrow{a} p'$ then either $a = \tau$ and $p'\mathcal{R}q$
   or $\exists z, q'$ such that $q \Longrightarrow z \xrightarrow{a} q'$ and $p'\mathcal{R}q'$.
2. If $p \xrightarrow{a} \sqrt{}$ then $\exists z, z'$ such that $q \Longrightarrow z \xrightarrow{a} z'$ with $\sqrt{}(z')$.
3. Respectively (1) and (2) with the role of p and q interchanged.

Definition 5. $\mathcal{R} \subset \mathcal{T}_u \times \mathcal{T}_u$ is an untimed *weak* bisimulation if whenever $p\mathcal{R}q$ then
1. If $p \xrightarrow{a} p'$ then either $a = \tau$ and $p'\mathcal{R}q$
   or $\exists z, z', q'$ such that $q \Longrightarrow z \xrightarrow{a} z' \Longrightarrow q'$ and $p'\mathcal{R}q'$.
2. If $p \xrightarrow{a} \sqrt{}$ then $\exists z, z'$ such that $q \Longrightarrow z \xrightarrow{a} z' \Longrightarrow \sqrt{}$.

3. Respectively (1) and (2) with the role of p and q interchanged.

We need the predicate $\surd$ in this paper to express that "τ-stuttering" afterwards is allowed. Hence, we obtain the law $X \cdot \tau = X$. For $* \in \{b, d, w\}$ we define $*$-bisimulation equivalence.

Definition 6. $p \leftrightarrow_* q$ iff there is a $*$-bisimulation relating p and q.

None of these equivalences is a congruence over $\mathcal{T}_u$. An untimed bisimulation is *rooted* w.r.t. p and q if it does not relate p and q with states which can be reached from either p or q in one or more steps. We obtain *rooted* $*$-bisimulation equivalences, denoted by $p \leftrightarrow_{r*} q$, by requiring that there is a rooted $*$-bisimulation relating p and q. For each $* \in \{b, d, w\}$ the equivalence $\leftrightarrow_{r*}$ is a congruence. We define the following τ-laws:

T1(M1)	$X \cdot \tau$	$=$	X
T2	$Z \cdot (\tau \cdot (X+Y) + X)$	$=$	$Z \cdot (X+Y)$
T3(M2)	$\tau \cdot X$	$=$	$\tau \cdot X + X$
T4(M3)	$a \cdot (\tau \cdot X + Y)$	$=$	$a \cdot (\tau \cdot X + Y) + a \cdot X$

T1 is Milner's first τ-law, T3 is his second and T4 is his third one. T2 is Van Glabbeek & Weijland's branching bisimulation law, note that A3+T1+T3 $\vdash$ T2.

If Θ is an equational theory over a set of terms $\mathcal{T}$ then $\Theta \vdash p = q$ denotes that there is a derivation in Θ which equalizes p with q. We define three theories, $\Theta_b = \text{BPA}_\delta + \text{T1} + \text{T2}, \Theta_d = \text{BPA}_\delta + \text{T1} + \text{T3}$ and $\Theta_w = \Theta_d + \text{T4}$. Each bisimulation equivalence can be axiomatized completely by its corresponding theory. More formally, for each $* \in \{b, d, w\}$ and $(p, q \in \mathcal{T}_u)$ we have

Theorem 7. $$p \leftrightarrow_{r*} q \quad \textit{iff} \quad \Theta_* \vdash p = q$$

In [GW91] the completeness is proven for branching bisimulation first. From this result the other completeness results can be found easily.

3 Adding Time

We follow Baeten & Bergstra ([BB91]) in adding time stamps to the atomic actions. We take a from $A_{\tau\delta}$ and t from $[0, \infty]$. Three auxiliary operators are added, they are appropriate in the axiomatization and they will be discussed later. The real time extension of $\text{BPA}_{\tau\delta}$ is denoted by $\text{BPA}\rho_{\tau\delta}$. The associated set of terms is denoted by $\mathcal{T}_\rho$.

$$p ::= a(t) \mid p_1 + p_2 \mid p_1 \cdot p_2 \mid t \gg p \mid p \gg t \mid p@t$$

We assume that $\gg$ and @ bind the strongest and that + binds the weakest. Thus

$$t \gg p \cdot q + p \text{ is parsed as } ((t \gg p) \cdot q) + p$$

Syntactic equivalence is denoted by $\equiv$. We apply the following abbreviations:

$$X \underline{\gg} t \text{ abbreviates } X \gg t + X@t$$
$$t \underline{\gg} X \text{ abbreviates } t \gg X + X@t$$

The term $\delta(0)$ is abbreviated by δ, it denotes the process which can not do any action nor can it idle to any point in time.

The time stamps are interpreted absolute, thus from the start time zero. Hence, some parts of a process term may be inaccessible since they are not allowed anymore by the elapse of time caused by preceding actions:

$$a(2) \cdot (b(1) + c(3)) = a(2) \cdot c(3)$$
$$a(2) \cdot b(1) \qquad = a(2) \cdot \delta$$

The process $t \gg p$ is that part of p which starts after t, the process $p \gg t$ is that part of p which starts before t. Finally, $p@t$ is that part of p which starts exactly at t and $p \ggeq t$ is that part of p which starts at or before t. (In the original paper [BB91] the operator $p \gg t$ is part of ACPρ and not of BPAρ_δ, but we need this operator to define the τ-laws later on. The operator $p@t$ and the abbreviations $p \ggeq t$ and $t \ggeq X$ are new in the context of BPAρ_δ. The notation @ occurs as well [Wan90])

The semantics of a process term is again a labeled transition system, the action rules and strong bisimulation equivalence have been taken from [Klu91a]. The labels are now timed actions (in $A_{\tau\delta}$). The states are terms. The root state of the transition system of a term p is the state p. We have the following transitions:

$$a(r) \xrightarrow{a(r)} \surd, \quad a(r) + p \xrightarrow{a(r)} \surd \quad \text{and} \quad a(r) \cdot p \xrightarrow{a(r)} r \gg p$$

The right most transition shows us that the course of time is encoded by $r \gg$ .... The action rules are given in Table 3 at the end of the paper. This calculus contains *time stops* or timed deadlocks; the process $\delta(t)$ may be regarded as the process with an unspecified behavior at time t. A timed deadlock at t will always be avoided if something else is possible at or after t. Hence we have $X + \delta(t) = X$ whenever X can idle till t. The terminology time stop can be found as well in ([Jef91]).

We define strong bisimulation for timed transition systems.

Definition 8. Strong Bisimulation
$\mathcal{R} \subset T_\rho \times T_\rho$ is a *strong bisimulation* if whenever $p\mathcal{R}q$ then
1. $p \xrightarrow{a(r)} p'$ implies $\exists q'$ such that $q \xrightarrow{a(r)} q'$ and $p'\mathcal{R}q'$.
2. $p \xrightarrow{a(r)} \surd$ implies $q \xrightarrow{a(r)} \surd$.
3. Respectively (1) and (2) with the role of p and q interchanged.

Definition 9. Strong Bisimulation Equivalence
$p \leftrightarrow_s q$ iff there is a strong bisimulation $\mathcal{R}$ such that $p\mathcal{R}q$.

Strong bisimulation equivalence is completely axiomatized by the theory BPAρ_δ ([Klu91a]). In the sequel we use the *ultimate delay* of a term p, denoted by $U(p)$. Intuitively, the ultimate delay of p is the upperbound of points in time which can be reached by p by idling only. Moreover, we have the *latest starttime* $L(p)$ for each time closed process; it is the maximum of points in time at which p can perform an (initial) action. The ultimate delay originates from [BB91]. It is similar to the *system delay* of Moller & Tofts [MT90]. Both $U(p)$ and $L(p)$ can be defined inductively ([Klu91a], [Klu92]).

4 Branching Bisimulation and Time

4.1 Rooted Timed Branching Bisimulation Equivalence

In untimed branching bisimulation it is allowed that a τ-transition on one side is mimicked with no transition at all at the other side if this τ-transition does not determine a choice. Take for example $a \cdot \tau \cdot b \leftrightarrow_{rb} a \cdot b$.

We have a function *time* which assigns to every process term the course of time, its inductive definition can be found in [Klu92]. It is guaranteed that if $p \xrightarrow{a(r)} p'$ then $time(p') = r$. If p may evolve into p' in zero or more τ actions we write $p \Longrightarrow_t p'$ where $t = time(p')$. Thus $p \Longrightarrow_{time(p)} p$ and if $p \xrightarrow{\tau_1(t_1)} p_1 \dots \xrightarrow{\tau(t_n)} p_n$ then $p \Longrightarrow_{t_n} p_n$.

In our real time context it does make sense to relate only states in which the time is passed equally. Formally for a (timed) branching bisimulation $\mathcal{R}$ we require that $p\mathcal{R}q$ implies $time(p) = time(q)$. In the first and the second clause of the definition of untimed branching bisimulation the "intermediate" state z in the transition system of q must be related to p. In real time it will always be the case that the time in z is greater than the time in q, hence in stead of $p\mathcal{R}_u z$ we have $(t \gg p)\mathcal{R}z$ where $t = time(z)$.

For an untimed branching bisimulation $p\mathcal{R}_u q$ and $p \xrightarrow{\tau} p'$ might imply that $p'\mathcal{R}_u q$. A (timed) branching bisimulation $\mathcal{R}$ must increase the time in q accordingly to the transition $p \xrightarrow{\tau(r)} p'$, resulting in $p'\mathcal{R}(r \gg q)$. Some care is needed, since it must be guaranteed that q is able to idle till r, which is formalized by $U(q) > r$.

In timed branching bisimulation no "τ-stuttering" is allowed afterwards; we enforce that two bisimilar process terms terminate at the same points in time.

Definition 10. Timed Branching Bisimulation
$\mathcal{R} \subset T_\rho \times T_\rho$ is a *branching bisimulation* if whenever $p\mathcal{R}q$ then

1. $p \xrightarrow{a(r)} p'$ implies
   - either $a = \tau$, $\exists q'$ with $U(q') > r$ and $\exists t$ such that $q \Longrightarrow_t q'$ and $(t \gg p)\mathcal{R}q'$ and $p'\mathcal{R}(r \gg q')$.
   - or $\exists z, q'$ and $\exists t$ such that $q \Longrightarrow_t z \xrightarrow{a(r)} q'$ and $(t \gg p)\mathcal{R}z$ and $p'\mathcal{R}q'$.
2. $p \xrightarrow{a(r)} \surd$ implies $\exists z$ and $\exists t$ such that $q \Longrightarrow_t z \xrightarrow{a(r)} \surd$ and $(t \gg p)\mathcal{R}z$.
3. Respectively (1) and (2) with the role of p and q interchanged.

One may argue that it would be better not to mention the structure of the states (which are just terms in our context) in a definition of timed branching bisimulation. In [Klu92] an operational semantics with idle steps is given according to Baeten & Bergstra ([BB91]). In that semantics idle steps are considered, denoted by $q \xrightarrow{\iota(r)} q'$. Instead of the requirement $U(q) > r$ and $p'\mathcal{R}(r \gg q)$ we then have the requirement $q \xrightarrow{\iota(r)} q'$ and $p'\mathcal{R}q'$.

Definition 11. Timed Branching Bisimulation Equivalence
$p \leftrightarrow_b q$ iff there is a branching bisimulation $\mathcal{R}$ such that $p\mathcal{R}q$.

Again, this equivalence is not a congruence:

$$a(2) + b(3) \quad \leftrightarrow_b \quad a(2) + \tau(2) \cdot b(3)$$

$$d(2) \cdot b(3) \leftrightarrow_s d(2) \cdot (a(2) + b(3)) \not\leftrightarrow_b d(2) \cdot (a(2) + \tau(2) \cdot b(3)) \leftrightarrow_s d(2) \cdot \delta$$

And we need a *rootedness* condition as usual.

Definition 12. Rooted Timed Branching Bisimulation Equivalence
$p \leftrightarrow_{rb} q$ iff

1. $p \xrightarrow{a(r)} p'$ implies $\exists q'$ such that $q \xrightarrow{a(r)} q'$ and $p' \leftrightarrow_b q'$.
2. $p \xrightarrow{a(r)} \surd$ implies $q \xrightarrow{a(r)} \surd$.
3. Respectively (1) and (2) with the role of p and q interchanged.

This rootedness condition is stricter than the untimed one. From a root it acts as if it were strong bisimulation equivalence. As soon as an action (from A_τ) is executed τ-transitions can not be distinguished anymore from idlings.

One could think of a machine which is not yet active at time zero, only when an action is executed it becomes active and it remains active until termination.

Following this idea it makes sense to distinguish idling and internal activity from the root state; the process $a(2) + b(3)$ can be inactive in between 2 and 3 while $a(2) + \tau(2) \cdot b(3)$ cannot.

Proposition 13. *$\leftrightarrow_{rb}$ is a congruence for* $\mathrm{BPA}\rho_{\delta\tau}$

4.2 One Law for the Silent Step

A typical identity is the following

$$a(1) \cdot (a(2) + b(3)) \quad \leftrightarrow_{rb} \quad a(1) \cdot (a(2) + \tau(2) \cdot b(3))$$

The choice for the $b(3)$ is determined at time 2, with or without $\tau(2)$ in front of it. In general, if p can idle till t we may split p into two parts, one which starts before or at a time t (denoted by $p \gtrsim t$) and the other part which starts after t (denoted by $t \gg p$). Whenever $t < U(p)$ we have $X \quad \leftrightarrow_s \quad X \gtrsim t + t \gg X \quad \leftrightarrow_b \quad X \gtrsim t + \tau(t) \cdot X$

In order to have a corresponding equivalence in $\leftrightarrow_{rb}$ we may put it in a context $a(r) \cdot (..)$ where $r < t$. We can express this algebraically by the following real time τ-law (AT stands for *Absolute Time*).

ATT1	$r < t < U(X)$	:	$a(r) \cdot X = a(r) \cdot (X \gtrsim t + \tau(t) \cdot X)$

In [Klu91b] this law was formulated as follows:

ATT1'	$r < t < U(X) \wedge U(Y) \leq t$	:	$a(r) \cdot (\tau(t) \cdot X + Y) = a(r) \cdot (t \gg X + Y)$

The following examples show the need for the condition $r < t < U(X)$, without this condition ATT1 would identify $a(1) \cdot \tau(1) \cdot b(3)$ with $a(1) \cdot b(3)$ and $a(1) \cdot \tau(3) \cdot b(3)$ with $a(1) \cdot b(3)$.

$$a(1) \cdot \delta \quad \leftrightarrow_s a(1) \cdot \tau(1) \cdot b(3) \not\leftrightarrow_b a(1) \cdot b(3)$$

$$a(1) \cdot \tau(3) \cdot \delta \leftrightarrow_s a(1) \cdot \tau(3) \cdot b(3) \not\leftrightarrow_b a(1) \cdot b(3)$$

We close this section with one theorem.

Theorem Soundness and Completeness $p, q \in \mathcal{T}_\rho$

$$\mathrm{BPA}\rho_{\delta\tau} + \mathrm{ATT1} \vdash p = q \quad \Longleftrightarrow \quad p \leftrightarrow_{rb} q$$

5 Weak Bisimulation and Time

5.1 The need for idle transitions

An untimed weak bisimulation $\mathcal{R}_u$ relating p and q can match a transition $p \xrightarrow{a} p'$ with a sequence $q \Longrightarrow z \xrightarrow{a} z' \Longrightarrow q'$ such that p' and q' are related again. As opposed to branching bisimulation intermediate states in between q and q' do not have to be related with either p or p'.

We have timed rooted weak bisimulation ($\leftrightarrow_{rw}$) which relates

$$a(1) \cdot (\tau(2) \cdot b(3) + c(2)) \quad \text{and} \quad a(1) \cdot (\tau(2) \cdot b(3) + c(2)) + a(1) \cdot b(3)$$

as was to be expected by the untimed law (T4) $a(\tau X + Y) = a(\tau X + Y) + aX$.

It can be axiomatized completely by $\mathrm{BPA}\rho_\delta$ + ATT1 and the timed version of T4.

ATT4	$r < t < U(X)$	:	$a(r) \cdot (\tau(t) \cdot X + Y) = a(r) \cdot (\tau(t) \cdot X + Y) + a(r) \cdot (t \gg X)$

We obtain new identities over the calculus without τ as well, these are characterized by combining ATT1 and ATT4.

$r < t < U(X) \wedge U(Y) \leq t \quad a(r) \cdot (t \gg X + Y) = a(r) \cdot (t \gg X + Y) + a(r) \cdot (t \gg X)$

5.2 Weak Bisimulation is not a congruence over ACPρ

Let us introduce communication and encapsulation. We assume $c|\bar{c} = \tau$.
Thus $c(2)|\bar{c}(2) = \tau(2)$. Furthermore we disallow the individual action $c(2)$ to happen by the encapsulation $\partial_{\{c\}}$.

Take $p = a(1)\cdot(b(3)+c(2))$ and $q = a(1)\cdot(b(3)+c(2))+a(1)\cdot b(3)$. Then we have BPA$\rho_\delta$+ ATT4 $\vdash p = q$. But in a context they can be distinguished. In $\partial_{\{c\}}(p\|\bar{c}(2))$ at time 2 a communication of $c(2)$ with $\bar{c}(2)$ is forced since it is the only option for the whole process not to deadlock at 2. However, $\partial_{\{c\}}(q\|\bar{c}(2))$ has a deadlock at time 2.

$$\text{ACP}\rho \vdash \partial_{\{c\}}(p\|\bar{c}(2)) = a(1)\cdot\tau(2)$$
$$\text{ACP}\rho \vdash \partial_{\{c\}}(q\|\bar{c}(2)) = a(1)\cdot\tau(2) + a(1)\cdot\delta(2)$$

This counterexample is due to Jan Bergstra. Hence, weak bisimulation is not a congruence in ACPρ. Therefore, we think that weak bisimulation is not appropriate for extending with time, at least in the context of ACP and ACPρ. We will not study this timed version of weak bisimulation anymore in this paper.

6 Branching Bisimulation and Integration

The first four subsections are taken from ([Klu91a],[Fok91]).

6.1 Introduction

Integration is the alternative composition over a continuum of alternatives ([BB91]). So if an action a can happen somewhere in the interval $[1,2]$ we write $\int_{v\in[1,2]} a(v)$. The bounds of an interval may depend on time variables, for example $\int_{v\in[1,2]} a(v)\cdot \int_{w\in[v+1,v+2]} b(w)$. In this section we take a more restrictive view on integration than in [BB91], called *prefixed* integration ([Klu91a]). We require that every integral $\int_{v\in V}$ is directly followed by an action $a(v)$. Moreover, time variables are chosen from intervals of which the bounds are liniar expressions. Examples of terms which are not in this from are $\int_{v\in\{n|n \text{ is prime}\}} a(v)$ and $\int_{v\in\langle 1,2\rangle} a(1)\cdot b(v)$.

6.2 Bounds and Intervals

TVar denotes an infinite, countable set of *time variables*. Let $t \in [0,\infty]$, $r \in \langle 0,\infty\rangle$ and $v \in TVar$. The set *Bound* of *bounds*, with typical elements b, b_1, b_2, is defined by

$$b ::= t \mid v \mid b_1 + b_2 \mid b_1 \dot{-} b_2 \mid r\cdot b$$

where $\dot{-}$ denotes the monus function, i.e. if $t_0 \leq t_1$ then $t_0 \dot{-} t_1 = 0$. In the sequel $\{\!\!\langle$ and $\rangle\!\!\}$ are elements of $\{\langle,[\}$ and $\{\rangle,]\}$ respectively. An interval V is of the form $\{\!\!\langle b_1, b_2\rangle\!\!\}$ with b_1, b_2 bounds. For $b \in Bound$ the set of time variables occurring in b is denoted by $tvar(b)$. Of course $tvar(\{\!\!\langle b,c\rangle\!\!\}) = tvar(b)\cup tvar(c)$. The set of intervals is denoted by Int, its subset of *time-closed* intervals, i.e. intervals for which $tvar(V) = \emptyset$, is denoted by Int^{cl}. On Int two operators *sup*, *inf* are defined. Let $V = \{\!\!\langle t_0, t_1\rangle\!\!\}$

$$V \neq \emptyset \;\; inf(V) = t_0 \text{ and } sup(V) = t_1$$
$$V = \emptyset \;\; inf(V) = sup(V) = 0$$

A *condition* is a logical expression build from bounds, for example $v < 1$ or $2v \leq w+4$. We may consider expressions like $t \in V$ and $V = \emptyset$ as conditions depending on the bounds of V and whether they are closed or open. Similarly, we may consider $b < V$ as a condition which reduces to true if b is lower then every point in V. For example, $1 < [2,3]$ and $v < \langle v, v+1]$. Finally, we may write $V < W$ where V and

W are both intervals; meaning that it is always true that an arbitrary point in W is greater than every point in V. We require in this case as well that W is not empty.

For example $\langle v, v+1\rangle < \langle 2v, 10]$ abbreviates the condition $1 < v < 5$.

6.3 Process Terms

Let $a \in A_\delta$, $t \in \mathbb{R}^{\geq 0}$, $v \in TVar$, V an interval and b a bound. The set $\mathcal{T}_{\mathrm{BPA}\rho I_{\tau\delta}}$ of *process terms*, is abbreviated by $\mathcal{T}_I$.

$p ::= a(t) \mid \int_{v\in V} a(v) \mid \int_{v\in V}(a(v)\cdot p) \mid p_1+p_2 \mid p_1\cdot p_2 \mid b \gg p \mid p \gg b \mid p@b$

We use a *scope convention*, saying that if we do not write scope brackets, then the scope is as large as possible. Thus we write $\int_{v\in V} a(v)\cdot p$ for $\int_{v\in V}(a(v)\cdot p)$. We can define inductively the collection $FV(p)$ of free time variables appearing in a process term p, that is those variables which are not bound by any of the integrals in p. A term p with $FV(p) = \emptyset$ is called a *time-closed* term. The subset of time closed terms is denoted by $\mathcal{T}_I^{cl}$. We define $U(\int_{v\in V} a(v)) = U(\int_{v\in V}(a(v)\cdot p)) = sup(V)$.

6.4 The Semantics

The action rules for the integral are very simple, they express exactly that an integral is a continuum of alternatives.

$$\frac{r \in V}{\int_{v\in V} a(v) \xrightarrow{a(r)} \surd} \qquad \frac{r \in V}{\int_{v\in V} a(v)\cdot p \xrightarrow{a(r)} r \gg p[r/v]}$$

Definition 9 of strong bisimulation and Definition 11 branching bisimulation still apply when we restrict ourselves to time closed terms. In [Klu91a] it is proven that $\underline{\leftrightarrow}_s$ is a congruence over $\mathcal{T}_I^{cl}$. Moreover, it is proven that $\underline{\leftrightarrow}_s$ over $\mathcal{T}_I^{cl}$ is completely axiomatized by the theory $\mathrm{BPA}\rho I_{\tau\delta}$ which is given in Table 4. We extend the definition of $\underline{\leftrightarrow}_{rb}$ to $\mathcal{T}_I^{cl} \times \mathcal{T}_I^{cl}$ in the obvious way.

Proposition 14. *$\underline{\leftrightarrow}_{rb}$ is a congruence over $\mathcal{T}_I^{cl}$*

6.5 Extending the Law ATT1 with Integrals

We have to redefine the conditions for the timed τ laws, since they have to deal with intervals. So, ATT1 is generalized to ATTI1; the I in the name refers to *integration*. now:

$$\text{ATTI1 } V < W < U(X)\ \int_{v\in V} a(v)\cdot X = \int_{v\in V} a(v)\cdot(X \gg inf(W) + \int_{w\in W}\tau(w)\cdot X)$$

So, we have $a(1)\cdot\int_{v\in\langle 1,10]} b(v) = a(1)\cdot(\int_{v\in\langle 1,5]} b(v) + \tau(5)\cdot\int_{v\in\langle 5,10]} b(v))$

6.6 The Timed Branching Law

However, not all identities can be covered by that law, for example:

$$a(1)\cdot(\int_{v\in\langle 1,10\rangle}\tau(v)\cdot(b(10)+c(5)) + b(10)) \ \underline{\leftrightarrow}_{rb}\ a(1)\cdot(b(10)+c(5))$$

$$a(1)\cdot(\int_{v\in\langle 1,10\rangle}\tau(v)\cdot(\int_{w\in\langle 1,20]} b(w) + \int_{z\in\langle 0,10\rangle} c(z)) + \int_{w\in\langle 1,20]} b(w)) \ \underline{\leftrightarrow}_{rb}\ a(1)\cdot(\int_{w\in\langle 1,20]} b(w) + \int_{z\in\langle 0,10\rangle} c(z))$$

They all look like instances of the second (untimed) τ-law:

$$\text{T2}\quad Z\cdot(\tau\cdot(X+Y)+X) = Z\cdot(X+Y)$$

When adding time to this law we have to be very careful with the conditions on all the intervals. Consider the pairs of branching bisimilar terms of above and the following pairs below of non bisimilar terms. The transition system of the left hand side of the first pair has a deadlock at 10 caused by the possible execution of the τ at 10. The transition system of the right hand side of the first pair does not have such a deadlock. In the transition system of the left hand side of the second pair the choice for doing the b might be done at 10, while at the right hand side it is always done at 11.

$$a(1) \cdot (\textstyle\int_{v \in \langle 1,10]} \tau(v) \cdot (b(10) + c(5)) + b(10)) \qquad\qquad \not\underline{\leftrightarrow}_{rb}$$
$$a(1) \cdot (b(10) + c(5))$$

$$a(1) \cdot (\textstyle\int_{v \in \langle 1,10)} \tau(v) \cdot (\textstyle\int_{w \in \langle 1,20]} b(w) + \textstyle\int_{z \in \langle 0,11)} c(z)) + \textstyle\int_{w \in \langle 1,20]} b(w)) \not\underline{\leftrightarrow}_{rb}$$
$$a(1) \cdot (\textstyle\int_{w \in \langle 1,20]} b(w) + \textstyle\int_{z \in \langle 0,11)} c(z))$$

ATTI2 $V < W < U(X+Y) \wedge U(Y) \leq sup(W)$
$\int_{v \in V} a(v) \cdot ((\int_{w \in W} \tau(w)) \cdot (X+Y) + X) = \int_{v \in V} a(v) \cdot (X + inf(W) \gg Y)$

We have been a bit sloppy with the conditions like $U(Y) \leq inf(W)$ in the τ-laws, since the expressions $U(Y)$ and $inf(W)$ may contain the time variable v. For more details on reasoning with process terms and conditions over bounds, we refer to [Fok91] and [Klu91a],[Klu92].

We have the following main Theorem. All other completeness Theorems of this paper can be derived from this one.

Theorem Soundness and Completeness $p, q \in T_I^{cl}$
$$\text{BPA}\rho I_\delta + \text{ATTI1} + \text{ATTI2} \vdash p = q \quad\Longleftrightarrow\quad p \underline{\leftrightarrow}_{rb} q$$

The proof of the Theorem is rather involved. We give only a short sketch here, all details can be found in [Klu91a].

First we generalize our set of terms by allowing conditions in the terms. Then we take the conditional axioms of [Klu91a] and we formulate the τ-laws we have found here with conditionals as well. We take the construction of normal forms of [Fok91] and extend it with a few steps which correspond with the saturation and removals of certain τ transitions in the transition systems. In this way we obtain τ-normal forms. Then we can prove that for τ-normal forms $p \underline{\leftrightarrow}_b q$ implies that p and q are isomorphic. Finally we have to prove that for each term $\int_{v \in V} a(v) \cdot p$ (with $FV(p) \subseteq \{v\}$) we can split V into $V_1, ..., V_n$ and we can construct p_i such that
$$\textstyle\int_{v \in V} a(v) \cdot p = \sum_i \int_{v \in V_i} a(v) \cdot p_i$$
where $r \in V_i$ implies $p_i[r/v]$ is a τ-normal form.

7 Delay Bisimulation and Time

7.1 Rooted Delay Bisimulation Equivalence

Now we have studied branching bisimulation in detail it is easy to introduce delay bisimulation. In the untimed case, delay bisimulation can be found by taking branching bisimulation and relaxing one condition (see Figure 1). So, we take the definition of (timed) branching bisimulation and do similar. For the formal definition we refer to [Klu92]. We define *delay bisimulation* equivalence, $\underline{\leftrightarrow}_d$, and *rooted delay bisimulation*, $\underline{\leftrightarrow}_{rd}$, along the standard line.

Proposition 15. $\underline{\leftrightarrow}_{rd}$ *is a congruence over* T_I^{cl}

7.2 The Timed Delay Law

We obtain the following new identity:

$$a(1) \cdot (\int_{v \in \langle 1,10 \rangle} \tau(v) \cdot \int_{w \in \langle 1,10 \rangle} b(w) + c(11)) \leftrightarrow_{rd}$$
$$a(1) \cdot (\int_{v \in \langle 1,10 \rangle} \tau(v) \cdot \int_{w \in \langle 1,10 \rangle} b(w) + \int_{w \in \langle 1,10 \rangle} b(w) + c(11))$$

This identity looks like an instantiation of the (untimed) delay law $\tau \cdot X + X = \tau \cdot X$. We can apply this law only within an appropriate context, since the root condition here is stronger than in the untimed case. The delay law becomes in real time

ATTI3	$V < W \wedge U(X) \leq sup(W)$
$\int_{v \in V} a(v) \cdot ((\int_{w \in W} \tau(w)) \cdot X + Y) = \int_{v \in V} a(v) \cdot ((\int_{w \in W} \tau(w)) \cdot X + inf(W) \gg X + Y)$	

Note the condition $U(X) \leq sup(W)$ since:

$$a(1) \cdot (\int_{v \in \langle 1,10 \rangle} \tau(v) \cdot \int_{w \in \langle 1,11 \rangle} b(w) + c(11)) \not\leftrightarrow_{rd}$$
$$a(1) \cdot (\int_{v \in \langle 1,10 \rangle} \tau(v) \cdot \int_{w \in \langle 1,11 \rangle} b(w) + \int_{w \in \langle 1,11 \rangle} b(w) + c(11))$$

And we have the following Theorem.

Theorem Soundness and Completeness $p, q \in \mathcal{T}_I^{cl}$

$$\text{BPA}\rho I_\delta + \text{ATTI1} + \text{ATTI2} + \text{ATTI3} \vdash p = q \quad \Longrightarrow \quad p \leftrightarrow_{rb} q$$

Next, we have the following proposition which explains why we have not introduced delay bisimulation without integration. This fact may look surprising, but it can easily be seen from the fact that the axiom ATTI3, which characterizes those identities which are not branching bisimilar, is derivable when restricted to $\mathcal{T}_\rho$. Note that if $W = [t, t]$ then $U(X) \leq sup(W)$ implies $inf(W) \gg X = \delta(t)$.

Proposition 16. $p, q \in \mathcal{T}_\rho \qquad p \leftrightarrow_{rd} q \quad \Longleftrightarrow \quad p \leftrightarrow_{rb} q$

8 Conclusions

We have carried over several formalizations of abstraction from untimed process algebra into real time process algebra. The characterizing laws are very close to the untimed ones. Weak bisimulation is not very appropriate since it is not a congruence for ACPρ, the extension of real time BPA with parallelism and synchronization. We expect that branching bisimulation and delay bisimulation, are congruences for real time ACP. However, these proofs still have to be done.

The equivalence of (timed) rooted delay bisimulation was already proposed in ([Klu91a]). There it was solely motivated as a possible interpretation of abstraction in real time while in this paper it is more motivated as a real time formulation of an existing (untimed) equivalence. Moreover, in that paper a protocol verification has been carried out as well. The law ATTI1 is the only law which is used there, hence, this verification is sound for branching bisimulation as well.

A lot of work still has to be done. For example, recursion has to be studied for strong, branching and delay bisimulation. Other proposals for introducing the silent step into a timed process algebra can be found in Wang's thesis ([Wan90]) and Moller & Tofts ([MT92]). Both proposals concern timed extensions of CCS. Comparison is not so easy due to the subtle differences in syntax and semantics, but it definitely has to be done. For example, our conclusions depend on the fact that the transitions systems we study obey the property that every transition increases the time. Both

Wang and Moller & Tofts have two types of transitions, one type of transitions for passing the time and one for the actual execution of actions. Moreover, we do not have *maximal progress* as basic assumption in our semantics. Wangs work is based on *maximal progress* which means that the τ has priority over all external actions which can be executed at the very same moment. When we translate the pair of terms like $a(1) + b(2)$ and $a(1) + \tau(1) \cdot b(2)$ then the resulting pair is not weak bisimilar any more in any of the versions of timed CCS.

Finally, it has to be studied whether the bisimulation equivalences we have defined so far enable the verification of real time protocols. Vaandrager and Lynch ([VL92]) argue that in real time any bisimulation equivalence is too fine since no abstraction is possible from the timings of external actions.

Acknowledgements

Jos Baeten (Technical University Eindhoven) and Jan Bergstra (University of Amsterdam) are thanked for their encouraging and critical comments. Willem Jan Fokkink is thanked for the many technical discussions. Frits Vaandrager is thanked for some final improvements.

References

[BB91] J.C.M. Baeten and J.A. Bergstra. Real time process algebra. *Journal of Formal Aspects of Computing Science*, 3(2):142–188, 1991.

[BG87] J.C.M. Baeten and R.J. van Glabbeek. Another look at abstraction in process algebra. In *Proceedings* 14th *ICALP*, Karlsruhe, LNCS 267, pages 84–94. Springer-Verlag, 1987.

[BW90] J.C.M. Baeten and W.P. Weijland. *Process algebra*. Cambridge Tracts in Theoretical Computer Science 18. Cambridge University Press, 1990.

[Fok91] W.J. Fokkink. Normal forms in real time process algebra. Report CS-R9149, CWI, Amsterdam, 1991.

[GW91] R.J. van Glabbeek and W.P. Weijland. Branching time and abstraction in bisimulation semantics. Report CS-R9120, CWI, Amsterdam, 1991. An extended abstract of an earlier version has appeared in *Information Processing 89*, North-Holland, 1989.

[HM80] M. Hennessy and R. Milner. On observing nondeterminism and concurrency. In *Proceedings* 7th *ICALP*, LNCS 85, pages 299–309. Springer-Verlag, 1980. This is a preliminary version of *Algebraic laws for nondeterminism and concurrency.* JACM 32(1), pp. 137–161, 1985.

[Hoa85] C.A.R. Hoare. *Communicating Sequential Processes.* Prentice Hall International, 1985.

[Jef91] A. Jeffrey. Discrete timed CSP. Technical Report Memo 78, Chalmers University, Goteborg, 1991.

[Klu91a] A.S. Klusener. Completeness in real time process algebra. Report CS-R9106, CWI, Amsterdam, 1991. An extended abstract appeared in *Proceedings CONCUR 91*, Amsterdam, LNCS 527, pages 376–392. Springer-Verlag, 1991.

[Klu91b] A.S. Klusener. Abstraction in real time process algebra. Report CS-R9144, CWI, Amsterdam, 1991. An extended abstract appeared in *Proceedings of the REX workshop "Real-Time: Theory in Practice"*, LNCS 600, Springer-Verlag, 1991.

[Klu92] A.S. Klusener. The silent step in time. To appear, CWI, Amsterdam, 1992.

[Mil80] R. Milner. *A Calculus of Communicating Systems*, LNCS 92. Springer-Verlag, 1980.

[Mil81] R. Milner. Modal characterisation of observable machine behaviour. In *Proceedings CAAP 81*, LNCS 112, pages 25–34. Springer-Verlag, 1981.

[Mil83] R. Milner. Calculi for synchrony and asynchrony. *Theoretical Computer Science*, 25:267–310, 1983.

[Mil89] R. Milner. *Communication and concurrency.* Prentice Hall International, 1989.

[MT90] F. Moller and C. Tofts. A temporal calculus of communicating systems. In *Proceedings CONCUR 90,* Amsterdam, LNCS 458, pages 401–415. Springer-Verlag, 1990.

[MT92] F. Moller and C. Tofts. Behavioural abstraction in TCCS. To appear in *Proceedings ICALP 92,* Vienna, LNCS, Springer-Verlag, 1992.

[Plo81] G.D. Plotkin. A structural approach to operational semantics. Report DAIMI FN-19, Computer Science Department, Aarhus University, 1981.

[VL92] F.W. Vaandrager and N.A. Lynch. Action transducers and timed automata. 1992. This Volume.

[Wan90] Y. Wang. *A Calculus of Real Time Systems.* PhD thesis, Chalmers University of Technology, Göteborg, 1990.

Table 3. Action Rules ($a \in A_\tau$ and $r \in [0,\infty)$, $s \in [0,\infty]$)

$a(r) \xrightarrow{a(r)} \surd$	$\dfrac{r > s \quad p \xrightarrow{a(r)} p'}{s \gg p \xrightarrow{a(r)} p'}$ $\quad \dfrac{r > s \quad p \xrightarrow{a(r)} \surd}{s \gg p \xrightarrow{a(r)} \surd}$
$\dfrac{p \xrightarrow{a(r)} p'}{p+q \xrightarrow{a(r)} p'}$ $\quad \dfrac{p \xrightarrow{a(r)} \surd}{p+q \xrightarrow{a(r)} \surd}$	$\dfrac{r < s \quad p \xrightarrow{a(r)} p'}{p \gg s \xrightarrow{a(r)} p'}$ $\quad \dfrac{r < s \quad p \xrightarrow{a(r)} \surd}{p \gg s \xrightarrow{a(r)} \surd}$
$\dfrac{p \xrightarrow{a(r)} p'}{q+p \xrightarrow{a(r)} p'}$ $\quad \dfrac{p \xrightarrow{a(r)} \surd}{q+p \xrightarrow{a(r)} \surd}$	$\dfrac{p \xrightarrow{a(r)} p'}{p@r \xrightarrow{a(r)} p'}$ $\quad \dfrac{p \xrightarrow{a(r)} \surd}{p@r \xrightarrow{a(r)} \surd}$
$\dfrac{p \xrightarrow{a(r)} p'}{p \cdot q \xrightarrow{a(r)} p' \cdot q}$ $\quad \dfrac{p \xrightarrow{a(r)} \surd}{p \cdot q \xrightarrow{a(r)} r \gg q}$	$\dfrac{U(p) > L(p)}{p \xrightarrow{\delta(U(p))} \surd}$

$\dfrac{r \in V}{\int_{v \in V} a(v) \xrightarrow{a(r)} \surd}$ $\quad \dfrac{r \in V}{\int_{v \in V} a(v) \cdot p \xrightarrow{a(r)} r \gg p[r/v]}$

This article was processed using the LaTeX macro package with LLNCS style

Table 4. The Axiom System

INT0		$\int_{v\in[t,t]} a(v) = a(t)$
INT1	$V = V_0 \cup V_1$	$\int_{v\in V_0} P + \int_{v\in V_1} P = \int_{v\in V} P$
INT2		$\int_{v\in\emptyset} P = \delta$
INT3a		$\int_{v\in V}(a(v)) \cdot Y = \int_{v\in V} a(v) \cdot Y$
INT3b		$\int_{v\in V}(a(v)\cdot p) \cdot Y = \int_{v\in V} a(v) \cdot (p \cdot Y)$
INT4		$\forall t \in V\ X + P[t/v] = X \implies X + \int_{v\in V} P = X$
ATI1a		$a(0) = \delta$
ATI1b		$a(\infty) = \delta(\infty)$
ATI2a		$\int_{v\in V} \delta(v) = \delta(sup(V))$
ATI2b		$\int_{v\in V} \delta(v) \cdot p = \delta(sup(V))$
ATI3	$t \le sup(V)$	$\int_{v\in V} P + \delta(t) = \int_{v\in V} P$
ATI4		$\int_{v\in V} a(v) \cdot p = \int_{v\in V} a(v) \cdot (v \gg p)$
ATBI1		$t \gg \int_{v\in V} P = \int_{v\in V\cap\langle t,\infty]} P + \delta(t)$
ATB2		$t \gg (X+Y) = (t \gg X) + (t \gg Y)$
ATB3		$t \gg (X \cdot Y) = (t \gg X) \cdot Y$
ATBI4		$\int_{v\in V} P \gg t = \int_{v\in V\cap\langle 0,t\rangle} P$
ATB5		$(X+Y) \gg t = (X \gg t) + (X \gg t)$
ATB6		$(X \cdot Y) \gg t = (X \gg t) \cdot Y$
ATBI7		$\int_{v\in V} P@t = \int_{v\in V\cap\{t\}} P + \delta(t)$
ATB8		$(p+q)@t = p@t + q@t$
ATB9		$(p \cdot q)@t = (p@t) \cdot q$
BPAρ_δ		
ATTI1	$V < W < U(X)$	$\int_{v\in V} a(v) \cdot X = \int_{v\in V} a(v) \cdot (X \geqq inf(W) + \int_{w\in W} \tau(w) \cdot X)$
ATTI2	$V < W < U(X+Y)\ \wedge\ U(Y) \le sup(W)$	$\int_{v\in V} a(v) \cdot ((\int_{w\in W} \tau(w)) \cdot (X+Y) + X) = \int_{v\in V} a(v) \cdot (X + inf(W) \gg Y)$
ATTI3	$V < W\ \wedge\ U(X) \le sup(W)$	$\int_{v\in V} a(v) \cdot ((\int_{w\in W} \tau(w)) \cdot X + Y) = \int_{v\in V} a(v) \cdot ((\int_{w\in W} \tau(w)) \cdot X + inf(W) \gg X + Y)$

($p, q \in \mathcal{T}_{\rho I}$ with $FV(p+q) \subseteq \{v\}$, $X, Y \in \mathcal{T}_I^{cl}$ and P is of the form $a(v)$ or $a(v) \cdot p$)

Action Transducers and Timed Automata*

Frits Vaandrager[1] ** and Nancy Lynch[2]

[1] Department of Software Technology, CWI
P.O. Box 4079, 1009 AB Amsterdam, The Netherlands
fritsv@cwi.nl
[2] MIT Laboratory for Computer Science
Cambridge, MA 02139, USA
lynch@theory.lcs.mit.edu

Abstract. The *timed automaton model* of [13, 12] is a general model for timing-based systems. A notion of *timed action transducer* is here defined as an automata-theoretic way of representing operations on timed automata. It is shown that two timed trace inclusion relations are substitutive with respect to operations that can be described by timed action transducers. Examples are given of operations that can be described in this way, and a preliminary proposal is given for an appropriate language of operators for describing timing-based systems. Finally, justification is given for the definition of implementation based on inclusion of timed trace sets; this is done in terms of a notion of *feasibility* which says that a timed automaton cannot prevent the passage of time.

1 Introduction

The *timed automaton model* of [13, 12] is a general model for timing-based systems. It is intended as a basis for formal reasoning about such systems, in particular, for verification of their correctness and for analysis of their complexity. In [13, 12], we develop a full range of *simulation* proof methods for timed automata; in this paper, we continue the development by studying *process algebras* for the same model. Eventually, we envision using a combination of proof methods, perhaps even using several in the verification of single system.

A timed automaton is an automaton (or labelled transition system) with some additional structure. There are three types of actions: *time-passage actions*, *visible actions* and the special *internal action* τ. All except the time-passage actions are thought of as occurring instantaneously. To specify times, a dense time domain is used, specifically, the nonnegative reals, and no lower bounds are imposed on the times between events. Two notions of external behavior are considered. First, as the finite behaviors, we take the *finite timed traces*, each of which consists of a finite sequence of timed visible actions together with a final time. Second, as the

* This work was supported by ONR contracts N00014-85-K-0168 and N00014-91-J-1988, by NSF grant CCR-8915206, and by DARPA contract N00014-89-J-1988.

** Most of the work on this paper was done while the first author was employed by the Ecole des Mines, CMA, Sophia Antipolis, France.

infinite behaviors, we take the *admissible timed traces*, each of which consists of a sequence of timed visible actions that can occur in an execution in which time grows unboundedly.

The timed automaton model permits description of algorithms and systems at different levels of abstraction. We say that a "low-level" timed automaton A *implements* a "high-level" timed automaton B if the sets of finite and admissible timed traces of A are included in the corresponding sets of B. To justify this notion of implementation, we must argue that the timed trace sets for A are not trivial. We can do this by classifying the visible actions of A as *input actions* or *output actions*, as in the I/O automaton model of [11]. We then require A to be *input enabled*, i.e., willing to accept each input action in each state, and *I/O feasible*, which means that each finite execution can be extended to an admissible execution via an execution fragment that contains no input actions. If A is input enabled and I/O feasible, then it can generate an admissible execution for any "non-Zeno" pattern of inputs, as follows. A starts with an admissible execution containing no inputs. When an input arrives, A performs a transition labelled by that input and continues from the resulting state with another admissible execution fragment containing no inputs until the next input arrives, etc. Thus, A must have a rich set of admissible timed traces.

In the 'classical' untimed case, bisimulation equivalences have been reasonably successful as a notion of implementation between transition systems [5, 15]. Consequently, bisimulation equivalences have also been put forward as a central notion in many studies on real-time process algebras [4, 9, 16, 17, 23]. However, we do not believe that bisimulation equivalences will be very useful as implementation relations in the timed case. The problem is that bisimulation equivalences do not allow one to abstract in specifications from the often very complex timing behavior of implementations (see [9] for an example).

Often, the design or verification of an algorithm includes several levels of abstraction, each of which implements the one above it. Note that it is only necessary to show that the trace sets at the *lowest level* are nontrivial. Thus, only at the lowest level does one require the I/O classification with its accompanying properties of input enabledness and I/O feasibility. Fortunately, at this level, the two properties are generally quite easy to achieve, since they correspond to the receptive and non-Zeno nature of physical machines.

Since we believe that timed trace inclusion does form a good notion of implementation, we are interested in identifying operations on timed automata for which the timed trace inclusion relation is substitutive. This substitutivity is a prerequisite for the compositional verification of systems using timed automata. It should also enable verification of systems using a combination of compositional methods and methods based on levels of abstraction.

We represent operations by automaton-like objects that we call *action transducers*, rather than, for example, using SOS specifications [20]. The importance of transducers for process algebra and concurrency theory was first noted by Larsen and Xinxin [10], who introduced a certain type of transducer, which they called a *context system*, to study compositionality questions in the setting of process algebra. For an example of an action transducer, consider the operation $|||$ of interleaving parallel composition. It can be described by an automaton with a single state s and transitions (one for each action a):

$$s \xrightarrow[(1,a)]{a} s \quad \text{and} \quad s \xrightarrow[(2,a)]{a} s.$$

The left transition says that if the first argument performs an a-action the composition will also perform an a-action, while the right transition says that if the second argument performs an a-action the composition will also perform an a-action. Together, the transitions say that the automaton $A \parallel B$ can do an a-step whenever one of its arguments can do so. In the SOS approach, the same operator $\parallel$ can be described by inference rules (one for each action a):

$$\frac{x \xrightarrow{a} x'}{x \parallel y \xrightarrow{a} x' \parallel y} \quad \text{and} \quad \frac{y \xrightarrow{a} y'}{x \parallel y \xrightarrow{a} x \parallel y'}.$$

The two styles of describing operators, SOS and action transducers, are quite similar.

However, action transducers are more convenient for our purposes. First, although it is easy to see how SOS specifications determine automata, it is less clear how to regard them as defining *operations* on automata. For action transducers, this correspondence is more direct. Second, as noted by Larsen and Xinxin [10], action transducers are a convenient tool for studying compositionality questions, and their use tends to simplify proofs. Third, action transducers can easily be defined to allow multiple start states. Multiple start states have turned out to be useful in untimed automaton formalisms for concurrency such as the I/O automaton model, and we would like to include them. We do not know how to handle start states in the setting of SOS.

A major result of our paper is that the timed trace inclusion relation is substitutive with respect to all operations that can be described by our action transducers, provided they satisfy a number of conditions that concern the handling of internal and time-passage steps.

A condition that is required of our action transducers for timed automata is that time passes uniformly (i.e., at the same rate) for the transducer and all active holes, and it does not pass at all for the inactive holes. This uniformity condition (along with some other technical constraints) is used in our proof of substitutivity. We note that this time uniformity condition is not necessary for substitutivity to hold. For instance, the timed trace inclusion relation is substitutive for a "speedup" operation that doubles the rate at which its timed automaton argument operates. However, we do not have a clean generalization of the uniformity condition that applies to the speedup operation and still guarantees substitutivity.

Having proved substitutivity for a general class of operations, we describe many examples of specific operations that fall into this class. In our view, an appropriate language for describing timing-based systems should consist of a small number of basic operations, both timed and untimed, out of which more complex operations can be built. The basic and derived operations together should be sufficient to describe most interesting timing-based systems. As a starting point, we believe that such a language ought to include the basic untimed operations that are already well understood and generally accepted. Nicollin and Sifakis [18] describe a simple and general construction to transform any untimed operation into a timed one that behaves essentially the same and moreover does not use or constrain the time. By applying this construction to the well-known untimed operations, we obtain a collection of corresponding timed operations that we believe should be included in a real-time process language.

The untimed operations alone are not enough, however; a real-time process language also must include operations that constrain time explicitly. Of the many possibilities, we would like to identify only one or two that can be used for constructing all the others. For this purpose, we tentatively propose a *timer* operation, derived from Alur and Dill [2]. Using only this timer operation and untimed operations, we can construct the timeout construct of Timed CSP [21, 7], and the execution delay operation of ATP [17]. Also, because this timer operation is derived from [2], we are able to use it to define a minor variant of Alur and Dill's ω-automata. Moreover, we can also use it to define the timed automata of Merritt, Modugno and Tuttle [14]. All of this provides evidence that our timer (or something very similar) may be appropriate to use as the sole timed operation in a real-time process language.

The decidability and closure properties of Alur-Dill automata suggest that they can be regarded as a real-time analog of classical finite automata. In the untimed setting, a crucial characteristic of algebras like CCS is that they can easily describe finite automata. Thus by analogy, a natural requirement for a real-time process language is that it can easily describe Alur-Dill automata. Nicollin, Sifakis and Yovine [19] give a translation from ATP into Alur-Dill automata, but do not investigate the reverse translation. In fact it appears that, besides our language, only the BPAr$\rho\delta$I-language of Baeten and Bergstra [4] is sufficiently expressive to allow for a direct encoding of Alur-Dill automata.

As discussed earlier, the appropriateness of the timed trace inclusion relation as a notion of implementation depends upon the lowest level implementation satisfying the input enabling and I/O feasibility properties. In the final section of our paper, we address the question of how to ensure that these properties hold. Although the input enabling condition can be required explicitly, I/O feasibility will typically require a proof. Since we envision the lowest level implementation being described in terms of operators from our real-time language, the way we would like to prove I/O feasibility is by proving it for the basic components and arguing that it is preserved by each of the operators. But examples (e.g., one studied in [1]) show that I/O feasibility is *not* preserved, e.g., by parallel composition. Therefore, we propose a condition, which we call ***strong I/O feasibility***, which implies I/O feasibility, is preserved by operations expressed appropriately as timed action transducers, and appears to be satisfied by "low-level" automata, i.e., those that are close to physical machines. This condition can be used as the basis of a proof method for demonstrating I/O feasibility of implementations.

We present our definitions and results for timed systems by first presenting related definitions and results for untimed systems, and then building upon those wherever we can.

In summary, the main contributions of the paper are: (1) the definitions of action transducers and timed action transducers, (2) the substitutivity results for traces and timed traces, (3) the notion of strong I/O feasibility, (4) the presentation of a large number of interesting operators, timed and untimed, as action transducers, and (5) a preliminary proposal for a process language for timed systems. We see these all as pieces of a unified proof methodology for timed systems.

2 The Untimed Setting

We begin by describing action transducers for the untimed setting. Later, the concepts needed for the timed setting will be defined in terms of corresponding concepts for the untimed setting.

2.1 Automata and Traces

An (untimed) *automaton* A consists of:

- a set $states(A)$ of *states*,
- a nonempty set $start(A) \subseteq states(A)$ of *start states*,
- a set $acts(A)$ of *actions* that includes the *internal action* τ, and
- a set $steps(A) \subseteq states(A) \times acts(A) \times states(A)$ of *steps*.

We let s, s', u, u',.. range over states, and a,.. over actions. The set $ext(A)$ of *external actions* is defined by $ext(A) \triangleq acts(A) - \{\tau\}$. We write $s' \xrightarrow{a}_A s$ as a shorthand for $(s', a, s) \in steps(A)$. We will suppress the subscript A where no confusion is likely. An *execution fragment* of A is a finite or infinite alternating sequence $s_0 a_1 s_1 a_2 s_2 \cdots$ of states and actions of A, beginning with a state, and if it is finite also ending with a state, such that for all i, $s_i \xrightarrow{a_{i+1}} s_{i+1}$. An *execution* of A is an execution fragment that begins with a start state.

For $\alpha = s_0 a_1 s_1 a_2 s_2 \cdots$ an execution $trace(\alpha)$ is defined as the sequence obtained from $a_1 a_2 \cdots$ by removing all τ's. A finite or infinite sequence β of actions is a *trace* of A if A has an execution α with $\beta = trace(\alpha)$. We write $traces^*(A)$, $traces^\omega(A)$ and $traces(A)$ for the sets of finite, infinite and all traces of A, respectively. These notions induce three preorders on automata: we define $A \leq_* B \triangleq traces^*(A) \subseteq traces^*(B)$, $A \leq_\omega B \triangleq traces^\omega(A) \subseteq traces^\omega(B)$, and $A \leq B \triangleq traces(A) \subseteq traces(B)$.

2.2 Action Transducers

We now define a notion of *action transducer*, as an explicit representation of certain extensional operations on automata. We consider operations with a possibly infinite set of arguments. As placeholders for these arguments, an action transducer contains a set of *colors*. Sometimes we will find it useful to make several copies of an argument automaton.[3] [4] To this end a transducer is equipped with a set of *holes*, and a

[3] The idea of copying arguments of transducers is not present in the work of Larsen and Xinxin [10].

[4] Note that, since we always start copies of an argument automaton from a start state, our copying operations are different from those of Bloom, Istrail and Meyer [6], which also allow copying from intermediate states. As a consequence, the trace preorder is substitutive for our operations, whereas it is not substitutive in general for the operations of [6].

mapping that associates a color to each hole. The idea is that we plug into each hole the argument automaton for which the color of the hole serves as placeholder. As a useful analogy one can consider the way in which a term with free variables determines an operation on terms: here the variables play the role of colors, and the occurrences of variables serve as holes. As the rest of its "static" description, a transducer has an associated global set of actions, and, for each color, a local set of actions. The "dynamic" part of a transducer is essentially an automaton: a set of states, a nonempty set of start states, and a step relation. The elements of the step relation are 4-tuples of source state, action, trigger and target state. Here the trigger is a function that tells, for each hole, whether the argument automaton in that hole is supposed to idle, or whether it has to participate in the step, and if so by which action. Finally, each state of the transducer has an associated set of *active holes*, and these are the only ones that can participate in the steps from that state. Formally, an *(action) transducer* T consists of:

- a set $states(T)$ of *states*,
- a nonempty set $start(T) \subseteq states(T)$ of *start states*,
- a set $holes(T)$ of *holes*,
- for each state s, a set $active_T(s) \subseteq holes(T)$ of holes that are *active* in s,
- a set $colors(T)$ of *colors*,
- a map $c\text{-}map_T : holes(T) \rightarrow colors(T)$, the *coloring* of the holes,
- a set $acts(T)$ of *actions* that includes τ,
- for each color m, a set $acts_T(m)$ of *actions* that includes τ,
- a set $steps(T) \subseteq states(T) \times acts(T) \times triggers(T) \times states(T)$, where $triggers(T)$ is the set of maps $\eta : holes(T) \rightarrow (\bigcup_m acts_T(m) \cup \{\perp\})$ such that, for all i, $\eta(i) \in acts_T(c\text{-}map_T(i)) \cup \{\perp\}$. We require that if $(s', a, \eta, s) \in steps(T)$ and $\eta(i) \neq \perp$, then $i \in active_T(s')$.

We define the sets of *external actions* of T by $ext(T) \triangleq acts(T) - \{\tau\}$, and, for each m, $ext_T(m) \triangleq acts_T(m) - \{\tau\}$. We write $s' \xrightarrow[\eta]{a}_T s$ instead of $(s', a, \eta, s) \in steps(T)$. We call s' the *source* of the step, s the *target*, a the *action*, and η the *trigger*. Often we will suppress the subscript T. We often represent a trigger η by the set $\{(i, a) \mid \eta(i) = a \neq \perp\}$. An *execution fragment* of T is a finite or infinite alternating sequence $\alpha = s_0 a_1 \eta_1 s_1 a_2 \eta_2 s_2 \cdots$ of states, actions and triggers of T, beginning with a state, and if it is finite also ending with a state, such that for all i, $s_i \xrightarrow[\eta_{i+1}]{a_{i+1}} s_{i+1}$. An *execution* of T is an execution fragment that begins with a start state.

We view action transducers as a generalization of automata. Thus we will frequently identify an action transducer having an empty set of holes with its underlying automaton.

2.3 Combining Transducers and Automata

We now define the meaning of a transducer as an operation on automata.[5] First, define an *automaton assignment* for T to be a function ζ that maps each color m

[5] In fact, it is often useful to interpret transducers in a more general (and somewhat more complex) way, as operations on transducers.

of T to an automaton in such a way that $acts(\zeta(m)) = acts_T(m)$. Suppose ζ is an automaton assignment for T, and let Z be the composition $\zeta \circ c\text{-}map_T$ (so Z associates an automaton to each hole). Then $T(\zeta)$ is the automaton A given by:

- $states(A) = \{(s, z) \mid s \in states(T)$ and z maps holes i of T to states of $Z(i)\}$,
- $start(A) = \{(s, z) \mid s \in start(T)$ and z maps holes i of T to start states of $Z(i)\}$,
- $acts(A) = acts(T)$, and
- $steps(A)$ is the least relation such that

$$s' \xrightarrow[\eta]{a}_T s \wedge \forall i : [\text{if } \eta(i) = \perp \text{ then } z'(i) = z(i) \text{ else } z'(i) \xrightarrow{\eta(i)}_{Z(i)} z(i)]$$
$$\Rightarrow \quad (s', z') \xrightarrow{a}_A (s, z).$$

2.4 Substitutivity

Now we state our substitutivity result for untimed action transducers. A relation R on a class of automata $\mathcal{A}$ is *substitutive* for an action transducer T if for all automaton assignments ζ, ζ' for T with range $\mathcal{A}$,

$$\forall m \in colors(T) : \zeta(m)\ R\ \zeta'(m) \quad \Rightarrow \quad T(\zeta)\ R\ T(\zeta').$$

An action transducer T is *τ-respecting* if it satisfies the following constraints:

1. For each state s and for each hole i that is active in s, T contains a *clearing step*, i.e., a step $s \xrightarrow[\{(i,\tau)\}]{\tau} s$.
2. If $s' \xrightarrow[\eta]{a} s$ and $\eta(i) = \tau$, then $s' \xrightarrow[\eta]{a} s$ is a clearing step for s' and i.
3. Only finitely many holes participate in each step, i.e., if $s' \xrightarrow[\eta]{a} s$ then $\{i \mid \eta(i) \neq \perp\}$ is finite.

Theorem 1. *The relations $\leq_*$ and $\leq$ on automata are substitutive for all τ-respecting action transducers.*

In the full paper we give an example to show that $\leq_\omega$ is not substitutive, even for τ-respecting action transducers. The converse of Theorem 1 does not hold: there are many examples of non-τ-respecting action transducers for which $\leq_*$ and $\leq$ are substitutive.

2.5 Examples

We give examples of operations that can be expressed as action transducers; these examples include variants of most of the usual operations considered in (untimed) process algebras ([8, 15, 5]).

We first describe a number of conventions so that, in most cases, we do not have to specify the static part of transducers explicitly. Since we only use the ability to make copies of arguments in one of our operations, we adopt the convention that, unless otherwise specified, the sets of holes and colors are equal, and the coloring function is the identity. Often, the set of holes will be an initial fragment $\{1, \ldots, n\}$

of the natural numbers. All action transducers that we define are parametrized by the actions sets of their arguments. By default, there are no restrictions on these action sets. The global action set of a transducer can be obtained by taking the set of all actions that occur in steps of the transducer. Finally, the set of active holes of a state will be implicitly defined as the set of holes that participate in one of the steps starting from that state.

In our language, it is convenient to allow each external action to be structured as a nonempty finite set of *labels*. Sometimes, for uniformity, it will be convenient to identify τ with the empty set. For each transducer T we define $labels(T) \triangleq \bigcup ext(T)$. Similarly we define, for each color m of T, $labels_T(m) \triangleq \bigcup ext_T(m)$. Often we will denote the singleton set $\{l\}$ by the symbol l.

The simplest action transducer is STOP_H. It is parametrized by a set H of actions, has no holes, no colors, no steps, just a single state, which is also start state.

Transducer SKIP denotes the process that terminates successfully. The transducer starts in state s_1, does action $\{\surd\}$ and then stops in state s_2.

$$s_1 \xrightarrow[\emptyset]{\surd} s_2$$

Here $\surd$ is a special label denoting *successful termination*. In accordance with this terminology, our language has been designed in such a way that no further transitions are possible after a transition whose label contains $\surd$. Also, $\surd$ does not occur in the process language itself.

For any nonempty set a of labels with $\surd \notin a$, transducer a starts in state s_1, performs action $a \cup \{\surd\}$ and then stops in state s_2.

$$s_1 \xrightarrow[\emptyset]{a \cup \{\surd\}} s_2$$

Transducer ";" describes the binary operation of *sequential composition*. This operation runs its first argument up to successful termination and then runs its second argument. The transducer has two states s_1, s_2, of which s_1 is the start state, and steps (with $\emptyset$ denoting τ):

$$s_1 \xrightarrow[\{(1,a)\}]{a} s_1 \text{ for } \surd \notin a \in acts_;(1)$$

$$s_1 \xrightarrow[\{(1,a)\}]{a-\{\surd\}} s_2 \text{ for } \surd \in a \in acts_;(1)$$

$$s_2 \xrightarrow[\{(2,a)\}]{a} s_2 \text{ for } a \in acts_;(2)$$

Parametrized by an index set I, the *external choice operation* $\Box$ waits for the first external action of any of its arguments and then runs that argument. The transducer has distinct states s_i, for each $i \in I$, plus an additional state s, which is the only start state. The steps are:

$$s \xrightarrow[\{(i,\tau)\}]{\tau} s \text{ for } i \in I$$

$$s \xrightarrow[\{(i,a)\}]{a} s_i \text{ for } i \in I \wedge a \in ext_\Box(i)$$

$$s_i \xrightarrow[\{(i,a)\}]{a} s_i \text{ for } i \in I \wedge a \in acts_\Box(i)$$

Parametrized by an index set I, transducer $\sqcup$ takes the *disjoint union* of automata indexed by I. Operationally it behaves as the *internal choice operation* $\sqcap$ of CSP [8]: the operation runs one, nondeterministically chosen argument.[6] For each $i \in I$, the transducer has a distinct state s_i, which is also a start state, and steps:

$$s_i \xrightarrow[\{(i,a)\}]{a} s_i \text{ for } i \in I \wedge a \in acts_{\sqcup}(i)$$

Transducer $\|$, which is parametrized by a *finite* index set I, describes the operation of *parallel composition*. The transducer has a single state s, which is the start state, and steps:

$$s \xrightarrow[\{(i,\tau)\}]{\tau} s \text{ for } i \in I$$

$$s \xrightarrow[\eta]{a} s \quad \text{for } \emptyset \neq a \subseteq \bigcup_{j\in I} labels_{\|}(j) \wedge$$
$$\forall i : [\eta(i) = \perp \wedge a \cap labels_{\|}(i) = \emptyset] \vee [\eta(i) = a \cap labels_{\|}(i) \in ext_{\|}(i)]$$

We require that $\surd$ is either in the label set of all arguments or in the label set of none of them.

The postfix *hiding* operation $\backslash L$ hides all labels from the set L by removing them from the actions. The steps are (with $\emptyset$ denoting τ):

$$s \xrightarrow[\{(1,a)\}]{a-L} s \text{ for } a \in acts_{\backslash L}(1)$$

For each function f on labels such that $f(l) = \surd$ implies that $l = \surd$, we introduce a unary *relabeling* operation "f" that renames actions of its argument according to f. The steps are (with f lifted to sets of labels and $\emptyset$ denoting τ):

$$s \xrightarrow[\{(1,a)\}]{f(a)} s \text{ for } a \in acts_f(1)$$

Transducer $\wedge$ describes the binary *interrupt* operation of CSP. The transducer runs its first argument until the second argument performs an external action; after that the first argument is disabled and the second argument takes over. The transducer has two states s_1, s_2, of which s_1 is the start state, and steps:

$$s_1 \xrightarrow[\{(1,a)\}]{a} s_1 \text{ for } a \in acts_{\wedge}(1)$$

$$s_1 \xrightarrow[\{(2,\tau)\}]{\tau} s_1$$

$$s_1 \xrightarrow[\{(2,a)\}]{a} s_2 \text{ for } a \in ext_{\wedge}(2)$$

$$s_2 \xrightarrow[\{(2,a)\}]{a} s_2 \text{ for } a \in acts_{\wedge}(2)$$

In order to guarantee that after a $\surd$-step no other transitions are possible, we require that $\surd \notin labels_{\wedge}(1)$.

[6] We have tried to use CSP notation where possible. However, here we have not followed CSP since we found it confusing to use an intersection-like symbol for the operation of disjoint union.

Now we describe a construct that exploits the ability of transducers to copy their arguments. The construct is inspired by the process creation mechanism of the object-oriented language POOL-T [3]. Transducer $\mathsf{CREATE}(C, I, .)$ is parametrized by a set C of *classes* and a nonempty subset $I \subset C$ of *initial classes*. The transducer takes colors from C. Initially the transducer starts up a single instance of one of the initial classes. Then, each time some process does an action containing a label $new(X)$, for some class X, a new instance of X is created. Formally, the states of CREATE are finite multisets over C, i.e., functions $M : C \rightarrow \mathsf{N}$ that are 0 almost everywhere. The set of start states of the transducer is $\{\{X\} | X \in I\}$, it takes holes from the set $C \times \mathsf{N}^+$, and the color of hole (X, j) is X. The steps are (with $\emptyset$ denoting τ):

$$M \xrightarrow[\{((X,j),a)\}]{a - \{new(Y) | Y \in C\}} M \cup \{Y | new(Y) \in a\} \quad \text{for} \quad M(X) \geq j \wedge a \in acts_{\mathsf{CREATE}}(X)$$

The CREATE construct adds a lot of expressive power to our language, and plays a role similar to the recursion constructs in process algebras such as ACP [5], CCS [15] and CSP [8]. By an example we show how this construct allows us to specify arbitrary automata *inside* our language up to isomorphism. As a useful notation, define

$$a \,.\, X \triangleq a \cup \{new(X)\} \,;\, \mathsf{STOP}$$

for X a class and a a nonempty set of non-*new* labels. Now the expression[7]

$$\begin{aligned} SWITCH \triangleq \mathsf{CREATE}(&\{OFF, ON\}, \{OFF\}, \\ &OFF \mapsto sw\_on \,.\, ON \\ &ON \mapsto sw\_on \,.\, ON \,\Box\, sw\_off \,.\, OFF \;) \end{aligned}$$

denotes a finite automaton that describes an automatic switch off mechanism. The system allows a lamp to be switched on at any time; once it has been switched on, it can be switched off (automatically). In Section 3 we will come back to this example and show how we can add real-time constraints to make it more interesting.

Using the CREATE operation with a single argument, we can define the *looping* operation ω, which restarts its argument upon each successful termination.

$$(A)^\omega \triangleq \mathsf{CREATE}(\{X\}, \{X\}, f(A)),$$

where f relabels $\surd$ to $new(X)$ and leaves all other labels unchanged.

Note that all of the operations described in this subsection are τ-respecting. Therefore, it follows from Theorem 1 that the preorders $\leq_*$ and $\leq$ are substitutive for all these operations.

---

[7] We warn the reader that, even though the notation we use here is very close to the standard notation in process algebra for systems of recursion equations, the operational semantics is quite different in the case of nonlinear systems.

3 The Timed Setting

3.1 Timed Automata

We use a slight variant of the *timed automaton* model from [12].[8] A *timed automaton* A is an automaton whose set of actions is a superset of R^+, the set of positive reals. Actions from R^+ are referred to as *time-passage actions*. We let $d, d', \ldots$ range over time-passage actions and, more generally, over the set R of real numbers. The set of *visible* actions is defined by $vis(A) \triangleq ext(A) - \mathsf{R}^+$. We assume that a timed automaton satisfies the following axioms:

S1 If $s' \xrightarrow{d} s''$ and $s'' \xrightarrow{d'} s$, then $s' \xrightarrow{d+d'} s$.

S2 If $s' \xrightarrow{d} s$ then there exists a trajectory from s' to s of length d.

Here a *trajectory from s' to s of length d* is a function $w : [0, d] \rightarrow states(A)$ such that $w(0) = s'$, $w(d) = s$, and $w(d_1) \xrightarrow{d_2 - d_1} w(d_2)$ for all $d_1, d_2 \in [0, d]$ with $d_1 < d_2$.

3.2 Timed Traces

Suppose $\beta = a_0 a_1 a_2 \cdots$ is a trace of a timed automaton A. For each i, we define the *time of occurrence* t_i of event a_i by:

$$t_0 = 0$$
$$t_{i+1} = \textbf{if } a_i \in \mathsf{R}^+ \textbf{ then } t_i + a_i \textbf{ else } t_i$$

The *limit time* of β, notation $\beta.ltime$, is the smallest element of $\mathsf{R}^{\geq 0} \cup \{\infty\}$ larger than or equal to (i.e., the supremum of) all the t_i. The *timed trace t-trace*(β) associated with β is defined by:

$$t\text{-}trace(\beta) \triangleq (((a_0, t_0)(a_1, t_1)(a_2, t_2) \cdots) \lceil (vis(A) \times \mathsf{R}^{\geq 0}), \beta.ltime)$$

So *t-trace*(β) records the visible events of β paired with their time of occurrence, as well as the limit time of the sequence. The above definitions are lifted to executions in the obvious way. For α an execution of A, $\alpha.ltime \triangleq trace(\alpha).ltime$ and $t\text{-}trace(\alpha) \triangleq t\text{-}trace(trace(\alpha))$. Execution α is *admissible* if $\alpha.ltime = \infty$.

A pair p is a *timed trace* of A if it is the timed trace of some finite or admissible execution of A. Thus, we explicitly exclude the timed traces that originate from *Zeno* executions, i.e., infinite executions with a finite limit time. We write *t-traces*(A) for the set of all timed traces of A, *t-traces*$^*(A)$ for the set of *finite* timed traces, i.e., those that originate from a finite execution, and *t-traces*$^\infty(A)$ for the *admissible* timed traces, i.e., those that originate from an admissible execution. These notions induce three preorders on timed automata: $A \leq^t B \triangleq t\text{-}traces(A) \subseteq t\text{-}traces(B)$, $A \leq^t_* B \triangleq t\text{-}traces^*(A) \subseteq t\text{-}traces^*(B)$, and $A \leq^t_\infty B \triangleq t\text{-}traces^\infty(A) \subseteq t\text{-}traces^\infty(B)$. The kernels of these preorders are denoted by $\equiv^t$, $\equiv^t_*$ and $\equiv^t_\infty$, respectively.

[8] The difference is just the explicit indication of the amount of elapsed time in the time-passage action instead of using a *now* function that associates the current time to a state.

3.3 Timed Action Transducers

A *timed action transducer* T is a transducer with $\mathsf{R}^+ \subseteq acts(T)$ and, for all colors m, $\mathsf{R}^+ \subseteq acts_T(m)$. The sets of *visible* actions are defined by $vis(T) \triangleq ext(T) - \mathsf{R}^+$ and, for all m, $vis_T(m) \triangleq ext_T(m) - \mathsf{R}^+$. We assume that T satisfies the following axioms.

T1 If $s' \xrightarrow[\eta]{d} s$ then $\eta(i) = d$ for all $i \in active_T(s')$.

(As a consequence of axiom **T1**, the trigger of a d step is fully determined. This justifies our convention below to write $s' \xrightarrow{d} s$ instead of $s' \xrightarrow[\eta]{d} s$).

T2 If $s' \xrightarrow[\eta]{a} s$ and $\eta(i) = d$, then $a = d$.

T3 If $s' \xrightarrow{d} s$ then $active_T(s') = active_T(s)$.

We also assume that T satisfies the axioms **S1** and **S2** for timed automata (with triggers added implicitly according to the convention above).

An automaton assignment ζ for a timed action transducer T is called *timed* if it maps each color to a timed automaton.

Lemma 2. *Suppose T is a timed action transducer and ζ is a timed automaton assignment for T. Then $T(\zeta)$ is a timed automaton.*

3.4 Substitutivity

A timed action transducer T is *Zeno respecting* if for each execution $s_0 a_1 \eta_1 s_1 a_2 \eta_2 \cdots$ of T in which the sum of the time passage actions in $a_1 a_2 \cdots$ grows to ∞, and for each hole i, either the sum of the time passage actions in $\eta_1(i)\eta_2(i) \cdots$ grows to ∞ as well, or there exists an index n such that $\eta_m(i) = \perp$ for all $m \geq n$.

Transducers that are not Zeno respecting can turn a Zeno execution of an argument into a non-Zeno execution, by deactivating the argument infinitely often while advancing time. Since timed trace inclusion does not preserve Zeno traces, this means that $\leq^t$ will in general not be substitutive for such transducers. However, we do have the following result.

Theorem 3. *The relations $\leq^t_*$ and $\leq^t$ on timed automata are substitutive for all Zeno- and τ-respecting timed action transducers.*

3.5 Examples

We give examples of operations that can be expressed as timed action transducers.

Timed Transducers from Untimed Transducers An important collection of timed transducers can be obtained from untimed transducers. In this subsection we present a simple but important construction, essentially due to Nicollin and Sifakis [18], that transforms an untimed action transducer into a timed one, by inserting arbitrary time-passage steps. Suppose T is an (untimed) action transducer whose sets of actions are disjoint with R^+. Then the structure *patient*(T) is obtained from T by adding R^+ to all the action sets, and also adding, for each state s of T and each $d \in R^+$, a step

$$s \xrightarrow[\eta]{d} s, \quad \text{where, for all } i, \eta(i) = \textbf{if } i \in active_T(s) \textbf{ then } d \textbf{ else } \perp .$$

It is straightforward to check that *patient*(T) is indeed a timed action transducer.

If T is τ-respecting then *patient*(T) is not τ-respecting in general. For instance, the transducer $\Box$ for external choice over an infinite index set of Section 2.5 is τ-respecting, but its patient timed version is not. The following simple lemma characterizes the situations in which the *patient* operation preserves the property of being τ-respecting.

Lemma 4. *Suppose T is a τ-respecting action transducer whose sets of actions are disjoint with R^+. Then patient(T) is Zeno respecting. Moreover, patient(T) is τ-respecting iff in each state of T only finitely many holes are active.*

Except for external choice over an infinite index set, all the patient timed versions of the transducers of Section 2.5 are τ-respecting, by Lemma 4. Thus, by Theorem 3, the timed trace preorders $\leq^t_*$ and $\leq^t$ are substitutive for the *patient* variants of all the transducers of Section 2.5 except for infinitary external choice.

The timed transducers obtained by the *patient* construction turn out to be quite useful, so in the subsequent sections we will adopt the convention that T means *patient*(T) for any of the transducers of Section 2.5.

Timers We consider a set X of *timer variables*, ranged over by $x, y, \ldots$. The set of *timer constraints* ϕ is defined inductively by:

$$\phi ::= x < d \mid x = d \mid \phi \wedge \phi' \mid \neg\phi$$

Note that constraints such as true, $5 < 4$, $x \geq d$, $x \in [2,5)$ can be defined as abbreviations.

A *time assignment* $\xi : X \to R^{\geq 0}$ assigns a nonnegative real value to each timer variable. We say that a time assignment ξ for *satisfies* a timer constraint ϕ, notation $\xi \models \phi$, iff ϕ evaluates to true using the values given by ξ. We say that ϕ is a *tautology* iff for all time assignments ξ, $\xi \models \phi$. We say that ϕ is *satisfiable* iff there exists a time assignment ξ such that $\xi \models \phi$. We denote by $\phi[d/x]$ the formula obtained from ϕ by replacing all occurrences of x by d.

The transducer TIMER^x_d models the behavior of a timer called x. The argument $d \in R^{\geq 0} \cup \{\infty\}$ gives a *bound* beyond which time cannot proceed.[9] The state set of the transducer is $R^{\geq 0} \times (R^{\geq 0} \cup \{\infty\})$ and the initial state is $(0, d)$. There is just one argument, that is, one hole and one color. The argument can reset timer x at

[9] For simplicity, we do not consider timers with strict bounds. Such timers can be defined by parametrizing the transducer with an additional boolean that tells whether the time bound is strict or not. Alternatively, one can follow a suggestion of Abadi and Lamport [1], and introduce, as additional elements of the time domain, the set of all 'infinitesimally shifted' real numbers r^-, where $t \leq r^-$ iff $t < r$, for any reals t and r.

any moment via a label $x := 0$; similarly the upper bound can be modified via a label $x :\leq d$. Besides assigning values to the timer, an automaton can use timer constraints as labels to test the values of the various timers in whose scope it occurs. TIMER_d^x has, for instance, a step

$$(1,10) \xrightarrow[\{(1,8.5)\}]{8.5} (9.5,10), \quad \text{but not a step} \quad (1,10) \xrightarrow[\{(1,9.5)\}]{9.5} (10.5,10),$$

because that would violate the time bound. If $a = \{sw\_off, 9 \leq x, x \leq 10, x :\leq \infty\}$ and $b = \{sw\_off\}$, then TIMER_d^x also has a step

$$(9.5,10) \xrightarrow[\{(1,a)\}]{b} (9.5,\infty), \quad \text{but not a step} \quad (1,10) \xrightarrow[\{(1,a)\}]{b} (1,\infty),$$

since a contains a constraint $9 \leq x$.

In order to define the step relation formally it is convenient to define some auxiliary functions. For x a timer, $d_1, d_2 \in \mathsf{R}^+$ and a a set of labels, $\mathcal{A}(x, d_1, a)$ is obtained from a by first removing all labels $x := 0$ and $x :\leq d$ (for all d) then replacing each time constraint ϕ in a by $\phi[d_1/x]$, and finally removing all tautologies from the result. We say $\mathcal{A}(x, d_1, a)$ is *satisfiable* if all time constraints contained in it are satisfiable. We also define

$$\mathcal{V}(x, d_1, a) \triangleq \textbf{if } x := 0 \in a \textbf{ then } 0 \textbf{ else } d_1,$$
$$\mathcal{B}(x, d_2, a) \triangleq \textbf{if } \{d | x :\leq d \in a\} = \emptyset \textbf{ then } d_2 \textbf{ else } \min\{d | x :\leq d \in a\}.$$

Now the steps of TIMER_d^x can be defined by (with $\emptyset$ denoting τ):

$$(d_1, d_2) \xrightarrow[\{(1,d)\}]{d} (d_1 + d, d_2) \qquad \text{for } d_1 + d \leq d_2,$$
$$(d_1, d_2) \xrightarrow[\{(1,a)\}]{b} (\mathcal{V}(x, d_1, a), \mathcal{B}(x, d_2, a)) \text{ for } b = \mathcal{A}(x, d_1, a) \text{ satisfiable.}$$

The reader can easily check that TIMER_d^x is Zeno- and τ-respecting. Thus relations $\leq_*^t$ and $\leq^t$ are substitutive for this transducer.

Our definition of a timer is similar to the one proposed by Alur and Dill [2] for their timed ω-automata. However, instead of a Büchi style acceptance criterion we use bounds to specify *urgency*, i.e., properties that say that certain actions *must* occur at a certain time. We suppose that for some applications, it will be useful to have a more general definition. One can, for instance, extend the set of time constraints with formulas like $x + y < 1$, or allow for assignments of the form $x := y + 4$, or introduce operations that ask the timer to emit its current time. The important point here is that explicit timers constitute an important and useful construct in real-time process algebra. Our specific choice of operations is just an example, subject to modification.

A Timed Process Algebra We think that a reasonable proposal for a timed process algebra might include *only* (1) the timed transducers obtained by applying the *patient* operation to the untimed transducers discussed earlier, and (2) the *timers* described above. A justification for this is that these are sufficient to implement all the other timed operators we have thought of. In this section, we give some of these derivations.

Using a single timer, we can define the process WAIT d of Timed CSP [21, 7], which waits time d and then terminates successfully.

$$\mathsf{WAIT}\ d \triangleq \mathsf{TIMER}^x_d(x = d)$$

More generally, we can specify a process that terminates successfully after waiting some nondeterministically chosen time from the closed interval $[d_1, d_2]$.

$$\mathsf{WAIT}\ [d_1, d_2] \triangleq \mathsf{TIMER}^x_{d_2}(x \geq d_1)$$

Using a timer with deadline 0, we can restrict any automaton to its behavior at time 0. We define

$$\overline{A} \triangleq \mathsf{TIMER}^x_0(A)$$

The above construction is useful for defining the *timeout* construct of Timed CSP. For a given delay d this operator is defined, as in [7], by

$$A \overset{d}{\triangleright} B \triangleq (A \,\Box\, (\mathsf{WAIT}\ d\ ;\ \overline{abort}\ ;\ B)) \backslash \{abort\}$$

If, at time d, A has not performed any visible action, an interrupt occurs and automaton B is started. Note that we need the auxiliary label *abort* (not in the label set of A and B) to force the choice between A and B at time d.

The *execution delay* operator of ATP [17] is given by:

$$\begin{aligned} \lceil A \rceil^d(B) \quad &\triangleq \quad (\mathsf{TIMER}^x_d((A \wedge (\{abort, x = d\}\ ;\ B))\|C))\backslash\{abort, cancel\} \\ &\text{where } C = \{cancel, x :\leq \infty\}\ ;\ \mathsf{STOP} \,\Box\, \{abort, x :\leq \infty\}\ ;\ \mathsf{SKIP} \end{aligned}$$

$\lceil A \rceil^d(B)$ behaves as A until time d; at time d, A is interrupted and B is started. However, if A performs a special label *cancel*, then the interrupt is cancelled and A can continue to run forever. The auxiliary process C takes care that once A has done *cancel*, it can no longer be interrupted by B. Also C removes deadline d after a *cancel* or *abort* action. We assume that A and B do not have *abort* in their label set, nor any label referring to timer x. For simplicity we also assume that A does not contain the termination symbol $\surd$. The labels *cancel* and *abort* are hidden so that they cannot communicate with any other action. A minor difference between our execution delay operator and the one from ATP is that ours allows machine A to perform visible actions *at* time d.

As an illustration of the use of the timed operators presented thus far, we specify a timed version of the automatic switch-off mechanism we described in Section 2.5. The system allows a lamp to be switched on at any time; then between 9 and 10 time units after the last time the lamp has been switched on, it will be switched off.

$$\begin{aligned} &T\_SWITCH \triangleq \mathsf{TIMER}^x_\infty(\mathsf{CREATE}(\{OFF, ON\}, \{OFF\}, \\ &\qquad OFF \mapsto \{sw\_on, x := 0, x :\leq 10\}\ .\ ON \\ &\qquad ON \mapsto \{sw\_on, x := 0\}\ .\ ON \,\Box\, \{sw\_off, 9 \leq x \leq 10, x :\leq \infty\}\ .\ OFF\ )) \end{aligned}$$

Counterexamples Although the converse of Theorem 3 does not hold, our result appears to be quite sharp: for many examples of timed transducers that are not τ-respecting, the timed trace preorders are indeed not substitutive. The best example is the CCS-like choice operator + that plays a dominant role in many real-time process calculi (TCCS [16], the timed extension of CCS proposed in [23], ATP [17], and ACPρ [4]). This operator can be viewed as the *patient* version of the choice operator from CCS. Relation $\leq_*$ is not substitutive for + because

$$\mathsf{WAIT}\ 2 + \mathsf{WAIT}\ 1.5 \quad \not\leq^t_* \quad (\mathsf{WAIT}\ 1\ ;\ \mathsf{WAIT}\ 1) + \mathsf{WAIT}\ 1.5$$

The first process will always terminate at time 1.5, whereas the second process will always terminate at time 2.

4 Nontriviality of Implementations

As explained in the Introduction, we exclude trivial implementations by distinguishing input and output actions and requiring implementations to be input enabled and I/O feasible. In a process algebraic setting, a natural way to prove the property $\phi \triangleq$ *input enabled* $\wedge$ *I/O feasible*, is to describe a timed automaton purely in terms of operators that preserve some property ψ that implies ϕ. In this section we describe such a ψ, i.e., a property that implies input enabledness and I/O feasibility and that is preserved by an interesting class of operators.

We begin by defining *timed I/O automata* and *I/O feasibility.*

4.1 Timed I/O Automata

A *timed I/O automaton A* is an timed automaton together with a set $inp(A) \subseteq vis(A)$ of *input actions.* We require that input actions always be enabled, i.e., for each state s and for each $a \in inp(A)$ there is a step $s \xrightarrow{a}_A s'$.

We define the set $out(A)$ of *output actions* by $out(A) \triangleq vis(A) - inp(A)$, and the set $loc(A)$ of *locally controlled actions* by $loc(A) \triangleq out(A) \cup \{\tau\}$. A step $s' \xrightarrow{a} s$ is said to be *under control of A* if $a \in loc(A)$. By *t-AUT*(A) we denote the timed automaton that underlies timed I/O automaton A. Conversely, for each timed automaton A and $IN \subseteq vis(A)$ a set of actions that are enabled in each state of A, $IO_{IN}(A)$ denotes the corresponding timed I/O automaton.

Timed I/O automata constitute a variant of the I/O automata model of Lynch and Tuttle [11] within our timed setting.[10]

4.2 I/O Feasibility

A timed I/O automaton is *I/O feasible* if each finite execution can be extended to an admissible execution via an execution fragment that does not contain input actions. I/O feasibility strengthens the notion of *feasibility* used in [13, 12], which simply requires that each finite execution can be extended to an admissible one.

[10] They do not include the class structure, used in I/O automata to model fairness.

This means that all the results of [13, 12] for feasible automata are valid for I/O feasible automata. However, the requirement of feasibility is too weak to exclude trivial implementations, because it may be that a feasible implementation can only produce an admissible execution in case the environment provides certains inputs. Thus if the environment does not offer these inputs, a Zeno execution is produced. To illustrate the problem, we consider as a specification the timed switch of the previous section, with as additional "feature" that it can break down at any moment and stop functioning:

$$IO_{\{sw\_on, br\_down\}}(T\_SWITCH \wedge (br\_down\ ;\ (sw\_on \,\square\, br\_down)^{\omega}))$$

This specification is implemented by the following feasible timed I/O automaton:

$$IO_{\{sw\_on, br\_down\}}((\overline{sw\_on})^{\omega} \wedge (br\_down\ ;\ (sw\_on \,\square\, br\_down)^{\omega}))$$

But this is an undesirable implementation because, first, the machine will never turn the switch off, even if no breakdown occurs, and moreover, the machine will not allow time to advance unless there is an immediate breakdown. The notion of I/O feasibility excludes this type of implementation.

4.3 Timed I/O Transducers

An I/O automaton should always be willing to engage in input actions, because these are thought of as being under the control of the environment. When defining operations on I/O automata, it is natural to require, as a dual property, that the environment (i.e., the operation) cannot block output actions, as these are under the control of the automata. In the untimed setting the nonblocking condition for outputs can be motivated technically because it is needed to obtain substitutivity of the *quiescent* and *fair* trace preorders [22]. Interestingly, the nonblocking condition also has a technical motivation in the timed case, but a different one: it is needed to preserve I/O feasibility. To illustrate this, we consider a trivial example of two I/O feasible timed I/O automata, A and B, whose parallel composition is not I/O feasible.

$$A = IO_{\emptyset}((\mathsf{WAIT}\ 1\ ;\ \overline{tick})^{\omega}) \qquad B = IO_{\emptyset}(\mathsf{STOP}_{\{tick\}})$$

Machine A describes a perfect clock that perform a *tick* action after each unit of time. Machine B does nothing except allow time to pass. Both machines are trivially input enabled (since there are no inputs) and I/O feasible. Still, in the parallel composition $A \| B$, time cannot proceed past 1, because B refuses to synchronize with the output action of A. Time deadlocks like this can be avoided if we do not allow for contexts (environments) that can block output actions. This leads us to the following definition.

A *timed I/O transducer* T is a timed transducer together with a set $inp(T) \subseteq vis(T)$ and, for each m, a set $inp_T(m) \subseteq vis_T(m)$. Analogous to the automaton case, we define derived sets $out(T) \triangleq vis(T) - inp(T)$, $loc(T) \triangleq out(T) \cup \{\tau\}$, and, for each m, $out_T(m) \triangleq vis_T(m) - inp_T(m)$ and $loc_T(m) \triangleq out_T(m) \cup \{\tau\}$. We require

- *Input enabling.* For all states s and for all $a \in inp(T)$, T has a step $s \xrightarrow[\eta]{a} s'$, such that, for all i, $\eta(i) \in inp_T(c\text{-}map_T(i)) \cup \{\perp\}$.

- *No output blocking.* For all states s, all active holes i of s, and all actions $a \in out_T(c\text{-}map_T(i))$, T has a step $s \xrightarrow[\eta]{b} s'$, with $b \in loc(T)$, $\eta(i) = a$, and, for all $i' \neq i$, $\eta(i') \in inp_T(c\text{-}map_T(i')) \cup \{\perp\}$.

By $t\text{-}TRANS(T)$ we denote the timed action transducer that underlies timed I/O transducer T. We say that a step $s' \xrightarrow[\eta]{a} s$ is *under control of* T if $a \in loc(T)$ and for all i, $\eta(i) \in inp_T(c\text{-}map_T(i)) \cup \{\perp\}$.

The conditions of input enabling and no output blocking for timed action transducers are direct translations of similar constraints presented in [22]. An important example of a timed I/O transducer can be obtained by taking the patient version of the composition operation of the I/O automaton model [11]. In fact, this operation can be viewed as a special case of of our parallel composition operator with additional requirements on the action sets of the arguments. These requirements are: (1) all visible actions are singletons, (2) the sets of output actions of different arguments are disjoint. The input actions of the composition are then defined as the union of the input actions of the arguments minus the union of the output actions of the arguments.

Suppose T is a timed I/O transducer. A *timed I/O automaton assignment* for T is a function ζ that maps each color m of T to a timed I/O automaton in such a way that $inp(\zeta(m)) = inp_T(m)$. Let T be a timed I/O transducer and let ζ be a timed I/O automaton assignment for T. Then we define $T(\zeta)$ to be the structure A given by $t\text{-}AUT(A) \triangleq t\text{-}TRANS(T)(t\text{-}TRANS \circ \zeta)$ and $inp(A) \triangleq inp(T)$.

Lemma 5. *Let T be a timed I/O transducer and let ζ be a timed I/O automaton assignment for T. Then $T(\zeta)$ is a timed I/O automaton.*

4.4 Feasibility Revisited

It turns out that I/O feasibility is not preserved by parallel composition. Consider the following two timed I/O automata.[11]

$$
\begin{aligned}
FAST\_SWITCH \triangleq IO_{\{sw\_on\}}\ (\mathrm{CREATE}(&\{OFF, ON\}, \{OFF\},\\
&OFF \mapsto sw\_on\ .\ ON\\
&ON \mapsto sw\_on\ .\ ON \ \Box\ \underline{\overline{sw\_off\ .\ OFF}}\ ))
\end{aligned}
$$

$$
\begin{aligned}
FAST\_USER \triangleq \underline{IO_{\{sw\_off\}}}\ (\mathrm{CREATE}(&\{OFF, ON\}, \{OFF\},\\
&OFF \mapsto \underline{sw\_on\ .\ ON} \ \Box\ sw\_off\ .\ OFF\\
&ON \mapsto sw\_off\ .\ OFF\ ))
\end{aligned}
$$

The first expression denotes a fast version of the timed switch of Section 3: each time the switch gets turned on the machine turns it off immediately. The second expression describes a fast user of this switch, who initially turns the switch on, and then, each time it goes off, immediately turns it on again. Both systems are I/O

[11] A similar example is given by Abadi and Lamport [1].

feasible, but their parallel composition is not. The problem arises from the ability of both systems to respond immediately to a given input. One way to avoid this problem is to introduce a ***system delay constant***, as in in Timed CSP: a minimum amount of time that has to pass between each pair of locally controlled actions. However, this seems too drastic and artificial to us, and the extra constants would probably complicate the task of analyzing system timing behavior. Fortunately, system delays are not needed to preserve I/O feasibility. Instead the following conditions suffice.

A timed I/O automaton A is ***strongly I/O feasible*** if it is I/O feasible and time grows unboundedly in any execution that contains infinitely many steps that are under control of A. This generalizes to transducers as follows. A timed I/O transducer T is *I/O feasible* if each finite execution can be extended to an admissible execution via an execution fragment that contains only time-passage steps and steps that are under the control of the transducer. T is ***strongly I/O feasible*** if in addition:

1. In any execution that contains infinitely many steps that are under control of T, time grows unboundedly.
2. In any execution for which the union over all states of their sets of active holes is infinite, time grows unboundedly.

The second condition is needed to exclude situations in which, each time an input arrives, a new machine is activated to produce an immediate output. We have the following result:

Theorem 6. *Suppose T is a strongly I/O feasible timed I/O transducer and ζ is a timed I/O automaton assignment for T that maps each color of T to a strongly I/O feasible timed I/O automaton. Then $T(\zeta)$ is a strongly I/O feasible timed I/O automaton.*

Note that the timer construct proposed above is *not* strongly I/O feasible, because it does not permit time to pass beyond its upper bound. However, the timed I/O transducer for parallel composition is strongly I/O feasible.

As a topic for future research it remains to find a collection of strongly I/O feasible I/O transducers that is sufficiently expressive to describe implementations of real-time systems. Ideally, these action transducers would be defined in terms of the action transducers of Section 3. Process algebraic and/or assertional proof techniques could then be applied to prove that a system described in terms of feasible I/O transducers implements a system described in terms of the higher-level primitives of Section 3.

References

1. M. Abadi and L. Lamport. An old-fashioned recipe for real time. In *Proceedings of the REX Workshop "Real-Time: Theory in Practice"*, LNCS 600. Springer-Verlag, 1992. To appear.
2. R. Alur and D.L. Dill. Automata for modeling real-time systems. In *Proceedings 17th ICALP*, LNCS 443, pages 322–335. Springer-Verlag, July 1990.
3. P. America. POOL-T — A parallel object-oriented language. In A. Yonezawa and M. Tokoro, editors, *Object-Oriented Concurrent Systems*. MIT Press, 1986.

4. J.C.M. Baeten and J.A. Bergstra. Real time process algebra. *Journal of Formal Aspects of Computing Science*, 3(2):142–188, 1991.
5. J.C.M. Baeten and W.P. Weijland. *Process Algebra.* Cambridge Tracts in Theoretical Computer Science 18. Cambridge University Press, 1990.
6. B. Bloom, S. Istrail, and A.R. Meyer. Bisimulation can't be traced: preliminary report. In *Conference Record of the* 15th *ACM Symposium on Principles of Programming Languages*, pages 229–239, 1988.
7. J. Davies and S. Schneider. An introduction to Timed CSP. Technical Monograph PRG-75, Oxford University Computing Laboratory, Programming Research Group, August 1989.
8. C.A.R. Hoare. *Communicating Sequential Processes.* Prentice-Hall International, Englewood Cliffs, 1985.
9. A.S. Klusener. Abstraction in real time process algebra. Report CS-R9144, CWI, Amsterdam, October 1991.
10. K.G. Larsen and L. Xinxin. Compositionality through an operational semantics of contexts. In *Proceedings* 17th *ICALP*, LNCS 443, pages 526–539. Springer-Verlag, July 1990.
11. N.A. Lynch and M.R. Tuttle. Hierarchical correctness proofs for distributed algorithms. In *Proceedings of the* 6th *Annual ACM Symposium on Principles of Distributed Computing*, pages 137–151, August 1987. A full version is available as MIT Technical Report MIT/LCS/TR-387.
12. N.A. Lynch and F.W. Vaandrager. Forward and backward simulations – part II: Timing-based systems, 1992. In preparation.
13. N.A. Lynch and F.W. Vaandrager. Forward and backward simulations for timing based systems. *Proceedings of the REX Workshop "Real-Time: Theory in Practice"*, LNCS 600. Springer-Verlag, 1992. To appear.
14. M. Merritt, F. Modugno, and M. Tuttle. Time constrained automata. In *Proceedings CONCUR 91*, LNCS 527, pages 408–423. Springer-Verlag, 1991.
15. R. Milner. *Communication and Concurrency.* Prentice-Hall International, Englewood Cliffs, 1989.
16. F. Moller and C. Tofts. A temporal calculus of communicating systems. In *Proceedings CONCUR 90*, LNCS 458, pages 401–415. Springer-Verlag, 1990.
17. X. Nicollin and J. Sifakis. The algebra of timed processes ATP: Theory and application (revised version). Technical Report RT-C26, LGI-IMAG, Grenoble, France, November 1991.
18. X. Nicollin and J. Sifakis. An overview and synthesis on timed process algebras. In *Proceedings CAV 91*, LNCS 575, pages 376–398. Springer-Verlag, 1992.
19. X. Nicollin, J. Sifakis, and S. Yovine. From ATP to timed graphs and hybrid systems. *Proceedings of the REX Workshop "Real-Time: Theory in Practice"*, LNCS 600. Springer-Verlag, 1992. To appear.
20. G.D. Plotkin. A structural approach to operational semantics. Report DAIMI FN-19, Computer Science Department, Aarhus University, 1981.
21. G.M. Reed and A.W. Roscoe. A timed model for communicating sequential processes. *Theoretical Computer Science*, 58:249–261, 1988.
22. F.W. Vaandrager. On the relationship between process algebra and input/output automata. In *Proceedings* 6th *Annual Symposium on Logic in Computer Science*, pages 387–398. IEEE Computer Society Press, 1991.
23. Wang Yi. Real-time behaviour of asynchronous agents. In *Proceedings CONCUR 90*, LNCS 458, pages 502–520. Springer-Verlag, 1990.

This article was processed using the LaTeX macro package with LLNCS style

Compositional Verification of Probabilistic Processes

Kim G. Larsen & Arne Skou
Aalborg University, Denmark *

Abstract

We introduce a simple calculus of probabilistic processes and we apply it as basis for an initial investigation of compositional verification for probabilistic processes. In particular we study the problem of decomposing logical specifications with respect to operators of the calculus. This study identifies a new probabilistic logic, which is needed in order to support decomposition. Complete axiomatizations are offered for both the calculus and the logic.

1 Introduction

When specifying real-life parallel systems, there is often a need to express requirements on the *stochastic* behaviour of a system and its individual components. For example, when specifying a communication medium we cannot hope to acquire a completely error-free device. Realistically, we have to be content as long as the probability of the medium loosing messages is sufficiently small. Therefore it is desirable as well as necessary to extend traditional specification formalisms such as logics, nets or algebras with notions of *probabilistic behaviour*. For *process algebras*, the work on such extensions has just begun recently [GJS90, LS89, HJ90, Chr90, CSZ91, JL91], and in this paper we make an initial investigation of the important question of *compositionality* for probabilistic process algebras.

The study of compositionality becomes important in a top-down design approach. In this approach, we want to decompose specifications of a composed system into sufficient and necessary specifications af the components of the system. Typically the composed system is of the form $C(X_1, \ldots, X_n)$ where C is a derived operator (context) of the given process algebra.

Now, assuming that the component processes $X_1 \ldots X_n$ are yet to be designed, it is desirable to decompose a given specification S for the combined system into individual subspecifications $S_1 \ldots S_n$ for the (unknown) subcomponents. This will allow the future design of these components to be carried out independently of one another. The decomposition should be sound in the following sense:

$$\textit{Whenever } X_1 \text{ sat } S_1, \ldots, X_n \text{ sat } S_n \textit{ then } C(X_1, \ldots, X_n) \text{ sat } S$$

Moreover — while maintaining soundness in the above sense — we want the specifications $S_1 \ldots S_n$ to be as weak as possible as this will leave the possible implementations of $X_1 \ldots X_n$ as open as possible. For classical process calculi (such as CCS) and logical specification formalisms (such as

*Address: Dep. of Math. and Comp. Sc., Aalborg University, Fredrik Bajersvej 7, 9220 Aalborg, Denmark. Telephone: +45 98158522. Email: kgl@iesd.auc.dk. This work has been supported in part by the Danish Natural Science Research Council through the DART project.

Hennessy Milner Logic [HM85]) decomposition results with the above properties have been obtained during the last few years [LX90, AW91].
For *probabilistic processes* the notion of compositionality is somewhat less understood. Only a few proposals have been made for probabilistic calculi [GJS90, HJ90], and the question of decomposition and related logics [LS89] has not been investigated previously. This motivates the main results of the present paper, where we define (and axiomatize) a new calculus of probabilistic processes and identify (and axiomatize) a new probabilistic modal logic which we prove decomposable with respect to the derived unary operators of the proposed calculus.

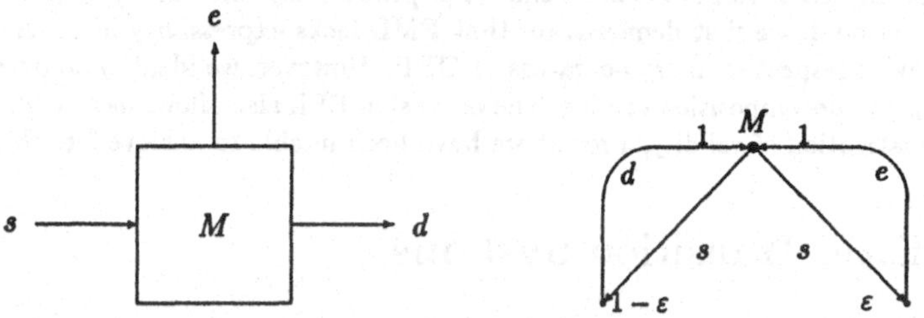

Figure 1: A faulty medium

To further motivate the potential practical relevance of such decomposition results, consider a faulty medium M which can receive a signal from a sender through a port s. Having received the signal, the medium may either deliver the signal correctly through a port d, or it may report a failure to the sender through port e. Assuming that the probability of entering the error reporting state is known to be ε, we can model the medium by a probabilistic transition system (see figure 1), where the arrows are decorated with transition probilities. Informally such a medium has the property 'Whenever an s is observed, the medium may with probability $1-\varepsilon$ enter a state, where d can be observed', or more formally using the probabilistic version of Hennessy–Milner Logic presented in [LS89]

$$M \text{ sat } \langle s\rangle_{\geq 1-\varepsilon}\langle d\rangle_{\geq 1}\text{tt}$$

Now wanting to increase the transmission reliability (to $1-\delta$ say) we may try to add an extra medium X in parallel with M. The desired property of the combined medias may then be expressed as 'After observing an s on both media, the probability for *both* of them having entered an error reporting state must not exceed δ'. Assuming that we can model the duplication of media by a SCCS-like parallel combinator $\times$ [Mil83], we may formulate the requirement as

$$M \times X \text{ sat } \langle ss\rangle_{\leq\delta}\langle ee\rangle_{1}\text{tt} \tag{1}$$

The important decomposition question is then: How faulty may the unknown medium X behave without damaging the desired property above? That is, can we find a property $G^{\times}_{M,F}$ (derived from M, the property $F = \langle ss\rangle_{\leq\delta}\langle ee\rangle_{1}\text{tt}$ and the parallel combinator $\times$) such that

$$X \text{ sat } G^{\times}_{M,F} \text{ if and only if } M \times X \text{ sat } F$$

In the present paper we answer this question from 3 different angles: First we demonstrate that the probabilistic version of Hennessy–Milner Logic introduced in [LS89] does *not* posses sufficient expressivity in order to define G in general (in contrast to the classical non-probabilistic case). To gain sufficient expressivity we introduce an extended version of the logic, which enables the desired decomposition, and finally we provide an axiomatization of the new logic that may be applied to synthesize the unknown medium X from the component property G.

Section Overview

In section 2 we review from [LS89] the basic notions of (reactive) probabilistic transition system, probabilistic bisimulation and probabilistic modal logic (PML). In section 3 we present a simple probabilistic calculus (CPP) together with its axiomatization for finite processes. Also, we show that it is possible for any CPP term to express the set of probabilistic bisimilar processes as a single PML formula. In section 4, we first demonstrate that PML lacks expressivity in order to support decompositionality with respect to unary operators of CPP. However, we identify an extended logic EPL, for which we prove decomposition results. The extension EPL also allows us to offer (in section 5) a complete axiomatization of validity, a result we have been unable to achieve for PML itself.

2 Probabilistic Transition Systems

As mentioned in the introduction, we want to study reactive systems where non-deterministic choices are described probabilistically. In order to define the semantics of such systems, we adopt the well-known notion of labeled transition systems and extend it with information about the transition probabilities. More precisely, we assume that in each system state, a set of actions (observations, experiments) are available to an external observer, and whenever an observer attempts to observe a given action, the system changes its state according to some probability distribution.

We have chosen a reactive viewpoint on probabilistic processes in the sense that there is a probability distribution for *each* possible action (and state).

The above intuition is formalized through the following notion of a probabilistic transition system:

Definition 2.1 *A probabilistic transition system is a structure $\mathcal{P} = (Pr, Act, \pi)$, where Pr is a set of processes (or states), Act is the set of actions that the processes may perform, and π is a transition probability function $\pi : Pr \times Act \times Pr \longrightarrow [0,1]$ such that for each $P \in Pr$ and $a \in Act$:*

$$\sum_{P' \in Pr} \pi(P,a,P') = 1 \text{ or } \sum_{P' \in Pr} \pi(P,a,P') = 0$$

indicating the possible next states and their probabilities after P has performed the action a.

In the above definition, $\pi(P,a,P') = \mu$ should be read as "P can perform the action a and with probability μ become the process P' afterwards". In particular, the case $\sum_{P' \in Pr} \pi(P,a,P') = 0$ should be read as "P cannot perform the action a". We introduce the following notation:

$$\begin{array}{ll} P \xrightarrow{a}_{\mu} P' & \text{whenever } \pi(P,a,P') = \mu \\ P \xrightarrow{a} P' & \text{whenever } \pi(P,a,P') > 0 \\ \pi(P,a,S) & \text{for } \sum_{P' \in S} \pi(P,a,P'),\ S \subseteq Pr \\ P \xrightarrow{a}_{\mu} S & \text{for } \pi(P,a,S) = \mu\ S \subseteq Pr \\ P \xrightarrow{a} & \text{for } \pi(P,a,Pr) = 1 \\ P \not\xrightarrow{a} & \text{for } \pi(P,a,Pr) = 0 \end{array}$$

This notation shows that a given probabilistic transition system may be interpreted as an ordinary labeled transition system where the actual probabilities are ignored. For such systems there exist a variety of proposals for semantic equivalence (or ordering), one of these being the well-known notion of bisimulation which enjoys a number of pleasant properties like fixedpoint proof technique and polynomial decidability (for finite state systems). For probabilistic processes, the notion of bisimulation equivalence has been extended [LS89] to the following notion of probabilistic bisimilarity:

Definition 2.2 *Let $\mathcal{P} = (Pr, Act, \pi)$ be a probabilistic transition system. Then a* probabilistic bisimulation *$\equiv_{\mathcal{P}}$, is an equivalence on Pr such that whenever $P \equiv_{\mathcal{P}} Q$, then the following holds:*[1]

$$\forall a \in Act. \forall S \in Pr/\equiv_{\mathcal{P}}.\ \pi(P, a, S) = \pi(Q, a, S)$$

Two processes P and Q are said to be probabilistic bisimilar *in case (P, Q) is contained in some probabilistic bisimulation. We write $P \sim_p Q$ in this case.*

So, two processes are probabilistic bisimilar just in case they have the same transitions and furthermore have identical accumulated probabilities for transitions to sets of equivalent processes.
The notion of probabilistic bisimilarity has been shown to identify the *limit* of the probabilistic testing framework of [LS89]: two processes are probabilistic bisimilar just in case they induce the same probability distribution on the observation set of *any* test. Furthermore the notions of probabilistic– and non– probabilistic bisimilarity have been related in [BM89] where it is shown that two processes are bisimilar, just in case there exists a probability function $\pi : Pr \times Act \times Pr \longrightarrow [0, 1]$, which makes them probabilistic bisimilar.
Similar to the simple modal logic, Hennessy-Milner Logic (or HML), which has been proved to be adequate for bisimulation [HM85, Pnu85], there exists a probabilistic counterpart, Probabilistic Modal Logic (or PML), being adequate for probabilistic bisimulation [LS89]:

$$F ::= \mathrm{tt} \mid F \wedge G \mid \neg F \mid \langle a \rangle_\mu F$$

The interpretation of this probabilistic modal logic is relative to a probabilistic transition system $\mathcal{P} = (Pr, Act, \pi)$. More precisely, the semantics of a formula F is a set of probabilistic processes $[\![F]\!]$ defined inductively as follows:

i) $[\![\mathrm{tt}]\!] = Pr$	*iii)* $[\![\neg F]\!] = Pr \backslash [\![F]\!]$
ii) $[\![F \wedge G]\!] = [\![F]\!] \cap [\![G]\!]$	*iv)* $[\![\langle a \rangle_\mu F]\!] = \{P \mid \pi(P, a, [\![F]\!]) \geq \mu\}$

In particular, a process satisfies a property $\langle a \rangle_\mu F$ if the sum of probabilities of transitions leading to F–satisfying states is not less than μ. Note, that the modalities of standard Hennessy–Milner Logic are derivable as follows: $\langle a \rangle \mathrm{tt} \equiv \langle a \rangle_\mu \mathrm{tt}$ for any $\mu > 0$; $[a]F \equiv \langle a \rangle_1 F \vee \neg \langle a \rangle \mathrm{tt}$, and $\langle a \rangle F \equiv \neg [a] \neg F$. We shall also in this paper apply the abbreviations $F \vee G \equiv \neg(\neg F \wedge \neg G)$ and $\Delta_a \equiv \neg(\langle a \rangle_1 \mathrm{tt})$. We shall write $P \models F$ in case $P \in [\![F]\!]$.
In [LS89] it has been shown that PML is adequate for probabilistic bisimulation, provided the transition system satisfies the so–called *minimal probability assumption*, that is, any value $\mu = \pi(P, a, P')$ is either zero or above a given minimal value $\varepsilon > 0$. This condition corresponds to the well–known image–finiteness condition for non–probabilistic processes.

Theorem 2.3 *Let $\mathcal{P} = (Pr, Act, \pi)$ be a probabilistic transition system satisfying the minimal probability assumption. Then two processes are probabilistic bisimilar, just in case they satisfy the same* PML *formulae.*

In chapter 3 we shall show that PML is also expressive with respect to probabilistic bisimulation in the sense of [Pnu85].

[1] $Pr/\equiv_{\mathcal{P}}$ denotes the equivalence classes of $\equiv_{\mathcal{P}}$

3 A Calculus of Probabilistic Processes

In order to enable the study of compositional verification of probabilistic processes, we introduce in this section a simple language for defining and combining probabilistic transition systems. The definition of such a language is also of interest in itself, because only a few proposals have been made so far [GJS90, HJ90]. Here we restrict ourselves to an SCCS-like calculus without recursion:

Definition 3.1 *Let Act be a set of actions satisfying $Act \times Act \subseteq Act$. Then the calculus af probabilistic processes* (CPP) *over a given set of actions Act has the following syntax*

$$P ::= 0 \mid a.P \mid P \oplus_\mu Q \mid P \times Q \mid P \lceil S$$

where $\mu \in]0,1[$, $a \in Act$ and $S \subseteq Act$.

In the above definition, 0 denotes (as usual) the completely inactive process, whereas $a.P$ is a process being able to perform the action a and thereafter behave like P. So in terms of the corresponding probabilistic transition system $\mathcal{P} = (Pr, Act, \pi)$, we have $\pi(0, b, P) = 0$ for all actions b and processes P. This is also the case for $\pi(a.P, b, P')$, except for the single case $\pi(a.P, a, P) = 1$. In the case of parallel composition, we have chosen a generalized[2] CSP-like semantics, i.e. the process $P \times Q$ can only make a transition $c = (a, b)$ when both components P and Q can perform the corresponding action components. When $P \times Q$ is able to do a transition, we consider the contributing transitions from P and Q to be chosen independently; that is, the resulting transition probabilities for $P \times Q$ are obtained as products of the probabilities for the contributing transitions.

Consider now the sum-construct $P \oplus_\mu Q$. In the semantics of this construct we want μ to indicate the probability used to resolve (in a probabilistic manner) alternative *internal* choices between P and Q. Hence, if both P and Q can do a particular action a, μ defines the probability of choosing among P's a-transitions, and $(1-\mu)$ defines the probability of choosing among the a-transitions of Q. On the other hand, if only P can do an a, an attempt to observe a will always (due to our reactive viewpoint) enable one of P's possible a-transitions; that is, the choice between P and Q will be independent of the particular μ-value.

The above intuitions are formalised through the following CPP semantics:

Definition 3.2 *Let Act be a set of actions. Then the language* CPP *defines a probabilistic transition system, where Pr is the the set of all* CPP*-terms and where the probability function π is defined by structure of terms as follows:*

$$\pi(0, a, P) = 0 \text{ for } a \in Act \text{ and for } P \in Pr$$

$$\pi(a.P, b, Q) = \begin{cases} 1 & \text{if } P = Q, b = a \\ 0 & \text{otherwise} \end{cases}$$

$$\pi(P \times Q, c, R) = \begin{cases} \pi(P, a, P') \cdot \pi(Q, b, Q') & ; c = (a, b) \text{ and } R = P' \times Q' \\ 0 & ; \text{otherwise} \end{cases}$$

$$\pi(P \oplus_\mu Q, a, R) = \begin{cases} \mu \cdot \pi(P, a, R) + (1-\mu) \cdot \pi(Q, a, R) & \text{if } P \xrightarrow{a}, Q \xrightarrow{a} \\ \pi(P, a, R) & \text{if } P \xrightarrow{a}, Q \not\xrightarrow{a} \\ \pi(Q, a, R) & \text{if } P \not\xrightarrow{a}, Q \xrightarrow{a} \\ 0 & \text{otherwise} \end{cases}$$

$$\pi(P \lceil S, a, R) = \begin{cases} \pi(P, a, P') & ; a \in S \text{ and } R = P' \lceil S \\ 0 & ; \text{otherwise} \end{cases}$$

---

[2] An approximation to the traditional CSP parallel operator can be obtained by combining $\times$ and $\lceil$.

Recall that in the semantic equation for $P \oplus_\mu Q$, $P \xrightarrow{a}$ is an abbreviation for $\pi(P, a, Pr) = 1$. Also note the first equation for $\pi(P \oplus_\mu Q, a, R)$, which ensures well-definedness of π in the case where both P and Q have an a-transition to the same term (like e.g. $a.\mathbf{0} \oplus_{.3} a.\mathbf{0}$)[3].
It is straightforward to verify that the above definition indeed defines a probabilistic transition system i.e. either $\pi(P, a, Pr) = 1$ or $\pi(P, a, Pr) = 0$ for any $P \in Pr$ and $a \in Act$.
A question which is relevant for compositional verification of CPP–processes is whether probabilistic bisimilarity is preserved by various constructs. The following theorem confirmes that this indeed is the case:

Theorem 3.3 *Probabilistic bisimulation is a congruence with respect to the combinators of* CPP.

For the axiomatization we need to establish syntactically the initial actions of processes. This is achieved by the following function.

Definition 3.4 *The function $In : Pr \longrightarrow 2^{Act}$ is defined as follows by structure of terms*[4]*:*

$$\begin{array}{lcl} In(\mathbf{0}) & = & \emptyset \\ In(a.P) & = & \{a\} \end{array} \qquad \begin{array}{lcl} In(P \oplus_\mu Q) & = & In(P) \cup In(Q) \\ In(P \times Q) & = & In(P) \cdot In(Q) \\ In(P\lceil S) & = & In(P) \cap S \end{array}$$

When $In(P) \cap In(Q) = \emptyset$ we shall apply the notation $P \perp Q$. Also in this case we shall use the notation $P + Q$ for $P \oplus_\mu Q$, reflecting that the probability μ in this case does not effect the semantics.

It is easily verified that the above predicate indeed captures precisely the possible first actions. That is, for any CPP–defined process P and action a, $a \in In(P)$ if and only if $P \xrightarrow{a}$.

Theorem 3.5 *For finite processes, the axiom system $\mathcal{A}$ in figure 2 is sound and relative complete (relative to arithmetic on $[0,1]$) with respect to probabilistic bisimulation.*

The soundness–part of the above theorem is quite straightforward and omitted here. For the completeness–part we use the traditional technique of *normalforms*. In order to define normalforms below, we introduce the notation:

$$\sum_{i=1}^{n} [\mu_i] a.E_i \tag{2}$$

whenever $\sum \mu_i = 1$ and $\mu_i > 0$ for all i. This notation abbreviates $\mathbf{0}$ when $n = 0$ and otherwise:

$$a.E_1 \oplus_{\mu_1} \left(\sum_{i=2}^{n} \left[\frac{\mu_i}{1 - \sum_{j<i} \mu_j} \right] a.E_i \right)$$

This notation is useful as it directly describes a terms' initial transitions as well as their probabilities. In particular for a term P of the form (2) we have $P \xrightarrow{a}_{\mu_i} E_i$.

Definition 3.6 *Let P be a* CPP*–process. Then P is said to be on normalform if P has the form:*

$$P = \sum_{a \in B} \sum_{i=1}^{n_a} [\mu_{a,i}] a.N_{a,i}$$

where $B \subseteq Act$, $N_{a,i} \not\sim_p N_{a,j}$ for $j \neq i$ and where all terms $N_{a,i}$ are themselves on normalform (note that $N = \sum_{a \in \emptyset} S_a = \mathbf{0}$ is a normalform).

[3]This is in contrast to [GJS90], where the individual transitions are indexed in order to separate transitions to identical terms. We find our definition more intuitive and certainly much simpler.
[4]For sets A and B, $A \cdot B$ denotes the set $\{(a, b) \mid a \in A, b \in B\}$.

$$
\begin{array}{lll}
\textbf{A1} & P \oplus_\mu P = P & \\
\textbf{A2} & P \oplus_\mu 0 = P & \\
\textbf{A3} & P \oplus_\mu Q = Q \oplus_{1-\mu} P & \\
\textbf{A4} & P \oplus_\mu (Q \oplus_{\frac{\nu}{\nu+\rho}} R) = (P \oplus_{\frac{\mu}{\mu+\nu}} Q) \oplus_{\mu+\nu} R & \begin{cases} \mu+\nu+\rho = 1 \\ In(P) = In(Q) = In(R) \end{cases} \\
\textbf{A5} & P \oplus_\mu Q = P \oplus_\nu Q \quad (= P + Q) & In(P) \cap In(Q) = \emptyset \\
\textbf{A6} & (P+Q) \oplus_\mu (R+S) = P \oplus_\mu R + Q \oplus_\mu S & \begin{cases} P \perp Q, R \perp S \\ P \perp S, Q \perp R \end{cases} \\
\textbf{A7} & 0 \times P = 0 & \\
\textbf{A8} & P \times (Q \times R) = (P \times Q) \times R & \\
\textbf{A9} & P \times (Q \oplus_\mu R) = (P \times Q) \oplus_\mu (P \times R) & \\
\textbf{A10} & a.P \times b.Q = (a,b).(P \times Q) & \\
\textbf{A11} & 0 \lceil S = 0 & \\
\textbf{A12} & (P+Q) \lceil S = P \lceil S + Q \lceil S & \\
\textbf{A13} & (a.P) \lceil S = \begin{cases} a.(P \lceil S) & a \neq 0 \\ 0 & \text{otherwise} \end{cases} &
\end{array}
$$

Figure 2: The Axiom System $\mathcal{A}$.

As expected it may be shown that any CPP process P may be transformed into a normalform term N_P such that P and N_P are provably equivalent in $\mathcal{A}$. We are now ready to sketch the completeness–part of Theorem 3.5.

Proof of completeness for $\mathcal{A}$:

Assume that $P \sim_p Q$. Now we first transform P and Q into provably equivalent normalform terms T_P and T_Q:

$$T_P = \sum_{a \in B_P} \sum_{i=1}^{n_a} [\mu^P_{a,i}] a.N^P_{a,i} \qquad T_Q = \sum_{a \in B_Q} \sum_{i=1}^{m_a} [\mu^Q_{a,i}] a.N^Q_{a,i}$$

Now it suffices to show that $T_P \sim_p T_Q$ implies $\mathcal{A} \vdash T_P = T_Q$ which we do by induction on the maximum depth of T_P and T_Q.

For the base case the proof is trivial. For the induction step we must have $B_P = B_Q$ and also $\sum_{i=1}^{n_a} [\mu^P_{a,i}] a.N^P_{a,i} \sim_p \sum_{i=1}^{m_a} [\mu^Q_{a,i}] a.N^Q_{a,i}$ for arbitrary $a \in B_P$. For arbitrary $a \in B_P$ the number of summand in these two terms (n_a and m_a) must be equal and also they must span identical equivalence classes. We may therefore arrange (using $\mathcal{A}$) the summands such than $N^P_{a,i} \sim_p N^Q_{a,i}$ for all i. For this arrangement we must also have $\mu^P_{a,i} = \mu^Q_{a,i}$ for all i. Now appealing to the Induction Hypothesis we obtain $\mathcal{A} \vdash N^P_{a,i} = N^Q_{a,i}$ and hence — using rules for substitution — $\mathcal{A} \vdash T_P = T_Q$. □

In section 2 (theorem 2.3) we recalled the adequacy result (from [LS89]) for the probabilistic modal logic PML. However, using the terminology of [Pnu85], PML is also *expressive* with respect to probabilistic bisimulation in the sense that we can express as a single PML formula the exact equivalence class denoted by any CPP term. First we introduce the notion of characteristic property for normalform CPP terms.

Definition 3.7 *Let N be a CPP-process on normalform, i.e.:*

$$N = \sum_{a\in B}\sum_{i=1}^{n_a}[\mu_{a,i}]a.N_{a,i}$$

Then we define the characteristic property C_N inductively as follows [5]*:*

$$C_N = \bigwedge_{a\in B} C_N^a \wedge \bigwedge_{a\notin B}[a]\mathrm{ff} \quad \textit{where} \quad C_N^a = \bigwedge_{I\subseteq\{1\dots n_a\}}\left[\langle a\rangle_{\mu_{a,I}}\left(\bigvee_{i\in I} C_{N_{a,i}}\right)\right]$$

where $\mu_{a,I} = \sum_{i\in I}\mu_{a,i}$ and $\bigwedge_{a\in\emptyset} F_a = \mathrm{tt}$.

We state without proof the expected characterization theorem:

Theorem 3.8 *Let M and N be CPP-processes in normalform. Then $M \models C_N$ if and only if $M \sim_p N$.*

4 Decomposing Formulae

We now address the question of decomposing logical specifications with respect to unary operators (or contexts) of CPP. That is, for any unary operators O of CPP we want to find a specification transformer $\mathcal{W}_o$ such that for any specification S and process P the following holds:

$$O(P) \models S \quad \text{if and only if} \quad P \models \mathcal{W}_o(S) \tag{3}$$

Hence, $\mathcal{W}_o(S)$ expresses the necessary and sufficient requirement to a component process P in order that the combined process $O(P)$ satisfies S. (In [Lar90] such transformers are called "wips" for weakest inner property-transformers).

For classical, non-probabilistic process algebras it has been shown in [LX90, Lar91] that Hennsessy Milner Logic does support decomposition in the sense that the logic is closed under all $\mathcal{W}_o$. Put differently, given any HML specifiation S and any unary process algebra operator O, $\mathcal{W}_o(S)$ is *expressible* in HML.

In the present probabilistic setting we lack a similar decomposition property. In particular as we shall demonstrate below, the logic PML (though adequate and expressive with respect to probabilistic bisimilarity) is simply not expressive enough to be closed under $\mathcal{W}_o$-transformers when O range over unary CPP operators. However, we propose an extension of PML for which the decomposition property will be shown to hold.

Before introducing this extended logic, we present two illustrating examples. In particular, we present an example demonstrating that PML does not support decomposition with respect to the parallel operator of CPP.

Example 4.1 Consider the following correctness assertions which involves the probabilistic sum-construct:

$$a.0 \oplus_\mu X \models \langle a\rangle_\nu\langle b\rangle_1\mathrm{tt} \tag{4}$$

$$a.b.0 \oplus_\mu X \models \langle a\rangle_\nu\langle b\rangle_1\mathrm{tt} \tag{5}$$

---

[5]We are here using as an additional assumption that Act is finite. Otherwise we would have to add atomic propositions I_B for any $B \subseteq_{\text{fin}} Act$ indicating that the initial actions of a process is within B.

In both cases the sufficient and necessary requirements to X will depend on a relationship between μ and ν. Some reflection will lead to the following equivalent formulations:

$$X \models \begin{cases} \mathrm{ff} & ; \ (1-\mu) < \nu \\ \langle a \rangle_{\frac{\nu}{(1-\mu)}} \langle b \rangle_1 \mathrm{tt} & ; \ \textit{otherwise} \end{cases} \tag{6}$$

$$X \models \begin{cases} \mathrm{tt} & ; \ \mu \geq \nu \\ \Delta_a \vee \langle a \rangle_{\frac{(\nu-\mu)}{(1-\mu)}} \langle b \rangle_1 \mathrm{tt} & ; \ \textit{otherwise} \end{cases} \tag{7}$$

That is, the assertion (4) is true if and only if the assertion (6) is true and likewise for the assertions (5) and (7). For the equivalence between (4) and (6) observe that no process can satisfy (4) if $(1-\mu) < \nu$, as $\pi(a.0 \oplus_\mu X, a, [\![\langle b \rangle_1]\!]\mathrm{tt}) \leq (1-\mu) \cdot \pi(X, a, [\![\langle b \rangle_1]\!]\mathrm{tt}) \leq (1-\mu)$.
Also for the equivalence between (5) and (7) note that any process will satisfy (5) when $\mu \geq \nu$, as $\pi(a.b.0 \oplus_\mu X, a, [\![\langle b \rangle_1 \mathrm{tt}]\!]) \geq \mu \cdot \pi(a.b.0, a, [\![\langle b \rangle_1 \mathrm{tt}]\!]) = \mu$.
It is also clear that (5) will be satisfied for any process X which can not perform a as in this case $\pi(a.b.0 \oplus_\mu X, a, [\![\langle b \rangle_1 \mathrm{tt}]\!]) = \pi(a.b.0, a, [\![\langle b \rangle_1 \mathrm{tt}]\!]) = 1$. □

Example 4.2 Let R be the probabilistic process $(a.R_1 \oplus_{\rho_1} a.R_2)$, where $R_1 = b.0$ and $R_2 = c.0$. Consider the following correctness assertion

$$R \times X \models \langle aa \rangle_\mu (\langle bb \rangle_1 \mathrm{tt} \vee \langle cc \rangle_1 \mathrm{tt}) \tag{8}$$

beeing equivalent to

$$\pi(R \times X, aa, [\![\langle bb \rangle_1 \mathrm{tt} \vee \langle cc \rangle_1 \mathrm{tt}]\!]) \geq \mu$$

We now calculate (letting $\rho_2 = 1 - \rho_1$)

$$\begin{aligned}
&\pi(R \times X, aa, [\![\langle bb \rangle_1 \mathrm{tt} \vee \langle cc \rangle_1 \mathrm{tt}]\!]) \\
&= \sum \{\pi(R \times X, aa, R' \times X') \mid R' \times X' \models \langle bb \rangle_1 \mathrm{tt} \vee \langle cc \rangle_1 \mathrm{tt}\} \\
&= \sum \{\pi(R, a, R') \cdot \pi(X, a, X') \mid R' \times X' \models \langle bb \rangle_1 \mathrm{tt} \vee \langle cc \rangle_1 \mathrm{tt}\} \\
&= \rho_1 \cdot \sum \{\pi(X, a, X') \mid R_1 \times X' \models \langle bb \rangle_1 \mathrm{tt} \vee \langle cc \rangle_1 \mathrm{tt}\} \\
&\quad + \rho_2 \cdot \sum \{\pi(X, a, X') \mid R_2 \times X' \models \langle bb \rangle_1 \mathrm{tt} \vee \langle cc \rangle_1 \mathrm{tt}\} \\
&= \rho_1 \cdot \sum \{\pi(X, a, X')) \mid X' \models \langle b \rangle_1 \mathrm{tt}\} + \rho_2 \cdot \sum \{\pi(X, a, X')) \mid X' \models \langle c \rangle_1 \mathrm{tt}\} \\
&= \rho_1 \cdot \pi(X, a, [\![\langle b \rangle_1 \mathrm{tt}]\!]) + \rho_2 \cdot \pi(X, a, [\![\langle c \rangle_1 \mathrm{tt}]\!])
\end{aligned}$$

Now, assume $X \xrightarrow{a}_s [\![\langle b \rangle_1 \mathrm{tt}]\!]$ and $X \xrightarrow{a}_t [\![\langle c \rangle_1 \mathrm{tt}]\!]$. Then (8) becomes equivalent to $\rho_1 s + \rho_2 t \geq \mu$. That is, for X to satisfy the assertion of (8) the probabilities of *two types* of transitions of X should satisfy a certain relationship. Clearly, this type of property is *not* expressible in PML, and hence PML *does not* support decomposition of specifications with respect to $\times$. □

Having shown that PML does not support decomposition for CPP, we now increase the expressive power of our logic resulting in a new logic, EPL by replacing the modality $\langle a \rangle_\mu F$ with a more general construct:

$$F \ ::= \ \mathrm{tt} \mid F \wedge G \mid \neg F \mid [\langle a \rangle_{x_1} F_1, \ldots, \langle a \rangle_{x_n} F_n \text{ where } \phi(x_1, \ldots, x_n)]$$

Here $x_1, \ldots, x_n$ are formal variables ranging over and binding probabilities and ϕ denotes an n-ary predicate over $[0,1]$. The semantics of the generalized modality is the following:

$$\begin{aligned}
&P \in [\![[\langle a \rangle_{x_1} F_1, \ldots, \langle a \rangle_{x_n} F_n \text{ where } \phi(x_1, \ldots, x_n)]]\!] \\
&\Leftrightarrow \ \phi(\pi(P, a, [\![F_1]\!])/x_1, \ldots, \pi(P, a, [\![F_n]\!])/x_n)
\end{aligned}$$

To see that the new modality indeed provides a generalization note:

$$\langle a\rangle_\mu F \equiv [\langle a\rangle_x F \text{ where } x \geq \mu]$$

Now, we claim that the extended probabilistic modal logic leads to a specification formalism EPL that *does support* decomposition:

We define four EPL transformers corresponding to the four typical unary contexts of CPP: $a.[\,]$, $P \oplus_\mu [\,]$, $P \times [\,]$ and $[\,]\lceil S$.

Definition 4.3 *Let $a \in Act$, $\mu \in]0,1[$, $S \subseteq Act$ and P be a process. We then define EPL–transformers $\mathcal{W}_{a.}$, $\mathcal{W}_{P\oplus_\mu}$, $\mathcal{W}_{P\times}$ and $\mathcal{W}_{\lceil S}$ inductively as follows:*

$$\begin{array}{lcl} \mathcal{W}_o(\mathrm{tt}) & = & \mathrm{tt} \\ \mathcal{W}_o(F \wedge G) & = & \mathcal{W}_o(F) \wedge \mathcal{W}_o(G) \\ \mathcal{W}_o(\neg F) & = & \neg \mathcal{W}_o(F) \end{array}$$

where O ranges over $a.$, $P\oplus_\mu$, $P\times$ and $\lceil S$.
For $F = [\langle a\rangle_{x_1} F_1, \ldots, \langle a\rangle_{x_n} F_n$ where $\phi(x_1,\ldots,x_n)]$ we define

$$\mathcal{W}_{b.}(F) = \begin{cases} \mathrm{tt}\ ; b \neq a, \phi(\overline{0}) \\ \mathrm{ff}\ ; b \neq a, \neg\phi(\overline{0}) \\ \mathrm{ff},\ b = a\ \Upsilon = \emptyset \\ \bigvee_{\nu\in\Upsilon}(\bigwedge_{\nu_i=1} F_i \wedge \bigwedge_{\nu_i=0} \neg F_i), \text{ otherwise} \end{cases}$$

where Υ denotes the set of tuples $\overline{\mu} = (v_1,\ldots,v_n)$, where $\phi(\overline{\mu})$ holds and v_i takes the value 0 or 1.

$$\mathcal{W}_{P\oplus_\mu}(F) = \begin{cases} F & ; P \stackrel{a}{\not\longrightarrow} \\ \Delta_a \vee G_{P,F} & ; P \stackrel{a}{\longrightarrow}\ \&\ P \models F \\ G_{P,F} & ; P \stackrel{a}{\longrightarrow}\ \&\ P \not\models F \end{cases}$$

where $G_{P,F} = [\langle a\rangle_{\overline{y}} \overline{F}$ where $\phi(\ldots,\mu\cdot\pi(P,a,[\![F_i]\!]) + (1-\mu)\cdot y_i,\ldots)]$. [6]
In the following definition of $\mathcal{W}_{P\times}$ [7]*, we assume that the action a in the formula F has the structure $a = (b,c)$.*

$$\mathcal{W}_{P\times}(F) = \begin{cases} \mathrm{tt} & ; P \stackrel{b}{\not\longrightarrow}\ \&\ \phi(\overline{0}) \\ \mathrm{ff} & ; P \stackrel{b}{\not\longrightarrow}\ \&\ \neg\phi(\overline{0}) \\ \bigwedge_{\substack{1\leq i\leq M \\ 1\leq j\leq n}} \langle c\rangle_{y_{i,j}} \mathcal{W}_{P_i\times}(F_j) & \\ \quad \text{where } \phi(\sum_i \mu_i\cdot y_{i,1},\ldots,\sum_i \mu_i\cdot y_{i,n}) & ; P \stackrel{b}{\longrightarrow} \end{cases}$$

where in the last equation $\{P_1 \ldots P_n\}$ constitutes the b-derivatives of P and $P \stackrel{b}{\longrightarrow}_{\mu_i} P_i$.

$$\mathcal{W}_{\lceil S}(F) = \begin{cases} \mathrm{tt} & ; a \notin S\ \&\ \phi(\overline{0}) \\ \mathrm{ff} & ; a \notin S\ \&\ \neg\phi(\overline{0}) \\ [\langle a\rangle_{\overline{x}} \overline{\mathcal{W}_S(F)} \text{ where } \phi(\overline{x})] & ; a \in S \end{cases}$$

---

[6] We apply vector notation $\langle a\rangle_{\overline{y}}\overline{F}$ for the tuple $\langle a\rangle_{y_1}F_1,\ldots,\langle a\rangle_{y_n}F_n$.
[7] Sometimes we shall use the notation $[(\langle a\rangle_{x_1}F_1 \wedge \ldots \wedge \langle a\rangle_{x_n}F_n$ where $\phi(x_1\ldots x_n)]$ as an alternative to $[(\langle a\rangle_{x_1}F_1,\ldots \langle a\rangle_{x_n}F_n$ where $\phi(x_1\ldots x_n)]$.

We have proved that the above defined transformers yield the following decomposition result:

Theorem 4.4 *Let O range over $a.$, $P\oplus_\mu$, $P\times$ and $\lceil S$ with $a \in Act$, $\mu \in]0,1[$, $S \subseteq Act$ and P a process. Then the following holds for all processes Q and all EPL specifications F:*

$$O(Q) \models F \text{ if and only if } Q \models \mathcal{W}_o(F)$$

Example 4.5 Let us now review the motivating example of a faulty medium from section 1. The requirement to an extra medium X is expressed as

$$M \times X \models [\langle ss\rangle_x \langle ee\rangle_1 \mathrm{tt} \text{ where } x \leq \delta]$$

By applying the EPL–transformer for parallel composition we obtain

$$\begin{aligned} &\mathcal{W}_{M\times}([\langle ss\rangle_x \langle ee\rangle_1 \mathrm{tt} \text{ where } x \leq \delta]) \\ &= [\langle s\rangle_x \mathcal{W}_{M_1\times}(\langle ee\rangle_1 \mathrm{tt}), \langle s\rangle_y \mathcal{W}_{M_2\times}(\langle ee\rangle_1 \mathrm{tt}) \text{ where } x\cdot\varepsilon + y\cdot(1-\varepsilon) \leq \delta] \\ &= [\langle s\rangle_x(\langle e\rangle_z \mathrm{tt} \text{ where } z \geq 1), \langle s\rangle_y \mathrm{ff} \text{ where } x\cdot\varepsilon + y\cdot(1-\varepsilon) \leq \delta] \end{aligned}$$

where $M = M_1 \oplus_\epsilon M_2$, $(M_1 = s.e.M, M_2 = s.d.M)$.
So we have obtained

$$M \times X \models \langle ss\rangle_x \langle ee\rangle_1 \mathrm{tt} \text{ where } x \leq \delta$$

if and only if

$$X \models [\langle s\rangle_x(\langle e\rangle_z \mathrm{tt} \text{ where } z \geq 1), \langle s\rangle_y \mathrm{ff} \text{ where } x\cdot\varepsilon + y\cdot(1-\varepsilon) \leq \delta]$$

In the next section we shall demonstrate how this decomposed property may be reduced even further by applying the complete axiomatization of EPL. □

5 Axiomatization of EPL

In this section we present an axiomatization of EPL complete relative to assertions about n-ary predicates over $[0,1]$.
We shall fell free to combine such predicates using boolean connectives with their standard interpretation. We shall also use the following standard abbreviations in EPL:

$$\begin{array}{llllll} F \vee G & \equiv^{def} & \neg(\neg F \wedge \neg G) & \mathrm{ff} & \equiv^{def} & \neg \mathrm{tt} \\ F \Rightarrow G & \equiv^{def} & \neg F \vee G & F \Leftrightarrow G & \equiv^{def} & (F \Rightarrow G) \wedge (G \Rightarrow F) \end{array}$$

As standard, we say that a formula F is *valid* in case $P \models F$ for all probabilistic processes P of any probabilistic transition system. We say a formula F is *satisfiable* if $P \models F$ for *some* probabilistic process P of some probabilistic transition system.
In figure 3 we present the proof system $\mathcal{E}$ for validity of EPL–formulae. The proof system is sound and (relative) complete as stated below:

Theorem 5.1 *Let F be an EPL–formula. Then F is valid if and only if $\mathcal{E}, \mathcal{O} \vdash F$, where $\mathcal{O}$ is an oracle for inclusions between n–ary predicates over $[0,1]$.*

Given that the soundness of most of the rules are fairly obvious we concentrate on the problem of (relative) completeness. For this we need the concept of a normal form formula:

AXIOMS

E1	All propositional tautologies
E2	$\vdash [\langle a\rangle_x \mathrm{tt} \text{ where } x = 1 \vee x = 0]$
E3	$\vdash [\langle a\rangle_x \mathrm{ff} \text{ where } x = 0]$
E4	$\vdash [\langle a\rangle_x F \text{ where } 0 \leq x \leq 1]$
E5	$\vdash [\langle a\rangle_x F, \langle a\rangle_y (F \wedge \neg G), \langle a\rangle_z (F \wedge G) \text{ where } x = y + z]$
E6	$\vdash \neg[\langle a\rangle_{\overline{x}} \overline{F} \text{ where } \phi(\overline{x})] \Leftrightarrow [\langle a\rangle_{\overline{x}} \overline{F} \text{ where } \neg\phi(\overline{x})]$
E7	$\vdash \big([\langle a\rangle_{\overline{x}} \overline{F} \text{ where } \phi(\overline{x})] \wedge [\langle a\rangle_{\overline{x}} \overline{F} \text{ where } \psi(\overline{x})]\big)$ $\Leftrightarrow [\langle a\rangle_{\overline{x}} \overline{F} \text{ where } \phi(\overline{x}) \wedge \psi(\overline{x})]$
E8	$\vdash [\langle a\rangle_{\overline{x}} \overline{F} \text{ where } \phi(\overline{x})] \Leftrightarrow [\langle a\rangle_{\overline{y}} \overline{F} \text{ where } \phi(\overline{y})]$
E9	$\vdash [\langle a\rangle_{x_1} F_1, \ldots, \langle a\rangle_{x_n} F_n \text{ where } \phi(\overline{x})]$ $\Leftrightarrow [\langle a\rangle_{x_{\sigma_1}} F_{\sigma_1}, \ldots, \langle a\rangle_{x_{\sigma_n}} F_{\sigma_n} \text{ where } \phi(\overline{x})]$ σ **is an arbitrary permutation**
E10	$\vdash [\langle a\rangle_y G, \langle a\rangle_{\overline{x}} \overline{F} \text{ where } \phi(\overline{x})]$ $\Leftrightarrow [\langle a\rangle_{\overline{x}} \overline{F} \text{ where } \phi(\overline{x})]$
E11	$\vdash [\langle a\rangle_{\overline{x}} \overline{F} \text{ where ff}] \Leftrightarrow \mathrm{ff}$
E12	$\vdash [\langle a\rangle_x F, \langle a\rangle_y F \text{ where } x = y]$

RULES

Cons
$$\frac{\models \phi(\overline{x}) \Rightarrow \psi(\overline{x})}{\vdash (\langle a\rangle_{\overline{x}} \overline{F} \text{ where } \phi(\overline{x})) \Rightarrow (\langle a\rangle_{\overline{x}} \overline{F} \text{ where } \psi(\overline{x}))}$$

MP
$$\frac{(\vdash F) \wedge (\vdash F \Rightarrow G)}{\vdash G}$$

Subst
$$\frac{\vdash F \Leftrightarrow G}{\vdash \mathcal{O}(F) \Leftrightarrow \mathcal{O}(G)} \qquad \mathcal{O} \text{ a formula context}$$

Figure 3: The Axiom System $\mathcal{E}$.

Definition 5.2 *An* EPL*–formula H is said to be on* disjunctive normal form *in case it has the form*

$$H = \bigvee_i \left[\langle a_i\rangle_{x_1^i} F_1^i, \ldots, \langle a_i\rangle_{x_{n_i}^i} F_{n_i}^i \text{ where } \phi_i(\overline{x})\right]$$

and F_j^i is on disjunctive normal form for all i, j.

As expected (and easily proved) any formula may be translated into a provably equivalent formula on disjunctive normal form.
As natural extensions of axioms **E2** and **E5** we have proved the following *Partioning Lemma:*

Lemma 5.3 *Let $G_1, \ldots, G_n$ $(n \geq 1)$ be* EPL*-formulae and let for $A \subseteq \{1 \ldots n\}$:*

$$\mathcal{G}_A^n = \bigwedge_{i \in A} G_i \wedge \bigwedge_{j \notin A} \neg G_j$$

Then the following are provable as theorems in $\mathcal{E}$:

$$\vdash \left[\bigwedge_A \langle a\rangle_{y_A} \mathcal{G}_A^n \wedge \langle a\rangle_{x_1} G_1 \text{ where } x_1 = \sum\{y_a | 1 \in A\}\right]$$

$$\vdash \left[\bigwedge_A \langle a\rangle_{y_A} \mathcal{G}_A^n \text{ where } \sum_A y_A = 1 \vee \sum_A y_A = 0\right]$$

Now, completeness of our axiomatization follows if we can show that *consistency* of a formula[8] implies *satisfiability*. We only consider formulae on disjunctive normal form of the form $F = [\langle a\rangle_{\overline{x}} \overline{F} \text{ where } \phi(\overline{x})]$. From the semantics of our logic the following observation is obvious:

$$F = [\langle a\rangle_{\overline{x}} \overline{F} \text{ where } \phi(\overline{x})] \text{ is satisfiable} \Rightarrow \text{that } \phi(\overline{x}) \text{ is satisfiable.}$$

where "$\phi(\overline{x})$ is satisfiable" simply means that $\phi(\overline{\mu})$ holds for some vector $\overline{\mu}$ of probabilities. Now, assume that we can transform F into a *provably equivalent* formula $H = [\langle a\rangle_{\overline{y}} \overline{\mathcal{J}} \text{ where } \psi(\overline{y})]$ satisfying the following stronger condition:

$$H = [\langle a\rangle_{\overline{y}} \overline{\mathcal{J}} \text{ where } \psi(\overline{y})] \text{ is satisfiable} \Leftrightarrow \psi(\overline{y}) \text{ is satisfiable.}$$

then it may easily be shown that consistency of F implies satisfiability of F.
As in the Partitioning Lemma let

$$\mathcal{J}_A = \bigwedge_{i \in A} F_i \wedge \bigwedge_{j \notin A} \neg F_j$$

for any $A \subseteq \{1 \ldots n\}$. Then let

$$H^\dagger = \left[\bigwedge_{i=1}^{n} \langle a\rangle_{x_i} F_i \wedge \bigwedge_{A \subseteq \{1..n\}} \langle a\rangle_{y_A} \mathcal{J}_A \text{ where } \begin{array}{c} x_i = \sum\{y_A | i \in A\} \\ \phi(\overline{x}) \\ 0 \leq y_a \leq 1 \\ \sum_A y_A = 1 \vee \sum_A y_A = 0 \end{array}\right]$$

Now conjuncting F (using **E7**) with the two Partitioning Lemmas and with axiom **E4** it may be shown that F and $H^\dagger$ are provably equivalent in $\mathcal{E}$. Now let $\mathcal{A}$ be the subsets A of $\{1 \ldots n\}$ for which $\mathcal{J}_A$ is consistent (or – using an Induction Hypothesis – is satisfiable). Then clearly $\mathcal{J}_B$ is inconsistent whenever $B \notin \mathcal{A}$ — and hence using **Subst** and **E3** it follows that:

$$\vdash [\langle a\rangle_{y_B} \mathcal{J}_B \text{ where } y_B = 0]$$

[8] *F is consistent* if not $\vdash \neg F$.

whenever $B \notin \mathcal{A}$. Now conjuncting this fact (using **E7**) to $H^\dagger$ for each $B \notin \mathcal{A}$, simplifying the new condition and applying **E10** yields a desired formula H:

$$H = \left[\bigwedge_{A \in \mathcal{A}} \langle a \rangle_{y_A} \mathcal{J}_A \text{ where } \begin{array}{c} 0 \leq y_A \leq 1 \\ \sum_{A \in \mathcal{A}} y_A = 1 \\ \phi(\sum\{y_A | i \in A\}/x_i) \end{array} \right]$$

which is provably equivalent to F. It is easy to see that satisfiability of H coincides with satisfiability of its condition. Furthermore, given processes satisfying $\mathcal{J}_A$ when $A \in \mathcal{A}$ (appealing to an Induction Hypothesis) it is easy to construct a probabilistic process satisfying H. In this sense our completeness proof is a constructive one.

Example 5.4 Once again reconsider the example of the faulty medium. The requirement to the unknown extra medium X has in Example 4.5 been determined to be:

$$X \models [\langle s \rangle_x (\langle e \rangle_z \text{tt where } z \geq 1), \langle s \rangle_y \text{ff where } x \cdot \varepsilon + y \cdot (1 - \varepsilon) \leq \delta]$$

Now using the axioms **E2**, **E3**, **E7** and **E10** together with the Cons–rule of the proof system $\mathcal{E}$ we can simplify this requirement to the following:

$$X \models [\langle s \rangle_x \langle e \rangle_1 \text{tt where } x \cdot \varepsilon \leq \delta]$$

□

6 Conclusions and Further Work

In this paper we have introduced a proposal for a reactive calculus of probabilistic processes, and we have identified a new probabilistic logic proved to support decomposition with respect to the derived unary operators of the calculus. We have proved that the logic is expressive with respect to probabilistic bisimulation, and we have also offered complete axiomatizations for the calculus and the logic.

From the expressiveness and the axiomatizations it follows easily that probabilistic bisimulation is decidable for finite CPP processes. Using the Characteristic Property Construction, the Decomposition Result together with the constructions of the completeness result from the axiomatization of EPL, we may obtain a method for solving equations of the form $M \times X \sim_p N$, where M and N are normal forms.

It is clear, that we have only made a very preliminary investigation of the extension of process algebra with probabilistic features. In general, there is a great need for attempting to redo the classical work [Mil89, Hoa85, BK84, Hen88] on process algebras with the addition of probabilities.

The introduction of the new probabilistic logic (EPL) has been dictated by the need for decomposition. However, it is not clear that we need all the expressive power of EPL in order to obtain this phenomenon. It would be interesting to investigate the existence of weaker logics (with respect to expressibility) also supporting decomposition. Also, to what extent does EPL support decomposition with respect to other probabilistic calculi? Here (from experiments not reported here) we are confident that EPL will do quite well.

The interest in probabilistic processes is not new. Pnueli and Zuck [PZ86] and Vardi [Var85] have studied the formal verification of probabilistic programs based on temporal logic; Kozen has given a semantic interpretation of probabilistic programs using generalizations of Scott Domains [Koz81], and introduced a Probabilistic Propositional Dynamic Logic [Koz83] useful for calculating the expected running time of probabilistic programs; Jones and Plotkin have studied probabilistic power domains [JP89].

The extension of process algebras with probability distributions has been studied by Giacalone, Jou and Smolka [GJS90] who introduced (and axiomatized) a probabilistic (generative) version of SCCS. Christoff [Chr90] has worked on testing equivalences for probabilistic processes, and Hansson and Jonsson [HJ90] have related time and probabilities. Smolka et.al. [vGSST90] have studied the relationship between the reactice and the generative models.

The notions of bisimulation and Hennessy–Milner Logic have been generalized to probabilistic versions by Larsen and Skou [LS89, Sko89] together with the related adequacy result. They also introduced a testing framework for probabilistic processes much like Hypothesis Testing from classical statistics. Bloom and Meyer [BM89] have shown an interesting relationship between non–probabilistic and probabilistic bisimulation. Fagin, Halpern and Megiddo have axiomatized a more restricted probabilistic logic [FHM88] than the one we introduce in this paper.

References

[AW91] H Andersen and G Winskel. Compositional Checking of Satisfaction. In *Proceedings of CAV '91*, 1991.

[BK84] J. Bergstra and J. Klop. Process algebra for synchronous communication. *Information and Control*, 60:109–137, 1984.

[BM89] B. Bloom and A. Meyer. A remark on bisimulation between probabilistic processes. In A. Meyer and M. Taitslin, editors, *Logic at Botik '89*, volume 363 of *Lecture Notes in Computer Science*, pages 26–40, 1989.

[Chr90] I. Christoff. *Tesing Equivalences for Probabilistic Processes.* PhD thesis, Uppsala University, 1990.

[CSZ91] R. Cleaveland, S. A. Smolka, and A. Zwarico. Testing Preorders for Probabilistic Processes. To appear in Proceedings of ICALP '92, 1991.

[FHM88] R. Fagin, J. Y. Halpern, and N. Megiddo. A Logic for Reasoning about Probabilities. In *Proceedings of LICS '88*, pages 410–421, 1988.

[GJS90] A. Giacalone, C. Jou, and S. Smolka. Algebraic Reasoning for Probabilistic Concurrent Systems. In *Proceedings of Working Conference on Programming Concetps and Methods.* IFIP TC2, 1990.

[Hen88] M. Hennessy. *Algebraic Theory of Processes.* MIT Press, 1988.

[HJ90] H. Hansson and B. Jonsson. A Calculus for Communicating Systems with Time and Probabilities. In *Proc. 11th IEEE Real-Time Systems Symposium*, Orlando, Florida, 1990.

[HM85] M. Hennessy and R. Milner. Algebraic laws for nondeterminism and concurrency. *Journal of the Association for Computing Mathinery*, pages 137-161, 1985.

[Hoa85] C. Hoare. *Communicating Sequential Processes.* Prentice-Hall, 1985.

[JL91] B. Jonsson and K. G. Larsen. Specification and Refinement of Probabilistic Processes. In *Proceedings of LICS '91*, 1991.

[JP89] C. Jones and G. Plotkin. A Probabilistic Powerdomain of Evaluations. In *Proceedings of 4th LICS*, 1989.

[Koz81] D. Kozen. Semantics of Probabilistic Programs. *JCSS*, 22, 1981.

[Koz83] D. Kozen. A Probabilistic PDL. In *Proceedings 10th ACM POPL*, 1983.

[Lar90] K.G. Larsen. Ideal specification formalism = expressivity + compositionality + decidability + testability + ... *Lecture Notes in Computer Science*, 458, 1990. Invited paper at CONCUR'90.

[Lar91] K. G. Larsen. The Expressive Power of Implicit Specifications. In *Proceedings of ICALP '91*, 1991.

[LS89] K.G. Larsen and A. Skou. Bisimulation through probabilistic testing: Preliminary report. In *Proceedings 16'th ACM POPL*, 1989. Full version in Information and Computation vol 94 no 1, 1991.

[LX90] K. G. Larsen and L. Xinxin. Compositionality through an Operarional Semantics of Contexts. In *Proceedings of ICALP '90*, 1990. Full version in Journal of Logic Computation, vol. 1 no. 6, pp 761–795, 1991.

[Mil83] R. Milner. Calculi for synchrony and asynchrony. *Theoretical Computer Science*, 25:267–310, 1983.

[Mil89] R. Milner. *Communication and Concurrency*. Prentice-Hall, 1989.

[Pnu85] A. Pnueli. Linear and branching structures in the semantics and logics of reactive systems. In W. Braur, editor, *ICALP '85*, volume 194 of *Lecture Notes in Computer Science*, pages 15–32, 1985.

[PZ86] A. Pnueli and L. Zuck. Verification of Multiprocess Probabilistic Protocols. *Distributed Computing*, 1, 1986.

[Sko89] A. Skou. *Validation of Concurrent Processes, with emphasis of testing*. PhD thesis, Aalborg University Centre, 1989.

[Var85] M. Y. Vardi. Automatic Verification of Probabilistic Concurrent Finite-State Programs. In *Proceedings 26th IEEE FOCS*, 1985.

[vGSST90] R. van Glabbeek, B. Steffen, S.A. Smolka, and C.M.N. Tofts. Reactive, Generative, and Stratified Models of Probabilistic Processes. In *Proceedings of LICS '90*, 1990.

Axiomatizing Probabilistic Processes: ACP with Generative Probabilities
(Extended Abstract)

J.C.M. Baeten
Dept. of Math and Computing Science, Eindhoven University of Technology
P.O. Box 513, 5600 MB Eindhoven, The Netherlands

J.A. Bergstra
Programming Research Group, University of Amsterdam
P.O. Box 41882, 1009 DB Amsterdam, The Netherlands

S.A. Smolka
Department of Computer Science, SUNY at Stony Brook
Stony Brook, NY 11794-4400, U.S.A.

Abstract

This paper is concerned with finding complete axiomatizations of probabilistic processes. We examine this problem within the context of the process algebra ACP and obtain as our end-result the axiom system $pr\mathrm{ACP}_I^-$, a probabilistic version of ACP which can be used to reason algebraically about the reliability and performance of concurrent systems. Our goal was to introduce probability into ACP in as simple a fashion as possible. Optimally, ACP should be the homomorphic image of the probabilistic version in which the probabilities are forgotten.

We begin by weakening slightly ACP to obtain the axiom system ACP_I^-. The main difference between ACP and ACP_I^- is that the axiom $x+\delta = x$, which does not yield a plausible interpretation in the generative model of probabilistic computation, is rejected in ACP_I^-. We argue that this does not affect the usefulness of ACP_I^- in practice, and show how ACP can be reconstructed from ACP_I^- with a minimal amount of technical machinery.

$pr\mathrm{ACP}_I^-$ is obtained from ACP_I^- through the introduction of probabilistic alternative and parallel composition operators, and a process graph model for $pr\mathrm{ACP}_I^-$ based on *probabilistic bisimulation* is developed. We show that $pr\mathrm{ACP}_I^-$ is a sound and complete axiomatization of probabilistic bisimulation for finite processes, and that $pr\mathrm{ACP}_I^-$ can be homomorphically embedded in ACP_I^- as desired.

Our results for ACP_I^- and $pr\mathrm{ACP}_I^-$ are presented in a modular fashion by first considering several subsets of the signatures. We conclude with a discussion about the suitability of an internal probabilistic choice operator in the context of $pr\mathrm{ACP}_I^-$.

1 Introduction

It is intriguing to consider the notion of probability (or probabilistic behavior) within the context of process algebra: a formal system of algebraic, equational, and operational techniques for the specification and verification of concurrent systems. Through the introduction of probabilistic measures, one can begin to analyze — in an algebraic fashion — "quantitative" aspects of concurrency such as reliability, performance, and fault tolerance.

In this paper, we address this problem in terms of complete axiomatizations of probabilistic processes within the context of the axiom system ACP [BK84]. ACP models an asynchronous merge, with synchronous communication, by means of arbitrary interleaving. It uses an additional constant δ, which plays the role of *NIL* from CCS [Mil80] (CCS is a predecessor of ACP). The key axioms for δ are:

$x + \delta = x$	A6
$\delta \cdot x = \delta$	A7

The process δ represents an unfeasible option; i.e. a task that cannot be performed and therefore will be postponed indefinitely. The interaction with merge (parallel composition) is as follows:

$$x \parallel \delta = x \cdot \delta$$

(This is not provable from ACP but for each closed process expression p we find that ACP $\vdash p \parallel \delta = p \cdot \delta$.) Now δ represents deadlock according to the explanation of [BK84].

Our goal is to introduce probability into ACP in as simple a fashion as possible. Optimally we would like ACP to be the homomorphic image of the probabilistic version in which the probabilities are forgotten. To this end, we develop a weaker version of ACP called ACP_I^-. This axiom system is just a minor alteration expressing almost the same process identities on finite processes. The virtues of this weaker axiom system are as follows:

(i) ACP_I^- does not imply $x + \delta = x$. In fact, this axiom has often been criticized as being non-obvious for the interpretation δ=deadlock=inaction.

(ii) $\text{ACP}_I^- + \{x + \delta = x\}$ implies the same identities on finite processes as ACP (but it is slightly weaker on identities between open processes).

(iii) ACP_I^- has for all practical purposes the same expressiveness as ACP. I.e., if one can specify a protocol in ACP, this can be done just as well in ACP_I^-.

(iv) ACP_I^- allows a probabilistic interpretation of +, and for this reason we need it as a point of departure for the development of a probabilistic version of ACP.

We introduce probability into ACP_I^- by replacing the operators for alternative and parallel composition with probabilistic counterparts to obtain the axiom system $pr\text{ACP}_I^-$. Probabilistic choice in $pr\text{ACP}_I^-$ is of the *generative* variety, as defined in [vGSST90], in that a single probability distribution is ascribed to all alternatives. Consequently, choices involving possibly *different* actions are resolved probabilistically. In contrast, in the *reactive* model of probabilistic computation [LS89, vGSST90], a separate distribution is associated with each action, and choices involving different actions are resolved nondeterministically.

A property of the generative model of probabilistic computation is that, unlike the reactive model, the probabilities of alternatives are conditional with respect to the set of actions offered by the environment. A more detailed comparison of the reactive and generative models can be found in [vGSST90]. There the *stratified* model is also considered and it is shown that the generative model is an abstraction of the stratified model and the reactive model is an abstraction of the generative model.

Previous work on probabilistic process algebra [LS89, GJS90, vGSST90, Chr90, BM89, JL91, CSZ92] has has been primarily of an operational/behavioral nature. Three exceptions, however, are [JS90, Tof90, LS92]. In [JS90], a complete axiomatization of generative probabilistic processes built from a limited set of operators (NIL, action prefix, probabilistic alternative composition, and tail recursion) are provided, while in [Tof90], axioms for synchronously composed "weighted processes" are given. A complete axiomatization of an SCCS-like calculus with reactive probabilities is presented in [LS92].

Summary of Technical Results

We have obtained the following results toward our goal of finding complete axiomatizations of probabilistic processes.

- We first present the axiom system ACP_I^-, our point of departure from ACP. Its development is modular beginning with BPA (consisting of process constants, alternative composition, and sequential composition), to which we add a merge operator to obtain PA. Finally, a communication merge operator, the constant δ, and an auxiliary *initials* operator I are added to PA to obtain ACP_I^-. In each case, we present a process graph model based on bisimulation and prove that the system is a sound and complete axiomatization of bisimulation for finite processes.

- We show in a technical sense, how ACP can be reconstructed from ACP_I^- through the reintroduction of the axiom A6.
- The axiom systems *pr*BPA, *pr*PA, and *pr*ACP_I^- for probabilistic processes are considered next. In each case, we present a process graph model based on *probabilistic bisimulation*, Larsen and Skou's [LS89] probabilistic extension of strong bisimulation, and prove that the system is a sound and complete axiomatization of probabilistic bisimulation for finite probabilistic processes.
- Connections between ACP_I^- and its probabilistic counterpart are then explored. We show that ACP_I^- is the homomorphic image of *pr*ACP_I^- in which the probabilities are forgotten. This result is obtained for both the graph model — the homomorphism preserves the structure of the bisimulation congruence classes, and the proof theory — the homomorphic image of a valid proof in *pr*ACP_I^- is a valid proof in ACP_I^-.
- We show that certain technical problems arise when a probabilistic internal choice operator is added to *pr*ACP_I^-, and argue that a state operator should be introduced to remedy the situation.

The structure of the rest of this paper is as follows. Section 2 presents the equational specifications BPA and ACP_I^-, and their accompanying process graph models and completeness results. Section 3 treats the probabilistic versions of these axiom systems, namely, *pr*BPA and *pr*ACP_I^-. The homomorphic derivability of ACP_I^- from *pr*ACP_I^- is the subject of Section 4, and, finally, Section 5 concludes. Note that we do not treat internal or τ-moves in this paper, so we stay within the setting of concrete process algebra.

Due to space limitations, all proofs of results are either omitted or sketched; the full proofs appear in [BBS92]. Also, we have eliminated from this extended abstract the sections on the axiom system PA and its probabilistic counterpart *pr*PA, and the section concerning probabilistic internal choice.

2 A Weaker Version of ACP

In this section we present the equational theory ACP_I^-, which, as described in Section 1, will be our point of departure for a probabilistic version of ACP. The main difference between ACP and ACP_I^- is that the axiom $x+\delta = x$, which does not yield a plausible interpretation in the generative model of probabilistic computation, is rejected in ACP_I^-. We begin with the theory BPA (Basic Process Algebra).

2.1 BPA

The signature $\Sigma(\mathrm{BPA}(A))$ consists of one sort P (for processes) and three types of operators: constant processes a, for each atomic action a, the sequential composition (or sequencing) operator '$\cdot$', and the alternative composition (or nondeterministic choice) operator '$+$'. The set of all constants is denoted by A, and is considered a parameter to the theory.

$$\Sigma(\mathrm{BPA}(A)) = \{a : \to \mathrm{P} | a \in A\} \cup \{+ : \mathrm{P} \times \mathrm{P} \to \mathrm{P}\} \cup \{\cdot : \mathrm{P} \times \mathrm{P} \to \mathrm{P}\}$$

The axiom system $\mathrm{BPA}(A)$ is given by:

$x + y = y + x$	A1
$(x + y) + z = x + (y + z)$	A2
$x + x = x$	A3
$(x + y) \cdot z = x \cdot z + y \cdot z$	A4
$(x \cdot y) \cdot z = x \cdot (y \cdot z)$	A5

Note the absence of the axiom $x \cdot (y + z) = x \cdot y + x \cdot z$ which does not hold in our process graph model.

Definition 2.1 *A process graph g is a triple $< V, r, \longrightarrow >$ such that*

- *V is the set of* nodes *(vertices) of g*
- *$r \in V$ is the* root *of g*
- *$\longrightarrow \subseteq V \times A \times V$ is the* transition relation *of g*

The *endpoints* of g are those nodes devoid of outgoing transitions and represent successful termination. We often write $v \xrightarrow{a} v'$ to denote the fact that $(v, a, v') \in \longrightarrow$. We denote by $\mathcal{G}$ the family of all process graphs. Bisimulation, due to Milner and Park, is the primary equivalence relation we consider on process graphs.

Definition 2.2 *Let $g_1 =< V_1, r_1, \longrightarrow_1 >$, $g_2 =< V_2, r_2, \longrightarrow_2 >$ be two process graphs. A* bisimulation *between g_1 and g_2 is a relation $\mathcal{R} \subseteq V_1 \times V_2$ with the following properties:*

- $\mathcal{R}(r_1, r_2)$
- *$\forall v \in V_1$, $w \in V_2$ with $\mathcal{R}(v, w)$:*
  $\forall a \in A$ and $v' \in V_1$,
  if $v \xrightarrow{a}_1 v'$ then $\exists w' \in V_2$ with $\mathcal{R}(v', w')$ and $w \xrightarrow{a}_2 w'$
- *and* vice versa *with the roles of v and w reversed.*

Graphs g_1 and g_2 are said to be bisimilar, *written $g_1 \leftrightarrows g_2$, if there exists a bisimulation between g_1 and g_2.*

The operators from $\Sigma(\text{BPA}(A))$ are defined on the domain of (root-unwound) process graphs in the standard way (e.g., [BW90]). For example, letting $g_1 = < V_1, r_1, \longrightarrow_1 >$, $g_2 = < V_2, r_2, \longrightarrow_2 >$, we have that $g_1 \cdot g_2$ is obtained by appending a copy of g_2 at each endpoint of g_1. In detail, $g_1 \cdot g_2$ is given by $< V_1 \times V_2, (r_1, r_2), \longrightarrow >$ where $(q_1, q_2) \xrightarrow{a} (q_1', q_2')$ if either

- $q_1 \xrightarrow{a}_1 q_1'$ and $q_2 = q_2' = r_2$
- $q_2 \xrightarrow{a}_2 q_2'$ and $q_1 = q_1'$ is an endpoint

In the setting of BPA, $\leftrightarrows$ is a congruence and BPA(A) constitutes a sound and complete axiomatization of process equivalence in $\mathcal{G}/\leftrightarrows$ for finite processes.

Theorem 2.1 ([BW90])

1. $\mathcal{G}/\leftrightarrows \models \text{BPA}(A)$
2. *For all closed expressions p,q over $\Sigma(\text{BPA}(A))$:*
   $\mathcal{G}/\leftrightarrows \models p = q \implies \text{BPA}(A) \vdash p = q.$

2.2 ACP without A6

The equational system $\text{ACP}_I^-(A)$ treats the operators of BPA(A) as well as the new constant δ, representing deadlock; a *communication merge* operator $|$ describing the result of a communication between any two atomic actions; a *merge* operator $\|$ representing the interleaved composition of two process which additionally admits the possibility of communication; a *left merge* operator $\lfloor\!\lfloor$ which is the same as $\|$ but always starts with the "left" process; and a family of restriction operators ∂_H, $H \subseteq A$. We will also need an auxiliary operator I that defines the initial actions (the *initials*) that a process can perform.

Letting $A_\delta = A \cup \{\delta\}$, the signature of $\text{ACP}_I^-(A)$ extends that of BPA(A) as follows:

$$\Sigma(\mathrm{ACP}_I^-(A)) = \Sigma(\mathrm{BPA}(A)) \cup \{\delta :\to \mathbf{P}\} \cup \{\|: \mathbf{P}\times\mathbf{P}\to\mathbf{P}\} \cup \{\lfloor: \mathbf{P}\times\mathbf{P}\to\mathbf{P}\} \cup \{| : \mathbf{P}\times\mathbf{P}\to\mathbf{P}\} \cup \{\partial_H : \mathbf{P}\to\mathbf{P} \mid H\subseteq A\} \cup \{I : \mathbf{P}\to 2^{A_\delta}\}$$

It is convenient to define the communication merge operator as a binary commutative and associative function on atomic actions; i.e., $| : A_\delta \times A_\delta \to A_\delta$,. In order to axiomatize $|$ as a function on processes (rather than on elements of A_δ) we define the characteristic predicate $\overline{A_\delta}$ of A_δ in the usual way:

$$\overline{A_\delta}(x) = \bigvee_{a\in A_\delta} (x = a)$$

We require $|$ to be total and this is captured by the following axiom:[1]

$\forall a,b \in \mathbf{P}\ \overline{A_\delta}(a) \wedge \overline{A_\delta}(b) \implies \exists c \in \mathbf{P}\ \overline{A_\delta}(c) \wedge a\|b = c$	C0

The axioms of $\mathrm{ACP}_I^-(A)$ are now given. In this system, a,b,c range over A_δ, and $\cap$,$\cup$ are used on 2^{A_δ} without further specification.

BPA(A) +

$\delta \cdot x = \delta$	A7

+

C0 +

$a\|b = b\|a$	C1
$(a\|b)\|c = a\|(b\|c)$	C2
$\delta\|a = \delta$	C3

+

$x \parallel y = x \lfloor y + y \lfloor x + x\|y$	CM1
$a \lfloor x = a \cdot x$	CM2
$(a\cdot x) \lfloor y = a(x \parallel y)$	CM3
$(x + y) \lfloor z = (x \lfloor z) + (y \lfloor z)$	CM4
$a\|(b\cdot x) = (a\|b)\cdot x$	CM5
$(a\cdot x)\|b = (a\|b)\cdot x$	CM6
$(a\cdot x)\|(b\cdot y) = (a\|b)\cdot(x \parallel y)$	CM7
$(x+y)\|z = x\|z + y\|z$	CM8
$x\|(y+z) = x\|y + x\|z$	CM9

+

$I(a) = \{a\}$	I1
$I(x\cdot y) = I(x)$	I2
$I(x+y) = I(x)\cup I(y)$	I3

[1] Axiom C0 is often replaced by choosing a total function $\gamma : A_\delta \times A_\delta \to A_\delta$ and having all identities of the graph of γ as axioms: $a|b = \gamma(a,b)$. In this way, γ becomes a parameter to the theory (see, e.g., [BW90]).

$$
\begin{array}{ll}
a \in H \implies \partial_H(a) = \delta & \text{D1} \\
a \notin H \implies \partial_H(a) = a & \text{D2} \\
I(x) \subseteq H \cup \{\delta\} \implies \partial_H(x+y) = \partial_H(y) & \text{D3.1} \\
I(x+y) \cap (H \cup \{\delta\}) = \emptyset \implies \partial_H(x+y) = \partial_H(x) + \partial_H(y) & \text{D3.2} \\
\partial_H(x \cdot y) = \partial_H(x) \cdot \partial_H(y) & \text{D4}
\end{array}
$$

Comments: $\mathrm{ACP}_I^-(A)$ differs from ACP by the absence of A6 and the presence of the weaker axioms D3.1-2 instead of D3: $\partial_H(x+y) = \partial_H(x) + \partial_H(y)$. Note that it is within axioms D3.1-2 where the auxiliary operator I comes into play. We give an example to illustrate the new axiom system.

$$
\begin{aligned}
\partial_{\{c\}}(a + (b+c)) &= \partial_{\{c\}}(c + (a+b)) & \text{(by A1 and A2)} \\
&= \partial_{\{c\}}(a+b) & \text{(by D3.1)} \\
&= \partial_{\{c\}}(a) + \partial_{\{c\}}(b) & \text{(by D3.2)} \\
&= a + b & \text{(by D2 twice)}
\end{aligned}
$$

Our graph model for $\mathrm{ACP}_I^-(A)$ is standard (see, e.g., [BW90]) with the exception of the restriction operator. This operator removes all edges labeled with actions from the set of restricted actions H. It also removes δ-edges, which it must do to ensure the soundness of D3.1. In case a node with at least one outgoing edge has all its edges removed, a new δ-edge, to the "dead" state v_δ, is added. Formally, $\partial_H(g_1)$ is given by $< V_1 \cup \{v_\delta\}, r_1, \longrightarrow >$ where $v_\delta \notin V_1$ and

$$
\longrightarrow \; = \{(v, a, v') \in \longrightarrow_1 \mid a \notin H \cup \{\delta\}\} \cup
$$

$$
\{(v, \delta, v_\delta) \mid \#(\{(v, a, v') \in \longrightarrow_1 \mid a \in H \cup \{\delta\}\}) = \#(\{(v, a, v') \in \longrightarrow_1 \mid a \in A_\delta\}) \geq 1\}
$$

Here $\#(S)$ is equivalent notation for $|S|$, for S a set; i.e. $\#$ is the cardinality function on sets.

The interpretation of δ as deadlock requires a new definition of bisimulation in which a weaker condition is imposed on δ-edges. The resulting relation, which we call a δ-bisimulation, is the same as in Definition 2.2 on non-δ edges. Otherwise, if $\mathcal{R}$ is a δ-bisimulation and $\mathcal{R}(v, w)$, then:

if $v \xrightarrow{\delta}_1 v'$, for some v', then $w \xrightarrow{\delta}_2 w'$, for some w'

and *vice versa* with the roles of v and w reversed. The resulting equivalence is denoted $\underline{\leftrightarrow}_\delta$ and can be shown to be a congruence in the context of $\mathrm{ACP}_I^-(A)$. That $\mathrm{ACP}_I^-(A)$ is a sound and complete axiomatization of $\underline{\leftrightarrow}_\delta$ for finite processes is given by the following.

Theorem 2.2

1. $\mathcal{G}/\underline{\leftrightarrow}_\delta \models \mathrm{ACP}_I^-(A)$

2. For all closed expressions p,q over $\Sigma(\mathrm{ACP}_I^-(A))$:

$$
\mathcal{G}/\underline{\leftrightarrow}_\delta \models p = q \implies \mathrm{ACP}_I^-(A) \vdash p = q.
$$

Proof sketch: The proof is by a normal form reduction and relies on the completeness of BPA (Theorem 2.1). We first define a *basic term* as one constructed from the constants A_δ, alternative composition, and (non-δ) action prefixing. Note that a basic term is a $\mathrm{BPA}(A_\delta)$ term. A term rewriting system, $\mathrm{RACP}_I^-(A)$, based on $\mathrm{ACP}_I^-(A)$ is introduced such that a normal form of the system is a $\mathrm{BPA}(A_\delta)$ term in which all occurrences of communication merge, merge, left-merge, and restriction have been eliminated. $\mathrm{RACP}_I^-(A)$ is shown to be *strongly normalizing* by transforming a reduction sequence π of $\mathrm{RACP}_I^-(A)$ into a valid reduction sequence of $\mathrm{RACP}(A)$ [BK84]. Finally, a normal form of $\mathrm{RACP}_I^-(A)$ is shown to be a basic term, and by the completeness of $\mathrm{BPA}(A_\delta)$ we are done. $\square$

2.2.1 Connections Between ACP and ACP_I^-

Let $\mathbf{A}$ be the usual bisimulation model for ACP(A), and let $\mathbf{A}^- = \mathcal{G}/\leftrightarrow_\delta$ be the bisimulation model for $\text{ACP}_I^-(A)$. Then for p,q closed expressions over $\Sigma(\text{ACP}(A))$ we have the following results.

1. $\mathbf{A}^- \models p = q \implies \mathbf{A} \models p = q$. This implies that $\mathbf{A}^-$ can be homomorphically embedded in $\mathbf{A}$ using the identity mapping.

2. $\mathbf{A} \models p = q \implies \mathbf{A}^- \models \partial_\emptyset(p) = \partial_\emptyset(q)$. This implies that $\mathbf{A}$ can be homomorphically embedded in $\mathbf{A}^-$ using the homomorphism $\varphi : \mathbf{A} \longrightarrow \mathbf{A}^-$, such that $\varphi(x) = \partial_\emptyset(x)$

3. $\text{ACP}(A) \vdash p = q \implies \text{ACP}_I^-(A) + \{x + \delta = x\} \vdash p = q$

4. $\text{ACP}_I^-(A) \vdash \partial_\emptyset(x + \delta) = \partial_\emptyset(x)$

3 A Probabilistic Version of ACP

Our discussion of probabilistic ACP will proceed in a manner similar to before. We will present the axioms systems prBPA(A) and $pr\text{ACP}_I^-(A)$, the probabilistic counterparts of BPA(A) and $\text{ACP}_I^-(A)$, and prove completeness in a graph model based on probabilistic bisimulation.

3.1 Probabilistic BPA

As usual, $(0,1)$ denotes the open interval of the real line $\{r \in \Re \mid 0 < r < 1\}$, and $[0,1]$ denotes the closed interval of the real line $\{r \in \Re \mid 0 \leq r \leq 1\}$. The signature $\Sigma(pr\text{BPA}(A))$ over the sort $pr\mathbf{P}$ (for probabilistic processes) is given by:

$$\begin{aligned}\Sigma(pr\text{BPA}(A)) \;&=\; \{a : \to pr\mathbf{P} | a \in A\} \cup \{+_p : pr\mathbf{P} \times pr\mathbf{P} \to pr\mathbf{P} \mid p \in (0,1)\} \cup \\ &\quad\; \{\cdot : pr\mathbf{P} \times pr\mathbf{P} \to pr\mathbf{P}\}\end{aligned}$$

The operator $+$ has been replaced by the family of operators $+_p$, for each probability p in the interval $(0,1)$, and is now called *probabilistic alternative composition*. Intuitively, the expression $x +_p y$ behaves like x with probability p and like y with probability $1-p$. Probabilistic alternative composition is *generative* [vGSST90] in that a single distribution (viz. the discrete probability distribution $\{p, 1-p\}$) is associated with the two alternatives x and y. As mentioned in Section 1, these probabilities are conditional with respect to the set of actions permitted by the environment. This will become clear in Section 3.2 with the introduction of the restriction operator ∂_H in the setting of probabilistic ACP.

We have the following axioms for prBPA(A):

$x +_p y = y +_{1-p} x$	prA1
$x +_p (y +_q z) = (x +_{p/(p+q-pq)} y) +_{p+q-pq} z$	prA2
$x +_p x = x$	prA3
$(x +_p y) \cdot z = x \cdot z +_p y \cdot z$	prA4
$(x \cdot y) \cdot z = x \cdot (y \cdot z)$	prA5

As for BPA(A), we consider process graphs, with labels from A, as a model for prBPA(A). Additionally, a probability distribution will be ascribed to each node's outgoing transitions.

Definition 3.1 *A* probabilistic process graph g *is a triple* $< V, r, \mu >$ *such that* V *and* r *are as in Definition 2.1 and* $\mu : (V \times A \times V) \to [0,1]$, *the* transition distribution function *of* g, *is a total function satisfying the following* stochasticity condition*:*

$$\forall v \in V \quad \sum_{\substack{a \in A, \\ v' \in V}} \mu(v,a,v') \in \{0,1\}$$

Intuitively, $\mu(v,a,v') = p$ means that, with probability p, node v can perform an a-transition to node v'. A node in a stochastic probabilistic process graph performs some transition with probability 1, unless it is an endpoint. When $\mu(v,a,v') > 0$ we say that v' is *reachable* from v and the notion of reachability extends to sequences of transitions in the natural way. We denote by $pr\mathcal{G}$ the family of all probabilistic process graphs.

The notion of strong bisimulation for nondeterministic processes has been extended by Larsen and Skou [LS89] to reactive probabilistic processes in the form of *probabilistic bisimulation*. Here we define probabilistic bisimulation on generative processes and to do so we first need to lift the definition of the transition distribution function as follows:

$$\mu : (V \times A \times 2^V) \longrightarrow [0,1] \text{ such that } \mu(v,a,S) = \sum_{v' \in S} \mu(v,a,v')$$

Intuitively, $\mu(v,a,S) = q$ means that node v, with total probability q, can perform an a-transition to some node in S.

Definition 3.2 **([LS89])** *Let* $g_1 = <V_1, r_1, \mu_1>$, $g_2 = <V_2, r_2, \mu_2>$ *be probabilistic process graphs. A* probabilistic bisimulation *between* g_1 *and* g_2 *is an equivalence relation* $\mathcal{R} \subseteq V_1 \times V_2$ *with the following properties:*

- $\mathcal{R}(r_1, r_2)$
- $\forall v \in V_1,\ w \in V_2$ *with* v *reachable from* r_1, w *reachable from* r_2, *and* $\mathcal{R}(v,w)$*:*
  $\forall a \in A,\ S \in (V_1 \cup V_2)/\mathcal{R},\quad \mu_1(v,a,S) = \mu_2(w,a,S)$

Graphs g_1 *and* g_2 *are* probabilistically bisimilar, *written* $g_1 \leftrightarroweq^{pr} g_2$, *if there exists a probabilistic bisimulation between* g_1 *and* g_2.

Intuitively, two nodes are probabilistically bisimilar if, for all actions in A, they transit to probabilistic bisimulation classes with equal probability. Note the somewhat subtle use of recursion in the definition.

We now define the operators of prBPA(A) on the domain of probabilistic process graphs. For this purpose, it is convenient to assume that probabilistic process graphs are acyclic with respect to transitions of non-zero probability (we consider only finite processes anyway) and that the root is not an endpoint. For the remainder of Section 3, let $g_1 = <V_1, r_1, \mu_1>$, $g_2 = <V_2, r_2, \mu_2>$ be probabilistic process graphs satisfying these assumptions such that $V_1 \cap V_2 = \emptyset$.

Definition 3.3 *The operators* $a \in A$, $+_p$, *and* $\cdot$ *are defined on* $pr\mathcal{G}$ *as follows:*

$a \in A$: *The process graph for each of these constants is given by* $<\{r_a, v\}, r_a, \mu_a>$, *where* $\mu_a(r_a, a, v) = 1$ *is the only transition with non-zero probability.*

$g_1 +_p g_2$: *is given by* $<V_1 \cup V_2 \cup \{r\}, r, \mu>$ *where* $r \notin V_1 \cup V_2$ *and*

$$\begin{array}{ll}
\mu(r,a,v') = p \cdot \mu_1(r_1,a,v') & \textit{if } v' \in V_1 \\
\mu(r,a,v') = (1-p) \cdot \mu_2(r_2,a,v') & \textit{if } v' \in V_2 \\
\mu(v,a,v') = \mu_1(v,a,v') & \textit{if } v, v' \in V_1 \\
\mu(v,a,v') = \mu_2(v,a,v') & \textit{if } v, v' \in V_2 \\
\mu(v,a,v') = 0 & \textit{otherwise}
\end{array}$$

The case for $g_1 \cdot g_2$ is analogous to the nonprobabilistic case. Note that as a consequence of this definition, and the fact that probabilistic process graphs are acyclic, transitions from r_1 and r_2 (some of which may be ascribed non-zero probabilities) are not reachable from the root r of $g_1 + g_2$.

We have that $\leftrightarroweq^{pr}$ is a congruence in prBPA(A).

Proposition 3.1 *If $g_1 \leftrightarrow^{pr} g_2$, then $g +_p g_1 \leftrightarrow^{pr} g +_p g_2$, $g \cdot g_1 \leftrightarrow^{pr} g \cdot g_2$, and $g_1 \cdot g \leftrightarrow^{pr} g_2 \cdot g$.*

The graph model for prBPA(A) is now given by $pr\mathcal{G}/\leftrightarrow^{pr}$ and we prove that prBPA(A) constitutes a sound and complete axiomatization of process equivalence in $pr\mathcal{G}/\leftrightarrow$ for finite processes.

Theorem 3.1

1. $pr\mathcal{G}/\leftrightarrow^{pr} \models pr\text{BPA}(A)$

2. For all closed expressions s,t over $\Sigma(pr\text{BPA}(A))$*:*

$$pr\mathcal{G}/\leftrightarrow^{pr} \models s = t \implies pr\text{BPA}(A) \vdash s = t.$$

Proof sketch: The proof is again by a normal form reduction. We first introduce the notation

$$\sum_{i=1}^{n}[p_i]x_i$$

with $\sum p_i = 1$ and $p_i > 0$ for all i. So, in particular, when $n = 1$, $p_1 = 1$. This notation abbreviates nested probabilistic alternative composition expressions as follows:

$$\sum_{i=1}^{1}[p_i]x_i = x_1 \quad \text{and} \quad \sum_{i=1}^{n+1}[p_i]x_i = x_1 +_{p_1} \left(\sum_{i=1}^{n}\left[\frac{p_{i+1}}{1-p_1}\right]x_{i+1}\right)$$

This *summation form* notation is useful as it directly reflects the transition structure of the probabilistic process graph underlying the nested probabilistic alternative composition. That is, the process graph of the summation form $\sum[p_i]x_i$ will have a probability-p_i transition from its root to the node representing the root of the process graph of x_i.

A *probabilistic basic term* is then defined to be a summation form whose summands are either constants from A_δ or of the form $a \cdot t$, where $a \in A$ and t itself is a probabilistic basic term. Furthermore, the summands of a probabilistic basic term are required to be pairwise probabilistically bisimulation inequivalent.

We next show that the axioms of prBPA(A) are sufficient to prove a closed prBPA(A) term t equal to a probabilistic basic term. The proof is in two steps. First the term rewriting system corresponding to prBPA(A) axioms prA4 and prA5 is used to transform t into a term in which the only occurrences of sequential composition are of the action-prefixing variety. Secondly, the constraint that the summands be pairwise inequivalent is met by using axioms prA1, prA2, and prA3 to group together and, in the process, compute the total probability assumed by a summand in a probabilistic basic term. Completeness is then proved by induction on the maximum depth of the probabilistic basic terms for the given s and t. □

3.2 Probabilistic ACP

The signature of $pr\text{ACP}_I^-(A)$ extends that of prBPA(A):

$$\begin{aligned}\Sigma(pr\text{ACP}_I^-(A)) = \Sigma(pr\text{BPA}(A)) &\cup \{\delta :\rightarrow pr\mathbf{P}\} \cup \{I : pr\mathbf{P} \rightarrow 2^{A_\delta}\} \cup \\ &\{|_{r,s} : pr\mathbf{P} \times pr\mathbf{P} \rightarrow pr\mathbf{P} \mid r,s \in (0,1)\} \cup \{\|_{r,s} : pr\mathbf{P} \times pr\mathbf{P} \rightarrow pr\mathbf{P} \mid r,s \in (0,1)\} \cup \\ &\{⌊\!⌊_{r,s} : pr\mathbf{P} \times pr\mathbf{P} \rightarrow pr\mathbf{P} \mid r,s \in (0,1)\} \cup \{\partial_H : pr\mathbf{P} \rightarrow pr\mathbf{P} \mid H \subseteq A\}\end{aligned}$$

Thus, for each of the operators $|$, $\|$, and $⌊\!⌊$ we have a family of operators, each indexed by two probabilities from the interval $(0, 1)$. These operators work intuitively as follows. Consider for example the merge operator. In the expression $x \|_{r,s} y$, a communication between x and y occurs with probability $1-s$, and an autonomous move by either x or y occurs with probability s. Given that an autonomous move occurs, it comes from x with probability r and from y with probability $1 - r$.

The treatment of the communication merge is exactly analogous to the situation in the nonprobabilistic case (Section 2.2). The "totality" axiom C0 now becomes:

$$\forall a,b \in pr\mathbf{P}\ \overline{A_\delta}(a) \wedge \overline{A_\delta}(b) \implies \exists c \in pr\mathbf{P}\ \forall r,s \in (0,1)\ \overline{A_\delta}(c) \wedge a\,|_{r,s}\, b = c \qquad pr\text{C0}$$

The axioms of $pr\text{ACP}_I^-(A)$ are as follows. In this system, a,b,c range over A_δ, and I has functionality $I : pr\mathbf{P} \to 2^{A_\delta}$.

$pr\text{BPA}(A)$ +

$$\delta \cdot x = \delta \qquad pr\text{A7}$$

+

$pr\text{C0}$ +

$a\,\vert_{r,s}\, b = b\,\vert_{(1-r),s}\, a$	prC1
$(a\,\vert_{r,s}\, b)\,\vert_{u,v}\, c = a\,\vert_{r,s}\,(b\,\vert_{u,v}\, c)$	prC2
$\delta\,\vert_{r,s}\, a = \delta$	prC3

+

$x \parallel_{r,s} y = ((x \mathbin{\lfloor\!\lfloor}_{r,s} y) +_r (y \mathbin{\lfloor\!\lfloor}_{(1-r),s} x)) +_s (x\,\vert_{r,s}\, y)$	prCM1
$a \mathbin{\lfloor\!\lfloor}_{r,s} y = a \cdot y$	prCM2
$(a \cdot x) \mathbin{\lfloor\!\lfloor}_{r,s} y = a \cdot (x \parallel_{r,s} y)$	prCM3
$(x +_p y) \mathbin{\lfloor\!\lfloor}_{r,s} z = (x \mathbin{\lfloor\!\lfloor}_{r,s} z) +_p (y \mathbin{\lfloor\!\lfloor}_{r,s} z)$	prCM4
$a\,\vert_{r,s}\,(b \cdot x) = (a\,\vert_{r,s}\, b) \cdot x$	prCM5
$(a \cdot x)\,\vert_{r,s}\, b = (a\,\vert_{r,s}\, b) \cdot x$	prCM6
$(a \cdot x)\,\vert_{r,s}\,(b \cdot y) = (a\vert_{r,s}b) \cdot (x \parallel_{r,s} y)$	prCM7
$(x +_p y)\,\vert_{r,s}\, z = x\,\vert_{r,s}\, z +_p y\,\vert_{r,s}\, z$	prCM8
$x\,\vert_{r,s}\,(y +_p z) = x\,\vert_{r,s}\, y +_p x\,\vert_{r,s}\, z$	prCM9

+

$I(a) = \{a\}$	prI1
$I(x \cdot y) = I(x)$	prI2
$I(x +_p y) = I(x) \cup I(y)$	prI3

+

$a \in H \implies \partial_H(a) = \delta$	prD1
$a \notin H \implies \partial_H(a) = a$	prD2
$I(x) \subseteq H \cup \{\delta\} \implies \partial_H(x +_p y) = \partial_H(y)$	prD3.1
$I(x + y) \cap (H \cup \{\delta\}) = \emptyset \implies \partial_H(x +_p y) = \partial_H(x) +_p \partial_H(y)$	prD3.2
$\partial_H(x \cdot y) = \partial_H(x) \cdot \partial_H(y)$	prD4

To define the graph model for $pr\text{ACP}_I^-(A)$, we need to introduce a "normalization factor" to be used in computing conditional probabilities in a restricted process.

Definition 3.4 *Let $g = <V, r, \mu>$ be a probabilistic process graph. Then, for $v \in V$, the normalization factor of v with respect to the set of actions $H \subseteq A$ is given by*

$$\nu_H(v) = 1 - \sum\{\mu(v,a,v') | a \in H \cup \{\delta\}, v' \in V\}$$

Intuitively, $\nu_H(v)$ is the sum of the probabilities of those transitions from v that remain after restricting by the set of actions H.

Definition 3.5 *The operators δ, $\|_{r,s}$, $\lfloor_{r,s}$, $|_{r,s}$ and ∂_H, $H \subseteq A$, are defined on prG as follows:*

δ: *The process graph of δ is given by Definition 3.3, treating δ as a normal atomic action; i.e., the graph of δ is $<\{r_\delta, v\}, r_\delta, \mu_\delta>$, where $\mu_\delta(r_\delta, \delta, v) = 1$ is the only transition with non-zero probability.*

$g_1 \|_{r,s} g_2$: *is given by $<V_1 \times V_2, (r_1, r_2), \mu>$ where for all $a \in A_\delta$, $v_1, v_1' \in V_1$, $v_2, v_2' \in V_2$*

$$\mu((v_1,v_2),a,(v_1',v_2)) = \begin{cases} r \cdot s \cdot \mu_1(v_1,a,v_1') & \text{if } v_2 \text{ not an endpoint} \\ \mu_1(v_1,a,v_1') & \text{otherwise} \end{cases}$$

$$\mu((v_1,v_2),a,(v_1,v_2')) = \begin{cases} (1-r) \cdot s \cdot \mu_2(v_2,a,v_2') & \text{if } v_1 \text{ not an endpoint} \\ \mu_2(v_2,a,v_2') & \text{otherwise} \end{cases}$$

$$\mu((v_1,v_2),a,(v_1',v_2')) = (1-s) \cdot \sum_{b,c:\, b|_{r,s} c = a} \mu_1(v_1,b,v_1') \cdot \mu_2(v_2,c,v_2')$$

$g_1 \lfloor_{r,s} g_2$: *is given by $<V_1 \times V_2, (r_1, r_2), \mu>$ where for all $a \in A_\delta$, $v_1, v_1', \in V_1$, $v_2, v_2', \in V_2$*

- $\mu((r_1,r_2),a,(v_1',r_2)) = \mu_1(r_1,a,v_1')$
- *if $v_1 \neq r_1$ or $v_2 \neq r_2$*

$$\mu((v_1,v_2),a,(v_1',v_2)) = \begin{cases} r \cdot s \cdot \mu_1(v_1,a,v_1') & \text{if } v_2 \text{ not an endpoint} \\ \mu_1(v_1,a,v_1') & \text{otherwise} \end{cases}$$

$$\mu((v_1,v_2),a,(v_1,v_2')) = \begin{cases} (1-r) \cdot s \cdot \mu_2(v_2,a,v_2') & \text{if } v_1 \text{ not an endpoint} \\ \mu_2(v_2,a,v_2') & \text{otherwise} \end{cases}$$

$$\mu((v_1,v_2),a,(v_1',v_2')) = (1-s) \cdot \sum_{b,c:\, b|_{r,s} c = a} \mu_1(v_1,b,v_1') \cdot \mu_2(v_2,c,v_2')$$

- *if $v_2' \neq r_2$* $\quad \mu((r_1,r_2),a,(v_1',v_2')) = 0$

$g_1 |_{r,s} g_2$: *is given by $<V_1 \times V_2, (r_1, r_2), \mu>$ where for all $a \in A_\delta$, $v_1, v_1' \in V_1$, $v_2, v_2' \in V_2$*

- $\mu((r_1,r_2),a,(v_1',v_2')) = \sum_{b,c:b|_{r,s} c=a} \mu_1(r_1,b,v_1') \cdot \mu_2(r_2,c,v_2')$
- *if $v_1 \neq r_1$ or $v_2 \neq r_2$*

$$\mu((v_1,v_2),a,(v_1',v_2)) = \begin{cases} r \cdot s \cdot \mu_1(v_1,a,v_1') & \text{if } v_2 \text{ not an endpoint} \\ \mu_1(v_1,a,v_1') & \text{otherwise} \end{cases}$$

$$\mu((v_1,v_2),a,(v_1,v_2')) = \begin{cases} (1-r) \cdot s \cdot \mu_2(v_2,a,v_2') & \text{if } v_1 \text{ not an endpoint} \\ \mu_2(v_2,a,v_2') & \text{otherwise} \end{cases}$$

$$\mu((v_1,v_2),a,(v_1',v_2')) = (1-s) \cdot \sum_{b,c:\, b|_{r,s} c = a} \mu_1(v_1,b,v_1') \cdot \mu_2(v_2,c,v_2')$$

- *if $(v_1' \neq r_1$ and $v_2' = r_2)$ or $(v_1' = r_1$ and $v_2' \neq r_2)$* $\quad \mu((r_1,r_2),a,(v_1',v_2')) = 0$

$\partial_H(g_1)$: *is given by $<V, r, \mu>$ where $V = V_1 \cup \{v_\delta\}$, v_δ a new endpoint not in V_1, $r = r_1$, and for all $a \in A$, $v_1, v_1' \in V_1$*

$$\mu(v_1,a,v_1') = \begin{cases} 0 & \text{if } a \in H \\ \mu_1(v_1,a,v_1')/\nu_H(v_1) & \text{otherwise} \end{cases}$$

$$\mu(v_1,\delta,v_1') = 0$$

$$\mu(v_1,a,v_\delta) = 0 \text{ if } a \neq \delta$$

$$\mu(v_1,\delta,v_\delta) = \begin{cases} 0 & \text{if } \exists a \notin H \cup \{\delta\}, v_1' \in V : \mu_1(v_1,a,v_1') > 0 \\ 1 & \text{otherwise} \end{cases}$$

Note the careful treatment of endpoints in the above definition—e.g., in a merge, if one process terminates, the other continues with its original, unweighted probability—and of transitions from the root (r_1, r_2) in $g_1 \mathbin{⌊\!\!⌊}_{r,s} g_2$ and $g_1 |_{r,s} g_2$—e.g., in a left merge, transitions emanating from the root that start with g_2 are given probability 0.

As in the nonprobabilistic case, the presence of δ-edges requires a new definition of probabilistic bisimulation. A *δ-probabilistic bisimulation* is the same as in Definition 3.2 with the additional clause

$$\mu_1(v, \delta, V_1) = \mu_2(w, \delta, V_2)$$

That is, δ-probabilistically bisimilar nodes must perform the action δ with the same total probability, without regard to where the δ-transitions lead. The resulting equivalence is denoted $\leftrightarrows_\delta^{pr}$.

Theorem 3.2

1. *$pr\mathcal{G}/\leftrightarrows_\delta^{pr} \models pr\mathrm{ACP}_I^-(A)$*
2. *For all closed expressions p,q over $\Sigma(pr\mathrm{ACP}_I^-(A))$:*

$$pr\mathcal{G}/\leftrightarrows_\delta^{pr} \models p = q \implies pr\mathrm{ACP}_I^-(A) \vdash p = q.$$

Proof sketch: The proof is analogous to the completeness proof of $\mathrm{ACP}_I^-(A)$.

- The definition of a probabilistic basic term uses $+_p$ instead of $+$ and the term rewriting system $pr\mathrm{RACP}_I^-(A)$ uses the probabilistic counterparts of the rules in $\mathrm{RACP}_I^-(A)$.
- $pr\mathrm{RACP}_I^-(A)$ is strongly normalizing: take a $pr\mathrm{RACP}_I^-(A)$ reduction and erase all probability subscripts. One obtains a valid $\mathrm{RACP}_I^-(A)$ reduction.
- The proof that a probabilistic normal form is also a probabilistic basic term proceeds as before – no rule in $pr\mathrm{RACP}_I^-(A)$ is conditional with respect to any probability. By the completeness of $pr\mathrm{BPA}(A_\delta)$ we are done.

□

4 ACP_I^- as an Abstraction of $pr\mathrm{ACP}_I^-$

In this section we demonstrate that $\mathrm{ACP}_I^-(A)$ can be considered an abstraction of $pr\mathrm{ACP}_I^-(A)$ at both the level of the graph model and at the level of the equational theory. For the former, we exhibit a homomorphism ϕ from probabilistic process graphs to nonprobabilistic process graphs that preserves the structure of the bisimulation congruence classes. For the latter, we exhibit a homomorphism Φ from $pr\mathrm{ACP}_I^-(A)$ terms to $\mathrm{ACP}_I^-(A)$ terms that preserves the validity of equational reasoning.

Definition 4.1 *Let $g = \langle V, r, \mu \rangle$ be a probabilistic process graph. Then $\phi(g) = \langle V, r, \longrightarrow \rangle$ has the same states and start state as g and $\longrightarrow$ is such that*

$$v_1 \xrightarrow{a} v_2 \iff \mu(v_1, a, v_2) > 0$$

Proposition 4.1 *Let g_1, g_2 be probabilistic process graphs.*

$$\begin{aligned}
\phi(a) &= a, a \in A_\delta \\
\phi(g_1 \cdot g_2) &= \phi(g_1) \cdot \phi(g_2) \\
\phi(g_1 +_p g_2) &= \phi(g_1) + \phi(g_2) \\
\phi(g_1 |_{r,s} g_2) &= \phi(g_1) \mid \phi(g_2) \\
\phi(g_1 \|_{r,s} g_2) &= \phi(g_1) \parallel \phi(g_2) \\
\phi(g_1 \mathbin{⌊\!\!⌊}_{r,s} g_2) &= \phi(g_1) \mathbin{⌊\!\!⌊} \phi(g_2) \\
\phi(\partial_H(g_1)) &= \partial_H(\phi(g_1))
\end{aligned}$$

Proposition 4.2 *The homomorphism ϕ preserves the structure of the bisimulation congruence classes. That is,*

$$g_1 \leftrightarrow_\delta^{pr} g_2 \implies \phi(g_1) \leftrightarrow_\delta \phi(g_2)$$

The converse of this result is clearly not true, e.g., $a + b \leftrightarrow_\delta b + a$ but $a +_{\frac{1}{2}} b \not\leftrightarrow_\delta^{pr} b +_{\frac{1}{3}} a$. Thus, the graph model $\mathcal{G}/\leftrightarrow_\delta$ of $\text{ACP}_I^-(A)$ is strictly more abstract than the probabilistic graph model $pr\mathcal{G}/\leftrightarrow_\delta^{pr}$ of $pr\text{ACP}_I^-(A)$.

The homomorphism $\Phi : \mathcal{L}(pr\text{ACP}_I^-(A)) \longrightarrow \mathcal{L}(\text{ACP}_I^-(A))$ from $pr\text{ACP}_I^-(A)$ terms to $\text{ACP}_I^-(A)$ terms is defined as follows:

$$\begin{aligned}
&\Phi(a) = a, a \in A_\delta \\
&\Phi(x) = x \\
&\Phi(x \cdot y) = \Phi(x) \cdot \Phi(y) \\
&\Phi(x +_p y) = \Phi(x) + \Phi(y) \\
&\Phi(x \mid_{r,s} y) = \Phi(x) \mid \Phi(y) \\
&\Phi(x \parallel_{r,s} y) = \Phi(x) \parallel \Phi(y) \\
&\Phi(x \lfloor_{r,s} y) = \Phi(x) \lfloor \Phi(y) \\
&\Phi(\partial_H(x)) = \partial_H(\Phi(x))
\end{aligned}$$

The following proposition states that any valid proof of $pr\text{ACP}_I^-(A)$ can be mapped into a valid proof of $\text{ACP}_I^-(A)$ using the homomorphism Φ.

Proposition 4.3 *Let t_1, t_2 be terms of $pr\text{ACP}_I^-(A)$, i.e., $t_1, t_2 \in \mathcal{L}(pr\text{ACP}_I^-(A))$.*

$$\frac{pr\text{ACP}_I^-(A) \vdash t_1 = t_2}{\text{ACP}_I^-(A) \vdash \Phi(t_1) = \Phi(t_2)}$$

Proof sketch: The proof is by induction on the length of the $pr\text{ACP}_I^-(A)$ proof, using the observation that, for every $pr\text{ACP}_I^-(A)$ axiom of the form $c \implies t_1 = t_2$, its homomorphic image $\Phi(c) \implies \Phi(t_1) = \Phi(t_2)$ is an $\text{ACP}_I^-(A)$ axiom. Here c is a possibly empty condition on the validity of the $pr\text{ACP}_I^-(A)$ axiom, and the fact that $\Phi(c)$ is equal to the condition of the corresponding $\text{ACP}_I^-(A)$ axiom means that no axiom of $pr\text{ACP}_I^-(A)$ is conditional on a probability appearing within an $pr\text{ACP}_I^-(A)$ term. □

Note that the converse of the result does not hold, e.g., $a+b = b+a$ but $a +_{\frac{1}{2}} b \neq b +_{\frac{1}{3}} a$. Thus, $\text{ACP}_I^-(A)$ is a strictly more abstract theory than $pr\text{ACP}_I^-(A)$.

5 Conclusions

In this paper, we have presented complete axiomatizations of probabilistic processes within the context of the process algebra ACP. Given that axiom A6 of ACP ($x + \delta = x$) does not have a plausible interpretation in the generative model of probabilistic computation, we introduced the somewhat weaker theory ACP_I^-, in which A6 is rejected. ACP_I^- is, in essence, a minor alteration of ACP expressing almost the same process identities on finite processes.

Our end-result is the axiom system $pr\text{ACP}_I^-$, which can be seen as a probabilistic extension of ACP_I^- for generative probabilistic processes. In particular, ACP_I^- is homomorphically derivable from $pr\text{ACP}_I^-$. As desired, we showed that $pr\text{ACP}_I^-$ constitutes a complete axiomatization of Larsen and Skou's probabilistic bisimulation for finite processes.

Several directions for future work can be identified. First, we are interested in adding certain important features to the model, such as recursion and unobservable τ actions. Secondly, we desire also to completely axiomatize the *reactive* and *stratified* models of probabilistic processes [vGSST90]. In the stratified model, which is well-suited for reasoning about probabilistic "fair" scheduling, distinctions are made between processes based on the branching structure of their purely probabilistic choices. We conjecture that by eliminating axiom *pr*A2 (probabilistic alternative composition is not associative in the stratified model!) and slightly modifying *pr*D3.2, the desired axiomatization can be obtained.

Acknowledgements

The authors gratefully acknowledge Rob van Glabbeek, Chi-Chiang Jou, and Bernhard Steffen for valuable discussions.

References

[BBS92] J. C. M. Baeten, J. A. Bergstra, and S. A. Smolka. *Axiomatizing Probabilistic Processes: ACP with Generative Probabilities.* Technical Report, Dept. of Math and Computing Science, Technical University of Eindhoven, Eindhoven, The Netherlands, 1992.

[BK84] J. A. Bergstra and J. W. Klop. Process algebra for synchronous communication. *Information and Computation*, 60:109–137, 1984.

[BM89] B. Bloom and A. R. Meyer. A remark on bisimulation between probabilistic processes. In Meyer and Tsailin, editors, *Logik at Botik*, Springer-Verlag, 1989.

[BW90] J. C. M. Baeten and W. P. Weijland. *Process Algebra. Cambridge Tracts in Computer Science 18*, Cambridge University Press, 1990.

[Chr90] I. Christoff. *Testing Equivalences for Probabilistic Processes.* Technical Report DoCS 90/22, Ph.D. Thesis, Department of Computer Science, Uppsala University, Uppsala, Sweden, 1990.

[CSZ92] R. Cleaveland, S. A. Smolka, and A. E. Zwarico. Testing preorders for probabilistic processes. In *Proceedings of the 19th ICALP*, July 1992.

[GJS90] A. Giacalone, C.-C. Jou, and S. A. Smolka. Algebraic reasoning for probabilistic concurrent systems. In *Proceedings of Working Conference on Programming Concepts and Methods*, IFIP TC 2, Sea of Gallilee, Israel, April 1990.

[JL91] B. Jonsson and K. G. Larsen. Specification and refinement of probabilistic processes. In *Proceedings of the 6th IEEE Symposium on Logic in Computer Science*, Amsterdam, July 1991.

[JS90] C.-C. Jou and S. A. Smolka. Equivalences, congruences, and complete axiomatizations for probabilistic processes. In J. C. M. Baeten and J. W. Klop, editors, *Proceedings of CONCUR '90*, pages 367–383, Springer-Verlag, Berlin, 1990.

[LS89] K. G. Larsen and A. Skou. Bisimulation through probabilistic testing. In *Proceedings of 16th Annual ACM Symposium on Principles of Programming Languages*, 1989.

[LS92] K. G. Larsen and A. Skou. Compositional verification of probabilistic processes. In *Proceedings of CONCUR '92*, Springer-Verlag Lecture Notes in Computer Science, 1992.

[Mil80] R. Milner. *A Calculus of Communicating Systems.* Volume 92 of *Lecture Notes in Computer Science*, Springer-Verlag, Berlin, 1980.

[Tof90] C. M. N. Tofts. A synchronous calculus of relative frequency. In J. C. M. Baeten and J. W. Klop, editors, *Proceedings of CONCUR '90*, pages 467–480, Springer-Verlag, Berlin, 1990.

[vGSST90] R. J. van Glabbeek, S. A. Smolka, B. Steffen, and C. M. N. Tofts. Reactive, generative, and stratified models of probabilistic processes. In *Proceedings of the 5th IEEE Symposium on Logic in Computer Science*, pages 130–141, Philadelphia, PA, 1990.

Embeddings Among Concurrent Programming Languages (Preliminary Version)

Ehud Shapiro

Department of Applied Mathematics and Computer Science
The Weizmann Institute of Science
Rehovot 76100, Israel

Abstract

We relate several well-known concurrent programming languages by demonstrating mappings among them that are homomorphic with respect to parallel composition and preserve fully-abstract semantic distinctions. The results are presented within a general framework for language comparison based on language embeddings, and complement our earlier negative results presented within this framework. Together, the positive and negative results induce a nontrivial preordering on the family of concurrent programming languages that quite often coincides with previous intuitions on the "expressive power" of these languages.

Our results reveal interesting connections between hitherto unrelated concurrent languages and models of concurrency, and provide evidence to the viability of the theory of structural simplicity as a formal framework for the study of programming language expressiveness.

1 Introduction

Concurrent programming enjoys a proliferation of languages but suffers from the lack of a general method of language comparison. In particular, concurrent (as well as sequential) programming languages cannot be usefully distinguished based on complexity-theoretic considerations, since most of them are Turing-complete. Nevertheless, differences between programming languages matter, else we would not have invented so many of them.

In a previous paper [34] we have developed a general method for comparing programming languages based on their algebraic (structural) simplicity, and, using this method, achieved separation results among many well-known concurrent languages. The method calls for investigating the existence of mappings among languages that preserve certain aspects of their syntactic and semantic structure. It is based on associating with each programming language a first-order structure that includes programs (or transition systems) as objects, program composition operations as functions, and a set of semantic relations. Given two programming languages L, L', with associated structures S and S', we say that L is *simpler than* L' if there is a homomorphic mapping of S into S', but not vice versa. We call such a mapping a *language embedding.*

In [34] we investigated concurrent languages with parallel composition as the sole composition operation and with semantic relations derived from the observable equivalence, and showed that some concurrent languages cannot be embedded in others. In this paper we complement the negative results with positive results: we demonstrate embeddings among various well-known concurrent programming languages and models. Together, the positive and negative results induce a nontrivial preordering on the family of concurrent programming languages that quite often coincides with previous intuitions on the "expressive power" of these languages.

The contributions of this paper are in:

- Uncovering novel relationships between well-known concurrent languages and models of concurrency, as summarized in Figure 1.
- Providing evidence to the viability of the theory of structural simplicity as a formal framework for the study of programming language expressiveness.
- Identifying a class of transition systems with parallel composition that can adequately capture the operational semantics of all the concurrent language and models investigated.

The rest of the paper is organized as follows. Section 2 presents the method of language comparison. Section 3 describes transition systems, their parallel composition and their observables. Section 4 lists the concurrent languages we compare. Section 5 contains a menagerie of language embeddings that justify some of the positive results shown in Figure 1. The negative results shown in the figure were proved in [34]. Section 6 discusses related work and Section 7 concludes the paper.

2 The Method of Language Comparison

2.1 The Framework: Structural Simplicity

We present a slightly generalized version of the framework presented in [34]. With every programming language or abstract machine model that we study we associate a first-order structure as follows.

Definition 1 (First-Order Structure). A *partial algebra* is a pair $\langle A; F \rangle$, where A is a non-empty set and F is a set of finitary (possibly partial) operations on A.

A *first-order structure* is a triple $\langle P; F; R \rangle$ where $\langle P; F \rangle$ is a partial algebra and R is a set of relations on P.

The first-order structure associated with every language has programs with composition operations as algebras, and semantic relations which generally include (the complement of) a semantic equivalence relation. In our comparison method we investigate the existence of language embeddings between the compared languages. A language embedding is a homomorphic mapping between the associated structures.

Definition 2 (Algebra Homomorphism). Let $PA_1 = \langle A_1; F_1 \rangle$ and $PA_2 = \langle A_2; F_2 \rangle$ be two partial algebras. A *homomorphism* ε of PA_1 into PA_2 is a mapping satisfying that for every $a_1, \ldots, a_n \in A_1$ and $f \in F_1$ such that $f(a_1, \ldots, a_n)$ is defined, $f\varepsilon(a_1\varepsilon, \ldots, a_n\varepsilon)$ is defined and satisfies $(f(a_1, \ldots, a_n))\varepsilon = f\varepsilon(a_1\varepsilon, \ldots, a_n\varepsilon)$.

Definition 3 (Structure Homomorphism). Let $S = \langle P; F; R \rangle$ and $S' = \langle P'; F'; R' \rangle$ be two first-order structures. A *homomorphism* ε of S into S' is a mapping satisfying:

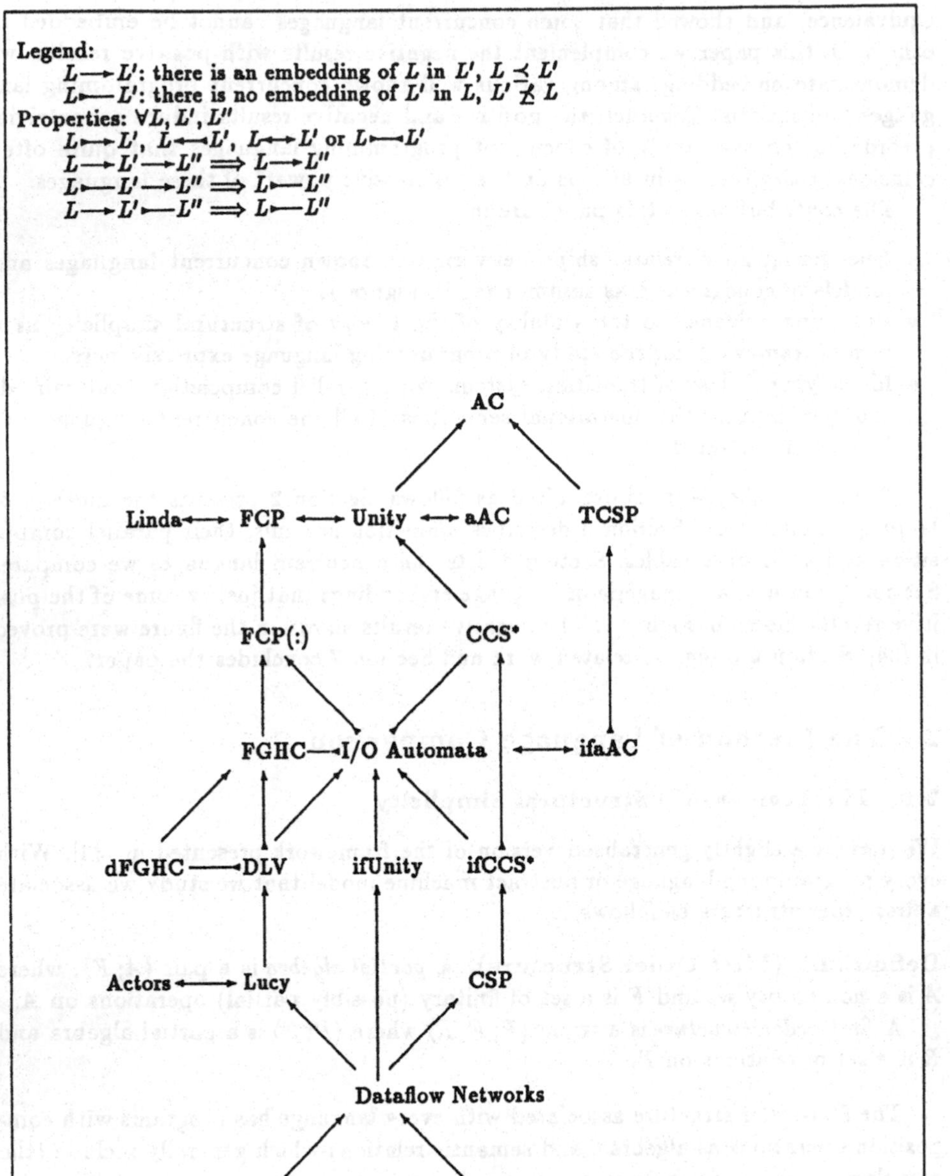

Fig. 1. Embeddings among concurrent programming languages with parallel composition

1. ε is a homomorphism of $\langle P; F\rangle$ into $\langle P'; F'\rangle$ (as partial algebras).
2. For every $a_1, \ldots, a_n \in P$ and every n-ary relation $r \in R$, if $r(a_1, \ldots, a_n)$ holds in S then $r\varepsilon(a_1\varepsilon, \ldots, a_n\varepsilon)$ holds in S' (for every n).

We would like language embeddings to be defined for languages with different sets of operations. Furthermore, we would like embeddings to allow realizing an operation in the source language using an operation in the closure of operations in the target language. We therefore define the following:

Definition 4 (Closure). Let $\langle A; F\rangle$ be an algebra. The *closure* of F, denoted by $\overline{F}$, is the set of all finitary operations that correspond to "terms with variables" over F and A (the elements of A serve as nullary operations over A).
For a structure $S = \langle P; F; R\rangle$, the closure of S is the structure $\overline{S} = \langle P; \overline{F}; R\rangle$.

Definition 5 (Structure Embedding). Let S and S' be two first-order structures. A *structure embedding* of S into S' is a homomorphism of S into $\overline{S'}$.

Given two languages L and L', with associated structures S and S', respectively, we refer to a structure embedding of S into S' also as a *language embedding* of L into L'.

The notion of structure embeddings induces a preorder $\preceq$ on any family of structures. Given two structures S and S', we say that $S \preceq S'$ if and only if there is a structure embedding of S in $\overline{S'}$. We use $S \sim S'$ as a notation for $S \preceq S'$ & $S' \preceq S$, and $S \prec S'$ as a notation for $S \preceq S'$ & $S' \not\preceq S$. We say that S is *as simple as* S' if $S \preceq S'$ and that S is *simpler than* S' if $S \prec S'$.

To prove that $S \preceq S'$, one should introduce a structure homomorphism of S in $\overline{S'}$. To prove that $S \not\preceq S'$, one should find a structure property $\mathcal{P}$ and apply the following observation:

Observation 1 (Separation Scheme) *Let S and S' be first-order structures. If there exists a structure property $\mathcal{P}$ such that:*

1. *$\mathcal{P}$ is satisfied by S, but not by any homomorphic image of S in $\overline{S'}$.*
2. *$\mathcal{P}$ is preserved by structure homomorphisms.*

then $S \not\preceq S'$.

2.2 The Structure of A Concurrent Programming Languages

To compare different concurrent programming languages, defined using different formalisms and within different semantic frameworks, we ignore some aspects of the languages and make some unifying assumptions. The main simplification comes from considering transition systems rather than programs. This approach allows us to ignore the concrete syntax of the different programming languages. It also allows us to compare concurrent programming languages with abstract models of concurrency, which are given directly as transition systems with a rule for parallel composition.

We assume that any concurrent programming language L is provided with a set of transition systems P, a parallel composition operation $\|$ over transition systems, and a (possibly partial) function Ob that associates with each transition system $p \in P$ a set of observable behaviors, called ***observables.***

Here and in the following equality (or equivalence) of expressions means that one side is defined iff the other side is, and if both are defined then they are equal (or equivalent).

We require that $Ob(p \parallel q) = Ob(q \parallel p)$ for every $p, q \in P$, and that there is a program $c \in P$, called a ***trivial program***, such that $Ob(p \parallel c) = Ob(p)$ for every $p \in P$.

We consider two semantic equivalence relations over L. The ***observable equivalence*** relation $\simeq$, defined to be the the kernel[1] of Ob, and the ***fully-abstract congruence***, $\cong$, induced by Ob and $\parallel$ [6]. The fully-abstract congruence is the unique largest equivalence relation satisfying $p \cong q \Longrightarrow p \simeq q$ and $p \cong p'$ & $q \cong q' \Longrightarrow (p \parallel q) \cong (p' \parallel q')$.

The notions of sound, faithful, and fully-abstract embeddings introduced in [34] can all be obtained as special cases of the notion of structure embedding, by taking the set of semantic relations to be an appropriate combination of the observable equivalence, the fully-abstract congruence, and/or their complements.

Notation: For the remaining of this paper L denotes a language with programs (transition systems) P, parallel composition $\parallel$, observable function Ob, observable equivalence $\simeq$, and fully-abstract congruence $\cong$, and L' denotes a language with the same component names but with a prime ($'$) suffix.

In this paper we associate with each programming language L we consider the first-order structure $(P; \parallel; \ncong)$. When there is no confusion we identify the language with its associated structure. We note that in [34] we studied sound embeddings, which preserve the complement of the observable equivalence rather than the complement of the fully-abstract congruence. It is easy to see that an embedding that preserves observable distinctions also preserves fully-abstract semantic distinctions:

Proposition 6. *Let $L = (P; \parallel; \ncong)$ and $L' = (P'; \parallel'; \ncong')$ be structures with $\ncong$ and $\ncong'$ being the complement of the fully abstract congruences induced by the observable functions Ob and Ob', respectively, and let $\varepsilon : P \rightarrow P'$ be a mapping satisfying $Ob(p) \neq Ob(q) \Longrightarrow Ob'(p\varepsilon) \neq Ob'(q\varepsilon)$. Then $p \ncong q \Longrightarrow p\varepsilon \ncong' q\varepsilon$.*

The negative results in [34] for embeddings preserving observable distinctions hold also for the more general class of embeddings investigated here. This more general class was chosen here since the fully-abstract congruence is often insensitive to minor variations in the definition of the observables, and hence provides a more robust basis for comparison. In addition, it allows greater freedom in designing embeddings, as shown by the following proposition, which strengthens the previous one.

Proposition 7. *Let $L = (P; \parallel; \ncong)$ and $L' = (P'; \parallel', \ncong')$ be two structures and let $\varepsilon : L \rightarrow L'$ be an embedding. Assume that for every $p, q \in L$ such that $Ob(p) \neq Ob(q)$ there is a program $c' \in L'$ such that $Ob'(p\varepsilon \parallel' c') \neq Ob'(q\varepsilon \parallel' c')$. Then $p \ncong q \Longrightarrow p\varepsilon \ncong' q\varepsilon$ for every $p, q \in L$.*

[1] The kernel of a function f is the equivalence relation $\equiv$ satisfying $a \equiv b$ iff $f(a) = f(b)$.

The fully-abstract congruence relation often has no convenient characterization. Proposition 7 is useful in that it allows establishing that the embedding preserves the fully-abstract congruence without referring to it explicitly.

3 Transition Systems With Parallel Composition

3.1 Basics

Definition 8. Assume a given set $\mathcal{S}$ of states and a set $\mathcal{A}$ of actions. A ***transition system over*** $\mathcal{S}$ ***and*** $\mathcal{A}$ is a pair $\langle I, T\rangle$ where $I \subseteq \mathcal{S}$ is the set of ***initial states*** and T, the ***transition relation***, is a subset of $\mathcal{S} \times \mathcal{A} \times \mathcal{S}$.

We do not incorporate fairness constraints in our transition systems.

A *computation prefix* is a sequence $s_0 \xrightarrow{a_0} s_1 \xrightarrow{a_1} s_2 \ldots$ of alternating states and actions, which begins with an initial state, ends with a state if it is finite, and for each i the triple $s_i \xrightarrow{a_i} s_{i+1}$ is in the transition relation. A *computation* is a computation prefix which is either infinite or ends in a state from which there are no transitions.

Definition 9. We say that a transition system $\langle I, T\rangle$ over A and S is *computable* if the function $\eta : S \rightarrow 2^{A \times S}$ defined by $\eta(s) = \{\langle a, s'\rangle | s \xrightarrow{a} s' \in T\}$ is computable.

In this paper we consider only computable transition systems.

For each set of transition systems $\mathcal{T}$ over $\mathcal{S}$ and $\mathcal{A}$ we define the parallel composition of transition systems $\|^{\mathcal{T}}$ using two auxiliary binary operations, state composition $\|^{\mathcal{S}}$ and action composition $\|^{\mathcal{A}}$. Both operations may be partial. We lift both to sets and define $S_1 \|^{\mathcal{S}} S_2 = \{s_1 \|^{\mathcal{S}} s_2 \mid s_i \in S_i \text{ and } s_1 \|^{\mathcal{S}} s_2 \text{ is defined}\}$, for any $S_1, S_2 \subseteq \mathcal{S}$. We omit the superscripts $\mathcal{T}$, $\mathcal{S}$, and $\mathcal{A}$ when they are clear from the context.

We define $\langle I, T\rangle = \langle I_1, T_1\rangle \parallel \langle I_2, T_2\rangle$ to have initial states $I = I_1 \parallel I_2$ and transitions $s_1 \parallel s_2 \xrightarrow{a_1 \| a_2} s'_1 \parallel s'_2$ if T_i has the transition $s_i \xrightarrow{a_i} s'_i$ for $i = 1, 2$ and all three components $s_1 \parallel s_2$, $a_1 \parallel a_2$, and $s'_1 \parallel s'_2$ are defined.

3.2 The Four Variants

Transition systems with parallel composition come in four basic variants:

1. **Global Actions:**
   - State composition is cartesian product.
   - Action composition is defined by $a \parallel a = a$ for every $a \in \mathcal{A}$.

   Global actions are employed in I/O Automata and in Theoretical CSP.
2. **Binary Actions:**

   The set of actions $\mathcal{A}$ includes a special action τ and admits a complementation operation satisfying $\overline{a} \neq a$ and $\overline{\overline{a}} = a$ for every $a \in \mathcal{A} \setminus \{\tau\}$.
   - State composition is cartesian product.
   - Action composition is defined by $a \parallel \overline{a} = \tau$ for every $a \in \mathcal{A} \setminus \{\tau\}$ and $a \parallel \tau = \tau \parallel a = a$ for every $a \in \mathcal{A}$.

   Binary actions are employed in CCS.

3. **Binary Channels:**
There is a function $c : \mathcal{S} \to 2^{\mathcal{A}}$ associating with each state a subset of $\mathcal{A}$, representing the channel ports "owned" by a process in that state. Transitions preserve c, i.e., $s \xrightarrow{a} s' \in T \Longrightarrow c(s) = c(s')$. State composition, when defined, satisfies $c(s_1 \parallel s_2) = c(s_1) \cup c(s_2)$.
   - State composition satisfies $s_1 \parallel s_2 = \langle s_1, s_2 \rangle$ if $c(s_1) \cap c(s_2) = \emptyset$.
   - Action composition is defined as in a binary actions language.

   Binary channels are employed in Occam and in CSP.
4. **Shared Store:**
States consist of two components $\mathcal{S} = \mathcal{Z} \times \mathcal{D}$, a local state taken from $\mathcal{Z}$ and a shared store value taken from $\mathcal{D}$. Actions have the form $\mathcal{A} = \mathcal{D} \times \{\uparrow, \downarrow\}$. Transitions labeled $d^{\uparrow}$ are called *active* and those labeled $d^{\downarrow}$ are called *passive*. We use $T^{\uparrow}$ and $T^{\downarrow}$ to denote the active and passive subsets of T, respectively.
We require that a transition $\langle s, d \rangle \xrightarrow{\ell} \langle s', d' \rangle$ satisfy $s \neq s'$ if it is active and $s = s'$ if it is passive. The intuition is that active transitions are carried out by the process and passive transitions by the environment.
The requirement that active transitions change internal state is no restriction since a process spinning in one internal state can be modeled by a process flipping between two internal states.
Since these rules allow reconstructing the set of actions from the set of states and the transition label from its source and target states, both can be omitted when describing a shared-store transition system.
   - State composition is defined by $s_1 \parallel s_2 = \langle s_1, s_2 \rangle$ for states $s_1 = \langle z_1, d_1 \rangle$, $s_2 = \langle z_2, d_2 \rangle$ if $d_1 = d_2 = d$. The resulting state $\langle \langle z_1, d \rangle \langle z_2, d \rangle \rangle$ is written more concisely as $\langle z_1, z_2; d \rangle$.
   - Action composition is defined by $d^{\uparrow} \parallel d^{\downarrow} = d^{\downarrow} \parallel d^{\uparrow} = d^{\uparrow}$ and $d^{\downarrow} \parallel d^{\downarrow} = d^{\downarrow}$ for every $d \in \mathcal{D}$.

   Shared-store transition systems are used for Dataflow Networks, Actors, the concurrent logic languages, Linda, Unity, and AC.
In most shared-store languages "output never blocks". This property is captured by:

Definition 10. A set of shared-store transition systems $\mathcal{T}$ is *asynchronous* if for every two transition systems $\langle I_1, T_1 \rangle$, $\langle I_2, T_2 \rangle \in \mathcal{T}$ and any state $\langle z_1, z_2; d \rangle \in \mathcal{S}$ if $\langle z_1; d \rangle \to \langle z_1'; d' \rangle \in T_1^{\uparrow}$ then $\langle z_2; d \rangle \to \langle z_2; d' \rangle \in T_2^{\downarrow}$.

All the shared-store languages that we investigate, except AC, are asynchronous. Some shared-store languages are "loosely coupled" [24] in the sense that active transitions from two concurrent processes sharing the same store can always commute. This property is captured by:

Definition 11. A set of shared-store asynchronous transition systems $\mathcal{T}$ is *loosely coupled* if for every two transition systems $\langle I_1, T_1 \rangle$, $\langle I_2, T_2 \rangle \in \mathcal{T}$ and any state $\langle z_1, z_2; d \rangle \in \mathcal{S}$ if $\langle z_1; d \rangle \to \langle z_1'; d_1 \rangle \in T_1^{\uparrow}$ and $\langle z_2; d \rangle \to \langle z_2'; d_2 \rangle \in T_2^{\uparrow}$ then there is $d' \in D$, uniform in d_1 and d_2, such that $\langle z_1; d_2 \rangle \to \langle z_1'; d' \rangle \in T_1^{\uparrow}$ and $\langle z_2; d_1 \rangle \to \langle z_2'; d' \rangle \in T_2^{\uparrow}$. Otherwise we say that $\mathcal{T}$ is *tightly coupled*.

Note that if there is a partial order on $\mathcal{D}$ such that active transitions never decrease the shared store then the requirement that the store value d' be uniform in d_1 and d_2 in the definition above induces a least upper bound operation on D.

Dataflow Networks, Actors, the concurrent logic languages Lucy, DLV, and FGHC are loosely coupled. They have natural partial orderings and least upper bound operations which are consistent with our definitions.

Linda, Unity, AC, and the concurrent logic languages FCP(:) and FCP are tightly coupled.

3.3 Observables

For each set of transition systems $\mathcal{T}$ we define a (possibly partial) function Ob that associates with each computation c of a transition $T \in \mathcal{T}$ an observable outcome. For a transition system T, we define $Ob(T)$, the ***observables of*** T, to be $\{Ob(c) \mid c$ is a computation of T and $Ob(c)$ is defined$\}$.

The observable outcome $Ob(c)$ of a computation c of an action-based transition system is a pair consisting of a termination mode and a trace, which is the sequence of actions of the transitions it employed, with internal actions removed.

The observable outcome of a shared-store transition system T is defined for computations in which the environment made no transitions, i.e., for computations of $T^{\dagger}$. It consists of a termination mode and an abstraction of the limit value of the shared store. Both the termination mode and the abstraction used are language specific.

4 Concurrent Programming Languages

We present a catalogue of the languages we consider.

1. IOA: I/O Automata [21] employ global actions over a set of actions A. Each transition system T has an associated set of "output" actions $T^A \subseteq A$, and for any state s, any $a \in A \setminus T^A$, and some s' has a transition $s \xrightarrow{a} s'$. The parallel composition of two transition systems T_1 and T_2 is defined iff $T_1^A \cap T_2^A = \emptyset$.
2. CCS* [22, 23]. CCS consists of processes employing dynamic binary actions. Parallel composition for CCS is totally defined.
   Since the official semantic equivalence for CCS cannot be derived naturally from observable outcomes of computations we refer to the language that we study as CCS*. It is closely related to the language EPL [9].
3. SV: Processes communicating by reading from and writing to shared variables [20] (*e.g.*, Unity [5]). Their transition systems are state-based, asynchronous and tightly-coupled with the shared store being the vector of values of the shared variables.
4. AC [7]: Processes communicating via a shared store that takes values from a partially ordered set. The value of a store can only be increased in the ordering.
5. TCSP: Theoretical CSP [13]. TCSP employs global actions. Its parallel composition operation is totally defined.

6. **DN: Nondeterministic Dataflow Networks (cf. [17] for the original deterministic model, and [15, 26] for fully-abstract semantics for nondeterministic dataflow networks). Dataflow networks operate on streams, which are two-port asynchronous FIFO channels with an internal unbounded buffer. Transition systems for dataflow networks employ a shared store that records channel histories. They are asynchronous and loosely coupled.**
   In DN, CSP, Occam, and DLV each transition system has an associated set of input and output ports. In these languages the parallel composition of two transition systems is defined iff they do not own a common port.
   The store abstraction used in the observables of T hides the history of the internal channels of T.
7. **CSP (cf. [12] for the original definition and [2] for an investigation of various sublanguages). CSP processes communicate using synchronous channels, which are two-port unbuffered channels in which sending and receiving must occur simultaneously. Transition systems for CSP are action based and employ static binary synchronization. Closure conditions ensure that a transition system cannot preview the value to be received on an input channel.**
8. **Occam [14]. Occam is similar to CSP. The main difference is that nondeterministic choices in CSP can be guarded by both input and output actions, whereas in Occam they can be guarded by input actions only. This is reflected in the transition system for sequential programs by the requirement that if from a given state an "output" transition is enabled, then no other transition is enabled at that state.**
9. **Actors [10]. The Actors model consists of processes, called actors, communicating via mailboxes. A mailbox is a many-to-one asynchronous channel, without a guarantee for order of message arrival. Each actor has a unique identifier and may "know" the identifiers of other actors, which are said to be its acquaintances. An actor a can send a message to b's mailbox only if b is an acquaintance of a. A message may include actor identifiers known to the sender, allowing the receiver to become acquainted with these actors as well.**
   Actors are modeled using state-based transition systems, whose shared store records the messages sent to each mailbox.
   The parallel composition of two transition systems is defined iff they do not share a common actor.
   The abstraction used in the observables hides all internal communication and leaves visible only messages sent to outside actors.
10. **Lucy [18] is a concurrent constraint language designed to be comparable to actors. Its only compound structure is a many-to-one bag and only output ports to a bag can be sent in messages. Its design is successful in the sense that from a transition system point of view it is essentially identical to actors.**
11. **DLV: Processes communicating via directed logic variables [19] (e.g. Doc [11] and a subset of Janus [30]). A directed logic variable is a two-port communication channel that can transmit at most one message, which may contain embedded ports. Unlike Actors and Lucy, both input and output ports can be sent in messages. A state-based transition system for directed logic variables is given in [19].**
    The store abstraction hides the values given to internal channels.

12. FGHC: Flat Guarded Horn Clauses [36]. Processes communicating using shared logic variable, which are single-assignment variable that take terms, which may contain variables, as values. The shared store consists of a set of equality constraints. Here and in the following concurrent logic languages parallel composition is totally defined. The store abstraction projects the final constraint on the initial variables.
13. FCP: Flat Concurrent Prolog [31, 32]. (See [38] for a definition of FCP(:,?), the variant used here.) A concurrent logic language employing atomic unification and read-only variables.
14. FCP(:) [29, 38]. A variant of FCP with atomic unification but no read only variables. (Initially presented with a different syntax under the name FCP($\downarrow$, $|$)[28].)
15. Linda [4]. Processes communicating by inserting and deleting tuples in a shared Tuple Space. Transition systems for Linda are state-based, with the shared store consisting of sets of tuples.

In addition, we are interested in subsets of some languages. These are defined specifically for each language, but have some properties in common:

- *dL*: The deterministic subset of *L*.
- *ifL*: An interference-free subset of a language *L*.

The deterministic subset *dL* of *L* is the set of *L* programs in which each sequential process is deterministic, has at most one transition emanating from each state. Note that *dL is* closed under the parallel composition operation of *L* and that the parallel composition of deterministic processes may exhibit nondeterministic behavior.

In some languages parallel composition is only partially defined to avoid naming conflicts, to ensure that communication is one-to-one, or to allow only a single writer per variable. A common aspect of these restrictions is captured by the following property, called interference freedom (a slightly more general definition of this property is given in [34]).

Definition 12. A program $p \in L$ is *trivial* if for every $q \in L$ such that $(p \parallel q) \in L$, $(p \parallel q) \cong (q \parallel p) \cong q$.

We say that L is *interference free* if for every nontrivial program $p \in L$, $(p \parallel p) \notin L$.

Some languages investigated here, e.g. Dataflow Networks, Occam, CSP, Actors, Lucy, DLV, and I/O-automata, are interference free. In addition to the languages mentioned here we investigate the following interference-free sublanguages:

16. ifCCS*: The sublanguage of CCS* in which the parallel composition of processes is allowed if and only if they employ a disjoint set of actions.
17. ifSV: The single-writer sublanguage of SV (called, in the case of Unity, the read-only sublanguage).
18. ifAC: The sublanguage of AC in which parallel composition of two transition systems T_1, T_2 is allowed if and only if the store augmentations of their active subsets $T_1^{\uparrow}$, $T_2^{\uparrow}$ are disjoint.

We also investigate two other sublanguages of AC:

19. aAC, the asynchronous sublanguage of AC.
20. ifaAC, the interference-free sublanguage of aAC.

5 A Menagerie of Language Embeddings

We now present justifications for some of the arrows in Figure 1.

For a set A, 2^A denotes the set of finite subsets of A, A^* denotes the set of finite sequences over A, and Λ denotes the singleton set containing the empty sequence. For sequences x and y over A (where elements of A are identified with singleton sequences) $x \diamond y$ denotes the concatenation of y to x if x is finite, and is undefined otherwise.

5.1 Embedding Interference-Free CCS* in I/O Automata

In this section we relate two abstract models of concurrency, CCS* and I/O Automata. There is no embedding of CCS* in I/O Automata, as shown in [34], since the latter is interference free whereas the former isn't. Here we show, however, that ifCCS*, the interference-free subset of CCS*, can be embedded in I/O Automata. This subset of CCS is interesting since it is rich enough to embed CSP [12]. The technique we use was inspired by an embedding of CSP in FCP [27, 32].

For a set A, let D_A denote the set of sequences over $A \cup 2^A$. We call an element of 2^A an *offer* and an element of D_A a *ready trace*.

Definition 13. Outstanding offer, enabled action, admissible ready trace
Let $d = r_1, r_2, \ldots$ be a ready trace. An offer r_i is *outstanding in d* if for no action a s.t. $\overline{a} \in r_i$, $a = r_j$, $j \geq i$.

An action a is *enabled in d* if $\overline{a} \in r_i$ for some outstanding offer r_i in d.

We say that d is *admissible* if every action $a = r_i$ in d is enabled in the ready trace $d_i = r_1, r_2, \ldots, r_{i-1}$.

Let p be a CCS program with transition system $\langle I, T\rangle$ over states S and actions A. We define $p\epsilon$ to be the I/O Automaton with states $\langle S \times 2^A \times D_A\rangle$, actions D_A, and transition system $\langle I \times \emptyset \times \Lambda, T'\rangle$, where T' is defined as follows:

Let $s \in S$ be a state with $s \xrightarrow{a_i} s_i$ for $1 \leq i \leq n$ being the only transitions from s in T, and let $\mathbf{a} = \{a_1, a_2, \ldots, a_n\}$. Then for every ready trace d and every i, $1 \leq i \leq n$, T' has the following transitions:

Output transitions:

- **Offer** (Offers $\mathbf{a}$ if no a_i is enabled):
  $\langle s, \emptyset, d\rangle \xrightarrow{\mathbf{a}} \langle\langle s, \mathbf{a}, d \diamond \mathbf{a}\rangle$, where no a_i is enabled in d, $1 \leq i \leq n$.
- **Act$_i$** (Do a_i if it is enabled):
  $\langle s, \emptyset, d\rangle \xrightarrow{a_i} \langle s_i, \emptyset, d \diamond a_i\rangle$ if a_i is enabled in d.

Input transitions:

- **Accept$_i$** (Accept choice of a_i if $\overline{a_i}$ was in a previous offer made by self):
  $\langle s, \mathbf{a}, d\rangle \xrightarrow{a_i} \langle s_i, \emptyset, d\rangle$, $1 \leq i \leq n$.
- **Observe**: In addition, for any state s, offer $\mathbf{a}$, and ready trace d, we have:
  $\langle s, \mathbf{a}, d\rangle \xrightarrow{a} \langle s, \mathbf{a}, d \diamond a\rangle$
  where either a is an action different from a_i for all $1 \leq i \leq n$ or a is an offer.

Note that only admissible ready traces can be generated by a computation. Note also that if the actions that can be taken by $p, q \in CCS^*$ are disjoint (implying that their parallel composition in ifCCS* is defined) then the actions that can be taken by $p\epsilon, q\epsilon$ are also disjoint, implying that the parallel composition of $p\epsilon, q\epsilon \in IOA$ is defined.

We claim that if $p, q \in CCS$ are observably distinct then so are $p\epsilon$ and $q\epsilon$. This is shown via a mapping from computations of the I/O automaton to computations of the corresponding CCS program, which maps an $\mathbf{Act}_i$ and an $\mathbf{Accept}_i$ transition to a CCS transition synchronizing on a_i and maps **Offer** and **Observe** to stutter.

Proposition 14. $ifCCS^* \preceq IOA$.

5.2 Embedding I/O Automata in Unity and in ifaAC

Let p be an I/O Automaton with transition system $\langle I, T\rangle$ over states S and actions A. Then $\epsilon : IOA \rightarrow Unity$ maps p to the Unity program $p\epsilon$ with the transition system $\langle I', T'\rangle$ over S' and A', where $S' = Z \times A^*$, $Z = S \times A^*$, and $I' = I \times \Lambda \times \Lambda$. That is, states of a Unity program have the structure $\langle s, d; d'\rangle$, where s is the internal state, d is the value of a local variable and d' is the value of the shared variable. T' has the following transitions, for every state $s \in S$ and sequence $d \in A^*$:

- **Output:** $\langle s, d; d\rangle \rightarrow \langle s', d \diamond a; d \diamond a\rangle$ if there is an output transition $s \xrightarrow{a} s' \in T$.
- **Observe:** $\langle s, d; d'\rangle \rightarrow \langle s, d; d \diamond a\rangle$ for every internal state (s, d), store value d' and $a \in A$.
- **Input:** $\langle s, d; d'\rangle \rightarrow \langle s', d \diamond a; d'\rangle$ if $d' = d \diamond a \diamond d''$ for some action a and sequence d'' and T has an input transition $s \xrightarrow{a} s'$.

We claim that if $p, q \in IOA$ are observably distinct then so are $p\epsilon$ and $q\epsilon$.

Proposition 15. $IOA \preceq Unity$.

It turns out that the same mapping ϵ constitutes an embedding of I/O Automata in AC, and that the image of the embedding is actually included in ifaAC, the interference-free, asynchronous sublanguage of AC.

Proposition 16. $IOA \preceq ifaAC$.

5.3 Embedding ifaAC in I/O Automata

Let p be an $ifaAC$ program with transition system $\langle S, I, T\rangle$, with $S = Z \times D$. Define $aug(T) = \{(d, d') | \langle s; d\rangle \rightarrow \langle s'; d'\rangle \in T\}$. Then for any $T \in ifaAC$, $T\epsilon$ is the I/O Automaton $\langle S, I, T'\rangle$, where the output actions are $aug(T^{\uparrow})$, the input actions are $aug(T^{\downarrow})$, and T' has transitions $\langle s; d\rangle \xrightarrow{(d,d')} \langle s'; d'\rangle$ for every $\langle s; d\rangle \rightarrow \langle s'; d'\rangle \in T$.

It is easy to see that active transitions of T are mapped to output transitions of the I/O automaton $T\epsilon$ and passive transitions of T to input transitions of $T\epsilon$.

The interference freedom of T means that $T_1\epsilon \parallel T_2\epsilon$ is defined in IOA whenever $T_1 \parallel T_2$ is defined in ifaAC. The asynchrony of ifaAC ensures that $T\epsilon$ has an input transition for every action $a \in aug(T\uparrow)$.

Proposition 17. $ifaAC \preceq IOA$.

Corollary 18. $IOA \sim ifaAC$.

5.4 Embedding I/O Automata in Theoretical CSP

I/O Automata and TCSP are similar in using global actions. The differences between these models is that in I/O Automata one process outputs an action and the others must input it. In TCSP all processes that can make an action must choose to make it in order for the action to take place.

However, it turns out that the identity mapping from IOA to TCSP happens to be an embedding in spite of this difference. The interference freedom of IOA ensures that even if its transition system is interpreted as a TCSP transition system, a necessary and sufficient condition for an action to take place is that the unique process "owning" the action takes it.

This may justify the view that, ignoring syntax and fairness, IOA is a special case of TCSP.

Proposition 19. *$IOA \preceq TCSP$.*

5.5 Embedding I/O Automata in FCP(:) and Unity in FCP(:,?)

The programming techniques used here to embed I/O Automata in FCP(:) and of Unity in FCP(:,?) are explained in [32].

Let p be an I/O Automaton with transition system $\langle I, T\rangle$ over states S and actions A. Then $\varepsilon : IOA \to FCP(:)$ maps p to the FCP(:) program $p\varepsilon$ with the clause

```
p(S,As) :- S=s, As=[a|As'] : true | p(s',As').
```

for every input transition $s \xrightarrow{a} s' \in T$, and the clause

```
p(S,As) :- S=s : As=[a|As'] | p(s',As').
```

for every output transition $s \xrightarrow{a} s' \in T$.

The embedding $\varepsilon : Unity \to FCP(:,?)$ maps a Unity program p with transition system $\langle I, T\rangle$, $S = Z \times D$ to the FCP(:,?) program with the clause:

```
p(S,Vs) :- Vs=[v|Vs'] : true | p(S,Vs').
```

for every passive transition $\langle s; v\rangle \to \langle s; v'\rangle \in T$, and the clause:

```
p(S,Vs) :- S=s : As=[V?|Vs'] | V=v, p(s',Vs').
```

for every active transition $\langle s; v\rangle \to \langle s'; v'\rangle \in T$.

In both cases the resulting programs are infinite in general, but by assumption on the computability of the transition system for the source programs they are each equivalent to a finite program.

5.6 Embedding CCS* in aAC

The embedding $\epsilon : CCS^* \to aAC$ is very similar to the embedding of ifCCS* in I/O Automata. It takes a CCS program p with transition system $\langle I, T\rangle$ over A and S to the aAC transition system $\langle I', T'\rangle$ over $S' = S \times D_{\mathcal{A}}$, where $D_{\mathcal{A}}$ is the domain of ready traces over A defined above, $I' = I \times \Lambda$, and T' having the following transitions:

For every state $s \in S$, where $s \xrightarrow{a_i} s_i$ for $1 \le i \le n$ are the only transitions from s in T, $\mathbf{a} = \{a_1, a_2, ..., a_n\}$, and every ready trace d:

- **Offer** (Offers **a** if no a_i is enabled): $\langle s, \emptyset; d\rangle \to \langle s, d \diamond \mathbf{a}; d \diamond \mathbf{a}\rangle$ if no a_i is enabled in d.
- **Act$_i$** (Do a_i if it is enabled): $\langle s, \emptyset; d\rangle \to \langle s_i, \emptyset; d \diamond a_i\rangle$ if a_i is enabled in d, for $1 \le i \le n$.
- **Accept$_i$** (A silent transition that changes the internal state if one of the actions offered by the process was taken by the environment): $\langle s, d'; d\rangle \to \langle s_i, \emptyset; d\rangle$ if $d = d' \diamond d''$ and $\overline{a_i}$ occurs in d'', for $1 \le i \le n$.
- **Input**: In addition, for every internal state (s, d'), store value d and every single-element legal store augmentation x, we have: $\langle s, d'; d\rangle \to \langle s, d'; d \diamond x\rangle$

The effect of this implementation is that every synchronous CCS transition is realized by three AC transitions: Offer, Act and Accept.

We claim that if $p, q \in CCS^*$ are observably distinct then there is an aAC program c for which $p\epsilon \parallel c$ and $q\epsilon \parallel c$ are observably distinct in aAC.

Proposition 20. $CCS^* \preceq aAC$.

6 Other Known Embeddings and Non-Embeddings

6.1 Known Embeddings

Various embeddings are known from the literature (although not necessarily under this name, and not always proven formally within our framework). An embedding of Dataflow Networks in Theoretical CSP is shown by Josephs, Hoare, and Jifeng [16]. The same method applies to embedding Dataflow Networks in CSP. It cannot, however, be used to embed DN in Occam, since Occam cannot implement unbounded buffers. An embedding of Dataflow Networks in Occam is given by de Boer and Palamidessi [1]. An embedding of Dataflow in ifUnity (i.e., the read-only subset of Unity) is given by Chandi and Misra [5]. It seems that another method employed there can be used to embed CSP in ifUnity.

The technique used by Milner to embed the value passing calculus of CCS in pure CCS [23] can be applied to embedding CSP in ifCCS*. There is a straightforward embedding of Dataflow Networks in concurrent logic languages. Even the weakest languages in the family, those employing directed logic variables, admit this embedding. An embedding of Actors in Lucy is given by Saraswat et al. [18]. There is a simple embedding of Lucy's bags in terms of Directed Logic Variables. There is a known embedding of directed logic variables in FGHC, which has been used in an actual implementation of a subset of Janus in FGHC. A straightforward embedding

of FGHC in FCP is given in [32]. That survey includes other embeddings among concurrent logic languages which are not considered here. An embedding of Linda in FCP is described in [33]. Embeddings of Dataflow Networks, Shared-Variable languages, and concurrent logic languages in the model AC were studied by Gaifman, Maher, and Shapiro [7].

6.2 Non Embeddings

SV in ifSV **Single-Writer Shared Variables (ifSV) are interference-free, whereas Shared Variables aren't. This precludes embedding ifSV in SV by our earlier results [34]. An implementation ϵ of SV in ifSV that realizes a shared variable by a process that manages it is not an embedding, since if p is a program that writes on some shared variable $p\epsilon$ must write on some shared variable. Assuming $(p \parallel p)\epsilon = p\epsilon \parallel p\epsilon$ contradicts the interference-freedom of ifSV.**

SV in IOA **I/O Automata are interference-free, whereas Shared Variables aren't. This precludes embedding SV in IOA [34]. In particular the implementation of shared variables in I/O Automata, shown by Goldman and Lynch [8], is not an embedding, since it is not homomorphic with respect to parallel composition. However, ifSV, the single-writer sublanguage of SV, can be embedded in I/O Automata according to the same principles pf [8].**

SV in CCS **Milner [23] shows an implementation of a shared variable language in CCS. The embedding is homomorphic in a very restricted sense, which does not cover SV. Concurrent processes can communicate only via variables declared outside their scope. They can interact with the outside world only via two fixed variables, which are identical for all programs. Hence independent programs cannot be composed in a meaningful way. Attempting to extend the embedding to allow 'open' composition, by allowing arbitrary external variable names, would not work since it is not clear with which agent to associate the process implementing a variable. Having multiple copies of a variable process would not be a sound embedding. The embedding of a shared-variable language with "open" composition in CCS is possible if the language is restricted to be single-writer, i.e., to ifSV. In this case the embedding associates the variable process with the (unique) agent that writes on it.**

CSP vs. CCS **Brookes [3] describes an embedding of CSP in CCS-like synchronization trees. Of course there is also an embedding of CCS in synchronization trees. However, it seems that from these two embeddings one cannot reconstruct an embedding of CSP in CCS, or vice versa.**

6.3 Other Related Work

A review of some previous work on the subject was given in [34]. Two more recent references not mentioned there are [25, 37]. Mitchell [25] introduces a notion of language translation which is similar to the notion of fully-abstract embeddings

described in [34], and uses it to demonstrate that Lisp is not universal in the sense that it cannot embed any language having an abstraction context, i.e., a context whose application to observably distinct programs may result in equivalent programs.

Vaandrager [37], following earlier work by de Simone [35], uses structural operational semantics to compare I/O Automata to other algebraic models of concurrency. We have yet to understand the relationship of this line of research to ours.

7 Conclusion and Future Work

The work presented here is preliminary in many ways:

- There are many missing arrowheads in Figure 1. Each represents an open problem. Our tentative conjectures as to which direction the missing arrowheads will point are reflected by the relative height of the languages in the figure.
- Many important languages and models of concurrency are missing from our figure. Incorporating them into the framework is a job that remains to be done.
- We have not considered operators other than ||. In particular, hiding in some languages is incorporated in parallel composition and is a separate operator in others. Allowing hiding as an additional operator seems a natural extension.
- We have ignored fairness in our treatment. Incorporating the different fairness criteria into the comparison may refine our understanding of the different models.

Acknowledgements

Comments by Frank de Boer, David Harel, Haim Gaifman, Joost Kok, Elia Lalovich, Yael Moscowitz, Catuscia Palamidessi, Amir Pnueli, Philippe Schnoebelen, Joseph Sifakis, and Igor Ulyanovsky are gratefully acknowledged. Part of this work was done while the author was visiting ICOT, the Institute for New Generation Computer Technology.

References

1. de Boer, F.S., and Palamidessi, C., Embedding as a Tools for Language Comparison: On the CSP Hierarchy, *Proc. of CONCUR'91*, LNCS, Springer, 1991.
2. Bougé, L., On the Existence of Symmetric Algorithms to Find Leaders in Networks of Communicating Sequential Processes, *Acta Informatica*, 25, pp. 179–201, 1988.
3. Brookes, S.D., On the Relationship of CCS and CSP, *Proc. of 10th Colloq. on Automata, Languages and Programming*, Lecture Notes in Computer Science 154, Springer-Verlag, pp.83-96, 1983
4. Carriero, N., and Gelernter, D., Linda in Context, *Comm. ACM*, 32(4), pp. 444–458, 1988.
5. Chandy, K.M., and Misra, J., *Parallel Program Design*, Addison-Wesley, 1988.
6. Gaifman, H., and Shapiro, E., Fully Abstract Compositional Semantics for Logic Programs, *Proc. ACM Symposium on Principles of Programming Languages*, pp. 134–142, 1989.
7. Gaifman, H., Maher, M.J., and Shapiro, E., Replay, Recovery, Replication, and Snapshots of Nondeterministic Concurrent Programs, *Proc. PODC'91*, ACM,1991.

8. Goldman, K.J., and Lynch, N.A., Modelling Shared State in a Shared Action Model, *Proc. 5th Annual Symposium on Logic in Computer Science*, pp. 450–463, IEEE, 1990.
9. Hennessy, M., *Algebraic Theory of Processes*, MIT Press, 1988.
10. Hewitt, C., A Universal, Modular Actor Formalism for Artificial Intelligence, *Proc. International Joint Conference on Artificial Intelligence*, 1973.
11. Hirata, M., Programming Language Doc and its Self-Description, or, $X = X$ is Considered Harmful, *Proc. 3rd Conference of Japan Society of Software Science and Technology*, pp. 69–72, 1986.
12. Hoare, C.A.R., Communicating Sequential Processes, *Comm. ACM*, 21(8), pp. 666–677, 1978.
13. Hoare, C.A.R., *Communicating Sequential Processes*, Prentice Hall, New Jersey, 1985.
14. INMOS Ltd., *OCCAM Programming Manual*, Prentice-Hall, New Jersey, 1984.
15. Jonsson, B., A Fully Abstract Trace Model for Dataflow Networks, 16th *Annual ACM Symposium on Principles of Programming Languages*, pp. 155–165, 1989.
16. Jifeng H., Josephs, M.B., and Hoare, C.A.R., A Theory for Synchrony and Asynchrony, *Proc. of IFIP TC 2 Working Conf. on Programming Concenpts and Methods*, Sea of Gallilee, Israel, April, 1990.
17. Kahn, G., The Semantics of a Simple Language for Parallel Programming, *Information Processing 74*, pp. 993–998, North-Holland, 1977.
18. Kahn, K., and Saraswat, V.A., Actors as a Special Case of Concurrent Constraint (Logic) Programming, Xerox Technical Report, 1990.
19. Kleinman, A., Moscowitz, Y., Pnueli, A., and Shapiro, E., Communication with Directed Logic Variables, *Proc. of ACM POPL*, 1991.
20. Lynch, N.A., and Fischer, M.J., On Describing the Behavior and Implementation of Distributed Systems, *TCS*, 13, pp. 17–43, 1981.
21. Lynch, N.A., and Tuttle, M.R., Hierarchical Correctness Proofs for Distributed Algorithms, *Proc. ACM Symposium PODC'87*, 1987.
22. Milner, R., *A Calculus of Communicating Systems*, Lecture Notes in Computer Science 92, Springer-Verlag, 1980.
23. Milner, R., *Communication and Concurrency*, Prentice-Hall, 1989.
24. Misra, J., Loosely-Coupled Processes, in E.H.L. Aarts, J. van Leeuwen, M. Ram (eds.), *Proc. of PARLE'91: Parallel Architectures and Languages Europe*, Vol. 2, LNCS 506, Springer, pp.1-26, 1989.
25. Mitchell, J., On Abstraction and the Expressive Power of Programming Languages, *Proc. of the Int'l Conference on Theoretical Aspects of Computer Science*, Sendai, Japan, LNCS 526, Springer, pp.290-310, 1991.
26. Russel, J.R., Full Abstraction for Nondeterministic Dataflow Networks, *Proc. 30th IEEE FOCS*, pp.170-175, 1989.
27. Safra, S., Partial Evaluation of Concurrent Prolog and its Implications, M.Sc. Thesis, Technical Report CS86-24, Dept. of Computer Science, Weizmann Institute of Science, 1986.
28. Saraswat, V.A., Partial Correctness Semantics for CP[↓,—,&], *Proc. 5th Conference on Foundations of Software Technology and Theoretical Computer Science*, New-Delhi, LNCS 205, Springer, pp.347-368, 1985.
29. Saraswat, V.A., A Somewhat Logical Formulation of CLP Synchronisation Primitives, in Bowen, K., and Kowalski, R.A. (eds.), *Proc. 5th International Conference Symposium on Logic Programming*, pp. 1298–1314, MIT Press, 1988.
30. Saraswat, V.A., Kahn, K., and Levy, J., Janus: A Step Towards Distributed Constraint Programming, *Proc. 1990 North American Conference on Logic Programming*, S. Debray and M. Hermenegildo (Eds.), MIT Press, 1990.

31. Shapiro, E. (Editor), *Concurrent Prolog: Collected Papers*, Vols. 1 & 2, MIT Press, 1987.
32. Shapiro, E., The Family of Concurrent Logic Programming Languages, *ACM Computing Surveys* 21(3), pp. 412–510, 1989.
33. Shapiro, E., Embedding Linda and other joys of concurrent logic programming, Technical Report CS89-07, Department of Computer Science, The Weismann Institute of Science, Rehovot, 1989.
34. Shapiro, E., Separating Concurrent Languages with Categories of Language Embeddings, *Proc. STOC'91*, ACM, pp.198-208, 1991.
35. De Simone, R., Higher-Level Synchronising Devices in MEIJE-SCCS, *Theoretical Computer Science*, Vol. 37, pp.245-267, 1985.
36. Ueda, K., *Guarded Horn Clauses*, Ph.D. Thesis, Information Engineering Course, University of Tokyo, Tokyo, 1986.
37. Vaandrager, F.W., On the Relationship Between Process Algebra and Input/Output Automata, Proc. *LICS'91*, IEEE, pp.387-398, 1991.
38. Yardeni, E., Kliger, S., and Shapiro, E., The Languages FCP(:) and FCP(:,?), *J. New Generation Computing*, 7, pp. 89–107, 1990.

Logic of Trace Languages
(Extended Abstract)

Alexander Rabinovich
IBM Research Division
T.J. Watson Research Center
P.O. Box 218
Yorktown Heights, NY 10598

Abstract

Usually, laws established in process calculi have the format of equations and/or inequations between process-terms. Though such set of laws captures important properties of the underlying algebra it cannot reveal some basic *logical* properties of the algebra. From the logical point of view, the consequence relation associated with an algebra is much more fundamental than the set of laws valid in it. That is why in this paper our main concern is about consequence relation which provides the answer to questions in the following format (the formalization is in terms of sequents): what terms are equal under the assumption that some other pairs of terms are equal. We compare two algebras: algebra of linear trace languages and algebra of relations. The fundamental operations in trace algebra are *synchronization* (parallel composition) of two trace languages, *nondeterministic choice* and *hiding* of a port in a language. The corresponding operations in relational algebra are *join*, *union* and *projection*. We show that these algebras have the same laws, i.e. two terms have the same meaning in all trace interpretations iff they have the same meaning in all relational interpretations. Moreover, we show that these algebras have the same consequence relations. We embed both algebras into first order logic and through this embedding obtain sound and complete proof systems for reasoning about the consequence relations in these algebras.

1. Introduction

1.1. Goals and Contribution

Synchronization, nondeterministic choice and hiding are fundamental operations in concurrency. The main goal of this paper is to investigate the algebra of these operations in a framework which is more general than that which is traditionally used in process algebras. We will deal here with the simplest model of concurrent behavior which does not take into account neither branching nor causality. The behavior of a system is represented by a linear trace language. For this simplest model one can observe that synchronization (notations $\|$), nondeterministic choice (notations $+$) and hiding obey laws similar to well known logical laws for conjunction, disjunction and existential quantifier. For example:

- Synchronization is idempotent, commutative and associative, i.e. $L\|L = L,\ L_1\|L_2 = L_2\|L_1,\ (L_1\|L_2)\|L_3 = L_1\|(L_2\|L_3)$.
- Distributivity of synchronization over nondeterministic choice: $L_1\|(L_2 + L_3) = (L_1\|L_2) + (L_1\|L_3)$.

Besides of being interesting observations such facts are also useful for applications. For example, the proof of the generalized Kahn principle in [10] and proofs of modularity for nets of processes in [11] are based on such observations.

The set of laws valid in an algebra captures many algebraic properties. But this set is not enough to reveal some basic logical properties of an algebra. For example, in the algebra of processes the equality of process Pr_1 and of process *hide* x *in* Pr_2 implies that the processes Pr_1 and *hide* x *in* $Pr_1\|Pr_2$ are also equal. But this cannot be derived from the set of laws which are valid in this algebra. From the logical point of view, the notion of consequence relation (see [1]) associated with an algebra is much more fundamental than the set of laws which are valid in it. Very roughly speaking, the consequence relation of an algebra captures the pattern of reasoning appropriate for the algebra. Knowledge of the consequence relations provides the answer to such questions: what terms are equal under the assumption that some other pairs of terms are equal.

Consequence relation is formalized through valid sequents. A sequent is an expression of the form $A_1, \cdots A_n \rightarrow B_1 \cdots B_k$, where A_i, B_j are formulas. A sequent is valid in an algebra iff whenever its antecedents formulas $A_1, \cdots A_n$ hold at least one of its succedents B_j holds. The consequence relation associated with an algebra is just the set of all sequents valid in the algebra.

Terms are used as notations for processes; as formulas one usually takes equations between terms, and in cases when there is a partial order on the elements of the algebra it is more convenient to work with inequations between terms.

As mentioned above, we consider only a subset of the usual repertoire of operations; in particular, prefix operation and recursion are omitted. Nevertheless, even synchronization and hiding are enough for many applications, in particular, for net based models of concurrency. Actually, applications for net models provided the initial motivation for this research.

Usually the behavior of nets is described in an operational way but in all cases we know the behavior of a net can be obtained from the behavior of its components using two operations: *synchronization* (notations $\|$) and *hiding* (a port).

When a semantical model and a specification language are chosen, the central question is to characterize when two systems are equivalent. Often in net theory, some components of the nets are underspecified, but one knows that two nets at certain nodes contain the same component. Therefore, from schematological perspective, the question of equivalence is modified as follows:

> ***characterize when two nets are equivalent for arbitrary interpretations of their components.***

This leads to the next question:

> ***assume that we consider only interpretations under which net N_1 is equivalent to N_1', N_2 is equivalent to N_2', ... N_k is equivalent to N_k'. Can we conclude from this that N is equivalent to N' under these particular interpretations?***

Recall that the behavior of a net is obtained from the behavior of its components through synchronization and hiding. Hence, the first question may be reduced to the question of equivalence of two terms constructed by these operations. The second question is reduced to the question of equality of two terms under the assumption that some other pairs of terms are equal, i.e. to the validity of an appropriate sequent.

We provide a complete and sound calculus for the consequence relation associated with linear trace algebra. It turns out that there is nothing incidental in the observation about similarity between synchronization, choice and hiding on the one hand and conjunction, disjunction and existential quantifier on the other hand. Roughly speaking our main theorem states that their corresponding consequence relations are the same.

Before proceeding to more detailed survey of the paper let us contrast it with other works in the field. Unlike others we investigate the simplest model of concurrency which ignores branching and causalities. On the other hand, despite the fact that in logic the consequence relation plays much more important and fundamental role than the set of valid sentences, to our knowledge, this is the first time that the consequence relation is investigated for a model of concurrency. More technical, but still important, are the following two points: (1) Though synchronization (parallel composition) is one of the most fundamental operations in concurrency, when it comes to axiomatization it is usually eliminated through expansion theorem [5, 7] and it is reduced to other operations such as prefixing, choice, left and right merge, etc. In such axiomatizations the properties of synchronization are disguised. (2) Unlike in this paper, completeness theorems which appear in the literature are very weak. Namely, they state that an (in)equation between *closed* terms is provable iff it is valid. Only in the works of Milner [8, 9] and Groote [4] versions of completeness theorem for open terms are considered.

1.2. Structure of the Paper

We compare two algebras: algebra of linear trace languages and algebra of relations. The fundamental operations in the trace algebra are synchronization (parallel composition) of two trace languages, nondeterministic choice and hiding of a port in a language. The corresponding operations in relational algebra are join, union and projection. We show that these algebras have the same laws, i.e., two terms have the same meaning in all trace interpretations iff they have the same meaning in all relational interpretations. Moreover, we show that these algebras have the same consequence relations.

Let us describe now the structure of the paper, some results and underlying ideas.

Section 2 describes two algebras: (1) algebra of relations with operations join, union and projection and (2) algebra of trace languages with operations synchronization, choice and hiding.

A linear trace language consists of the sets of all runs of a concurrent system. Each trace language L is equipped with the alphabet of actions which are available in L. Runs of L use actions from this alphabet but not necessarily all of them.

Section 3 begins with the syntactical definition of our language $\mathcal{L}$ which has three syntactical categories: typed terms, formulas and sequents. Two semantics for $\mathcal{L}$ are provided. The first one is relational semantics which assigns *port relations* to $\mathcal{L}$-terms. Ports of such relations are the same as attributes for database relations; the type of a term is the set of ports (attributes) of the corresponding relation in this semantics. The second semantics assigns trace languages to terms; and the type of a term provides information about the alphabet of the corresponding trace language.

Terms of $\mathcal{L}$ are constructed from typed variables by three constructors which are interpreted in relational semantics as join, union and projection; in trace semantics these constructors are interpreted as synchronization, choice and hiding.

As it usually happens in programming languages semantics, inequational reasoning is more simple than the equational one. This is one of the reasons that formulas of $\mathcal{L}$ are inequations between terms. A formula $t_1 \subseteq t_2$ is valid in one of the semantics above if the meaning of t_1 is a subset of the meaning of t_2. Hence, in a well formed formula $t_1 \subseteq t_2$ terms t_1 and t_2 have the same type.

Recall that a sequent is an expression of the form $A_1, \cdots A_n \rightarrow B_1 \cdots B_k$, and it holds in a semantics iff whenever all of its antecedent formulas $A_1, \cdots A_n$ hold at least one of its succedent formula B_j holds. The consequence relation associated with an algebra is just the set of all sequents which hold in the algebra.

Here is the main theorem of the paper

Theorem 1.1: *A single conclusion sequent $A_1, \cdots A_n \rightarrow B$ is valid in trace algebra iff it is valid in relational algebra.*

The fact that trace validity of a sequent implies its relational validity is more simple and we give a sketch of the proof in section 4.

Section 5 describes an embedding of $\mathcal{L}$ into a proper subset of first order logic. This embedding transforms relationally valid sequents into first order valid sequents. Moreover, sequents which are not valid relationally are transformed into sequents which are not valid in first order logic. This transformation gives us a way to construct a complete proof system for relational semantics from any complete proof system of first order logic.

In section 6 we describe the plan of the proof that relational validity of a sequent implies its trace validity; we point to difficulties which arise here. In this way together with section 4 we accomplish the proof of theorem 1.1. Section 7 contains concluding remarks and states some open problems.

2. Two Algebras

Let J be a non-empty set. Each element of J is called a port. Finite sets of ports are called types. a, b, c will be metavariables which range over ports; Σ_0, $\Sigma_1, \cdots$ are metavariables which range over types. We describe in this section two algebras: algebra of relations and algebra of trace languages. Objects in these algebras will have a type. In relational algebra ports have the role similar to the role of sorts in many-sorted logic. But unlike relations in many sorted logic, every relation can have at most one argument of a given sort. In trace algebra a type of relation provides an information not only which communications can happen, but also which communications cannot happen.

2.1. Algebra of Relations

Let D be a set. A **port relation** of type Σ over D is a subset of D^{Σ}. Given $x \in D^{\Sigma}$ and $x_1 \in D^{\Sigma_1}$. Assume that Σ_1 is a subset of Σ and for every port p in Σ_1 the equality $x(p) = x_1(p)$ holds. In this case we say that x_1 is the **projection** of x on Σ_1. For a port relation R of type Σ its projection on Σ_1 is the relation of type Σ_1 which consists of the projection of elements of R on Σ_1.

Let R_1 be a relation of type Σ_1 and R_2 be a relation of type Σ_2. The **join** of R_1 and R_2 is the relation of type $\Sigma_1 \cup \Sigma_2$ defined as follows: $x \in R_1 \& R_2$ if the projection of x on Σ_1 is in R_1 and the projection of x on Σ_2 is in R_2.

For a relation R of type Σ the relation $\exists a.R$ has the type $\Sigma - \{a\}$ and it is the projection of R on this type.

For relations R_1, R_2 of the same type Σ the relation $R_1 \cup R_2$ has type Σ and it is the **union** of R_1 and R_2.

Example 1: *$(a_1 \rightsquigarrow d_1, \ldots, a_n \rightsquigarrow d_n)$ is the notation for the element of $D^{\{a_1, \ldots a_n\}}$ which maps a_i to d_i for $i = 1, \ldots, n$. Consider port relations R_1, R_2, R_3 over set $\{0,\ 1\}$ of types $\{a_1,\ a_2\}$, $\{a_1,\ a_2\}$ and $\{a_1,\ a_3\}$ respectively which are defined as follows:*

$$R_1 = \{(a_1 \rightsquigarrow 0, a_2 \rightsquigarrow 1)\}$$
$$R_2 = \{(a_1 \rightsquigarrow 1, a_2 \rightsquigarrow 0),\ (a_1 \rightsquigarrow 1, a_2 \rightsquigarrow 1)\}$$
$$R_3 = \{(a_1 \rightsquigarrow 1, a_3 \rightsquigarrow 0),\ (a_1 \rightsquigarrow 0, a_3 \rightsquigarrow 1)\}$$

Then

$R_1 \& R_2$ is the empty relation of type $\{a_1,\ a_2\}$;

$$R_1 \& R_3 = \{(a_1 \rightsquigarrow 0,\ a_2 \rightsquigarrow 1,\ a_3 \rightsquigarrow 1)\};$$
$$R_2 \& R_3 = \{(a_1 \rightsquigarrow 1,\ a_2 \rightsquigarrow 0,\ a_3 \rightsquigarrow 0),\ (a_1 \rightsquigarrow 1,\ a_2 \rightsquigarrow 1,\ a_3 \rightsquigarrow 0)\};$$
$$\exists a_1.R_3 = \{(a_3 \rightsquigarrow 0),\ (a_3 \rightsquigarrow 1)\};$$
$$\exists a_2.R_3 = \{(a_1 \rightsquigarrow 1, a_3 \rightsquigarrow 0),\ (a_1 \rightsquigarrow 0, a_3 \rightsquigarrow 1)\};.$$
$$\exists a_2.R_2 = \{(a_1 \rightsquigarrow 1)\}.$$

Remark. In database terminology (see [6]) ports are called attributes; types are called relational schemes; join is denoted by $\bowtie$; projection of R on Σ is denoted by $\pi_\Sigma(R)$.

2.2. Algebra of Trace Languages

Let T^* denotes the set of all strings over alphabet T. As usual the concatenation of strings $s_1,\ s_2 \in T^*$ will be denoted by $s_1 s_2$; we say that s_1 is a prefix of s_2 if there is a string s_3 such that $s_1 s_3 = s_2$. A (linear) trace language over alphabet T is a pair $< S,\ T >$, where S is a subset of T^*. For a language L its alphabet is denoted by $alph(L)$ and its strings denoted by $trace(L)$.

Below we will consider trace languages over structured alphabets. Such an alphabet is the cartesian product of set Σ of ports and set D of values. Notation $< a,\ d >$ is used for elements of an alphabet $\Sigma \times D$. (Intuitevely an action $< a,\ d >$ corresponds to the transmission of the value d over the port a.) If a language L has an alphabet $\Sigma \times D$ then we say that L has type Σ. Assertion that L has type Σ will be written as $L:\ \Sigma$. We consider the following operations on languages:

Synchronization ($\|$).

$$alph(L_1 \| L_2) = alph(L_1) \cup alph(L_2)$$
$$s \in trace(L_1 \| L_2) \text{ iff for } i = 1, 2$$
$$s|alph(L_i) \in trace(L_i)$$

where $s|A$ is the notation for the string one gets from s by deleting all events which are not in A.

Hiding. *hide* p in L results in the language over $alph(L) - \{p \times D\}$; its strings are obtained from the strings of L through the deletion of all occurrences of actions on the port p.

Choice. For languages L_1 and L_2 of the same type Σ, language $L_1 + L_2$ has type Σ; the language $L_1 + L_2$ consists of all the strings which belong to at least one of the languages $L_1,\ L_2$.

Whenever trace languages will occur in the same context we will always assume that their alphabet is over the same set of values D. Under this assumption, it is clear that if $L_1:\ \Sigma_1$ and $L_1:\ \Sigma_2$ then $L_1 \| L_2:\ \Sigma_1 \cup \Sigma_2$ and *hide* a in $L:\ \Sigma - \{a\}$.

Example 2: *Consider languages L_1, L_2, L_3 of types $\{a,\ b\}$, $\{a,\ c\}$, $\{b,\ c\}$ respectively which contain the following traces:*

$$trace(L_1) = \{<a,1><a,0><b,0>,\ <a,0><a,1><b,1>\}$$
$$trace(L_2) = \{<a,1><a,0><c,0>\}$$
$$trace(L_3) = \{<c,1><b,1>\}$$

Below are the traces of some languages resulting by applying synchronization and hiding to L_1, L_2 and L_3.

$L_1 \| L_2 = \{<a,1><a,0><b,0><c,0>,\ <a,1><a,0><c,0><b,0>\}$

$L_1 \| L_3 = \{<a,0><a,1><c,1><b,1>,\ <a,0><c,1><a,1><b,1>,$
$<c,1><a,0><a,1><b,1>\}$

$L_2 \| L_3$ is the empty language of type $\{a,\ b,\ c\}$

hide a in L_1 : $\{b\}$ and trace(hide a in L_1) $= \{<b,0>,\ <b,1>\}$

hide b in L_1 : $\{a\}$ and trace(hide b in L_1) $= \{<a,0><a,1>,\ <a,1><a,0>\}$

hide c in $L_1 = L_1$

Remark: The role of alphabet. The type of language is very important in the definition of synchronization. For languages L_1, L_2 with same set of strings the languages $L_1 \| L_3$ and $L_2 \| L_3$ usually have not only different types but also different sets of strings as in the following

Example 3: *Consider languages L_1, L_2 and L_3 over unary set D of values. a, b, c will be used below as ports and as elements of alphabet $\Sigma \times D$. Assume that L_1 and L_2 contain only one string ab, and have respectively types $\{a,\ b,\ c\}$ and $\{a,\ b\}$. Let L_3 be of type $\{a,\ c\}$ and contain only one string ac. Then $L_1 \| L_3$ and $L_2 \| L_3$ have the same type $\{a,\ b,\ c\}$; $L_1 \| L_3$ contains no strings, but $L_2 \| L_3$ contains two strings abc and acb.*

3. Syntax and Semantics of $\mathcal{L}$

In 3.1 we define language $\mathcal{L}$ which has three syntactical categories: typed terms, formulas and sequents. Two semantics for $\mathcal{L}$ are provided. The first one is relational semantics (section 3.2) which assigns port relations to $\mathcal{L}$-terms. The second semantics (section 3.3) assigns trace languages to terms.

Terms of $\mathcal{L}$ are constructed from typed variables by three constructors which are interpreted in relational semantics as join, union and projection; in trace semantics these constructors are interpreted as synchronization, choice and hiding.

3.1. Syntax

Definition 1: *The alphabet of $\mathcal{L}$ consists of the following sets of symbols:*

- *Ports.*
- *An infinite list of variables X_1^{Σ}, $X_2^{\Sigma}, \dots$ for each type Σ.*
- *Constructors: $\wedge$, $\exists$, $\subseteq$, $\vee$.*
- *Auxiliary symbols: '(', ')', '$\rightarrow$' and ','.*

Below we define inductively terms and their types. Metavariables t_0, $t_1 \cdots$ range over terms; metavariables Σ_0, $\Sigma_1 \cdots$ range over types. An assertion that t has a type Σ will be written as $t : \Sigma$.

Definition 2: *Terms and their types.*

- *Variable X^{Σ} is a term of type Σ.*
- *If t_1 and t_2 are terms of types Σ_1 and Σ_2 respectively, then $(t_1 \wedge t_2)$ is a term of type $\Sigma_1 \cup \Sigma_2$.*
- *If t_1 and t_2 are terms of type Σ, then $(t_1 \vee t_2)$ is a term of type Σ.*
- *If t is a term of type Σ then $\exists a.t$ is a term of type $\Sigma - \{a\}$.*

A formula is an expression of the form $t_1 \subseteq t_2$, where t_1, t_2 are terms of the same type. A, B, $C \cdots$ will be metavariables ranging over formulas.

A sequent is an expression of the form $A_1, A_2, \cdots A_n \rightarrow B_1, \cdots B_m$, where $A_1, A_2, \cdots A_n, B_1, \cdots B_m$ are formulas. $A_1, A_2, \cdots A_n$ are the antecedents of the sequent and $B_1, \cdots B_m$ are the succedents of the sequent.

3.2. Relational Semantics of $\mathcal{L}$ terms

A relational environment η is a map from the variables to port relations which respects types, i.e. if X is of type Σ then $\eta(X)$ is a subset of D^{Σ}. For a term t of type Σ its relational semantics in an environment η (notation - $[|t|]_{\eta}^{rel}$) is a subset of D^{Σ} defined as follows:

Definition 3: *(Relational semantics for $\mathcal{L}$-terms)*

- *If t is a variable X then $[|t|]_{\eta}^{rel} = \eta(X)$*
- $[|t_1 \wedge t_2|]_{\eta}^{rel}$ *is* $[|t_1|]_{\eta}^{rel} \& [|t_2|]_{\eta}^{rel}$.
- $[|\exists a.t|]_{\eta}^{rel}$ *is* $\exists a.[|t|]_{\eta}^{rel}$.
- $[|t_1 \vee t_2|]_{\eta}^{rel}$ *is* $[|t_1|]_{\eta}^{rel} \cup [|t_2|]_{\eta}^{rel}$.

3.3. Trace Semantics of $\mathcal{L}$ terms

A trace environment η is a map from the variables to trace languages which respects types, i.e. if X is of type Σ then $\eta(X)$ is a trace language of type Σ. For a term t of type Σ its trace semantics in an environment η (notation - $[|t|]_\eta^{trace}$) is a trace language of type Σ defined as follows:

Definition 4: *(Trace semantics for $\mathcal{L}$-terms)*

- *If t is a variable X then $[|t|]_\eta^{trace} = \eta(X)$*
- $[|t_1 \wedge t_2|]_\eta^{trace}$ *is* $[|t_1|]_\eta^{trace} \parallel [|t_2|]_\eta^{trace}$.
- $[|\exists a.t|]_\eta^{trace}$ *is hide a in* $[|t|]_\eta^{trace}$
- $[|t_1 \vee t_2|]_\eta^{trace}$ *is* $[|t_1|]_\eta^{trace} + [|t_2|]_\eta^{trace}$.

3.4. Semantics of Formulas and Sequents

Definition 5: *Let A be a formula $t_1 \subseteq t_2$.*

- *We say that a relational environment η satisfies the formula A if the relation $[|t_1|]_\eta^{rel}$ is a subset of the relation $[|t_2|]_\eta^{rel}$. Similarly, we say that a trace environment η satisfies A if the trace language $[|t_1|]_\eta^{trace}$ is a subset of the trace language $[|t_2|]_\eta^{trace}$.*
- *A formula A is relationally (trace) valid if every relational (trace) environment satisfies it.*
- *An environment satisfies a sequent if it satisfies one of the succedents of the sequent or if it does not satisfy one of the antecedents of the sequent.*
- *A sequent is relational (trace) valid if every relational (trace) environment satisfies it.*

From the above definition it follows that for both relational and trace semantics an environment satisfies a formula A iff it satisfies the sequent $\rightarrow A$.

4. Trace Validity Implies Relational Validity

Theorem 4.1: *If a sequent is trace valid then it is relationally valid.*

To prove that every trace valid sequent is relationally valid we define in the full paper an injective map ν from port relations to trace languages, which increases with respect to inclusion. We then show that ν is a homomorphism i.e.

1. $\nu(R_1 \& R_2) = \nu(R_1) \| \nu(R_2)$

2. $\nu(\exists a.R) = hide\ a\ in\ \nu(R)$

3. $\nu(R_1 \cup R_2) = \nu(R_1) + \nu(R_2)$

The fact that ν is an injective and an increasing homomorphism implies

Claim 4.2: *Let η be a relational environment and η' be trace environment defined as follows: $\eta'(X) = \nu(\eta(X))$. Then*

1. $\nu([|t|]^{rel}_{\eta}) = [|t|]^{trace}_{\eta'}$

2. $x \in [|t|]^{rel}_{\eta}$ *iff* $\nu(x) \subseteq [|t|]^{trace}_{\eta'}$

3. *η satisfies formula A in relational semantics iff η' satisfies formula A in trace semantics.*

4. *η satisfies sequent S in relational semantics iff η' satisfies sequent S in trace semantics.*

5. *If a sequent S is valid in trace semantics then it is valid in relational semantics.*

Claim 4.2 (5) implies theorem 4.1.

5. Embedding of $\mathcal{L}$ into First Order Logic

Relational semantics of $\mathcal{L}$ is tightly connected with semantics of first order logic. Below we give an embedding (transformation) of terms and formulas of $\mathcal{L}$ into first order formulas. In this transformation, port names will play a role of individual variables and to each variable X_i of $\mathcal{L}$ there corresponds a predicate variable R_i in the first order logic. To a term of type $\Sigma = \{a_1, \cdots a_n\}$ there corresponds a formula with the free variables $\{a_1, \cdots a_n\}$. The embedding is straightforward. Here is

Definition 6: (Transformation) *Fix a bijection ψ between variables of $\mathcal{L}$ and predicate variable such that if $X : \Sigma$ then $\psi(X)$ has arity $|\Sigma|$. Assume that the set of $\mathcal{L}$-ports is the same as the set of first order individual variables and fix a linear ordering $\prec$ on this set.*

- *For each variable X_i of type $\Sigma = \{a_1, \prec a_2, \cdots \prec a_n\}$ there corresponds an atomic formula $R_i(a_1, \cdots, a_n)$, where $R_i = \psi(X_i)$*

- *A term is transformed into first order formula by replacing all variables by the corresponding atomic formulae.*

- *A formula $t_1 \subseteq t_2$ is transformed into first order formula by transforming t_i into first order formulae and replacing $\subseteq$ by implication $\Rightarrow$.*

Example 4: $\exists c.\ (X_1^{\{a,\ b\}} \wedge X_2^{\{a,\ c\}}) \subseteq (\exists c.\ X_2^{\{a,\ c\}}) \wedge X_1^{\{a,\ b\}}$ *is transformed into* $\exists c.\ (R_1(a,\ b) \wedge R_2(a,\ c)) \Rightarrow (\exists c.\ R_2(a,\ c)) \wedge R_1(a,\ b)$

Now given a relational environment η. Consider an interpretation $\overline{\eta}$ of first order predicate defined as follows: $\overline{\eta}(R_i)(d_1, \cdots, d_n) = true$ iff the function which maps a_j into d_j for $j = 1 \cdots n$ is in $\eta(X_i)$

Claim 5.1: *Let t be a term and η be a relational environment. Let $\bar{t}$ and $\overline{\eta}$ be the corresponding first order formula and first order interpretation. Assume that t is of type $\Sigma = \{a_1, \cdots a_n\}$ and ρ is any first order valuation which maps a_j to $d_j \in D$ for $j = 1, \cdots n$. Then ρ satisfies $\bar{t}$ in the interpretation $\overline{\eta}$ iff the element of D^{Σ} which maps a_j to $d_j \in D$ for $j = 1, \cdots n$ is in $[|t|]_{\eta}^{rel}$.*

From the claim it follows

Corollary 5.2: *(Faithfulness of the transformation)*

1. *A relational environment η satisfies an $\mathcal{L}$ formula t iff the transformation of t holds in the interpretation $\overline{\eta}$.*
2. *A relational environment η satisfies a sequent S iff the sequent obtained by universal closure of the formulas which correspond to formulas of S holds in the interpretation $\overline{\eta}$.*

Remark Complete proof systems for relational semantics. Claim 5.2.2 reduces relational validity of $\mathcal{L}$ to the validity of first order formulas. Therefore, from any complete proof system G for first order logic a procedure can be constructed which accepts relationally valid $\mathcal{L}$-sequents. Here is a description of such a procedure.

$Proc_G$(input: an $\mathcal{L}$ sequent $A_1, \ldots A_n \rightarrow B_1, \ldots B_m$).

1. Let first order formulas $\bar{A}_i$, $\bar{B}_j$ be obtained from $\mathcal{L}$ formulas A_i, B_j as in definition 6.
2. if formula $(\bar{A}_1^{\forall} \wedge \bar{A}_2^{\forall} \wedge \cdots \wedge \bar{A}_n^{\forall}) \Rightarrow (\bar{B}_1^{\forall} \vee \bar{B}_2^{\forall} \vee \cdots \vee B_m^{\forall})$ is provable in G then the sequent $A_1, \ldots A_n \rightarrow B_1, \ldots B_m$ is relationally valid.[1].

It is clear that if G is a complete proof system for first order logic then $Proc_G$ accepts exactly relationally valid sequents.

6. Relational Validity Implies Trace Validity

Let us summarize what we proved till now. We know that (1) trace validity implies relational validity; (2) we have complete proof systems for relational validity. Therefore, to show that trace validity coincides with relational validity it is enough to check that the inference rules of a complete proof system for relational semantics are sound for trace semantics. It turns out that the Gentzen sequent calculus [2] is most appropriate for these purposes.

[1] We use notations $C^{\forall}$ for the universal closure of the formula C

To show that a proof system is sound for trace semantics we have to verify that if the premises of the inference rules are trace valid, then their conclusions are trace valid. But here arises a new obstacle. Note that $\mathcal{L}$ was mapped to a proper subset of first order logic. As a result of this it might happen that the premises of an inference rule are images of $\mathcal{L}$-sequents but the conclusion is not the image of any $\mathcal{L}$-sequent. So trace semantics does not assign any meaning to this sequent. One can hope that a proof system has the following *property:*

(*) if the image $\bar{S}$ of an $\mathcal{L}$-sequent is provable then there exists a proof of $\bar{S}$ which contains only the images of $\mathcal{L}$-sequents.

This property would allow to consider only such instances of inference rules in which premises and conclusions are images of $\mathcal{L}$ sequents. The Gentzen sequent calculus has many nice characterizations which guarantee that it is a conservative extension of its subcalculus. This is one of the main reasons of its choice for our proofs. But nevertheless, the property (*) does not hold for the Gentzen sequent calculus.

To overcome these difficulties another language $\mathcal{L}'$ which contains $\mathcal{L}$ is introduced in the full paper. Trace semantics is provided for $\mathcal{L}'$ sequents. For single conclusion sequents the trace semantics of $\mathcal{L}'$ conservatively extends the trace semantics of $\mathcal{L}$. $\mathcal{L}'$ sequents are transformed into a proper subset of first order logic sequents. But now, the property (*) mentioned above holds for $\mathcal{L}'$-sequents and the proofs in the Gentzen calculus. The proof of this appeals to quite deep proof theoretical results and will appear in the full paper.

From Gentzen calculus one can extract a sequent calculus for $\mathcal{L}'$, and verify the soundness of its inference rules for $\mathcal{L}'$.

Since this calculus is complete for relational semantics it follows that every relational valid sequent is provable. Since this calculus is sound for $\mathcal{L}'$ trace semantics it follows that every provable $\mathcal{L}'$ sequent is trace valid. Therefore, every single conclusion $\mathcal{L}$ sequent which is relationally valid is trace valid. Together with theorem 4.1 this completes the proof of equivalence of relational and trace validity for single conclusion $\mathcal{L}$-sequents (theorem 1.1).

7. Concluding Remarks and Some Open Problems

7.1. Prefix Closed Languages (Linear Processes)

We modeled concurrent systems by trace languages. Each string in a language represents a 'complete' run. Sometimes all runs of a concurrent system are considered. In this case semantical objects will be trace languages which have the following prefix closeness property:

> If the trace language L contains a string s then L also contains all prefixes of s.

Below, prefix closed trace languages are called (linear) processes. Processes are closed under synchronization, choice and hiding. One of the natural questions is to characterize the algebra of processes. Since, processes are trace languages every trace valid sequent is process valid. Therefore, according to theorems 1.1 every relationally valid single conclusion sequent is process valid. Hence one can safely apply first order logic to find laws in the process algebra and to reason about process validity of (single conclusion) sequents. Is every process valid sequent relationally valid? The following example shows that this is not the case.

Example 5: *(A process valid sequent which is not relationally valid.)*

$\exists a.\ X^{\{a\}} \subseteq \exists a.\ Y^{\{a\}}, \exists a.\ Y^{\{a\}} \subseteq \exists a.\ X^{\{a\}} \rightarrow \exists a.\ Y^{\{a\}} \subseteq \exists a.\ (X^{\{a\}} \wedge Y^{\{a\}})$

The sequent is valid for processes because its antecedents hold if either both X and Y are the empty languages of type a or they contain an empty string ϵ. In the first case the languages $\exists a.\ Y^{\{a\}}$ and $\exists a.\ (X^{\{a\}} \wedge Y^{\{a\}})$ are empty. In the second case they consist of the empty string.

The corresponding first order sequent is: $\exists a.\ R_1(a) \Rightarrow \exists a.\ R_2(a), \exists a.\ R_2(a) \Rightarrow \exists a.\ R_1(a) \rightarrow \exists a.\ R_2(a) \Rightarrow \exists a.\ (R_1(a) \wedge R_2(a))$. *It is clear that it is not valid.*

We would like to know the answer to the following problems:

- Is it possible to add some sequents to the Gentzen calculus such that a single conclusion sequent is process valid iff it is provable?
- Does there exist process valid laws which are not relationally valid?
- Is it decidable whether a single conclusion sequent is process valid?

7.2. Decidability

It is clear that the validity of $\mathcal{L}$ formulas is decidable. We don't know whether validity of $\mathcal{L}$-sequents is decidable.

7.3. Conclusion

We hope that our paper will stimulate the study of consequence relations for more subtle concurrent models. These consequence relations are quite different from our simplest case (for example, almost in all other models synchronization is not idempotent). We believe that knowledge of consequence relations will highlight merits and demerits of concurrent models. Some of these models are based on the same intuition and only different (and not always justified) technical decisions lead to the diversity. The study of the corresponding consequence relations might justify these decisions and even suggest other more natural models.

The previous discussion highlights the way from concurrent models to their consequence relations and to logic. Another challenge is to go in the opposite direction and to provide 'concurrent' models for logics which claim to have a flavor of parallelism. Among such logics one of the most important is linear logic [3].

Acknowledgments

Many ideas of the paper have arisen in numerous discussions with Boris A. Trakhtenbrot. Thanks also to Arnon Avron, Joram Hirshfield and Albert R. Meyer for many stimulating discussions and helpful comments.

References

[1] A. Avron. Simple Consequence Relations. In *Information and Computation*, volume 92, 1991

[2] G. Gentzen. Investigations into logical deduction. In *The collected works of Gerhard Gentzen*, (M. Szabo, Ed.), North-Holland, Amsterdam, 1969.

[3] J.-Y. Girard. Linear Logic. In *Theoretical Computer Science*, 50(1):1-102, 1987.

[4] J. F. Groote. A new strategy for proving ω-completeness with applications in process algebra. In *Proceedings of CONCUR 90*, volume 458 of *Lect. Notes in Computer Science*. Springer Verlag, 1990.

[5] M. Hennessy. *Algebraic Theory of Processes*. MIT Press, 1988.

[6] P. .C. Kanellakis. *Elements of Relational Database Theory*. In *Handbook of Theoretical Computer Science*, North-Holland, 1989.

[7] R. Milner. *Communication and Concurrency*. Prentice-Hall International, Englewood Cliffs, 1989.

[8] R. Milner. A complete proof system for a class of regular behaviors. In *JCSS* , volume 28, pp 439-466,1984.

[9] R. Milner. A complete axiomatisation for observational Congruence of finite state behaviors. In *Information and Computation*, volume 81, pp 227-247, 1989.

[10] A. Rabinovich and B. A. Trakhtenbrot. Communication among relations. In *International Conference on Automata, Languages and Programming*, volume 443 of *Lect. Notes in Computer Science*. Springer Verlag, 1990.

[11] A. Rabinovich and B. A. Trakhtenbrot. On Nets, Algebras and Modularity. In *International Conference on Theoretical Aspects of Computer Software*, to appear in *Lect. Notes in Computer Science*. Springer Verlag, 1991.

Multiway Synchronization Verified with Coupled Simulation

Joachim Parrow* and Peter Sjödin

Swedish Institute of Computer Science, Box 1263, S-16428 Kista, Sweden
and
Department of Computer Systems, Uppsala University, Sweden

Abstract. We consider the problem of implementing multiway synchronization in a distributed environment providing only binary asynchronous communication. Our implementation strategy is formulated as a transformation on transition systems and we give a distributed algorithm for multiway synchronization. Correctness assertions and proofs are based on a new method: *coupled simulations*. The coupled simulation equivalence is weaker than observation equivalence and stronger than testing equivalence and combines some of their advantages. Like observation equivalence (and unlike testing) it is established through case analysis over single transitions. Like testing equivalence (and unlike observation) it allows an internal choice to be distributed onto several internal choices. The latter is particularly important when relating our distributed implementations to their specifications.

1 Introduction

In this paper we will formally verify a strategy for implementing multiway synchronizations between processes in an environment supporting only binary asynchronous communication. This type of environment is common in physically distributed systems, and multiway synchronization is a communication primitive that has attained much interest [1, 3, 4, 7, 8, 9, 10, 16]. We will here concentrate on the verification methodology, i.e. the formulation and proof of the relevant correctness properties.

Our point of departure will be a *specification* of a system of n "client" processes, written $P_1 \parallel \cdots \parallel P_n$. A process P_i can perform two kinds of actions: *synchronization actions* involve multiway synchronizations over ports with other processes, and *external actions* are interactions between a single process and the environment of the system. An example specification with four client processes is depicted in Fig. 1. The lines connecting the processes represent multiway synchronization capabilities; thus a is a synchronization between all four processes, b is between P_1, P_2 and P_3, and c is between P_3 and P_4. A process can only participate in one synchronization at a time, so there may be conflicts between different enabled synchronizations. Suppose that the specification in Fig. 1 is in a state where all three multiway synchronizations are enabled. Since there is one process that is involved in all synchronizations, namely P_3, only one of the synchronizations will take place. The choice of what synchronization takes place is nondeterministic.

* Partly funded by NUTEK as part of Esprit Basic Research Action 3006: Concur.

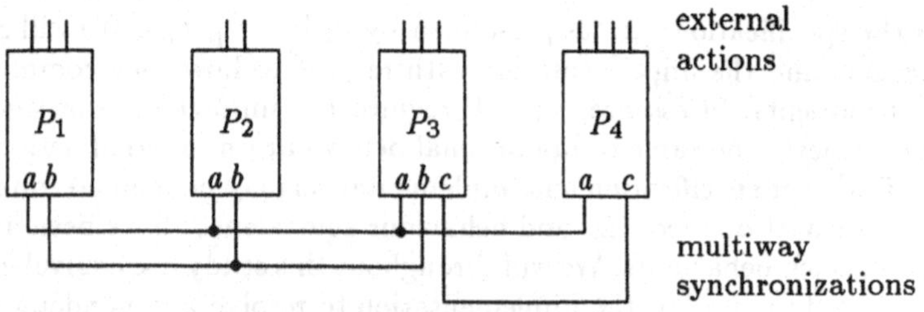

Fig. 1. A specification with four processes.

The first step in formulating a general strategy for implementing multiway synchronization is to replace multiway synchronizations by binary interactions with a monolithic *central synchronizer*. We call the resulting system a *central implementation*, and the central implementation of the specification $P_1 \parallel \cdots \parallel P_n$ is written $\mathcal{C}(P_1, \ldots, P_n)$. Each client process P_i here communicates, on actions req_i, its intentions to participate in synchronizations and receives confirmations about scheduled interactions on actions $conf_i$. A central implementation of a specification is achieved by transforming each client process so that it uses req_i and $conf_i$ actions rather than multiway synchronization actions, and by adding the central synchronizer as an extra process with no external actions. The central implementation of the specification in Fig. 1 is depicted in Fig. 2.

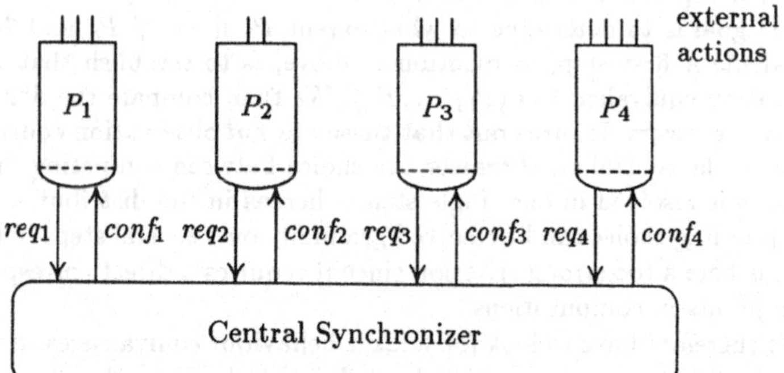

Fig. 2. A central implementation.

Traditionally, correctness proofs for multiway synchronization algorithms [2, 6, 10, 14, 15] amount to proving particular safety properties such as deadlock freedom and particular liveness properties such as "an enabled synchronization must eventually be scheduled for execution". But our implementation strategy admits a more general correctness assertion, namely that the implementation behaves as prescribed by the specification (i.e., as prescribed by $P_1 \parallel \cdots \parallel P_n$). We will compare the specification and the implementation with respect to how they communicate with their environments. In essence, we will require the implementation and specification to have exactly the same computational behaviour on external actions.

Both our specification and implementation may be nondeterministic, and various implementation preorders and behaviour equivalences have been suggested for comparing such behaviours. We will throughout this study use equivalences, which means that we do not allow the implementation to resolve any nondeterminism present in the specification; all of the specified behaviours must be implemented. When comparing $P_1 \parallel \cdots \parallel P_n$ with $\mathcal{C}(P_1, \ldots, P_n)$ we will use observation equivalence [12]. This is stronger than e.g. testing equivalence [5] (at least for the class of convergent systems which is of interest for this paper), so equivalent processes cannot be separated by any amount of testing. Furthermore observation equivalence can be shown by establishing weak bisimulations, a convenient proof technique applicable directly on transition systems.

A central implementation is not realistic if the client processes execute on physically distributed computers and communicate over an asynchronous network. Therefore we replace the central synchronizer with a *distributed synchronizer* which in turn consists of mediators and ports. There is one *mediator* for each client process; the mediator can be thought of as representing the client process in negotiations to establish multiway synchronizations. There is one *port* for each multiway synchronization action in the specification; the port establishes that all required clients participate in the synchronization. Mediators and ports interact asynchronously, each such interaction is always between one mediator and one port. The distributed implementation of n processes is written $\mathcal{D}(P_1, \ldots, P_n)$. Figure 3 contains the distributed implementation corresponding to Fig. 1 and 2.

Our goal is to determine to what extent $P_1 \parallel \cdots \parallel P_n$ and $\mathcal{D}(P_1, \ldots, P_n)$ are equivalent. A first step, as mentioned above, is to establish that $P_1 \parallel \cdots \parallel P_n$ is observation equivalent to $\mathcal{C}(P_1, \ldots, P_n)$. We then compare the distributed and central synchronizers. It turns out that these are *not* observation equivalent; the reason is that in the central synchronizer the choice between competing multiway synchronizations is resolved in one single step, whereas in the distributed synchronizer the corresponding choice can be resolved gradually over several steps. Observation equivalence is here a too strong criterion since it requires a direct correspondence between choice points in computations.

We therefore have to look for weaker behaviour equivalences. An interesting candidate is testing equivalence, which implies that both synchronizers satisfy the same tests on the external actions. Unfortunately, in order to establish testing equivalence one must do a case analysis on all possible computations (this is in contrast to observation equivalence which can be established through a case analysis over all single computation steps, a more manageable task). We will therefore introduce a new equivalence which we call *coupled simulation* equivalence, $=_{cs}$. For convergent systems this lies between testing and observation equivalence. The main point is that coupled simulation equivalence can be established by case analysis over single computation steps and yet does not require an exact correspondence between choice points in computations. We will establish that the distributed and central synchronizers are coupled simulation equivalent, and that the context of the synchronizers

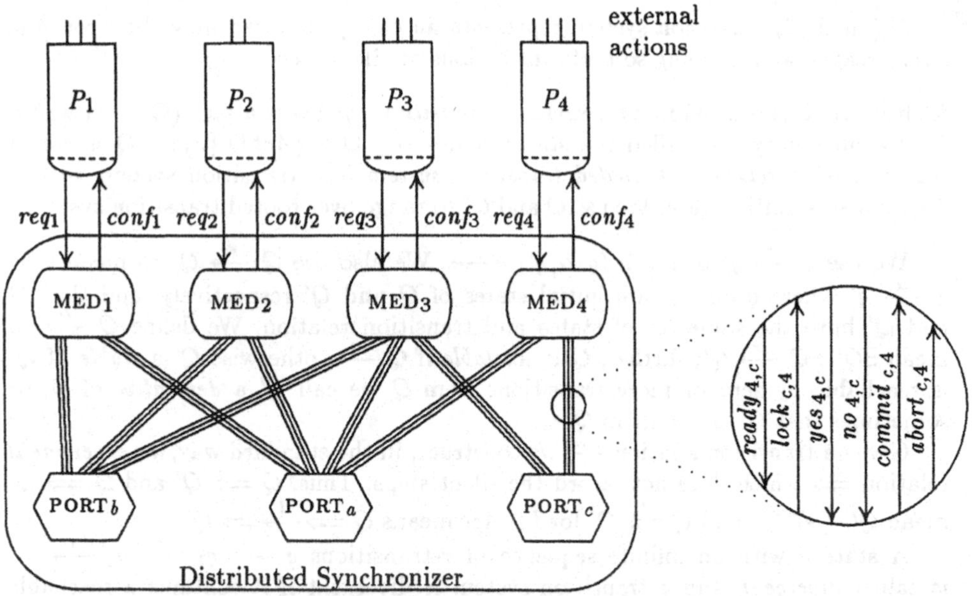

Fig. 3. A distributed implementation. The mediator for client i appears as MED_i and the port for action l as PORT_l. A port and a mediator interact over six actions (*ready*, ..., *abort*).

preserves this equivalence. Thus we can conclude $P_1 \parallel \cdots \parallel P_n =_{\text{cs}} \mathcal{D}(P_1, \ldots, P_n)$. In order to use coupled simulations we must first establish convergence of the involved processes. This result is also interesting in itself since it implies absence of livelocks in the implementation.

In summary, this paper addresses the problem of verifying systems where decisions on a high level of abstraction are regarded as atomic but in the implementation are made distributedly and incrementally.

The rest of the paper is structured as follows. In Sect. 2 we define transition systems, equivalences and parallel composition formally. Section 3 contains the definition and verification of central implementations. In Sect. 4 we define distributed synchronizers and prove them equivalent to the corresponding central synchronizers. Section 5 contains a brief summary, ideas for further work, and related work on verifications of multiway synchronization.

2 Preliminaries

2.1 Transition Systems

We assume a set of *actions* Act ranged over by k, l. The *silent step* τ is not in Act, and we use α to range over $Act \cup \{\tau\}$. A silent step represents an internal computation step of an entity.

A (labelled) transition system represents an entity that can move between different states, and in doing so performs actions or silent steps:

Definition 1 (transition system). A *transition system* is a pair $(\mathbf{Q}, \longrightarrow)$ where $\mathbf{Q}$ is a nonempty set, called the *states*, and $\longrightarrow \subseteq \mathbf{Q} \times (Act \cup \{\tau\}) \times \mathbf{Q}$ is called the *transition relation*. A *rooted transition system* is a transition system with a distinguished initial state. We use Q and Q' to range over rooted transition systems.

We use $q \xrightarrow{\alpha} q'$ to mean $(q, \alpha, q') \in \longrightarrow$. We also use $Q \xrightarrow{\alpha} Q'$ to mean that $q \xrightarrow{\alpha} q'$ where q and q' are initial states of Q and Q' respectively, and that Q and Q' have the same set of states and transition relation. We define $Q \xrightarrow{\alpha}$ to mean $\exists Q' : Q \xrightarrow{\alpha} Q'$; further, Q is *unstable* if $Q \xrightarrow{\tau}$, otherwise Q is *stable*. If Q' is reachable by zero or more transitions from Q we call Q' a *derivative* of Q, or sometimes that Q' is a *state* in Q.

Given a transition relation $\longrightarrow$ we construct, in the standard way, an *experiment* relation $\Longrightarrow$ which does not record the silent steps. Thus, $Q \Longrightarrow Q'$ and $Q \stackrel{\tau}{\Longrightarrow} Q'$ mean $Q(\xrightarrow{\tau})^* Q'$, and $Q \stackrel{l}{\Longrightarrow} Q'$ for $l \in Act$ means $Q \Longrightarrow \xrightarrow{l} \Longrightarrow Q'$.

A state q with an infinite sequence of τ-transitions $q \xrightarrow{\tau} q_1 \xrightarrow{\tau} q_2 \xrightarrow{\tau} \cdots$ is called *divergent*, and a transition system is divergent if it contains a reachable divergent state, otherwise it is *convergent*.

2.2 Equivalence Relations

Definition 2 (observation equivalence). A binary relation $\mathcal{S}$ on transition systems is called a *(weak) simulation* if

$$\text{whenever } Q\mathcal{S}R \text{ and } Q \xrightarrow{\alpha} Q', \text{ then } \exists R' : R \stackrel{\alpha}{\Longrightarrow} R' \text{ and } Q'\mathcal{S}R' .$$

The relation $\mathcal{S}$ is a (weak) *bisimulation* if both $\mathcal{S}$ and $\mathcal{S}^{-1}$ are weak simulations. Two transition systems are *observation equivalent*, written $\approx$, if their initial states are related by a weak bisimulation.

Definition 3 (coupled simulation equivalence). A pair $(\mathcal{S}_1, \mathcal{S}_2)$ of binary relations on transition systems is called a *coupled simulation* if

1. If $Q\mathcal{S}_1R$ then both of:
   (a) $Q \xrightarrow{\alpha} Q'$ implies $\exists R' : R \stackrel{\alpha}{\Longrightarrow} R'$ and $Q'\mathcal{S}_1R'$;
   (b) If Q is stable then $Q\mathcal{S}_2R$.
2. If $Q\mathcal{S}_2R$ then both of:
   (a) $R \xrightarrow{\alpha} R'$ implies $\exists Q' : Q \stackrel{\alpha}{\Longrightarrow} Q'$ and $Q'\mathcal{S}_2R'$;
   (b) If R is stable then $Q\mathcal{S}_1R$.

Two transition systems are *cs-equivalent*, written $=_{cs}$, if their initial states are related by both elements of a coupled simulation.

Thus, $(\mathcal{S}_1, \mathcal{S}_2)$ is a coupled simulation if both $\mathcal{S}_1$ and $\mathcal{S}_2^{-1}$ are simulations which are "coupled" at the stable states.

Proposition 4. $\approx \subset =_{cs}$.

The inclusion is strict as demonstrated by the following example, where $Q \not\approx R$ and $Q =_{cs} R$.

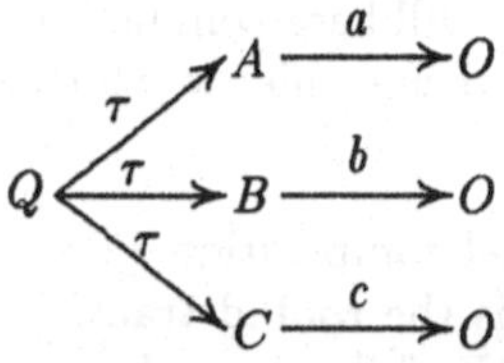

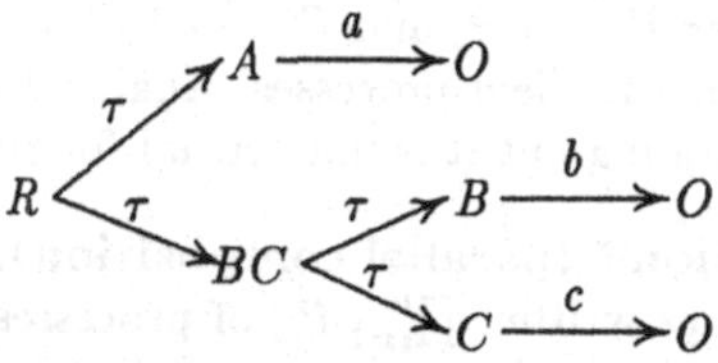

This example demonstrates the significant difference between $=_{cs}$ and $\approx$; the choice to preempt a must in the left-hand transition system imply a choice between b and c, while in the right-hand system there is an extra state, namely BC, where a is preempted but the choice between b and c has not been resolved. We have established the general result that $=_{cs}$ on convergent systems implies testing equivalence (the simple proof of this is beyond the scope of the present paper). $=_{cs}$ shares with $\approx$ the nice property that it can be established by simulations which involve case analysis over individual transitions; this is in contrast with testing equivalence which involves case analysis over sequences of transitions. We will thus use both $\approx$ and $=_{cs}$ as practical proof methods for convergent systems.

Proposition 5. $=_{cs}$ is an equivalence relation on convergent transition systems.

We omit the straightforward proof. $=_{cs}$ is not transitive on divergent transition systems; we will however only use it on convergent systems.

2.3 Processes and Parallel Composition

Definition 6 (process). A *process* P is a rooted transition system with an associated *synchronization sort*, $L(P)$, which is a set of actions such that each state Q of P is either

1. *waiting*, meaning that $\forall \alpha : Q \xrightarrow{\alpha}$ implies $\alpha \in L(P)$, or
2. *external*, meaning that $\neg \exists \alpha \in L(P) : Q \xrightarrow{\alpha}$.

The synchronization sort of a process will be significant only when processes are composed in parallel. Intuitively, when P occurs in a parallel composition then the actions in $L(P)$ represent synchronizations with other entities in the composition, while actions not in $L(P)$, which we call *external* for P, are interactions with the environment outside the parallel composition. We restrict attention to processes where external and internal interactions never compete, thus a state must either be waiting or external. (This is not a vital restriction, our protocols and proofs can be extended to allow competition between internal and external synchronizations, see [17].)

It will sometimes be convenient to identify a process with its underlying transition system. Thus, if P and P' have transition systems Q and Q' respectively we use $P \xrightarrow{\alpha} P'$ to mean that $Q \xrightarrow{\alpha} Q'$ and that P and P' have the same synchronization sort. Similarly we will also use $P \overset{\alpha}{\Longrightarrow} P'$ and $P \xrightarrow{\alpha}$ on processes, and say that a process is divergent if its transition system is divergent. $P =_{cs} P'$ will mean that $L(P) = L(P')$ and $Q =_{cs} Q'$.

A process or transition system P is called *synchronization deterministic* (or *s-deterministic* for short) if for all derivatives P' and all actions $l \in L(P)$ it holds that whenever $P' \xrightarrow{l} P_1$ and $P' \xrightarrow{l} P_2$ then $P_1 = P_2$. We will here confine interest to s-deterministic client processes. Again, this is a technical convenience which simplifies our reasoning but it is not crucial for the results.

Definition 7 (parallel composition). The parallel composition $P_1 \parallel \cdots \parallel P_n$, sometimes written $\prod_{i=1}^{n} P_i$, of processes $P_1, \ldots, P_n$ is the rooted transition system whose states and transitions can be generated from the following rules:

1. For all $i, 1 \leq i \leq n$ and all $\alpha \notin L(P_i)$: if $P_i \xrightarrow{\alpha} P_i'$ then
$$P_1 \parallel \cdots \parallel P_i \parallel \cdots \parallel P_n \xrightarrow{\alpha} P_1 \parallel \cdots \parallel P_i' \parallel \cdots \parallel P_n \ .$$
2. For all l: if $\forall i : (l \in L(P_i)$ implies $P_i \xrightarrow{l} P_i')$, then
$$P_1 \parallel \cdots \parallel P_n \xrightarrow{\tau} P_1' \parallel \cdots \parallel P_n'$$
(where P_j' is defined to be P_j whenever $l \notin L(P_j)$).

Clause 2 here corresponds to an internal synchronization at l; this involves all processes with l in their sort and results in a silent step. Note that our parallel composition operator is not associative—formally, the result of a parallel composition is a rooted transition system and not a process. We will need the following simple results:

Lemma 8. If $P =_{cs} P'$ then $P_1 \parallel \cdots \parallel P_n \parallel P =_{cs} P_1 \parallel \cdots \parallel P_n \parallel P'$.

Proof: Omitted.

Lemma 9. If $P =_{cs} P'$ and both P and $P_1 \parallel \cdots \parallel P_n \parallel P'$ are convergent, then also $P_1 \parallel \cdots \parallel P_n \parallel P$ is convergent.

Proof: Omitted.

3 Central Implementation

3.1 Implementation Transformation

Let P be a client process and i its index, i.e. a natural number, in a specification. Then $\mathcal{R}_i(P)$ is a process which, instead of participating in multiway synchronizations on actions in $L(P)$, interacts with a synchronizer on the following actions:

$req_i(L)$ where L is a set of actions; this is a request from the process to the synchronizer to participate in any action in L.

$conf_i(l)$ where l is an action; this is a confirmation from the synchronizer to the process that a multiway synchronization on l has been scheduled.

Note that the index of the client (the number i) is part of the actions; this ensures that all interactions will be binary, namely between one client process and the synchronizer.

Definition 10 (implementation transformation). Let P be any process and i a natural number. The processes $\mathcal{R}_i(P)$ and $\mathcal{W}_i(P)$ are defined to have the following transitions:

1. If $P \xrightarrow{\alpha} Q$ and $\alpha \notin L(P)$ then $\mathcal{R}_i(P) \xrightarrow{\alpha} \mathcal{R}_i(Q)$.
2. If P is waiting and $L = \{l : P \xrightarrow{l}\}$ then $\mathcal{R}_i(P) \xrightarrow{req_i(L)} \mathcal{W}_i(P)$.
3. If $P \xrightarrow{l} P'$ then $\mathcal{W}_i(P) \xrightarrow{conf_i(l)} \mathcal{R}_i(P')$.

Further, $L(\mathcal{R}_i(P))$ and $L(\mathcal{W}_i(P))$ are both defined to be

$$\{conf_i(l) : l \in L(P)\} \cup \{req_i(L) : L \subseteq L(P)\} .$$

The purpose of $\mathcal{R}_i$ is to implement a small protocol for interaction with the synchronizer. If P is in a waiting state where a set of actions L are enabled (i.e., $P \xrightarrow{l}$ for all l in L), then L represents the multiway synchronizations in which P wants to participate. So the transformed processes $\mathcal{R}_i(P)$ can do the action $req_i(L)$, signifying "Process i requests to do a synchronization in L", and thereby reach state $\mathcal{W}_i(P)$. This state can continue with a $conf_i(l)$ action, signifying that "the multiway synchronization on l has been achieved". The transformation leaves external action unaffected.

The notation "$req_i(L)$" and "$conf_i(l)$" suggests that l and L are "data" transferred in the interaction. There is no need to introduce data transfer in a formal way; the actions are atomic and there is one such action for each appropriate l and L.

3.2 Central Synchronizer

The central synchronizer is a monolithic entity which interacts with the transformed client processes. A state of the central synchronizer is written CSYNCH(RQ, CF) where both RQ and CF are sets of offers. An *offer* is a pair (i, l) where $l \in L(P_i)$. The parameters mean:

RQ: If $(i, l) \in RQ$ then client process i has requested to participate in a synchronization on l.

CF: If $(i, l) \in CF$ then a synchronization on l has been scheduled and the synchronizer shall confirm this to process i.

Definition 11 (central synchronizer). The *central synchronizer* for a sequence $P_1, \ldots, P_n$ of processes is a process with synchronization sort

$$\bigcup_i \{conf_i(l) : l \in L(P_i)\} \cup \{req_i(L) : L \subseteq L(P_i)\} .$$

The states of the central synchronizer are of the form CSYNCH(RQ, CF), and the transitions from CSYNCH(RQ, CF) to CSYNCH(RQ', CF') are given by the following table:

precondition	action	RQ'	CF'
$\neg\exists l : (i, l) \in RQ \cup CF$	$req_i(L)$	$RQ \cup \{(i, l) : l \in L\}$	$-$
$S \subseteq RQ \wedge Cand(S)$	τ	$RQ - \{(i, l) : \exists k : (i, k) \in S\}$	$CF \cup S$
$(i, l) \in CF$	$conf_i(l)$	$-$	$CF - \{(i, l)\}$

In this and subsequent similar tables each line corresponds to a set of transitions, namely the transitions which satisfy the precondition. The columns "RQ'" and "CF'" give the resulting state $\text{CSYNCH}(RQ', CF')$; the symbol "–" means that the value does not change. In the second line, $Cand(S)$ means that the set of offers S is a ***candidate synchronization***, i.e. that all offers in S have the same action l and $S = \{(i, l) : l \in L(P_i)\}$.

The initial state of the central synchronizer is $\text{CSYNCH}(\{\}, \{\})$ and we will abbreviate it as CSYNCH_0.

The first transition type in the table corresponds to process i requesting synchronizations on actions in L. The second type corresponds to the selection of a multiway synchronization at some particular action l. This is an internal step within the synchronizer, which selects (among the offers in RQ) a "candidate synchronization", i.e. a set of offers where all required processes—processes with l in the synchronization sort—are present. The third type corresponds to the synchronizer confirming to process i that the synchronization on l has taken place.

Definition 12 (central implementation). The *central implementation* of processes $P_1, \ldots, P_n$ is written $\mathcal{C}(P_1, \ldots, P_n)$ and is defined to be

$$\mathcal{R}_1(P_1) \parallel \cdots \parallel \mathcal{R}_n(P_n) \parallel \text{CSYNCH}_0$$

(where CSYNCH_0 is the initial state of the central synchronizer for $P_1, \ldots, P_n$).

We have established two correctness properties of the central implementation (the proofs are omitted for lack of space). The first is:

Theorem 13. For s-deterministic processes $P_1, \ldots, P_n$:

$$P_1 \parallel \cdots \parallel P_n \approx \mathcal{C}(P_1, \ldots, P_n) .$$

This guarantees that both sides can mimick each other indefinitely as far as external actions are concerned. However, it does not guarantee anything about silent steps; in particular it does not imply divergence freedom. So our second correctness property is:

Theorem 14. For s-deterministic processes $P_1, \ldots, P_n$: if $P_1 \parallel \cdots \parallel P_n$ is convergent, then $\mathcal{C}(P_1, \ldots, P_n)$ is convergent.

4 Distributed Implementation

4.1 Overview

The distributed synchronizer for processes $P_1, \ldots, P_n$ consists of ***mediators***, one for each process, and ***ports***, one for each action $l \in \bigcup_i L(P_i)$. Mediator i synchronizes with process P_i via req_i and $conf_i$ actions. Ports and mediators communicate via ***messages***. Messages are transferred asynchronously from a sender process to a receiver process. We model this as (two-way) synchronizations where the sender and receiver are represented as indices of actions and where receivers are always prepared to accept a message (i.e. receptions are enabled in all states). For example, the action $ready_{i,l}$ represents the transmission of a *ready* message from mediator i to port l. Thus for mediator i this action corresponds to a transmission, and for port l it corresponds to a reception.

Port l is responsible for establishing multiway synchronizations at l. The algorithm for establishing synchronizations is a *two-phase locking* algorithm, and the following messages are used in communication between mediators and ports.

$ready_{i,l}$ is an indication from mediator i to port l that process P_i has requested, by a $req_i(L)$ synchronization where $l \in L$, to synchronize at l.
$lock_{l,i}$ is a request from port l to lock mediator i.
$yes_{i,l}$ is a positive response, from mediator i to port l, to a lock request.
$no_{i,l}$ is a negative response, from mediator i to port l, to a lock request.
$commit_{l,i}$ is an indication, from port l to mediator i, that P_i has been scheduled for a multiway synchronization at l.
$abort_{l,i}$ is an indication, from port l to mediator i, that a multiway synchronization at l could not be established.

Informally, the algorithm works as follows. After a $req_i(L)$ synchronization mediator i sends to each port $l \in L$ a $ready_{i,l}$ message. When a port l has received sufficient *ready* messages to determine that a multiway synchronization is possible between a set of processes I, it can initiate the locking procedure.

In the first phase, port l attempts to lock each mediator in I. To lock a mediator i, the port sends a *lock* message and then awaits a reply. If the mediator replies by a *yes* message, the mediator is thereby locked and the port continues to lock the next mediator. When all mediators have been locked, the second phase of the locking procedure is commenced: to each mediator $i \in I$, a $commit_{l,i}$ message is sent. When mediator i has received a $commit_{l,i}$ message, it signals to process P_i that it participated in a multiway synchronization at l, by a $conf_i(l)$ synchronization.

It can happen that a mediator i cannot be locked for a synchronization at a port l. This happens, for instance, when P_i has already participated in another multiway synchronization. Then mediator i rejects the lock request from l by responding with a *no* message. If port l at that point has managed to lock some mediators, it must unlock them by sending an *abort* message to each of them.

A mediator can only be locked by one port at a time. Should a mediator receive lock requests while being locked, the mediator does not respond until it receives a *commit* or *abort*. If the mediator receives a *commit* (an indication that a synchronization was established), the mediator rejects all other lock requests (by sending *no* messages). Should the mediator receive an *abort* (an indication that a synchronization could not be established), the mediator may respond with a *yes* to any delayed lock request. To prevent deadlocks, mediators are locked in the order defined by their indices, so whenever a port should lock mediators i and j, and $i < j$, it will lock i before j.

4.2 Mediators

A state of a mediator for process i is written $\text{MED}_i(R, Q, V, c)$ where R, Q and V are sets of actions and c is an action or the special symbol $\perp$. The parameters have the following meanings:

R is the indices of ports to which *ready* messages should be sent. If the last request from the mediator's client was a $req_i(L)$-synchronization, then $R \subseteq L$.
Q is the indices of ports that have sent *lock* requests to the mediator and to which the mediator has not yet responded. If $l \in Q$, then a $lock_{l,i}$-synchronization has taken place and either a $yes_{i,l}$- or a $no_{i,l}$-synchronization will take place.
V is the valid actions, i.e. the actions the client currently is prepared to do. If $V \neq \{\}$ then the last request from the client process was $req_i(V)$.
c is the action to which the mediator has committed itself, that is, to which the mediator has sent a *yes* message and from which the mediator expects to receive a *commit* or an *abort* message. If the mediator has not committed itself, then c is $\bot$.

If c is different from $\bot$ and V is empty, then the mediator has received a *commit* message and should synchronize with its process by a *conf* action.

Definition 15 (mediator). Let K be a synchronization sort and i a natural number. Then the *mediator* for i and K is a process with synchronization sort

$$\bigcup_{l \in K} \{ready_{i,l}, lock_{l,i}, yes_{i,l}, no_{i,l}, commit_{l,i}, abort_{l,i}\} .$$

The states of a mediator are of the form $\text{MED}_i(R, Q, V, c)$, and the transitions from $\text{MED}_i(R, Q, V, c)$ to $\text{MED}_i(R', Q', V', c')$ are given by the following table:

precondition	action	R'	Q'	V'	c'
$c = \bot \wedge V = \{\}$	$req_i(L)$	L	–	L	–
$l \in R$	$ready_{i,l}$	$R - \{l\}$	–	–	–
–	$lock_{l,i}$	–	$Q \cup \{l\}$	–	–
$l \in Q \wedge l \in V \wedge c = \bot$	$yes_{i,l}$	–	$Q - \{l\}$	–	l
$l \in Q \wedge l \notin V$	$no_{i,l}$	–	$Q - \{l\}$	–	–
–	$commit_{l,i}$	$\{\}$	–	$\{\}$	–
–	$abort_{l,i}$	–	–	–	$\bot$
$V = \{\} \wedge c \neq \bot$	$conf_i(c)$	–	–	–	$\bot$

The initial state is $\text{MED}_i(\{\}, \{\}, \{\}, \bot)$, i.e. all action sets are empty and c is $\bot$.

Note that actions corresponding to asynchronous receptions have the trivial precondition "–" which always holds.

4.3 Ports

A state of a port for multiway synchronization action l is written $\text{PORT}_l(T, N, w, Y, A)$ where T, N, Y and A are sets of process indices, i.e. natural numbers, and w is a process index or the special symbol $\bot$. The parameters have the following meanings:

T is the indices of mediators that have sent *ready* messages to the port, but are not currently involved in a locking procedure of the port.
N is the indices of mediators that have sent *ready* messages to the port, and that the port has decided to lock for a synchronization, but has not yet locked.

w is the index of the mediator that the port is waiting for. If $w = i$ then the port has sent a *lock* message to mediator i, and is waiting for a a *yes* or a *no* message from i. In states where the port is not waiting for any mediator, $w = \bot$.

Y is the indices of mediators that have responded with *yes* messages to lock requests, and thereby have committed themselves to a synchronization at l. The mediator will eventually send a *commit* or an *abort* message to each mediator in Y.

A is the indices of mediators that have responded with *yes* messages to lock requests, and to which the mediator should send *abort* messages.

If N is non-empty, or w is different from $\bot$, then the port is in the first phase of the locking procedure. If Y is non-empty, N is empty and $w = \bot$, then the port is in the second phase of the locking procedure and should send *commit* messages to all mediators in Y.

Definition 16 (port). Let l be an action and I a set of natural numbers. Then the *port* for l and I is a process with synchronization sort

$$\bigcup_{i \in I} \{ready_{i,l}, lock_{l,i}, yes_{i,l}, no_{i,l}, commit_{l,i}, abort_{l,i}\} .$$

The states of a port are of the form $\text{PORT}_l(T, N, w, Y, A)$, and the transitions from $\text{PORT}_l(T, N, w, Y, A)$ to $\text{PORT}_l(T', N', w', Y', A')$ are collected in the following table:

precondition	action	T'	N'	w'	Y'	A'
$-$	$ready_{i,l}$	$T \cup \{i\}$	$-$	$-$	$-$	$-$
$Cand(\{(i,l) : i \in T\}) \wedge N = \{\} \wedge w = \bot$	τ	$\{\}$	T	$-$	$-$	$-$
$N \neq \{\} \wedge w = \bot \wedge i = least(N)$	$lock_{l,i}$	$-$	$N - \{i\}$	i	$-$	$-$
$-$	$yes_{i,l}$	$-$	$-$	$\bot$	$Y \cup \{i\}$	$-$
$-$	$no_{i,l}$	$T \cup N$	$\{\}$	$\bot$	$\{\}$	Y
$N = \{\} \wedge w = \bot \wedge i \in Y$	$commit_{l,i}$	$-$	$-$	$-$	$Y - \{i\}$	$-$
$i \in A$	$abort_{l,i}$	$T \cup \{i\}$	$-$	$-$	$-$	$A - \{i\}$

Here $least(N)$ gives the least element of the set of natural numbers N. The initial state is $\text{PORT}_l(\{\}, \{\}, \bot, \{\}, \{\})$, i.e. all index sets are empty and w is $\bot$.

4.4 Distributed Synchronizer

Definition 17 (distributed synchronizer). The *distributed synchronizer* for a sequence $P_1, \ldots, P_n$ of processes is a process with synchronization sort

$$\bigcup_i \{conf_i(l) : l \in L(P_i)\} \cup \{req_i(L) : L \subseteq L(P_i)\} .$$

Let $\widetilde{L} = \bigcup_i L(P_i)$. The states of the distributed synchronizer are of the form

$$\prod_{i=1}^{n} \text{MED}_i(R_i, Q_i, V_i, c_i) \; || \; \prod_{l \in \widetilde{L}} \text{PORT}_l(T_l, N_l, w_l, Y_l, A_l)$$

where $\text{MED}_i(R_i, Q_i, V_i, c_i)$ is the mediator for i and $\widetilde{L}$ and $\text{PORT}_l(T_l, N_l, w_l, Y_l, A_l)$ is the port for l and $\{1, \ldots, n\}$. The initial state of the distributed synchronizer, written DSYNCH_0, is the parallel composition of the initial states of its mediators and ports.

Definition 18 (distributed implementation). The ***distributed implementation*** of processes $P_1, \ldots, P_n$ is written $\mathcal{D}(P_1, \ldots, P_n)$ and is defined to be

$$\mathcal{R}_1(P_1) \parallel \cdots \parallel \mathcal{R}_n(P_n) \parallel \text{DSYNCH}_0$$

(where DSYNCH_0 is the initial state of the distributed synchronizer for $P_1, \ldots, P_n$).

So $\mathcal{D}(P_1, \ldots, P_n)$ differs from $\mathcal{C}(P_1, \ldots, P_n)$ only in that DSYNCH_0 replaces CSYNCH_0.

4.5 Correctness Properties

Theorem 19. For s-deterministic processes $P_1, \ldots, P_n$:

$$P_1 \parallel \cdots \parallel P_n \quad =_{\text{cs}} \quad \mathcal{D}(P_1, \ldots, P_n) \ .$$

To prove this theorem it suffices, in view of Proposition 4, Lemma 8 and Theorem 13, to establish:

Lemma 20. $\text{CSYNCH}_0 =_{\text{cs}} \text{DSYNCH}_0$.

To prove this lemma we need a coupled simulation between the central and the distributed synchronizer:

Definition 21. Let $CS = \text{CSYNCH}(RQ, CF)$ range over derivatives of CSYNCH_0 for a sequence of processes $P_1, \ldots, P_n$ and

$$DS \quad = \quad \prod_{i=1}^{n} \text{MED}_i(R_i, Q_i, V_i, c_i) \parallel \prod_{l \in \widetilde{L}} \text{PORT}_l(T_l, N_l, w_l, Y_l)$$

range over derivatives of DSYNCH_0 for the same process sequence. Let *Reqs* and *Confs* be the following functions from derivatives of DSYNCH_0 to offer sets:

$$\begin{aligned} \mathit{Reqs}(DS) &\stackrel{\text{def}}{=} \{(i,l) : l \in V_i \wedge \neg(\exists k : i \in Y_k \wedge N_k = \{\} \wedge w_k = \bot)\}, \\ \mathit{Confs}(DS) &\stackrel{\text{def}}{=} \{(i,l) : (c_i = l \wedge V_i = \{\}) \vee (i \in Y_l \wedge N_l = \{\} \wedge w_l = \bot)\} \ . \end{aligned}$$

Let $\mathcal{S}_1$ and $\mathcal{S}_2$ be relations between derivatives of CSYNCH_0 and DSYNCH_0 as follows:

$$\begin{aligned} \mathcal{S}_1 &= \left\{ (CS, DS) : \begin{pmatrix} RQ = \mathit{Reqs}(DS) \wedge CF = \mathit{Confs}(DS) \wedge \\ \forall i, l : R_i = N_l = Y_l = A_l = \{\} \wedge w_l = \bot \end{pmatrix} \right\}, \\ \mathcal{S}_2 &= \{(CS, DS) : RQ = \mathit{Reqs}(DS) \wedge CF = \mathit{Confs}(DS)\} \ . \end{aligned}$$

Obviously $(\text{CSYNCH}_0, \text{DSYNCH}_0) \in \mathcal{S}_1 \cap \mathcal{S}_2$. So in order to prove Lemma 20 we have to prove:

Lemma 22. $(\mathcal{S}_1, \mathcal{S}_2)$ is a coupled simulation.

The second correctness criterion guarantees that the distributed implementation is convergent:

Theorem 23. For s-deterministic processes $P_1, \ldots, P_n$: if $P_1 \parallel \cdots \parallel P_n$ is convergent, then $\mathcal{D}(P_1, \ldots, P_n)$ is convergent.

To prove this theorem it is enough, by Lemma 9 and Theorem 14, to establish the following lemma:

Lemma 24. DSYNCH_0 is convergent.

We finally give an example that demonstrates why the distributed synchronizer and the central synchronizer are not observation equivalent. Assume a specification with three client processes P_1, P_2, and P_3 and three possible (two-way) synchronizations: P_1 and P_2 can synchronize with each other at both a and b, and P_2 and P_3 can synchronize with each other at c. Suppose all three clients have requested to synchronize, then all three synchronizations are enabled but since P_2 must participate in every synchronization only one can occur.

The central synchronizer selects, by an internal transition, one of the three possible synchronizations and then confirms the selected synchronization to the clients. Hence any state in the central synchronizer is either before the choice has been made, in which case all three synchronizations are still possible, or after the choice, in which case only the chosen synchronization is possible.

In the distributed synchronizer, on the other hand, the choice may be resolved incrementally by first excluding the possibility of an a synchronization (cf. the example in Section 2.2). Consider the state where ports a and b have requested to lock mediator 1, and port c has requested to lock mediator 2. If mediator 1 sends a $yes_{1,b}$ to port b then synchronizations on b and c are still possible. But a synchronization on port a can only happen if b relinquishes its claim on mediator 1 (through $abort_{b,1}$); this only happens if mediator 2 refuses b ($no_{2,b}$), and the only possible reason for such a refusal is that c has been selected ($yes_{2,c}$), whence a c synchronization (and not a) will occur.

5 Conclusions

We have presented a new way to formulate and prove correctness properties of distributed algorithms through a new equivalence relation $=_{cs}$, coupled simulation equivalence (cs-equivalence), on transitions systems. It is weaker than observation equivalence but stronger than e.g. testing equivalence, and it is motivated by its practical applicability; it can, in contrast to testing equivalence, be established by case analysis over individual transitions. Although it is not suitable for divergent transition systems (systems with infinite sequences of internal transitions), we believe that the cs-equivalence is useful, for instance for implementation-oriented verifications where divergence often is regarded as harmful (a divergent system has potential "livelocks").

We have demonstrated the use of the cs-equivalence by verifying a strategy for implementing multiway synchronization. We started from a specification of a set of processes, composed in parallel, that communicate by multiway synchronizations.

The specification was transformed into an implementation consisting of another set of processes and a synchronizer which are composed in parallel. We first defined a *central* synchronizer, such that the implementation with the central synchronizer is convergent and observation equivalent to its specification. We then defined a *distributed* synchronizer, which is convergent and cs-equivalent to the central synchronizer. Cs-equivalence is preserved by parallel composition, wherefore we conclude that the implementation with the distributed synchronizer is cs-equivalent to the specification.

The fact that we have used an equivalence (and not an asymmetric implementation preorder such as the testing preorder) means that our implementation strategy does not exclude any synchronization possibilities in the specification. Thus the implementation will inherit all global properties such as fairness or deadlock freedom from the specification.

Other proposals for distributed algorithms for multiway synchronization can be found in [2, 6, 10, 11, 14, 15]. We have given priority to verification and thus chosen a simple algorithm, so the algorithm is not optimal, in terms of the number of messages required to establish a synchronization, and there are several possible improvements of the algorithm. For a discussion on this topic we refer to [17], where one of the authors presents and verifies a strategy for generating distributed implementations from LOTOS specifications, including process creation and communication with environment.

Mitchell [13] gives a transformation from CCS expressions to distributed implementations with a distributed algorithm to implement the CCS (two-way) synchronization. A result similar to ours was obtained; the implementation is not observation equivalent to the transformed CCS expression. The distributed algorithm is a busy-wait algorithm and, consequently, it causes divergence in the implementation so testing equivalence does not hold either. Mitchell therefore introduces *weak-must* testing equivalence, which is insensitive to certain kinds of divergence, but concludes that its lack of "bisimulation style" proof technique makes it difficult to use.

We have left the modal logic characteristics and algebraic theory of the cs-equivalence as future work. The distributed algorithm we have proposed can be improved in several ways; we believe that the results of such improvements would be (at least) cs-equivalent to the algorithm given in this paper, and therefore also correct according to our criteria. Our investigation could also consider distributed algorithms for multiway synchronization proposed by others.

References

1. R. J. R. Back and R. Kurki-Suonio. Distributed cooperation with action systems. *ACM Transactions on Programming Languages and Systems*, 10(4):513–554, 1988.
2. R. Bagrodia. Process synchronization: Design and performance evaluation of distributed algorithms. *IEEE Transactions on Software Engineering*, 15(9):1053–1065, September 1989.
3. T. Bolognesi and E. Brinksma. Introduction to the ISO specification language LOTOS. In P. H. J. van Eijk, C. A. Vissers, and M. Diaz, editors, *The Formal Description Technique LOTOS*, pages 23–73. North-Holland, 1989.

4. A. Charlesworth. The multiway rendezvous. *ACM Transactions on Programming Languages and Systems*, 9(2):350–366, July 1987.
5. R. De Nicola and M. C. B. Hennessy. Testing equivalences for processes. *Theoretical Computer Science*, 34(1,2):83–133, 1984.
6. P. Eklund. Synchronizing multiple processes in common handshakes. Reports on Computer Science and Mathematics 39, Åbo Akademi, Finland, 1984.
7. M. Evangelist, N. Francez, and S. Katz. Multiparty interactions for interprocess communication and synchronization. *IEEE Transactions on Software Engineering*, 15(11):1417–1426, November 1989.
8. N. Francez and I. R. Forman. Superimposition for interacting processes. In J. C. M. Baeten and J. W. Klop, editors, *Proceedings of CONCUR '90*, volume 458 of *Lecture Notes in Computer Science*, pages 230–245. Springer-Verlag, 1990.
9. N. Francez, B. Hailpern, and G. Taubenfeld. Script: A communication abstraction mechanism. *Science of Computer Programming*, 6(1):35–88, January 1986.
10. Y.-J. Joung and S. A. Smolka. Coordinating first-order multiparty interactions. In *Proc. 17th Annual ACM Symposium on Principles of Programming Languages*, pages 209–220, Orlando, Florida, January 1991.
11. D. Kumar. An implementation of N-party synchronization using tokens. In *Proc. 10th International Conference on Distributed Computing Systems*, pages 320–327. IEEE, 1990.
12. R. Milner. *Communication and Concurrency*. Prentice Hall, 1989.
13. K. Mitchell. *Implementations of Process Synchronization and their Analysis*. PhD thesis, Department of Computer Science, University of Edinburgh, July 1986.
14. M. H. Park and M. Kim. A distributed synchronization scheme for fair multi-process handshakes. *Information Processing Letters*, 34(3):131–138, April 1990.
15. S. Ramesh. A new and efficient implementation of multiprocess synchronization. In *Parallel Architectures and Languages Europe*, volume 259 of *Lecture Notes in Computer Science*, pages 387–401. Springer-Verlag, 1987.
16. S. Ramesh and S. L. Mehndiratta. A methodology for developing distributed programs. *IEEE Transactions on Software Engineering*, SE-13(8):967–976, August 1987.
17. P. Sjödin. *From LOTOS Specifications to Distributed Implementations*. PhD thesis, Department of Computer Systems, Uppsala University, Uppsala, Sweden, 1991.

This article was processed using the LaTeX macro package with LLNCS style

Programming in a General Model of Synchronization

Steven M. German

GTE Laboratories Incorporated
Computer and Intelligent Systems Laboratory
40 Sylvan Road, Waltham, Ma. 02154
german@gte.com

Abstract. We propose a programming formalism that provides multiprocess synchronization and priorities. As in CCS and CSP, processes communicate by executing pairs of complementary actions. Processes are labelled transition systems, where the labels are formed by combining atomic actions with the operators $\wedge, \neg$. Intuitively, $a \wedge b$ expresses multiple synchronization on actions a, b, while $\neg e$ expresses that the actions specified by e cannot be performed in the current environment.

The operational semantics is based on a notion of *stratification*, like that in logic programming. Stratification requires a partial order on actions, which is related to priority. We allow systems to place different priorities on actions at different states.

The goals of the language are: to serve as a high level executable specification language for concurrent programs, and to enable mechanical reasoning techniques such as model checking to be applied to specifications. We identify a large class of specifications that are efficiently executable, and present a simple synchronization algorithm.

The language is being used to design multiprocess systems for controlling telephone switching. We describe progress towards the goal of applying model checking to complex software systems.

1 Introduction

Languages based on synchronous atomic actions have been widely used to specify, reason about, and implement concurrent systems. In CSP [Ho78] and Occam [In84] and process calculi such as CCS [Mi80], synchronization is between pairs of cooperating processes. More recently, there has been a trend to explore multiprocess synchronization. Also, there has been increasing interest in formalizing other aspects of synchronization, such as relative priorities between processes and queueing behavior.

We are developing a language for specifying concurrent systems which attempts to synthesize all of these aspects of synchronization into a simple common framework. The language is intended to be used for formally specifying and reasoning about concurrent systems. We are also investigating efficient algorithms for executing the language, making it suitable for rapid prototyping of concurrent systems.

As in CCS, the language is based on asynchronous processes that synchronize by performing complementary *actions*. A process is a labelled transition system, where the labels describe actions performed by individual processes. In our language, the labels allowed on transitions are generated from the set of visible actions by the operators $\wedge, \neg$. A conjuction of visible actions, $a_1 \wedge \ldots \wedge a_n$, is called a *word*; intuitively, a word specifies that a transition involves multiple synchronization in which all of the actions in the word are performed. A word must be matched by the labels of other processes in the synchronization, in such a way that every action a that appears is matched by the complementary action $\overline{a}$.

Negation in labels specifies that a transition can occur only if the actions appearing under the negation are not possible in the environment of the process. We use negation to express relative priorities between processes. The state transition semantics with negation in labels is defined formally by providing a partial ordering on the actions. This approach is related to *stratification* in logic programming [Ap90].

In order to provide a convenient programming language, the notation can be enriched by providing local variables and other operations on actions. An action can be *indexed* by a variable. Intuitively, an indexed action corresponds to a set of actions, with the choice of the action to perform determined by the value of the variable. It is also possible to define notions of FIFO queue behavior on actions, using the operators $\wedge, \neg$.

This work is intended as a step towards the goal of automatically producing distributed implementations of multiprocess systems directly from high-level specifications. The specification language is intended to be expressive enough to define advanced communication services [Ge91a]. For these applications, it seems to be necessary to express a combination of kinds of synchronization, including multiprocess handshakes, priorities, and queueing. At the same time, the language is intended to permit the use of mechanical verification techniques such as temporal logic *model checking* [CES86]. Thus we want it to be fairly easy to extract finite-state models from specifications written in the language. This is discussed in [Ge91a].

The organization of the rest of this paper is as follows. Section 2 discusses related previous work. Section 3 gives the syntax of processes and presents an introductory example. Section 4 defines a state transition semantics based on stratification. Section 5 discusses the problem of efficiently determining when processes can synchronize. In Section 6, we discuss language extensions and use of the language to specify new communications services.

2 Comparison with Previous Work

Multiprocess synchronization has been extensively studied. Previous programming languages with multiprocess synchronization include *Action Systems* [BK88], *Scripts* [FHT86], and *IP* [FF90]. Aspects of our language are closely related to a number of previous programming formalisms, but our approach appears to be one of the first languages with asynchronous processes to provide multiprocess synchronization and priorities, and to do this in a way that is amenable to mechanical verification.

The specification language is closely related to work in the area of *process algebra*, although it is not currently presented in the form of an algebra. A number of process algebras have multiprocess synchronization, for instance *CSP* [Ho85], *Meije* [Si85], and *LOTOS* [BB87]. Notions of priority have also been studied in the context of process algebra [CH88, GSST90, SS90, Ca91, CW91]. To the best of knowledge, the present paper is one of the first to describe an asynchronous language that combines multiprocess synchronization and priorities. In Section 4, our approach to priority is compared with that of [CW91, Ca91].

A number of works in the area of process algebra focus on general frameworks for defining algebras with negative premises [BIM88, Gr89, BG90]. Algebras can use transition rules with negative premises to define priorities and queueing. Some care is needed to assure that systems with negative premises are well defined. The papers [Gr89, BG90] introduce stratification as a way to give meaning to transition system specifications in structured operational semantics. The semantic definition given in this paper differs from [Gr89, BG90] in two ways. First, we present the semantics of our language directly in terms of a transition system model, rather than as a set of rules. In our approach, it is necessary to find a certain type of ordering for the actions in a given system of processes. Given the existence of such an ordering, it is straightforward to show that our model is well defined.

Our main concern is in developing a particular language which we believe to be useful for modelling practical concurrent systems, rather than in developing general frameworks for such languages. A second difference is that our stratification is an ordering of the actions appearing in a particular program in the language. In contrast, in [Gr89, BG90], stratification is an ordering of the semantic rules of a system of structured operational semantics. It would be interesting to investigate whether the approaches of [BIM88, Gr89, BG90] could be used to provide a structured operational semantics for our language.

3 Syntax and an Example

Processes are state-transition systems defined in a CCS-like notation. In syntax definitions, [*term*]* denotes a sequence of zero or more strings of the form *term*. A *process-definition* has the form

$$\begin{array}{l} \textit{process-definition} ::= \textbf{process}\ \textit{pname} = \{\textit{transition-set}\ [;\textit{transition-set}]^*\} \\ \textit{transition-set} ::= \textit{State} \Rightarrow \textit{plabel.State}\ [+\ \textit{plabel.State}]^* \end{array}$$

where *pname* is the name of the process being defined and each occurrence of *State* is a name of a process state. A *transition-set* expresses the transitions from a given process state; summation with + represents the nondeterministic choice of CCS. The first process state mentioned in a definition is taken as the *initial* state. A *plabel*, or primitive label, expresses the synchronization that occurs on a transition. The syntax of primitive labels is as follows,

$$\begin{array}{l} \textit{pword} ::= (\textit{vis}\ [\wedge\ \textit{vis}]^*) \\ \textit{negated-pword} ::= \neg\, \textit{pword} \\ \textit{plabel} ::= \tau \mid (\textit{pword}\ [\wedge\ \textit{negated-pword}]^*) \end{array}$$

where each occurrence of *vis* is a visible action. A word is *simple* if it has just one occurrence of a visible action and is *compound* otherwise. Note that a primitive label can contain only one *pword*, but it can have any number of *negated-pwords*. Where no ambiguity results, we may drop parentheses. An action appears *negatively* in a primitive label if it appears within a negated expression, and appears *positively* if it has an appearance that is not in a negated expression.

Coordination for three committees

$\text{Committee}_A = \{\text{Person}_1, \text{Person}_2\}$

$\text{Committee}_B = \{\text{Person}_2, \text{Person}_3, \text{Person}_4\}$

$\text{Committee}_C = \{\text{Person}_1, \text{Person}_2, \text{Person}_5\}$

can be modelled by processes as shown below.

---

process $Person_2$ =

$\{Idlepers \Rightarrow \tau.Readypers;$

$Readypers \Rightarrow enter_{A,2}.Meeting + enter_{B,2}.Meeting + enter_{C,2}.Meeting;$

$Meeting \Rightarrow done_2.Idlepers\}$

process $Committee_A$ =

$\{Idlecom \Rightarrow (\overline{enter_{A,1}} \wedge \overline{enter_{A,2}}).Meetingcom;$

$Meetingcom \Rightarrow (\overline{done_1} \wedge \overline{done_2}).Idlecom\}$

Fig. 1. Committee Coordination

As an example, consider the Committee Coordination problem [CM88, Chapter 14]. It is desired to synchronize the activity of a group of persons, each of whom can belong to a number of committees. A committee can meet only if all of its members are ready to participate. A person cannot participate in more than one meeting at a time. When not in a meeting, a person can pursue other activities.

First, we will consider a specification that uses multiprocess synchronization. The specification can be easily extended, for instance, to put different priorities on the actions of entering different committees. While it is nontrivial to design a distributed algorithm for a problem such as Committee Coordination, the specification language is convenient for specifying such problems, and for reasoning about and executing the specifications.

Figure 1 shows membership in Committee_A, Committee_B, and Committee_C. Each person and committee can be modelled by a process; the figure shows the processes $Person_2$ and $Committee_A$. Process $Person_n$ performs the action $enter_{Com,n}$ to enter Committee_{Com} and $done_n$ to leave a committee.[1] An action of the form $act_1 \wedge \ldots \wedge act_n$ represents the simultaneous occurrance of $act_1 \ldots act_n$.

Suppose now that different committee meetings have different priorities. For example, assume Person_2 assigns the ordering Committee_A, Committee_B, Comittee_C from highest to lowest priority. This can be modelled by changing the transitions from the *Readypers* state of process $Person_2$.

$$Readypers \Rightarrow enter_{A,2}.Meeting + (enter_{B,2} \wedge \neg enter_{A,2}).Meeting + (enter_{C,2} \wedge \neg enter_{A,2} \wedge \neg enter_{B,2}).Meeting$$

Intuitively, a transition labelled $a \wedge \neg b \wedge \neg c$ can be taken if another process is ready to do $\overline{a}$, but no process can perform $\overline{b}$ or $\overline{c}$. By using $\wedge, \neg$ in this way, it is possible to express any partial ordering of priorities on actions. For a program to be well

[1] In [CM88], each person can leave a meeting autonomously, while the definition in Figure 1 uses exit synchronization. It is straightforward to express autonomous exit in our language.

defined, priorities assigned to actions by different processes must be consistent in a sense that is discussed in the next section. It is not necessary for all processes to give identical priorities to a given action. Since the language contains both multiparty synchronization and priorities, it leads naturally to the idea of assigning priority not just to single actions, but to simultaneous occurrences of sets of actions. The label $a \wedge \neg(b \wedge c)$ expresses readiness to synchronize on the action a, provided it is not possible to *simultaneously* perform b, c. This differs from $a \wedge \neg b \wedge \neg c$, where a is enabled provided neither b, c can be performed individually.

4 Transition System Semantics

This section presents an interleaving semantics based on labelled transition systems. A *process* consists of a set of states, a set of transitions, and an initial state. A *transition* consists of a source state, a label, and a successor state. A transition is said to be *outgoing* from its source state, and *incoming* to its successor state. Labels are generated from a set of actions, *Act*. There is a distinguished action, $\tau \in Act$, which is the *invisible* action. All other actions are *visible*. Visible actions have complements; the complement of the action a is $\overline{a}$. The complement of $\overline{a}$, or $\overline{\overline{a}}$, is equal to a.

A *concurrent system* is an ordered set of processes. To avoid undue complexity, we currently work only with systems having a single level of parallel composition. It would be interesting to see if language can be extended to have embedded levels of parallelism. The sets of states (and hence of transitions) of processes in a concurrent system are required to be non-overlapping.

A *global state* of a concurrent system $S = \{p_1, \ldots, p_n\}$, where p_i is a process, is an ordered set $\{s_1, \ldots, s_n\}$, where s_i is a state of p_i. An *environment* of the system S is an ordered set $E = \{T_1, \ldots, T_n\}$, where T_i is a set of outgoing transitions of a state of p_i. The interleaving semantics of a system is a structure $(\mathcal{S}, \mathcal{I}, \mathcal{R})$, where $\mathcal{S}$ is the set of global states, $\mathcal{I} \in \mathcal{S}$ is the initial global state, and $\mathcal{R} \subseteq \mathcal{S} \times \mathcal{S}$ is the transition relation. The transition relation is defined as follows: we first define the notion of an *enabled set of transitions* with respect to an environment. If a set of transitions is an enabled set in the environment of a certain global state of a system, then the set of transitions can occur together as an atomic action. This brings the system to a new global state in which the source state of each transition in the set is replaced with the successor state.

In this paper, we consider closed systems, where a process can only perform actions that are matched by complementary actions of other processes.

We will develop the definition of enabled set in two steps: first we consider environments with multiparty synchronization, and then we add negated actions.

To begin, we consider environments where transitions have *negation-free* labels. A negation-free label is either τ or a set of visible actions appearing positively. Since more than one process may have a transition containing a given primitive word, the primitive words appearing in a set of transitions form a multiset. Let us say that a multiset W of primitive words is a *matching set* provided:

(POSMATCH) Each action a appears in W as many times as $\overline{a}$ appears.

Under this definition, a computation step corresponds to a matching multiset of atomic actions, because an action can appear more than once.

We can now define enabled sets in environments with negation-free labels. A set of transitions Tr in an environment E, where E is negation-free, is enabled iff the following properties hold:

(E.1) Tr contains no more than one transition of each process.

(E.2) Tr is minimal; it does not properly contain an enabled set.

(E.3) The multiset of *pwords* (i.e. words appearing positively) in Tr is a matching set.

This notion of enabling generalizes the CCS case, and provides a highly expressive form of multiple-process synchronization. The minimality condition E.2 has the effect of saying that a computation step can only be executed by a single group of interacting processes. Note that Tr is enabled if it contains a single transition with label τ (in E.3, Tr contains no positive words). Also, enabling does not depend on the entire environment in negation-free environments. We will use the environment to define matching for transitions with negations.

A negated word appearing in a label has the effect of inhibiting the transition. Intuitively, if a transition of a process has a negated word and it is possible for the process to perform the actions in the negated word, then the transition is inhibited.

In order to give semantics to transitions with negated words, we use a notion of *stratification*. It will be possible to determine if certain actions are possible without considering any negations. This forms the lowest level of the stratification. Once the actions that are possible on a given level have been defined, these actions can appear in negated form in transitions on the next level, and we can determine which actions are possible on this level.

Let $\equiv$ be an equivalence relation on visible actions, and $\prec$ be a strong (irreflexive, transitive) ordering on the equivalence classes of $\equiv$. Intuitively, $\equiv$ relates actions that can occur together in a multiparty synchronization, while $\prec$ expresses priorities. We extend $\prec$ to actions a, b by defining $a \prec b$ iff $[a] \prec [b]$. We write $a \preceq b$ iff $a \prec b$ or $a \equiv b$. If α and β are words or transitions, then we say that $\alpha \prec \beta$ (resp. $\preceq$) if for every action a in α and b in β, $a \prec b$ (resp. $\preceq$) holds.

We say that $(\prec, \equiv)$ is a stratification of a set of labels L if the following conditions are satisfied:

(S.1) For all visible actions a, $a \equiv \bar{a}$.

(S.2) If there is some primitive word (either positive or negated) appearing in L which contains both a and b, then $a \equiv b$.

(S.3) If there is a label in L such that action a appears negated and action b appears positively, then $a \prec b$.

The first two conditions say that any two actions that can occur in the same step of a multiparty synchronization are in the same equivalence class. Condition S.3 requires all of the labels in L to assign a consistent preference on actions. Note that there is only one ordering, which applies at all process states. Later in this section, we discuss the problem of modelling systems where different states assign conflicting priorities to actions. Also note that $\equiv, \prec$ are not defined uniquely by conditions (S), but that when such relations exist, there are always unique minimal relations satisfying the conditions.

An important point is that nondeterministic choice with $+$ does not impose constraints on $\equiv$ or $\prec$. For instance, it is possible to have a system where process p_1 has transitions labelled $a.State1 + b.State2$, while process p_2 has transitions labelled

$a.State3 + (b \wedge \neg a).State4$. Process p_1 has no preference between actions a, b while p_2 will do action b only if it cannot do a. Both of these transition-sets are consistent with the ordering $a \prec b$. The reader can easily check the example in Section 3 to see that the transitions of all processes are stratifiable, provided there are no conflicts in priority. A conflict would arise if $Person_i$ assigns $enter_{X,i} \prec enter_{Y,i}$, while $Person_j$ assigns $enter_{Y,j} \prec enter_{X,j}$. This would be a conflict because for each committee X, the set of all actions of the form $enter_{X,i}$ must be contained in one equivalence class (S.1, S.2).

Henceforth, we will assume that $(\prec, \equiv)$ is a stratification of the set of all labels appearing in all processes.

We will use induction on the ordering $\prec$ to define the enabling condition for sets of transitions with negations. Consider a system $S = \{p_1, \ldots, p_n\}$. We need to define what it means for a primitive word w to be enabled for a process p_i in an environment E of S. Intuitively, this means that the process p_i can perform the actions in w by interacting with the other processes in the environment. Formally, we say that w is enabled for process p_i in $E = \{T_1, \ldots, T_n\}$ if there is a set containing the transition $State \Rightarrow w.State'$ which is enabled with respect to the environment $E[(T_i \cup \{State \Rightarrow w.State'\})/T_i]$, where $State$ is the source state of the transitions in T_i, $State'$ is a new state name, and $E[S/T_i]$ is the environment formed from E by replacing T_i with S.

Now, we say that a set of transitions Tr is enabled in an environment E, where the labels are arbitrary *plabels*, iff conditions E.1 - E.4 hold. The first three conditions are the ones used in the definition of enabling in negation-free environments.

(E.1) Tr contains no more than one transition of each process.

(E.2) Tr is minimal; it does not properly contain an enabled set.

(E.3) The multiset of *pwords* (i.e. words appearing positively) in Tr is a matching set.

(E.4) $\forall t \in Tr$, $\forall$ negated words nw in t, if t is a transition of p_i and $nw = \neg(w)$ then w is not enabled for p_i in E.

Condition E.4 tests whether it is possible for the process p_i to perform the actions in a negated word nw appearing in its transition t. Intuitively, to determine whether a word w is enabled for a process in an environment, it is only necessary to consider transitions with positive words $\preceq w$. Since the ordering $\prec$ is well founded, the definition eventually reduces to considering transitions without negations, using only E.1 - E.3.

We now mention two necessary properties that the definitions have.

Proposition 1 *It is well defined whether a set of transitions Tr is enabled in an environment E.* (See Appendix for Proof.)

The second property is that enabling depends only on the transition labels, i.e. it is invariant with respect to different stratifications.

Proposition 2 *The enabled sets of transitions are defined uniquely with respect to any stratifiable set of transition labels.* (Simple proof, omitted due to space limit.)

To conclude this section, we discuss the problem of representing systems in which different transitions assign conflicting priorities. Consider an example of a *server* process that can perform either a, b in state S. The server can also silently enter states Sa, Sb in which it can only perform one of a or b.

$$\textbf{process } Server = \{S \Rightarrow a.S + b.S + \tau.Sa + \tau.Sb;\ Sa \Rightarrow a.Sa;\ Sb \Rightarrow b.Sb\}$$

We would like to model two *client* processes. Process *Clienta* prefers to do $\overline{a}$, but will do $\overline{b}$ if the server does not let it do $\overline{a}$. Process *Clientb* has the priorities reversed. One way to represent the clients would be with the transitions

$$\textbf{process } Clienta = \{Ca \Rightarrow \overline{a}.Ca + (\overline{b} \wedge \neg\overline{a}).Ca\}$$

$$\textbf{process } Clientb = \{Cb \Rightarrow \overline{b}.Cb + (\overline{a} \wedge \neg\overline{b}).Cb\}$$

The language presented thus far does not permit the two clients to be represented in this way, because taken together, the transitions are not stratifiable. However, the processes can be modelled by splitting action a into a pair of actions a_1, a_2. In a state where the server offers a, the new version of the server offers both a_1, a_2. The clients will perform a_1 when they prefer to do a, and will do a_2 when doing an a but they really preferred to have done b. (Action b is treated similarly). In this representation, we would write the server and the clients as follows:

$$\textbf{process } Server\text{-}split = \{T \Rightarrow a_1.T + a_2.T + b_1.T + b_2.T + \tau.Ta + \tau.Tb;\ Ta \Rightarrow a_1.Ta + a_2.Ta;\ Tb \Rightarrow b_1.Tb + b_2.Tb\}$$

$$\textbf{process } Clienta\text{-}split = \{Da \Rightarrow \overline{a}_1.Da + (\overline{b}_2 \wedge \neg\overline{a}_1).Da\}$$

$$\textbf{process } Clientb\text{-}split = \{Db \Rightarrow \overline{b}_1.Db + (\overline{a}_2 \wedge \neg\overline{b}_1).Db\}$$

In this version, we can determine the enabling of a_1, b_1 first, and then use this information to determine whether a_2, b_2 are enabled. While there are now more names for the actions, we have represented the priority structure on the state transitions of the system.

Of previous approaches to priority in CCS-like systems, our work is perhaps most related to that of [Ca91, CW91]. These works define operators that give priority to certain actions, depending on the environment. [Ca91, CW91] introduce an asymmetry between actions and their complements; complemented actions are not permitted in the priority operators. This is a somewhat unnatural restriction, running counter to the idea that both a process and its environment may have priorities. However, it allows some systems to be written without splitting actions.

We now show how to extend our language to permit systems such as *Server*, *Clienta*, *Clientb*. Say that a visible action is *free* if it only appears as the entire label of a transition, an action is *co-free* if its complement is free, and a visible action is *ordinary* if it is neither free nor co-free. In the example, a, b are free and $\overline{a}, \overline{b}$ are co-free. Let us say that $(\equiv, \prec)$ is an *f-stratification* of a set of labels L if $\equiv$ is a relation on the ordinary actions such that S.1, S.2 hold for all ordinary actions in L, and $\prec$ is an ordering on the equivalence classes of $\equiv$ such that S.3 is satisfied for any ordinary a, b. This definition allows the co-free actions to appear without being restricted by $\equiv, \prec$. With this extension, our language essentially includes that of [CW91]. It is not difficult to see that Propositions 1 and 2 continue to hold for sets of labels that are f-stratifiable.

5 Efficient Execution

In this section, we turn attention to the problem of executing a system of processes. Ultimately, we seek concurrent and distributed algorithms to execute the language. However, as a first step, it is necessary to understand the sequential complexity of execution.

Consider the problem of testing for enabled sets of transitions: given a system of processes $\{p_1, \ldots, p_n\}$ and an ordered set $\{T_1, \ldots, T_n\}$, where T_i is a set of transitions of process p_i, determine whether $T = T_1 \cup \ldots \cup T_n$ contains an enabled set. In [JS92], this problem is studied for a wide variety of languages with multiprocess synchronization. In the terminology of [JS92], our systems with *negation-free* transitions would be called a port-based language with guards in format DNF (disjunctive normal form). Here, $+$ is regarded as disjunction. Actions are called *ports* if they can be performed by more than one process. It is shown in [JS92] that the problem of testing for enabled sets in port-based languages is NP-complete, even in the very restrictive case that the transition set of each process is a single transition mentioning the $\wedge$ operator, i.e. no process mentions the nondeterministic choice $+$ operator. However, the problem becomes computationally easier if we restrict attention to so-called channel-based systems.

A *channel* is an abstraction that provides communication between two processes. We can model a channel by assigning one action in a complementary pair of actions to each of the processes connected by the channel. These actions may not appear in any other processes. This motivates the following definition: A system of processes is a *channel system* if no visible action appears in more than one process. Under this definition, an action may appear both positively and negatively in the same process.

Channel systems occur naturally in our specification language. Often, the initial specification of a system will involve an undetermined number of processes, and actions will be ports. If a configuration is then desired with a fixed number of processes, it is possible to disjunctively expand the port-based specification into a channel-based one by assigning a different action to each process.

In [JS92], a hierarchy of special cases of the channel-based synchronization problem is studied. Recall that a primitive word is *compound* if it contains more than one action (combined with $\wedge$). The set of transitions from a process state is said to be in DNF^i if it does not contain any negations and it can be written as $t_1 + \cdots + t_n$, where at most i of the terms are compound. A system is said to be in DNF^i if for every process state s, the set of transitions from s has this form. It is shown in [JS92] that the problem of testing for enabled sets is NP-complete for channel-based systems that are DNF^i, $i \geq 2$. On the other hand, there is an efficient algorithm for determining synchronization of channel-based systems in DNF^1. For a set T of transitions, let $||T||$ be the sum of the lengths of the labels in T.

Theorem [JS92] *Given a set T of transitions of a channel system in DNF^1, there is an algorithm to determine if T contains a matching set in $O(n||T||)$ time, where n is the number of processes in the system.*

We now extend this result to systems with negative actions. Say that the set of transitions from a process state is in DNFN^i, or DNF^i *with negations,* if it can be written as a sum of terms where at most i terms involve compound positive words. Note that this form permits any number of terms to contain negated words with $\wedge$.

A system is in DNFN^i if for every process state s, the set of transitions from s has this form.

To present an algorithm for determining if there is a possible synchronization, we need names for the actions corresponding to a channel. Let c^i_{ij} denote the action that is performed by process i when it communicates with process j, and let c^j_{ij} be the complementary action. In this notation, c^i_{ij} and c^i_{ji} denote the same action. To obtain a unique form, we can write actions with the subscripts in increasing order.

Algorithm Test takes as input an ordered set $\{T_1, \ldots, T_n\}$, where T_i is a set of transitions for process i, and determines whether there is an enabled set. The system is a channel system in DNFN^1. The algorithm first applies procedure Reduce to obtain a system without any negated words. Then it uses techniques of [JS92] to test whether there is an enabled set. To simplify the exposition, we define the algorithm for stratifiable systems without τ transitions; the extension to f-stratifiable systems with τ is straightforward.

We use the following functions on transitions to define the algorithm. For a transition t, $Act(t)$ is the set of actions appearing in t, $pword(t)$ is the positive word in the label of t, and $nwords(t)$ is the set of negated words appearing in t. We write $label(t)$ for the label of t.

In Procedure Reduce, line 4 checks whether the actions in a negated word nw contained in T_i can be performed by synchronizing with the other processes. This check is made by calling Enabled with a system in which T_i has been replaced with a new transition. The new transition is constructed by the function *pospart*. Given a negated word $nw = \neg(pw)$, $pospart(nw)$ denotes a single transition with the positive label pw. In lines 5-6, Reduce either removes t from T_i or replaces t with a transition whose label is just the positive word of t.

Line 2 of Test checks whether there is a matching pair of single-action labels. If there is no such pair, then any enabled set must contain a transition with compound labels. The ith process can have at most one compound label in T_i. Line 4 checks whether there is an enabled set containing this label. Procedure Enabled takes advantage of the NDNF^1 format to determine if there is an enabled set containing T_l. C is a set of compound labels and R is variable ranging over sets of actions. Intuitively, R is a set of actions that must be performed to complement actions in C. For any i, $R\uparrow_i$ denotes the set of actions in R of the form c^i_{ij}, i.e. actions that can be performed by process i. In line 4, it is necessary to add a compound label from T_i to C if $|R\uparrow_i| > 1$ or $R\uparrow_i = \{c^i_{ik}\} \wedge c^i_{ik} \notin pwords(T_i)$, where $pwords(T_i)$ is the set of positive words in T_i. If no command satisfying the conditions in line 6 can be added, then there is no enabled set. If the test in line 4 is false, then $C \cup R$ is an enabled set, which the procedure returns in line 10. Procedure Enabled can be implemented to run in time $O(||T||)$ time, where $T = \bigcup_{i=1}^{n} T_i$ [JS92].

Theorem 1 *Given a set T of labels of a channel system in NDNF^1, there is an algorithm to determine if T contains an enabled set in $O(n||T|| + w||T||)$ time, where n is the number of processes in the system and w is the number of negated words.*

Proof. Reduce calls Enabled once for each negated word, thus its cost is $O(w||T||)$. Lines 2-3 of Test can be implemented to run in time $O(||T||)$, e.g. by presorting the commands $c^i_{i,j}$ in T_i on index j. Finally, lines 4-6 of Test take time $O(n||T||)$, because there are at most n compound commands. □

Algorithm Test
Input: $\{T_1, \ldots, T_n\}$
1. $\{T_1, \ldots, T_n\} \leftarrow$ Reduce($\{T_1, \ldots, T_n\}$)
2. if $\exists i, j$, $s \in T_i$, $t \in T_j$ such that $pword(s) = c^i_{ij}$ & $pword(t) = c^j_{ij}$ then return("True")
3. for $i = 1, \ldots, n$ if $\exists t \in T_i$ such that $pword(t)$ is compound, then do
    begin
4. evaluate Enabled(i, $\{T_1, \ldots, T_n\}[\{t\}/T_i]$)
5. if Enabled returns non-empty then return("Yes")
    end
6. return("No")

Procedure Reduce($\{T_1, \ldots, T_n\}$)
Input: $T_1, \ldots, T_n$, where T_i is set of transitions of process i.
Output: reduced set of transitions by evaluating all negated words.
1. do step 2 for each equivalence class Ψ of actions, starting with minimal elements.
   Process equivalence class Ψ when all $\Upsilon \prec \Psi$ have been processed.
2. if Ψ is not minimal and not marked as done, then do lines 3 - 6
3. for each i, $1 \leq i \leq n$, and $t \in T_i$ such that $[pword(t)] = \Psi$ do
    begin
4. for each $nw \in nwords(t)$ do Enabled(i, $\{T_1, \ldots, T_n\}[\{pospart(nw)\}/T_i]$)
5. if Enabled returns non-empty in Step 4 at least once, then $T_i \leftarrow T_i - \{t\}$
6. else $T_i \leftarrow (T_i - \{t\}) \cup t[label(t) \leftarrow pword(t)]$
    end
7. mark level Ψ as reduced.
    end
8. return($\{T_1, \ldots, T_n\}$)

Procedure Enabled(l, $\{T_1, \ldots, T_n\}$)
Input: $1 \leq l \leq n$, sets of transitions $T_1, \ldots, T_n$. T_l is a singleton.
   T_l and all other transitions on the same level are negation-free.
Output: determines if there is an enabled set containing T_l.
1. let $t \in T_l$
2. $R \leftarrow \{c^j_{jk},\ c^k_{jk} \in Act(t)\}$
3. if $pword(t)$ is compound then $C \leftarrow \{pword(t)\}$ else $C \leftarrow \emptyset$
4. while $\exists i$, $|R\uparrow_i| > 1 \vee (R\uparrow_i = \{c^i_{ik}\} \wedge c^i_{ik} \notin pwords(T_i))$ do
    begin
5. let j be the value of index i in line 4
6. if $\exists \beta \in T_j$ satisfying (1) $R\uparrow_j \subseteq Act(\beta)$ and (2) $\forall k$, $c^j_{jk} \in (Act(\beta) - R\uparrow_j)$
   implies that C does not include a transition label from T_k
    then begin
7. $C \leftarrow C \cup \{pword(\beta)\}$
8. $R \leftarrow (R - Act(\beta)) \cup \{c^k_{jk},\ c^j_{jk} \in (Act(\beta) - R)\}$
    end
9. else return($\emptyset$)
    end
10. return($C \cup R$)

6 System Modelling and Model Checking

In this section, we describe how the language is being used to model and verify concurrent systems for controlling telephone switching. Such systems may involve a very large number of interacting processes. This application uses multiparty synchronization and negated actions. A selection of other examples of synchronization problems and solutions in the language can be found in [Ge91b].

Some widely used formal languages for specifying software systems are based on processes communicating through message buffers. For instance, the languages SDL [CCITT85] and Estelle [BD87], which are used in the telecommunications industry, are of this type. In contrast, we are applying the concept of shared atomic handshakes directly to specification of complex software systems. We feel that this approach has advantages in simplifying system design and making it easier to apply formal reasoning techniques such as model checking. By freeing system designers from having to specify asynchronous message passing in their high level designs, it removes a common source of errors. In practice, when software designers use a langauge based on message passing, the specifications often implement synchronous handshakes, but in an application-dependent way. This suggests that synchronizing on atomic actions will not add too much overhead.

One experiment in modelling and verification is described in [Ge91a]. The 911 Emergency Operator service is a typical communication service. In such services, many calls are active at the same time, which creates a potentially unbounded degree of concurrency. The calls synchronize with system processes that represent resources such as data terminals and operator lines. If a requested resource is busy, a call may perform some less-preferred action and wait in a queue. Thus it is often necessary to specify priorities.

In order to capture as much of a system as possible in the specification language, an extended language has been defined [Ge91b]. The extended language includes some conventional features such as local variables in processes, boolean guards on transitions, and value passing during synchronization. The extended language also defines *indexed actions* and *queued actions*. An indexed action has one or more subscripts. In a process definition, the action can be subscripted by the values of local variables, and the indexed action defines a distinct atomic action for each value of the subscripts. Indexed actions are useful for specifying systems where a potentially unbounded number of similar processes are active concurrently.

In the extended language, the semantics of a finite system of processes can involve an infinite set of actions. Stratification can be extended to handle an infinite set of actions by requiring the ordering to be well-founded.

Queued actions are a form of queueing that can be defined in terms of the $\wedge, \neg$ operators. A process can be either a client or a server with respect to a given queued action. The queued action appears in transition labels in a client in much the same way as other actions. However, semantically, in each instance that a client reaches a state with outgoing transitions on a queued action, we treat it as though the client has a transition with a unique new action. These actions are called queued variants. The server accepts requests from the clients by placing a priority ordering on the queued variants, i.e. its highest priority is the perform the oldest queued variant, followed by the next oldest etc. In client processes, a queued action can be combined with other actions with the $\wedge, \neg$ operators. This allows the client to use multiparty synchronization, priorities, and non-deterministic choice on queued events. Details of how we use stratification to define the semantics are given in [Ge91b].

Two model checking programs have been implemented for the subset of the language limited to CCS synchronization. The first program performs reachability analysis to construct models for the EMC model checker [CES86]. In [Ge91a], we describe how the 911 service can be modelled as finite-state processes in this language subset. Using the temporal logic CTL and the model checker, certain correctness properties were verified for a correct model. Also, the falsify of these properties was demonstrated in an incorrect model.

Recently, experiments were started using a model checker based on Binary Decision Diagrams (BDDs) [BCMDH90]. This system, which was developed by David Long at Carnegie Mellon University, can potentially handle much larger systems. Further experiments are needed to evaluate this approach. It would also be interesting to see if the BDD approach can be extended to larger subsets of the specification language.

7 Discussion

We have introduced a state-transition language for processes with multiparty synchronization and priorities. The design of the language has been influenced by modelling a range of practical examples, including some telephone switching. The full language has high expressive power, and is potentially expensive to execute. However, we have identified a significant subset of the language where synchronization can be tested efficiently. Initial work indicates that it may be possible to verify large designs using temporal logic model checking based on BDDs, but this remains to be determined in practice. We believe that the algorithm of Section 5 may be adaptable for implementation on shared-memory multiprocessors. It may also be possible to modify existing concurrent algorithms for multiparty synchronization to handle the negated actions of the language.

Acknowledgments

It is a pleasure to acknowledge conversations with Robert Rubin, Anita Chow, Jim Walker, Scott Smolka, Andrew Uselton, and Yuh-Jzer Joung. The referees made useful comments on a draft of the paper.

References

[Ap90] K.R. Apt, Logic programming, in *Handbook of Theoretical Computer Science, Volume B, Formal Models and Semantics*, Chapter 10, North Holland, 1990.

[BK88] R.J. Back and R. Kurki-Suonio, Distributed cooperation with action systems, *ACM Transactions on Programming Languages and Systems*, 10, 4 (1988) pp. 513-544.

[BIM88] B. Bloom, S. Istrail, A. Meyer, Bisimulation can't be traced: preliminary report, in *Proceedings of Fifteenth Annual ACM SIGACT-SIGPLAN Symposium on Principles of Programming Languages*, January 1988, pp. 229-239.

[BG90] R. Bol and J.F. Groote, The meaning of negative premises in transition system specifications, Report CS-R9054, Centre for Mathematics and Computer Science, Amsterdam, The Netherlands, 1990, and in *Proceedings of* 18th *ICALP*, 1991, pp. 481-494.

[BB87] T. Bolognesi and E. Brinksma, Introduction to the ISO specification language LOTOS, *Computer Networks and ISDN Systems*, 14 (1987), pp. 25-59.

[BD87] S. Budkowski and P. Dembinski, An Introduction to Estelle: A Specification Language for Distributed Systems. *Computer Networks and ISDN Systems* 14 (1987), pp. 3-23.

[BCMDH90] J.R. Burch, E.M. Clarke, K.L. McMillan, D.L. Dill, L.J. Hwang, Symbolic model checking: 10^{20} states and beyond, in *Proceedings of Fifth Annual IEEE Symposium on Logic in Computer Science*, June 1990, pp. 428-439.

[Ca91] J. Camilleri, A conditional operator for CCS, in *Proceedings of Concur '91*, Amsterdam, August 1991.

[CCITT85] CCITT/SGXI Recommendation Z101 to Z104, Functional Specification and Description Language, 1985.

[CW91] J. Camilleri and G. Winskel, CCS with priority choice, in *Proceedings of the Sixth Annual IEEE Symposium on Logic in Computer Science*, July 1991.

[CM88] K.M. Chandy and J. Misra, *Parallel program design: a foundation*, Addison-Wesley, 1988.

[CES86] E.M. Clarke, E.A. Emerson, and A.P. Sistla, Automatic verification of finite-state concurrent systems using temporal logic specifications, *ACM Transactions on Programming Languages and Systems*, 8, 2, April 1986, pp. 244-263.

[CH88] R. Cleaveland and M. Hennessy, Priorities in process algebras, in *Proceedings of the Third Annual IEEE Symposium on Logic in Computer Science*, July 1988, pp. 193-202.

[FF90] N. Francez and I.R. Forman, Interacting processes: a language for coordinated distributed programming, invited paper for Jerusalem Conference on Information Technology, Jerusalem, 1990.

[FHT86] N. Francez, B. Hailpern, and G. Taubenfeld, Script: a communication abstraction mechanism, *Science of Computer Programming*, 6, 1, January 1986, pp. 35-88.

[Ge91a] S.M. German, Rapid Prototyping and Verification of Communication Services, GTE Laboratories TM-0369-01-91-152, January 1991.

[Ge91b] S.M. German, A Language for Specifying Synchronization, GTE Laboratories, October 1991.

[GSST90] R. van Glabbeek, S. Smolka, B. Steffen, and C. Tofts, Reactive, generative, and stratified models of probabilistic processes, in *Proceedings of 5th Annual IEEE Symposium on Logic in Computer Science*, June 1990, pp. 130-141.

[Gr89] J.F. Groote, Transition system specifications with negative premises, Report CS-R8950, Centre for Mathematics and Computer Science, Amsterdam, The Netherlands, December 1989, and in J.C.M. Baeten and J.W. Klop, eds., *Proceedings of CONCUR 90*, Springer-Verlag LNCS 458, pp. 332-341.

[Ho78] C.A.R. Hoare, Communicating sequential processes, *Communications of the ACM*, 21, 8, August 1978, pp. 666-677.

[Ho85] C.A.R. Hoare, *Communicating Sequential Processes*, Prentice Hall, 1985.

[In84] INMOS Ltd., *Occam Programming Manual*, Prentice Hall, 1984.

[JS92] Yuh-Jzer Joung and S.A. Smolka, A Comprehensive Study of the Complexity of Multiparty Interaction, in *Proceedings of POPL 1992*.

[Mi80] R. Milner, *A Calculus of Communicating Systems*, LNCS 92, Springer Verlag, 1980.

[Si85] R. de Simone, Higher-level synchronizing devices in Meije-SCCS, *Theoretical Computer Science*, 37 (1985) pp. 245-267.

[SS90] S.A. Smolka and B. Steffen, Priority as extermal probability, in *Proceedings of Concur '90*, August 1990, Springer-Verlag *Lecture Notes in Computer Science* 458, pp. 456-466.

Appendix

Proposition 1 *It is well defined whether a set of transitions Tr is enabled in an environment E.*

Proof. There are two cases. For the first case, Tr contains visible actions appearing positively. In the other case, we show that the proposition holds if Tr is a set of τ transitions.

For the first case, begin by observing that if Tr is an enabled set, then all of the actions appearing positively in Tr are contained in one equivalence class of $\equiv$. Suppose to the contrary, that Tr is enabled, but contains actions a, b appearing positively, where $a \not\equiv b$. By S.2, all of the actions in a primitive word are equivalent. Let $Tr[a]$ (resp. $Tr[b]$) be the set of transitions in Tr with positive words equivalent to a (resp. b). One can easily see that $Tr[a]$ and $Tr[b]$ must each satisfy E.1, E.3, and E.4. It follows that Tr cannot be minimal. This contradicts the assumption that Tr is enabled. We conclude that if Tr is enabled, all of the actions that appear positively in Tr are in the same equivalence class of $\equiv$.

For any set of actions A, let $[A]$ denote the equivalence class of $\equiv$ that contains A, if such an equivalence class exists. Also, if T is a set of transitions, let $pos(T)$ denote set of all actions appearing positively in T.

By the remark above, if $[pos(Tr)]$ does not exist, then Tr cannot be an enabled set with visible positive actions. Thus we will prove the proposition by induction on the position of $[pos(Tr)]$ in the ordering $\prec$. For the base case, assume that $[pos(Tr)]$ is a minimal element of the ordering. Then Tr contains no negated actions. In this case, it is well defined whether Tr is an enabled set, because this is defined by E.1 - E.3.

Now assume by induction that for any set of transitions Tr' and any environment E if $[pos(Tr')] \prec [pos(Tr)]$ then it is well defined whether Tr' is enabled in E. Suppose that nw appears as a negative word in Tr. Then it must be the case that $nw \prec [pos(Tr)]$. By the inductive hypothesis, the condition for testing negative words (E.4) in E is well defined. It follows that it is well defined whether Tr is enabled in E.

For the second case, suppose that Tr contains only transitions of the form $\tau \wedge \neg(w_1) \wedge \ldots \wedge \neg(w_m)$. Because of the minimality condition, only singleton sets of this form can be enabled. By the first case, it is well defined whether condition E.4 holds in an environment for each of the w_i. Thus it is well defined whether Tr is enabled in E. □

Proposition 2 *The enabled sets of transitions are defined uniquely with respect to any stratifiable set of transition labels.*

Proof. Let L be a set of transition labels, let $\prec_0, \equiv_0$ be the unique mininal relations that form a stratification, and let $enabled_0$ be the relation that holds for enabled sets of transitions under this stratification. If $\prec_1, \equiv_1$ is any other pair of relations that forms a stratification and $enabled_1$ is the corresponding enabled relation, we show that $enabled_1$ agrees with $enabled_0$.

For any set of actions A, let $[A]_i$, $i = 1, 2$ be the equivalence class of $\equiv_i$ containing A, if one exists. Note that if $[A]_0$ exists then so does $[A]_1$, because $\equiv_0$ is a minimal relation. Recall from the proof of Proposition 1 that if a set of transitions Tr is enabled under a given stratification, then $[pos(Tr)]$ exists for the equivalence relation of the stratification. It follows that for any Tr, if $[pos(Tr)]_1$ does not exist, then both $enabled_0(Tr), enabled_1(Tr)$ are false.

So, it is sufficient to prove the proposition for Tr such that $[pos(Tr)]_1$ exists. First, consider the case that Tr has no negated actions. In this case, it is clear from E.1 - E.3 that $enabled_0(Tr)$ iff $enabled_1(Tr)$.

Now, proceed by induction on the ordering $\prec_1$. For the induction case, assume that for all Tr', $[pos(Tr')]_1 \prec_1 [pos(Tr)]_1$ implies $enabled_0(Tr')$ iff $enabled_1(Tr')$. If a negative word $nw = \neg(pw)$ appears in Tr then $nw \prec_1 [pos(Tr)]_1$. Thus in condition E.4, $enabled_0$ must agree with $enabled_1$ for any set of transitions containing a transition with the label pw. It follows that $enabled_0(Tr)$ must agree with $enabled_1(Tr)$. The proof for τ transitions is similar. □

Operational and Compositional Semantics of Synchronous Automaton Compositions

Florence Maraninchi

Laboratoire de Génie Informatique
IMAG-Campus, B.P. 53X, 38041 Grenoble Cedex – FRANCE
email: maraninchi@imag.fr

Abstract: The state/transition paradigm has been used extensively for the description of event-driven, parallel systems. However, the lack for hierarchic structure in such descriptions usually prevents us from using this paradigm in a real programming language. We propose the Argos language for reactive systems. The basic components of a program are input/output-labeled transition systems verifying *reactivity* (a property similar to input-enabling in IO-automata). The composition operations (parallel composition and *refinement*, providing hierarchy) are based upon the synchronous broadcast mechanism of Esterel. We define the language formally in an algebraic framework, and give an operational semantics. The main result is the *compositionality* of the semantics; we prove that the bisimulation of models induces an equivalence which is a congruence for the operators we propose. An interesting point is the way we introduce hierarchy in a compositional way.

1 Introduction

The problem of *specifying*, *programming* and *verifying reactive* systems, together with the definition of appropriate development environments, is still an important research problem [Har87, HP85, SYN], the following being widely accepted. There exist appropriate design methods, programming languages and environments for *transformational* systems (or, at least, for systems which are mainly transformational, like compilers). This is not yet the case for *reactive* systems, like digital watches, and all kinds of real time process controllers. Moreover, the need for formal verification methods and tools is even more crucial for reactive systems than for transformational ones, because strong reliability requirements are associated to them.

The family of *synchronous* languages and formalisms has been a very important contribution to the domain. The synchronous approach is the mathematical foundation for the Esterel [BG88], Lustre [CHPP87], Signal [GBBG85] languages and for the algebra ATP [NRSV90]. To some extent, Statecharts [Har87] are also synchronous. Synchronous languages or formalisms make the assumption that the reaction time of the system is always zero, hence the outputs are produced simultaneously with the inputs. It has been demonstrated that such systems compose very well, and are easier to describe and verify than asynchronous ones. Moreover, it allows to deal with functional and timing correctnesses of a system separately [SYN].

Esterel, Lustre and ATP are compiled into labeled transition systems which constitute the *models* of programs. Verification may be performed by reducing and comparing these models, modulo a given equivalence relation. This is not the case for Statecharts. However, the state/transition paradigm is sometimes the most appropriate.

We propose the Argos language [Mar89, Mar91a], which is a state/transition-based synchronous language, compiled into labeled transition systems, and to which existing verification methods and tools can be applied. A program is either an automaton, or an object of the form $P_1 \text{OP } P_2$, where P_1 and P_2 are programs and OP is an operator. The main operators are: the parallel composition, and an operator called *refinement* able to express watchdogs, interruptions, exceptions, termination of subprograms, etc.

The *semantics* of the language is syntax-directed. It describes the reactive behaviour of a program by defining a function $\mathcal{S}$ such that: $\mathcal{S}(\text{automaton}) = \text{automaton}$ and $\mathcal{S}(P_1 \text{op} P_2) = \mathcal{F}_{\text{op}}(\mathcal{S}(P_1), \mathcal{S}(P_2))$.

The reactive behaviours we want to build from Argos programs are denoted by *reactive* and *deterministic* input/output-labeled transition systems. However, this representation does not capture the exact meaning of a reactive behaviour, and an equivalence relation has to be added. We call a reactive behaviour an equivalence class of this relation.

An input/output-labeled transition system is *reactive* if, in any state, for any input configuration, the reaction of the system can be defined, i.e. there exists at least one transition. This transition may be the representation of a "null reaction", i.e. the system does not change states, and generates no output; but it *must* be defined. The notion is similar to that of *input enabling* in IO-automata [LT89]. A system is *deterministic* if, for any state, for any input configuration, there exists at most one transition.

Reactivity is required in all cases. Determinism is not so important as reactivity. It is required for a *programming* language only. For a *specification* language, non-determinism is a convenient way to describe and verify a (possibly infinite) set of behaviours, in a concise manner. The Argos semantics can be extended to support *intrinsic* non-determinism in basic components; however, the operators do not *introduce* non-determinism.

The Semantic Function and its Required Properties. Let $\mathcal{P}$ denote the set of programs. It is defined syntactically. Let $\mathcal{I}olts$ denote the set of input/output-labeled transition systems (a transition label describes an elementary reaction of the system: an input condition, and the simultaneous output caused by this input). $\mathcal{I}olts_{dr} \subsetneq \mathcal{I}olts$ is the set of reactive and deterministic systems. We denote by $\sim \subsetneq \mathcal{I}olts \times \mathcal{I}olts$ the equivalence of models. The semantics of the language associates an element of $\mathcal{I}olts_{dr}$ to an element of $\mathcal{P}$. This function is defined in two phases. First, we define a function $\mathcal{S}' \quad : \quad \mathcal{P} \longrightarrow \mathcal{I}olts$, to capture the intuition of the synchronous behaviour of programs.

Extensive work [BG88, Hui91] about synchronous formalisms has shown that *causality problems are intrinsic to synchronous semantics.* From a technical point of view, causality problems are related to the fact that the synchronous semantics $\mathcal{S}'$ is such that: $\exists P \in \mathcal{P}, \mathcal{S}'(P) \in \mathcal{I}olts - \mathcal{I}olts_{dr}$. (This is even stronger: there exist programs whose semantics are not reactive, and there exist programs whose semantics are not deterministic.) We define the set of *incorrect* programs $\mathcal{P}_i \subseteq \mathcal{P}$ by: $P \in \mathcal{P}_i \iff \mathcal{S}'(P) \notin \mathcal{I}olts_{dr}$. In a practical programming environment, such programs have to be rejected. Hence the semantic function is defined by:

$$\mathcal{S} \;:\; \mathcal{P} \longrightarrow \mathcal{I}olts_{dr} \cup \{\text{ ERROR}\}$$

$$P \in \mathcal{P}_i \iff \mathcal{S}(P) = \text{ERROR}, \quad P \notin \mathcal{P}_i \implies \mathcal{S}(P) = \mathcal{S}'(P)$$

We define the equivalence $\sim_{dr}$ of z, z' in $\mathcal{I}olts_{dr} \cup \{\text{ERROR}\}$ by:

$$z \sim_{dr} z' \iff \begin{cases} z = z' = \text{ERROR or} \\ z, z' \in \mathcal{I}olts_{dr} \text{ and } z \sim z' \end{cases}$$

We first require *expressivity* of the language. The reactive behaviours are exactly the equivalence classes of $\sim_{dr}$, and the language should be such that:

$$\forall \mathcal{M} \in \mathcal{I}olts_{dr}, \exists P \in \mathcal{P} \text{ s.t. } \mathcal{S}(P) \sim_{dr} \mathcal{M}$$

We also require modularity. The equivalence of models induces an equivalence of programs: $P_1 \equiv P_2 \iff \mathcal{S}(P_1) \sim_{dr} \mathcal{S}(P_2)$

The language is *modular* if the semantics is *compositional*, i.e. if $\equiv$ is a *congruence* for the Argos operators: $\forall P_1, P_2, Q \in \mathcal{P}, P_1 \equiv P_2 \implies P_1 \text{ op } Q \equiv P_2 \text{ op } Q$.

We use strong bisimulation as the $\sim$ relation.

Structure of the Paper. Section 2 gives an informal description of the operators, using a graphical syntax. The reader should refer to [Mar89, Mar91a] for further details. Section 3 defines the language formally, and we prove the required properties of the operational semantics. Related work is summarized in section 4. Finally we conclude and give some hints about the complete Argos language and its programming environment in section 5.

2 Description of the Language Constructs

2.1 Automata

When the system is simple enough, one describes it using a single automaton. An automaton has a unique initial state and it has to be deterministic.

The transitions are labeled by an input part (a conjunction of events and negations of events) and an output part (a set of events), separated by a slash (example: $a.\bar{b}.c/e, f, g$). We consider *saturated* automata: there exists a set I called the input set of the automaton, such that the input parts of the transition labels contain all the elements of I. A concrete syntax usually provides abbreviations to avoid this.

2.2 Parallel Composition and Local Events

An example: the Three-bit a Counter. Figure 1-(a) describes a modulo 8 a-counter. Dashed lines are used to give a *parallel* structure to a program, as in Statecharts. The surrounding box, with the indication "b, c", defines the scope in which events b and c can be used: they are *local* to this scope. Consequently, the only input of the program is a, and its only output is e. The global behaviour is

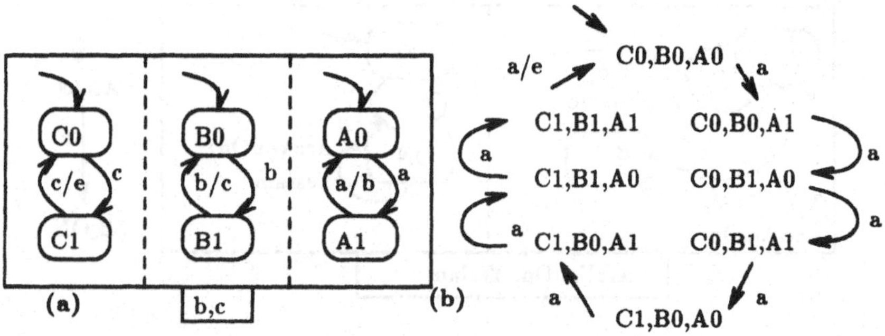

Figure 1: The three-bit **a** counter

defined by: the global initial state is **C0,B0,A0** ; when it has reacted to input **a** n times, the program is in state $\mathtt{C}_k,\mathtt{B}_j,\mathtt{A}_i$, where $i + 2j + 4k = n \bmod 8$; the event **e** is output every 8 **a**'s. This behaviour is achieved by connecting three one-bit counters. The first one (A) reacts to external input **a**, and triggers the second one (B) with event **b**, every two **a**'s. The second one, reacting to **b**, triggers the third one (C) with **c**, every two **b**'s. The third one outputs **e** every two **c**'s. Figure 1-(b) gives the global behaviour of the counter.

The scope of some events being restricted, the counter does not output **b** and **c** to the global environment, but only to those components which belong to their scope. On the other hand, it cannot take inputs **b** and **c** in the global environment: they have to be output somewhere else in the scope. The communication mechanism is the *synchronous broadcast.* A component X can always output events; any component in their scope, which is able to react to them, must do so.

Status of Local Events During Global Reactions. The intuitive presentation of the counter explains the global transitions by giving *chains* of automaton transitions. The first one is triggered by external inputs, and each of the subsequent ones is triggered by, either inputs or already emitted internal events. This is not the exact definition of the synchronous broadcast.

Figure 2 illustrates a program whose transitions cannot be explained by chains of transitions in the different components. (This example is known as the "instantaneous dialogue" in Esterel). On reacting to x, in state **A**, the first component outputs the question **AreYouOn**. The other one, if **ON**, outputs the answer **YesIam**. Hence the first component reaches **B** on input x if and only if the second one is **ON**.

In order to determine the reaction of the program to input x, when in state **A,ON**, one has to decide whether the local events **AreYouOn** and **YesIam** participate to the reaction. They do so if and only if they are emitted during the reaction.

The status of the local events is given by the solution of a set of boolean equations, where event names are used as boolean variables. the equations associated to the dialogue, for state **A,ON**, are: $\{$ **YesIam** $=$ **AreYouOn**, **AreYouOn** $= x \wedge$ **YesIam** $\vee x \wedge \neg$ **YesIam** $\}$ which expresses that: **YesIam** will be emitted only if the loop in the second component is taken, i.e. if **AreYouOn** participates to the reaction; **AreYouOn** may be

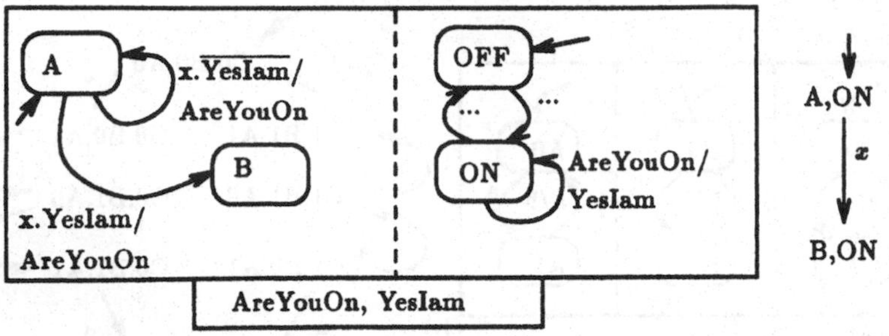

Figure 2: Determining the behaviour of a program

emitted in two cases (they are two transitions sourced in A which output it).

For input x, the variable x has to be replaced by TRUE, and the system has a unique solution: both YesIam and AreYouOn are true. Hence the global reaction is made of: the loop in the second component; the transition from A to B in the first one.

Typical Incorrect Programs. In the introduction, we mentioned that, if the global behaviour of a program happens to be non reactive or non deterministic, an error has to be reported: the program is *incorrect*. This happens when the set of boolean equations has several solutions or no solution at all.

Consider a program P_1 (resp. P_2) with two local events a, b and an input i, where one parallel component has a transition $X \xrightarrow{i.a/b} X'$ and another component has a transition $Y \xrightarrow{i.b/a} Y'$ (resp. $Y \xrightarrow{i.\bar{b}/a} Y'$). The sets of equations associated to the state X,Y are: $\mathcal{E}_1 = \{a = b \wedge i, b = a \wedge i\}$ for P_1, $\mathcal{E}_2 = \{a = \neg b \wedge i, b = a \wedge i\}$ for P_2. For input i, $\mathcal{E}_1$ has two solutions: $a = b =$ TRUE and $a = b =$ FALSE. Conversely, $\mathcal{E}_2$ has no solution. In the first case, there is no good reason to choose either solution, and keeping both would introduce non-determinism in the global behaviour (even if we accept "*intrinsic*" non-determinism in basic components, the interpretation of the communication should not introduce non-determinism). In the second case, the reaction of the program to input x when in state X,Y cannot be defined. The behaviour would be non reactive. Both programs are rejected.

2.3 The Refinement Operation

Unlike parallel composition, the refinement is not *symmetric*. One of the components, which has to be an automaton, is the *controller*, and the other ones are *controlled* by it. The controller may start and kill them.

Consider an automaton A, and suppose we place the description of subprograms P_i in its states. Then, define the effect of such an operation by the following rules. 1) When a state is entered, the subprogram inside is *started*, in its initial state; 2) when a state is left, the subprogram inside is *killed*. All information about its current state is lost. These rules induce a notion of *active* subprogram: the subprogram inside

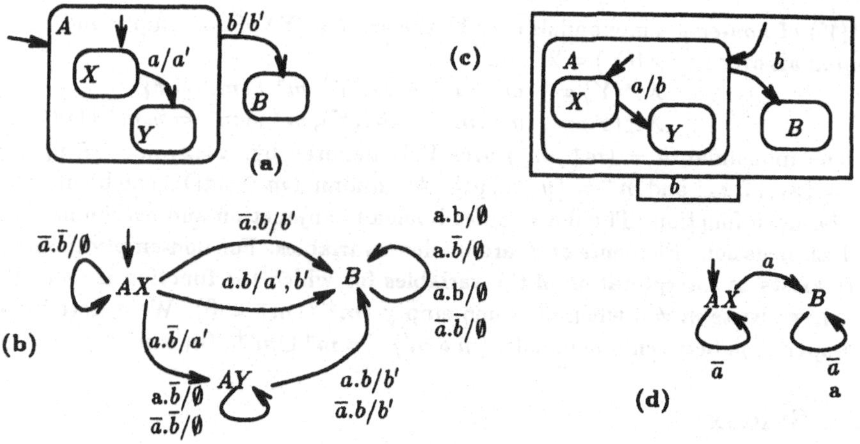

Figure 3: Refinement and self-termination mechanism

state i is said to be *active* when and only when the automaton is in state i.

Now, observe Fig. 3, where we focus on one state of the automaton. The behaviour of program (a) can be defined intuitively by considering that the two components of the refinement operation are actually *parallel* components. In the initial state, when events a and b occur simultaneously, the two components react simultaneously; *but*, by taking the b-transition, the A,B-component leaves state A, and then the X,Y-component state becomes irrelevant. The global behaviour is (b).

In the previous example, b comes from the global environment of the program, and thus the transition triggered by b behaves like an *interruption* of the refining subprogram. This interruption is not *preemptive*, in the sense that the interrupted subprogram may react at the instant when it is killed (the reaction to $a.b$ outputs b' *and* a'). In the complete Argos language, an *inhibition* operator is provided, for several purposes. A program P inhibited by an event a reacts exactly as P would do, when a is absent; when a is present, it has only null reactions. Interrupting P can be made preemptive if P is inhibited by a, and the interrupting transition outputs a.

Now, suppose b comes from the refining subprogram itself (see Fig. 3-c). The a/b transition can be used to describe a *self-termination* of the program inside state A, which broadcasts this information with event b. The controller may decide to kill the XY-component, on receiving b, as done by the AB-component. The behaviour of program (c) is (d).

3 Formal Definition of the Language

3.1 Notations

We denote by $\mathcal{A}$ the finite alphabet of *events*: $\mathcal{A} = \{a, b, c, ...\}$

For a subset Y of $\mathcal{A}$, we define: the set $\mathcal{M}(Y)$ of *monomials* over Y; the set

$\mathcal{M}^\star(Y)$ of *non-empty* monomials over Y; the set $\mathcal{M}^\star_c(Y)$ of non-empty and *complete* monomials over Y: $\mathcal{M}(Y) = 2^Y \times 2^Y$

$$\mathcal{M}^\star(Y) = \{(m^+, m^-) \in \mathcal{M}(Y), m^+ \cap m^- = \emptyset\}$$
$$\mathcal{M}^\star_c(Y) = \{(m^+, m^-) \in \mathcal{M}(Y), m^+ \cap m^- = \emptyset, m^+ \cup m^- = Y\}$$

The monomial $m = (m^+, m^-)$ over Y is denoted by: $x_1.x_2.....x_n.\overline{y_1}.\overline{y_2}.....\overline{y_p}$ if $m^+ = \{x_1, ..., x_n\}$ and $m^- = \{y_1, ..., y_p\}$. A monomial $m \in \mathcal{M}(Y)$ can be interpreted as a boolean function. The dot denotes boolean conjunction and overlining denotes boolean negation. Elements of Y are boolean variables. For non-empty monomials, there exists an interpretation of the variables for which the function is true. In the sequel, we write $m \neq 0$ when m is non-empty ($m^+ \cap m^- = \emptyset$). We denote by $\bullet$ the AND-operation between monomials: $(m \bullet m')^+ = m^+ \cup m'^+$, $(m \bullet m')^- = m^- \cup m'^-$.

3.2 Syntax

An *automaton* is a tuple $A = (Q, q_0, T, I)$. Q is a set of *states*, $q_0 \in Q$ is the *initial* state, and T is the set of *transitions*. The set $I \neq \emptyset, I \subseteq \mathcal{A}$, is called the *input set* of the automaton. $T \subseteq Q \times \mathcal{M}^\star_c(I) \times 2^{\mathcal{A}} \times Q$. A transition $t = (q_s, m, o, q_t) \in T$ has three parts: the *source* state q_s, the *target* state q_t, and the *label* (m, o). m is the input part of the label, and o is the output part.
An automaton $A = (Q, q_0, T, I)$ is said to be *deterministic* iff the following holds:

$$\forall q \in Q, \quad \forall m \in \mathcal{M}^\star_c(I), \quad |\{(q, m, o, q') \in T\}| \leq 1$$

The set of programs is denoted by $\mathcal{P}$. It is defined syntactically in two phases: we give a context-free grammar first, and then add some contextual constraints. A program $P \in \mathcal{P}$ is of the form:

$$P ::= \mathbf{A}_A(R_1, ..., R_n) \mid P \| P \mid \overline{P^Y}. \qquad R ::= P \mid \text{NIL}.$$

Each automaton $A = (Q, q_0, T, I)$ defines a parameterized refinement operator $\mathbf{A}_A()$, whose arity is $|Q|$. To relate a state in Q to its refining program, we identify the set Q to the interval $[1..|Q|]$. Then $\mathbf{A}_A(R_1, ..., R_n)$ is a program made with the automaton A, the state i being refined by R_i. The R_i may be programs or the special value NIL, which means that the state is not refined. $\|$ denotes parallel composition; overlining denotes local event declaration ($Y \subseteq \mathcal{A}$).

In order to express the contextual constraints, we define the function $\mathcal{I}$ (resp. $\mathcal{O}$) which gives the inputs (resp. the outputs) of a program. It is extended to automata and to NIL.

$$\mathcal{I}(\text{NIL}) = \emptyset \qquad \mathcal{I}(P_1 \| P_2) = \mathcal{I}(P_1) \cup \mathcal{I}(P_2)$$
$$\mathcal{O}(\text{NIL}) = \emptyset \qquad \mathcal{O}(P_1 \| P_2) = \mathcal{O}(P_1) \cup \mathcal{O}(P_2)$$
$$\mathcal{I}(\mathbf{A}_A(R_1, ...R_n)) = \mathcal{I}(A) \cup \textstyle\bigcup_{i=1}^{i=n} \mathcal{I}(R_i) \qquad \mathcal{I}(\overline{P^Y}) = \mathcal{I}(P) - Y$$
$$\mathcal{O}(\mathbf{A}_A(R_1, ...R_n)) = \mathcal{O}(A) \cup \textstyle\bigcup_{i=1}^{i=n} \mathcal{O}(R_i) \qquad \mathcal{O}(\overline{P^Y}) = \mathcal{O}(P) - Y$$

$\mathcal{I}((Q, q_0, T, I)) = I$, $\quad \mathcal{O}((Q, q_0, T, I)) = \{a, \exists q_s \in Q, m \in \mathcal{M}^\star_c(I), o \in 2^{\mathcal{A}}, q_t \in Q, \text{ s.t. } (q_s, m, o, q_t) \in T, a \in o\}$

If $P_1, P_2 \in \mathcal{P}$ then $P_1 \| P_2 \in \mathcal{P}$. If $P \in \mathcal{P}$ and $Y \subseteq \mathcal{I}(P) \cup \mathcal{O}(P)$, then $\overline{P^Y} \in \mathcal{P}$.
if (Q, q_0, T, I) is deterministic and $\forall i \in [1, n]$, $R_i \in \mathcal{P}$ or $R_i = \text{NIL}$ then $\mathbf{A}_{(Q,q_0,T,I)}(R_1, ..., R_n) \in \mathcal{P}$.

A program P is said to be a *main* program if $\mathcal{I}(P) \cap \mathcal{O}(P) = \emptyset$, which means that an event which is used both as input and as output has to be declared local to the appropriate part of the program.

3.3 Semantics

The semantics of Argos is given by defining the function $\mathcal{S} : \mathcal{P} \longrightarrow \mathcal{I}olts_{dr} \cup \{\text{ERROR}\}$. $\mathcal{I}olts$ is the set of *input/output-labeled transition systems* of the form: (Q, q_0, T, I), where Q is a set of states, $q_0 \in Q$ is the initial state, $I \subseteq \mathcal{A}$ is the set of inputs, $T \subseteq Q \times \mathcal{M}_c^*(I) \times 2^{\mathcal{A}} \times Q$ is the set of transitions. Note that the input part of the labels is a *complete* monomial over I (hence I is a redundant information). We define the *degree* of a transition system, for a given state and a given monomial by:

Definition 3.1 *(degree of a system, reactivity and determinism)*
For $q \in Q, m \in \mathcal{M}_c^*(I), \deg(q,m) = |\{(q,m,o,q') \in T\}|$.
The system is deterministic *iff* $\forall q \in Q, \forall m \in \mathcal{M}_c^*(I), \deg(q,m) \leq 1$.
It is reactive *iff* $\forall q \in Q, \forall m \in \mathcal{M}_c^*(I), \deg(q,m) \geq 1$. □

$\mathcal{I}olts_{dr} \subseteq \mathcal{I}olts$ is the subset of deterministic and reactive elements. The model of a correct program P is $\mathcal{S}(P) = (\textit{states}(P), \textit{init}(P), \textit{trans}(P), \mathcal{I}(P))$. The model of an incorrect program is the special value ERROR.

$\mathcal{I}(P)$ is defined in section 3.2. The potential states of the input/output-labeled transition system $\mathcal{S}(P)$ have the form:

$$PS ::= \mathbf{A}_{AS}(RS_1, ..., RS_n) \mid PS \| PS \mid \overline{PS^Y}. \quad RS ::= PS \mid \text{NIL}.$$

AS is an automaton state of the form (Q, q_0, q_c, T, I) (an automaton with additional information about the *current* state q_c). A state of P is obtained by specifying a *current* state in each of its automaton components. The initial state of P is obtained by specifying the initial state for each of its automata:

$$\textit{init}(\text{NIL}) = \text{NIL}, \ \textit{init}(\overline{P^Y}) = \overline{\textit{init}(P)^Y}, \ \textit{init}(P\|P') = \textit{init}(P)\|\textit{init}(P')$$
$$\textit{init}(\mathbf{A}_{(Q,q_0,T,I)}(R_1, ...R_n)) = \mathbf{A}_{(Q,q_0,q_0,T,I)}(\textit{init}(R_1), ..., \textit{init}(R_n))$$

$\textit{states}(P)$ is defined as the set of reachable states from $\textit{init}(P)$, according to $\textit{trans}(P)$. $\textit{trans}(P)$ is defined by giving a set of Plotkin-style rules. In the sequel, we write $q \xrightarrow{m/o} q'$ when $(q, m, o, q') \in \textit{trans}(P), q, q'$ having the form defined above. Errors are reported by adding a special ERROR state to the potential states of a program. This state is a *sink* state. The model of a program P has to be considered as the special value ERROR if and only if it contains the state ERROR.

Defining the Set of Transitions

Parallel composition and local events have been presented together. However, the abstract syntax makes use of two operators. The behaviour associated to a parallel program is a kind of synchronous product of the two behaviours: transitions triggered by the same input are taken together, and their outputs are added (see rule [P]). If two components communicate with each other, an input i of the one is an output of the other. In this case, the global behaviour contains transitions where i appears both as input and as output. In a *main* program, i has to be declared local somewhere; then the effects of the communication are taken into account, and i is hidden (see rules [I1], [I2]).

All operators propagate the errors reported in their components. Errors are *detected* with the local event declaration only.

Parallel Composition.
$P = P_1 \| P_2$. If q_1 (resp. q_2) denotes a global state of program P_1 (resp. P_2), the following rule computes the behaviour of $P_1 \| P_2$.

$$\frac{m \in \mathcal{M}_c^\star(\mathcal{I}(P)), \quad q_1 \xrightarrow{m[\mathcal{I}(P_1)]/o_1} q_1', \quad q_2 \xrightarrow{m[\mathcal{I}(P_2)]/o_2} q_2'}{q_1 \| q_2 \xrightarrow{m/o_1 \cup o_2} q_1' \| q_2'} \quad \text{[P]}$$

$m[I]$ is the restriction of m to I, defined by: $m[I]^+ = m^+ \cap I, m[I]^- = m^- \cap I$.
$q_1' \| q_2'$ may be the ERROR state according to: $\text{ERROR} \| p = p \| \text{ERROR} = \text{ERROR} \| \text{ERROR} = \text{ERROR}$.

Local Event Declaration.
Consider a program $\overline{P^Y}$. The behaviour of P has been computed without taking into account the fact that the events in Y are local to it. For each state q and each input $m \in \mathcal{M}_c^\star(\mathcal{I}(P))$, it contains a *unique* (see section 3.4) transition $q \xrightarrow{m/o} q'$ which summarizes the reactions to m of the basic components. m can be written $m = m' \bullet mi$, where $mi \in \mathcal{M}_c^\star(Y \cap \mathcal{I}(P))$ and $m' \in \mathcal{M}_c^\star(\mathcal{I}(P) - Y)$. q' and o can be considered as functions of (q, m). We write $o = \text{OUT}(q, m), q' = \text{TARG}(q, m)$ when $q \xrightarrow{m/o} q'$. The outputs in o illustrate the effect of making the assumption that mi is the correct status of local events during the reaction of P to the input m'.

This information is sufficient to apply the mechanism which has been presented informally in section 2, without building the sets of equations explicitly. It is easy to show that a status of local events, represented by a monomial mi, corresponds to a solution of the boolean equations if and only if: $mi^+ \subseteq o$ and $mi^- \cap o = \emptyset$, which can be read as follows: all events supposed to participate to the reaction are indeed emitted during the reaction; conversely, all events supposed not to participate are not emitted.

An error has to be reported when no status is correct, or several status are. Rule [I1] expresses that, if all the transitions in P can indeed be computed without errors, and only one of them has a correct status, then it constitutes the global reaction:

$$I = \mathcal{I}(P), J = \mathcal{I}(\overline{P^Y}), M = \mathcal{M}_c^\star(Y \cap I)$$

$$\frac{\begin{array}{c} m \in \mathcal{M}_c^\star(J), \ (\forall mi \in M, \ \text{TARG}(q, m \bullet mi) \neq \text{ERROR}) \ \wedge \\ (\exists! mi_O \in M \text{ s.t. } mi_O^+ \subseteq \text{OUT}(q, m \bullet mi_O) \wedge mi_O^- \cap \text{OUT}(q, m \bullet mi_O) = \emptyset) \end{array}}{\overline{q^Y} \xrightarrow{m/\text{OUT}(q, m \bullet mi_0) - Y} \overline{\text{TARG}(q, m \bullet mi_0)}^Y} \quad \text{[I1]}$$

If one of the transitions in P reports an error, or there does not exist a unique correct status, an error is reported:

$$\frac{\begin{array}{c} m \in \mathcal{M}_c^\star(J), \ (\exists mi \in M, \text{TARG}(q, m \bullet mi) = \text{ERROR}) \quad \vee \\ |\{mi \in M \text{ s.t. } mi^+ \subseteq \text{OUT}(q, m \bullet mi) \wedge mi^- \cap \text{OUT}(q, m \bullet mi) = \emptyset\}| \neq 1 \end{array}}{\overline{q^Y} \xrightarrow{m/} \text{ERROR}} \quad \text{[I2]}$$

Refinement.
Consider a program $P = \mathbf{A}_{(Q,q_0,T,I)}(R_1, ..., R_n)$ and $J = \mathcal{I}(P)$. The behaviour of P is defined by rules [R1], [R2]. The transition triggered by m and sourced in $\mathbf{A}_{(Q,q_0,q_c,T,I)}(R_1, ..., R_n)$ can be built in two ways. [R1]: if there exists a transition triggered by m (i.e., with input m[I]) and sourced in the current state q_c of the automaton, then the automaton and the active refining subprogram R_c react together; their outputs o_1, o_2 are added; the automaton changes states (from q_c to q_d); all the refining subprograms are reinitialized. [R2]: if there exists no such transition in the automaton, the active refining subprogram reacts alone, from R_c to R'_c; the automaton does not change states.

$$\frac{m \in \mathcal{M}_c^\star(J), (q_c, m[I], o_1, q_d) \in T, \quad R_c \xrightarrow{m[\mathcal{I}(R_c)]/o_2} R'_c}{\mathbf{A}_{(Q,q_0,q_c,T,I)}(R_1, ..., R_n) \xrightarrow{m/o_1 \cup o_2} \mathbf{A}_{(Q,q_0,q_d,T,I)}(reinit(R_1), ..., reinit(R'_c), ..., reinit(R_n))} \quad \text{[R1]}$$

Where $reinit(PS) = init(prog(PS))$ and $prog(PS)$ (the program to which the PS state belongs) is obtained by removing the current-state information from PS.

Notice that, in the global state $\mathbf{A}_{(Q,q_0,q_d,T,I)}(reinit(R_1), ..., reinit(R_n))$, the automata which belong to the programs R_i are all in their initial state. For the ones belonging to R_d, this corresponds to the fact that R_d is started in its initial state; for the other ones, it ensures that the states of $P = \mathbf{A}_{(Q,q_0,T,I)}(R_1, ..., R_n)$ are built in a unique form: the states of the automata in the $R_{i \neq d}$ are semantically irrelevant in a global state of the form $\mathbf{A}_{(Q,q_0,q_d,T,I)}(reinit(R_1), ..., reinit(R_n))$, and they are always set to the initial one.

$$\frac{m \in \mathcal{M}_c^\star(J), \quad \nexists(q_c, m[I], o_1, q_d) \in T, \quad R_c \xrightarrow{m[\mathcal{I}(R_c)]/o} R'_c}{\mathbf{A}_{(Q,q_0,q_c,T,I)}(R_1, ..., R_n) \xrightarrow{m/o} \mathbf{A}_{(Q,q_0,q_c,T,I)}(R_1, ..., R'_c, ..., R_n)} \quad \text{[R2]}$$

The global target states can be the ERROR state, according to:
$reinit(\text{ERROR}) = \text{ERROR}$
if $\exists i \in [1, n], R_i = \text{ERROR}$ then $\mathbf{A}_{(Q,q_0,q_c,T,I)}(R_1, ..., R'_c, ..., R_n) = \text{ERROR}$.
The case $R_c = \text{NIL}$ is integrated in rule [R2] thanks to the axiom: $\text{NIL} \xrightarrow{0/\emptyset} \text{NIL}$, where 0 denotes the *null* monomial. (Note that the null monomial is *non-empty*, as defined in 3.1).

3.4 Properties of the Semantics

Property 3.1 *(Expressivity)*
$\forall \mathcal{M} \in \mathcal{I}\text{olts}_{dr}, \exists P \in \mathcal{P} \text{ s.t. } \mathcal{S}(P) \sim_{dr} \mathcal{M}$ □

Proof: Very simple. $\mathcal{S}(\mathbf{A}_{\mathcal{M}}(\text{NIL}, ..., \text{NIL})) = \mathcal{M}$. ■

Property 3.2 *(Determinism and reactivity, error detection and propagation)*
$\forall P, P_1, P_2, R_1, ..., R_n \in \mathcal{P}$,

$\mathcal{S}(P) = \text{ERROR} \implies \mathcal{S}(\overline{P^Y}) = \text{ERROR}$
$\mathcal{S}(P_1) = \text{ERROR} \vee \mathcal{S}(P_2) = \text{ERROR} \iff \mathcal{S}(P_1 \| P_2) = \text{ERROR}$
$\exists i \in [1, n], \mathcal{S}(R_i) = \text{ERROR} \iff \mathcal{S}(\mathbf{A}_{(Q,q_0,T,I)}(R_1, ..., R_n)) = \text{ERROR}$

$\forall P \in \mathcal{P}, \mathcal{S}(P) \in \mathcal{I}olts_{dr} \cup \{\text{ERROR}\}$
(We denote by DRE(P) *the property* $\mathcal{S}(P) \in \mathcal{I}olts_{dr} \cup \{\text{ERROR}\}$*).* □

Proof: By structural induction on $\mathcal{P}$, we prove: $\forall P \in \mathcal{P}, \text{DRE}(P)$. For each operator, we show that error-detection and propagation are done correctly.
NIL: $\mathcal{S}(\text{NIL}) = (\{\text{NIL}\}, \text{NIL}, \{\text{NIL} \xrightarrow{0/\emptyset} \text{NIL}\}, \emptyset) \in \mathcal{I}olts_{dr}$
Parallel composition: Errors: Assume $\mathcal{S}(P_1) = \text{ERROR}$. This means that the ERROR state is reachable in the model of P_1. The key idea is that the reachable states of P_1 are still reachable in $P_1 \| P_2$, according to rule [P] (this can be proved easily by induction on the length of paths from the initial state to a given reachable state of P_1; the result is based upon the fact that, in order to build the transitions of $P_1 \| P_2$, we need *all* the transitions of the P_i): there exists a state q_2 of P_2 such that $\text{ERROR} \| q_2 = \text{ERROR}$ is reachable in the model of $P_1 \| P_2$. Hence, $\mathcal{S}(P_1 \| P_2) = \text{ERROR}$. (The proof is similar if $\mathcal{S}(P_2) = \text{ERROR}$). Conversely, it is clear from rule [P] that: $\mathcal{S}(P_1) \neq \text{ERROR} \wedge \mathcal{S}(P_2) \neq \text{ERROR} \implies \mathcal{S}(P_1 \| P_2) \neq \text{ERROR}$.
Determinism and reactivity: Now, consider $P = P_1 \| P_2$, a state $q = q_1 \| q_2$ of P, and a monomial $m \in \mathcal{M}_c^\star(\mathcal{I}(P))$. This transition can only be made with a transition triggered by $m[\mathcal{I}(P_1)]$ sourced in q_1, and a transition triggered by $m[\mathcal{I}(P_2)]$ sourced in q_2. Assume, by induction, that $\text{DRE}(P_1), \text{DRE}(P_2)$.
Then, if $\mathcal{S}(P_1) = \text{ERROR}$ or $\mathcal{S}(P_2) = \text{ERROR}$, then $\mathcal{S}(P) = \text{ERROR}$, hence $\text{DRE}(P)$.
Else, $\mathcal{S}(P_1) \neq \text{ERROR}$ and $\mathcal{S}(P_2) \neq \text{ERROR}$, and they are deterministic and reactive: the above transitions exist and are unique. Hence the m transition sourced in q exists and is unique, for all m. This proves $\mathcal{S}(P) \in \mathcal{I}olts_{dr}$, i.e. $\text{DRE}(P)$.
Refinement: Errors: the proof is similar to the proof for parallel composition. The key idea for propagation is that the reachable states of the refining programs are still reachable in the refined program, provided that the states of the automaton are all reachable from the initial one. Conversely, it is clear from rules [R1] and [R2] that no error is detected by the refinement operator: all errors in the composed program come from errors in the components.
Determinism and reactivity: Consider $P = \mathbf{A}_{(Q,q_0,T,I)}(R_1, ..., R_n)$, a monomial $m \in \mathcal{M}_c^\star(\mathcal{I}(P))$ and a state $q = \mathbf{A}_{(Q,q_0,q_c,T,I)}(R_1, ..., R_c, ..., R_n)$ of P. Assume $\text{DRE}(R_i)$ or $R_i = \text{NIL}, \forall i \in [1, n]$. The automaton is not necessarily reactive.
First case: the transition triggered by $m[I]$ sourced in q_c exists (and is unique) in the automaton. By rule [R1] we get: either $\exists i \in [1, n], \mathcal{S}(R_i) = \text{ERROR}$, and then $\mathcal{S}(P) = \text{ERROR}$, or $\mathcal{S}(R_i) \in \mathcal{I}olts_{dr}, \forall i \in [1, .n]$, and then the transition triggered by m sourced in q exists and is unique. Hence $\text{DRE}(P)$.
Second case: similar, with rule [R2].
Local events: Errors: Assume that the ERROR state is reachable in $\mathcal{S}(P)$, from a state q. $\overline{q^Y}$ is not necessarily reachable in $\mathcal{S}(\overline{P^Y})$, *but* if it is not, that is because an error is reached between the initial state of $\mathcal{S}(\overline{P^Y})$ and $\overline{q^Y}$. So, if $\overline{q^Y}$ is reachable, rule [I2] ensures that the ERROR state will be reachable too. If it is not, the ERROR state has been reached "before". In all cases, $\mathcal{S}(\overline{P^Y}) = \text{ERROR}$.

Determinism and reactivity: Consider $P_1 = \overline{P^Y}$, a state $q_1 = \overline{q^Y}$ of P_1 and a monomial $m \in \mathcal{M}_c^\star(\mathcal{I}(P_1))$. Assume $\text{DRE}(P)$.
If $\mathcal{S}(P) = \text{ERROR}$, then $\mathcal{S}(P_1) = \text{ERROR}$.

If $\mathcal{S}(P) \neq$ ERROR then, either the transition triggered by m and sourced in $\overline{q^Y}$ can be built in a unique way (rule [I1]), or $\mathcal{S}(P_1) =$ ERROR. Hence DRE$(\overline{P^Y})$. ∎

Let us define the equivalence of models by:
$(Q_1, q_{01}, T_1, I_1) \sim (Q_2, q_{02}, T_2, I_2) \iff I_1 = I_2$ and (Q_1, q_{01}, T_1) and (Q_2, q_{02}, T_2) are *bisimilar*.
Two transition systems (Q_1, q_1, T_1) and (Q_2, q_2, T_2) are *bisimilar* if and only if there exists a relation $\mathcal{R} \subseteq Q_1 \times Q_2$ such that:

$$(q_1, q_2) \in \mathcal{R} \implies \begin{cases} q_1 \xrightarrow{m/o} q_1' & \implies \exists q_2' \text{ s.t. } q_2 \xrightarrow{m/o} q_2' \text{ and } (q_1', q_2') \in \mathcal{R}. \\ q_2 \xrightarrow{m/o} q_2' & \implies \exists q_1' \text{ s.t. } q_1 \xrightarrow{m/o} q_1' \text{ and } (q_1', q_2') \in \mathcal{R}. \end{cases}$$

The relation $\sim_{dr}$ on $\mathcal{I}olts_{dr} \cup \{\text{ERROR}\}$ (cf. introduction) is defined by:

$$z \sim_{dr} z' \iff \begin{cases} z = z' = \text{ERROR or} \\ z, z' \in \mathcal{I}olts_{dr} \text{ and } z \sim z' \end{cases}$$

and the equivalence of programs by: $P_1 \equiv P_2 \iff \mathcal{S}(P_1) \sim_{dr} \mathcal{S}(P_2)$

Property 3.3 *(Modularity)*
$\equiv$*is a* congruence *for the Argos operators, i.e.* : $\forall P_1, P_2, Q, R_1, ..., R_n, R_i, R_i' \in \mathcal{P}, Y \subseteq \alpha$,

$$\begin{aligned} P_1 \equiv P_2 &\implies P_1 \| Q \equiv P_2 \| Q, \quad \overline{P_1{}^Y} \equiv \overline{P_2{}^Y} \\ R_i \equiv R_i' &\implies \mathbf{A}_{(Q, q_0, T, I)}(R_1, ..., R_i, ..., R_n) \equiv \mathbf{A}_{(Q, q_0, T, I)}(R_1, ..., R_i', ... R_n) \end{aligned}$$ □

Proof: We give the proof for parallel composition, and only sketch it for the other operators.
Parallel composition:

Assume $P_1 \equiv P_2$, i.e. $\begin{cases} \mathcal{S}(P_1) = \mathcal{S}(P_2) = \text{ERROR or} & (1) \\ \mathcal{S}(P_1) \neq \text{ERROR}, \mathcal{S}(P_2) \neq \text{ERROR}, \mathcal{S}(P_1) \sim \mathcal{S}(P_2) & (2) \end{cases}$

Case (1): if $\mathcal{S}(P_1) =$ ERROR and $\mathcal{S}(P_2) =$ ERROR then $\mathcal{S}(P_1 \| P) = \mathcal{S}(P_2 \| P) =$ ERROR (from property 3.2), hence $P_1 \| P \equiv P_2 \| P$.
case (2): Else, assume $\mathcal{S}(P_i) = (Q_i, q_i, T_i, I_i), i \in \{1, 2\}$. $P_1 \equiv P_2$ implies $I_1 = I_2$ and there exists a bisimulation relation $\mathcal{R}$ between (Q_1, q_1, T_1) and (Q_2, q_2, T_2). We know from property 3.2 that $\mathcal{S}(P_i \| P) \neq$ ERROR. Assume $\mathcal{S}(P_i \| P) = (Q_i', q_i', T_i', I_i'), i \in \{1, 2\}$. We have: $I_1' = I_1 \cup \mathcal{I}(P) = I_2 \cup \mathcal{I}(P) = I_2'$ and we can define a relation $\mathcal{R}'$ between (Q_1', q_1', T_1') and (Q_2', q_2', T_2') which is a bisimulation:
$\mathcal{R}' \in Q_1' \times Q_2'$ is such that: $(q_1 \| q, q_2 \| q) \in \mathcal{R}' \iff (q_1, q_2) \in \mathcal{R}$.
Indeed, assume $q_1 \| q \xrightarrow{m/o} p'$ (t_1). We prove $\exists p'', q_2 \| q \xrightarrow{m/o} p'', (p', p'') \in \mathcal{R}'$.
The transition t_1 comes from the transitions $q_1 \xrightarrow{m[\mathcal{I}(P_1)]/o_1} q_1'$, $q \xrightarrow{m[\mathcal{I}(P)]/o'} q'$, with $o = o_1 \cup o'$, $p' = q_1' \| q'$. Then $P_1 \equiv P_2$ implies $\exists q_2', q_2 \xrightarrow{m[\mathcal{I}(P_1)]/o_1} q_2'$ and $(q_1', q_2') \in \mathcal{R}$ (and $\mathcal{I}(P_1) = \mathcal{I}(P_2)$).
Hence $q_2 \| q \xrightarrow{m/o_1 \cup o'} q_2' \| q'$, by rule [P], and $(q_1' \| q', q_2' \| q') \in \mathcal{R}'$ since $(q_1', q_2') \in \mathcal{R}$.
The other direction is symmetrical.
Refinement:

The idea is the same as for the parallel composition. If $\mathcal{S}(R_i) = \mathcal{S}(R_i') = \text{ERROR}$ then $\mathcal{S}(\mathbf{A}_{(Q,q_0,T,I)}(R_1,...,R_i,...,R_n)) = \mathcal{S}(\mathbf{A}_{(Q,q_0,T,I)}(R_1,...,R_i',...,R_n)) = \text{ERROR}$, hence these two programs are congruent. Else, $\mathcal{S}(R_i){\sim}\mathcal{S}(R_i')$, and the corresponding bisimulation relation is $\mathcal{R}$. Then $\mathcal{R}'$ is a bisimulation between the states of $\mathbf{A}_{(Q,q_0,T,I)}(R_1,...,R_i,...,R_n)$ and the states of $\mathbf{A}_{(Q,q_0,T,I)}(R_1,...,R_i',...R_n)$, where $(\mathbf{A}_{(Q,q_0,q_c,T,I}(R_1,...R_i,...,R_n), \mathbf{A}_{(Q,q_0,q_d,T,I}(R_1,...R_i',...,R_n)) \in \mathcal{R}' \iff (d = c = i \wedge (R_i, R_i') \in \mathcal{R}) \vee (d = c, d \neq i)$

<u>*Local events:*</u>

If $\mathcal{S}(P_1) = \mathcal{S}(P_2) = \text{ERROR}$, then $\mathcal{S}(\overline{P_1{}^Y}) = \mathcal{S}(\overline{P_2{}^Y}) = \text{ERROR}$ hence these two programs are congruent.

Else, $\mathcal{S}(P_1){\sim}\mathcal{S}(P_2)$. Assume $\mathcal{R}$ is the bisimulation relation between the states of P_1 and the states of P_2. The relation $\mathcal{R}'$ between the states of $\overline{P_1{}^Y}$ and the states of $\overline{P_2{}^Y}$ is a bisimulation:

$$(q_1', q_2') \in \mathcal{R}' \iff \begin{cases} q_1' = \overline{q_1{}^Y}, q_2' = \overline{q_2{}^Y}, (q_1, q_2) \in \mathcal{R} \text{ or} \\ q_1' = q_2' = \text{ERROR} \end{cases}$$

This proves: either $\mathcal{S}(\overline{P_1{}^Y}) = \mathcal{S}(\overline{P_2{}^Y}) = \text{ERROR}$ or $\mathcal{S}(\overline{P_1{}^Y}) \neq \text{ERROR}$ and $\mathcal{S}(\overline{P_2{}^Y}) \neq \text{ERROR}$ and they are bisimilar. Hence $\overline{P_1{}^Y} \equiv \overline{P_2{}^Y}$. ■

4 Related Work

Argos is inspired from Statecharts, and from the synchronous family of languages (Lustre, Esterel).

The refinement operator is clearly related to the notion of termination in Esterel. We proved in [Mar91a] that all the termination modes of Esterel can be translated into Argos refined programs, by adding appropriate control to the components (see the self-termination mechanism in section 2 for instance). On the other hand, the exact error-detection mechanism presented in this paper is possible only because there are no variables.

The differences between Argos and Statecharts are due to several reasons.

In Argos, a program is a collection of automata, since level-crossing arrows are suppressed. A program is clearly decomposed into *subprograms*. Compositionality of the semantics allows modular development of programs. On the contrary, the syntax-directed semantics of Statecharts decompose a program into smaller syntactic parts, which cannot be viewed as subprograms (ex: the syntax of [HGdR88], whose operators are connection, juxtaposition, etc.) Compositionality does not make sense with a global semantics. With a syntax-directed one, it does not allow modular development of programs when the operators deal with pictures rather than subprograms. The differences between Argos and Statecharts, concerning the use of the state/transition paradigm, have been discussed in [Mar91b].

A second difference is the way causality problems are dealt with. Argos adopts the point of view of Esterel, where the programs for which causality problems occur must be rejected, or, at least, detected. In practise, the non-causal situations are not wanted by the designer, and he must be at least warned of their presence. The most interesting semantics of Statecharts, concerning these problems, is [PS88]. It is global. The semantics computes only the transitions which can be explained by a chain of transitions (see example in section 2); instantaneous dialogue cannot be expressed. A minimality criterion is taken into account in order to compute *causal* reactions only, but it is impossible to detect the problems. In the typical example where a and b are local to $i.a/b \| i.b/a$, the *causal* reaction is the one where a and b are absent. If the program is $i.\bar{a}/b \| i.\bar{b}/a$, the two possible reactions to i are considered to be *causal*. When there is no possible reaction, as in $i.\bar{a}/b \| i.b/a$, no error is reported; the global behaviour is *not* reactive.

5 Complements and conclusion

Argos is currently used as the basis for the Argonaute verification environment, in which the user can edit Argos programs, visualize them, simulate their behaviour, etc. The compiler produces labeled transition systems (LTS) according to various formats which allow the connection of Argonaute to existing tools.

The main point is the connection to Aldébaran [Fer90]. This tool allows efficient comparison and minimization of LTS according to various equivalence or congruence relations (observational equivalence [Mil80],bisimulation [Par81], ...). Argonaute is also connected to the Sahara [Ghe92] interface generator of the Esterel environment.

We have presented the set of operators which constitutes the basis of Argos. Argos is based upon the state/transition paradigm, it is synchronous, and it provides the user with powerful constructs like the parallel composition and the *refinement*. The refinement is able to express watchdogs, interruptions, exceptions, terminations of subprograms, etc. The complete Argos language is slightly different from the version presented here. A graphical and a textual concrete syntaxes allowing abbreviations in the automaton descriptions are provided. Some other operators, like the inhibition one (see section 2) are defined, and non-determinism is allowed in basic components.

The main result is the modularity of the language. It is due to the fact that the operational semantics is compositional. The bisimulation of the models induces an equivalence relation between programs which is a *congruence* for the operators we presented. The main difficulty is the expression of the refinement operation in a compositional framework. It is achieved by considering the automata as *operators*. Each n-state automaton can be used as a kind of "sequencing" n-ary operator.

References

[BG88] G. Berry and G. Gonthier. *The ESTEREL Synchronous Programming Language: Design, Semantics, Implementation.* Technical Report, 842, INRIA, 1988.

[CHPP87] P. Caspi, N. Halbwachs, D. Pilaud, and J. Plaice. LUSTRE, a declarative language for programming synchronous systems. In *14th Symposium on Principles of Programming Languages*, january 1987.

[Fer90] J.C. Fernandez. An implementation of an efficient algorithm for bisimulation equivalence. *Science of Computer Programming*, 13(2-3), may 1990.

[GBBG85] P. Le Guernic, A. Benveniste, P. Bournai, and T. Gauthier. *Signal: A Data Flow Oriented Language for Signal Processing.* Technical Report, IRISA report 246, IRISA, Rennes, France, 1985.

[Ghe92] G. Gherardi. *Sahara: un environnement de mise au point graphique pour les programmes Esterel (in preparation).* Thesis, Université de Nice, 1992.

[Har87] D. Harel. Statecharts : a visual approach to complex systems. *Science of Computer Programming*, 8:231–275, 1987.

[HGdR88] C. Huizing, R. Gerth, and W.P. de Roever. Modelling statecharts behaviour in a fully abstract way. In 13th *CAAP*, LNCS 299, Springer Verlag, April 22-24 1988.

[HP85] D. Harel and A. Pnueli. On the development of reactive systems. In *Logic and Models of Concurrent Systems, NATO Advanced Study Institute on Logics and Models for Verification and Specification of Concurrent Systems*, NATO ASI series F, Springer Verlag, 1985.

[Hui91] C. Huizing. *Semantics of reactive systems: comparison and full abstraction.* thesis, Eindhoven University, 1991.

[LT89] N. A. Lynch and M. R. Tuttle. *An Introduction to Input/Output Automata.* CWI-Quaterly 3, 1989.

[Mar89] F. Maraninchi. Argonaute, graphical description, semantics and verification of reactive systems by using a process algebra. In *Workshop on Automatic Verification Methods for Finite State Systems*, LNCS 407, Springer Verlag, june 1989.

[Mar91a] F. Maraninchi. *Argos : a Graphical Synchronous Language for the Description of Reactive Systems.* Spectre Report C29, LGI-IMAG, Grenoble, march 1991.

[Mar91b] F. Maraninchi. The argos language: graphical representation of automata and description of reactive systems. In *IEEE Workshop on Visual Languages*, october 1991.

[Mil80] R. Milner. A calculus of communication systems. In *LNCS 92*, Springer Verlag, 1980.

[NRSV90] X. Nicollin, J.L. Richier, J. Sifakis, and J. Voiron. ATP: an algebra for timed processes. In *IFIP Working Conference on Programming Concepts and Methods*, april 1990.

[Par81] D. Park. Concurrency and automata on infinite sequences. In *5th GI-Conference on Theoretical Computer Science*, Springer Verlag, 1981. LNCS 104.

[PS88] A. Pnueli and M. Shalev. *What is in a Step.* Technical Report, Dept. of Applied Mathematics and Computer Science, The Weizmann Institute of Science, Israel, may 1988.

[SYN] Another look at real-time programming. IEEE proceedings, 1991.

Towards a Theory of Actor Computation

(Extended Abstract)

Gul Agha
University of Illinois
agha@cs.uiuc.edu

Ian A. Mason
Stanford University
iam@cs.stanford.edu

Scott Smith
Johns Hopkins University
scott@cs.jhu.edu

Carolyn Talcott
Stanford University
clt@sail.stanford.edu

Abstract

In this paper we present preliminary results of a rigorous development of the actor model of computation. We present an actor language which is an extension of a simple functional language, and provide a precise operational semantics for this extension. Our actor systems are open distributed systems, meaning we explicitly take into account the interface with external components in the specification of an actor system. We define and study various notions of equivalence on actor expressions and systems. We show that the usual tripartite family of testing equivalence relations collapses to two in the presence of fairness. We define a notion of operational bisimulation as a tool for establishing equivalence under fairness assumptions, and illustrate its use.

1. Introduction

The actor model of computation was originally proposed by Hewitt [5]. Actors are self-contained, concurrently interacting entities of a computing system. They communicate via message passing which is asynchronous and fair. Actors can be dynamically created and the topology of actor systems can change dynamically. The actor model is a primitive model of computation, but nonetheless easily expresses a wide range of computation paradigms. It directly supports encapsulation and sharing, and provides a natural extension of both functional programming and object style data abstraction to concurrent open systems [1, 2].

The main features of an open distributed system are that new components can be added, existing components can be replaced, and interconnections can be changed, largely without disturbing the functioning of the system. Components have no control over the components with which they might be connected. The behavior of a component is locally determined by its initial state and the history of its interactions with the environment through its interface. The internal state of a component must only be accessible through operations provided by the interface. Since the actor model has a built-in notion of local component and interface, it is a natural model to use as a basis for a theory of open distributed computation.

The semantics we define combines the message passing features of the primitive actor model with an applicative functional language for describing individual actor behaviors. This provides a simple yet expressive language with a manageable formal semantics. In our formalization we make explicit the notion of open system. An actor system is a collection of individually named concurrently executing actors, plus two collections of actor names that define the interface to the environment. The *receptionists* ρ are names of actors within the system that external components may freely interact with; all other actors in the system are local and thus inaccessible from outside. The *external* actors χ are names of actors that are outside this actor system but to which messages may be directed. Each system is a self-contained entity, and we define operations for composing systems to form larger systems.

Most of the research in the area of parallel language design has either been practical but with very limited formal basis, or has been formal and theoretical but at the expense of realism. Our choices and approach are motivated by a desire to bridge the gap between theory and practice. The theory we develop is intended to be useful for justifying program transformations for real languages, and to formalize intuitive arguments and properties used by programmers.

Following the tradition of [9, 7] we adopt an operational interpretation of actor systems. Actor behavior is defined by a transition relation on configurations. Each configuration is a symbolic instantaneous representation of an actor system with respect to some idealized observer [1], and a transition on configurations maps configurations to possible future ones. Two actor expressions/systems are said to be observationally equivalent if they give rise to the same observations, suitably defined, inside all observing contexts. This notion is closely related to testing equivalence [4]. Observational equivalence provides a semantic basis for developing sound transformation rules.

Some highlights of this paper include the following. The operational semantics extends that of the embedded functional language in such a way that the equational theory of the functional language is preserved. We define a notion of open actor configuration which makes explicit the interface to the environment. As a first step towards an algebra of operations on open system components, we define a composition operator on configurations. An important aspect of the actor model is the fairness requirement: message delivery is guaranteed, and individual actor computations are guaranteed to progress. This makes the computation model more realistic: many intuitively correct equations fail in the absence of fairness. Although fairness makes some aspects of reasoning more complicated, it simplifies others. We prove that in the presence of fairness, the three standard notions of observational equivalence collapse to two. Finally, we give a simple bisimulation principle that allows equivalences to be established in the presence of fairness.

The remainder of this abstract is organized as follows. §2 gives the syntax and operational semantics of our actor language. §3 define various notions of equivalence and state their basic properties. §4 defines a notion of operational bisimulation as a sound approximation to operational equivalence and illustrates its use. §5 contains some concluding remarks.

Notation.

We use the usual notation for set membership and function application. Let Y, Y_0, Y_1 be sets. We specify meta-variable conventions in the form: let y range over Y, which should be read as: the meta-variable y and decorated variants such as y', y_0, ..., range over the set Y. Y^n is the set of sequences of elements of Y of length

n. Y^* is the set of finite sequences of elements of Y. $\bar{y} = [y_1, \ldots, y_n]$ is the sequence of length n with ith element y_i. (Thus $[]$ is the empty sequence.) $u * v$ denotes the concatenation of the sequences u and v. $\mathbf{P}_\omega[Y]$ is the set of finite subsets of Y. $\mathbf{M}_\omega[Y]$ is the set of (finite) multi-sets with elements in Y. $[Y_0 \to Y_1]$ is the set of total functions, f, with domain Y_0 and range contained in Y_1. We write $\mathrm{Dom}(f)$ for the domain of a function and $\mathrm{Rng}(f)$ for its range. $Fmap[Y_0, Y_1]$ is the set of finite maps from Y_0 to Y_1.

2. A Simple Lambda Based Actor Language

Actors are self-contained, components of a computing system that communicate by asynchronous message passing. Message delivery is guaranteed (fairness). In response to a message an actor can send messages to actors that it knows about, and create new actors. It can also change its state/behavior. This change will be in effect when the next message is received by the actor, and the only time an actor's local state changes is when the actor changes it in response to a message. (Local cause for local effect principle).

Our actor language is an extension of the call-by-value lambda calculus that includes (in addition to arithmetic primitives and structure constructors, recognizers, and destructors) primitives for creating and manipulating actors. An actor's behavior is described by a closure which embodies the code to be executed when a message is received, and the local store (values bound to free variables). The actor primitives are: **send** (for sending messages); **become** (for changing behavior); and **newadr** and **initbeh** (for actor creation). $\mathtt{send}(a, v)$ creates a new message with receiver a and contents v and puts the message into the message delivery system. $\mathtt{become}(b)$ clones an anonymous actor to carry out the rest of the current computation, alters the behavior of the actor executing the become to b, and frees that actor to accept another message. The cloned actor may send messages or create new actors in the process of completing its computation, but will never receive any messages as its address can never be known. $\mathtt{newadr}()$ creates a new (uninitialized) actor and returns its address. $\mathtt{initbeh}(a, b)$ initializes the behavior of a newly created actor with address a to be b. An uninitialized actor can only be initialized by the actor which created it. Without this restriction composability of actor systems is problematic, as it would permit an external actor to initialize an internally created actor. The allocation of a new address and initialization of the actor's behavior have been separated in order to allow an an actor to know its own address. This is a weak form of synchronization and would not be necessary if message sending were synchronous. An alternative would be to have built into the semantics that every actor knows its own name, as is done in many actor and object-oriented languages. See [1] for intuitions behind these constructs.

2.1. Syntax

We take as given countable sets $\mathbb{X}$ (variables) and $\mathbb{A}\mathrm{t}$ (atoms). $\mathbb{F}_n$ is the set of primitive operations of rank n and $\mathbb{F} = \bigcup_{n \in \mathbb{N}} \mathbb{F}_n$. We assume $\mathbb{A}\mathrm{t}$ contains $\mathtt{t}, \mathtt{nil}$

for booleans, as well as integers. $\mathbb{F}$ contains arithmetic operations, branching br (rank 3), pairing ispr, pr, 1st, 2nd (ranks 1, 2, 1, 1), and actor primitives newadr, initbeh, send, and become (ranks 0, 2, 2, 1). The sets of expressions, $\mathbb{E}$, value expressions, $\mathbb{V}$, and contexts (expressions with holes), $\mathbb{C}$, are defined inductively as follows.

Definition ($\mathbb{E}$, $\mathbb{V}$, $\mathbb{C}$):

$$\mathbb{V} = \mathbb{A}t \cup \mathbb{X} \cup \lambda\mathbb{X}.\mathbb{E} \cup \mathtt{pr}(\mathbb{V}, \mathbb{V})$$

$$\mathbb{E} = \mathbb{A}t \cup \mathbb{X} \cup \lambda\mathbb{X}.\mathbb{E} \cup \mathtt{app}(\mathbb{E}, \mathbb{E}) \cup \mathbb{F}_n(\mathbb{E}^n)$$

$$\mathbb{C} = \mathbb{A}t \cup \mathbb{X} \cup \lambda\mathbb{X}.\mathbb{C} \cup \mathtt{app}(\mathbb{C}, \mathbb{C}) \cup \mathbb{F}_n(\mathbb{C}^n) \cup \{\varepsilon\}$$

We let x, y, z range over $\mathbb{X}$, v range over $\mathbb{V}$, e range over $\mathbb{E}$, and C range over $\mathbb{C}$. $\lambda x.e$ binds the variable x in the expression e. We write $\mathrm{FV}(e)$ for the set of free variables of e. We write $e\{x := e'\}$ to denote the expression obtained from e by replacing all free occurrences of x by e', avoiding the capture of free variables in e'. Contexts are expressions with holes. We use ε to denote a hole. $C[\![e]\!]$ denotes the result of replacing any holes in C by e. Free variables of e may become bound in this process. let, if and seq are the usual syntactic sugar, seq being a sequencing primitive.

A simple actor behavior b that expects its message to be an actor address, sends the message 5 to that address, and becomes the same behavior, may be expressed using a definable call-by-value fixed-point combinator rec (cf. [7]) as follows.

$$b = \mathtt{app}(\mathtt{rec}, \lambda y.\lambda x.\mathtt{seq}(\mathtt{become}(y), \mathtt{send}(x, 5)))$$

An expression that would create an actor with this behavior and send it some other actor address a is

$$e = \mathtt{let}\{x := \mathtt{newadr}()\}\mathtt{seq}(\mathtt{initbeh}(x, b), \mathtt{send}(x, a)).$$

The behavior of a sink, an actor that ignores its messages and becomes this same behavior, is defined by

$$\mathtt{sink} = \mathtt{rec}(\lambda b.\lambda m.\mathtt{become}(b)).$$

2.2. Reduction Semantics for Open Configurations

We give the semantics of actor expressions by defining a transition on open configurations. Open configurations describe actor systems in which addresses of some (but not necessarily all) of the actors are known to the outside world. These actors are called receptionists. An open configuration may also know addresses of some actors in the outside world. These actors are called external actors. The sets of receptionists and external actors are the interface of an actor system to its environment. They specify what actors are visible and what actor connections must be provided for the system to function. The set of receptionists may grow and the set of required external connections may shrink as the system evolves. In addition, an open configuration contains an actor map and a multi-set of pending messages. An actor map is a finite map from actor addresses to actor states. An actor state is either uninitialized (having been newly created by an actor, a) written $(?_a)$; ready to accept a message, written (b) where b is its behavior, a lambda abstraction; or busy executing e, written $[e]$, here e represents the actor's current (local) processing state. A message contains the address of the actor to whom it is sent and the message contents. The contents can be any value constructed from atoms and actor addresses using constructors.

Lambda abstractions and constructions containing lambda abstractions are not allowed to be communicated in messages. There are two reasons for this restriction. Firstly, allowing lambda abstractions to be communicated in values violates the actor principle that only an actor can change its own behavior, because a `become` in a lambda message may change the receiving actor behavior. Secondly, if lambda abstractions are communicated to external actors, there is no reasonable way to control what actor addresses are actually exported. This has unpleasant consequences in reasoning about equivalence, amongst other things. This restriction is not a serious limitation since the address of an actor whose behavior is the desired lambda abstraction can be passed in a message. Thirdly, if lambda abstractions can be communicated in messages then syntactic extensions to the language that involve transformations such as CPS can not be done on a per actor basis, since it would requires transformation of code that might arrive in a message. We classify transitions as internal or external. The internal transitions of a configuration are:

(1) an actor executing a step of its current computation;

(2) an actor initializing the behavior of a newly created actor; and

(3) acceptance of a message by an actor not currently busy computing.

The transitions of class (1) involve a single actor. They may be purely internal (a λ-transition), or messages may be sent, or a new actor may be created. The transitions of class (2) involve two actors, and the initialized actor becomes ready to accept a message. The transitions of class (3) involve an actor and a message. The message is consumed and the actor becomes busy. In addition to the internal transitions of a configuration, there are transitions that correspond to interactions with external agents:

(4) arrival of a message to a receptionist from the outside; and

(5) passing a message out to an external actor.

We assume that we are given a countable set $\mathbb{A}\mathrm{d}$ of actor addresses. To simplify notation, we identify $\mathbb{A}\mathrm{d}$ with $\mathbb{X}$. This pun is useful for two reasons: it allows us to use expressions to describe actor states and message contents; and it allows us to avoid problems of choice of names for newly created actors by appealing to an extended form of alpha conversion. (See [7] for use of this pun to represent reference cells.)

Definition ($\mathrm{c}\mathbb{V}$, $\mathbb{A}\mathrm{s}$, $\mathbb{M}$): The set of communicable values, $\mathrm{c}\mathbb{V}$, the set of actor states, $\mathbb{A}\mathrm{s}$, and the set of messages, $\mathbb{M}$, are defined as follows.

$$\mathrm{c}\mathbb{V} = \mathbb{A}\mathrm{t} \cup \mathbb{X} \cup \mathtt{pr}(\mathrm{c}\mathbb{V}, \mathrm{c}\mathbb{V}) \qquad \mathbb{A}\mathrm{s} = (?_{\mathbb{X}}) \cup (\mathbb{L}) \cup [\mathbb{E}] \qquad \mathbb{M} = \langle \mathbb{X} \Leftarrow \mathrm{c}\mathbb{V} \rangle$$

We let cv range over $\mathrm{c}\mathbb{V}$.

Definition (Actor Configurations): An actor configuration with actor map, α, multi-set of messages, μ, receptionists, ρ, and external actors, χ, is written

$$\left\langle\!\left\langle \alpha \mid \mu \right\rangle\!\right\rangle_{\chi}^{\rho}$$

where $\rho, \chi \in \mathbf{P}_\omega[\mathbf{X}]$, $\alpha \in Fmap[\mathbf{X}, \mathbf{As}]$, and $\mu \in \mathbf{M}_\omega[\mathbf{M}]$. Further, it is required that, letting $A = \mathrm{Dom}(\alpha)$, the following constraints are satisfied:

(0) $\rho \subseteq A$ and $A \cap \chi = \emptyset$,

(1) if $a \in A$, then $\mathrm{FV}(\alpha(a)) \subseteq A \cup \chi$, and if $\alpha(a) = (?_{a'})$, then $a' \in A$,

(2) if $<a \Leftarrow v> \in \mu$, then $\mathrm{FV}(v) \cup \{a\} \subseteq A \cup \chi$.

We let κ range over actor configurations. A configuration in which both the receptionist and external actor sets are empty is said to be *closed*. For closed configurations we may omit explicit mention of the empty sets. The actor map portion of a configuration is presented as a list of actor states each subscripted by the actor address which is mapped to this state. $\alpha, (b)_a$ denotes the map α' such that $\mathrm{Dom}(\alpha') = \mathrm{Dom}(\alpha) \cup \{a\}$, $\alpha'(a) = (b)$, and $\alpha'(a') = \alpha(a)$ if $a' \neq a$. Similarly for other states subscripted with addresses. We use _ to denote a fresh address whose actual name we do not care about. Such addresses refer to actors not known to any other actors (anonymous actors). In a configuration, there may be multiple occurrences of actor states with address represented by _. These are in fact distinct, and simply reflect that the choice of address is irrelevant.

The set of possible computations of an actor configuration is defined in terms of the transition relation $\mapsto$ on configurations. To describe the internal transitions other than message receipt, an expression is decomposed into a reduction context filled with a redex. Reduction contexts are expressions with a unique hole, that play the role of continuations in abstract machine models of sequential computation. We have defined the decomposition to correspond to a left-most, outer-most, call-by-value evaluation order, thus preserving the semantics of the embedded functional language. Decomposition of non-value expressions is unique. Thus, locally computation is deterministic.

Definition ($\mathbb{E}_{\mathrm{rdx}}$,$\mathbb{R}$): The set of redexes, $\mathbb{E}_{\mathrm{rdx}}$, and the set of reduction contexts, $\mathbb{R}$, are defined by

$$\mathbb{E}_{\mathrm{rdx}} = \mathrm{app}(\mathbb{V}, \mathbb{V}) \cup (\mathbb{F}_n(\mathbb{V}^n) - \mathrm{pr}(\mathbb{V}, \mathbb{V}))$$

$$\mathbb{R} = \{\varepsilon\} \cup \mathrm{app}(\mathbb{R}, \mathbb{E}) \cup \mathrm{app}(\mathbb{V}, \mathbb{R}) \cup \mathbb{F}_{n+m+1}(\mathbb{V}^n, \mathbb{R}, \mathbb{E}^m)$$

We let R range over $\mathbb{R}$.

Redexes can be split into two classes: purely functional and actor redexes. Reduction rules for the purely functional case are given by a relation $\overset{\lambda}{\mapsto}$ on expressions. They correspond to the usual operational semantics for the purely functional fragment of our actor language and we omit them from this abstract. The actor redexes are: `newadr()`, `initbeh(a, b)`, `become(b)`, and `send(a, v)`.

Definition ($\mapsto$): The single-step transition relation $\mapsto$, on actor configurations is the least relation satisfying the following conditions.

$$\mathrm{f}(a) \qquad e \overset{\lambda}{\mapsto} e' \Rightarrow \left\langle\!\left\langle \alpha, [e]_a \mid \mu \right\rangle\!\right\rangle_{\chi}^{\rho} \mapsto \left\langle\!\left\langle \alpha, [e']_a \mid \mu \right\rangle\!\right\rangle_{\chi}^{\rho}$$

$$\texttt{n}(a,a) \qquad \left\langle\!\!\left\langle \alpha, [R[\![\texttt{newadr()}]\!]]_a \mid \mu \right\rangle\!\!\right\rangle^{\rho}_{\chi} \mapsto \left\langle\!\!\left\langle \alpha, [R[\![a']\!]]_a, (?_a)_{a'} \mid \mu \right\rangle\!\!\right\rangle^{\rho}_{\chi} \qquad a' \text{ fresh}$$

$$\texttt{c}(a) \qquad \left\langle\!\!\left\langle \alpha, [R[\![\texttt{initbeh}(a',b)]\!]]_a, (?_a)_{a'} \mid \mu \right\rangle\!\!\right\rangle^{\rho}_{\chi} \mapsto \left\langle\!\!\left\langle \alpha, [R[\![\texttt{nil}]\!]]_a, (b)_{a'} \mid \mu \right\rangle\!\!\right\rangle^{\rho}_{\chi}$$

$$\texttt{b}(a) \qquad \left\langle\!\!\left\langle \alpha, [R[\![\texttt{become}(b)]\!]]_a \mid \mu \right\rangle\!\!\right\rangle^{\rho}_{\chi} \mapsto \left\langle\!\!\left\langle \alpha, [R[\![\texttt{nil}]\!]]_{\_}, (b)_a \mid \mu \right\rangle\!\!\right\rangle^{\rho}_{\chi}$$

$$\texttt{s}(a,a',cv) \qquad \left\langle\!\!\left\langle \alpha, [R[\![\texttt{send}(a',cv)]\!]]_a \mid \mu \right\rangle\!\!\right\rangle^{\rho}_{\chi} \mapsto \left\langle\!\!\left\langle \alpha, [R[\![\texttt{nil}]\!]]_a \mid \mu, \texttt{<}a' \Leftarrow cv\texttt{>} \right\rangle\!\!\right\rangle^{\rho}_{\chi}$$

$$\texttt{r}(a,cv) \qquad \left\langle\!\!\left\langle \alpha, (b)_a \mid \texttt{<}a \Leftarrow cv\texttt{>}, \mu \right\rangle\!\!\right\rangle^{\rho}_{\chi} \mapsto \left\langle\!\!\left\langle \alpha, [\texttt{app}(b,cv)]_a \mid \mu \right\rangle\!\!\right\rangle^{\rho}_{\chi}$$

$$\texttt{o}(a,cv) \qquad \left\langle\!\!\left\langle \alpha \mid \texttt{<}a \Leftarrow cv\texttt{>}, \mu \right\rangle\!\!\right\rangle^{\rho}_{\chi} \mapsto \left\langle\!\!\left\langle \alpha \mid \mu \right\rangle\!\!\right\rangle^{\rho'}_{\chi}$$

where $\rho' = \rho \cup (\mathrm{FV}(cv) \cap \mathrm{Dom}(\alpha))$ and $a \in \chi$

$$\texttt{i}(a,cv) \qquad \left\langle\!\!\left\langle \alpha \mid \mu \right\rangle\!\!\right\rangle^{\rho}_{\chi} \mapsto \left\langle\!\!\left\langle \alpha \mid \mu, \texttt{<}a \Leftarrow cv\texttt{>} \right\rangle\!\!\right\rangle^{\rho}_{\chi \cup (\mathrm{FV}(cv) - \mathrm{Dom}(\alpha))}$$

provided $a \in \rho$ and $\mathrm{FV}(cv) \cap \mathrm{Dom}(\alpha) \subseteq \rho$

Note that in the last four rules the message contents are restricted to be communicable values. $\overset{*}{\mapsto}$ is the transitive reflexive closure of $\mapsto$. The configurations reachable from a given configuration κ are those configurations κ' such that $\kappa \overset{*}{\mapsto} \kappa'$.

The transitions are labelled to facilitate some technical definitions. We write $\kappa_0 \xrightarrow{l} \kappa_1$ if $\kappa_0 \mapsto \kappa_1$ according to the rule encoded by l. We say l is enabled in configuration κ if there is some κ' such that $\kappa \xrightarrow{l} \kappa'$.

Definition (Computation trees and paths): If κ is a configuration, then we define $\mathcal{T}(\kappa)$ to be the set of all finite sequences of labeled transitions of the form $[\kappa_i \xrightarrow{l_i} \kappa_i' \mid i < n]$ for some $n \in \mathbb{N}$ such that $\kappa_0 = \kappa$ and $\kappa_i' = \kappa_{i+1}$ for $i < n-1$. We call such sequences *nodes* and let ν range over nodes. We order nodes of a tree by the subtree relation: $\nu_0 < \nu_1$ iff ν_0 is below (properly extends) ν_1. A computation path for κ is a maximal linearly ordered set of nodes in $\mathcal{T}(\kappa)$. Note that a computation path can also be regarded as a (possibly infinite) sequence of transitions. We let π range over computation paths and use $\mathcal{T}^{\infty}(\kappa)$ to denote the set of all such π for $\mathcal{T}(\kappa)$.

We now rule out those computations that are unfair, i.e. those that either starve out a particular actor computation, or keep a message queued forever when the receiving actor is either external or has infinitely often been ready to receive a message.

Definition (Fair computation paths): A computation path $\pi = [\nu_i \xrightarrow{l_i} \nu_{i+1} \mid i \in I]$ in the computation tree $\mathcal{T}(\kappa)$ is fair if each enabled transition eventually happens or becomes permanently disabled. That is, if l is enabled in κ_i then $\kappa_j \xrightarrow{l} \kappa_{j+1}$ for some $j \geq i$, or l has the form $\texttt{r}(a,cv)$ and for some $j \geq i$ a is busy and never again becomes ready to accept a message. For a configuration κ we define $\mathcal{T}_{\mathrm{f}}^{\infty}(\kappa)$ to be the subset of $\mathcal{T}^{\infty}(\kappa)$ that are fair.

Note that finite computation paths are fair, since all of the enabled transitions must have happened.

Actor systems compose well, as indicated by the following definition and theorem.

Definition (Composition of Open Configurations): Two open configurations $\kappa_i = \left\langle\!\left\langle \alpha_i \mid \mu_i \right\rangle\!\right\rangle^{\rho_i}_{\chi_i}$, $i < 2$ are composable if $\mathrm{Dom}(\alpha_0) \cap \mathrm{Dom}(\alpha_1) = \emptyset$. The composition $\kappa_0 \cup \kappa_1$ is defined by

$$\kappa_0 \cup \kappa_1 = \left\langle\!\left\langle \alpha_0 \cup \alpha_1 \mid \mu_0 \cup \mu_1 \right\rangle\!\right\rangle^{(\rho_0 \cup \rho_1)-(S_0 \cup S_1)}_{(\chi_0 \cup \chi_1)-(S_0 \cup S_1)}$$

where $S_0 = \chi_0 \cap \rho_1$ and $S_1 = \chi_1 \cap \rho_0$.

Theorem (Composition of Open Configurations): There exists a binary operation $\mathcal{M}$ on computation trees such that if κ_i are composable configurations then

$$\mathcal{T}(\kappa_0 \cup \kappa_1) = \mathcal{M}(\mathcal{T}(\kappa_0), \mathcal{T}(\kappa_1))$$

where S_0, S_1, and $\kappa_0 \cup \kappa_1$ are as above.

In brief, the operation $\mathcal{M}$ merges pairs of computations that have matching i/o transitions for those external actors of one system that are (identified with) receptionists of the other system. Note that this theorem fails if arbitrary actors are allowed to initialize the behavior of newly created actors.

3. Notions of Equivalence for Actors

Two forms of equivalence are given, one for expressing the equivalence of actor expressions, and another for expressing the equivalence of actor configurations. We base our notion of equivalence on the now classic *operational equivalence* of [9]. For the deterministic functional languages of the sort Plotkin studied, this equality is defined as follows. Two program expressions are said to be equivalent if they behave the same when placed in any observing context, where an observing context is some complete program with a hole, such that all of the free variables in the expressions being observed are captured when the expressions are placed in the hole. The notion of "behave the same" is (for deterministic functional languages) typically equi-termination, i.e. either both converge or both diverge.

3.1. Equivalence of actor expressions

We first define equivalence of actor expressions, the equivalence of actor configurations will be defined later. The first step is to find proper notions of "observing context" and "behave the same" in an actor setting. For actor expressions, the analogue of observing context is an observing actor configuration that contains an actor with a hole in which the expression to be observed is placed. Since termination is not particularly relevant for actor configurations, we instead introduce an observer primitive, **event** and observe whether or not in a given computation, **event** is executed. This approach is similar to that used in testing equivalence for CCS [4]. Since the language is nondeterministic, three different observations may be made instead of two: either **event** occurs for all possible executions, it occurs in some executions but not others, or it never occurs.

Formally, the language of observing contexts is obtained by introducing a new 0-ary primitive operator, **event**. We extend the reduction relation $\mapsto$ by adding the following rule.

$$\mathrm{e}(a) \quad \left\langle\!\left\langle \alpha, \,[R[\![\mathtt{event}()]\!]]_a \mid \mu \right\rangle\!\right\rangle^{\rho}_{\chi} \mapsto \left\langle\!\left\langle \alpha, \,[R[\![\mathtt{nil}]\!]]_a \mid \mu \right\rangle\!\right\rangle^{\rho}_{\chi}$$

For an expression e, the observing configurations are configuration contexts of the form $\langle\!\langle \alpha, [C[\![\,]\!]]_a \mid \mu \rangle\!\rangle$ over the extended language, such that filling the hole in $C[\![\,]\!]$ with e results in a closed configuration. (Let $\mathbb{K}$ be the set of configuration contexts (configurations with holes), and let K range over $\mathbb{K}$.

We observe **event** transitions in the fair computation paths. We say that a computation path succeeds (s) if an **event** transition occurs in it, otherwise it fails (f). $obs(\pi)$ is the s/f observation of a single complete computation π, and $Obs(\kappa)$ is the set of observations possible for a closed actor configuration.

Definition (observations): Let κ be a configuration of the extended language, and let $\pi = [\kappa_i \xrightarrow{l_i} \kappa_{i+1} \mid i \in I]$ be a fair computation path, i.e. $\pi \in \mathcal{T}_f^{\infty}(\kappa)$. Define

$$obs(\pi) = \begin{cases} \mathrm{s} & \text{if } (\exists \kappa_0, \kappa_1, a)(\kappa_0 \xrightarrow{\mathrm{e}(a)} \kappa_1 \in \pi) \\ \mathrm{f} & \text{otherwise} \end{cases}$$

$$Obs(\kappa) = \begin{cases} \mathrm{s} & \text{if } (\forall \pi \in \mathcal{T}_f^{\infty}(\kappa))(obs(\pi) = \mathrm{s}) \\ \mathbf{sf} & \text{if } (\exists \pi \in \mathcal{T}_f^{\infty}(\kappa))(obs(\pi) = \mathrm{s}) \text{ and } (\exists \pi \in \mathcal{T}_f^{\infty}(\kappa))(obs(\pi) = \mathbf{f}) \\ \mathbf{f} & \text{if } (\forall \pi \in \mathcal{T}_f^{\infty}(\kappa))(obs(\pi) = \mathbf{f}) \end{cases}$$

The natural notion of operational equivalence is that equal observations are made in all closing configuration contexts. It may be desirable in some cases to consider using a weaker equality, however. An **sf** observation may be considered as good as an s observation, and a new equivalence arises if these observations are equated. Similarly, an **sf** observation may be as bad as an **f** observation. We may thus define the following three equivalences.

Definition ($\cong_{1,2,3}$):

(1) $e_0 \cong_1 e_1$ (testing or convex or Plotkin or Egli-Milner) iff $Obs(K[\![e_0]\!]) = Obs(K[\![e_1]\!])$ for all closing configuration contexts K

(2) $e_0 \cong_2 e_1$ (must or upper or Smyth) iff $Obs(K[\![e_0]\!]) =_{(\mathrm{sf}=\mathrm{f})} Obs(K[\![e_1]\!])$ for all closing configuration contexts K

(3) $e_0 \cong_3 e_1$ (may or lower or Hoare) iff $Obs(K[\![e_0]\!]) =_{(\mathrm{sf}=\mathrm{s})} Obs(K[\![e_1]\!])$ for all closing configuration contexts K

where $x =_{(o=o')} y$ iff $x = y$ or $x, y \in \{o, o'\}$.

Note that may-equivalence ($\cong_3$) depends only on the computation trees, not on the choice of paths admitted as computations, because all **events** are observed after some finite amount of time. This means it is independent of whether or not fairness is required. Since fairness sometimes makes proving equivalences more difficult, it is useful that may-equivalence can always be proved ignoring the fairness assumption. The other two equalities are sensitive to choice of paths admitted as computations, in particular when fairness is required, as in our model, $\cong_2$ is in fact the same as $\cong_1$. In models without the fairness requirement, they are distinct. In either case, $\cong_3$ is distinct from $\cong_1$ and $\cong_2$.

Theorem (partial collapse):

(1=2) $e_0 \cong_2 e_1$ iff $e_0 \cong_1 e_1$

(1-3) $e_0 \cong_1 e_1$ implies $e_0 \cong_3 e_1$

(3-1) $e_0 \cong_3 e_1$ does not imply $e_0 \cong_1 e_1$

Proof (partial collapse):

2-1 $\cong_1$ implies $\cong_2$ follows from the definitions. The key to showing that $\cong_2$ implies $\cong_1$ is the observation that if $Obs(K[\![e_0]\!]) = \mathbf{f}$ and $Obs(K[\![e_1]\!]) = \mathbf{sf}$ it is always possible to construct a K^* such that $Obs(K^*[\![e_0]\!]) = \mathrm{s}$, and $Obs(K^*[\![e_1]\!]) = \mathbf{sf}$. To see this, suppose that K satisfies the hypothesis. Form K' by replacing all occurrences of **event**() in K by **send**(a, **nil**) for some fresh variable a. Let K^* be obtained by adding to K' a message $<a \Leftarrow \mathtt{t}>$ and an actor a where a has the following behavior: If a receives the message t, it executes **event**() and becomes a sink, and if a receives the message nil, it just becomes a sink. Recall that a sink is an actor that ignores its message and becomes a sink. We claim K^* is the desired observing context. If $K[\![e_0]\!]$ never executes **event**(), then in any fair complete computation, the t message will be received by a, so $K^*[\![e_0]\!]$ will always execute **event**(). If $K[\![e_1]\!]$ executes **event**() in some computation, then in the corresponding computations for $K^*[\![e_1]\!]$, sometimes nil will be received by a before t is received and sometimes it won't, hence $K^*[\![e_1]\!]$ will execute **event**() in some computations, but not in all. $\Box_{2-1}$

1-3 from the definitions. $\Box_{1-3}$

3-1 We construct expressions e_0, e_1 such that $e_0 \cong_3 e_1$, but $\neg(e_0 \cong_2 e_1)$. Let e_0 create an actor that sends a message (say nil) to an external actor a and becomes a sink, and let e_1 create an actor that may or may not send a message nil to a depending on a coin flip (there are numerous methods of constructing coin flipping actors), and also then becomes a sink. Let K be an observing configuration context that with an actor a that executes **event** just if nil is received. Then $Obs(K[\![e_0]\!]) = \mathrm{s}$ but $Obs(K[\![e_1]\!]) = \mathbf{sf}$, so $\neg(e_0 \cong_2 e_1)$. To show that $e_0 \cong_3 e_1$, show for arbitrary K that some path in the computation of $K[\![e_0]\!]$ contains an event iff some path in the computation of $K[\![e_1]\!]$ contains an event. This is easy, because when e_1's coin flip indicates nil is sent, the computation proceeds identically to e_0's computation. $\Box_{3-1}$

$\Box$

Hereafter, $\cong$ (operational equivalence) will be used as shorthand for either $\cong_1$ or $\cong_2$. A possibly useful analogy is that $\cong_3$ corresponds to partial correctness and $\cong$ corresponds to total correctness.

The fairness requirement is critical in the proof of **(2-1)**. For example in CCS, where fairness is not assumed, no such collapse of $\cong_2$ to $\cong_1$ occurs. So, although fairness complicates some aspects of the theory, it simplifies others. If we omitted the fairness requirement we could make more $\cong$-distinctions between actors. For example, let a_0 be a sink. Let a_1 be an actor that also ignores its messages and becomes the same behavior, but it continues executing an infinite loop. The infinite looping actor could starve out the rest of the configuration, but in the presence of fairness no such starvation can occur, so the two are equal.

Since our reduction rules preserve the evaluation semantics of the embedded functional language, many of the equational laws for this language (cf. [11]) continue to hold in the full actor language. For example, operational equivalence is a congruence and the laws concerning lambda abstraction and application continue to hold.

Theorem (lambda laws):

(cong) $e_0 \cong e_1 \Rightarrow C[\![e_0]\!] \cong C[\![e_1]\!]$

(betav) $\mathtt{let}\{x := v\}e = (\lambda x.e)(v) \cong e\{x := v\}$

(app) $e_0(e_1) \cong (\lambda f.f(e_1))(e_0) = \mathtt{let}\{f := e_0\}f(e_1)$

(cmps) $f(g(e)) \cong (\lambda x.f(g(x)))(e) = (f \circ g)(e)$

(id) $\mathtt{let}\{x := e\}x = (\lambda x.x)(e) \cong e$

The proof of this theorem uses the notion of bisimulation (cf. the next section).

3.2. Equivalence of actor configurations

Equivalence is now defined for open actor configurations $\left\langle\!\left\langle \alpha \mid \mu \right\rangle\!\right\rangle^{\rho}_{\chi}$. As with actor expressions, we wish to close the open configuration by adding observers. This produces a notion of equivalence for actor configurations that is closely connected with equivalence of actor expressions.

Definition (Closing an Actor Configuration): A closing of an actor configuration $\kappa = \left\langle\!\left\langle \alpha \mid \mu \right\rangle\!\right\rangle^{\rho}_{\chi}$ is defined to be an actor configuration $\kappa' = \left\langle\!\left\langle \alpha' \mid \mu' \right\rangle\!\right\rangle^{\chi}_{\rho}$, in the extended language, composable with κ.

Definition ($\cong_s$): $\kappa_0 = \left\langle\!\left\langle \alpha_0 \mid \mu_0 \right\rangle\!\right\rangle^{\rho}_{\chi} \cong_s \left\langle\!\left\langle \alpha_1 \mid \mu_1 \right\rangle\!\right\rangle^{\rho}_{\chi} = \kappa_1$ iff $Obs(\kappa_0 \cup \kappa') = Obs(\kappa_0 \cup \kappa')$ for all actor configurations κ' closing κ_j, $j < 2$.

Theorem ($\cong$ / $\cong_s$): If $e_0 \cong e_1$ and $e_0' \cong e_1'$, then

$$\left\langle\!\left\langle \alpha, [C[\![e_0]\!]]_a, (\lambda x.C'[\![e_0']\!])_{a'} \mid \mu \right\rangle\!\right\rangle^{\rho}_{\chi} \cong_s \left\langle\!\left\langle \alpha, [C[\![e_1]\!]]_a, (\lambda x.C'[\![e_1']\!])_{a'} \mid \mu \right\rangle\!\right\rangle^{\rho}_{\chi}.$$

Note that while two closed configurations (configurations that have no receptionists and no external actors) cannot be distinguished by any external observation, two closed expressions can be distinguished by a behavior context that makes use of the values returned.

4. Operational Bisimulations

Given two computation trees $\mathcal{T}(\kappa_0)$ and $\mathcal{T}(\kappa_1)$ of actor configurations κ_0 and κ_1, we define the notion of an operational bisimulation, $R \subseteq \mathcal{T}(\kappa_0) \times \mathcal{T}(\kappa_1)$, in such a way as to ensure that if two computation trees, $\mathcal{T}(\kappa_0)$ and $\mathcal{T}(\kappa_1)$, are operationally bisimilar, then $\kappa_0 \cong_s \kappa_1$. We view operational bisimulation as a proof technique, not as an alternative notion of equivalence. To keep the notation somewhat under control, we shall treat a simple case in this paper. A computation tree is *non-expansive* iff the set of receptionists never increases. We say a configuration is *non-expansive* iff its computation tree is. In what follows we restrict our attention to non-expansive trees. Extending the results to expansive configurations poses only notational complications (e.g identifying newly created actors), and the non-expansive case suffices to prove operational equivalences.

The definition of an operational bisimulation requires a little notation. Firstly, an $R \subseteq \mathcal{T}(\kappa_0) \times \mathcal{T}(\kappa_1)$, naturally extends to an $R \subseteq \mathcal{T}^\infty(\kappa_0) \times \mathcal{T}^\infty(\kappa_1)$ as follows. For $\pi_i \in \mathcal{T}^\infty(\kappa_i)$ for $i < 2$

$$\pi_0 R \pi_1 \quad \text{iff} \quad (\forall \nu_0' \in \pi_0)(\forall \nu_1' \in \pi_1)(\exists \nu_0'' \in \pi_0)(\exists \nu_1'' \in \pi_1)(\nu_0'' < \nu_0' \wedge \nu_1'' < \nu_1' \wedge \nu_0'' R \nu_1'')$$

(Recall that $\nu < \nu'$ is the subtree relation on nodes.) Secondly, for $\nu = [\kappa_i \xrightarrow{l_i} \kappa_i' \mid i < n]$, the i/o restriction $(\nu)_{\mathrm{io}}$, is defined to be $(l_0)_{\mathrm{io}} * (l_1)_{\mathrm{io}} * \ldots * (l_{n-1})_{\mathrm{io}}$, where

$$(l)_{\mathrm{io}} = \begin{cases} [\mathrm{o}(a, cv)] & \text{if } l = \mathrm{o}(a, cv) \\ [\mathrm{i}(a, cv)] & \text{if } l = \mathrm{i}(a, cv) \\ [\mathrm{e}()] & \text{if } (\exists a)(l = \mathrm{e}(a)) \\ [] & \text{otherwise} \end{cases}$$

Definition (operational bisimulation): Given two *non-expansive* actor configurations, κ_0 and κ_1, we say that a relation, $R \subseteq \mathcal{T}(\kappa_0) \times \mathcal{T}(\kappa_1)$, is an operational bisimulation iff the following conditions hold:

(1) $[] R []$

(2) $(\forall \nu_0 \nu_1)(\nu_0 R \nu_1 \Rightarrow (\forall \nu_0' < \nu_0)(\exists \nu_1')(\nu_0' R \nu_1' \wedge \nu_1' \leq \nu_1))$

(3) $(\forall \nu_0 \nu_1)(\nu_0 R \nu_1 \Rightarrow (\forall \nu_1' < \nu_1)(\exists \nu_0')(\nu_0' R \nu_1' \wedge \nu_0' \leq \nu_0))$

(4) $(\forall \nu_0' \nu_1')(\nu_0' R \nu_1' \Rightarrow (\nu_0')_{\mathrm{io}} = (\nu_1')_{\mathrm{io}})$

(5) If $\pi_0 R \pi_1$, then π_0 is *fair* iff π_1 is *fair*.

A few simple consequences of this definition should be made clear. Suppose that

$$\kappa_i = \left\langle\!\!\left\langle\, \alpha_i \mid \mu_i \,\right\rangle\!\!\right\rangle^{\rho_i}_{\chi_i} \qquad i < 2$$

are operationally bisimilar. Then $\rho_0 = \rho_1$. The same can not quite be said for χ_0 and χ_1. Since there may be members of these sets that never ever receive any messages. Modulo this sort of garbage, these two sets must be the same.

Lemma (bisimilar composition): Suppose that κ_i are are operationally bisimilar via R, and κ is a composable configuration (i.e its actors are disjoint from those in κ_i). Then there is a relation $R' \subseteq \mathcal{M}(\mathcal{T}(\kappa_0), \mathcal{T}(\kappa)) \times \mathcal{M}(\mathcal{T}(\kappa_1), \mathcal{T}(\kappa))$ which makes $\mathcal{M}(\mathcal{T}(\kappa_0), \mathcal{T}(\kappa))$ operationally bisimilar to $\mathcal{M}(\mathcal{T}(\kappa_1), \mathcal{T}(\kappa))$.

The following is the most important consequence of bisimulation.

Theorem (opeq): If two non-expansive computation trees, $\mathcal{T}(\kappa_0)$ and $\mathcal{T}(\kappa_1)$, are operationally bisimilar, then $\kappa_0 \cong_s \kappa_1$.

Proof (opeq): Suppose that $\kappa_i = \left\langle\!\left\langle \alpha_i \mid \mu_i \right\rangle\!\right\rangle^{\rho}_{\chi}$, $i < 2$ are operationally bisimilar via R and that κ is a closing configuration. Then by the composition theorem we have that $\mathcal{T}(\kappa_i \cup \kappa) = \mathcal{M}(\mathcal{T}(\kappa_i), \mathcal{T}(\kappa))$, and by the bisimilar composition theorem we have that $\mathcal{T}(\kappa_0 \cup \kappa)$ is operationally bisimilar to $\mathcal{T}(\kappa_1 \cup \kappa)$ via R^J. This is easily seen to imply that $Obs(\kappa_0 \cup \kappa) = Obs(\kappa_1 \cup \kappa)$ $\Box_{\mathbf{opeq}}$

4.1. Example: Removal of Message Indirection

To illustrate how bisimulations can be used to prove two open configurations equivalent, we apply a transformation that removes indirection in message transmission. We begin with system 0, a two actor system with one receptionist r and one reference to an external actor x. The receptionist r takes requests for transforming data, applies some operation, f, and sends the result to the other internal actor a. This actor applies a second operation g and sends the result back to the receptionist, who passes it on, unchanged, to the external actor. This system is transformed into system 1, in which the second actor returns its results directly to the external actor. Equivalence is proved by establishing an operational bisimulation between the two systems, in which related nodes of the computation trees are constrained to be 'in step'. This approach also works for transformations such as fusion or splitting of internal actors [12]. We outline the construction of the bisimulation informally. We call the original system S_0 and the transformed system S_1. We let K_0, K_1 be the initial configurations of the S_0, S_1, respectively. The actors r and a each have only two "interesting" states:

- B_r – the initial behavior of r,
- $B_r[m]$ – the state in which r is processing message m,
- B^i_a – the initial behavior of a, in S_i,
- $B^i_a[m]$ – the state in which a is processing message m, in S_i.

The difference between the behaviors of a in the two systems is the address for sending replies (r in S_0 and x in S_1). When r receives a request or reply message m it sends a message and returns to its initial state. The message sent is $<a \Leftarrow f(v)>$ if m is a request with data v, and $<x \Leftarrow m>$ if m is a reply. When a receives a request message m with data v it sends a reply message with data $g(v)$ and returns to its initial state. In S_0 the reply is sent to r and in S_1 the reply is sent to x.

Now we describe when a S_0 configuration is in step with aS_1 configuration. To do this we define a redirection map X on messages. If m is a reply message with receiver r then $X(m)$ is a reply message with the same data as m and receiver x. Configurations κ_j reachable from K_j are said to be in step if a is in the same state in both configurations and one of the following four cases holds.

(1) the state of r is the same in both configurations, and the pending messages of κ_1 are the image under X of the pending messages of κ_0.

(2) r is processing a reply message m in κ_0 and waiting in κ_1, and the pending messages of κ_1 are the image under X of the pending messages of κ_0 plus a reply m to x.

(3) r is waiting in κ_0 and is processing a request message m in κ_1, and the pending messages of κ_1 are the image under X of the result of removing m from the pending messages of κ_0.

(4) r is processing a reply message m_0 in κ_0 and is processing a request message m_1 in κ_1, and the pending messages of κ_1 are the result of adding a reply m_0 to x to the image under X of the result of removing m_1 from the pending messages of κ_0.

The bisimulation $\mathcal{R}$ is now essentially determined by the clauses (1–4) of the definition and the requirement that $\mathcal{R}$ related configurations must be in step.

5. Discussion

This extended abstract outlines a first step towards a general theory for specification, interconnection, and transformation of components of actor systems. The full paper fills in details including a full set of transition rules and proofs of theorems. Our next task is to develop a logic for specifying components of actor systems, methods for verifying that programs implementing components meet their specifications, and methods for refining specifications into implementations. In addition, we plan to develop methods for modularizing specifications and combining components to build complex systems from simpler systems.

We contrast our work with three related efforts: CSP / Occam [6], the π-calculus [8], and Concurrent ML [3]. The CSP / Occam model is very restrictive. Occam assumes a fixed interconnection topology of processes, supports only static storage allocation, and disallows recursive procedures.

Many of the aims Milner and others had in developing the π-calculus are the same as ours, namely to formulate a language for concurrent computation that allows treatment of data channels as first-class objects, and furthermore for which an algebraic theory may be developed. Equally important, however, are how our aims differ. We aim for a model that can be regarded as a realistic model of a real language, not just an abstract calculus.

We believe realistic models must incorporate fairness assumptions, otherwise the model is impoverished by a collection of starving processes that have been enabled and in any realistic implementation would not be starving. Realistic models also must account for the inherently open nature of distributed systems, but the π-calculus only partially accounts for this: it is impossible to extrude a local port name to an external process. In addition, both CCS and the π-calculus treat senders and receivers uniformly, meaning there can be multiple receivers. However, this means a local receiving process can be corrupted by an external process that also receives on the same port. Lack of locality in these model may cause problems similar to those encountered in denotational models of reference/block structure in higher order languages.

We also contrast our work with more practical development of π-calculus-style communications primitives found in Berry, Milner, and Turner's effort [3], and Reppy's CML [10]. For our purposes we equate these presentations. Even though these theories are more realistic because the languages include other constructs such as functions and atomic data, they still do not incorporate fairness, they still have no theory of open systems, and they still suffer from the drawbacks of a unified treatment of senders and receivers alluded to above. Furthermore, neither of these presentations makes any attempt at developing an equational theory and reasoning principles as we do here.

Aknowledgements

This research was partially supported by DARPA contracts N00039-84-C-0211 and NAG2-703, and NSF grants CCR-8917606, CCR-8915663, and CCR-91-090070 by DARPA and NSF joint contract CCR 90-07195, by ONR contract N00014-90-J-1899, and by the Digital Equipment Corporation.

6. References

[1] G. Agha. *Actors: A Model of Concurrent Computation in Distributed Systems*. MIT Press, Cambridge, Mass., 1986.

[2] Gul Agha. Concurrent object-oriented programming. *Communications of the ACM*, 33(9):125–141, September 1990.

[3] D. Berry, R. Milner, and D.N. Turner. A semantics for ML concurrency primitives. In *Conference record of the 19th annual ACM symposium on principles of programming languages*, pages 105–129, 1992.

[4] R. de Nicola and M. C. B. Hennessy. Testing equivalences for processes. *Theoretical Computer Science*, 34:83–133, 1984.

[5] C. Hewitt. Viewing control structures as patterns of passing messages. *Journal of Artificial Intelligence*, 8(3), 1977.

[6] C. A. R. Hoare. *Communicating Sequential Processes*. Prentice-Hall, 1985.

[7] I. A. Mason and C. L. Talcott. Equivalence in functional languages with effects. *Journal of Functional Programming*, 1:287–327, 1991.

[8] R. Milner, J. G. Parrow, and D. J. Walker. A calculus of mobile processes, parts i and ii. Technical Report ECS-LFCS-89-85, -86, Edinburgh University, 1989.

[9] G. Plotkin. Call-by-name, call-by-value and the lambda-v-calculus. *Theoretical Computer Science*, 1:125–159, 1975.

[10] J. H. Reppy. An operational semantics of first-class synchronous operations. Technical Report TR 91-1232, Cornell University, 1991.

[11] C. L. Talcott. A theory for program and data specification. In *Design and Implementation of Symbolic Computation Systems, DISCO'90*, volume 429 of *Lecture Notes in Computer Science*. Springer-Verlag, 1990. full version to appear in TCS special issue.

[12] A. Yonezawa. *ABCL: An Object-Oriented Concurrent System*. MIT Press, Cambridge Mass., 1990.

Authors Index

Lecture Notes in Computer Science

For information about Vols. 1–544
please contact your bookseller or Springer-Verlag

Vol. 545: H. Alblas, B. Melichar (Eds.), Attribute Grammars , Applications and Systems. Proceedings, 1991. IX, 513 pages. 1991.

Vol. 546: O. Herzog, C.-R. Rollinger (Eds.), Text Understanding in LILOG. XI, 738 pages. 1991. (Subseries LNAI).

Vol. 547: D. W. Davies (Ed.), Advances in Cryptology – EUROCRYPT '91. Proceedings, 1991. XII, 556 pages. 1991.

Vol. 548: R. Kruse, P. Siegel (Eds.), Symbolic and Quantitative Approaches to Uncertainty. Proceedings, 1991. XI, 362 pages. 1991.

Vol. 549: E. Ardizzone, S. Gaglio, F. Sorbello (Eds.), Trends in Artificial Intelligence. Proceedings, 1991. XIV, 479 pages. 1991. (Subseries LNAI).

Vol. 550: A. van Lamsweerde, A. Fugetta (Eds.), ESEC '91. Proceedings, 1991. XII, 515 pages. 1991.

Vol. 551:S. Prehn, W. J. Toetenel (Eds.), VDM '91. Formal Software Development Methods. Volume 1. Proceedings, 1991. XIII, 699 pages. 1991.

Vol. 552: S. Prehn, W. J. Toetenel (Eds.), VDM '91. Formal Software Development Methods. Volume 2. Proceedings, 1991. XIV, 430 pages. 1991.

Vol. 553: H. Bieri, H. Noltemeier (Eds.), Computational Geometry - Methods, Algorithms and Applications '91. Proceedings, 1991. VIII, 320 pages. 1991.

Vol. 554: G. Grahne, The Problem of Incomplete Information in Relational Databases. VIII, 156 pages. 1991.

Vol. 555: H. Maurer (Ed.), New Results and New Trends in Computer Science. Proceedings, 1991. VIII, 403 pages. 1991.

Vol. 556: J.-M. Jacquet, Conclog: A Methodological Approach to Concurrent Logic Programming. XII, 781 pages. 1991.

Vol. 557: W. L. Hsu, R. C. T. Lee (Eds.), ISA '91 Algorithms. Proceedings, 1991. X, 396 pages. 1991.

Vol. 558: J. Hooman, Specification and Compositional Verification of Real-Time Systems. VIII, 235 pages. 1991.

Vol. 559: G. Butler, Fundamental Algorithms for Permutation Groups. XII, 238 pages. 1991.

Vol. 560: S. Biswas, K. V. Nori (Eds.), Foundations of Software Technology and Theoretical Computer Science. Proceedings, 1991. X, 420 pages. 1991.

Vol. 561: C. Ding, G. Xiao, W. Shan, The Stability Theory of Stream Ciphers. IX, 187 pages. 1991.

Vol. 562: R. Breu, Algebraic Specification Techniques in Object Oriented Programming Environments. XI, 228 pages. 1991.

Vol. 563: A. Karshmer, J. Nehmer (Eds.), Operating Systems of the 90s and Beyond. Proceedings, 1991. X, 285 pages. 1991.

Vol. 564: I. Herman, The Use of Projective Geometry in Computer Graphics. VIII, 146 pages. 1992.

Vol. 565: J. D. Becker, I. Eisele, F. W. Mündemann (Eds.), Parallelism, Learning, Evolution. Proceedings, 1989. VIII, 525 pages. 1991. (Subseries LNAI).

Vol. 566: C. Delobel, M. Kifer, Y. Masunaga (Eds.), Deductive and Object-Oriented Databases. Proceedings, 1991. XV, 581 pages. 1991.

Vol. 567: H. Boley, M. M. Richter (Eds.), Processing Declarative Kowledge. Proceedings, 1991. XII, 427 pages. 1991. (Subseries LNAI).

Vol. 568: H.-J. Bürckert, A Resolution Principle for a Logic with Restricted Quantifiers. X, 116 pages. 1991. (Subseries LNAI).

Vol. 569: A. Beaumont, G. Gupta (Eds.), Parallel Execution of Logic Programs. Proceedings, 1991. VII, 195 pages. 1991.

Vol. 570: R. Berghammer, G. Schmidt (Eds.), Graph-Theoretic Concepts in Computer Science. Proceedings, 1991. VIII, 253 pages. 1992.

Vol. 571: J. Vytopil (Ed.), Formal Techniques in Real-Time and Fault-Tolerant Systems. Proceedings, 1992. IX, 620 pages. 1991.

Vol. 572: K. U. Schulz (Ed.), Word Equations and Related Topics. Proceedings, 1990. VII, 256 pages. 1992.

Vol. 573: G. Cohen, S. N. Litsyn, A. Lobstein, G. Zémor (Eds.), Algebraic Coding. Proceedings, 1991. X, 158 pages. 1992.

Vol. 574: J. P. Banâtre, D. Le Métayer (Eds.), Research Directions in High-Level Parallel Programming Languages. Proceedings, 1991. VIII, 387 pages. 1992.

Vol. 575: K. G. Larsen, A. Skou (Eds.), Computer Aided Verification. Proceedings, 1991. X, 487 pages. 1992.

Vol. 576: J. Feigenbaum (Ed.), Advances in Cryptology - CRYPTO '91. Proceedings. X, 485 pages. 1992.

Vol. 577: A. Finkel, M. Jantzen (Eds.), STACS 92. Proceedings, 1992. XIV, 621 pages. 1992.

Vol. 578: Th. Beth, M. Frisch, G. J. Simmons (Eds.), Public-Key Cryptography: State of the Art and Future Directions. XI, 97 pages. 1992.

Vol. 579: S. Toueg, P. G. Spirakis, L. Kirousis (Eds.), Distributed Algorithms. Proceedings, 1991. X, 319 pages. 1992.

Vol. 580: A. Pirotte, C. Delobel, G. Gottlob (Eds.), Advances in Database Technology – EDBT '92. Proceedings. XII, 551 pages. 1992.

Vol. 581: J.-C. Raoult (Ed.), CAAP '92. Proceedings. VIII, 361 pages. 1992.

Vol. 582: B. Krieg-Brückner (Ed.), ESOP '92. Proceedings. VIII, 491 pages. 1992.

Vol. 583: I. Simon (Ed.), LATIN '92. Proceedings. IX, 545 pages. 1992.

Vol. 584: R. E. Zippel (Ed.), Computer Algebra and Parallelism. Proceedings, 1990. IX, 114 pages. 1992.

Vol. 585: F. Pichler, R. Moreno Díaz (Eds.), Computer Aided System Theory – EUROCAST '91. Proceedings. X, 761 pages. 1992.

Vol. 586: A. Cheese, Parallel Execution of Parlog. IX, 184 pages. 1992.

Vol. 587: R. Dale, E. Hovy, D. Rösner, O. Stock (Eds.), Aspects of Automated Natural Language Generation. Proceedings, 1992. VIII, 311 pages. 1992. (Subseries LNAI).

Vol. 588: G. Sandini (Ed.), Computer Vision – ECCV '92. Proceedings. XV, 909 pages. 1992.

Vol. 589: U. Banerjee, D. Gelernter, A. Nicolau, D. Padua (Eds.), Languages and Compilers for Parallel Computing. Proceedings, 1991. IX, 419 pages. 1992.

Vol. 590: B. Fronhöfer, G. Wrightson (Eds.), Parallelization in Inference Systems. Proceedings, 1990. VIII, 372 pages. 1992. (Subseries LNAI).

Vol. 591: H. P. Zima (Ed.), Parallel Computation. Proceedings, 1991. IX, 451 pages. 1992.

Vol. 592: A. Voronkov (Ed.), Logic Programming. Proceedings, 1991. IX, 514 pages. 1992. (Subseries LNAI).

Vol. 593: P. Loucopoulos (Ed.), Advanced Information Systems Engineering. Proceedings. XI, 650 pages. 1992.

Vol. 594: B. Monien, Th. Ottmann (Eds.), Data Structures and Efficient Algorithms. VIII, 389 pages. 1992.

Vol. 595: M. Levene, The Nested Universal Relation Database Model. X, 177 pages. 1992.

Vol. 596: L.-H. Eriksson, L. Hallnäs, P. Schroeder-Heister (Eds.), Extensions of Logic Programming. Proceedings, 1991. VII, 369 pages. 1992. (Subseries LNAI).

Vol. 597: H. W. Guesgen, J. Hertzberg, A Perspective of Constraint-Based Reasoning. VIII, 123 pages. 1992. (Subseries LNAI).

Vol. 598: S. Brookes, M. Main, A. Melton, M. Mislove, D. Schmidt (Eds.), Mathematical Foundations of Programming Semantics. Proceedings, 1991. VIII, 506 pages. 1992.

Vol. 599: Th. Wetter, K.-D. Althoff, J. Boose, B. R. Gaines, M. Linster, F. Schmalhofer (Eds.), Current Developments in Knowledge Acquisition - EKAW '92. Proceedings. XIII, 444 pages. 1992. (Subseries LNAI).

Vol. 600: J. W. de Bakker, C. Huizing, W. P. de Roever, G. Rozenberg (Eds.), Real-Time: Theory in Practice. Proceedings, 1991. VIII, 723 pages. 1992.

Vol. 601: D. Dolev, Z. Galil, M. Rodeh (Eds.), Theory of Computing and Systems. Proceedings, 1992. VIII, 220 pages. 1992.

Vol. 602: I. Tomek (Ed.), Computer Assisted Learning. Proceedings, 1992. X, 615 pages. 1992.

Vol. 603: J. van Katwijk (Ed.), Ada: Moving Towards 2000. Proceedings, 1992. VIII, 324 pages. 1992.

Vol. 604: F. Belli, F.-J. Radermacher (Eds.), Industrial and Engineering Applications of Artificial Intelligence and Expert Systems. Proceedings, 1992. XV, 702 pages. 1992. (Subseries LNAI).

Vol. 605: D. Etiemble, J.-C. Syre (Eds.), PARLE '92. Parallel Architectures and Languages Europe. Proceedings, 1992. XVII, 984 pages. 1992.

Vol. 606: D. E. Knuth, Axioms and Hulls. IX, 109 pages. 1992.

Vol. 607: D. Kapur (Ed.), Automated Deduction – CADE-11. Proceedings, 1992. XV, 793 pages. 1992. (Subseries LNAI).

Vol. 608: C. Frasson, G. Gauthier, G. I. McCalla (Eds.), Intelligent Tutoring Systems. Proceedings, 1992. XIV, 686 pages. 1992.

Vol. 609: G. Rozenberg (Ed.), Advances in Petri Nets 1992. VIII, 472 pages. 1992.

Vol. 610: F. von Martial, Coordinating Plans of Autonomous Agents. XII, 246 pages. 1992. (Subseries LNAI).

Vol. 611: M. P. Papazoglou, J. Zeleznikow (Eds.), The Next Generation of Information Systems: From Data to Knowledge. VIII, 310 pages. 1992. (Subseries LNAI).

Vol. 612: M. Tokoro, O. Nierstrasz, P. Wegner (Eds.), Object-Based Concurrent Computing. Proceedings, 1991. X, 265 pages. 1992.

Vol. 613: J. P. Myers, Jr., M. J. O'Donnell (Eds.), Constructivity in Computer Science. Proceedings, 1991. X, 247 pages. 1992.

Vol. 614: R. G. Herrtwich (Ed.), Network and Operating System Support for Digital Audio and Video. Proceedings, 1991. XII, 403 pages. 1992.

Vol. 615: O. Lehrmann Madsen (Ed.), ECOOP '92. European Conference on Object Oriented Programming. Proceedings. X, 426 pages. 1992.

Vol. 616: K. Jensen (Ed.), Application and Theory of Petri Nets 1992. Proceedings, 1992. VIII, 398 pages. 1992.

Vol. 617: V. Mařík, O. Štěpánková, R. Trappl (Eds.), Advanced Topics in Artificial Intelligence. Proceedings, 1992. IX, 484 pages. 1992. (Subseries LNAI).

Vol. 618: P. M. D. Gray, R. J. Lucas (Eds.), Advanced Database Systems. Proceedings, 1992. X, 260 pages. 1992.

Vol. 619: D. Pearce, H. Wansing (Eds.), Nonclassical Logics and Information Proceedings. Proceedings, 1990. VII, 171 pages. 1992. (Subseries LNAI).

Vol. 620: A. Nerode, M. Taitslin (Eds.), Logical Foundations of Computer Science – Tver '92. Proceedings. IX, 514 pages. 1992.

Vol. 621: O. Nurmi, E. Ukkonen (Eds.), Algorithm Theory – SWAT '92. Proceedings. VIII, 434 pages. 1992.

Vol. 622: F. Schmalhofer, G. Strube, Th. Wetter (Eds.), Contemporary Knowledge Engineering and Cognition. Proceedings, 1991. XII, 258 pages. 1992. (Subseries LNAI).

Vol. 623: W. Kuich (Ed.), Automata, Languages and Programming. Proceedings, 1992. XII, 721 pages. 1992.

Vol. 624: A. Voronkov (Ed.), Logic Programming and Automated Reasoning. Proceedings, 1992. XIV, 509 pages. 1992. (Subseries LNAI).

Vol. 625: W. Vogler, Modular Construction and Partial Order Semantics of Petri Nets. IX, 252 pages. 1992.

Vol. 626: E. Börger, G. Jäger, H. Kleine Büning, M. M. Richter (Eds.), Computer Science Logic. Proceedings, 1991. VIII, 428 pages. 1992.

Vol. 628: G. Vosselman, Relational Matching. IX, 190 pages. 1992.

Vol. 629: I. M. Havel, V. Koubek (Eds.), Mathematical Foundations of Computer Science 1992. Proceedings. IX, 521 pages. 1992.

Vol. 630: W. R. Cleaveland (Ed.), CONCUR '92. Proceedings. X, 580 pages. 1992.